AF848031

Upwelling Systems of the World

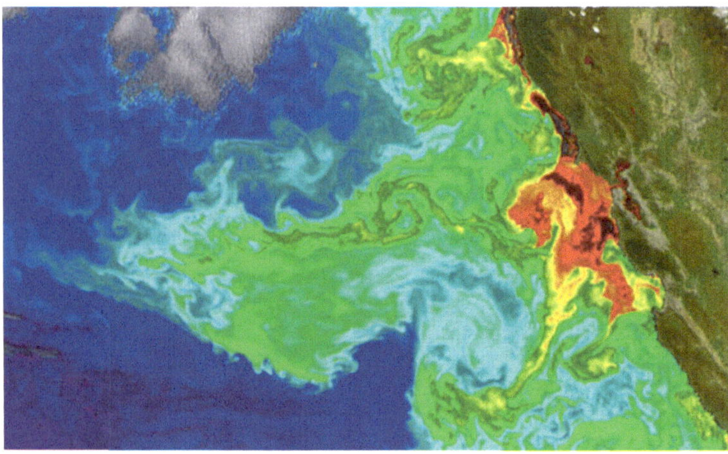

Phytoplankton blooms in an upwelling area in the Pacific Ocean off the California coast. *Image source* NASA http://visibleearth.nasa.gov/view.php?id=4317 [accessed 2/06/2016]

Jochen Kämpf · Piers Chapman

Upwelling Systems of the World

A Scientific Journey to the Most Productive Marine Ecosystems

Springer

Jochen Kämpf
Flinders University
Adelaide, SA
Australia

Piers Chapman
Texas A&M University
College Station, TX
USA

ISBN 978-3-319-42522-1 ISBN 978-3-319-42524-5 (eBook)
DOI 10.1007/978-3-319-42524-5

Library of Congress Control Number: 2016945937

© Springer International Publishing Switzerland 2016
This work is subject to copyright. All rights are reserved by the Publisher, whether the whole or part of the material is concerned, specifically the rights of translation, reprinting, reuse of illustrations, recitation, broadcasting, reproduction on microfilms or in any other physical way, and transmission or information storage and retrieval, electronic adaptation, computer software, or by similar or dissimilar methodology now known or hereafter developed.
The use of general descriptive names, registered names, trademarks, service marks, etc. in this publication does not imply, even in the absence of a specific statement, that such names are exempt from the relevant protective laws and regulations and therefore free for general use.
The publisher, the authors and the editors are safe to assume that the advice and information in this book are believed to be true and accurate at the date of publication. Neither the publisher nor the authors or the editors give a warranty, express or implied, with respect to the material contained herein or for any errors or omissions that may have been made.

Printed on acid-free paper

This Springer imprint is published by Springer Nature
The registered company is Springer International Publishing AG Switzerland

Preface

To early explorers and fishermen, the ocean seemed to be limitless, teeming with vast quantities of fish and other food organisms. However, as people got to know the ocean better, they realized that not all regions were the same. Large portions of the oceans in fact contained little marine life, while other regions, particularly along certain coasts, were much more productive. The most productive regions were found along the west coast of the main continents, in what are now known as *eastern boundary currents*, and these regions, which account for only about 1 % of the global ocean, produce about 20 % of the global fish catch. The four main eastern boundary systems are those off California/Oregon/Washington in the North Pacific, Peru and Chile in the South Pacific, off northwest Africa and Portugal in the North Atlantic, and off South Africa and Namibia in the South Atlantic. These upwelling systems have long provided large quantities of fish and are also known to support seabirds and mammals such as whales and fur seals.

We now know that a number of other upwelling systems exist throughout the global ocean, some of which are year-round features, whereas others occur on a seasonal basis. Recently, a number of reviews of individual systems have appeared in the scientific literature, some concentrating on physics and chemistry, others on biology, but we do not know of any consolidated text that covers all of them. Because of their importance in global productivity, biogeochemical cycles and food-web dynamics under exposure to global climate change, we believe that such an interdisciplinary book covering all important upwelling systems of the word is needed to describe their similarities and differences. We hope that this book will fill the gap and that you, the reader, will enjoy this scientific journey to the most productive ecosystems of the world.

Writing a book always takes a lot longer than anticipated, and this is particularly true of scientific books. While the World Wide Web makes it relatively easy to find information, it also complicates matters because of the enormous number of research papers that have been written about the different upwelling systems.

Undoubtedly we may have missed papers that some of you regard as being of supreme importance, but we have tried our best to cover all the major advances in the four major eastern boundary currents and give a good overview of the other upwelling regions. We welcome any suggestions you may have to improve this book for future editions.

Adelaide, Australia Jochen Kämpf
College Station, USA Piers Chapman
May 2016

Contents

About the Authors

Jochen Kämpf is an Associate Professor of Oceanography at the School of the Environment at Flinders University, Adelaide, Australia. Including the discovery of several important coastal upwelling regions, his previous research covered a broad range of subjects from small-scale convective mixing in polar regions, the circulation of inverse estuaries, suspended sediment dynamics and turbidity currents to canyon-flow interactions. He also published two textbooks on hydrodynamic modelling at Springer.

Piers Chapman is a Professor in the Department of Oceanography at Texas A&M University, where he currently works on the physics and chemistry of the Gulf of Mexico, concentrating on the low-oxygen environment that forms each year in summer. He worked for many years in South Africa, has published widely on the Benguela upwelling system, and has been on over 50 research cruises totalling almost three years at sea.

Chapter 1
Preliminaries

Abstract The distinct distributions of sunlight, nutrients and oxygen are fundamental to the way marine life forms in the ocean. This chapter describes these and the significant role that coastal upwelling plays in the cycles of nutrients, carbon and marine life. We also give an overview of early scientific expeditions that led to the discovery of upwelling systems and long-term interdisciplinary monitoring programs that significantly improved our understanding of upwelling processes.

Keywords Upwelling · Large marine ecosystems · Nutrient and carbon cycles · Marine food webs · General description · Early expeditions

You can't do anything about the length of your life,

but you can do something about its width and depth

Evan Esar (1899–1995)

(Taken from the *Stone Cutters' Journal*: *1922–1924*, Volumes 37–39)

1.1 Introduction

Although certain oceanic life forms, such as the microbes that support unique deep-sea biomes around hydrothermal vents, obtain their energy requirements from methane or hydrogen sulphide, almost all marine ecosystems owe their existence to light-induced photosynthesis carried out by phytoplankton (floating marine plants) and phototropic bacteria confined to the upper 50–100 m of the water column, the *euphotic zone*. As a by-product, these organisms produce oxygen as a by-product, another requirement for most life forms. The associated conversion of inorganic to organic carbon, also called *carbon fixation*, allows marine organisms to grow and reproduce. The rate of carbon fixation is strongly controlled by the availability of nutrients (e.g., nitrogen, phosphorous, silica), which are supplied to the euphotic

© Springer International Publishing Switzerland 2016
J. Kämpf and P. Chapman, *Upwelling Systems of the World*,
DOI 10.1007/978-3-319-42524-5_1

zone via currents bringing up high-nutrient water from below (upwelling), vertical mixing, continental runoff of sediment-laden waters via rivers or groundwater seepage, and to some extent by atmospheric dust deposition.

Before we discuss the fundamentals of light, nutrient and oxygen distributions in the oceans, which are essential to the understanding of physical-biological interactions in upwelling regions, we introduce the reader to the globalized concept of Large Marine Ecosystems (LMEs), a management framework that defines and ranks marine regions according to their annual *gross primary productivity*, i.e., the rate of conversion of CO_2 to organic carbon per unit surface area. These estimates are based on satellite-data algorithms, which accounts for the rather wide limits put on their classification. Many of the more productive LMEs are affiliated with upwelling processes, and it is scientifically useful to study and compare individual upwelling regions in the global context.

1.2 Large Marine Ecosystems

The World Summit on Sustainable Development, convened in Johannesburg in 2002, recognized the importance for coastal nations to work towards the sustainable development and use of ocean resources. Participating world leaders agreed to pursue four main targets:

(i) to achieve substantial reductions in land-based sources of pollution,
(ii) to introduce an ecosystems approach to marine resource assessment and management,
(iii) to designate a network of marine protected areas, and
(iv) to maintain and restore fish stocks to maximum sustainable yield levels.

This effort has led to the definition of Large Marine Ecosystems (LMEs), which are grouped into three different productivity categories according to their annual gross primary productivity (Sherman and Hempel 2008):

- **Class I**, high productivity (>300 g C/m^2/yr)
- **Class II**, moderate productivity (150–300 g C/m^2/yr), and
- **Class III,** low (<150 g C/m^{2}yr) productivity.

Despite some shortcomings of this methodology (e.g., the value of its productivity depends on the definition of the spatial extent of an ecosystem), this method produces a global map of significant marine ecosystems (Fig. 1.1) used as a scientific basis in international negotiations.

Regardless of their productivity level, all LMEs require a supply of nutrients [nitrogen (N), phosphorous (P), silica (Si) and smaller concentrations of elements such as iron (Fe)] to support their productivity. In some regions nutrients are supplied by large rivers/estuaries. In other regions, the nutrient supply comes from the ocean's interior, including the seabed, either through vertical movement of the

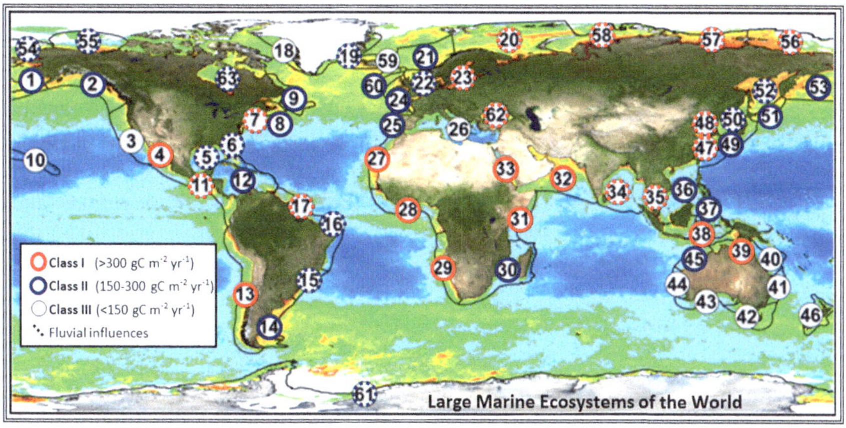

Fig. 1.1 Classification of Large Marine Ecosystems (LMEs) of the world ocean. Displayed are SeaWIFS chlorophyll-*a* distributions and the locations of most LMEs. The LMEs of the Arctic Ocean and Hudson Bay are not shown. Fluvial influences refer to continental runoff including estuarine ecosystems. Source of background image: http://lme.edc.uri.edu/ [accessed on 4 April 2016]

water column (i.e., upwelling) or by the vertical stirring of nutrient-enriched sub-surface water towards the surface. This book focusses on these upwelling regions, most of which are classified as individual LMEs. Before we can begin to describe how upwelling systems resemble or differ from each other, however, we need to consider some basic controls of life in the sea.

1.3 Life in the Ocean

The ocean acts as a giant mixing bowl, and contains varying concentrations of all the elements that make up the Earth as a result of erosion and dissolution of the Earth's surface by the action of water, ice, wind, and waves. Although we do not know how life on Earth began, it is possible that it started either in the ocean or in brackish pools that provided the necessary chemical building blocks, some form of substrate such as clay particles on which complex molecules can be built up, and energy in the form of heat and light that can catalyze the necessary reactions (e.g., Cairns-Smith 1982). A more recent, alternative suggestion is that life started in small droplets, with reactions occurring at the droplet surface (Fallah-Araghi et al. 2014).

When life began, some 3.5 billion years ago, the Earth's atmosphere was similar in composition to gases emitted today by volcanoes and was made up of water vapour, methane, hydrogen, carbon dioxide, nitrogen and ammonia; there was no oxygen. In a classic experiment in 1953, two scientists at the University of Chicago, Stanley Miller and Harold Urey, showed that a mixture of these gases, when subjected to electrical sparks meant to simulate lightning, could form numerous organic compounds including basic biochemical building blocks such as amino acids and simple sugars (Miller 1953; Miller and Urey 1959).

Over millions of years these earliest, simple life forms, which are known from fossil bacteria in marine rocks from this epoch, expanded until they probably occupied much of the ocean, but it took over a billion years until some of these species began to produce oxygen, about 2.5 billion years ago, thus allowing more complex life forms to develop. By about 1.8 billion years ago, the oxygen content of the atmosphere and ocean was high enough that most of these early life forms were killed off. It is thought, however, that bacteria known as *archeobacteria,* found today in regions such as subsea hot vents where volcanic gases escape into the ocean and oxygen concentrations are very low, are descended from these earliest life forms.

The multitudinous variety of marine life seen today includes bacteria, plants (e.g., sea grasses, seaweeds, and phytoplankton), and animals (e.g., zooplankton, crustacea, fish, birds and mammals). All these species interact through complex food webs, the dynamics of which are based on the uptake of simple inorganic chemicals and their conversion to more complex organic material by bacteria and phytoplankton, known as *autotrophs* because they produce their own organic matter. These are fed on by *heterotrophic* organisms that require pre-manufactured organic matter.

The following sections in this chapter describe the basic properties that control organic matter production in the ocean, including how they vary in space and time. From this, the reader will learn about important phenomena including the euphotic zone, oxygen-minimum zones, nutrient limitation, saturation horizons of carbonate minerals, oceanic carbon pumps and the relevance of the physical process of oceanic upwelling to marine ecosystems and the carbon cycle. From this it becomes clear that the study of upwelling is intrinsically interdisciplinary in nature, involving interactions between physical, chemical and biological environments that are interconnected by complex biogeochemical cycles and food web dynamics and that are also affected by human interference (Fig. 1.2).

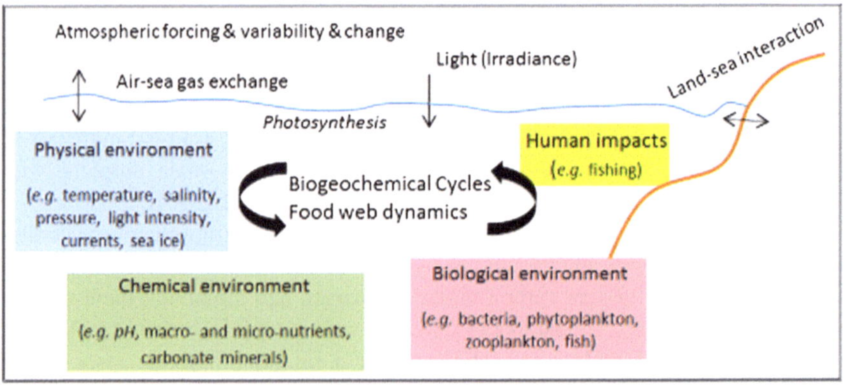

Fig. 1.2 The general interdisciplinary nature of processes influencing marine life

1.4 Basics of Marine Ecology

1.4.1 Types of Marine Life Forms

Open-ocean marine life forms can be loosely classified according to size. Bacteria are the smallest organisms, while organism size generally increases through phytoplankton and zooplankton to fish and mammals. There are also many benthic organisms that live attached to hard surfaces or within the bottom sediments, such as crustaceans, mollusks, worms, and coelenterates, as well as macroalgae (commonly called seaweeds), and filamentous algae, and even some true plants with roots, such as eelgrass, that photosynthesize near the seabed as long as the water is clear enough to allow light penetration.

Bacteria play extremely important roles in the marine food web through:

(i) controlling decomposition of dead organisms, which recycles nutrients, including iron and other trace elements, in the water column and sediment for reuse by other marine organisms, and
(ii) producing organic carbon (and oxygen) via photosynthesis by cyanobacteria. The microbes themselves can also serve as food for many larger organisms, especially filter feeders.

Prior to the discovery of the microbial loop, the classic view of marine food webs was one of a linear chain from phytoplankton to nekton. Generally, marine bacteria were not thought to be significant consumers of organic matter (including carbon), although they were known to exist. However, the view of a marine pelagic food web was challenged during the 1970s and 1980s by Pomeroy (1974) and Azam et al. (1983), who suggested the alternative pathway of carbon flow from bacteria to protozoans to metazoans. Early on it was recognised that bacteria play a substantial role comparable to that of the primary producers in terms of element cycling in the water column (Kirchman et al. 1982; Williams 1981). For more details, see the review by Fenchel (2008).

Phytoplankton (floating marine plants) are photosynthesizing microscopic organisms that inhabit the upper sunlit layer of almost all oceans. They include the important groups of diatoms, which require silica to construct their internal skeletons, coccolithophores (Fig. 1.3), which use calcium carbonate for the same purpose, and dinoflagellates, which possess flagellae that help them move rather than being totally dependent on currents, as well as the much smaller pico- and nanoplankton (see Fig. 1.4). Most phytoplankton cells are denser than seawater (Mann and Lazier 1996), hence they tend to sink in the water column except where an upward movement of water prevents it. Average sinking rates in quiescent water are between about 0.1–10 m/day, depending largely on cell size. To overcome sinking, which is important for continued growth and reproduction, plankton have devised various strategies. Dinoflagellates use their flagellae for active, energy-consuming self-propulsion that can oppose gravitational sinking. Other members of the

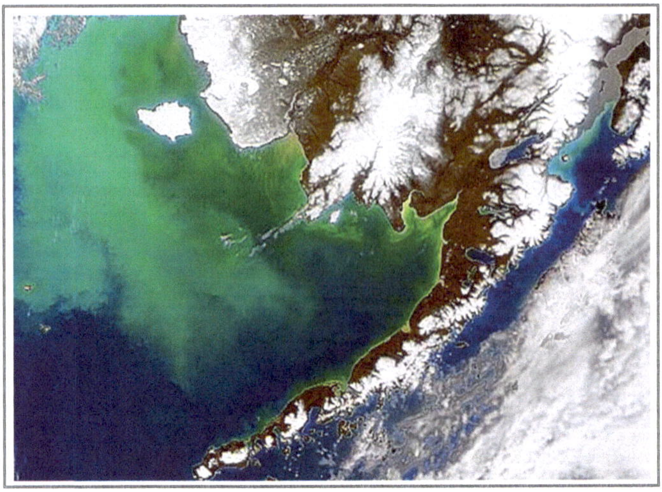

Fig. 1.3 Coccolithophore bloom (the pale green colour) in the Bering Sea off southwest Alaska on April 25, 1998. *Image source* NASA http://www.afsc.noaa.gov/Quarterly/AMJ2014/amj14featurelead.htm [accessed on 4 April 2016]

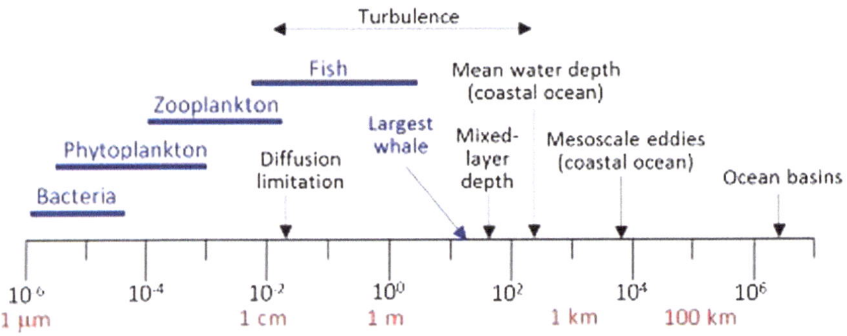

Fig. 1.4 Physical and biological length scales of oceanic processes and marine organisms. Adapted from Mann and Lazier (1996)

phytoplankton can modify their buoyancy, becoming, at least for sometime, positively buoyant, particularly through the formation of gas vacuoles. Others can exploit the turbulence in the mixed layer, using it to stay suspended for longer times. Additionally, some species, especially those with spiny outgrowths or those forming long chains, use their shape to reduce sinking rates by increasing their surface area.

The zooplankton (drifting marine animals) span a range of sizes from small *protozoans* (sizes 10–50 μm, up to 1 mm) to large *metazoans* (sizes 0.001 to >1 m). Ecologically important protozoan zooplankton groups include the *foraminiferans*, which like coccolithophores have calcium carbonate tests, and *radiolarians*, which use silica. These two groups, along with coccolithphores and diatoms, have, over geologic time, contributed enormous amounts of silica and calcium carbonate to deep ocean sediments.

Important metazoan zooplankton include *cnidarians* such as *jellyfish*, *crustaceans* such as *copepods* and *krill*; *chaetognaths* (arrow worms); *molluscs* such as *pteropods*; and *chordates* (animals with dorsal nerve cords) such as *salps*and *juvenile fish*. Within these groups are *holo-planktonic* organisms whose complete lifecycle lies within the plankton (e.g., the protozoans and jellyfish), as well as *mero-planktonic* organisms that spend part of their lives in the plankton during larval stages before graduating to either the nekton (swimming marine animals) such as copepods and fish, or a sessile, benthic existence (i.e., attached to the seafloor), such as sea anemones and many molluscs. Although zooplankton are primarily transported by ambient water currents, many have self-propulsion abilities, used to avoid predators or to increase prey encounter rates (Mann and Lazier 1996), and many species are known to migrate hundreds of meters vertically on a daily basis, staying at depth during the day and coming up towards the surface at night.

Diel vertical migration of both marine zooplankton was first described in the 1920s (see the review by Lewis (1954)). Behaviour adaptation and locomotive abilities play an important role for many zooplankton species and larval fish (Mann and Lazier 1996). Selective vertical migration whether diel, seasonal or ontogenetic (i.e. dependant on stage of life cycle) helps the species to conserve energy, locate food, retain a certain location or to move to other locations (see the review by Lampert (1989)). In the four major upwelling regions, which are characterised by poleward undercurrents, vertical migration between the equatorward surface flow and the undercurrent is particularly important and enhances the potential of self-recruitment (Carr et al. 2007).

Zooplankton feed on the bacterial component of plankton (bacterio-plankton), phytoplankton, other smaller zooplankton, detritus (marines now) and even nektonic (swimming) organisms (e.g., jellyfish eat fish). As a result, zooplankton are primarily found in surface waters where food resources (phytoplankton or other zooplankton) are abundant.

One can also define marine organisms as *photo-autotrophs*, or *heterotrophs* (Sigman and Hain 2012), depending on how they acquire the energy they need for basic living processes. Photo-autographs harvest light as an energy source to convert inorganic carbon to organic forms during photosynthesis, which also produces oxygen. Marine photoautotrophs include cyanobacteria, phytoplankton, algae, and marine plants. Heterotrophs, which include all other groups including bacteria as well as more complex single-and multi-celled zooplankton, *nekton*, and the *benthos*, utilize either the organic carbon produced by phototrophic organisms as an energy source, or, in the case of certain specialized organisms found near deep-sea vents, carbon produced by bacteria that can use volcanic gases such as hydrogen sulphide rather than carbon dioxide.

Gross primary production refers to the total rate of organic carbon production by autotrophs, while *respiration* refers to the energy-yielding oxidation of organic carbon back to carbon dioxide. The basic equation is the same in both cases, although the two mechanisms operate in reverse. The chemical reaction reads:

$$\text{Energy} + \text{nutrients} + 6CO_2 + 6H_2O \leftrightarrow C_6H_{12}O_6 + 6O_2 \tag{1.1}$$

where CO_2 is carbon dioxide, H_2O is the water molecule, energy comes from solar radiation, $C_6H_{12}O_6$ is a sugar molecule and O_2 is the oxygen molecule. Thus, production results in a decrease in carbon dioxide with production of oxygen, while respiration uses up oxygen to break down organic matter. Respiration occurs continuously, while production in surface ocean waters can only take place during daylight.

Net primary production is gross production minus the autotrophs' own rate of respiration; it is thus the rate at which the full metabolism of cyanobacteria and phytoplankton produce biomass (Bender et al. 1987). Thus, to estimate gross production, we have to measure respiration as well. About half of the Earth's primary production occurs in the ocean, and half of this is carried out by cyanobacteria. It is important to stress that primary production is not a measure of the growth rate of phytoplankton; this is measured by the rate of change of biomass, and depends on the original population size.

Secondary production typically refers to the growth rate of heterotrophic biomass. Only a small fraction of the organic matter ingested by heterotrophic organisms is used for growth, the majority being respired back to dissolved inorganic carbon and nutrients that can be reused by autotrophs. Therefore, secondary production in the ocean is small when compared to net primary production, and decreases each step up the trophic ladder from phytoplankton to zooplankton to small fish to larger fish. As a rule of thumb, there is generally about a factor of 10 difference in production at each stage of a trophic ladder, so that to produce 1 kg of zooplankton-eating fish requires 10 kg of zooplankton or 100 kg of phytoplankton.

1.4.2 Controls of the Marine Food Web

The controls of marine ecosystems can be characterized as (i) bottom-up, (ii) top-down, or (iii) wasp-waist. The *bottom-up control* of the ecosystem is driven by nutrient supply to the primary producers. If the nutrient supply is increased, e.g., when upwelling winds start to blow, the resulting increase in production of autotrophs is propagated through the food web and all of the other trophic levels will respond to the increased availability of food. Conversely, a less favourable physical environment, such as occurs off the coast of Peru during an El Niño, leads to a decrease in phytoplankton abundance, which in turn has a negative impact on the abundance of the zooplankton. The decrease of the zooplankton population similarly causes a decrease in the abundance of small forage fish, which itself leads to a decrease in the abundance of higher trophic level predators, such as tuna, seals, cetaceans and birds.

The *top-down control* implies that predation and grazing by higher trophic levels on lower trophic levels controls ecosystem function. An increase in predators will result in fewer grazers, and such a decrease in grazers will result in turn in more primary producers because fewer of them are being eaten by the grazing organisms. Thus the control of population numbers and overall productivity "cascades" from the top levels of the food chain down to the bottom trophic levels.

Wasps have a very characteristic appearance, with a very narrow stalk-like waist and a prominent thorax (chest) and abdomen. The term *wasp-waist control* is used for ecosystems in which small plankton-eating fish (such as sardines), called *forage fish*, control both higher and lower trophic levels. A wasp-waist ecosystem structure exhibits a mixture of the two methods of population control: a top-down control for zooplankton and a bottom-up control of upper trophic level predators. A decrease in the forage fish abundance affects the abundance of the predators negatively. The same decrease in abundance of the prey fish reduces the predation on zooplankton, which increase in abundance. A more abundant zooplankton population increases grazing pressure and leads to a diminishing phytoplankton abundance. Wasp-waist marine ecosystems usually occur in coastal upwelling regions, where small pelagic fish such as anchovy and sardines are dominant species. This does not, however, prevent larger demersal species, such as hake, from playing a major role in the ecosystem, and several upwelling systems also have large populations of *myctophid* (lanternfish) species that inhabit the offshore mid-water zone.

1.4.3 Spatial and Temporal Scales

Each species of marine organism has its own individual timescales (e.g. life span, duration of larval phase, etc.), length scales (e.g. body size) and locomotive abilities that determine possible levels of response and adjustment to physical processes occurring in the sea (Fig. 1.4). Depending on size and type, marine organisms are subject to different levels of physical interactions with their environment. The feeding of organisms less than a few millimeters in size, for instance, is strongly controlled by diffusive processes and *diffusion limitation*. Phytoplankton cells are suspended motionless in the water column and tend to consume nutrients in the water around them at a rate that is determined by how fast nutrients can diffuse towards them. Thus, while phytoplankton can grow quickly in calm water, unless there is sufficient turbulence that continually replenishes the nutrients in the immediate vicinity of the cells, growth will stop relatively quickly as nutrient concentrations are depleted. One way to overcome this limitation is to generate movement relative to the water, and some phytoplankters overcome this limitation by sinking to increase their nutrient uptake (see Mann and Lazier 1996).

Physical processes in the coastal ocean involve a large range of spatial and temporal scales. Turbulent stirring elements, called *vortices*, range in size from a few centimeters to tens of meters, whereas *mesoscale eddies* have horizontal diameters ~100 km in the open ocean and ~5–20 km in shelf seas. Temporal variations occur on the scales of vertical turbulence (including regular waves) and internal waves (seconds to minutes), semidiurnal and diurnal tides (~12–24 h), the daily sunlight cycle (24 h), weather events and associated coastal upwelling events (2–10 days), seasonal processes (e.g. warming-cooling cycle), and on multi-year scales (climate variability; e.g., El Niño events). All of these time and length scales can affect oceanic primary and secondary production.

Life spans and sizes of marine organisms are typically proportional to each other (Fig. 1.5). While the life span of a large marine mammal, such as a blue whale, may be close to 100 years, those of fish are more like 1–10 years, and zooplankton may complete a generation in a few days or weeks. Phytoplankton have doubling times on the order of days and bacteria of hours (Mann and Lazier 1996). In addition, each marine species has a certain population size and genetic diversity. Attached to this are population-related timescales characteristic of population variations and genetic evolution. Biological-physical interactions in coastal upwelling systems can encompass the entire range of spatial and temporal scales discussed above.

As an example, phytoplankton blooms typically last about 5–10 days, while the copepods that prey on them have life cycles of around 25–40 days, and the fish that eat the copepods even longer (Hutchings 1992). Thus, there is a basic mismatch between the life cycle of predator and prey. In order for copepods to survive for long enough to develop and reproduce, they need to take advantage of areas where

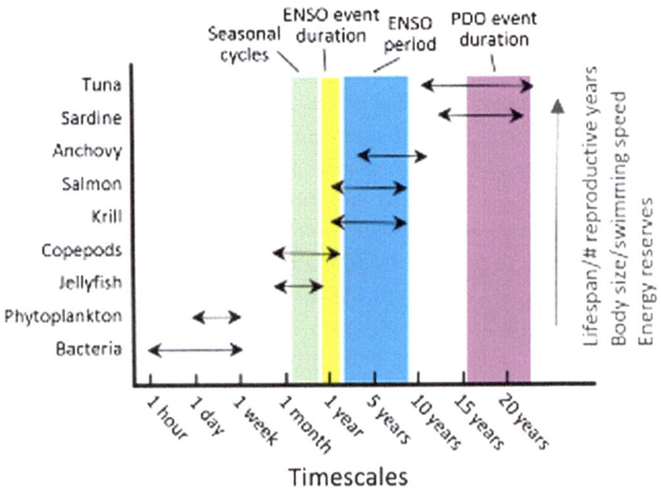

Fig. 1.5 Biological timescales of key marine species in the context of seasonal and climatic variability in the Pacific Ocean. See Chap. 3 for details on ENSO (El Niño Southern Oscillation) and the PDO (Pacific Decadal Oscillation)

the phytoplankton blooms are concentrated by physical effects and also make use of mechanisms that ensure that young copepods are not swept out of the productive region. This can include changing their depth in the water column by active swimming or sinking to make use of currents that bring them back to the area where the phytoplankton density is greatest. Similarly, small forage fish often spawn in one region that ensures the currents move their larvae into regions of high concentration of phytoplankton and zooplankton as they develop.

It should be highlighted that bacteria and phytoplankton respond strongly to weather events (storms and upwelling driven by synoptic coastal wind variations), whereas longer-lived species including krill and pelagic fish (e.g., salmon, anchovy, sardine and tuna) are also influenced by climate-scale variability such as El Niño events. Hence, there is a complex coupling between physical processes and food-web dynamics on a vast range of spatial and temporal scales. This fact makes the study of marine ecosystems inherently difficult and interdisciplinary in nature.

1.5 Light, Nutrients and Oxygen in the Sea

1.5.1 Photosynthesis

Most life in the sea, apart from specialized deep-sea communities that can survive by using sulphur or methane as a terminal electron acceptor, depends on photosynthesis, the conversion of inorganic to organic carbon, as the basis of marine food webs. As stated above, photosynthesis is carried out by cyanobacteria and phytoplankton, and sunlight provides the principal energy for photosynthesis in the sea according to chemical reaction (1.1). Photosynthesis takes place in specialized compartments within plant cells called *chloroplasts*, where light energy is absorbed by chemicals such as chlorophyll-*a* and used to power a series of chemical reactions, known as the Calvin-Benson cycle, that converts six molecules of CO_2 to glucose by splitting water molecules to produce the hydrogen needed to react with the CO_2. Oxygen, which also comes from the water molecules, is a side product of photosynthesis.

1.5.2 Light

Sunlight intensity decreases rapidly with depth as the water molecules absorb the sun's energy. Sunlight is made up of light of different colours, and each colour has a specific wavelength from red (longest) to violet (shortest) that corresponds inversely to an energy level. The longer wavelengths of light have less energy than shorter wavelengths and are absorbed first, so red light disappears quickly and eventually only blue or violet light is left. In the clearest ocean, light intensity is reduced to 1 % of the surface value at a depth of ~50–100 m, but this intensity can be reached at depths as shallow as 5–10 m in turbid coastal waters, where high levels of suspended material reduce light penetration. The 1 % level, called *euphotic depth*, is often used to define the base of the euphotic zone, the depth below which light intensity is too low for photosynthesis to take place. Mathematically, vertical attenuation of light intensity (I) with depth (z) can be described by the exponential equation:

$$I(z) = I_o \exp(-kz), \tag{1.2}$$

where I_o is the light intensity at the sea surface, and k is the attenuation coefficient, which depends strongly on turbidity (i.e. the concentration of particulate matter in the water column) (Fig. 1.6). Photosynthesis depends on the availability of light energy. Hence, the rate at which primary production of organic carbon can occur also decreases rapidly with depth from a maximum at the ocean surface. However, the rate of metabolic energy use by phytoplankton (*respiration*) varies little with depth, leading to the concept of a *compensation depth*—the depth at which phytoplankton production equals respiration (Gran and Trygve 1935). Below this depth, net phytoplankton production is not possible as respiration dominates over production. Essentially, the compensation depth is similar to the euphotic depth.

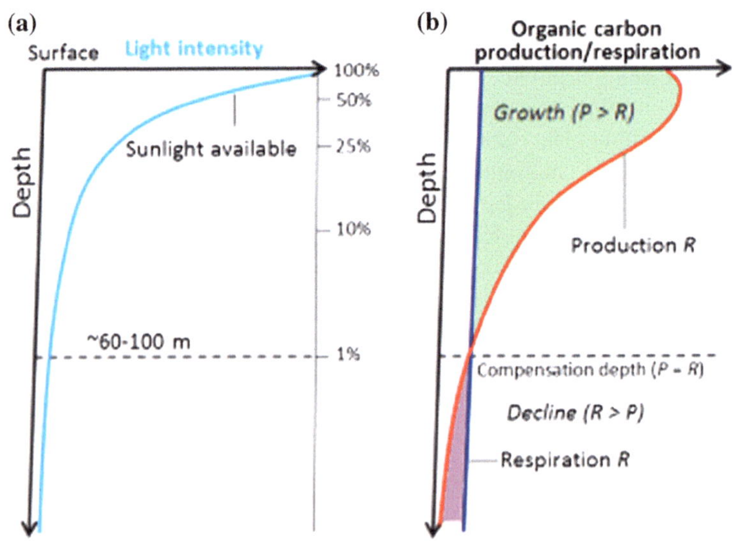

Fig. 1.6 Relation between daily averaged light intensity and production/respiration of organic carbon. The compensation depth is the depth at which production and respiration rate are equal. Adapted from Segar (2007)

1.5.3 Oxygen

The availability of dissolved oxygen (O_2) is vital for almost all marine life, with the exceptions of either bacteria that can use sulphur or methane or their associated vent communities. Because there is a continuous air-sea flux of gases between the atmosphere and ocean, there is typically plenty of dissolved oxygen in surface waters to support marine life. Below the surface in the ocean's interior, however, the oxygen level within a particular water mass depends on when the water was last at the surface, how much oxygen has been used up during remineralization of detritus (dead organic matter) as the water mass has moved along its path, and how much it has mixed with other water masses during its journey. Hence, the dissolved oxygen level at a given location depends on both ventilation age (time elapsed since the last contact of a water mass with the atmosphere) and oxygen utilization. Temperature is also important, as cold water can hold more dissolved oxygen than hot water.

As has long been understood, the combination of significant utilization and weak ventilation leads to a mid-depth *oxygen minimum* (Sverdrup 1938; Wyrtki 1962; Fig. 1.7). Shallower waters tend to have higher dissolved oxygen concentrations because they are better ventilated, despite higher rates of utilization, while deeper waters (below about 2,000 m depth) tend to contain more oxygen because of lower utilization rates and higher initial oxygen levels when they leave the surface, as they are derived from cold, high-latitude source waters in e.g., the Greenland and Norwegian Seas or around Antarctica, despite also having longer ventilation ages.

In most oceanic regions, oxygen-minimum layers are typically located at depths of between 400 and 1200 m near the base of the permanent thermocline (Keeling et al. 2010). In some regions however, such as the Northern Indian Ocean and most of the Pacific Ocean, the oxygen minimum is associated with critically low, potentially lethal oxygen levels. While these layers are typically located well below

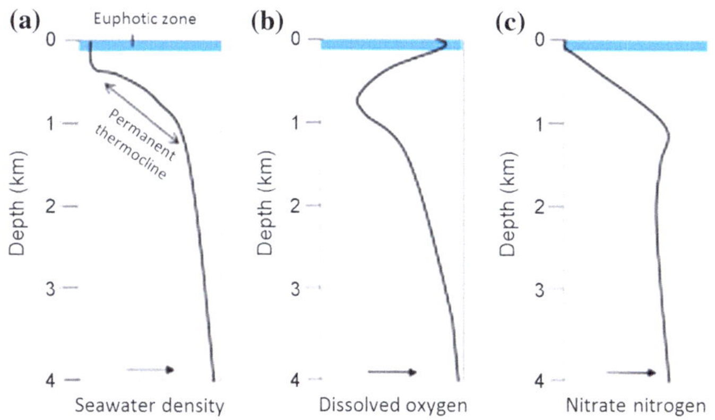

Fig. 1.7 Typical vertical profiles of **a** seawater density, **b** dissolved oxygen, and **c** nitrate nitrogen in the central Atlantic Ocean. Adapted from Segar (2007)

the euphotic zone in most regions of the oceans, they can come close to the sea surface (depths of 150–300 m) in the vicinity of the major coastal upwelling regions of the tropical eastern Pacific and Atlantic Oceans and may even rise to affect the euphotic zone (see Chap. 4).

The sensitivity of organisms, particularly macro-organisms, to changes in oxygen levels is variable. Most organisms are not very sensitive to oxygen levels as long as the concentrations remain above a certain threshold, but once the oxygen concentration falls below this threshold, the organism suffers from a variety of stresses, leading ultimately to death if the concentration stays too low for too long. Such low oxygen conditions are termed *hypoxic* or *anoxic* depending on whether the oxygen concentration is merely low or totally absent, and thresholds for hypoxia vary greatly between marine taxa, with fish and crustaceans tending to be the most sensitive (Fig. 1.8).

A typical threshold for hypoxia is ~60 μM/kg dissolved oxygen (Gray et al. 2002), which is equivalent to ~1.4 mL/L or 2 mg/L. Hypoxic zones are often called *dead zones* as the low oxygen concentrations can be lethal for many marine animals, particularly benthic or burrowing organisms that cannot move rapidly, and the area of coastal waters that is affected by hypoxia is increasing because of the runoff of nutrients from farming and other anthropogenic inputs (Breitburg et al. 2009; Diaz and Rosenberg 2008). The added nutrients fuel increased phytoplankton production, and when the phytoplankton die, their cells sink to the bottom and

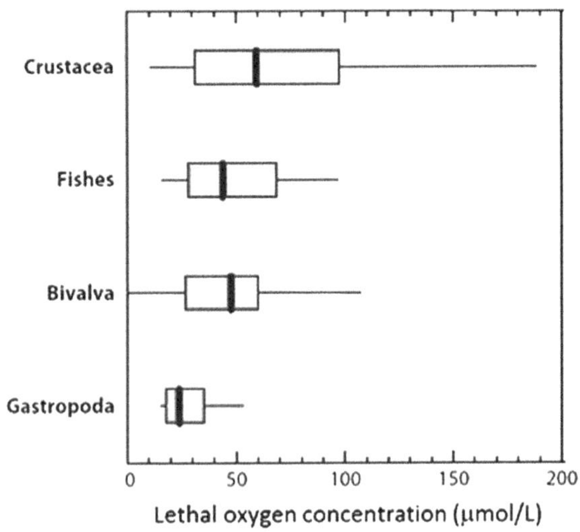

Fig. 1.8 Median lethal oxygen concentration for four different taxa. Boxes run from the lower (25 %) to the upper (75 %) quartile and also include the median. Redrawn after Vaquer-Sunyer and Duarte (2008). *Crustacea* form a very large group of arthropods which includes familiar animals such as crabs, lobsters, crayfish, shrimp, krill and barnacles. The species range in size from *Stygotantulus stocki* at 0.1 mm to the Japanese spider crab with a leg span of up to 3.8 m. *Bivalva* is a class of marine molluscs including clams, oysters, cockles, mussels and scallops. *Gastropoda* (gastropods) include snails and slugs of a large range of sizes

decompose, using up dissolved oxygen. This is particularly important in stratified water bodies, where there is a sudden change in density, as the resulting pycnocline prevents oxygen from mixing downwards from the upper, oxygen-rich layer into the bottom layer where decomposition is occurring. Hypoxic conditions, which may be permanent or temporary, depending on local mixing conditions, are typical of coastal upwelling regions (e.g, Monteiro et al. 2006).

1.5.4 Nutrients

A suite of chemicals, typically identified as *nutrients*, is required for phytoplankton production in the ocean. Broadly important nutrients include nitrogen (N), phosphorus (P), silicon (Si) and iron (Fe). All phytoplankton taxa have relatively uniform requirements for N and P, but only diatoms and radiolarian protozoa require silica. Plankton build their biomass with C:N:P ratios (C is carbon) of ~ 106:16:1, known as the *Redfield ratio* (Redfield 1958). Due to their high abundance in seawater, carbon and calcium, which is needed by the many organisms (including corals and crustaceans) that create calcium carbonate shells, are typically not listed among the nutrient elements.

Nutrient concentrations are generally low in the euphotic zone, because of their uptake during photosynthesis, high rates of biological utilization and gravitational settling of detritus. The remineralization of detritus as it sinks to greater depths brings nutrients back into solution (and uses up dissolved oxygen in the process). Hence, at some depth below the euphotic zone, usually around 1,000–1,500 m, the ocean is generally rich in dissolved nutrients and a nutrient maximum is found (Fig. 1.9b). This nutrient maximum corresponds to the midwater oxygen minimum (Fig. 1.9a). The transition zone between the base of the surface layer where nutrient concentrations are low and the depth of the nutrient maximum is called the *nutricline*. Similar changes in oxygen and nutrient concentrations occur as ocean water circulates between the different ocean basins. Thus as deep water moves from its formation region in the North Atlantic via the South Atlantic to the North Pacific Ocean, the dissolved oxygen concentrations slowly decrease and the concentrations of dissolved macro nutrients (N, P and Si) and carbon dioxide increase, so that the deep North Pacific has the highest overall dissolved nutrient concentrations (Fig. 1.9b).

Only a small fraction (<0.1 %) of sinking particulate matter reaches the seafloor in the open ocean (Martin et al. 1987). Hence, a large fraction of nutrients is recycled in the ocean interior. The physical process of upwelling (i.e., the upward movement of water parcels) plays a fundamental role in marine ecosystems as it lifts nutrient-enriched deeper water into the euphotic zone stimulating photosynthesis.

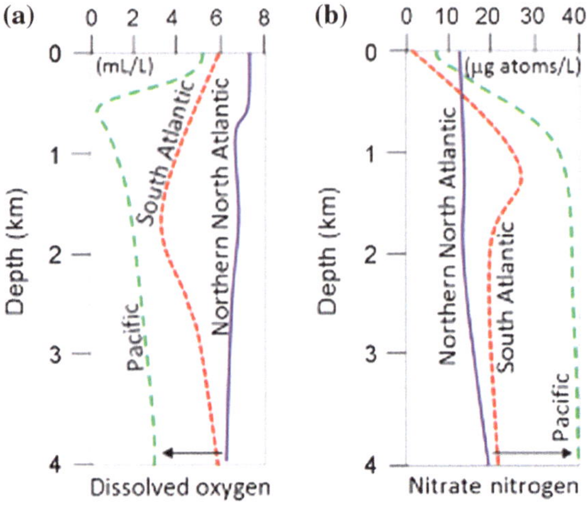

Fig. 1.9 Typical vertical distributions of dissolved oxygen and nitrate nitrogen (as an example of a key nutrient) in different oceanic regions. Adapted from Segar (2007)

1.5.5 Nutrient Limitation

The limitation of phytoplankton growth has traditionally been interpreted in the context of Liebig's *Law of the Minimum* (Sprengel 1828; Liebig 1855), which states that plant growth will be as great as allowed by the least available resource, the *limiting nutrient* that sets the productivity of the system (de Baar 1994). In the ocean, nitrogen, in the form of nitrate, is usually considered to limit production. This is in contrast to freshwater systems, where phosphorous is usually considered to be the limiting nutrient. However, recent studies in the eutrophic waters off the Louisiana shelf have shown that phosphorous can be limiting here at certain times (Sylvan et al. 2006; Quigg et al. 2011.)

In the 1930s the English biologist, Joseph Hart, speculated that the ocean's great "desolate zones" (areas apparently rich in nutrients, but lacking in plankton activity or other sea life) might simply be iron deficient (Weier 2001). Little further scientific discussion of this issue was recorded until the 1980s, when oceanographer John Martin renewed the controversy on the topic with his nutrient analyses of seawater. His studies indicated it was indeed a scarcity of the micronutrient iron that was limiting phytoplankton growth and overall productivity in these "desolate" regions, which came to be called "High Nutrient, Low Chlorophyll" (HNLC) zones (Martin and Fitzwater 1988; Boyd et al. 2007). These represent about 40–50 % of the areal extent of the world's oceans (Moore et al. 2002).

Iron limitation has been identified for the upwelling regions of the Humboldt Current (Hutchins et al. 2002); and the California Current (Hutchins et al. 1998; Chase et al. 2007). The role of external nutrient input, particularly iron, via atmospheric dust plumes became apparent from observations in the Canary Current upwelling system (Neuer et al. 2004). In upwelling regions, the continual importation of deeper waters, either along a "line feature" in areas where there is a belt of upwelling (such as Oregon), or more usually at upwelling centres, as found for example off California or in the Benguela region, means that this limitation is generally removed, and indeed, silica has been suggested as limiting production in the equatorial Pacific and off Peru (Dugdale et al. 1995). Silicon availability can also be a major limiting factor in polar regions (e.g., Nelson and Tréguer 1992).

While this view of nutrient limitation is powerful, interactions among nutrients and between nutrients and light can also control productivity. A simple but important example of this potential for "co-limitation" comes from polar regions, where oblique solar insolation combines with deep mixing of surface waters to yield low light levels. In such environments, higher iron supply can increase the efficiency with which phytoplankton capture light energy (Sunda and Huntsman 1997; Maldonado et al. 1999). More broadly, it has been argued that phytoplankton generally reside in a state of co-limitation by all the chemicals they require, including the many trace metal nutrients (Morel 2008).

1.5.6 Mechanisms Limiting Phytoplankton Blooms

As stated above, there are three key factors controlling primary production in the surface ocean. The first factor is the maximum light intensity and, hence, euphotic depth which underlies seasonal changes. Light intensity can be dramatically reduced in coastal regions by continental influences (e.g., sediment resuspension or sediment inputs from rivers).

The second factor is the depth of the surface mixed layer. Phytoplankton and other organic matter are vertically stirred throughout the surface mixed layer. Hence, the depth of surface mixed layer influences the relative time of growth that such organisms have when moving through the euphotic zone. The overall depth of the surface mixed layer depends on surface heat and freshwater fluxes and the magnitude of the wind stress, which are seasonally variable. The mixed-layer depth can vary on time scales from minutes to weeks under the influence of storm-induced mixing, internal waves and upwelling processes.

The third factor is the concentration level of nutrients within the euphotic zone, which depends on both physical processes (i.e., nutrient supply from mixed-layer deepening or upwelling) and biological processes (i.e., nutrient consumption for primary production). Nutrients become rapidly exhausted in the surface mixed layer via consumption unless there is an external nutrient source. This includes the upwelling process, which reduces the depth of the surface mixed layer and lifts elevated nutrient levels closer to the sea surface, and mixed-layer deepening from storms or thermohaline convection which entrains nutrient-enriched sub-pycnocline water into the surface mixed layer. Hence, light intensity, nutrient distributions and mixed layer depth all operate together to control primary production in the surface ocean. Dramatic reduction in oxygen levels during excessive algal growth can be

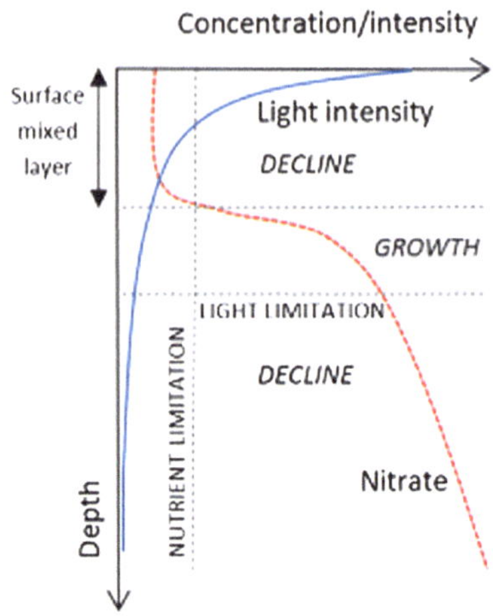

Fig. 1.10 Schematic of the situation in which light and nutrient limitations created a zone of sub-surface phytoplankton production confined to the base of the surface mixed layer

another controlling factor. Figure 1.10 shows the creation of a zone of sub-surface phytoplankton production which is confined by too low nutrient concentrations above and too low light intensity below.

1.5.7 Nutrient Regeneration

Benthic nutrient regeneration from ammonia was first explored and described by Dugdale and Goering (1967) and later by Eppley (1992) for the California Current upwelling system. This led to the introduction of the *f* ratio, the ratio between new and regenerated production, by Eppley and Peterson (1979). The *f* ratio plays a key role in the characterization of upwelling systems.

Nutrient regeneration plays a significant role in upwelling regions. It takes place primarily through two processes, bacterial regeneration at the sediment-water interface and in the water column, and by grazing activities of herbivores (Fig. 1.11 illustrates the regeneration process). The supply of nitrogen as dissolved nitrate allows us to differentiate between "new" production fueled by nitrate and "regenerated" production fueled by recycled ammonium and urea. In the open ocean, the *f*-ratio is generally about 0.1. In coastal upwelling regions, however, it can be as high as 0.8 (Laws 2004).

The increased productivity of upwelling regions results directly from the continuing availability of upwelled nitrate for new production, in contrast to other coastal regions or the open ocean that rely on much smaller quantities of recycled nitrogen. The fractions of regeneration attributable to herbivores and to bacterial action vary from region to region (Dugdale 1972). Early analyses of Dugdale and Goering (1970) show that regeneration of nitrogen and phosphorous by the anchoveta populations in the Peru upwelling system take place at such high rates

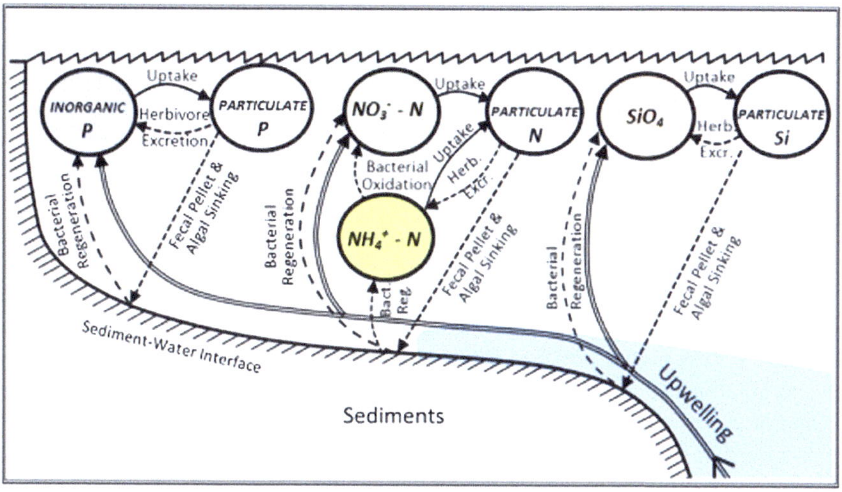

Fig. 1.11 The approximate pathways of phosphorous, nitrogen, and silica circulation, and biological uptake and regeneration in an upwelling region. Redrawn after Dugdale (1972)

that the anchoveta must be the dominant regenerators there. Direct silica regeneration was found to take place through anchoveta grazing activities at 10–20 % of the rate for nitrogen.

1.6 The Carbon Cycle and Oceanic Carbon Pumps

1.6.1 Overview

The initial source of carbon on Earth is outgassing of CO_2, stored in the mantle when the Earth was formed, from the Earth's interior at mid-ocean ridges or hotspot volcanoes. A second source is found at subduction-related volcanic arcs, and most CO_2 released at these subduction zones is derived from the metamorphism of sedimentary carbonate rocks subducting with the ocean crust. On geological timescales (millions of years), carbon is released into the atmosphere and ocean through the weathering of carbonate rocks such as limestone and via volcanic emissions. It returns as new rocks formed through sediment deposition.

On the much shorter timescale <100 years, carbon is exchanged between the atmosphere, the ocean and living and dead organisms, and air-sea gas exchange is the major process controlling carbon-dioxide fluxes across the sea surface. From the start of the industrial revolution in the mid-18th century, the atmospheric carbon budget has been substantially disturbed through human activities, such as fossil fuel combustion and cement manufacture, so that the pre-industrial atmospheric CO_2 concentration of about 270 ppm now exceeds 403 ppm and is continuing to increase.

Roughly 50 % of the CO_2 produced by human activities is taken up by the ocean, the remainder staying in the atmosphere where it contributes to global warming.

Atmospheric CO_2 enters the ocean via air-sea gas transfer. This transfer is a function of a transfer coefficient, called *piston velocity*, and the difference in partial gas pressures across the sea surface. Under the assumption that the thin surface skin of the ocean is fully saturated with a gas and applying Henry's and Fick's laws, the air-sea gas flux can be formulated as:

$$F = u\left(k_H P_{gas} - C_{ml}\right), \tag{1.3}$$

where u is the transfer coefficient, which is strongly controlled by wind speed, k_H is solubility, P_{gas} is partial pressure of the gas in the atmosphere, and C_{ml} is the gas concentration in the surface mixed layer of the ocean. Equation (1.3) is known as the *film model of gas exchange*. The direction of the gas flux depends on whether the dissolved gas in the mixed layer is under-saturated ($F > 0$), when CO_2 will enter the ocean from the atmosphere, or oversaturated ($F < 0$), when the gas will move in the opposite direction. As for oxygen, the solubility of carbon dioxide decreases with increasing sea temperature. Its saturation concentration value more than doubles as temperatures decrease from 24 °C (tropical regions) to 0 °C (polar regions). Hence, fully-saturated cold water can hold more CO_2 than warm water and the deep oceans have higher concentrations of CO_2 than the surface layers.

Within the ocean, the general carbon cycle is a complex process driven by both biogeochemistry and physics (Fig. 1.12). The oceanic carbon cycle can be described by different *carbon pumps*, each describing specific mechanisms that transfer carbon dioxide from the upper to the deep ocean or vice versa. These pumps are the *solubility pump* and the *biological pump*. The solubility pump is responsible for about 20 % of the vertical gradient in dissolved inorganic carbon in the ocean, while the remaining 80 % originates from the biological pump (Sarmiento et al. 1995). The solubility pump operates as a combination of:

(a) the temperature dependency of the solubility of carbon dioxide in seawater; i.e. under the same atmospheric conditions, cold water can dissolve more CO_2 than warm water before it reaches an equilibrium with the atmosphere, and

(b) oceanic flows that either export surface water to the ocean interior (called *oceanic subduction*) or bring deeper water back to the sea surface (upwelling).

One branch of the solubility pump, for example, is the deep circulation of the oceans driven by open-ocean convection in sub-polar regions of the North Atlantic Ocean. The other branch is the reverse process of upwelling in which CO_2 enriched deeper water is returned to the sea surface.

The *biologic pump* (responsible for 80 % of total carbon fixation in the ocean) describes vertical carbon transfers in the ocean associated with biochemical processes. The biologic pump comprises:

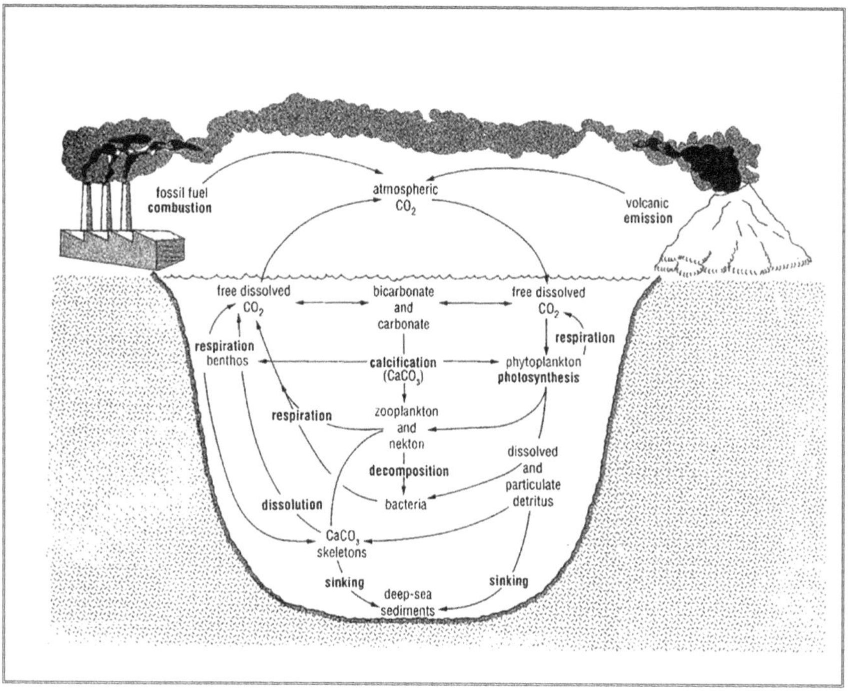

Fig. 1.12 The basic oceanic carbon cycle. Adapted from Lalli and Parsons (1993)

(a) the *organic carbon pump*, associated with primary production in the euphotic zone and remineralization of detritus at depths, and

(b) the *calcium carbonate counter pump*, associated with skeleton and shell formation in the surface ocean and the dissolution of calcareous particles at depth.

The biological pump starts with the conversion of inorganic carbon to organic forms. Some of the phytoplankton are remineralized when they die, but the major fraction is consumed by zooplankton and nekton, some of which also take up carbon dioxide directly to form calcium carbonate shells. Zooplankton faecal material and dead phytoplankton cells sink, transferring carbon into the deeper ocean, and remineralization continues throughout the water column. While a small portion (<0.1 %) of the carbon can eventually be preserved in ocean sediments, most is remineralized into carbon dioxide below 300 m depth, after which upwelling and the general circulation eventually brings it back to the surface layer as bicarbonate or carbonate ions, from where some returns to the atmosphere as carbon dioxide gas. Although only a small percentage in terms of the total mass of carbon at a given instant, over geological time the preservation of the carbonate skeletons of marine organisms is an extremely important component of the global carbon cycle, with about 1,000 times as much carbon sequestered in limestone or organic marine sediments as exists as free CO_2, bicarbonate or carbonate ions (Lalli and Parsons 1993).

When carbon dioxide from the atmosphere reacts with seawater (H_2O), it immediately forms a weak acid, carbonic acid (H_2CO_3), which in itself is chemically unstable. This acid further dissociates to form bicarbonate HCO_3^- (a base) and hydrogen ions H^+ (an acid):

$$CO_2 + H_2O \rightarrow H_2CO_3 \rightarrow HCO_3^- + H^+ \tag{1.4}$$

Excess hydrogen ions (H^+) react with carbonate ions (CO_3^{2-}) (seawater is naturally saturated with this base) to form further bicarbonate ions:

$$H^+ + CO_3^{2-} \rightarrow 2HCO_3^- \tag{1.5}$$

The acidity of the oceans is determined by the concentration of hydrogen ions; a greater amount results in more acidic conditions, represented by a lower pH.

The carbonate ions (CO_3^{2-}) and bicarbonate ions (HCO_3^-) can react with calcium ions, which are in excess in seawater, to form calcium carbonate ($CaCO_3$) which underpins skeleton and shell formation (also known as *calcification*) in marine organisms such as corals, shellfish and marine plankton (Feely et al. 2008). The main chemical reactions for the mineral formation and the dissolution of calcium carbonate ($CaCO_3$) are as follows:

$$CaCO_3 \leftrightarrow Ca^{2+} + CO_3^{2-} \tag{1.6}$$

$$CaCO_3 + H_2O + CO_2 \leftrightarrow Ca^{2+} + 2HCO_3^- \tag{1.7}$$

Calcium carbonate is formed as the reaction proceeds from right to left, and dissolved from left to right. In contrast to the organic carbon pump, the calcification process releases CO_2 back into the ambient seawater and dissolution of calcareous particles at depth takes up dissolved CO_2. The reaction of CO_2 with seawater to form bicarbonate and carbonate ions means that the resultant increase in gaseous seawater CO_2 concentration is smaller than the actual amount of carbon dioxide entering the seawater. This chemical reaction together with the gravitational export of detritus from the euphotic zone supports a continuous air-sea gas transfer of CO_2 into the ocean and is quantitatively the most important oceanic process contributing to the ocean as an overall carbon sink (Feely et al. 2008). The buffering capacity of seawater also implies that seawater maintains a slightly basic pH state within relatively narrow limits, despite the uptake of atmospheric CO_2, although this appears to be changing towards less basic conditions.

The calcification process depends critically on the availability of two specific carbonate minerals, aragonite and calcite. Aragonite is used by pteropods to construct their shells, while calcite is used by coccoliths and foraminifera. When seawater is supersaturated with these minerals, as is the case in all ocean surface waters at present, the formation of shells and skeletons will be favoured. Conversely, when seawater is under-saturated with respect to these minerals, the seawater becomes corrosive and the shells of calcifying organisms are increasingly prone to dissolution (Feely et al. 1988).

$CaCO_3$ becomes more soluble with decreasing temperature and increasing pressure and therefore ocean depth, creating a natural boundary known as the *saturation horizon* above which $CaCO_3$ can form, but below which it readily dissolves. The saturation horizon for aragonite varies spatially between 200 and 1500 m, while that for calcite varies between about 750–4300 m. The shallowest saturation depths for both chemical species are found in the northern, eastern and equatorial Pacific Ocean, in the northwestern Indian Ocean, and in the tropical eastern Atlantic Ocean. Overall, the aragonite saturation horizons in the North Pacific Ocean are much shallower than those in the North Atlantic Ocean. It is well known that deep waters of the North Pacific Ocean are the oldest in the world oceans. This means that the concentration of carbonate ions in the North Pacific Ocean is lowest here and this has dramatic consequences for the carbonate saturation states in this region.

As CO_2-driven ocean acidity increases (known as *ocean acidification*), carbonate ions are removed from the system in the carbonate buffering process, and the saturation horizon shallows towards surface waters. This shoaling of the saturation horizon reduces the habitat available for calcifying organisms reliant on the carbonate minerals and has implications for ecosystem productivity. On millennial timescales, the shoaling of these saturation horizons and subsequent dissolution of sedimentary carbonates is one of the major long-term buffering mechanisms by which the ocean's pH can be restored (Montenegro et al. 2007). The appearance of undersaturated, corrosive seawater on the continental shelf has been observed in the upwelling regions of California (Feely et al. 2008; Fig. 1.13) and is also indicated for the Peruvian-Chilean upwelling system (see Chap. 5).

As a consequence of the biological pump and the ability of cold water to dissolve more CO_2 than warmer water, the ocean contains about 60 times as much inorganic carbon as the atmosphere. The biological pump is an important control on atmospheric CO_2 concentrations. If it were not active, then the atmospheric CO_2 concentration would be about 550 ppm instead of its current 400 ppm. Conversely, if all the nutrients could be used to produce organic carbon, then the atmospheric carbon dioxide concentration would be only about 140 ppm.

Although carbon is needed for primary production, it is never limiting in the ocean, unlike nitrogen or phosphorous. Since remineralization and respiration take place at a much deeper level in the ocean than photosynthesis, there is a general transfer of carbon and nutrients from the surface to the deep ocean. This transfer will continue, in theory, until all the nutrients in the euphotic zone are used up. However, the continuous input of nutrients by rivers (and to a minor extent from the atmosphere) keeps replenishing the surface nutrient pool near the coasts, while the large-scale circulation of the ocean causes deep water to return to the surface layer, bringing with it high concentrations of nutrients and carbon. Both processes ensure that production in the euphotic zone can continue. Exceptions are interferences from waters that are either hypoxic and/or undersaturated with carbonate minerals.

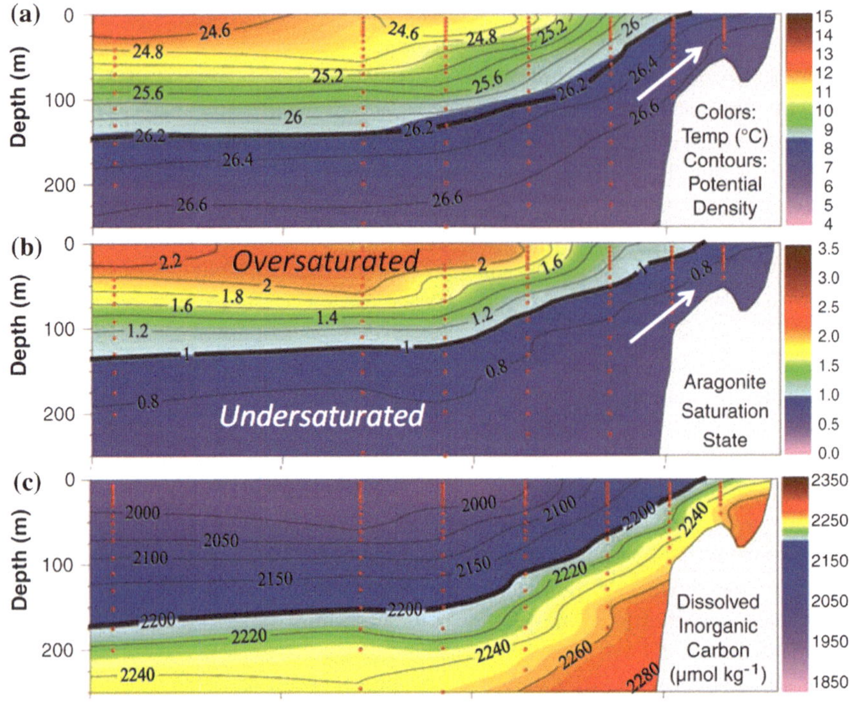

Fig. 1.13 Observations of **a** temperature/seawater density, **b** aragonite saturation state, and **c** dissolved inorganic carbon concentration during coastal upwelling on the Californian shelf. Taken from Feely et al. (2008)

1.6.2 The Role of Upwelling in the Carbon Cycle

Coastal upwelling systems are the "powerhouse" of phytoplankton production and marine productivity in the ocean. As a result, they play a disproportionately important role in the microbially mediated cycling of marine nutrients. These systems are characterized by strong natural variations in carbon dioxide concentrations, pH, nutrient levels and sea surface temperatures on both seasonal and interannual timescales (Capone and Hutchins 2013).

Upwelling systems can facilitate an efflux (outgassing) of CO_2 into the atmosphere when atmospheric heating reduces the solubility of CO_2. On the other hand, the enhanced primary production and export of particulate organic carbon (POC) to the seafloor implies that some of the upwelled CO_2 is retained and recycled within the ocean (Jiao et al. 2014). Hence, the biological pump plays an important role in the carbon cycle and air-sea gas fluxes in coastal upwelling regions. As stated above, in some upwelling regions upwelling events can induce the inflow of either hypoxic water and/or water that is undersaturated in carbonate minerals (particularly aragonite) onto the continental shelf. Global warming is predicted to lead to a shoaling (upward progression) of both oxygen-minimum zones (Keeling et al. 2010) and saturation horizons of carbonate minerals (aragonite) (Feely et al. 2012). As a result, substantial ecosystem modifications are to be expected in future as we continue to produce CO_2 from burning fossil fuels, in particular for the upwelling regions in the eastern tropical and subtropical Pacific Ocean.

1.7 Early Scientific Expeditions

Advancement of knowledge of coastal upwelling dynamics would not have been possible without extensive field surveys. Of the early global encircling voyages starting with the British *Challenger* expedition (1872–76), it was the *Discovery* and *Meteor* Expeditions (Fig. 1.14) in the South Atlantic that first considered the Benguela upwelling system (Wattenberg 1938; Hart and Currie 1960), while the second Danish *Galathea* expedition (1950–52) discovered enhanced biologic production in equatorial upwelling regions. This was first described by Steemann Nielsen (1952). Many other expeditions followed such as JOINT-I, which was the first major expedition of the Coastal Upwelling Ecosystems Analysis (CUEA) program, a project of the International Decade of Ocean Exploration (IDOE) office of the U.S. National Science Foundation. JOINT-I took place off the coasts of Mauretania and Spanish Sahara (now properly called Rio de Oro) from February to May 1974 and involved three U.S. research vessels, an aircraft from the National Center for Atmospheric Research, a Mauretanian research vessel, and ten shore-based meteorological stations along the Cape Blanc peninsula. First research findings were discussed in a special issue (Vol. 24, 1977) of Deep-Sea Research.

The first International Indian Ocean Expedition (IIOE) during years 1962–1965 was one of the greatest international, interdisciplinary oceanographic research efforts to explore the Indian Ocean. Forty oceanographic research vessels belonging to 13 countries surveyed the Indian Ocean and collected useful data in almost all disciplines in the marine sciences, leading to the publication of the first to the publication of the first oceanographic atlas of the region (Wyrtki et al. 1971; Wyrtki 1973) and many other legacies such as the first comprehensive description of the Somali Current coastal upwelling system (see Chap. 8). Data from IIOE still underpin current scientific research half a century later, such as research on upwelling in the Arafura Sea (see Sect. 8.2.3). These early expeditions set the foundation for our understanding of upwelling systems. Indeed, the Joint Global Ocean Flux Study (GOFS), beginning in 1984 in the U.S., was pivotal to the understanding of processes involved in the biological pump (Ducklow et al. 2001).

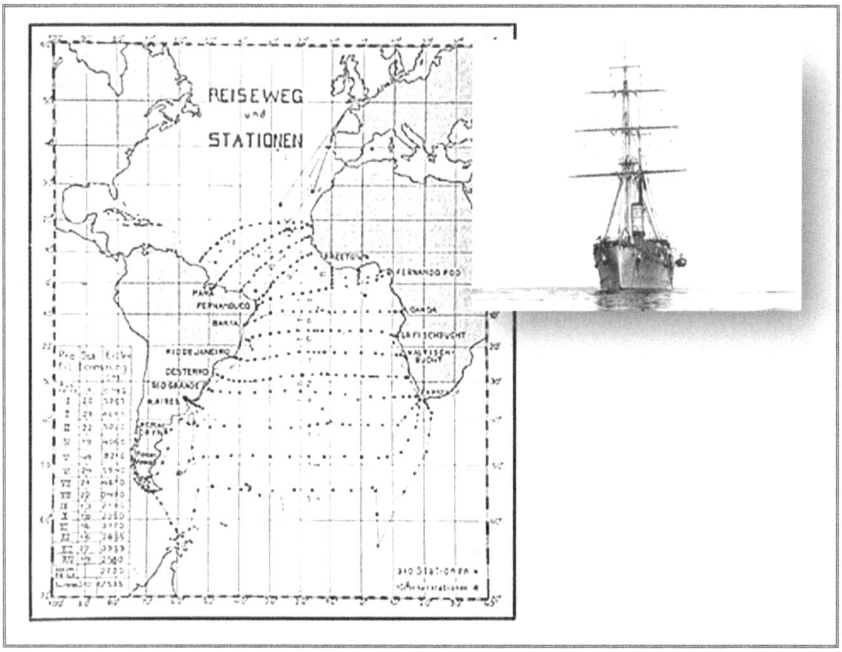

Fig. 1.14 Cruise plan, station map and photography of the *Meteor* expedition in the South Atlantic (1922–27) (from Spiess et al. 1932)

1.8 Long-Term Scientific Monitoring Programs

Worldwide there are only a few systematic long-term scientific monitoring programs that have measured oceanic, and in particular biological parameters, on periods >20 years. These programs include the California Cooperative Oceanic Fisheries Investigations (CalCOFI) initiated in 1941 (see Chap. 4), and the Continuous Plankton Recorder survey established in 1931. These were followed in the 1980s and 90s by establishment of time series programs such as the Hawaii Ocean Time-series (HOT), the Bermuda Atlantic Time-series Study (BATS), and the marine Long-Term Ecological Research (LTER) sites, although these are not all based in upwelling regions. The breakthrough in upwelling science was the availability of satellite data from the mid-1980s and, more recently, the ARGO float program and other modern technology such as oceanic gliders. Despite these monitoring programs and many snapshot field surveys, there is still a large gap in the understanding of most of the upwelling systems described in this book.

1.9 Summary

This chapter described fundamental properties and processes associated with the creation of marine life. The most important constraint is gravity which leads to sinking of particulate organic matter out of the euphotic zone. An effective recycling of nutrients is only possible due to relatively rapid decomposition of detritus by marine bacteria in the upper ocean and physical processes that lift (upwell) or entrain nutrient-enriched subsurface water back into the euphotic zone. These physical upwelling processes are detailed in the next chapter.

References

Azam, F., T. Fenchel, J.G. Field, J.S. Gray, L.A. Meyer-Reil, and F. Thingstad. 1983. The ecological role of water-column microbes in the sea. *Marine Ecology Progress Series* 10: 257–263.

Bender, M., K. Grande, K. Johnson, J. Marra, P.J.L. Williams, J. Sieburth, M. Pilson, C. Langdon, G. Hitchcock, J. Orchardo, C. Hunt, P. Donaghay, and K. Heinemann. 1987. A comparison of four methods for determining planktonic community production. *Limnology and Oceanography* 32: 1085–1098.

Boyd, P.W., T.D. Jickells, C.S. Law, S. Blain, E. Boyke, et al. 2007. Mesoscale iron enrichment experiments 1991–2005: synthesis and future directions. *Science* 315: 612–617.

Breitburg, D.L., D.W. Hondorp, L.A. Davias, and R.J. Diaz. 2009. Hypoxia, nitrogen, and fisheries: integrating effects across local and global landscapes. *Annual Review of Marine Sciences* 1: 329–349.

Cairns-Smith, A. 1982. *Genetic takeover and the mineral origins of life*. New York: Cambridge University Press.

Capone, D. G., and D.A. Hutchins. 2013. Microbial biogeochemistry of coastal upwelling regimes in a changing ocean. *Nature Geoscience*, 6, 711–717. doi:10.1038/ngeo1916.

Carr, S.D., X.J. Capet, J.C. McWilliams, J.T. Pennington, and F.P. Chavez. 2007. The influence of diel vertical migration on zooplankton transport and recruitment in an upwelling region: estimates from a coupled behavioral-physical model. *Fisheries Oceanography* 17: 1–15.

Chase, Z., P.G. Strutton, and B. Hales. 2007. Iron links river runoff and shelf width to phytoplankton biomass along the US West Coast. *Geophysical Research Letters* 34: L04607. doi:10.1029/2006GL028069.

De Baar, H.J.W. 1994. Von Liebig's law of the minimum and plankton ecology (1899–1991). *Progress in Oceanography* 33: 347–386.

Diaz, R.J., and R. Rosenberg. 2008. Spreading dead zones and consequences for marine ecosystems. *Science* 321: 926–929.

Ducklow, H.W., D.K. Steinberg, and K.O. Buesseler. 2001. Upper-ocean export and the biological pump. *Oceanography* 14: 50–58.

Dugdale, R.C. 1972. Chemical oceanography and primary productivity in upwelling regions. *Geoforum* 11: 47–61.

Dugdale, R.C., and J. Goering. 1967. Uptake of new and regenerated forms of nitrogen in primary productivity. *Limnology and Oceanography* 12: 196–206.

Dugdale, R. C., and J. J. Goering. 1970. Nutrient limitation and the path of nitrogen in Peru Current production. *Anton Bruun Report No.* 4, Texas A & M Press, p. 5.3–5.8.

Dugdale, R.C., F.P. Wilkerson, and H.J. Minas. 1995. The role of a silicate pump in driving new production. *Deep-Sea Research I* 42: 697–719.

Eppley, R.W. 1992. Chlorophyll, photosynthesis and new production in the Southern California Bight. *Progress in Oceanography* 30: 117–150.
Eppley, R.W., and B.J. Peterson. 1979. Particulate organic flux and planktonic new production in the deep ocean. *Nature* 282: 677–680.
Fallah-Araghi, A., K. Meguellati, J.-C. Baret, A. El Harrak, T. Mangeat, M. Karplus, S. Ladame, C.M. Marques, and A.D. Griffiths. 2014. Enhanced chemical synthesis at soft interfaces: a universal reaction-adsorption mechanism in microcompartments. *Physical Review Letters* 112: 028301. doi:10.1103/PhysRevLett112.028301.
Feely, R.A., R.H. Byrne, J.G. Acker, P.R. Betzer, C.-T.A. Chen, J.F. Gendron, and M.F. Lamb. 1988. Winter-summer variations of calcite and aragonite saturation in the north-east Pacific. *Marine Chemistry* 25: 227–241.
Feely, R.A., C.L. Sabine, J.M. Hernandez-Ayon, D. Lanson, and B. Hales. 2008. Evidence for upwelling of corrosive "acidified" water onto the continental shelf. *Science* 320: 1490–1492.
Feely, R.A., C.L. Sabine, R.H. Byrne, F.J. Millero, A.G. Dickson, R. Wanninkhof, A. Murata, L. A. Miller, and D. Greeley. 2012. Decadal changes in the aragonite and calcite saturation state of the Pacific Ocean. *Global Biogeochemical Cycles*, 26, GB3001. doi:10.1029/2011GB004157.
Fenchel, T. 2008. The microbial loop—25 years later. *Journal of Experimental Marine Biology and Ecology* 366: 99–103. doi:10.1016/j.jembe.2008.07.013.
Gran, H.H., and B. Trygve. 1935. A quantitative study of the phytoplankton in the Bay of Fundy and the Gulf of Maine (including observations on hydrography, chemistry and turbidity). *Journal of the Biological Board of Canada* 1(5): 279–467. doi:10.1139/f35-012.
Gray, J.S., R.S.S. Wu, and Y.Y. Or. 2002. Effects of hypoxia and organic enrichment on the coastal marine environment. *Marine Ecology Progress Series* 238: 249–279.
Hart, T.J., and R.I. Currie. 1960. The Benguela Current. *Discovery Reports* 31: 123–298.
Hutchings, L. 1992. Fish harvesting in a variable, productive environment-searching for rules or searching for exceptions? *South African Journal of Marine Science* 12: 297–318.
Hutchins, D.A., G.R. DiTullio, Y. Zhang, and K.W. Bruland. 1998. An iron limitation mosaic in the California upwelling regime. *Limnology and Oceanography* 43: 1037–1054.
Hutchins, D.A., C.E. Hare, R.S. Weaver, Y. Zhang, G.F. Firme, G.R. DiTullio, M.B. Alm, S.F. Riseman, J.M. Maucher, M.E. Geesey, C.G. Trick, G.J. Smith, E.L. Rue, J. Conn, and K.W. Bruland. 2002. Phytoplankton iron limitation in the Humboldt Current and Peru. *Limnology and Oceanography* 47: 997–1011.
Jiao, N., Y. Zhang, K. Zhou, Q. Li, M. Dai, J. Liu, J. Guo, and B. Huang. 2014. Revisiting the CO_2 "source" problem in ocean upwelling areas—acomparative study on eddy upwellings in the South China Sea. *Biogeosciences* 11: 2465–2475. doi:10.5194/bg-11-2465-2014.
Keeling, R.F., A. Körtzinger, and N. Gruber. 2010. Ocean deoxygenation in a warming world. *Annual Review of Marine Science* 2: 199–229.
Kirchman, D., H. Ducklow, and R. Mitchell. 1982. Estimates of bacterial growth from changes in uptake rates and biomass. *Applied Environmental Microbiology* 44: 1296–1307.
Lalli, C.M., and T.R. Parsons. 1993. *Biological oceanography: An introduction*. Oxford: Pergamon Press. 301 pp.
Lampert, W. 1989. The adaptive significance of diel vertical migration of zooplankton. *Functional Ecology* 3: 21–27.
Laws, E.A. 2004. New production in the equatorial Pacific: A comparison of field data with estimates derived from empirical and theoretical models. *Deep-Sea Research I* 51: 205–211.
Lewis, J.B. 1954. The occurrence and vertical distribution of the Euphausiacea of the Florida Current. *Bulletin of Marine Science of the Gulf and Caribbean* 4: 265–301.
Liebig, J.V. 1855. *Die Grundsätze der Agricultur-Chemie mit Rücksicht auf die in England angestellten Untersuchungen*. Braunschweig, Germany: Friedrich Vieweg and Sohn.
Mann, K.H., and J.R.N. Lazier. 1996. *Dynamics of marine ecosystems: Biological-physical interactions in the oceans*. Blackwell Science, 2nd edn, 394 pp.
Maldonado, M., M.C. Carmona, M.J. Uriz, and A. Cruzado. 1999. Decline in Mesozoic reef-building sponges explained by silicon limitation. *Nature* 401: 785–788.

Martin, J.H., and S.E. Fitzwater. 1988. Iron deficiency limits phytoplankton growth in the northeast Pacific subarctic. *Nature* 331: 341–343.
Martin, J.H., G.A. Knauer, D.M. Karl, and W.W. Broenkow. 1987. VERTEX: Carbon cycling in the northeast Pacific. *Deep Sea Research* 34: 267–285.
Miller, S.L. 1953. Production of amino acids under possible primitive earth conditions. *Science* 117: 528–529.
Miller, S.L., and H.C. Urey. 1959. Organic compound synthesis on the primitive earth. *Science* 130: 245–251.
Monteiro, P.M.S., A. Van Der Plas, V. Mohrholz, E., Mabille, A., Pascall and W. Joubert. 2006. Variability of natural hypoxia and methane in a coastal upwelling system: Oceanic physics or shelf biology? *Geophysical Research Letters*, 33(16): L16614. doi:10.1029/2006GL026234.
Montenegro, A., A.V. Brovkin, M. Eby, D. Archer, and A.J. Weaver. 2007. Long term fate of anthropogenic carbon. *Geophysical Research Letters* 34: L19707. doi:10.1029/2007GL030905.
Moore, J.K., S.C. Doney, D.M. Glover, and I.Y. Fung. 2002. Iron cycling and nutrient-limitation patterns in surface waters of the World Ocean. *Deep-Sea Research II* 49: 463–507.
Morel, F.M.M. 2008. The co-evolution of phytoplankton and trace element cycles in the oceans. *Geobiology* 6(3): 318–324. doi:10.1111/j.1472-4669.2008.00144.x.
Nelson, D.M., and P. Tréguer. 1992. On the role of silicon as a limiting nutrient to Antarctic diatoms: evidence from kinetic studies in the Ross Sea ice-edge zone. *Marine Ecology Progress Series* 80: 255–264.
Neuer, S., M.E. Torres-Padron, M.D. Gelado-Caballeo, M.J. Rueda, J. Hernandez-Brito, R. Davenport and G. Wefer. 2004. Dust deposition to the eastern subtropical North Atlantic gyre: Does ocean's biogeochemistry respond? *Global Biogeochemical Cycles*, 18, GB4020, 10. doi:10.1029/2004GB002228.
Quigg, A., J.B. Sylvan, A.B. Gustafson, T.R. Fisher, R.L. Oliver, et al. 2011. Going west: nutrient limitation of primary production in the northern Gulf of Mexico and the importance of the Atchafalaya River. *Aquatic Geochemistry* 17: 519–544.
Pomeroy, L.R. 1974. The oceans food web: a changing paradigm. *BioScience* 24: 409–504.
Redfield, A. 1958. The biological control of chemical factors in the environment. *American Scientist* 46: 205–221.
Sarmiento, J.L., R. Murnane, and C. LeQuere. 1995. Air-sea CO_2 transfer and the carbon budget of the North Atlantic. *Philosophical Transactions of the Royal Society London (B Biological Science)*, 348: 211–219.
Segar, D.A. 2007. *Introduction to ocean sciences*, 2nd ed. New York: W.W Norton & Company. 581 pp.
Sherman, K., and G. Hempel. eds. 2008. *The UNEP large marine ecosystem report: a perspective on changing conditions in LMEs of the world's Regional Seas*. UNEP Regional Seas Report and Studies No. 182. United Nations Environment Programme. Nairobi, Kenya.
Sigman, D.M., and M.P. Hain. 2012. The biological productivity of the ocean. *Nature Education* 3 (6): 1–16.
Sprengel, C. 1828. Von den Substanzen der Ackerkrume und des Untergrundes. *Journal für Technische und Ökonomische Chemie*, 2, 423–474; 3, 42–99, 313–352, and 397–421.
Spiess, F., et al. 1932. Wissenschaftliche Ergebnisse der Deutschen Atlantischen Expedition auf dem Forschungs-und Vermessungsschiff METEOR.
Steemann Nielsen, E. 1952. The use of radioactive carbon (C^{14}) for measuring organic production in the sea. *Journal du Conseil/Conseil Permanent International pour l'Exploration de la Mer* 8: 117–140.
Sunda, W.G., and S.A. Huntsman. 1997. Interrelated influence of iron, light and cell size on marine phytoplankton growth. *Nature* 39: 389–392.
Sverdrup, H.U. 1938. On the exploration of the oxygen minima and maxima in the oceans. *Journal du Conseil-Conseil Permanent International pour l'Exploration de la Mer* 13: 163–172.
Sylvan, J.B., Q. Dortch, D.M. Nelson, A.F. Brown, W. Morrison, et al. 2006. Phosphorus limits phytoplankton growth on the Louisiana shelf during the period of hypoxia formation. *Environmental Science and Technology* 43: 7548–7553.

Vaquer-Sunyer, R., and C.M. Duarte. 2008. Thresholds of hypoxia for marine biodiversity. In *Proceedings of the National Academy of Sciences of the United States of America*, 105: 15, 452–15, 457.
Wattenberg, H. 1938. Die Verteilung des Sauerstoffs im Atlantischen Ozean. *Wissenschaftliche Ergebnisse der Deutschen Atlantischen Expedition der Meteor* 1925–1927(9): 1–132.
Weier, J. (2001) John Martin (1935–1993): On the Shoulders of Giants. http://earthobservatory.nasa.gov/Features/Martin/. Accessed 29 March 2016.
Williams, P.J.E.B. 1981. Incorporation of microheterotrophic processes into the classical paradigm of the planktonic food web. *Kieler Meeresforschung* 5: 1–28.
Wyrtki, K. 1962. The oxygen minimum in relation to ocean circulation. *Deep-Sea Research* 9: 11–23.
Wyrtki, K. 1973. *Physical oceanography of the Indian Ocean: The biology of the Indian Ocean*, ed. B. Zeitzschel, 18–36. Springer, Berlin.
Wyrtki, K., E.B. Bennet, and D.J. Rochford. 1971. *Oceanographic Atlas of the International Indian Ocean expedition*. Washington, D.C.: National Science Foundation. 1971.

Chapter 2
The Functioning of Coastal Upwelling Systems

Abstract This chapter describes the general physics driving coastal upwelling, various upwelling mechanisms and indicators (e.g. upwelling index), the ecological response to coastal upwelling events, the location of significant upwelling regions, their general biogeochemistry, and various hypotheses for the high abundance of anchovies and sardines in coastal upwelling regions.

Keywords Upwelling · Physical mechanisms · Upwelling index · Biogeochemistry · Ecological response · Theories on fish production

How do you expect to communicate with the ocean,

when you can't even understand one another?

Stanislav Lem (1921–2006)

(Taken from *Solaris*, 1961, reproduced with permission)

2.1 The Physics of Coastal Upwelling

In general, upwelling refers to an upward movement of water parcels in the water column that is maintained over a reasonably long period (~ several days to weeks), long enough to lift water parcels over a vertical distance of ~ 100 m or more. Here we focus on the classical wind-driven coastal upwelling mechanism that creates the largest and most persistent upwelling regions in the world ocean. Other upwelling mechanisms are described in the last part of this section.

The main forms of wind-driven upwelling are (i) coastal upwelling, (ii) equatorial upwelling, and (iii) ice-edge upwelling (Fig. 2.1). The Earth's rotation and its associated effects, such as the Coriolis force, play a dominant role in the dynamics of upwelling in all three systems. Coastal upwelling owes its existence to the presence of both the coast as an impermeable lateral boundary and relatively

© Springer International Publishing Switzerland 2016
J. Kämpf and P. Chapman, *Upwelling Systems of the World*,
DOI 10.1007/978-3-319-42524-5_2

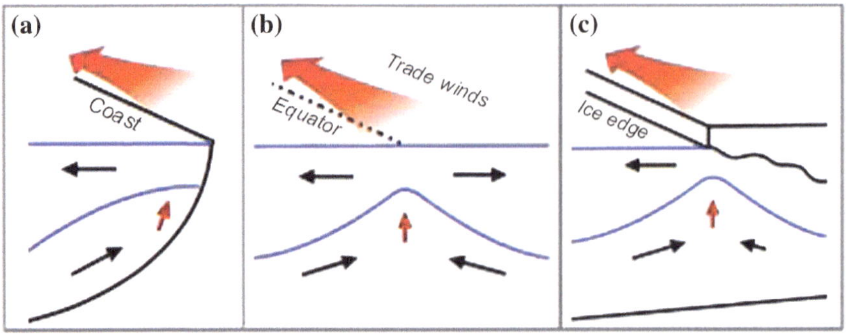

Fig. 2.1 Types of oceanic upwelling: **a** coastal upwelling, **b** equatorial upwelling, and **c** ice-edge upwelling. The *broad brown arrow* shows the direction of the prevailing wind relative to a coast, the equator, or an ice edge. In the case of the coast this shows the situation for the southern hemisphere, with the wind blowing towards the north

shallow water on the continental shelf. Equatorial upwelling is related to the fact that the Coriolis parameter (the constant of proportionality in the Coriolis force) changes sign across the equator. Although the Coriolis force vanishes at the equator it comes into full swing at relatively short distances (>50 km) from it. Due to this spatial variation of rotational effects, the dynamical role of the equator in the upwelling process is similar to that of coasts. Ice-edge upwelling is created via a substantial dampening of the effect of wind stresses on currents under the sea ice.

Depending on typical wind conditions in a region, coastal upwelling can be either a quasi-permanent feature in so-called *major* coastal upwelling systems or a seasonal feature in *seasonal* coastal upwelling systems. While upwelling can occur all along a straight coastline (e.g., off Oregon, see Chap. 4), since coastlines and seafloors are frequently irregular, wind-driven coastal upwelling events are

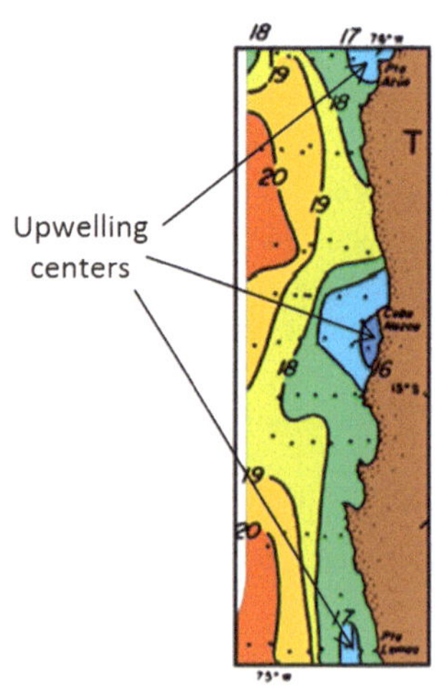

Fig. 2.2 In warm-water regions of the ocean, coastal upwelling of colder sub-surface water leads to a decrease in sea surface temperatures. This decrease need not be spatially uniform as it can occur at specific upwelling centres. The graph shows in-situ measured sea surface temperatures for the Peru-Chile upwelling system (from Tomczak and Godfrey 2003)

generally localized, and upwelling is not at all uniform. As a consequence, upwelling is more pronounced in certain regions, called *upwelling centres*, than in others (Fig. 2.2). Upwelling centres are often associated with strong frontal flows associated with an *upwelling jet* that breaks up into mesoscale (10–20 km in coastal waters) circular circulation patterns called *eddies*. While most of the primary productivity takes place inside and a short distance downstream of coastal upwelling centres, more quiescent regions adjacent to upwelling centres, called *upwelling shadows*, are important as spawning and nursery grounds for pelagic fish.

2.1.1 Description of the Upwelling Process

Seawater is largely incompressible. Hence, the vertical volume flux induced by upwelling is only possible if the same volume of water per unit time is moved away laterally from a region by a divergence of the horizontal flow. As a result, the trigger of coastal upwelling is the wind-induced offshore movement of surface water, which is replenished with water from below. In the case of equatorial upwelling, the prevailing trade winds induce net surface water movement away from the equator in both hemispheres, resulting in its replacement by sub-surface waters. Given that the upwelled water is generally heavier (denser) than the water it replaces, the associated increase in potential energy of the system requires an external energy source, which is provided by surface wind stresses.

Any dynamical process in the ocean lasting longer than a day, such as coastal upwelling, is controlled by rotational effects, i.e. the Coriolis force. A balance between the Coriolis force and the horizontal pressure-gradient force, known as the *geostrophic balance*, governs the dynamics of horizontal ocean currents in the ocean interior (i.e., at some distance from the sea surface and the seafloor). The geostrophic balance implies that horizontal currents strictly follow isobars (lines of constant pressure). Surface pressure anomalies and, hence, horizontal geostrophic flows extend throughout the entire water column, unless they are weakened by a dynamical adjustment of density interfaces in the ocean's interior, called *baroclinic compensation*. In addition to the geostrophic flow regime, frictional effects become relevant near vertical (and lateral) boundaries. Near vertical boundaries, the balance between friction and the Coriolis force creates departures of the flow from the geostrophic balance, known as *Ekman layers* (see next section for more details), first studied and described by Ekman (1905) (Fig. 2.3).

The general structure of the oceanic circulation in coastal upwelling situations consists of a coast-parallel geostrophic current, called an *upwelling jet*, which is deflected seaward in an Ekman layer near the surface and shoreward in an Ekman layer near the bottom (Fig. 2.4). The coast-parallel component of the wind stress induces offshore movement in the surface Ekman layer which operates to lower the coastal sea level by about ~5–10 cm until a dynamical equilibrium is reached. This

Fig. 2.3 Vagn Walfrid Ekman operating a current meter. Photography courtesy of University of Bergen

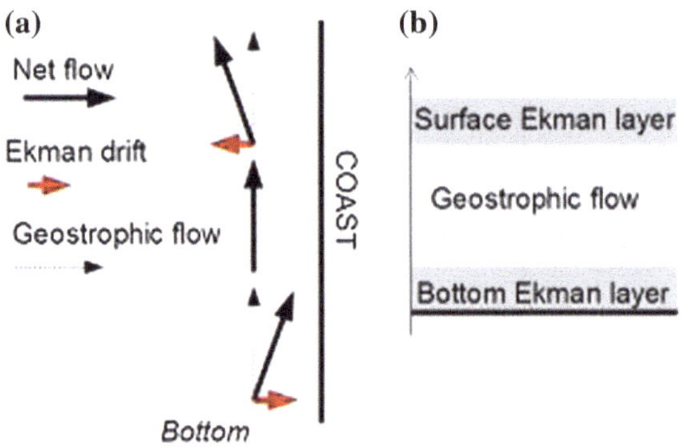

Fig. 2.4 The general dynamic structure of coastal upwelling. Panel **a** illustrates the horizontal flow structure in water near the surface, in the middle of the water column, and near the seafloor. Panel **b** illustrates the principal vertical structure of the ocean dynamics. In panel (**a**) the wind is blowing from the bottom of the figure towards the top

drop in sea level is sufficient to create a shoreward pressure-gradient force driving a swift geostrophic upwelling jet of 20–50 cm/s in speed. In turn, the geostrophic flow becomes subject to frictional effects of the seafloor. This creates a shoreward flow in the bottom Ekman layer, which is about 5–25 m thick. This near-bottom flow is the final agent of the upwelling process as it moves near-bottom water shoreward and, as it hits the coast, upward into the euphotic zone.

Some upwelling systems, such as the world's largest upwelling regions found off Peru/Chile, California, northwest Africa, and southwest Africa, exhibit nutrient-rich *undercurrents* at shelf-break depth acting as both a source for upwelled water and as a nutrient trap (Montecino et al. 2005).

2.1.2 *Wind Stress and Ekman Transport*

The magnitude of the frictional wind stress at the sea surface can be calculated from the formula:

$$\tau = \rho_{air} C_D W, \tag{2.1}$$

where $\rho_{air} \approx 1.28$ kg/m^3 is the air density, $C_D \approx 0.001$–0.002 is the wind drag coefficient (its value depends on turbulence levels in the lower atmosphere), and W is the horizontal wind speed at a reference height of 10 m above sea level. The direction of the wind stress vector is the same as that of the horizontal wind velocity.

Owing to rotational effects (i.e., the Earth's rotation) and the associated Coriolis force, the response of the surface ocean to wind stress is not straightforward. On timescales of a couple of days and longer, the frictional effect of a wind stress is confined to a surface Ekman layer having thickness of ~50–100 m (away from the equator). In the absence of other processes, oceanic currents in the Ekman layer are strictly horizontal but change direction with depth in a spiral fashion, known as the *Ekman spiral* (see Pond and Pickard 1983).

Dynamically more important than this vertical structure is the fact that the depth-averaged horizontal volume transport in a surface Ekman layer is at right angles to the wind direction; it is deflected 90° to the right (left) of the wind direction in the northern (southern) hemisphere (Fig. 2.5). This change of direction between the hemispheres is caused by the change of the direction of the Coriolis force at the equator. Moreover, theory reveals that the magnitude of the net volume transport in the surface Ekman layer M—called *Ekman transport* or *Ekman drift*—can be quantified as:

$$M = \frac{\tau}{\rho_{sea}|f|} \tag{2.2}$$

where τ is the magnitude of the wind stress in m/sec, $\rho_{sea} \approx 1026$ kg/m^3 is the average seawater density, and f is the Coriolis parameter. The Coriolis parameter exclusively depends on geographical latitude φ and can be calculated from $f = 4\pi/T_{earth} \sin(\varphi)$, where T_{earth} = 86,400 s is the period of the Earth's rotation. Hence, only information on wind stress and geographical latitude is required to determine surface Ekman

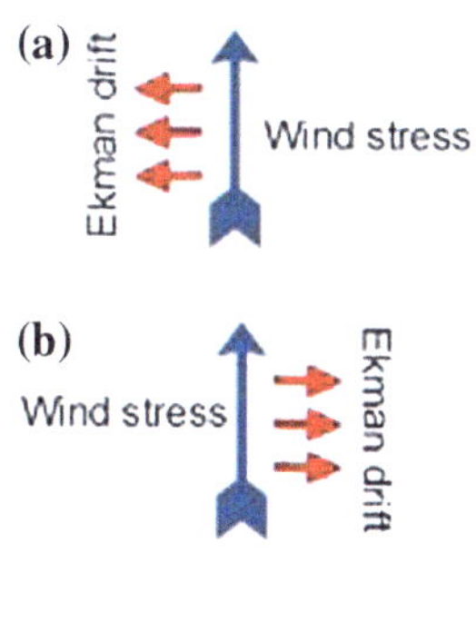

Fig. 2.5 The simple directional relationship between wind stress and Ekman drift in the **a** southern hemisphere and **b** the northern hemisphere

transports in the ocean, which can be done by landlubbers without the actual need of marine measurements. Note that M carries units of m^2/s which has to be interpreted as total volume transport (m^3/s) per unit width of the flow.

2.1.3 The Upwelling Index

The theory of Ekman layers (and geostrophic flows) (Jenkins and Bye 2006) is fundamental in the understanding of wind-driven coastal upwelling. We believe that Bakun (1973) was the first who initially described the methods used to calculate upwelling indices, presenting monthly, quarterly, and annual indices for 15 near-coastal positions along the North American west coast for the period 1946–71. Daily and weekly means of the upwelling indices at these standard west coast locations were published by Bakun (1975) for the period 1967–73 (Fig. 2.6). A third report in the series (Mason and Bakun 1986) summarized the monthly indices for 1946–71, along with daily and weekly means of the six-hourly indices for 1974–85 at six locations along the U.S. west coast (33–48° N).

Continental shelves are generally deep enough for the undisturbed development of a surface Ekman layer at some distance (typically a few kilometers) from the coast. Given that the coast acts as a barrier to flow, offshore Ekman drift follows from the coast-parallel component of the wind-stress vector. In the general case of any possible coastline orientation, this offshore Ekman transport can be calculated from:

$$UI = M\cos(\alpha^*) \tag{2.3}$$

where α^* is the relative angle between the wind direction and the actual coastline orientation. Coastline orientation is defined such that $\alpha^* = 0$ implies a wind blowing perfectly parallel to the coast with the coast on its left (right) in the

COASTAL UPWELLING INDICES, DAILY AND WEEKLY MEANS
DURING 1970 AT 60N, 149W

WEEK BEGINNING	DAILY VALUES SUN	MON	TUE	WED	THU	FRI	SAT	WEEKLY MEAN
6 JAN	-224	-290	-106	-61	-70	-81	-120	-150
11 JAN	-230	-318	-377	-274	-190	-68	-114	-226
18 JAN	-219	-324	-334	-162	-132	-118	-139	-200
25 JAN	-110	-47	-9	-39	-48	-258	-31	-75
1 FEB	-100	-436	-16	-51	-59	-48	-91	-117
8 FEB	-18	-131	-122	-28	-110	-135	-143	-99
15 FEB	-34	-38	-140	-114	-29	-5	-196	-69
22 FEB	-105	-14	110	-3	-19	-4	-38	-5
1 MAR	-8	-40	-65	-72	-74	-3	3	-36
8 MAR	-69	-88	-119	-110	-83	-47	-90	-82
15 MAR	-36	-1	-92	56	9	1	-10	-11
22 MAR	-25	2	-45	0	47	38	70	13

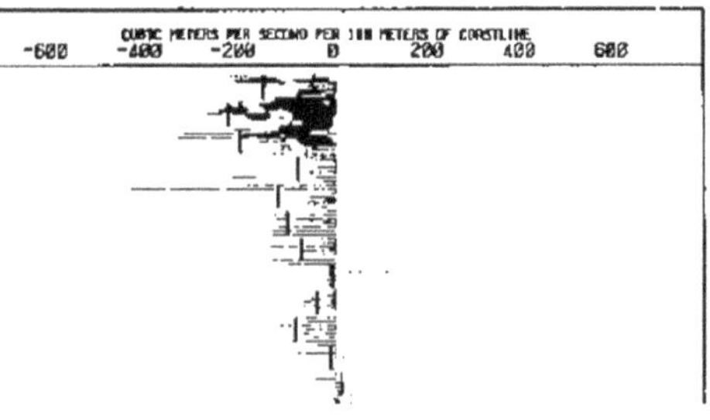

Fig. 2.6 Example of daily upwelling indices published in Bakun (1975)

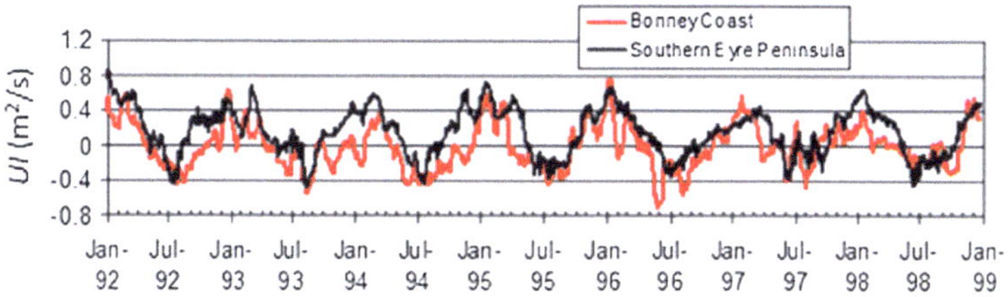

Fig. 2.7 Time-series of the window-averaged (30 days) upwelling index for two upwelling centres of the Great South Australian Coastal Upwelling System (from Kämpf et al. 2004). Indicated is the existence of a seasonal coastal upwelling system in which upwelling occurs during austral summer months

northern (southern) hemisphere. It should be pointed out that, due to volume conservation, *UI* is a direct measure of the vertical volume transport inherent with upwelling. The dimensional parameter *UI* is widely known as the *upwelling index*. The upwelling index is the offshore volume transport per unit length of the coastline. The latter is sometimes expressed in kilometres is calculated with reference to a distance of 1 km.

Using coastal wind data, the upwelling index is a simple method to explore and identify coastal upwelling activity in a region (Fig. 2.7 shows an example). Situations in which *UI* is negative imply an onshore Ekman drift pushing seawater against the coast and density interfaces downward in a process called *coastal downwelling*. Often, prevailing coastal winds create alternating upwelling and downwelling sequences associated with synoptic (3–10 days) variability of weather patterns.

2.1.4 Physical Timescales of the Upwelling Process

It is clear that it takes a certain time before sub-surface water is brought to the sea surface from its initial depth before upwelling starts. To estimate this time, let us consider a coastal ocean consisting of two distinct layers: a warm and lighter surface layer above a colder, denser and nutrient-rich bottom layer. The upper layer has a thickness of H. The density difference between the layers is $\Delta\rho$. The combined effect of magnitude and duration of upwelling-favourable wind stresses can be formulated in terms of a quantity called *wind impulse* (Ω), defined by:

$$\Omega = \frac{|f|}{H} \int_{event} UI\mathrm{d}t, \tag{2.4}$$

where H is the undisturbed thickness of the surface layer (Cushman-Roisin 1994). Depending on the magnitude of Ω for an upwelling event, the wind stress causes the

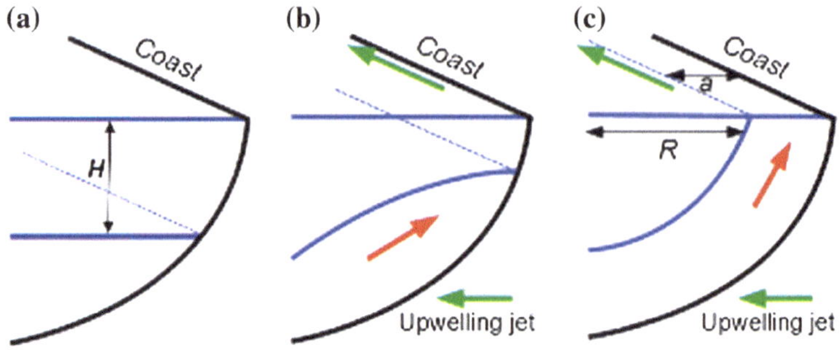

Fig. 2.8 Two possible outcomes of coastal upwelling after a longshore wind event of finite duration. Panel **a** shows the reference state. Panel **b** after a weak or brief wind event, the interface has upwelled but not to the point of reaching the surface. Panel **c** a strong or prolonged wind event causes the density interface to reach and intersect the sea surface, where it forms a front that may depart from the coast under the action of offshore Ekman drift. The latter case corresponds to a mature upwelling that favours biological activity. Adapted from Cushman-Roisin (1994)

density interface to rise towards the coast and it may or may not eventually reach the sea surface, leading to either *partial upwelling* or *full upwelling* (Csanady 1977; Fig. 2.8).

Partial upwelling results in a sloping density interface which does not reach the surface (Fig. 2.8b). Full upwelling, on the other hand, implies the formation of a surface *density front*, which is a narrow frontal zone across which seawater density changes rapidly, also referred to as *upwelling front* (Fig. 2.8c). The upwelling jet coincides with the frontal axis. Sea surface temperatures can vary by several degrees across such fronts. During the process of full upwelling, the upwelling front and the associated upwelling jet continue to move offshore under the action of the Ekman drift as long as the wind forcing continues to exist.

In order to estimate the duration of the partial upwelling phase and to determine the final offshore distance of the upwelling front, we assume that the coastline extends to infinity and that the coastal wind stress varies in time but not spatially. For such a simplified conceptual model, Cushman-Roisin (1994) derived a formula to estimate the distance (a) of the upwelling front from the coast (see Fig. 2.8c), given by:

$$a = \frac{\Omega}{|f|} - R \quad \text{with}\, R = \frac{\sqrt{g'H}}{|f|} \tag{2.5}$$

where R is the *internal deformation radius* (which is an estimate of the width of the upwelling front) with $g' = \Delta\rho/\rho_o\, g$ being reduced gravity ($g = 9.81$ m/s^2 is acceleration due to gravity). Equation (2.5) can also be used to estimate the time span full upwelling to develop from $\Omega = \sqrt{g'H}$ for $a = 0$.

If we furthermore express the wind impulse (2.5) by an average value of UI for an event of duration Δt, we obtain $\Delta t = RH/\langle UI \rangle$. Using realistic values

($R = 5$ km, $H = 100$ m, and $\langle UI \rangle = 1$ m^2/s) gives $\Delta t = 5$–6 days, which falls within the timescale of synoptic weather events. Hence, full upwelling does generally not occur for relatively short wind events of only 1–2 days or less in duration, but it may follow from a series of consecutive events. According to this simplified conceptual model an upwelling event of a duration of 10–12 days would move the density front a distance of $a = R = 5$ km offshore.

Does this imply that the upwelling front is moved to vast distances from the coast for quasi-continuous coastal upwelling ($\Omega \gg 1$) in major coastal upwelling systems? In contrast to the above conceptual model, the transition between partial and full upwelling (i.e., $a = 0$) in real upwelling systems occurs typically at a fixed location (rather than along the entire coast) and the width of the upwelling zone gradually increases with increasing downstream distance from this point. Based on the above example, an upwelling jet with a speed of, say, 0.2 m/s travels $\sim$200 km along the coast on a time scale of 10–12 days before it is moved 5 km offshore. Hence, coastal upwelling jets remain relatively close to the shore in smaller upwelling regions. On the other hand, coastal upwelling jets typically become dynamically unstable and break up into a turbulent field of mesoscale eddies. In some upwelling regions, this mechanism leads to a widening of the upwelling zone to several hundred kilometers.

2.1.5 Significance of Upwelling Jets

Coastal upwelling jets are an important part of the wind-driven circulation. On the global scale, such jets within major upwelling systems intensify equatorward flows on the eastern margin of subtropical gyres and thus play an important role in the ocean's meridional transport of heat and freshwater. As such, upwelling jets also transport nutrients and organic matter along the coast. While coastal upwelling centres are largely stationary features, the existence of coastal upwelling jets implies the formation of different biological zones; that is, spatially separated zones of maximum primary, secondary, and fish production form downstream the upwelling region. This separation occurs due to the difference in the time scales associated with the creation of phytoplankton, zooplankton and fish growth, e.g., 5 days for phytoplankton, 25 days for zooplankton, and 2–3 months for small pelagic fish (Hutchings et al. 2006).

On the other hand, upwelling jets are not smooth (laminar) flows. Like other frontal flows (e.g. western boundary currents) upwelling jets quickly become dynamically unstable on time scales of days to weeks and shed *mesoscale eddies*. Mesoscale eddies are circular flow patterns in which the geostrophic flow surrounds a high-pressure (low-pressure) centre associated with an elevated (depressed) sea level. It is the low-pressure eddies (i.e. *cyclones*) that largely sustain upwelled water in their centres. Eddies in the coastal ocean can have diameters as small as 10–20 km, in contrast to the more commonly known open-ocean eddies that have diameters of up to 300 km. As a result of eddy shedding, the width of the upwelling zone generally increases along the

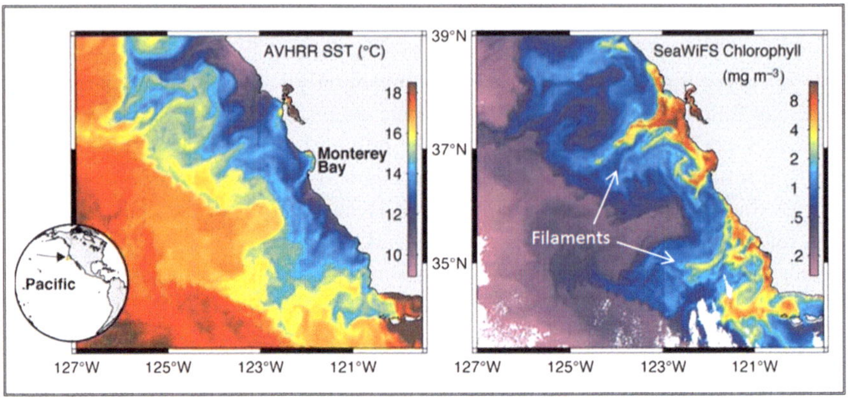

Fig. 2.9 Example of a fully developed eddy field and upwelling filaments seen in satellite images of sea surface temperature and chlorophyll-*a* concentrations in the Californian upwelling system (from Ryan et al. 2005)

coast in the direction of the upwelling jet. In addition, upwelling jets often form turbulent wakes in the lee of headlands or islands. Fully developed eddy fields exhibit specific pathways, called *filaments*, along which upwelled water is advected offshore (Fig. 2.9). Filaments, which can be quasi-stationary or transient features, generally operate as an export mechanism of organic matter to the open ocean (e.g. Arístegui et al. 1997). Eddies also operate to disperse properties of the upwelling centre (e.g. heat anomalies, organic matter and zooplankton and fish larvae) offshore.

As the upwelling front separates from the shore, so does the core of the geostrophic jet (see Fig. 2.8). Despite this, flows within the surface and bottom Ekman layers continue to exist inshore of the geostrophic jet. This implies that upwelling reaches into relatively stagnant coastal seawater that is not subject to the swift geostrophic flow of the upwelling jet. Seawater that has upwelled in this zone is not as rapidly "washed away" as it is in the upwelling jet and, hence, the nutrients contained within it are available for a longer period for biological use.

2.1.6 Coastal Upwelling Regimes

The upwelling process is only of ecological consequence if it actually carries nutrient-rich sub-surface water into the euphotic zone. Hence, the ecological effectiveness of the upwelling process critically depends on the properties of shelf bottom water. Nutrient concentrations in the sea generally increase with increasing water depth. Consequently, dynamical processes that move nutrient-richer seawater from the upper continental slope across the shelf break and onto the continental shelf can substantially enhance the biological productivity in an upwelling region. nutricline

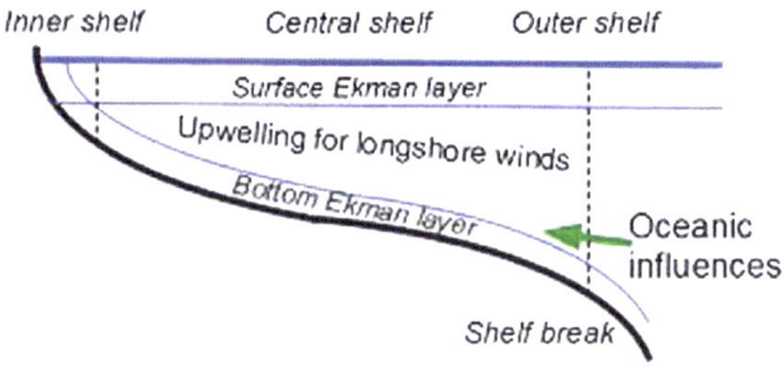

Fig. 2.10 Coastal upwelling regimes. Oceanic influences on the outer continental shelf can precondition shelf waters with nutrient-rich (in some cases, also oxygen-poor) water for subsequent wind-driven coastal upwelling events, which incorporate central regions of the shelf. The inner region of the continental shelf is defined by total water depths <50 m. It is a zone where surface and bottom Ekman layers interfere such that the wind-driven water movement is more aligned with the wind direction

Accordingly, the coastal upwelling dynamics can be grouped into different regimes (Fig. 2.10). Oceanic influences (e.g. the upward tilt of the large-scale nutricline on the eastern side of subtropical gyres, or localized onshore flows in submarine shelf-break canyons) are preconditions for coastal upwelling as they import nutrient-rich waters onto the shelf. The actual wind-driven coastal upwelling process takes place on the central parts of the continental shelf and follows from upwelling-favourable (i.e. coast-parallel) wind stresses. This upwelling excludes near-shore regions of the shelf in which the water is too shallow (<50 m) for the development of spatially separated surface and bottom Ekman layers. Here, the Ekman layers overlap and create water movements that are more aligned with the wind direction. Turbulence mixes waters from the main upwelling zone into nearshore waters.

2.1.7 *Indicators of Upwelling*

The upwelling index based on the wind stress (2.3) is commonly used as the primary indicator of coastal upwelling events. Another indicator of upwelling events is the lowering of the coastal sea level induced by offshore Ekman transport (Fig. 2.11). Nonetheless, the upwelling-induced signal is relatively weak ($\sim$5–20 cm) and low-pass filters have to be applied to extract this signal from other more dominant sea-level oscillations caused by wind waves, swell, tides and coastally trapped waves.

Since the 1970s, satellites have revolutionized our ability to see large-scale changes in the ocean over relatively short time periods. Satellite-derived sea surface temperatures are sufficiently accurate and of high-enough temporal resolution (<1 day) to identify upwelling events and their associated coastal upwelling centres. In warmer oceanic regions, where sea temperature decreases rapidly with depth, upwelling can readily be identified from sudden drops in sea surface temperatures

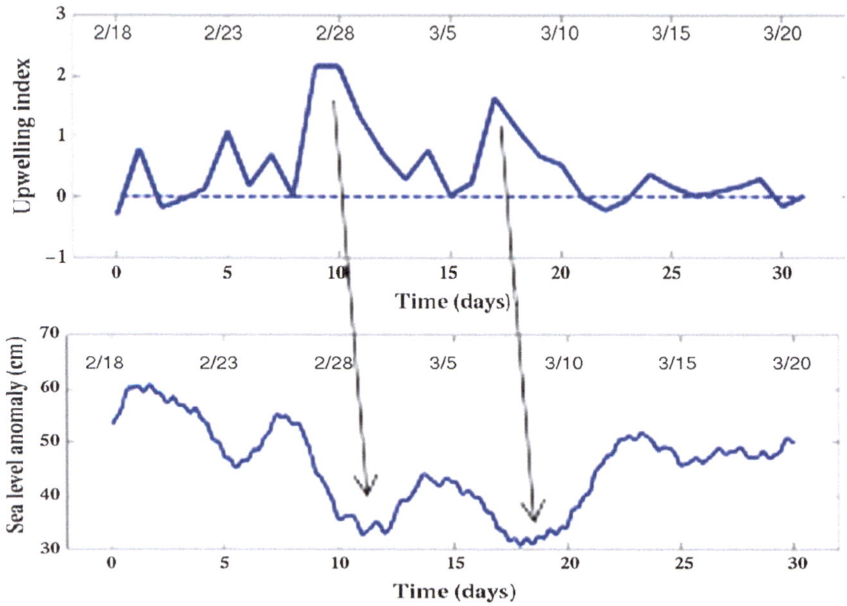

Fig. 2.11 Sea level response (cm) to upwelling-favourable winds for the Bonney upwelling region, Australia. Upwelling index is based on daily averaged wind data for Portland. A moving average window of 5 days in width is applied to sea level data of Portland's tide gauge. Unpublished data. For more information on this upwelling region, see Kämpf et al. (2004) and chap. 8

by >1 °C on the timescale of synoptic weather events (3–10 days). Elevated phytoplankton concentration associated with upwelling events can be inferred from Ocean Color (SeaWiFS) data (Fig. 2.12 shows examples).

Ocean colour data are based on standard blue/green ratio algorithms for chlorophyll-*a* retrieval, which can substantially overestimate phytoplankton blooms in coastal waters in the presence of coloured dissolved organic matter (CDOM or *Gelbstoff*) and suspended sediment (e.g., Garcia et al. 2006; Shanmugam 2011). The fluorescence line height (FLH) is a relative measure of the amount of radiance leaving the sea surface in the chlorophyll fluorescence emission band (e.g., Xing et al. 2007). The chlorophyll-*a* fluorescence peak near 685 nm has been proposed as an alternative option of chlorophyll retrieval (Neville and Gower 1977; Gordon 1979). Many researchers have studied the relationship between chlorophyll-*a* concentration and FLH, and have found a good correlation between these properties (Maritorena et al. 2000; Gower et al. 2004; Gower and King 2007; Kämpf 2015) for relatively low chlorophyll-*a* concentrations of <9 mg m^{-3}. Modifications are required at higher chlorophyll concentrations, where the fluorescence peak shifts to longer wavelengths (e.g., Gitelson 1993).

Ocean Color images often show upwelling filaments moving organic matter offshore (Fig. 2.12b). While these filaments and eddies can be generated locally, there are also known interferences from remotely created eddies, e.g., off the

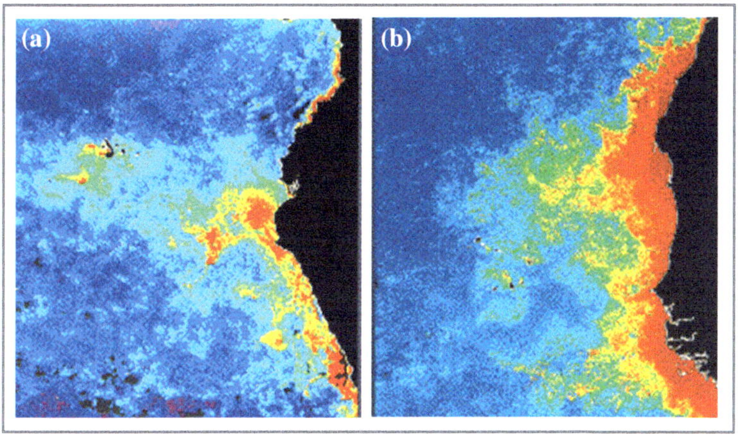

Fig. 2.12 Ocean Color as an indicator of coastal upwelling in waters off **a** Peru and **b** northwest Africa, which are among the most productive in the global ocean. *Red* (*blue*) *colours* represent waters of high (low) chlorophyll concentrations. The upwelling is reflected here as chlorophyll-rich bands along both coasts, shown in *red* and *yellow*, up to 100-km wide, and turbulent mixing plumes of productive waters extending ~500–1,000 km offshore. Note that island-induced upwelling generates highly productive regions around the Galapagos Archipelago (seen in Panel **a**). Source of images: NASA. http://disc.sci.gsfcnasa.gov/education-and-outreach/additional/science-ocus/space/ocdst_upwelling_dynamics.shtml [accessed 4 April 2016]

southwest coast of South Africa, where Agulhas eddies call pull shelf water several hundred kilometres into the open ocean and remove large quantities of fish eggs and larvae (Duncombe Rae et al. 1992). In contrast to sea surface temperatures, Ocean Color data can also identify upwelling events in colder oceanic regions, where upwelling-induced temperature anomalies are less pronounced. Other upwelling indicators are high abundances of zooplankton and fish, which can only be measured from sea-based surveys.

2.1.8 *Other Upwelling Mechanisms*

Apart from the wind-driven coastal upwelling process, there are a number of other physical mechanisms that can produce upwelling in the coastal ocean. These mechanisms are:

- Ekman pumping
- dynamic uplift (shelf-break upwelling),
- tidally-induced upwelling
- upwelling in the wake of islands
- topographically caused upwelling (e.g. in shelf-break canyons), and
- transient upwelling caused by the passage of coastally trapped waves.

Ekman pumping refers to the vertical adjustment of the pycnocline associated with a spatially varying horizontal wind-stress field, known as *wind-stress curl* (Sverdrup et al. 1942), which induces a divergence of horizontal Ekman transports in addition to boundary effects due to the existence of a coast. While the coastal upwelling process involves a cross-shelf transfer inherent with the dynamics of Ekman layers, Ekman pumping exclusively induces a vertical flow. Previous studies indicate that Ekman pumping contributes significantly (>25 %) to the vertical nutrient flux in major eastern boundary upwelling regions (Messié et al. 2009). The important difference between the effects of wind stress and wind stress curl is that Ekman pumping caused by the latter takes place considerably farther offshore, often at the shelf edge, and can occur at faster rates in terms of vertical transport than Ekman transport (Kraus and Businger 1994). Thus, it is separated from the inshore Ekman transport and can provide a second input of nutrients to the system.

Ekman pumping and its associated upward nutrient fluxes have been reported for the California Current (Checkley and Barth 2009; Rykaczewski and Checkley 2008), the Benguela Current (Bakun and Nelson 1991), the eastern Banda Sea (Gordon and Susanto 2001), and the upwelling system around Cabo Frio (Castelão and Barth 2006). Chavez and Messié (2009) suggest that this type of upwelling has different biochemical implications, given that it does not recruit shelf and slope waters that have interacted with the nepheloid layer, and hence brings lower concentrations of iron into the euphotic zone than classical upwelling closer inshore. As a consequence, Chavez and Messié (2009) suggest that the coastal upwelling process stimulates a food web of larger organisms (diatoms, copepods, krill) and accumulation of biomass, whereas the wind-stress-curl-induced upwelling farther offshore leads to relatively low biomass and ecosystem characteristics of high-nutrient-low-chlorophyll regions.

Dynamic uplift (also called *shelf-break upwelling*) is an upwelling mechanism that operates along the inshore edge of western boundary currents (where coastal winds are generally not upwelling favourable). Western boundary currents are geostrophic frontal flows associated with a steep rise of the thermocline from the ocean towards the coast. Being part of large oceanic gyres, their dynamics are largely independent of local wind conditions. On average, western boundary currents are located just offshore from the continental shelf, and their influence does not reach far into the coastal ocean, which establishes its own circulation dominated by local wind and thermohaline forcing. Changes in the current's position (e.g. caused by eddy formation) can lead to upwelling of deeper water onto the outer shelf. The process applies to all western boundary currents, and examples can be found in all oceans on both hemispheres (Tomczak and Godfrey 2003, see also chap. 9).

Tidally-induced "upwelling", also called *tidal pumping*, is a regional process in which oscillatory tidal flows regularly move nutrient-rich deeper water across the shelf break (Thompson and Golding 1981). Tidal pumping is most pronounced in regions of strong tidal flows in tidal channels between islands located along the shelf break. Although this process is oscillatory rather than steady (similar to the action of internal waves), it causes the regular appearance of nutrient-rich water on the outer shelf and therefore constitutes a form of shelf-break upwelling.

Fishermen have long known of the enhanced biological productivity in the vicinity of islands (see Fig. 2.10a). The island-wake effect was first scientifically documented by Doty and Oguri (1956). Nutrients, such as nitrate, can be 10 times higher in the wake compared to upstream values (Heywood et al. 1990), and density surfaces can shoal by as much as 100 m. Upwelling in the wake of islands in shallow seas can be explained by the well-known "tea cup effect" (Wolanski and Hamner 1988). Typically, flows around an island create a counter-rotating pair of eddies in the lee of the island. In an eddy, rotating in either direction, there is an inward pressure gradient supporting the centripetal acceleration of the fluid. Friction slows the flow near the bottom, reducing the centripetal acceleration. The pressure gradient remains, is unbalanced, and pushes the near-bottom fluid towards the centre of rotation. This flow convergence drives the upwelling and, in the case of a tea cup, concentrates the "tea leaves" in the cup's centre (Hasegawa et al. 2004).

Shelf-break canyons are submarine canyons cutting through the shelf break. Depending on their direction, ambient currents at shelf-break depth can trigger localized upwelling of deeper water from the upper continental slope onto the shelf (e.g., Hickey 1997; Allen and Durrieu de Madron 2009). This localized upwelling follows from ambient flows opposed to the propagation direction of topographic Rossby waves; that is, ambient flows that have shallower water (and the coastline) on their left (right) in the northern (southern) hemisphere. Topography Rossby waves are quasi-geostrophic flow disturbances created by water-column stretching/squeezing over topographic undulations. While these waves on their own can only propagate in one specific longshore direction, ambient flows can offset this propagation via the creation of standing wave patterns that consist of stationary but spatially alternating regions of onshore and offshore flows across the shelf break. As such, the canyon-upwelling process can be interpreted as the signature of a stationary topographic Rossby wave, that may induce localized onshore flows near the "crests" of this wave that forms downstream from the canyon (Kämpf 2012). Technically, localized upwelling in shelf-break canyons is also a form of shelf-break upwelling.

Coastally trapped waves are waves of intraseasonal variability (30–60 days periodicity) that attain the largest amplitude near the coast and induce transient upwelling/downwelling in conjunction with onshore/offshore flows across the shelf edge. These waves are often created by coastal winds, but they can also be generated by equatorial Kelvin waves impinging on the coast on the eastern side of the equator. Coastally trapped waves can travel several thousand kilometers along coastlines. After a strong upwelling event, the regional perturbation of the oceanic pressure field (comprising both sea-level and density anomalies) in the vicinity of an upwelling centre may initiate a coastally trapped wave propagating along the shelf break and inducing transient upwelling along its way. These have been observed in coastal tide gauge sea-level records along the coasts of Ecuador, Peru and Chile, as well as in currents measured at moorings along north-central Chile (Shaffer et al. 1997; Hormazabal et al. 2004).

2.1.9 Location of Significant Upwelling Regions

Most major and seasonal coastal upwelling systems can be inferred from surface wind maps (Fig. 2.13) when using the rule that upwelling is created by coast-parallel winds driving off-shore transport in the surface Ekman layer. This intuitive method identifies the major upwelling regions off the coasts of California, Peru, Chile, northwest and southwest Africa, and Portugal as well as their associated upwelling jets (i.e., California Current, Peru-Chile Current, Portugal Current, Canary and Benguela Current) (Fig. 2.14). These year-round upwelling regions are referred to as

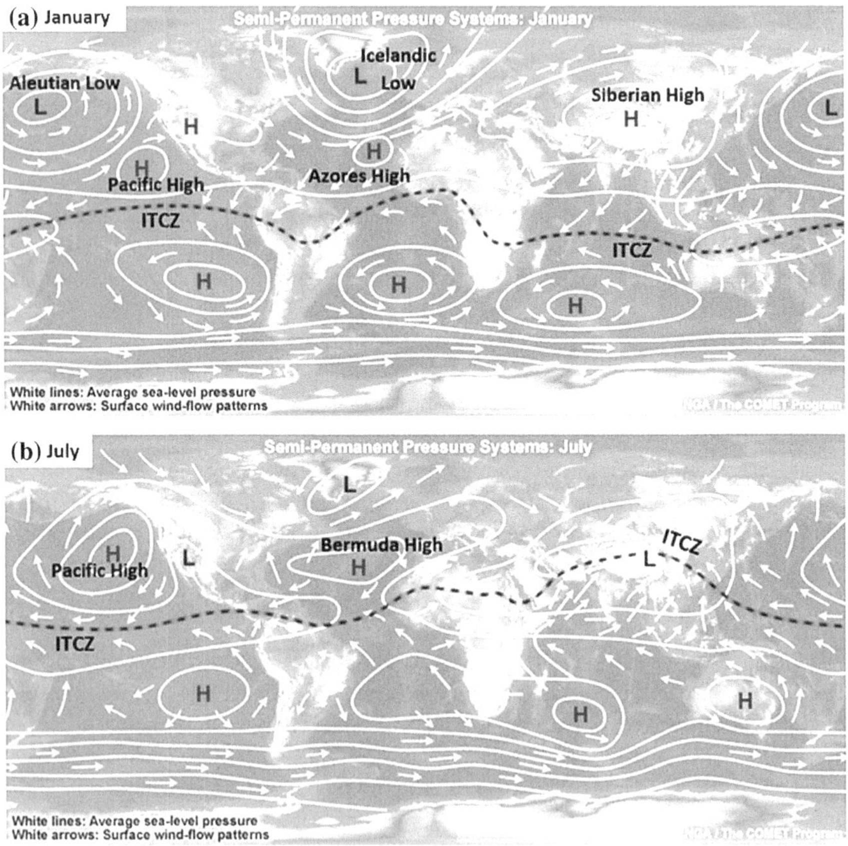

Fig. 2.13 Seasonal variations of surface wind patterns. Can you guess where coastal upwelling regions form? The source of this material is the COMET® Website at http://meted.ucar.edu/ of the University Corporation for Atmospheric Research (UCAR), sponsored in part through cooperative agreement(s) with the National Oceanic and Atmospheric Administration (NOAA), U.S. Department of Commerce (DOC). ©1997–2016 University Corporation for Atmospheric Research. All Rights Reserved

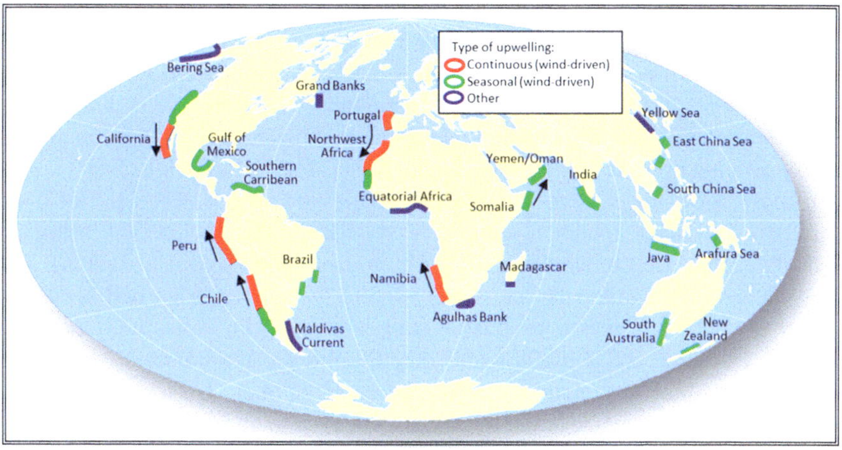

Fig. 2.14 Locations of significant coastal upwelling regions in the world ocean

major coastal upwelling regions, or *eastern boundary upwelling regions*. While the prevailing winds in these upwelling regions are equatorward, a distinct feature of these systems is the existence of poleward undercurrents that play significant roles in the distribution of nutrients and dissolved oxygen in these regions and also affect the lateral displacement of marine organisms undertaking diel vertical migration.

This method also identifies seasonal upwelling systems off the coasts of Somalia, Yemen/Oman, Sumatra, and in the South China Sea. There are also seasonal coastal upwelling systems along the southwest coast of India and the southern shelves of Australia, which are less apparent from the wind distributions shown in Fig. 2.13. For completeness, the map also includes other types of upwelling/highly-productive coastal regions such as the "Green Belt" of the Bering Sea, the Grand Banks, the Agulhas Bank and the shelf-break region of the Malvinas Current, in which upwelling is caused by dynamic uplift/shelf-break upwelling, and the Irish Sea and the Yellow Sea, where nutrient enrichment/concentration is induced by tidal mixing fronts.

2.2 The Biogeochemistry of Coastal Upwelling Systems

2.2.1 General Description

While the basics of nutrient cycling by phytoplankton and bacteria were described by Harvey (1928), and Redfield et al. (1963) described the ratios in marine phytoplankton of nutrient elements to carbon, much of our understanding of biogeochemical cycles in the ocean including the solubility and biological pumps was informed and motivated by Wallace Broecker's inference of the deep circulation from the distributions of chemical tracers in the deep ocean (Broecker and Peng 1982). This led to the international Joint Global Ocean Flux Study (JGOFS), which refined the understanding of processes involved in the biological pump (Ducklow et al. 2001). Broecker's work has since motivated many groundbreaking studies, including CO_2 flux studies in upwelling regions, and related concepts such as that of the *continental shelf pump*, first proposed by Tsunogai et al. (1999) for the East China Sea.

In most upwelling systems, the source of the upwelled water is the offshore intermediate waters, found mostly between 100–300 m below the seasonal pycnocline. Because the wind is such an important component of upwelling systems, the amount of nutrients that is delivered to the surface layer, and hence the theoretical maximum in new production and carbon export, varies with the wind stress. As discussed above, weak winds may lift only small volumes of water onto the shelf, and this water may not reach the euphotic zone. If the winds are too strong, however, it means that the phytoplankton may be unable to make use of the nutrients, either because they are advected out of the coastal zone, or because the increased turbulence means that the surface mixed layer is deepened to such an extent that "seed" cells are removed vertically from the euphotic layer. In either case the potential productivity is lost to the system.

Intermediate wind speeds are therefore needed for maximum production rates, and off California these have been estimated at 7–8 m/s (Ware and Thompson 1991), while in the northern Benguela, minimum wind speeds of about 5 m/s are required to initiate upwelling (Shannon 1985). As wind stress varies very considerably on both seasonal and shorter time-scales, so pulsing of upwelling and the resultant production is seen in all areas, and it is common for maximum production to occur following a wind relaxation, which stabilizes the upper water column and allows phytoplankton to use the nutrients that have been upwelled.

A feature common to all upwelling systems is an increase in nutrients and a decrease in dissolved oxygen on the shelf below the seasonal pycnocline. This occurs both because of the nutrients brought in with the upwelling water mass and through local decomposition and regeneration processes in the bottom water and sediments. A typical example is shown in Fig. 2.15, which shows temperature, oxygen, nitrate and silicate from a line of stations across St. Helena Bay in the southern Benguela system. The main mass of the upwelling water, which contains about 20 μM/kg of both nitrate and silicate, in this figure is from about 175 m depth, but considerably higher concentrations of nitrate and silicate were found at 50-m depth on the shelf, accompanied by reduced oxygen concentrations, which are typically <90 μM/kg during the summer upwelling season, compared to saturated oxygen concentrations of about 225-275 μM/kg for waters having the same temperature and salinity. While it might be expected that the recycled nitrogen appears as ammonia or urea, and sediments in upwelling regions are certainly reducing, nitrification is active in the water column below the euphotic zone so that much of the recycled nitrogen is reoxidized rapidly to nitrate, despite the lower oxygen concentrations. While studies prior to the 1990s assumed that this process was facilitated by nitrifying bacteria such as *Nitrosomonas* (Ward and Carlucci 1985),

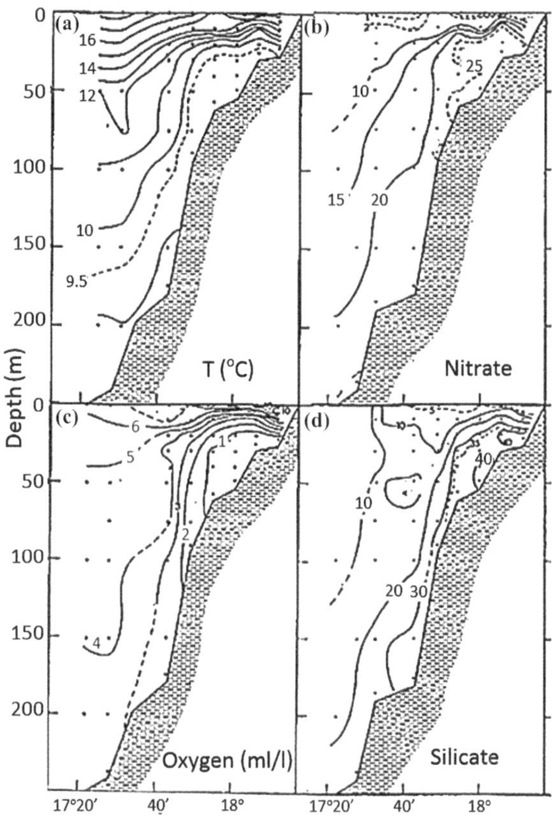

Fig. 2.15 **a** Temperature, **b** nitrate, **c** dissolved oxygen, **d** silicate in a section across the shelf off St. Helena Bay, South Africa (32° 35′ S, 17° 30′–18° 20′ E) in February 1979. Data from Bailey and Chapman (1985)

it is now accepted that Archaea, which are much more abundant than nitrifiying bacteria, play a large role in the initial oxidation of ammonia to nitrite, after which nitrite oxidizing bacteria convert the nitrite to nitrate (Zehr and Kudela 2011).

When productivity is very high, as is often the case in upwelling regions, the sinking and subsequent decomposition of the organic matter produced in the euphotic zone can lead to extremely low dissolved oxygen concentrations in the water column (<20 μM/kg). Because nitrate and nitrite have similar oxidation potentials to oxygen, their half-reactions during reduction can be used to provide microorganisms with very nearly the same amount of energy as they would gain from aerobic respiration. This switch from using oxygen to nitrate or nitrite as an electron acceptor can lead to nitrate reduction (when nitrate is converted to nitrite or ammonia) or even denitrification (when it is converted to nitrous oxide or free nitrogen gas) at oxygen concentrations below 5 μM/kg, and leads to considerably less nitrate being measured than would be expected from the standard 16:1 Redfield ratio of dissolved N:P.

An example, from the Peruvian upwelling system, is given in Fig. 2.16. In this case, a nitrite maximum can be seen at about 200 m depth, which corresponds to

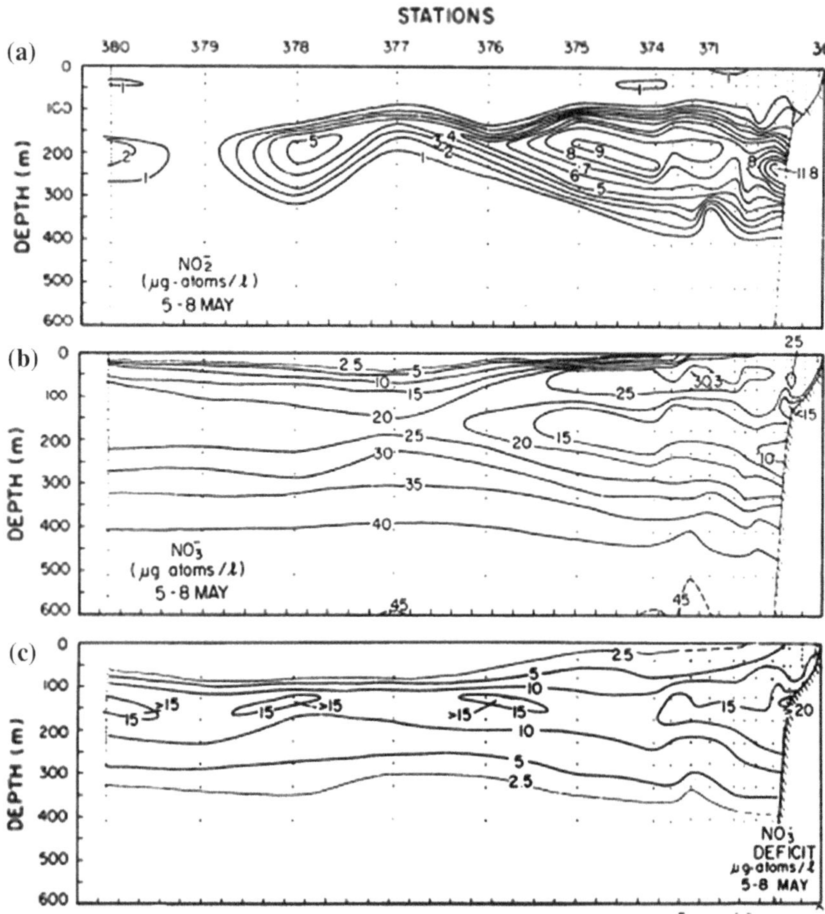

Fig. 2.16 Concentrations of **a** nitrite, **b** nitrate and **c** nitrate deficit (mmol/m^3) at 15° S off Peru (after Codispoti 1983). The nitrate deficit is the amount of nitrate or nitrite reduced to nitrogen gas

the site of maximum nitrate reduction and hence the zone in which nitrogen loss by conversion to gaseous nitrogen is also at a maximum. Oxygen concentrations at this depth are extremely low, with concentrations less than 10 μM/kg extending over 1,000 km from the coast into the subtropical South Pacific. Recently, Gruber and Sarmiento (1997) introduced the idea of expressing nitrogen deficits in terms of N^*, calculated as

$$N^* = [\text{Total inorganic nitrogen}] - 16\left[PO_4^{\ 3-}\right] + 2.9\,\mu\text{M/kg} \qquad (2.6)$$

where [Total inorganic nitrogen] equals the sum of nitrate, nitrite, and ammonia concentrations, so that modern values of nitrate deficit differ slightly from earlier numbers.

2.2.2 *Nitrogen Production by Anaerobic Oxidation of Ammonia*

Until the late 1990s, nitrogen reduction was assumed to proceed only via the pathways $NO_3^- \rightarrow NO_2^- \rightarrow NO \rightarrow N_2O \rightarrow N_2$ (heterotrophic denitrification) or $NO_3^- \rightarrow NO_2^- \rightarrow NH_4^+$ (dissimilatory nitrate reduction). However, following the identification of the direct production of nitrogen by anaerobic oxidation of ammonia (Anammox) by bacteria in sewage farms (Mulder et al. 1995; van der Graaf et al. 1995), it is now believed that this reaction, that is,

$$NH_4^+ + NO_2 \rightarrow N_2 + 2H_2O, \tag{2.7}$$

accounts for much of the loss of nitrogen from the ocean in eastern boundary regions (Lam and Kuypers 2011), both in the water column and within the sediments. Kuypers et al. (2005) showed that up to 5 mM N/m^2/day could be lost to the system via this route from the Benguela, and other researchers have shown similar losses from the eastern tropical South Pacific (Thamdrup et al. 2006; Hamersley et al. 2007). Good reviews of the complex interactions of different nitrogen species in upwelling systems and other low oxygen zones are given by Lam and Kuypers (2011) and Zehr and Kudela (2011).

Concentrations of dissolved phosphate and silicate are also enhanced in bottom waters over the shelf in coastal upwelling systems, in a similar way to nitrate. In these cases, however, the cycle is considerably simpler, as they are only found in one inorganic chemical form, although both phosphorous and nitrogen also exist as dissolved organic and particulate forms. Although the Redfield ratio of 16 atoms of nitrogen for every one of phosphate was originally defined from open-ocean North Atlantic data (Redfield et al. 1963), this ratio, or numbers close to it, is found to hold for upwelling regions as well (e.g., Bailey and Chapman 1985). Deviations to ratios less than 16 are assumed to occur because of denitrification (e.g., in the Humboldt Curent; Quiñones et al. 2010). However, phosphate concentrations can also vary because of adsorption/desorption reactions within the sediment as the system changes from oxic to anoxic (Suess 1976).

2.2.3 *The Role of Silica*

Silica is an essential component of diatom frustules, and diatoms are the most common phytoplankton species in coastal upwelling systems. The silica to nitrogen ratio in diatoms is usually accepted as being about 1, given their 16:16:1 Si:N:P atomic ratio. Because diatom cells are relatively large, with a small surface-to-volume ratio, they require a high supply of nutrients to enable them to grow (Chisholm 1992). Thus, these organisms are found in high-nutrient and turbulent regimes and transport carbon and silicon both to the seabed and to the open ocean as they sink, are incorporated into zooplankton faecal pellets, or are otherwise advected out of the coastal upwelling zone. This export results in a silica pump into the deep sea and potential silica limitation in the surface layer (Dugdale et al. 1995). The recycling of the element in the deep ocean is slow because it depends on the rate of dissolution of opaline silica following the decomposition of a protective layer of organic material. Thus its regeneration occurs considerably deeper in the water column that the regeneration of nitrogen and phosphorous and the ratio of silica to nitrate in the surface layer can be reduced to less than 1. In cases where silica is limiting, there is likely to be a shift from diatoms to non-siliceous algae (Officer and Ryther 1980; Egge and Aksnes 1992). This may have major effects on the local ecology, by encouraging harmful algal blooms (e.g., Smayda 1990), but such effects are usually of short duration.

2.2.4 *Upwelling and Carbon Fluxes*

The enhanced production and decomposition rates in coastal upwelling areas means that they are centres for carbon cycling. The observations show interannual variability caused by large-scale ocean effects, such as El Niño-La Niña (see Chap. 3), as well as considerable seasonal and regional variability, which depend to some extent on changes in offshore Ekman transport rates (Thomas et al. 2004). Overall, the world's eastern boundary current upwelling systems show theoretical maximum primary production rates of more than 3 g C/m^2/day. More information on seasonality and variability on other time scales will be given in the chapters on individual upwelling regions.

Primary production in all four regions feeds over into secondary production of zooplankton and fish. How much of this primary production ends up as secondary production varies considerably with season and region. The general transfer of carbon from the euphotic zone to the deep sea has been discussed in Chap. 1. However, within upwelling regions, considerable amounts of carbon are also recycled locally on the shelves either as dead phytoplankton or as zooplankton faecal pellets, so that sediments in some regions can contain more than 10 % percent organic carbon by weight (Bremner 1980; Dale et al. 2014). How much of this material is lost to the atmosphere following either oxidation to carbon dioxide or reduction to methane and dimethyl sulphide within the sediments remains unclear, but sulphide emissions are known from both the Benguela and Peru upwelling systems once oxygen levels in the water column drop to zero.

Whether eastern boundary systems are sinks or sources for carbon dioxide is equally unclear; Torres et al. (2002) suggest that northern Chile is a source, while Santana-Casiano et al. (2009) and Monteiro (2010) believe that the Benguela is a net small sink. In this case the sink in the southern Benguela outweighs the source in the north, with the boundary being at about 20° S, but both systems show considerable seasonal and interannual variability. Other eastern boundary systems also show high variability, depending on season, wind strength and direction, and there is presently little consensus on the issue.

2.3 The Ecology of Coastal Upwelling Systems

2.3.1 *Biological Response to Coastal Upwelling Events*

The biological response to coastal upwelling events in terms of primary and secondary production depends strongly on:

- Existing ambient conditions (illumination and available nutrient concentrations in shelf bottom waters);
- The intensity and duration of upwelling-favourable wind conditions;
- Biochemical factors (i.e. possible nutrient limitation, interaction with oxygen-minimum zones, development of harmful algae blooms and subsequent oxygen depletion),
- The availability of a seed population of phytoplankton (this may be in the form of resting spores on the bottom), and
- Complex food-web dynamics including human interference, such as the harvest of forage fish (sardines, anchovies, pilchards).

The upwelling process moves water and its nutrients from the bottom of the shelf into the surface mixed layer (Fig. 2.17). As nutrient-enriched water moves upward, it can lead to sub-surface phytoplankton production near the base of the euphotic zone, particularly during partial upwelling events, which cannot be detected from satellite measurements. Full upwelling, i.e., when the upwelled water intersects the surface, generally creates the strongest biological response, but the magnitude of the

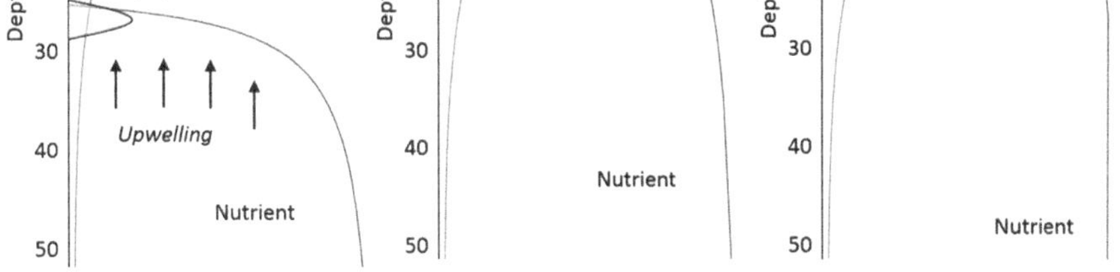

Fig. 2.17 Schematic of the response in phytoplankton production to upwelling of nutrient-rich sub-surface water. Shown is the typical summer situation for temperate waters in which thermal stratification triggers sub-surface phytoplankton production in a narrow zone near the base of the surface mixed layer. As the upwelling continues, this zone is moved upward in the water column to a level of greater light intensity, which increases production. The maximum possible production occurs for full upwelling

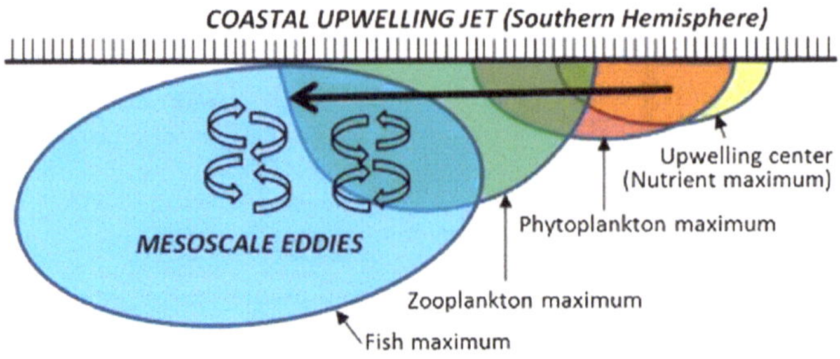

Fig. 2.18 Zonation of different trophic levels of the marine food web in upwelling regions (*top view*). Shown is the case for the southern hemisphere. The coastline is located along the top of the graph

response depends on whether the wind stress is maintained or relaxes to allow a phytoplankton bloom to develop.

The formation of a coastal upwelling centre implies that the nutrient source is a rather regional and localized feature. Given the formation of a coastal upwelling jet and biological response times, the zones of maximum phytoplankton concentration, maximum zooplankton abundance and fish abundance do not form at the same location. Instead, there is a natural zonation and the fish maximum is generally located farthest away from the nutrient maximum (Fig. 2.18). The distance between the centres of individual zones depends on the speed of the upwelling jet and integral biological response times (e.g., doubling time of phytoplankton cells). The effect of mesoscale eddies (created by the upwelling jet) is to increase the width of the upwelling zone with increasing distance from the upwelling centre. If, however, there is a natural physical feature, such as a headland or an eddy system confining the upwelled water, then the trophic succession can take place in the same area.

2.3.2 *The Significance of Upwelling Shadows*

Upwelling shadows, also known as *retention zones*, are relatively quiescent regions typically located downstream of upwelling centres in embayments that are sheltered from the swift and highly energetic upwelling jets (Fig. 2.19 shown an example). In addition to upwelling centres where nutrient fluxes into the euphotic zone fuel primary production, upwelling shadows form another important ecologic role in the food web dynamics of upwelling systems as these provide sheltered breeding and feeding grounds for fish species. The name "upwelling shadows" was introduced and described by Graham et al. (1992), who analyzed the zooplankton abundance in northern Monterey Bay, California, USA, although the idea of such retention zones trapping water close to the coast thus allowing blooms to develop had been considered earlier (e.g., Shannon et al. 1984; Taunton-Clark 1985).

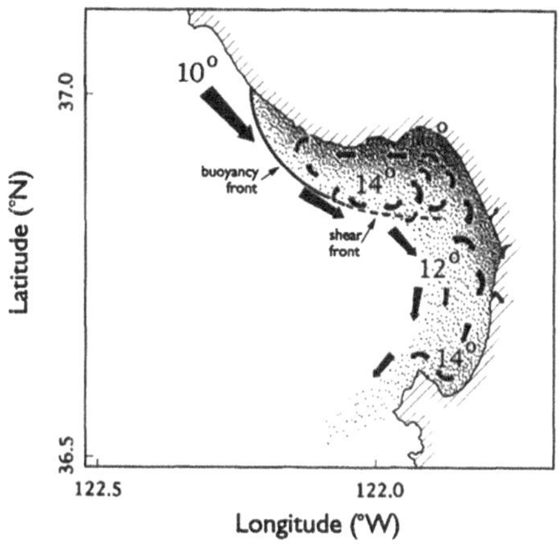

Fig. 2.19 Hypothetical circulation in the Monterey Bay upwelling shadow. Arrows indicate currrents and numbers show water temperatures. The shaded area is the retention zone where current speeds are reduced (from Graham and Largier 1997)

2.3.3 Timing and Duration of Phytoplankton Blooms

A major control on production is the timing and duration of individual phytoplankton blooms. These depend to a large extent on the transition from a well-mixed to a stratified water column, and lead to changes in the dominant species at different times during a bloom (Hutchings et al. 1995). As the bloom ages and nutrient concentrations decrease, there is a shift from small to large, colonial or chain-forming diatoms that use "new" nitrogen, and then to smaller species that use recycled nitrogen, while the importance of the bacterial loop in recycling carbon and nutrients from dead phytoplankton cells and chemical exudates increases.

Continuous upwelling-favourable winds ensure a constant supply of nutrients to the euphotic zone, but if the winds are too strong, then the increased turbulence means that phytoplankton cannot grow close to the upwelling region, but are moved well downstream and may be advected offshore and away from the coast. Productivity in many regions depends on the fact that upwelling is intermittent, so that phytoplankton growth can occur once the upwelling winds relax and the upper part of the water column stabilizes. This then allows time, typically about 5–10 days, for dense phytoplankton blooms to develop. Zooplankton generation times are considerably longer than this, of the order of 25 days, so moderate upwelling likely leads to the maximum populations of zooplankton and hence to increased forage fish stocks, Adult zooplankton can also migrate vertically to take advantage of the presence of undercurrents; this means they can stay in the vicinity of a bloom for long periods, even though larval and juvenile forms develop only in the surface layers (Hutchings et al. 1995).

As a result of the general unpredictability of coastal upwelling systems, the small forage fish typically exhibit serial spawning patterns. Instead of spawning only once a year, as many demersal fish do, they spawn multiple times at short intervals.

This allows them to take advantage of times when the conditions for survival and growth of their larvae are optimal, and minimizes the chance that all the larvae will be lost from the system through physical effects, such as eddies, predation, or starvation. Despite this adaptability, these fish do experience large changes in recruitment and biomass, depending on such factors as parent stock size, egg production, predation, availability of food resources, and physical conditions. The "window of opportunity" after hatching, during which time fish larvae can find a suitable food source, is probably only about 2 or 3 days (Hutchings et al. 1995), after which they exhaust the energy in their egg sacs and die.

Changes in the physical environment and the ecosystem biology, such as those described above, mean that feeding patterns can change from the generally accepted typical short food chain that leads from phytoplankton to zooplankton to forage fish to predatory fish. If thermal stratification is intense, so that upwelling and diatom blooms are reduced, then the plankton population changes to smaller species that use recycled nutrients, leading to an increase in smaller zooplankton and a reduction in the amount of primary production that goes to support fish biomass (Mann 1993). This then alters the whole food web as more recycling goes on in the water column and the general productivity of the system slows down. It may also mean that forage fish switch from eating zooplankton to feeding directly on phytoplankton cells, and/or lead to changes in species dominance, e.g., a switch from sardines to anchovy. Such coherent changes in dominance over decadal time scales, based on changes in the abundance of fish scales in laminated anoxic sediments, have been seen in several coastal upwelling regions (Baumgartner et al. 1992). Whether these changes are dependent on large-scale environmental variability, such as that caused by particularly intense El Niños, remains unclear (Arntz et al. 2006), as is the existence of long-range teleconnections between different upwelling regions, as suggested by Shannon and Nelson (1996).

2.4 Theories on High Fish Production

Because of the importance of recruitment to sustaining fish populations, there have been many studies on what conditions are required for maximum recruitment to occur. We give here brief descriptions of four examples that are applicable to upwelling systems.

2.4.1 Bakun's Triad

Bakun (1996) postulated that marine systems having high productivity of upper trophic levels have three key requirements: (i) Bottom-up enrichment, with high nutrient input and subsequent primary productivity; (ii) Aggregation through physically mediated processes that operate to concentrate food resources, and

(iii) Retention through physically-mediated processes that prevent the loss of weak swimmers before they can complete their life cycle.

Mackas et al. (2006) discussed this triad of requirements for the eastern boundary upwelling systems and found that all methodologies and indices agree that the highest nutrient supply and primary productivity are in these regions. Aggregation and retention are mainly associated with the ability of planktonic organisms, including larval fish and invertebrates to travel horizontally due to selective diel migration (taking advantage of flows of opposite directions at different depth levels), and the existence of upwelling shadow zones, where current movements are limited or form gyral systems that retain organisms in the area (see Mackas et al. 2006; and references therein).

2.4.2 *The "Optimal Environmental Window" Hypothesis*

The "optimal environmental window" hypothesis is based on observations of the variability of oceanographic conditions and forage fish populations (anchovy and sardines) in the large upwelling zones (Cury and Roy 1989). The basic concept is that the link between recruitment and upwelling intensity is very nonlinear. The best conditions for larval survival and hence the development of abundant anchovy and sardine populations occur following moderate upwelling intensities (Fig. 2.20) rather than very strong or weak intensities.

A gradual wind speed increase in the upwelling zone leads initially to increased production. However, too low a wind speed means the resulting reduced upwelling does not provide nutrient transport to the surface, hindering both phytoplankton and zooplankton development. This defines a minimal level of vertical turbulence promoting larval survival. Successful larval transition to the active feeding stage requires an intermediate level of turbulence, which regulates the frequency of encounter of the larval and their particulate microplankton food above the critical level (Rothschild and Osborn 1988).

As shown, this mechanism is critical for sardine, anchovy, cod, herring, pollock, and some other species of pelagic fishes (see Klyashtorin and Lyubushin 2007; and

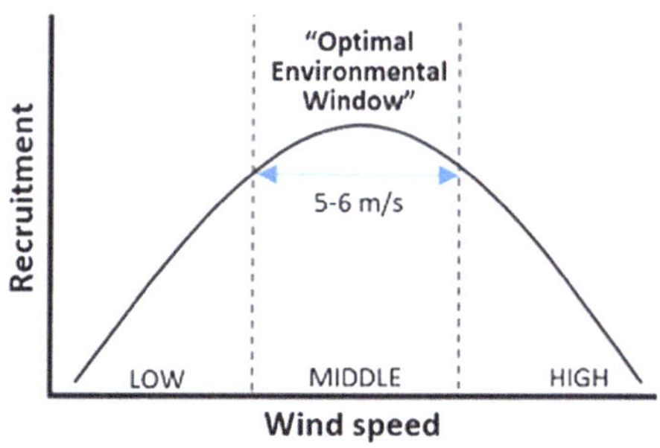

Fig. 2.20 Schematic of the "optimal environmental window". Based on data from Cury and Roy (1989)

references therein). However, with a further increase in wind speed the population growth is slowed, and maximum production requires an optimal wind speed of about 5–6 m/s. This wind speed range, which corresponds to average upwelling intensities and high recruitment, is labelled the "optimal environmental window". Further increase of wind speed and upwelling intensity increases offshore larval transport from the more favourable coastal environment into the open sea, where their mortality increases significantly (Parrish et al. 2000). In addition, excessive turbulent mixing disrupts the supportive local aggregates of small plankton required for the successful transition of fish larvae to their more mobile active feeding stage (Lasker 1978; Owen 1981).

2.4.3 Lasker's Hypothesis of a "Calm Ocean"

According to Lasker (1978), the microplankton required for initiation of active feeding by anchovy larvae comprises unicellular flagellate algae and copepod nauplii, which are irregularly distributed in the upper ocean, forming patches and often occurring as microlayers. Increased concentrations of anchovy are found in these microlayers, where they are observed to survive best. This layered structure is stable at low turbulence, but is disrupted by storms or strong winds, which cause intensive mixing of the surface layer. It has been found that a minimum 5-day stable zero-wind weather situation is required to form patches and microlayers of higher food organism concentrations. These periods of the so-called "calm ocean" that promote high survival of larvae were named "Lasker windows".

2.4.4 Cushing's "Match/Mismatch" Hypothesis

Based on the multi-year studies in the North Atlantic, Cushing (1990) formulated the underlying reasons for development of abundant recruitment of herring, cod, and other fish with drifting pelagic eggs. While the North Atlantic is not an upwelling system, the basic principles of this hypothesis can still apply. Since there is only a short season for phytoplankton and zooplankton development in this region, favourable conditions for high survival and growth of fish larvae are rather short-term. Ultimately, their survival, and thus breeding success, depends on the massive occurrence of copepod nauplii (the main food of larvae during their transition to active feeding) coinciding with the hatched larvae. When these events coincide, larval survival and the probability of a subsequent abundant generation developing increases. Without such a coincidence, recruitment decreases.

Seawater temperature strictly determines the rate of development of both eggs and larvae and the plankton that directly affect fish recruitment. Optimal vertical turbulence in the surface layer significantly accelerates zooplankton reproduction and promotes high survival of larvae. In contrast, high intensity wind mixing (regular spring storms, for example) destroys the oceans' upper layer structure and negatively affects survival of larvae, and thus the population recruitment.

2.5 Marine Food Web Structure in Coastal Upwelling Systems

Upwelling regions are important sources of marine productivity, and they attract hundreds of species at different trophic levels to create biodiversity hot spots. Many marine ecosystems typically contain a large number of species at the lower (e.g. planktonic) trophic levels. They also contain a substantial number of predatory fish, seabirds, or marine mammals that feed at the upper apex and near-apex trophic levels. However, in many of the highly productive ecosystems of the world, and particularly in upwelling regions, there tends to be a vital intermediate trophic level occupied by small, plankton-feeding pelagic fish that is typically dominated by only one (e.g. sardine), or at most a few, species (Bakun 1996).

For example, in South Africa's marine fauna, which is particularly rich and well documented, the three phyla mollusca, crustacea, and chordata are represented by 3062, 2333, and 2492 species, respectively (Gibbons et al. 1999). Among the 2000 marine fish species recorded, about 70 % are demersal, benthic, or reef species; in contrast, only 6.1 % large and 3.7 % small pelagic fish species are found (Fig. 2.21). Species diversity is relatively high at the bottom of the food chain (e.g. 429 copepods, 2262 gastropods) and at its top (92 species of marine birds and 41 species of marine mammals). This wasp-waist richness pattern appears to be a common characteristic of upwelling systems.

Food webs of eastern boundary current upwelling systems share various common structural characteristics, such as the dominance of pelagic over demersal consumers, the cell/body size and coarse-taxonomic-level community structure in both plankton and fish, and dominant life history adaptation strategies. A microbial loop consisting of heterotrophic bacteria, very small photosynthetic autotrophs, and unicellular microzooplankton is present in the plankton in all regions, but this is overlaid by another plankton food web consisting of medium-to-large photosynthetic autotrophs grazed by metazoans. For more details, see the reviews by Hutchings et al. (1995) and Mackas et al. (2006).

Apart from a high primary productivity by medium-to-large cell size diatoms, which is generally localized in time and space, patterns of productivity and consumption are leaky and time-space lagged (see Fig. 2.19). Mesozooplankton biomass is usually dominated by a few species of medium and large copepods, one or two euphausiid species, and one or more chaetognaths, all taking advantage of behavioural and life-history adaptations (Mackas et al. 2006). As noted previously, upwelling trophic levels typically have a wasp-waist structure (Cury et al. 2000), whereby the wasp-waist species (sardines) have localized spawning and nursery

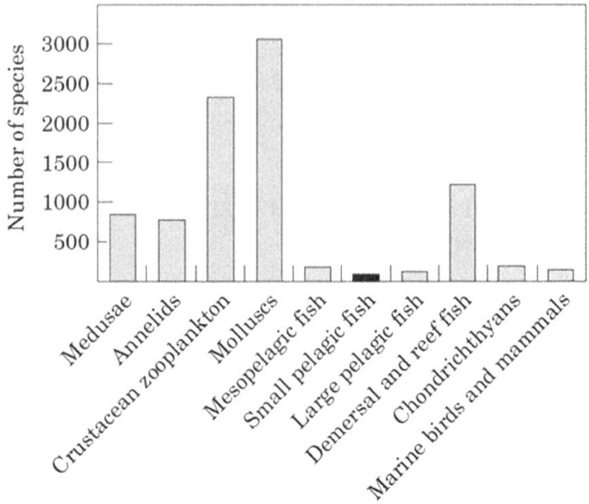

Fig. 2.21 Biodiversity of South Africa's marine fauna. Taken from Cury et al. (2000)

areas within the larger region and life spans that are short relative to decadal environmental variability. Dominant resident predators include hake and migratory fish (jack mackerel and tuna) that enter the upwelling system to feed, but spend part of their life offshore. Larger predators, such a fish, squid, seabirds and whales, survey the entire upwelling zone, targeting individual prey aggregations (Batchelder et al. 2002). The diversity of benthic macrofauna can be high, except where it is diminished in zones of hypoxia. Nearshore benthic species with meroplanktonic larval often rely on shoreward frontal displacement associated with upwelling relaxation in order to return to their settlement sites (Mackas et al. 2006).

The interaction of multiple time scales also influences behaviour of zooplankton and larval fish. In the Benguela Current, for example, various zooplankton and fish larvae undertake diel vertical migration, rising to the surface at night and descending into darker regions during the day. The usual explanation for this behaviour is predator-avoidance during daylight, but it has been shown that this behaviour interacts with the seasonal patterns of currents such that it retains the animals at a given latitude and distance offshore (e.g., Peterson et al. 1979; Peterson 1998). Moreover, some zooplankton and anchovy larvae use advanced techniques of diel vertical migration for moving onshore to more suitable spawning areas (see Mackas et al. 2006; and references therein).

2.6 Summary

This chapter described the functioning of the upwelling process in terms of physical and biogeochemical processes and marine food web dynamics, all closely interlinked and constrained by a large range of temporal and spatial scales and complex interactions. Generally, wind-driven coastal upwelling is the dominant upwelling mechanism creating highly productive coastal ecosystems and lucrative coastal fisheries. The next chapter details influences by the large-scale circulation, the natural variability of upwelling, and anthropogenic influences including eutrophication, global warming and exploitation of marine resources.

References

Allen, S.E., and X. Durrieu de Madron. 2009. A review of the role of submarine canyons in deep-ocean exchange with the shelf. *Ocean Science* 5: 607–620.

Arístegui, J., P. Tett, A. Hernández-Guerra, G. Basterretxea, M.F. Montero, K. Wild, P. Sangrà, S. Hernández-León, M. Cantón, J.A. García-Braun, M. Pacheco, and E.D. Barton. 1997. The influence of island-generated eddies on chlorophyll distribution: A study of mesoscale variation around Gran Canaria. *Deep-Sea Research I* 44: 71–96.

Arntz, W.E., V.A. Gallardo, D. Gutierrez, E. Isla, L.A. Levin, J. Mendo, C. Neira, G.T. Rowe, J. Tarazona and M. Wolff. 2006. El Niño and similar perturbation effects on the benthos of the Humboldt, California, and Benguela Current upwelling ecosystems. *Advances in Geosciences* 6: 243–265.

Bailey, G.W., and P. Chapman. 1985. The nutrient status of the St. Helena Bay region in February 1979. In *The South Africa Ocean Colour and Upwelling Experiment*, ed. L.V. Shannon, 125–145. Cape Town: Sea Fisheries Research Institute.

Bakun, A. 1973. Coastal upwelling indices, west coast of North America, 1946–71. U.S. Department of Commerce, *NOAA Technical Report*, NMFS SSRF-671, 103 p.

Bakun, A. 1975. Daily and weekly upwelling indices, west coast of North America, 1967–73. U.S. Department of Commerce, *NOAA Technical Report*, NMFS SSRF-693, 114 p.

Bakun, A. 1996. *Patterns in the Ocean: Ocean Processes and Marine Population Dynamics*. University of California Sea Grant, San Diego, California, USA, in cooperation with Centro de Investigaciones Biologicas de Noroeste, La Paz, Baja California Sur, Mexico, 323 pp.

Bakun, A., and C.S. Nelson. 1991. Wind stress curl in subtropical eastern boundary current regions. *Journal of Physical Oceanography* 21: 1815–1834.

Batchelder, H.P., et al. 2002. The GLOBEC northeast Pacific California Current System Program. *Oceanography* 15: 36–47.

Baumgartner, T.R., A. Soutar, and V. Ferreira-Bartrina. 1992. Reconstruction of the history of Pacific sardine and northern anchovy poulations over the past two millennia from sediments in the Santa Barbara Basin. *CalCOFI Reports* 33: 24–40.

Bremner, J.M. 1980. Physical parameters of the diatomaceous mud belt off South West Africa. *Marine Geology* 34: M67–M76.

Broecker, W.S., and T.H. Peng. 1982. *Tracers in the Sea*. Palisades, N. Y.: Lamont-Doherty Earth Observatory.

Castelão, R.M., and J.A. Barth. 2006. Upwelling around Cabo Frio, Brazil: The importance of wind stress curl. *Geophysical Research Letters* 33: L03602. doi:10.1029/2005GL025182.

Chavez, F.P., and M. Messié. 2009. A comparison of eastern boundary upwelling ecosystems. *Progress in Oceanography* 83(1–4): 80–96. doi:10.1016/j.pocean.2009.07.032.

Checkley, D.M., and J.A. Barth. 2009. Patterns and processes in the California Current System. *Progress in Oceanography* 83: 49–64. doi:10.1016/j.pocean.2009.07.028.

Chisholm, S.W. 1992. Phytoplankton size. In *Primary Productivity and Biogeochemical Cycles in the Sea*, ed. P.G. Falkowski, and A.D. Woodhead, 213–238. New York: Plenum.

Codispoti, L.A. 1983. On nutrient variability and sediments in upwelling regions. In *Coastal Upwelling, its Sediment Record. A. Response of the Sedimentary Regime to Present Coastal Upwelling*, ed. J. Thiede and E. Suess, 125–145. New York: Plenum Press.

Csanady, G.T. 1977. Intermittent "full" upwelling in Lake Ontario. *Journal of Geophysical Research* 82(1977): 397–419.

Cury, P., and C. Roy. 1989. Optimal environmental window and pelagic fish recruitment success in upwelling areas. *Canadian Journal of Fishery and Aquatic Sciences* 46: 670–680.

Cury, P., A. Bakun, R.J.M. Crawford, A. Jarre, R.A. Quinones, L.J. Shannon, and H.M. Verheye. 2000. Small pelagics in upwelling systems: Patterns of interaction and structural changes in "wasp-waist" ecosystems. *ICES Journal of Marine Science* 57: 603–618.

Cushing, D.H. 1990. Plankton production and year-class strength fish populations: An update of the match/mismatch hypothesis. *Advances in Marine Biology* 26: 249–293.

Cushman-Roisin, B. 1994. *Introduction to Geophysical Fluid Dynamics*. Upper Saddle River, N. J.: Prentice-Hall. 320 pp.

Dale, A.W., S. Sommer, U. Lomnitz, I. Montes, T. Treude, J. Gier, C. Hensen, M. Dengler, K. Stolpovsky, I.D. Bryant, and K. Wallmann. 2014. Organic carbon production, mineralization and preservation on the Peruvian margin. *Biogeosciences Discussions* 11: 13067–13126. doi:10.5194/bgd-11-13067-2014.

Doty, M.S., and M. Oguri. 1956. The island mass effect. *International Council for the Exploration of the Sea* 22: 33–37.

Ducklow, H.W., D.K. Steinberg, and K.O. Buesseler. 2001. Upper-ocean export and the biological pump. *Oceanography* 14: 50–58.

Dugdale, R.C., F.P. Wilkerson, and H.J. Minas. 1995. The role of a silicate pump in driving new production. *Deep-Sea Research I* 42: 697–719.

Duncombe Rae, C.M., F.A. Shillington, J.J. Agenbag, J. Taunton-Clark, and M.L. Grundlingh. 1992. An Agulhas ring in the South Atlantic Ocean and its interaction with the Benguela upwelling system. *Deep-Sea Research I* 39: 2009–2027.

Egge, J.K., and L. Aksnes. 1992. Silicate as regulating nutrient in phytoplankton competition. *Marine Ecology Progress Series* 83: 281–289.

Ekman, V.W. 1905. On the influence of the Earth's rotation on ocean currents. *Arkiv foer Matematik Astronomi och Fysik* 2: 11.

Garcia, V.M.T., S. Signorini, C.A.E. Garcia, and C.R. McClain. 2006. Empirical and semianalytical chlorophyll-*a* algorithms in the southwestern Atlantic coastal region (25–40° S and 60–45° W). *International Journal of Remote Sensing* 27: 1539–1562.

Gibbons, M.J., et al. 1999. The taxonomic richness of South Africa's marine fauna: A crisis at hand. *South African Journal of Science* 95: 8–12.

Gitelson, A. 1993. Algorithms for remote-sensing of phytoplankton pigments in inland waters. *Advances in Space Research* 13: 197–201.

Gordon, H.R. 1979. Diffuse reflectance of the ocean: The theory of its augmentation by chlorophyll-*a* fluorescence at 685 nm. *Applied Optics* 18: 1161–1166.

Gordon, A.L., and R.D. Susanto. 2001. Banda Sea surface layer divergence. *Ocean Dynamics* 52: 2–10.

Gower, J., and S. King. 2007. Validation of chlorophyll-*a* fluorescence derived from MERIS on the west coast of Canada. *International Journal of Remote Sensing* 28: 625–635.

Gower, J.F.R., L. Brown, and G.A. Borstad. 2004. Observation of chlorophyll-*a* fluorescence in west coast waters of Canada using the MODIS satellite sensor. *Canadian Journal of Remote Sensing* 30: 17–25.

Graham, W.M., and J.L. Largier. 1997. Upwelling shadows as nearshore retention sites: The example of northern Monterey Bay. *Continental Shelf Research* 17: 509–532.

Graham, W.M., J.G. Field, and D.C. Potts. 1992. Persistent 'upwelling shadows' and their influence on zooplankton distributions. *Marine Biology* 114: 561–570.

Gruber, N., and J.L. Sarmiento. 1997. Global patterns of marine nitrogen fixation and denitrification. *Global Biogeochemical Cycles* 11: 235–266.

Hamersley, M.R., G. Lavik, D. Woebken, J.E. Rattray, P. Lam, et al. 2007. Anaerobic ammonium oxidation in the Peruvian oxygen minimum zone. *Limnology and Oceanography* 52: 923–933.

Harvey, H.W. 1928. *Biological Chemistry and Physics of Sea Water*. Cambridge: Cambridge University Press. 194 pp.

Hasegawa, D., H. Yamazaki, R.G. Lueck, and L. Seuront. 2004. How islands stir and fertilize the upper ocean. *Geophysical Research Letters* 31: L16303. doi:10.1029/2004GL020143.

Heywood, K.J., E.D. Barton, and J.H. Simpson. 1990. The effects of flow disturbance by an oceanic island. *Journal of Marine Research* 48: 55–73.

Hickey, B.M. 1997. The response of a steep-sided, narrow canyon to time-variable wind forcing. *Journal of Physical Oceanography* 27: 697–726.

Hormazabal, S., G. Shaffer, and O. Leth. 2004. Coastal transition zone off Chile. *Journal of Geophysical Research* 109: C01021. doi:10.1029/2003JC001956.

Hutchings, L., G.C. Pitcher, T.A. Probyn, and G.W. Bailey. 1995. The chemical and biological consequences of coastal upwelling. In *Upwelling in the Ocean: Modern Processes and Ancient Records*, ed. C.P. Summerhayes, K.-C. Emeis, M.V. Angel, R.L. Smith, and B. Zeitschel, 65–81. Chichester: Wiley.

Hutchings, L., H.M. Verheye, J.A. Hugett, H. Demarcq, R. Cloete, R.G. Barlow, D. Louw, and A. da Silva. 2006. Variability of plankton with reference to fish variability in the Benguela Current large marine ecosystem: An overview. In *Benguela: Predicting a Large Marine Ecosystem*, ed. V. Shannon, G. Hempel, P. Manalotte-Rizzoli, C. Moloney, and J. Woods, 91–124. Amsterdam: Elsevier.

Jenkins, A.D., and J.A.T. Bye. 2006. Some aspects of the work of V.W. Ekman. *Polar Record* 42 (220): 15–22.

Kämpf, J. 2012. Lee effects of localized upwelling in a shelf-break canyon. *Continental Shelf Research* 42: 78–88. doi:10.1016/j.csr.2012.05.005.

Kämpf, J. 2015. Phytoplankton blooms on the western shelf of Tasmania: Evidence of a highly productive ecosystem. *Ocean Science* 11: 1–11. doi:10.5194/os-11-1-2015.

Kämpf, J., M. Doubell, D. Griffin, R.L. Matthews, and T.M. Ward. 2004. Evidence of a large seasonal coastal upwelling system along the southern shelf of Australia. *Geophysical Research Letters* 31: L09310. doi:10.1029/2003GLO19221.

Klyashtorin, L., and A. Lyubushin. 2007. *Cyclic Climate Changes and Fish Productivity*. Moscow: VNIRO. 223 pp.

Kraus, E.B., and J.A. Businger. 1994. *Atmosphere-Ocean Interaction*. Oxford University Press, 362 pp.

Kuypers, M.M.M., G. Lavik, D. Woebken, M. Schmid, B.M. Fuchs, R. Amann, B.B. Jorgensen, and M.S.M. Jetten. 2005. Massive nitrogen loss from the Benguela upwelling system through anaerobic ammonium oxidation. *Proceedings of the National Academy of Sciences* 102: 6478–6483.

Lam, P., and M.M.M. Kuypers. 2011. Microbial nitrogen cycling processes in oxygen minimum zones. *Annual Reviews of Marine Science* 3: 317–345.

Lasker, R. 1978. The relation between oceanographic conditions and larval anchovy food in the California Current: Identification of the factors leading to recruitment failure. *Rapports et procès-verbaux des réunions/Conseil permanent international pour l'exploration de la mer* 1073: 212–230.

Mackas, D.L., P.T. Strub, A. Thomas, and V. Montecino. 2006. Eastern regional ocean boundaries pan-regional overview. In *The Sea*, vol. 14a, ed. A.R. Robinson, and K. Brink, 21–60. Cambridge, Massachussetts: Harvard University Press.

Mann, K.H. 1993. Physical oceanography, food chains, and fish stocks: A review. *ICES Journal of Marine Science* 50: 105–119.

Maritorena, S., A. Morel, and B. Gentili. 2000. Determination of the fluorescence quantum yield by oceanic phytoplankton in their natural habitat. *Applied Optics* 39: 6725–6737.

Mason, J.E., and A. Bakun. 1986. Upwelling index update, U.S. west coast, 33 N-48 N latitude. U.S. Dep. Commer., NOAA Tech. Memo., NOAA-TM-NMFS-SWFC- 67, 81 p.

Messié, M., J. Ledesma, D.D. Kolber, R.P. Michisaki, D.G. Foley, and F.P. Chavez. 2009. Potential new production estimates in four eastern boundary upwelling systems. *Progress in Oceanography* 83(1–4): 151–158.

Montecino, V., P.T. Strub, F. Chavez, A.C. Thomas, J. Tarazona and T. Baumgartner. 2005. Biophysical interaction off western South America. *The sea, vol 14* In ed. A.R. Robinson and K.H. Brink, pp. 329–360. Harvard University Press.

Monteiro, P.M.S. 2010. The Benguela Current system. In *Carbon and Nutrient Fluxes in Continental Margins*, ed. K.-K. Liu, L. Atkinson, R. Quiñones, and L. Talaue-McManus, 29–44. Berlin: Springer.

Mulder, A., A.A. van der Graaf, L.A. Robertson, and J.G. Kuenen. 1995. Anaerobic ammonium oxidation discovered in a denitrifying fluidized bed reactor. *FEMS Microbiological Ecology* 16: 177–184.

Neville, R.A., and J.F.R. Gower. 1977. Passive remote sensing of phytoplankton via chlorophyll-*a* fluorescence. *Journal of Geophysical Research* 82: 3487–3493.

Officer, C.B., and J.H. Ryther. 1980. The possible importance of silicon in marine eutrophication. *Marine Ecology Progress Series* 3: 83–91.

Owen, R.W. 1981. Microscale plankton patchiness in the larval anchovy environment. *Rapports et procès-verbaux des réunions/ Conseil permanent international pour l'exploration de la mer* 178: 364–368.

Parrish, R.H., F.B. Schwing, and P. Mendelsson. 2000. Mid latitude wind stress: The energy source for climatic shifts in the North Pacific Ocean. *Fishery Oceanography* 9: 224–238.

Peterson, W. 1998. Life cycle strategies of copepods in coastal upwelling zones. *Journal of Marine Systems* 15(1–4): 313–326.

Peterson, W.T., C.B. Miller, and A. Hutchinson. 1979. Zonation and maintenance of copepod populations in the Oregon upwelling zone. *Deep-Sea Research* 26: 467–494.

Pond, S., and G.L. Pickard. 1983. *Introductory Dynamical Oceanography*, 2nd ed. Oxford: Pergamon Press. 329 pp.

Quiñones, R.A., M.H. Gutierrez, G. Daneri, D.G. Aguilar, H.E. Gonzales, and F.P. Chavez. 2010. The Humboldt Current system. In *Carbon and Nutrient Fluxes in Continental Margins*, ed. K.-K. Liu, L. Atkinson, R. Quiñones, and L. Talaue-McManus, 44–64. Berlin: Springer-Verlag.

Redfield, A.C., B.H. Ketchum, and F.A. Richards. 1963. The influence of organisms on the composition of seawater. In *The Sea*, vol. 2, ed. M.N. Hill, 26–77. New York: Wiley Interscience.

Rothschild, B.J., and T.R. Osborn. 1988. The effect of turbulence on planktonic contact rates. *Journal of Plankton Research* 10: 465–474.

Ryan, J.P., F.P. Chavez, and J.G. Bellingham. 2005. Physical-biological coupling in Monterey Bay, California: Topographic influences on phytoplankton ecology. *Marine Ecology Progress Series* 287: 23–32.

Rykaczewski, R.R., and D.M. Checkley. 2008. Influence of ocean winds on the pelagic ecosystem in upwelling regions. *Proceedings of the National Academy of Sciences of the United States of America* 105: 1965–1970. doi:10.1073/pnas.0711777105.

Santana-Casiano, J.M., M. Gonzales-Davil, and I.R. Ucha. 2009. Carbon dioxide fluxes in the Benguela upwelling system during winter and spring. A comparison between 2005 and 2006. *Deep-Sea Research II* 46: 533–541.

Shaffer, G., O. Pizarro, L. Djurfeldt, S. Salinas, and J. Rutllant. 1997. Circulation and low-frequency variability near the Chilean coast: Remotely forced fluctuations during the 1991–92 El Niño. *Journal of Physical Oceanography* 27: 217–230.

Shannon, L.V. 1985. The Benguela Ecosystem Part I: Evolution of the Benguela, physical features and processes. *Oceanography and Marine Biology Annual Reviews* 23: 105–182.

Shannon, L.V., and G. Nelson. 1996. The Benguela: Large-scale phenomena and processes and system variability. In *The South Atlantic: Present and Past Circulation*, ed. G. Wefer, W.H. Berger, G. Siedler, and D.J. Webb, 163–210. Berlin: Springer.

Shannon, L.V., P. Schlittenhardt, and S.A. Mostert. 1984. The Nimbus-7 CZCS experiment in the Benguela Current region off southern Africa, February 1980. 2. Interpretation of imagery and oceanographic implications. *Journal of Geophysical Research* 89: 4968–4976.

Shanmugam, P. 2011. A new bio-optical algorithm for the remote sensing of algal blooms in complex ocean waters. *Journal of Geophysical Research* 116: C04016. doi:10.1029/2010JC006796.

Smayda, T.J. 1990. Novel and nuisance phytoplankton blooms in the sea: Evidence for a global epidemic. In *Toxic Marine Phytoplankton*, ed. E. Graneli, B. Sundstrom, L. Edler, et al., 29–40. Amsterdam: Elsevier.

Suess, E. 1976. Nutrients near the depositional interface. In *The Benthic Boundary Layer*, ed. I.N. McCave, 57–79. London: Plenum Press.

Sverdrup, H.U., M.W. Johnson and R.H. Fleming. 1942. *The oceans, their physics, chemistry, and general biology*. Prentice-Hall Inc.

Taunton-Clark, J. 1985. The formation, growth and decay of upwelling tongues in response to the mesoscale wind field during summer. In *South African Ocean Colour and Upwelling Experiment*, ed. L.V. Shannon, 47–61. South Africa: Sea Fisheries Research Institute.

Thamdrup, B., T. Dalsgaard, M.M. Jensen, O. Ulloa, L. Farias, and R. Escribano. 2006. Anaerobic ammonium oxidation in the oxygen-deficient waters off northern Chile. *Limnology and Oceanography* 51: 2146–2156.

Thomas, A.C., P.T. Strub, M.-E. Carr, and R. Weatherbee. 2004. Comparisons of chlorophyll variability between the four major global eastern boundary currents. *International Journal of Remote Sensing* 25: 1443–1447.

Thompson, R.O.R.Y., and T.J. Golding. 1981. Tidally induced 'upwelling' by the Great Barrier Reef. *Journal of Geophysical Research* 86: 6517–6521.

Tomczak, M., and J.S. Godfrey. 2003. *Regional Oceanography: An Introduction*, 2nd edn. Daya Publishing House, 390 pp.

Torres, R., D. Turner, J. Rutllant, M. Sobarzo, T. Antezana, and H. Gonzalez. 2002. CO_2 outgassing off central Chile (31–30° S) and northern Chile (24–23° S) during austral summer 1997. The effect of wind intensity on the upwelling and ventilation of CO_2 rich waters. *Deep-Sea Research I* 49: 1413–1429.

Tsunogai, S., S. Watanabe, and T. Sato. 1999. Is there a "continental shelf pump" for the absorption of atmospheric CO_2? *Tellus Series B* 51: 701–712.

Van der Graaf, A.A., A. Mulder, P. de Bruijn, M.S.M. Jetten, L.A. Robertson, and J.G. Kuenen. 1995. Anaerobic oxidation of ammonium is a biologically mediated process. *Applied Environmental Microbiology* 61: 1246–1251.

Ward, B.B., and A.F. Carlucci. 1985. Marine ammonia- and nitrite-oxidizing bacteria serological diversity determined by immunofluorescence in culture and in the environment. *Applied Environmental Microbiology* 50: 194–201.

Ware, D.M., and R.E. Thompson. 1991. Link between long-term variability in upwelling and fish production in the northeast Pacific Ocean. *Canadian Journal of Fisheries and Aquatic Sciences* 48(12): 2296–2306.

Wolanski, E., and W.M. Hamner. 1988. Topographically controlled fronts in the ocean and their biological influence. *Science* 241: 177–181.

Xing, X.-G., D.-Z. Zhao, Y.-G. Liu, J.-H. Yang, P. Xiu, and L. Wang. 2007. An overview of remote sensing of chlorophyll fluorescence. *Ocean Science Journal* 42: 49–59. doi:10.1007/BF03020910.

Zehr, J.P., and R.M. Kudela. 2011. Nitrogen cycle of the open ocean: From genes to ecosystems. *Annual Reviews of Marine Science* 3: 197–225.

Chapter 3
Large-Scale Setting, Natural Variability and Human Influences

Abstract This chapter describes the large-scale oceanic circulation and how this influences upwelling systems, natural variability of physical factors shaping marine ecosystems and factors influencing the ecology of coastal upwelling systems including climate variability, climate change, commercial fishing, and various sources of pollution.

Keywords Upwelling · Large-scale setting · Water masses · Nutricline structure · Climate variability · Anthropogenic influences · Fishery impacts · Fisk stock variations

There is, one knows not what sweet mystery about the sea,
whose gently awful stirrings seem to speak of some hidden soul beneath.

Herman Melville (1819–1891)

(Taken from *Moby Dick*, 1851)

3.1 The Large-Scale Setting, Water Masses and Ventilation

3.1.1 Wind-Driven Circulation and Nutricline Structure

Upwelling systems are attached to and influenced by the large-scale oceanic circulation that sets the overall structure and depth of the nutricline. The wind-driven circulation of the ocean (Fig. 3.1) includes both year-round and seasonal current patterns. Year-round features include the subtropical and subpolar gyres, equatorial countercurrents, the Leeuwin Current and the Antarctic Circumpolar Current. The most important variable seasonal currents are found in the northern Indian Ocean, where monsoonal wind changes (not shown in Fig. 3.1) cause major current reversals. While these currents are important agents of meridional water, heat and freshwater transports, they are also reflected in the structure of the permanent

© Springer International Publishing Switzerland 2016

J. Kämpf and P. Chapman, *Upwelling Systems of the World*,
DOI 10.1007/978-3-319-42524-5_3

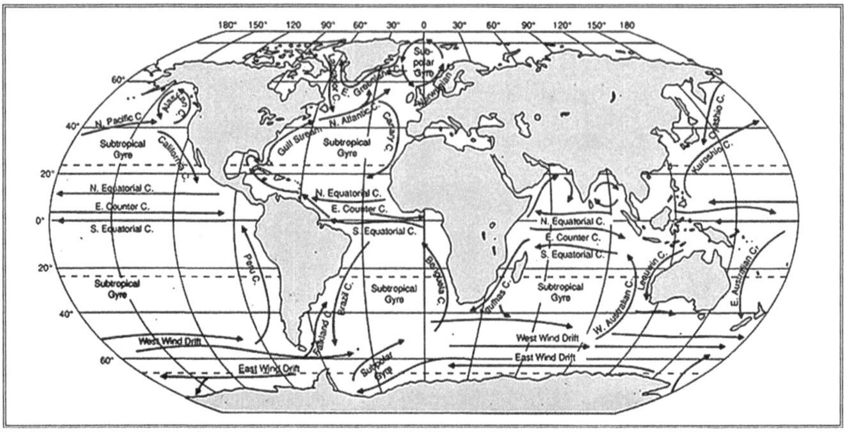

Fig. 3.1 The structure of the large-scale wind-driven circulation of the ocean. Note that the circulation during the southwest monsoon is shown in the Indian Ocean. Taken from Thurman and Trujillo (1999)

thermocline and nutricline. For example, all subtropical gyres are characterized by an eastward upward-doming of the thermocline and nutricline, which brings nutrient-rich water close to the euphotic zone near their eastern boundaries (Fig. 3.2). It is clear that this shoaling of the nutricline plays a key role in the high productivity of the big four coastal upwelling systems (see Chap. 10). Theories of the ventilation of the thermocline, which is intrinsically coupled with the distribution of the oxygen minimum zone (e.g., Brandt et al. 2010), were first developed by Luyten et al. (1983).

3.1.2 *Source Depth of Upwelled Water and Water Masses*

Upwelled nutrient-rich water typically comes from relatively shallow depths of 100–200 m. The properties (temperature, salinity, oxygen and nutrient levels) of water masses at that depth are often controlled by the large-scale circulation rather than by regional processes. Upwelled water in the four major eastern boundary coastal upwelling systems, for instance, is derived from complex current systems that involve relatively oxygen-poor, nutrient-rich water of equatorial undercurrents feeding into poleward undercurrents that follow the shelf break and provide source waters to the cross-shore upwelling circulation (see Chap. 10).

Shelf-break currents can also operate as a preconditioning process for the classical wind-driven upwelling mechanism. Examples are the nutrient-rich intermediate water of the Kuroshio Current which interacts with the shelf break of the East China Sea (see Sect. 8.2.2) and the Flinders Current that, as a result of its

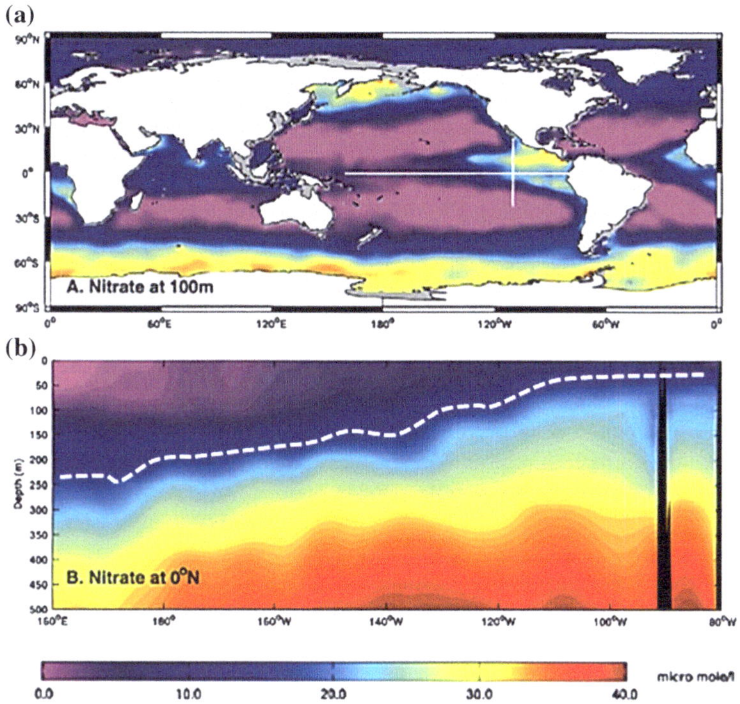

Fig. 3.2 Large-scale structure of the nutricline **a** at a depth of 100 m, and **b** along the equator in the Pacific Ocean. The *dashed white line* highlights the tilt of the 15 μM/L nitrate contour. Taken from Pennington et al. (2010)

interaction with shelf-break canyons at the shelf edge, brings nutrient-rich sub-surface water onto the southern shelves of Australia (see Sect. 8.2.4).

Water-mass analysis is a technique that allows us to identify the origins and pathways of individual water masses in the ocean (see Tomczak and Godfrey 2003). Individual water masses, which can be distinguished from each other through e.g., temperature/salinity or other chemical characteristics, are formed via air-sea exchanges of heat and freshwater in specific regions of the world ocean where they also receive their initial chemical signature via gas and particle transfers and biogeochemical processes in the surface mixed layer. In their formation regions, water masses are confined to the surface mixed layer from where they are pushed down or "subducted" into the ocean interior. Subduction takes place either in the form of density-driven gravity flows or is induced by external forcing such as the wind-induced convergence of Ekman flows in a region with outcropping isopyncals (see Tomczak and Godfrey 2003). As water masses in the ocean interior are transported away from their formation regions, they accrue dissolved nutrients from the remineralization of sinking detritus and their oxygen concentration drops (see Chap. 2).

Water masses formed by subduction in the open ocean that are relevant to upwelling regions are:

- Central Water, i.e. waters within the permanent thermocline and nutricline, and
- Intermediate Water, typically found below the permanent thermocline.

Central waters generally have relatively high nutrient levels, but nutrient concentrations vary significantly in the vertical and, to some extent, also laterally, due to the ventilation history of individual water masses. Figure 3.3 shows the distribution of different types of Central Water, which can be distinguished by their water-mass properties. There are also so-called 'Mode Waters' which refer to layers of nearly vertically homogeneous water found over a relatively large geographical area. After subduction, mode waters usually occur within or near the top of the permanent pycnocline, and hence are apparent through the contrast in stratification within the pycnocline waters. Mode waters are defined by the properties of water masses at their formation region. Mode waters can also contribute to the formation of Central Water (Hanawa and Talley 2001).

Intermediate Water is typically found somewhat deeper at depths <1000–1500 m, but, in some instances, western boundary currents such as the Kuroshio Current can dynamically lift this nutrient-rich water up across the shelf break and onto the continental shelf (Chen 1996). It is clear that the source depth of upwelled water plays an important role in the biogeochemical cycles in upwelling systems by determining the eventual concentrations of nutrients in the water that reaches the surface. Water mass analysis is also key to inferring changes in the large-scale circulation, which is indeed important for the understanding of changes in upwelling-induced nutrient fluxes.

3.1.3 Water Mass Properties of Upwelling Water

The proportions of different water masses that make up the water properties in a region can often be inferred from temperature and salinity data. In turn, this

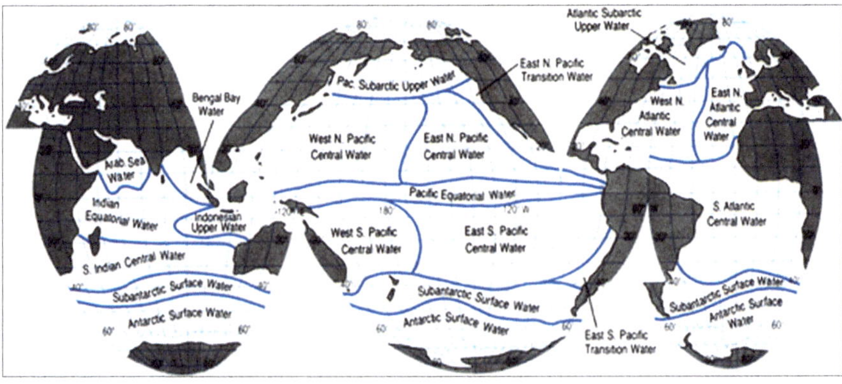

Fig. 3.3 Global distribution of upper water masses (0–500 m) (from Lalli and Parsons 1993)

information gives clues about chemical characteristics and the circulation of water masses, and the source depth from which the water is upwelled. In contrast to the open ocean, coastal ocean water-mass properties are markedly influenced by amplified seasonal cycles of temperature and salinity because of their shallow depths and continental influences such as river discharge or outflows of hypersaline water from inverse estuaries. Thus it is no surprise that measurements of temperature and salinity in coastal waters are highly variable and often difficult to interpret. Nevertheless, this task is easier for upwelling regions, given that upwelling events leave behind a pronounced temperature signature. In upwelling regions, in situ temperature (and salinity) measurements can give information on the source water properties including nutrient levels of upwelled water.

For instance, let us briefly consider the seasonal coastal upwelling system that develops on the southern shelves of Australia during austral summer months (Kämpf et al. 2004). Measurements undertaken during an upwelling event in the upwelling centre of the region indicate that upwelled water has temperatures <12 °C (Fig. 3.4a). This water can be traced back to an average depth of ~310 m (Fig. 3.4b) within the nutricline of Subantarctic Mode Water that the Flinders Current carries westward along the adjacent upper continental slope (Kämpf 2010).

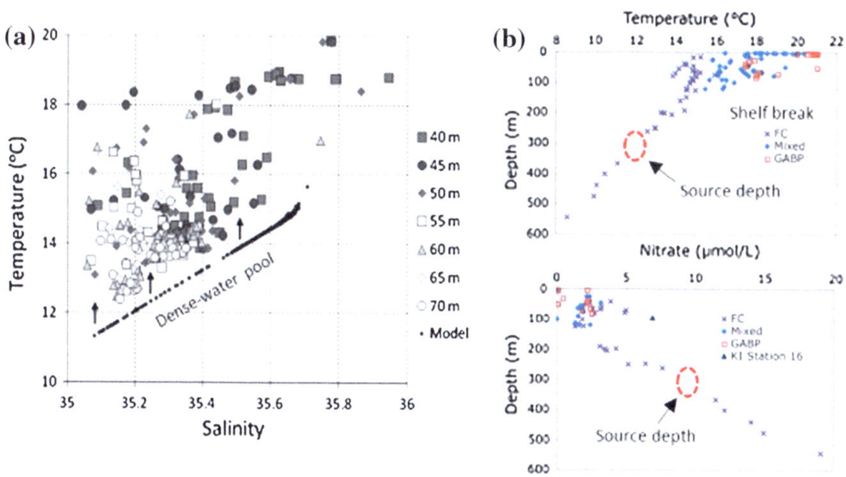

Fig. 3.4 Panel **a** Symbols are in situ measurements of temperature and salinity of bottom shelf waters during an upwelling event in the upwelling centre off the southern coast of the Eyre Peninsula, South Australia. Note how upwelling water tends to get warmer and saltier towards the surface. Coastal upwelling is fed by shelf waters formed in an adjacent dense-water pool, known as the Kangaroo Island Pool. The T-S properties of this pool (*black dots*) are derived from a hydrodynamic model application. *Source* Kämpf (2010). Panel **b** shows vertical profiles of temperature and nitrate measured in the open ocean adjacent to the upwelling centre. FC is the Flinders Current. Ellipses indicate the source depth and nutrient level of upwelled water. Modified from Richardson et al. (2009)

3.2 Seasonal Variability

If the surface ocean was at steady state, with an assumed constant nutrient supply and an unvarying surface mixed layer having the thickness of the euphotic zone, the production of phytoplankton in the surface ocean would simply follow the seasonal cycle of insolation. In reality, both density stratification, which affects the surface mixed layer thickness, and changes in nutrient availability interrupt this cycle. For example, a deeper surface mixed layer reduces the contact time of phytoplankton cells with the euphotic zone. On the other hand, nutrients become depleted over time in a shallow surface mixed layer in the absence of nutrient supply mechanisms such as wind-induced entrainment, upwelling or internal waves.

In the open North Atlantic Ocean, for instance, deep convective winter mixing brings nutrients to the upper ocean without significant consumption. The establishment of a shallow surface mixed layer with reduced turbulence during spring triggers a widespread phytoplankton spring bloom which recedes in the following months due to nutrient limitation (Fig. 3.5). A zooplankton maximum follows during early summer. Autumn storms and atmospheric cooling lead to the entrainment of nutrient-enriched water into the surface mixed layer, again triggering a transient phytoplankton bloom. This bloom recedes in winter months given reduced insolation and substantial convective deepening of the surface mixed layer.

In stark contrast to the two-peak bloom feature of the North Atlantic, the North Pacific and Arctic waters only experience a single phytoplankton bloom in late summer (Fig. 3.6). In the North Pacific, the lack of a spring diatom bloom may be due to iron deficiency (e.g., Gregg et al. 2003), while in the Arctic Ocean the problem is light limitation and persistently strong density stratification (e.g., Popova et al. 2010). In tropical waters, there is little seasonal variation in insolation or density stratification in the upper ocean. Hence, there are continuous levels of primary and secondary production year-round outside seasonal coastal upwelling systems (Cushing 1975). In shelf seas subject to nutrient input from upwelling or

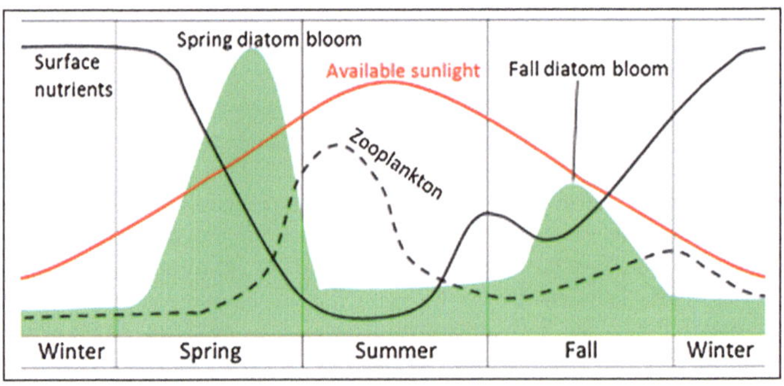

Fig. 3.5 Schematic of ambient conditions that trigger the development of spring and fall (autumn) blooms in the North Atlantic

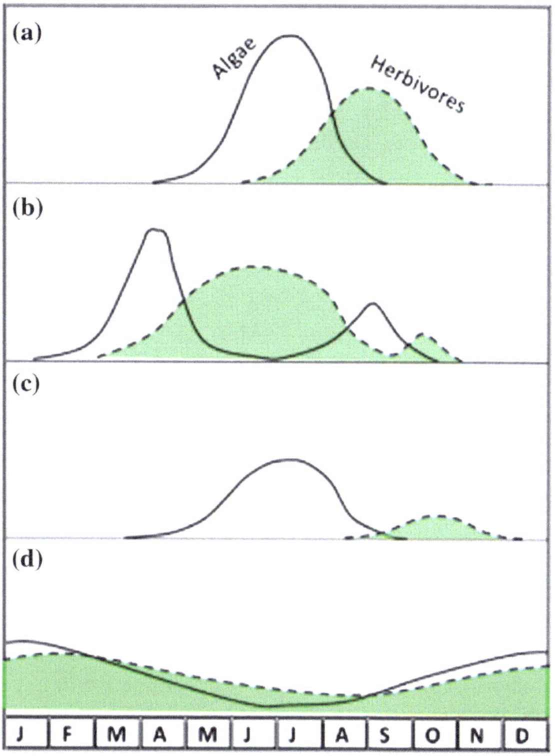

Fig. 3.6 Seasonal cycles of phytoplankton (algae) and zooplankton (herbivores) in different regions of the ocean. Modified from Cushing (1975). **a** Arctic. **b** North Atlantic. **c** North Pacific. **d** Tropical.

fluvial sources, seasonal cycles of phytoplankton concentration and zooplankton abundance can be dramatically different when compared to the ambient ocean.

3.3 Climate Variability and Climate Change

3.3.1 Modes of Climate Variability

Climate variables are properties of the climate system that influence climate from regional to global scales. Sea surface temperature, for instance, is a key oceanic climate variable in the surface heat-flux budget. How global climate varies depends on the way these climate variables change on timescales of 2–10 years, or even longer. Each zonal wind system of the general atmospheric circulation can be associated with a specific mode of climatic variability which is determined statistically from long-term records. For instance, two long-term climate oscillations have been recognized in the northern hemisphere, the *North Atlantic Oscillation* (NAO) and the *Arctic Oscillation*, also called Northern Annular Mode (NAM), that appear to be governed by changes in the boreal westerly winds on decadal-scale time periods. Changes in the trade winds on 2–5 year time-scales give rise to another identifiable oscillation, the *El Niño Southern Oscillation* (ENSO) , whereas variations of the austral Westerlies—the Roaring Forties—can be described by *Southern Annular Mode* (SAM), also known as the Antarctic Oscillation. Interactions between these climatic modes are referred to as *teleconnections*.

Because wind speed is driven essentially by atmospheric pressure differences, these climatic modes are mainly described by atmospheric indices (i.e., the sea level pressure difference between two selected regions). There is, however, another class of climatic modes, based on seawater temperatures, that reflects the ocean's internal dynamics. This includes the *Pacific Decadal Oscillation* (PDO) , the *Atlantic Multidecadal Oscillation* (AMO), the *Antarctic Circumpolar Wave* (AACW) and the *Indian Ocean Dipole* (IOD).

Oceanic climatic modes typically take place over longer time periods than the atmospheric modes, and their cycles are typically of the order of one or more decades. This reflects the much greater viscosity of the ocean compared with the atmosphere and the slower rates of ocean mixing processes. While there is considerable interest in whether variations in these indices have direct effects on upwelling systems, and if so, which, the intrinsic variability of ocean-atmosphere interactions, the multiplicity of potential interactions between different modes, and the scarcity of data from most regions makes it difficult to determine how large such effects are (e.g., Reason et al. 2006). Here, we only discuss those modes of variability that are likely to impact the main upwelling regions, so will omit those in the high latitudes.

The North Atlantic Oscillation (Walker and Bliss 1932; van Loon and Rogers 1978; Wallace and Gutzler 1981) is a mode of climatic variability that depends on modulation of the strength and storm tracks of the boreal Westerlies. Thus, it strongly influences the weather in North America and northern Europe. It is defined in terms of the pressure difference between the Azores (high pressure) and Iceland (low pressure), and positive NAO indices (large pressure differences) result in stronger winter storms with a more northerly track, leading to warm wet winters in Europe and cold, dry winters in northern Canada and Greenland.

The negative (low pressure difference) phase of the NAO results in fewer, weaker storms that result in a cold northern Europe and a wet Mediterranean. Although the index varies from year to year, it can become fixed in a positive or negative mode for several years; thus the period from 1955–1970 was consistently negative, while during 1970–2000 it was generally in the positive phase. The NAO also influences the formation of North Atlantic Deep Water—the key driver of the ocean's deep circulation—in the Greenland and Labrador Seas (e.g., Dickson et al. 1996), and biological production in the northern Atlantic through storm-induced interference with the spring bloom (e.g., Zhai et al. 2013). Nonetheless, the North Atlantic Oscillation has only little impact on the average intensity of coastal upwelling along the coasts of northwest Africa and Portugal (Narayan et al. 2010).

In the 1920s, Gerhard Schott and Erwin Schweigger in the 1920s (Cushman 2004) and later Klaus Wyrtki (1975) deserve recognition for interpreting oceanic features of the El Niño Southern Oscillation, which has a profound influence on upwelling off Peru. Later came the first descriptions of decadal-scale variability associated with the Pacific Decadal Oscillation (PDO) (e.g., Beamish 1993; Latif and Barnett 1994; Francis and Hare 1994).

The El Niño Southern Oscillation (Karoly 1989; Philander 1990; Burgers 1999; Barnett et al. 1999), of which the importance was first realized 500 years ago off Peru, is a climatic mode of variability that influences most of the equatorial Pacific Ocean and can at times extend into the Atlantic and Indian Oceans as well. It was named El Niño in association with the coming of the Christ child, as it was usually noticed in late December. The *Walker Circulation* (Walker 1923) is a vertical circulation cell in the atmosphere that is set up along the equator in the Pacific Ocean (Fig. 3.7a). This circulation consists of the trade winds blowing towards the west near the sea surface and an eastward return flow of air at higher levels. On average, warm and moist air rises over the *Warm Pool* in the western Pacific Ocean, where it creates tropical cumulus clouds and enhanced precipitation. The location of

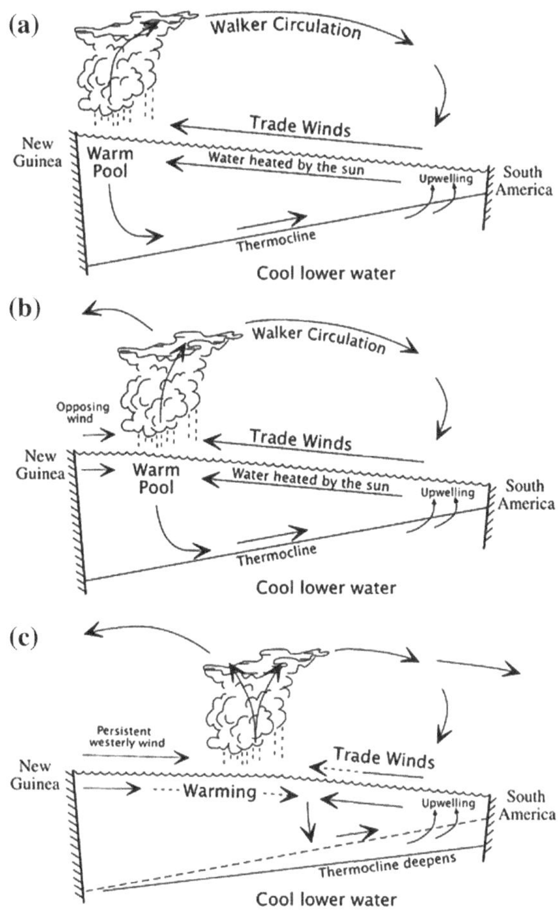

Fig. 3.7 Illustration of processes involved in the creation of an El Niño event. Panel **a** shows the average situation. Panels **b** and **c** display the onset and full development of the event. Courtesy of Billy Kessler, Pacific Marine Environmental Laboratory/NOAA. http://faculty.washington.edu/kessler/occasionally-asked-questions.html [accessed on 4 April 2016]

this thermal updraft region strongly depends on sea surface temperature, which modulates the flux of moisture (evaporation) into the atmosphere.

An El Niño event commences with a westerly wind burst in the western Pacific Ocean and a reversal of the trade winds in this sector (Fig. 3.7b). This usually creates a tropical cyclone pair in the lower atmosphere and an oceanic wave—an *equatorial Kelvin wave*—that travels eastward along the equator at a speed of 2–3 m/s (see Tomczak and Godfrey 2003). It takes ~20–30 days for the wave to cross the Pacific Ocean. This wave lowers the thermocline and suppresses equatorial upwelling along its path (Fig. 3.7c). As a consequence of this disturbance, the Warm Pool and thermal updraft region of the Walker Circulation is shifted eastward to the central equatorial Pacific, near the dateline, where it induces enhanced precipitation.

As the wave hits the west coast of central America, it turns into a pair of *coastal Kelvin waves* that propagate along the coasts and away from the equator both northward and southward. Like the equatorial wave, the coastal Kelvin waves operate to lower the thermocline along their path, which suppresses coastal upwelling of cold, nutrient-rich water off the coasts of Chile, Peru and California, even though the winds in the coastal region are still blowing towards the equator. Eventually the waves move back towards the centre of the Pacific Ocean as a pair of westward-propagating Rossby waves and the thermocline along the coast rises again to “average” conditions.

While ENSO events are triggered by the westerly wind bursts, the process vacillates irregularly between two phases, called El Niño and La Niña, which are the two ends of a continuum. El Niño is the warm phase, while La Niña is the cold phase. The La Niña phase implies an intensification of the regular trade winds (and enhanced equatorial upwelling) and precipitation in the western equatorial Pacific. The consequence of an El Niño event is a dramatic increase in sea surface temperatures in the eastern equatorial and tropical Pacific. Figure 3.8 shows an example of the temperature changes during the 1997/98 El Niño event, which is one of the strongest ENSO events in recorded history. Typically, ENSO events last from about 6–18 months but can be very important in terms of their effects on rainfall, agriculture, fisheries, and changes in biological populations, which can expand or contract by up to 1,000 km. Recent strong El Niño events occurred in 1957–58, 1982–83 and 1997–98, with weaker events in 1987–88, 1991–92, and 2014–15, although their occurrence appears to have become less regular since the 1982–83 event. A good introduction to ENSO and its effects on global oceanography is given by Philander (1990), while specific accounts of the 1982–83 and 1998–99 events are in Barber and Chavez (1983) and McPhaden (1999).

Similar El Niño type variability also exists in the Atlantic Ocean (see Tomczak and Godfrey 2003; Shannon et al. 1986). Although westerly wind bursts in Brazil are thought to control these Atlantic events, they do not seem to be linked directly to those in the Pacific, and have only been recorded intermittently, in 1934, 1950, 1963, 1984, and 1995 (Shannon et al. 1986; Gammelsrød et al. 1998), although additional warm events have also been recorded of lesser intensity, for example in 2001 and 2010 (Rouault et al. 2007; Bode et al. 2014; see Chap. 7).

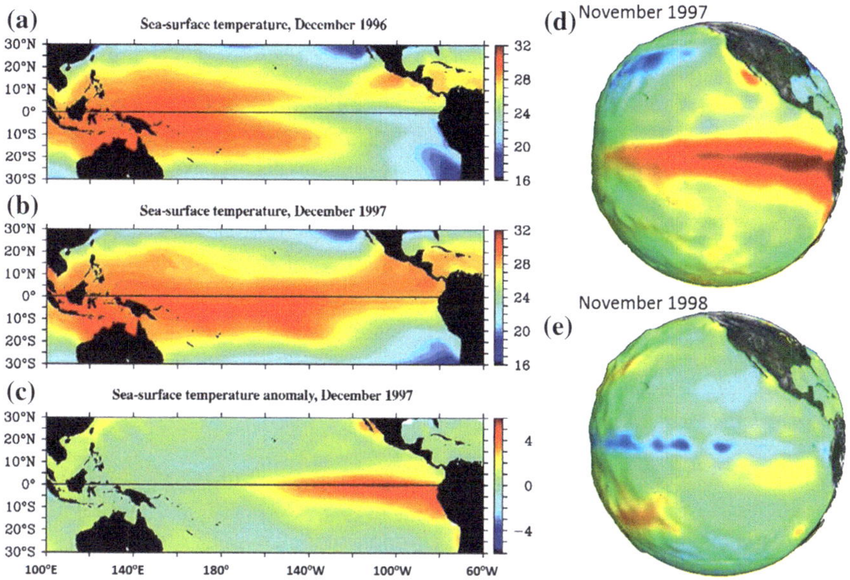

Fig. 3.8 Sea-surface temperature (SST) distributions in the equatorial Pacific Ocean for selected months and years. Panels **a** and **b** show the transition between the La-Niña phase and the El Niño phase of the ENSO in terms of actual SSTs. The impact of the late 1997 El Niño event is best seen from SST anomalies (Panel **c**), which displays the SST difference between panels **a** and **b**. Panels **d** and **e** show SST anomalies in late 1997 and 1998. Images courtesy of NASA

The intensity of the ENSO can be monitored by means of the atmosphere pressure difference at sea level between Darwin (Australia) and Tahiti, with the pressure at Darwin being low when the index is high, and vice versa. This gives the *Southern Oscillation Index* (SOI) (Fig. 3.9). There are also oceanic indices, such the Oceanic Niño Index (ONI), that record ENSO activity based on sea-surface-temperature anomalies in defined areas of the central equatorial Pacific Ocean. Some El Niño events have stronger impacts than others. For instance, the global flu pandemic of 1918/1919, which coincided with intense cold in the northeastern United States and a crippling drought in India, is now blamed at least partly on such an event (Giese et al. 2010), while the infamous 1997/1998 El Niño (McPhaden 1999) had far-reaching impacts with droughts in the western pacific islands and Indonesia as well as in Mexico and Central America and resulted in coral bleaching events in most tropical regions of the world ocean (Wilkinson and Hodgson 1999). The other recent strong El Niño, in 1982/1983, which was known at the time as the "El Niño of the century" led to devastating flooding across Bolivia, Cuba, Ecuador, northern Peru and in the U.S. states bordering the Gulf of Mexico, and spawned hurricanes in Hawaii and Tahiti, as well as droughts and fires throughout southern Africa, central America, Australia, India, Indonesia, and the Philippines (see, e.g., Gill and Rasmussen 1983; Glantz 1996; Rasmussen and Wallace 1983).

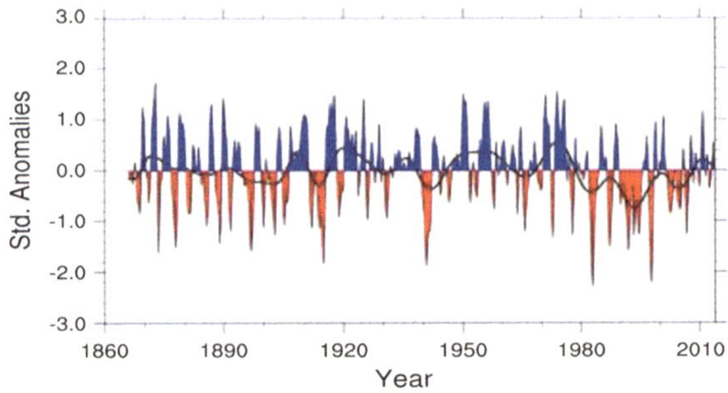

Fig. 3.9 Time series of the Southern Oscillation Index (SOI) . El Niño events correspond to negative values. Note that the unit of the y-axis is based on the standard deviation of the data, based on the normal distribution. Image source: http://www.cgd.ucar.edu/cas/catalog/climind/soi.html [accessed 4 April 2016]

The Pacific Decadal Oscillation (PDO) is the key mode of oceanic (as opposed to atmospheric) variability of the Pacific Ocean (e.g., Trenberth and Hurrell 1994), and is calculated as the first principal component of an EOF analysis of North Pacific sea surface temperature anomalies, based on a time series from 1900–1993. The recent study of Shakun and Shaman (2009) indicates that the PDO is an aftereffect of ENSO events; that is, the dominance of more frequent, stronger and longer La Niñas leads to the negative phase of the PDO, whereas more frequent, stronger and longer El Niños trigger the positive PDO mode. Accordingly, it not surprising that the phases of the PDO have a similar appearance in their SST anomaly distributions (Fig. 3.10).

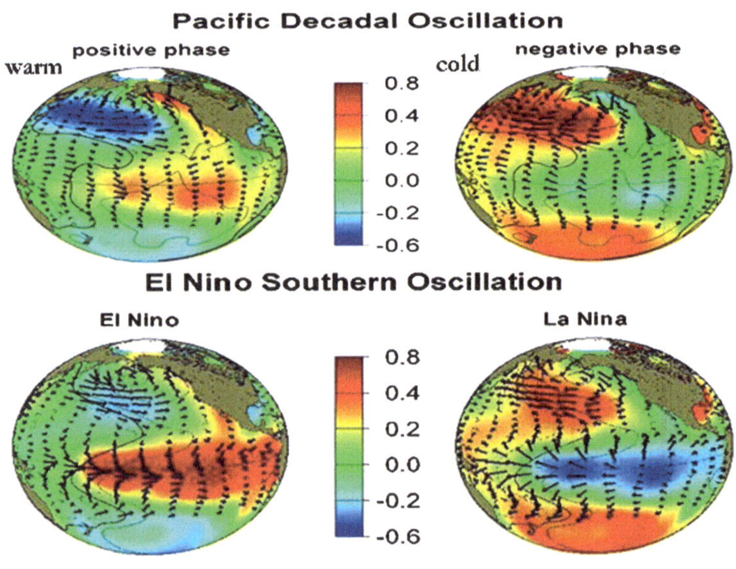

Fig. 3.10 Anomalies of sea surface temperatures (*colors*) and surface winds (*arrows*) for positive and negative phases of the Pacific Decadal and El Niño Southern Oscillation. Taken from http://jisao.washington.edu/pdo/graphics.html [accessed on 12 April 2016]

There is evidence of just two full PDO phases in the past century, with a step change at the end of each: "cool" PDO regimes prevailed from 1947–1976, whereas "warm" PDO regimes dominated from 1925–1946 and from 1977–1998 (Fig. 3.11). While efforts have been made to extend the records back to 1854 (from instrumental data) and even 1600 (based on tree rings), these show considerable inconsistency.

Major changes in northeast Pacific marine ecosystems have been correlated with phase changes in the PDO, with over 40 environmental variables, including local temperature and rainfall, changing simultaneously (Ebbesmeyer et al. 1991); warm eras have seen enhanced coastal ocean biological productivity in Alaska and inhibited productivity off the west coast of the US, while cold PDO eras have seen the opposite north-south pattern of marine ecosystem productivity (Mantua et al. 1997; Francis et al. 1998; Chavez et al. 2003). In particular, the longer-term periodicity of the PDO over the past 50–60 years is similar to that of sardines and anchovy populations reconstructed from lamina-based analysis of fish scales in samples of sediments in the Californian upwelling region (Baumgartner et al. 1992; see Chap. 4). There are also more distant effects, with South Pacific corals also showing similar variability to a PDO time-series.

More recently, another index, the *North Pacific Gyre Oscillation* (NPGO; Di Lorenzo et al. 2008), that uses the same data as for calculating the PDO but treats it differently, using the second EOF and principal component instead of the first, has been suggested as giving better agreement with additional parameters than just fish populations. Changes in the NPGO index are assumed to be driven by regional to basin-scale wind stress variability, thus affecting directly upwelling and horizontal advection and hence salinity and nutrient concentrations. Positive NPGO indices mean an increased flow in the North Pacific Current, leading to a stronger (upwelling favourable) California Current, although with increased downwelling off Alaska. Whether the PDO or NPGO is more important as regards changes in the California Current is presently uncertain, but they probably both exert effects at different times.

The Indian Ocean Dipole (Reverdin et al. 1986; Webster et al. 1999; Saji et al. 1999) is associated with sea-surface-temperature anomalies in the equatorial Indian Ocean. The Indian Ocean Dipole modulates the South Asian monsoon and has impacts on rainfall patterns in eastern central Africa, the lower reaches of the Persian Gulf and the Indonesian region. Both the Indian Ocean Dipole and ENSO

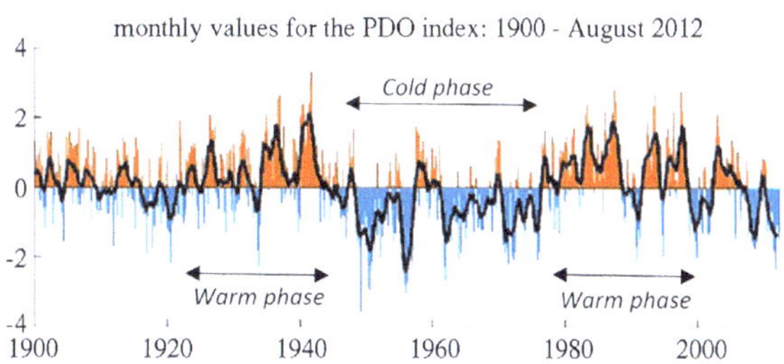

Fig. 3.11 Monthly values of the PDO index. Taken from http://jisao.washington.edu/pdo/graphics.html [accessed on 12 April 2016]

remotely influence coastal upwelling activity around the Indian Ocean through changes in large-scale wind patterns (e.g., Susanto et al. 2001).

The Atlantic Multidecadal Oscillation (AMO) is a multi-decadal mode of variability that affects sea surface temperature (SST) in the North Atlantic, and is based on the difference between the SST in the extra-tropical Atlantic and the mean global temperature. Several different ways of calculating the index exist (Schlesinger 1994; Trenberth and Shea 2005; van Oldenborgh et al. 2009; Guan and Nigam 2009). While the different calculations give different intensities to the positive and negative phases, the length of each phase is about 30–35 years, with positive phases from about 1925–1965 and since 2000, interspersed by a negative phase from 1965–2000 (Fig. 3.12; Trenberth and Shea 2005; van Oldenborgh et al. 2009). The warm (positive) phase is thought to bring droughts to the southwest and central U.S., but to increase rainfall over India and the Sahel, as well as increasing North Atlantic tropical cyclone activity, and hence overall hurricane activity (Chylek and Lesins 2008). The AMO is thought to have been occurring for at least 8,000 years (Knudsen et al. 2010), with external forcing from global warming trends since 1800 (Knudsen et al. 2014), but whether it has any effect on North Atlantic upwelling is unclear.

3.3.2 *Interference with Other Physical Processes*

Apart from interannual variability, upwelling in the eastern boundary currents is subject to other timescales of variability ranging from diurnal (tides), synoptic (mid-latitude weather patterns), intraseasonal (coastally trapped waves), interannual (e.g., ENSO) to interdecadal (e.g., PDO; see Mackas et al. (2006) for more details). While ENSO has a profound influence on upwelling in the eastern Pacific Ocean, a similar, but less dramatic variability is also seen in the equatorial Atlantic (Philander 1986; Gammelsrød et al. 1998).

The interaction of two or more processes with different timescales of variability can lead to unexpected results. For instance, one would anticipate that El Niño events simply lower the nutricline off Chile leading to reduced primary production

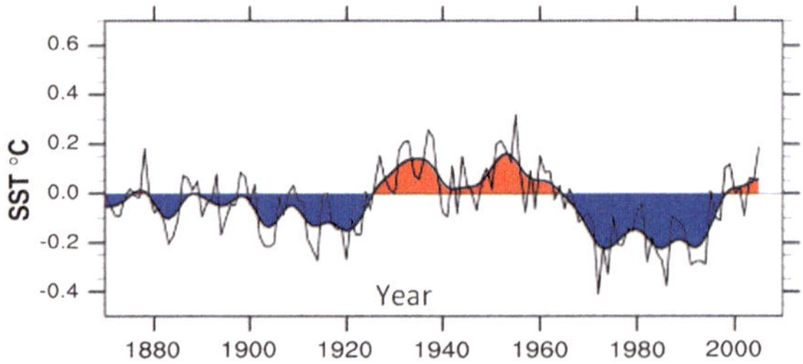

Fig. 3.12 Atlantic multidecadal index based on annual SST anomalies for the North Atlantic for 1870–2005, relative to 1901–1970 (°C) with the global mean SST removed to remove the bias from global warming. From Trenberth and Shea (2005)

and biomass. The opposite is expected for La Niña events. However, several observations contradict this expectation, at least on intermittent timescales. In reality, the depth of the nutricline is caused by the interference (superposition) of different processes, the main being signals from ENSO events, the seasonal cycle of wind forcing, intraseasonal coastally trapped waves (period 30–60 days) and synoptic upwelling-favourable wind events (timescale is 2–10 days). Hence, while El Niño events on their own lead to a downward displacement of the nutricline, other processes such as coastally trapped waves can partially eliminate the El Niño effect, at least transiently on intraseasonal timescales (Montecino et al. 2004), and synoptic wind-driven upwelling will still be able to carry nutrients into the euphotic zone.

3.3.3 Impacts of Climate Change

There can be direct or indirect impacts of climate change on the ecology of coastal upwelling systems. Direct impacts include changes in temperature differences between land and sea during summer that could possibly lead to an increased cross-boundary pressure gradient and hence amplify upwelling through strengthened alongshore geostrophic winds and offshore Ekman transport (Bakun 1990). Coastal wind patterns may also change, influencing the strength and frequency of upwelling events, but it is difficult to distinguish climate-change impacts with natural climate variability such as ENSO or the PDO.

Another likely direct climate-change impact is strengthening of the pycnocline, which leads to both the shoaling of oxygen minimum zones in the vicinity of coastal upwelling regions and causes a shallowing of the associated nutricline (Arntz et al. 2006; Gilly et al. 2013). Such shallowing has been seen in all the major eastern boundary systems as well as in the Arabian Sea and the Bay of Bengal (Stramma et al. 2008, 2010), leading to increased area and volume of the oxygen minimum zones associated with coastal upwelling, as well as reduced oxygen concentrations in the zones. This happens because the surface layer of the global ocean is warming, which means that it cannot contain as much oxygen as colder waters. As the additional heat sinks deeper into the ocean, it reaches layers that contribute directly to the water that is brought back to the surface during the upwelling process (Keeling et al. 2010).

While it seems likely that this will be an ongoing process, there may be some cyclic effects, as links to PDO cycles have been seen in the California Current (McClatchie et al. 2010; Deutsch et al. 2007) and the Gulf of Alaska (Whitney et al. 2007).

Indirect impacts include changes in biochemical processes, such as carbon, nitrogen or sulphur cycling, and food-web dynamics due to increased or decreased marine productivity, ocean acidification, and socio-economic factors associated with fisheries and exploitation of natural resources. Nevertheless, it is still far from clear how these changes will affect the productivity of upwelling regions. Increased upwelling may increase phytoplankton production, but continuous strong winds can reduce fisheries productivity as the plankton and fish larvae may be moved offshore more rapidly, thus being lost from the system rather than retained in spawning areas.

Changes in pycnocline and nutricline depths can also affect production; if they are brought closer to the surface, this means more nutrients are available for phytoplankton production, potentially feeding through into enhanced fish biomass, while the increased strength and reduced depth of the pycnocline may mean better retention of sedimenting material close to the coast (Hutchings et al. 1995). This, in turn, could lead to more extensive low oxygen zones, which may reduce fish production by decreasing the space available for demersal organisms, or increasing the rates of predation in the euphotic zone by higher trophic level predators, since there will be less volume for the prey organisms to inhabit.

It is likely that the most important changes will be linked to the combination of reduced oxygen concentration, higher water temperatures, reduction in pH, and changes in predator/prey relationships (Gruber 2011), and these will combine to affect pelagic, mesopelagic, demersal, and benthic organisms in different ways (Gilly et al. 2013). For example, Gobler et al. (2014) have shown that both hypoxia and increased acidification (lower pH) strongly reduce both larval survival and growth rates of scallops *Argopecten irradians* and clams *Mercenaria mercenaria* by more than 50 %, and that the combination of both stressors is more severe than either alone. Both of these organisms are important economically, suggesting that future harvests of these and similar shellfish could be severely impacted if climate change continues at its present rate.

3.4 Harmful Algal Blooms and Hypoxia

As discussed in Sect. 2.2, low oxygen conditions are a common feature of both coastal and open ocean upwelling systems. In extreme cases the oxygen concentrations can be depleted to zero (e.g., Chapman and Shannon 1985; Farias et al. 2004; Morrison et al. 1999), but despite this, these regions provide enormous quantities of high quality protein and sustain large-scale pelagic and demersal fisheries, and the organisms that live there are clearly able to survive the periodic fluctuations in oxygen concentration. This is in contrast to other coastal areas, such as the northern Gulf of Mexico, or the Baltic and Black Seas, where eutrophication is increasing and has been recorded as having deleterious effects on local organisms (Diaz and Rosenberg 2008; Rabalais et al. 2007).

Problems can arise, however, when oxygen concentrations in the inshore zones of upwelling regions are reduced suddenly. This can occur, for example, if there is a switch from upwelling to downwelling conditions following a bloom. In this case, the overall flow pattern along the coast changes from offshore to onshore, and this can bring water with very low oxygen concentrations into the surf zone. Demersal and benthic organisms can then be trapped in a narrow zone along the coast that still retains some oxygen. If the downwelling conditions persist, even the surf zone may become hypoxic, leading to mass mortalities. Typical examples of this occur in the Benguela system along the west coast of South Africa, when crayfish (*Jasus lalandii*) are known to walk out of the water during extreme downwelling conditions (e.g., Newman and Pollock 1974; Cockcroft et al. 2000). Such low oxygen occurrences can also affect estuaries, and deaths of benthic invertebrates and juvenile salmon have been shown to result from upwelling dynamics bringing upwelled water with low oxygen concentrations into the Columbia River and other estuaries on the Pacific northwest coast of the U.S (Grantham et al. 2004; Roegner et al. 2011).

A second possible reason for large-scale mortalities of fish and other marine organisms follows very high rates of phytoplankton production, when so-called "red tides" develop (Fig. 3.13). Red tides frequently occur in large upwelling shadows, e.g. in the Southern California Bight or off southern Peru and northern Chile, in conjunction with the development of thermal stratification (Ryan et al. 2008). Although these blooms may seem unpredictable, it should be remembered that the organisms that cause them are found as members of the normal phytoplankton population and are opportunistic only in the sense that they can produce dense blooms given the right conditions (Zingone and Wyatt 2005).

Red tides are not limited to upwelling regions, as they can also be triggered by eutrophication in other coastal areas, but do occur quite frequently each year following upwelling events once the wind relaxes and the upper mixed layer stabilizes again (Sordo et al. 2001; Pitcher and Nelson 2006). The result is that the upper mixed layer contains large quantities of nutrients, together with seeding spores that have been brought into the euphotic zone during the upwelling event. Typical red-tide organisms, such as the dinoflagellate *Karenia*, can form such dense blooms that they change the colour of the seawater, which despite their name can be brown, orange, purple or yellow as well as red, depending on the species involved. Some organisms, such as *Mesodinium rubrum*, *Prorocentrum micans* and *Ceratium furca*, are non-toxic, but reduce oxygen concentrations to very low levels when they start to decay at the end of the bloom, and it is the low oxygen conditions that result that cause the problem. Alternatively, the large number of plant cells can clog the gills

Fig. 3.13 Bloom of the non-toxic dinoflagellate *Noctiluca miliaris* near Cape Town. This organism is responsible for spectacular red tides and is also capable of luminescence at night. Downloaded from http://www.botany.uwc.ac.za/Envfacts/redtides/bloom.htm [accessed on 4 April 2016]

of other marine organisms, causing death by direct suffocation. They can affect fish, mollusks, and sea urchins, including species that live in the sediments.

Many species of dinoflagellates can produce chemicals that are toxic to other marine organisms. Species such as *Gonyaulax catanella* (presently *Alexandrium catanella* as a result of genetic typing) or *Karenia brevis*, produce potent neurotoxins, which are some of the most toxic compounds known. Filter-feeding organisms such as clams, mussels and oysters are particularly susceptible to red tide toxins because of their habit of filtering food particles from seawater, although abalone are also affected (Pitcher et al. 2001) and some species of dinoflagellate produce compounds that are toxic to fish as well. Filter feeding concentrates the toxins in the organisms' digestive systems so that it is passed on to predators, including humans, when the shellfish are eaten, and shellfish can continue to be toxic for several days or even weeks after the red tide has disappeared. Cooking only slightly reduces the potency of these toxins, which can poison humans in at least four different ways, and the toxins can also be transported as atmospheric aerosols (Pitcher and Calder 2000).

Paralytic shellfish poisoning (PSP) has been known since the 1700s and has caused numerous deaths globally, mainly as a result of people eating contaminated shellfish. The main toxin is saxitoxin, which initially causes tingling and numbness around the lips, tongue and fingertips, leading to general numbness through disruption of normal nerve functions, and can induce death from respiratory failure within less than 24 h. The other forms of poisoning are diarrhetic shellfish poisoning (DSP), where the toxic agent is okadaic acid, amnesiac shellfish poisoning (ASP) , caused by domoic acid, and neurotoxic shellfish poisoning (NSP). Symptoms of DSP, ASP, and NSP are generally much milder than for PSP, causing diarrhoea, nausea, and dizziness, although ASP also results in short-term memory loss (Pitcher and Calder 2000).

3.5 Exploitation of Marine Resources

3.5.1 Key Locations of Commercial Fisheries

Marine life is neither evenly nor randomly distributed across the oceans, since the environmental parameters and conditions that determine ocean productivity vary greatly in both time and space. While the primary and most important fishing grounds in the world are found on and along continental shelves within less than 200 nautical miles of the shores, their distribution is patchy and localized. In 2004, for instance, more than half of the marine landings were caught within 100 km off the coast at depths <200 m—on continental shelves—covering an area of less than 7.5 % of the world's oceans (e.g., Nellemann et al. 2008, Fig. 3.14). This is partly

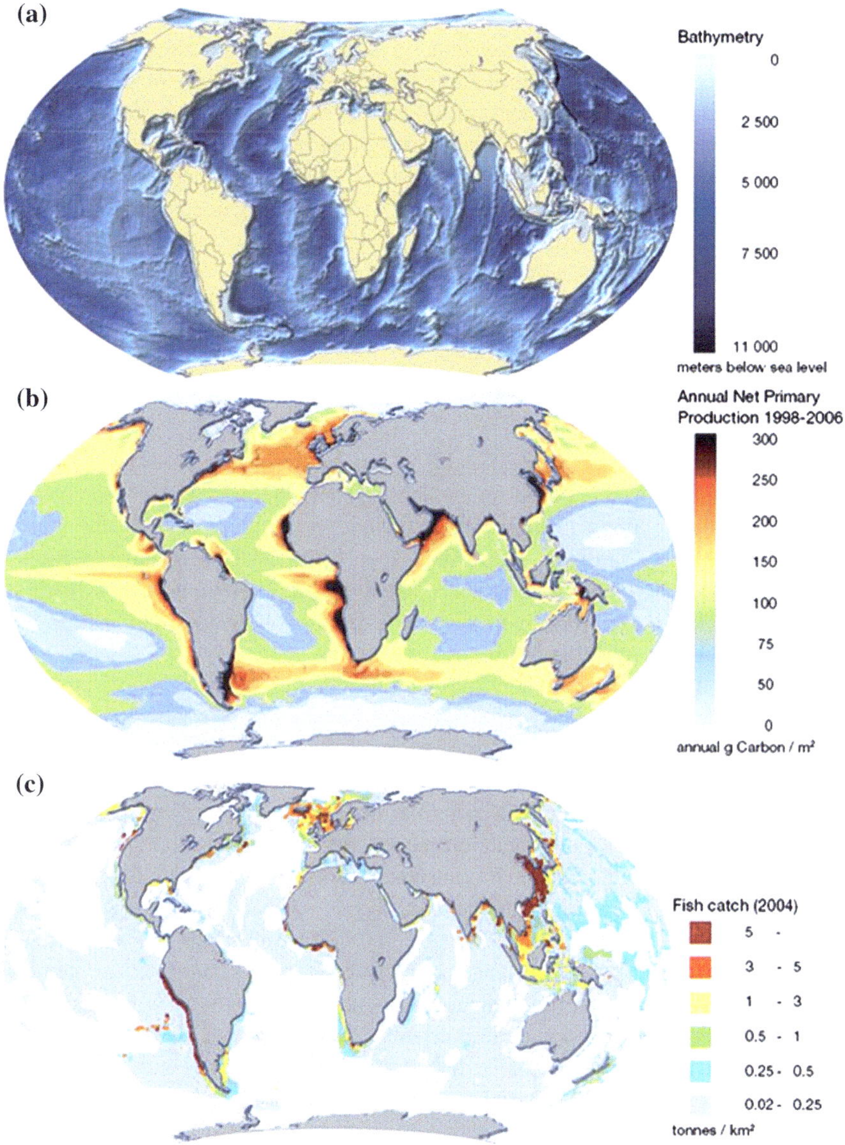

Fig. 3.14 Comparison between **a** ocean depth, **b** annual net primary production, and **c** global fish catches for 2004. Image source: GRID-Arendal, Nellemann et al. (2008). http://www.unep.org/pdf/InDeadWater_LR.pdf [accessed on 4 April 2016]. Reproduced with permission

for economic and logistical reasons, as it helps to have ports close to fishing grounds, but mainly because fishery hotspots are located in regions where the primary productivity is generally high. This includes coastal upwelling systems and localized shelf seas, such as the Irish Sea and the Yellow Sea, in which tidal mixing fronts support a rich marine ecosystem. In addition, the amount of fish caught depends on both demand and limitations such as catch quotas implemented in an attempt to avoid negative impacts on the resources due to overexploitation, also called *overfishing* Fishery data, however, are often incomplete and can be unreliable. Hilborn and Peterman (1996) identified key sources of uncertainty in fisheries stock assessments including the estimates of fish abundance, errors in fishery models, prediction of future conditions, and factors associated with particular economic, political and social conditions.

3.5.2 Variability of Forage Fish Stocks

Forage fish are defined as small pelagic fish which are preyed on by larger predators for food. Typical ocean forage fish feed near the base of the food chain on plankton, often by filter feeding. They include particularly fish of the family *Clupeidae* (herrings, sardines, shad, hilsa, menhaden, anchovies and sprats), but also other small fish, including halfbeaks, smelt such as capelin, goldband fusiliers, and young jack mackerel. Forage fish play a central role in the marine food-web dynamics of coastal upwelling systems.

While it is difficult to accurately estimate the size of fish stocks the variations of catches of different fish species over time may be taken as proxy of their temporal variability. Interestingly, the catches of forage fish and interlinked higher-trophic fish species exhibit substantial decadal variations, many of which are similar (Fig. 3.15). Two different types of temporal behaviour can be distinguished, with one group of species showing behaviour almost opposite to the other group (Bakun 1995; Klyashtorin 1998, 2001). For instance, while a number of fisheries (e.g., Pacific and Atlantic herring, Atlantic cod, South African sardine, and Peruvian anchovy) collapsed in the late 1960s and early 1970s, other fish species started to recover (e.g., Japanese, Californian and Peruvian sardines, and Pacific salmon). This alternation of sardine and anchovy stocks seems to be a worldwide phenomenon, not confined to upwelling systems (Schwartzlose et al. 1999).

Some of the variability of forage fish stocks (inferred from catch data) can be explained by natural climate variability linked to the long-term periodicities discussed in Sect. 3.2, and it is obvious that a fish species is most vulnerable when its stock size is at its lowest level. For instance, overfishing of Atlantic cod stocks in the 1980s during periods of naturally lower abundance has led to near extinction of the species on the east coast of Newfoundland. Overfishing during vulnerable periods has similarly affected fish stocks in upwelling regions at different times (Bakun 1993).

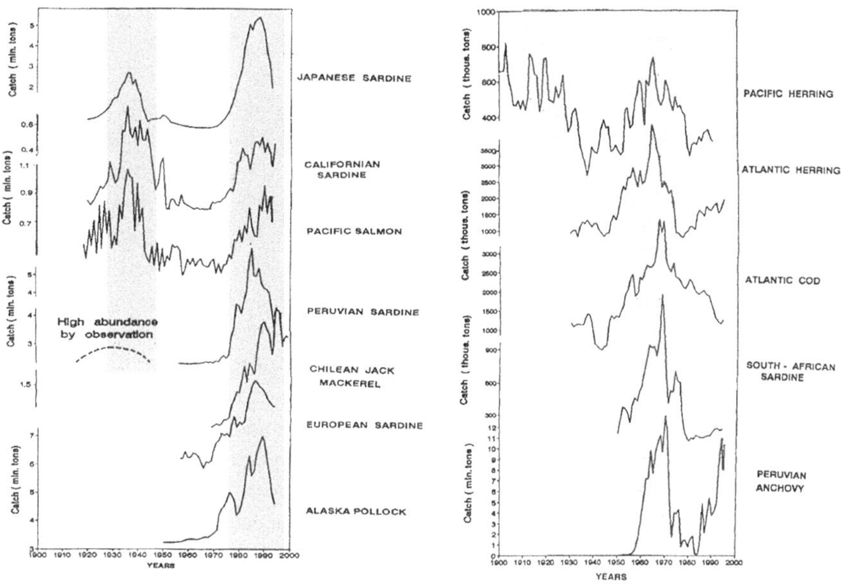

Fig. 3.15 Time series of catches of key pelagic fish species in the Atlantic and Pacific Oceans from 1900 to 2000. *Source* FAO, data from Klyashtorin (1998). http://www.fao.org/docrep/005/y2787e/y2787e00.pdf [accessed on 4 April 2016]

3.5.3 Overexploitation

Human interactions with ocean ecosystems have been drastically transformed in recent decades. New, technically sophisticated methods of fish capture and the production of edible fish products have led to an unprecedented level of productive capacity. Latest research reveals that every square mile of the world's oceans has been affected by anthropogenic forces, of which overfishing is a primary concern (Halpern et al. 2008). More than 90 % of the ocean's top predators have been overexploited and are ecologically threatened. Throughout the world, fisheries are in fast decline and aquaculture has become a major source of marine protein for human consumption (FAO 2007).

Over 30 years ago, in view of the overexploitation of many conventional fish stocks and the growing interest in harvesting new kinds of food from the sea, May et al. (1979) stressed the need for fisheries managers to take account of interactions among species. This appears even more important today because of the sustained increasing trend in the catches of pelagic fish since the 1950s (Fig. 3.14). Their contribution to the total marine fish catch has ranged from about 50 % in 1950 to over 64 % in 1994. Just six species (anchoveta, Atlantic herring, Japanese and South American pilchard, chub mackerel, capelin, and Chilean jack mackerel) among the 186 pelagic species exploited (FAO 1997) represent about half of the catch.

Pelagic fish stocks have been studied intensively, and environmental factors are recognized as a determinant of recruitment success. Food availability and physical processes pay a significant role in larval survival, which determines subsequent fish abundance. Records of scale deposition from anaerobic sediments show that large-amplitude fluctuations occurred even in the absence of any fishery (Baumgartner et al. 1992). As shown in Fig. 3.16, pelagic fish populations exhibit a substantial degree of global synchrony that is believed to be driven by global climatic teleconnections (Klyashtorin 1998). An earlier review of the collapse of 10 major fisheries on small pelagic species (Beverton 1990) concluded that fishing was the main cause of collapse in most, but not all, cases. Therefore, the impact of commercial fisheries on pelagic resources remains controversial. Another perspective, however, has been added recently. Using two-species competition models, Ferrière and Cazelles (1999) showed that phases of extremely low abundance followed by short peaks in abundance could arise in a simple community model as a result of competitive interactions within and between species.

Fisheries data assembled by the United Nations Food and Agriculture Organization (FAO) suggest that global marine fisheries catches increased steadily from 1950 to a maximum of 86 million tonnes in 1996, and have since slightly declined. A very recent study by Pauly and Zeller (2016), using a decade-long multinational 'catch reconstruction' project covering the Exclusive Economic Zones of the world's maritime countries and the High Seas from 1950 to 2010, and accounting for all fisheries, came to a different conclusion, however. Their reconstructed catch trajectories differ considerably from the national data submitted to the FAO and indicate that catch actually peaked at 130 million tonnes, and has been declining much more strongly since (Fig. 3.17).

The largest differences between reported and reconstructed catches were identified in the eastern Central Atlantic (i.e., the Canary Current Upwelling System), where about 70 % of the total catch is unreported. In other upwelling system, unreported catches account for ~20–40 % of the total value. While some fishery zones show relatively constant catch rates over the period 2000–2010, regions that contribute mostly to the reconstructed decline in total catches during this period are the northeast Atlantic and southeast Pacific. Overall, the data suggest that the world's fisheries have reached a maximum capacity from 1990 onwards.

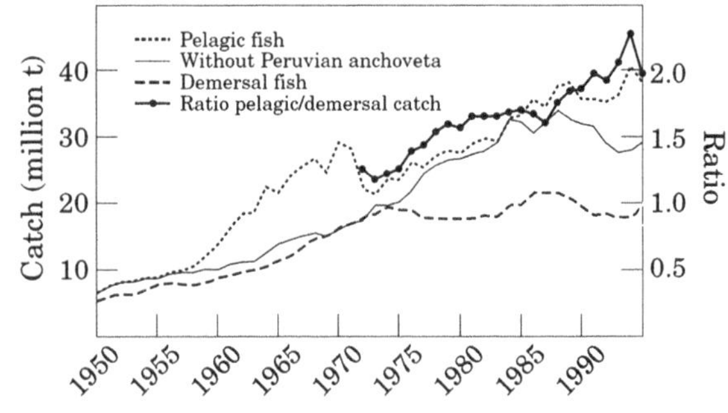

Fig. 3.16 World annual pelagic (with and without Peruvian anchoveta) and demersal marine fish catch from 1950 to 1995, and ratio between pelagic and demersal catches (from Cury et al. 2000)

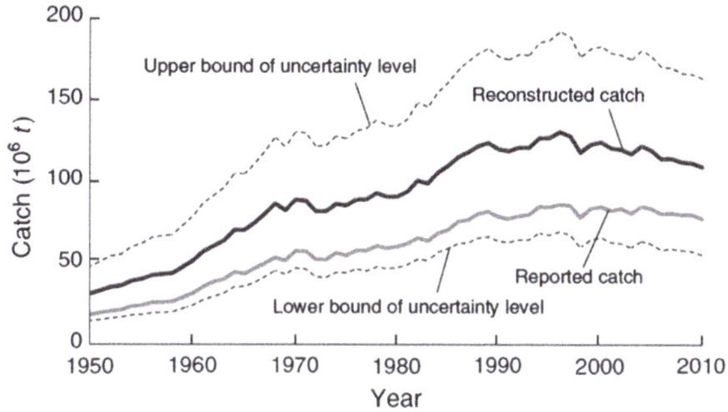

Fig. 3.17 Trajectories of reported and reconstructed marine fisheries catches 1950–2010 (from Pauly and Zeller 2016)

This stagnation has led to new marine technologies including land-based aquaculture and the construction of large-sized "super trawlers", which are over 130 m long (Fig. 3.18), use much larger nets than traditional fishing vessels and have fish factories aboard.

Both technologies are controversial. Aquaculture of marine species is based on catching large quantities of juvenile fish and forage feed (e.g. sardines) as a food source, which has impacts on the marine food web. For instance, there is a long-standing dispute between regulators, scientists and fisheries as to whether the Southern Bluefin Tuna (*Thunnus maccoyii*) is overexploited (see Sect. 8.2.4). On the other hand, it is certainly feared that mega catches from super trawlers overexploit regional fish stocks and controversy about the sustainability of increased fishing efforts continues. As an example, the FV *Margiris*, the world's second largest fishing boat, was originally authorized to catch a quota of 18,000 tonnes of jack mackerel and redbait along the southern shores of Australia. After protests against her use by environmental and fishing industry groups, the Australian government passed legislation prohibiting the trawler from fishing in Australian waters for 2 years.

Fig. 3.18 Example of a large-size super trawler. Taken from: http://www.greenpeace.org/australia/en/what-we-do/oceans/No-Super-Trawlers/ [accessed on 4 April 2016]

3.6 Summary

Upwelling regions are extremely important marine ecosystems providing large natural resources of fish to feed the world's population. Overall, it appears that the world's natural fish stocks have reached their limit in terms of providing food demanded by an increasing population, and climate variability in terms of El Niño events in the Pacific dramatically interrupts this food supply in some large upwelling systems. Controversial new technologies (super trawlers and aquaculture) have been developed to further increase fish production constrained by management measures such as catch quota. Unreported catches and unreliable catch data are severe problems, as is the global management of fish stocks. There is also evidence of large-scale multi-decadal shifts in forage fish stocks in the world oceans that in the Pacific Ocean have been linked to interdecadal oceanic variability such as the Pacific Decadal Oscillation. Impacts of global warming on fish productivity in upwelling regions cannot be predicted with any certainty so far, but significant changes in the food-web structure and functioning are expected.

References

Arntz, W.E., V.A. Gallardo, D. Gutiérrez, E. Isla, L.A. Levin, J. Mendo, C. Neira, G.T. Rowe, J. Tarazona, and M. Wolff. 2006. El Niño and similar perturbation effects on the benthos of the Humboldt, California, and Benguela Current upwelling ecosystems. *Advances in Geosciences* 6: 243–265.

Bakun, A. 1990. Global change and intensification of coastal upwelling. *Science* 247: 198–201.

Bakun, A. 1993. The California Current, Benguela Current, and Southwestern Atlantic Shelf Ecosystems: A comparative approach to identifying factors regulating biomass yields. In *Large Marine Ecosystems—Stress, Mitigation, and Sustainability*, ed. K. Sherman, L.M. Alexander, and B. Gold, 199–224. American Association for the Advancement of Science.

Bakun, A. 1995. A dynamical scenario for simultaneous "regime-scale" marine population shifts in widely separated LMEs of the Pacific. In *The large marine ecosystems of the Pacific Rim*, ed. K. Sherman, and Q. Tang, 47–72. Switzerland: IUCN, The World Conservation Union, Gland.

Barber, R.T., and F.P. Chavez. 1983. Biological consequences of El Nino. *Science* 222: 1203–1210.

Barnett, T., D. Pierce, M. Latif, D. Dommenget, and R. Saravanan. 1999. Interdecadal interactions between the tropics and midlatitudes in the Pacific basin. *Geophysical Research Letters* 26(5): 615–618.

Baumgartner, T., A. Soutar, and V. Ferreira-Bartrina. 1992. Reconstruction of the history of Pacific sardine and northern anchovy populations over the past two millennia from sediments of the Santa Barbara Basin, California. *California Cooperative Oceanic Fisheries Investigations Report* 33: 24–40.

Beamish, R.J. 1993. Climate and exceptional fish production off the west coast of North America. *Canadian Journal of Fisheries and Aquatic Sciences* 50: 2270–2291.

Beverton, R.J.H. 1990. Small marine pelagic fish and the threat of fishing; are they endangered? *Journal of Fish Biology* 37: 5–16.

Bode, M., A. Kreiner, A.K. van der Plas, D.C. Louw, R. Horaeb, H. Auel, and W. Hagen. 2014. Spatio-temporal variability of copepod abundance along the 20° S monitoring transect in the northern Benguela upwelling system from 2005 to 2011. *PLoS ONE* 9: e97738. doi:10.137/journal.pone.0097738.

Brandt, P., V. Hormann, A. Körtzinger, M. Visbeck, G. Krahmann, L. Stramma, R. Lumpkin, and C. Schmid. 2010. Changes in the ventilation of the oxygen minimum zone of the tropical North Atlantic. *Journal of Physical Oceanography* 40(8): 1784–1801.

Burgers, G. 1999. The El Niño stochastic oscillator. *Climate Dynamics* 15: 521–531.

Chapman, P., and L.V. Shannon. 1985. The Benguela ecosystem. Part II. Chemistry and related processes. *Oceanography and Marine Biology: An Annual Review* 21: 183–251.

Chavez, F., J. Ryan, S. Lluch-Cota, and C.M. Ñiquen. 2003. From anchovies to sardines and back: Multidecadal change in the Pacific Ocean. *Science* 299(5604): 217–221.

Chen, C.-T.A. 1996. The Kuroshio intermediate water is the major source of nutrients on the East China Sea continental shelf. *Oceanologica Acta* 19(5): 523–527.

Chylek, P., and G. Lesins. 2008. Multidecadal variability of Atlantic hurricane activity: 1851–2007. *Journal of Geophysical Research* 113: D22106. doi:10.1029/2008JD010036.

Cockcroft, A.C., D.S. Schoeman, G.C. Pitcher, G.W. Bailey, and D.L. Van Zyl. 2000. A mass stranding, or 'walk out' of west coast rock lobster, *Jasus lalandii*, in Elands Bay, South Africa: Causes, results, and implications. In *Biodiversity crisis and crustacea*, ed. J. von Vaupel Klein, and F.R. Schram, 673–688. A.A.: Balkema.

Cury, P., A. Bakun, R.J.M. Crawford, A. Jarre, R.A. Quinones, L.J. Shannon, and H.M. Verheye. 2000. Small pelagics in upwelling systems: patterns of interaction and structural changes in "wasp-waist" ecosystems. *ICES Journal of Marine Science* 57: 603–618.

Cushing, D.H. 1975. *Marine ecology and fisheries*. Cambridge, UK: Cambridge University Press. 278 pp.

Cushman, G.T. 2004. Enclave vision: Foreign networks in Peru and the internationalization of El Niño research during the 1920s. *Proceedings of the International Commission on History of Meteorology* 1: 1.

Deutsch, C., J.L. Sarmiento, D.M. Sigman, N. Gruber, and J.P. Dunne. 2007. Spatial coupling of nitrogen inputs and losses in the ocean. *Nature* 445: 163–167.

Diaz, R.J., and R. Rosenberg. 2008. Spreading dead zones and consequences for marine ecosystems. *Science* 321: 926–929.

Dickson, R., J. Lazier, J. Meincke, P. Rhines, and J. Swift. 1996. Long-term coordinated change in the convective activity of the North Atlantic. *Progress in Oceanography* 38: 241–295.

Di Lorenzo, E., N. Schneider, K.M. Cobb, P.J.S. Franks, K. Chhak, A.J. Miller, J.C. McWilliams, S.J. Bograd, H. Arango, E. Curchitser, T.M. Powell, and P. Riviere. 2008. North Pacific Gyre Oscillation links ocean climate and ecosystem change. *Geophysical Research Letters* 35: L08607. doi:10.1029/2007GL032838.

Ebbesmeyer, C.C., D.R. Cayan, D.R. Milan, F.H. Nichols, D.H. Peterson and K.T. Redmond. 1991. 1976 step in the Pacific climate: forty environmental changes between 1968–1975 and 1977–1984. In *Proceeedings of the 7th annual climate (PACLIM) workshop, April 1990*, ed. J. L. Betancourt and V.L. Tharp, 115–126. California Department of Water Resources, Interagency Studies Program Technical Report 26.

FAO. 1997. *The state of world fisheries and aquaculture*. Rome, Italy: FAO Fisheries Department. 126 pp.

FAO. 2007. *Fish and fishery products. World apparent consumption statistics based on food balance sheets (1961–2003)*, G. Laurenti (comp.). FAO Fisheries Circular, No. 821, Rev.8. Rome. 429 pp.

Farias, L., M. Graco, and O. Ulloa. 2004. Temporal variability of nitrogen cycling in continental shelf sediments of the upwelling ecosystem off central Chile. *Deep-Sea Research II* 51: 2491–2505.

Ferrière, R., and B. Cazelles. 1999. Universal power laws governing intermittent rarity in communities of interacting species. *Ecology* 80(5): 1505–1521.

Francis, R.C., and S.R. Hare. 1994. Decadal-scale regime shifts in the large marine ecosystems of the North-east Pacific: a case for historical science. *Fisheries Oceanography* 3: 279–291.

Francis, R.C., S.R. Hare, A.B. Hollowed, and W.S. Wooster. 1998. Effects of interdecadal climate variability on the oceanic ecosystems of the Northeast Pacific. *Fisheries Oceanography* 7: 1–21.

Gammelsrød, T., C.H. Bartholomae, D.C. Boyer, V.L.L. Filipe, and M.J. O'Toole. 1998. Intrusion of warm surface water along the Angolan-Namibian coast on February-March 1995: the 1995 Benguela Niño. *South African Journal for Marine Science* 19: 41–56.
Giese, B.S., G.P. Compo, N.C. Slowey, P.D. Sardeshmukh, J.A. Carton, S. Ray, and J.S. Whitaker. 2010. The 1918/19 El Niño. *Bulletin of the American Meteorological Society* 91: 177–183.
Gill, A.E., and E.M. Rasmussen. 1983. The 1982–83 climate anomaly in the equatorial Pacific. *Nature* 306: 229–234.
Gilly, W.F., J.M. Beman, S.Y. Litvin, and B.H. Robison. 2013. Oceanographic and biological effects of shoaling of the oxygen minimum zone. *Annual Reviews of Marine Science* 5: 393–420.
Glantz, M.H. 1996. *Currents of change: El Niño's impact on climate and society*. Cambridge University Press, 194 pp.
Gobler, C.J., E.L. DePasquale, A.W. Griffith, and H. Baumann. 2014. Hypoxia and acidification have additive and synergistic negative effects on the growth, survival, and metamorphosis of early life stage bivalves. *PLoS ONE* 9: e83648. doi:10.1371/journal.pone.0083648.
Grantham, B.A., F. Chan, K.J. Nielsen, D.S. Fox, J.A. Barth, A. Huyer, J. Lubchenco, and B.A. Menge. 2004. Upwelling-driven nearshore hypoxia signals ecosystem and oceanographic changes in the northeast Pacific. *Nature* 429: 749–754.
Gregg, W.W., P. Ginoux, P.S. Schopf, et al. 2003. Phytoplankton and iron: Validation of a global three-dimensional ocean biogeochemical model. *Deep-Sea Research II* 50: 3143–3169.
Gruber, N. 2011. Warming up, turning sour, losing breath: ocean biogeochemistry under global change. *Philosophical Transactions of the Royal Society of London A* 369: 1980–1996.
Guan, B., and S. Nigam. 2009. Analysis of Atlantic SST variability factoring interbasin links and the secular trend: clarified structure of the Atlantic Multidecadal Oscillation. *Journal of Climate* 22: 4228–4240.
Halpern, B.S., S. Walbridge, K.A. Selkoe, et al. 2008. A global map of human impact on marine ecosystems. *Science* 319: 948–952.
Hanawa, K., and L. D. Talley. 2001. Mode waters. In *Ocean circulation and climate, ed.* G. Siedler and J. Church, 373–386. International Geophysics Series, Academic Press.
Hilborn, R., and R.M. Peterman. 1996. The development of scientific advice with incomplete information in the context of the precautionary approach. *FAO Fisheries Technical Paper* 350: 77–97.
Hutchings, L., G.C. Pitcher, T.A. Probyn, and G.W. Bailey. 1995. The chemical and biological consequences of coastal upwelling. In *Upwelling in the ocean: Modern processes and ancient records*, ed. C.P. Summerhayes, K.-C. Emeis, M.V. Angel, R.L. Smith, and B. Zeitschel, 65–81. Chichester: Wiley.
Kämpf, J. 2010. On the preconditioning of coastal upwelling in the eastern Great Australian Bight. *Journal of Geophysical Research*, 115: C12071, 11 pp. doi:10.1029/2010JC006294.
Kämpf, J., M. Doubell, D. Griffin, R.L. Matthews, and T.M. Ward. 2004. Evidence of a large seasonal coastal upwelling system along the southern shelf of Australia. *Geophysical Research Letters* 31: L09310. doi:10.1029/2003GLO19221.
Karoly, D. 1989. Southern Hemisphere circulation features associated with El Nino-Southern Oscillation events. *Journal of Climate* 2: 1239–1252.
Keeling, R.F., A. Kortzinger, and N. Gruber. 2010. Ocean deoxygenation in a warming world. *Annual Reviews of Marine Science* 2: 199–229.
Klyashtorin, L.B. 1998. Long-term climate change and main commercial fish production in the Atlantic and Pacific. *Fisheries Research* 37: 115–125. doi:10.1016/S0165-7836(98)00131-3.
Klyashtorin, L.B. 2001. Climate change and long-term fluctuations of commercial catches: the possibility of forecasting. *FAO Fisheries Technical Paper*, 410 p.
Knudsen, M.F., M.-S. Seidenkrantz, B.H. Jacobsen, and A. Kuijpers. 2010. Tracking the Atlantic multidecadal oscillation through the last 8,000 years. *Nature Communications* 2: 178. doi:10.1038/ncomms1186.

Knudsen, M.F., B.H. Jacobsen, M.-S. Seidenkrantz, and J. Olsen. 2014. Evidence for external forcing of the Atlantic multidecadal oscillation since termination of the Little Ice Age. *Nature Communications* 5: 3323. doi:10.1038/ncomms4323.

Lalli, C.M., and T.R. Parsons. 1993. *Biological oceanography: An introduction*. Oxford: Pergamon Press. 301 pp.

Latif, M., and T.P. Barnett. 1994. Causes of decadal climate variability over the North Pacific and North America. *Science* 266: 634–637.

Luyten, J., J. Pedlosky, and H.M. Stommel. 1983. The ventilated thermocline. *Journal of Physical Oceanography* 13: 292–309.

Mackas, D. L., P.T. Strub, A. Thomas and V. Montecino. 2006. Eastern ocean boundaries. In *The sea*, vol. 14a, ed. A.R. Robinson and K.H. Brink. Harvard University Press.

Mantua, N.J., S.R. Hare, Y. Zhang, J.M. Wallace, and R.C. Francis. 1997. A Pacific decadal climate oscillation with impacts on salmon. *Bulletin of the American Meteorological Society* 78: 1069–1079.

May, R.M., J.R. Beddington, C.W. Clark, S.J. Holt, and R.M. Laws. 1979. Management of multispecies fisheries. *Science* 205: 267–277.

McClatchie, S., R. Goericke, R. Cosgrove, G. Auad, and R. Vetter. 2010. Oxygen in the Southern California Bight: multidecadal trends and implications for demersal fisheries. *Geophysical Research Letters* 37: L19602.

McPhaden, M.J. 1999. Genesis and evolution of the 1997–98 El Niño. *Science* 283: 950–954.

Montecino, V., P.T. Strub, F. Chavez, A. Thomas, J. Tarazona and T. Baumgartner. 2004. Bio-physical interactions off western South America. In *The Global coastal ocean, interdisciplinary studies and syntheses. the sea*, vol. 14, ed. A.R. Robinson and K.H. Brink, Chapter 10. Harvard University Press.

Morrison, J.M., L.A. Codispoti, S.L. Smith, K. Wishner, C. Flagg, W.D. Gardner, S. Gaurin, S.W. A. Naqvi, V. Manghnani, L. Prosperie, and J.S. Gundersen. 1999. The oxygen minimum zone in the Arabian Sea during 1995. *Deep-Sea Research II*, 46: 1903–1931.

Narayan, N., A. Paul, S. Mulitza, and M. Shulz. 2010. Trends in coastal upwelling intensity during the late 20th century. *Ocean Science Discussion* 7: 335–360.

Nellemann, C., S. Hain and J. Alder. 2008. In Dead Water—Merging of climate change with pollution, over-harvest, and infestations in the world's fishing grounds. UNEP, Arendal, Norway, http://www.unep.org/pdf/InDeadWater_LR.pdf. Accessed 4 Apr 2016.

Newman, G.G., and D.E. Pollock. 1974. A mass stranding of rock lobsters *Jasus lalandii* (H. Milne Edwards, 1837) at Elands Bay, South Africa (Decapoda, Palinuridia). *Crustaceana*, 26: 1–4.

Pauly, D., and D. Zeller. 2016. Catch reconstructions reveal that global marine fisheries catches are higher than reported and declining. *Nature Communications* 7: 10244. doi:10.1038/ncommc10244.

Pennington, J.T., G.E. Friederich, C.G. Castro, C.A. Collins, W.W. Evans, and F.P. Chavez. 2010. The northern and central California coastal upwelling system. In *Carbon and nutrient fluxes in continental margins*, ed. K.-K. Liu, L. Atkinson, R. Quiñones, and L. Talaue-McManus, 65–78. Berlin: Springer-Verlag.

Philander, S.G.H. 1986. Unusual conditions in the tropical Atlantic Ocean in 1984. *Nature* 322: 236–238.

Philander, S.G.H. 1990. *El Niño, La Niña, and the Southern Oscillation*. San Diego: Academic Press. 283 pp.

Pitcher, G.C., and D. Calder. 2000. Harmful algal blooms of the southern Benguela current: A review and appraisal of monitoring from 1989 to 1997. *South African Journal of Marine Science* 22: 255–271.

Pitcher, G.C., and G. Nelson. 2006. Characteristics of the surface boundary layer important to the development of red tide on the southern Namaqua shelf of the Benguela upwelling system. *Limnology and Oceanography* 51: 2660–2674.

Pitcher, G.C., J.M. Franco, G.J. Doucette, C.L. Powell, and A. Mouton. 2001. Paralytic shellfish poisoning in the abalone *Haliotis midae* on the west coast of South Africa. *Journal of Shellfish Research* 20: 895–904.

Popova, E.E., A. Yool, A.C. Coward, Y.K. Aksenov, S.G. Alderson, B.A. de Cuevas, and T.R. Anderson. 2010. Control of primary production in the Arctic by nutrients and light: Insights from a high resolution ocean general circulation model. *Biogeosciences* 7: 3569–3591. doi:10.5194/bg-7-3569-2010.

Rabalais, N.N., R.E. Turner, B.K. Sen Gupta, D.F. Boesch, P. Chapman, and M.C. Murrell. 2007. Hypoxia in the northern Gulf of Mexico: Does the science support the plan to reduce, mitigate and control hypoxia? *Estuaries and Coasts* 30: 753–772.

Rasmussen, E.M., and J.M. Wallace. 1983. Meteorological aspects of the El Niño/Southern Oscillation. *Science* 222: 1195–1202.

Reason, C.J.C., P. Florenchie, M. Rouault and J. Veitch. 2006. 10 Influences of large scale climate modes and Agulhas system variability on the BCLME region. In *Large marine ecosystems*, ed. V. Shannon, G. Hempel, P. Malanotte-Rizzoli, C. Moloney and J. Woods, Elsevier, 2006, vol. 14, pp. 223–238. doi:10.1016/S1570-0461(06)80015-7.

Reverdin, G., D.L. Cadet, and D. Gutzler. 1986. Interannual displacement of convection and surface circulation over the equatorial Indian Ocean. *Quarterly Journal of the Royal Meteorological Society* 112: 43–67.

Richardson, L.E., T.K. Kyser, N.P. James and Y. Bone. 2009. Analysis of hydrographic and stable isotope data to determine water masses, circulation, and mixing in the eastern Great Australian Bight. *Journal of Geophysical Research*, 114 (C10016). doi:10.1029/2009JC005407.

Roegner, G.C., J.A. Needoba, and A.M. Baptista. 2011. Coastal upwelling supplies oxygen-depleted water to the Columbia River estuary. *PLoS ONE* 6: e18672. doi:10.137/journal.pone.0018672.

Rouault, M., S. Illig, C. Bartholomae, C.J.C. Reason, and A. Bentamy. 2007. Propagation and origin of warm anomalies in the Angola Benguela upwelling system in 2001. *Journal of Marine Systems* 68: 473–488.

Ryan, J.P., J.F.R. Gower, S.A. King, W.P. Bissett, A.M. Fischer, R.M. Kudela, Z. Kolber, F. Mazzillo, E.V. Rienecker, and F.P. Chavez. 2008. A coastal ocean extreme bloom incubator. *Geophysical Research Letters* 35: L12602. doi:10.1029/2008GL034081(12).

Saji, N.H., B.N. Goswami, P.N. Vinayachandran, and T. Yamagata. 1999. A dipole mode in the tropical Indian Ocean. *Nature* 401: 360–363.

Schlesinger, M.E. 1994. An oscillation in the global climate system of period 65–70 years. *Nature* 367: 723–726.

Schwartzlose, R.A., J. Alheit, A. Bakun, T. Baumgartner, R. Cloete, R.J.M. Crawford, W.J. Fletcher, Y. Green-Ruiz, E. Hagen, T. Kawasaki, D. Lluch-Belda, S.E. Lluch-Cota, A.D. MacCall, Y. Matsuura, M.O. Nevares-Martinez, R.H. Parrish, C. Roy, R. Serra, K.V. Shust, N.M. Ward, and J.Z. Zuzunaga. 1999. Worldwide large-scale fluctuations of sardine and anchovy populations. *South African Journal of Marine Science* 21: 289–347.

Shakun, J., and J. Shaman. 2009. Tropical origins of North and South Pacific decadal variability. *Geophysical Research Letters* 36: L19711. doi:10.1029/2009GL040313.

Shannon, L.V., A.J. Boyd, G.B. Brundrit, and J. Taunton-Clark. 1986. On the existence of an El Niño-type phenomenon in the Benguela system. *Journal of Marine Research* 44: 495–520.

Sordo, I., E.D. Barton, J.M. Cotos, and Y. Pazos. 2001. An inshore poleward current in the NW of the Iberian Peninsula detected from satellite images, and its relation with *G. catenatum* and *D. acuminata* blooms in the Galician Rias. *Estuarine and Coastal Shelf Science* 53: 787–799.

Stramma, L., G.C. Johnson, J. Sprintall, and V. Mohrholz. 2008. Expanding oxygen-minimum zones in the tropical oceans. *Science* 320: 655–658.

Stramma, L., S. Schmidtko, L.A. Levin, and G.C. Johnson. 2010. Ocean oxygen minima expansions and their biological impacts. *Deep-Sea Research I* 57: 587–595.

Susanto, R.D., A.L. Gordon, and Q. Zheng. 2001. Upwelling along the coast of Java and Sumatra and its relation to ENSO. *Geophysical Research Letters* 28: 1599–1602.

Thurman, H., and A. Trujillo. 1999. *Essentials of Oceanography*. Upper Saddle River, N. J.: Prentice Hall.

Tomczak, M., and J.S. Godfrey. 2003. *Regional oceanography: An introduction*. 2nd edn. Daya Publishing House, 390 pp.

Trenberth, K.E., and J.W. Hurrell. 1994. Decadal atmosphere-ocean variations in the Pacific. *Climate Dynamics* 9: 303–319.

Trenberth, K.E., and D.J. Shea. 2005. Atlantic hurricanes and natural variability in 2005. *Geophysical Research Letters,* 33. doi:10.1029/2006GL026894.

van Loon, H., and J.C. Rogers. 1978. The seesaw in winter temperatures between Greenland and northern Europe. Part I: General description. *Monthly Weather Review* 106: 296–310.

van Oldenborgh, G.J., L.A. te Raa, H.A. Dijkstra, and S.Y. Philip. 2009. Frequency- or amplitude-dependent effects of the Atlantic meridional overturning on the tropical Pacific Ocean. *Ocean Science* 5: 293–301.

Walker, G.T. 1923. Correlation in seasonal variations of weather, VIII: a preliminary study of world weather. *Memoires of the Indian Meteorological Department, Calcutta* 24(4): 75–131.

Walker, G.T., and E.W. Bliss. 1932. World weather V. *Memoires of the Royal Meteorological Society, London* 4(36): 53–84.

Wallace, J.M., and D.S. Gutzler. 1981. Teleconnections in the potential height field during the Northern Hemisphere winter. *Monthly Weather Review* 109: 784–812.

Webster, P.J., A.M. Moore, J.P. Loschnigg, and R.R. Leben. 1999. Coupled oceanic-atmospheric dynamics in the Indian Ocean during 1997–98. *Nature* 401: 356–360.

Whitney, F.A., H.J. Freeland, and M. Robert. 2007. Persistently declining oxygen levels in the interior waters of the eastern subarctic Pacific. *Progress in Oceanography* 75: 179–199.

Wilkinson, C., and G. Hodgson. 1999. Coral reefs and the 1997–1998 mass bleaching and mortality. *Nature and Resources* 35: 16–25.

Wyrtki, K. 1975. El Niño-the dynamic response of the equatorial Pacific Ocean to atmospheric forcing. *Journal of Physical Oceanography* 5(4): 572–584.

Zhai, L., T. Platt, C. Tang, S. Sathyendranath, and A. Walne. 2013. The response of phytoplankton to climate variability associated with the North Atlantic Oscillation. *Deep Sea Research II* 93: 159–168.

Zingone, A., and T. Wyatt. 2005. Harmful algal blooms: keys to the understanding of phytoplankton ecology. In *The sea, Volume 13. The Global Coastal Ocean: Multiscale Interdisciplinary Processes*, ed. A.R. Robinson and K.H. Brink, 867–926. Harvard University Press.

Chapter 4
The California Current Upwelling System

Abstract In this chapter, we discuss the California Current upwelling system along the west coast of North America, from its southern point in Baja California to where it meets the cyclonic gyre in the Gulf of Alaska. The southern part of the system is very well studied, thanks to the CALCOFI program that has been ongoing since 1949, but new research results during the past decade have changed our views of the way the system operates considerably.

Keywords Upwelling · Oceanography · California current · CALCOFI program · Hypoxia · Marine ecology · Biogeochemistry · Ecological regime shifts · Fisheries · Fish stock variations · Climate-change impacts

What is it about the sea? Is it because it's there?

Jerry Pinto

(Taken from *Em and The Big Hoom*, 2012, reproduced with permission)

4.1 Introduction

The 2700 km California Current System (CCS) along the western coastline of North America stretches from the U.S.-Canadian border at the Strait of Juan de Fuca near 48° N to the tip of the Baja California peninsula, at 23° N, and can also include parts of Vancouver Island as far as 52° N (Mackas 2006; Figs. 4.1 and 4.2). It therefore includes the coastlines of Washington, Oregon and California, as well as northern Mexico. The coastline in the northern region is generally straight and oriented roughly north-south, but there are major bends at Cape Mendocino (40° N) and Point Conception (34° N). South of Point Conception the southern California Bight contains both islands and deep offshore basins, unlike the rest of the coast, although these disappear off Baja California. Thus, the CCS can be divided into four main regions: Vancouver Island-Cape Mendocino, Cape Mendocino-Point Conception, the California Bight, and Baja California.

© Springer International Publishing Switzerland 2016

J. Kämpf and P. Chapman, *Upwelling Systems of the World*,
DOI 10.1007/978-3-319-42524-5_4

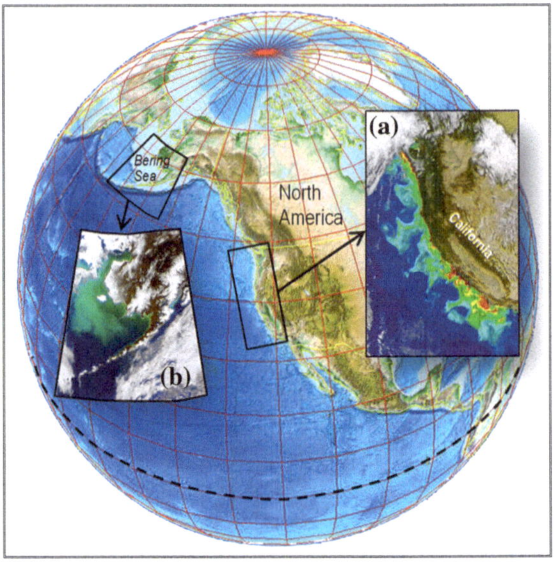

Fig. 4.1 Upwelling systems in the North Pacific revealed by SeaWiFs chlorophyll-*a* data (*source* NASA). Panel **a** shows the California Current Upwelling system and panel **b** the "green belt" of the Bering Sea (see Chap. 9 for more details). Background image: Google Earth with ETOPO 1 bathymetry layer

The current flow, which forms the eastern extremity of the North Pacific gyre is essentially southwards (equatorward) all along the coast during summer in a band about 100 km wide, although there can be northward (poleward) flow at the surface off Washington and Vancouver Island depending on the volume of freshwater flowing from the Fraser and Columbia Rivers, which are the only large freshwater inputs in the region. The shelf is generally narrow and deep, although intersected by canyons, and the strong wind stress forcing leads to large scale alongshore variability, including coastal-trapped waves that can propagate over long distances, particularly in the northern part of the CCS with its straight coastline. Currents offshore have different time and space scales to those on the shelf, and filaments can be pulled several hundred kilometers from the coast (Hickey 1998). In non-summer months, northward flow, known as the *Davidson Current*, is generally found from Point Conception to Vancouver Island (Hickey 1979; Chelton 1984). At the northern end, the system interacts with the sub-polar Alaskan Gyre, while at the southern end the California Current moves offshore under the influence of the trade winds to form the westward flowing North Equatorial Current.

As is typical of the four major upwelling systems, the hinterland in the lower latitudes of the CCS is semi-desert. This occurs because the trade winds contain little moisture after their passage across the American continent, and results in a low-pressure system over the coastal region. The offshore Ekman transport lowers sea level along the coast, producing a zonal pressure gradient, an equatorward geostrophic surface current about 100 km wide, and coastal upwelling in a narrow zone close to the coast. Below the surface there is a meridional pressure gradient near the shelf break that sets up a poleward undercurrent as described in Chap. 2.

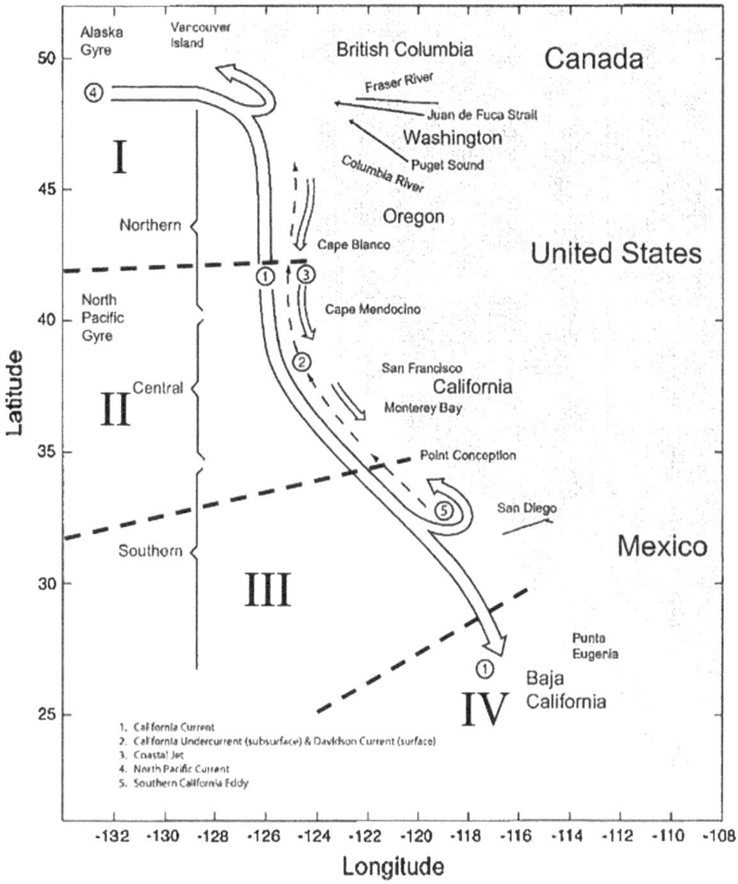

Fig. 4.2 Map of the California Current system showing major regions, currents and locations mentioned in the text. Modified from Checkley and Barth (2009)

Despite the relatively inhospitable terrain in the southern part of the region, the California Current is classified as a region of moderate to high productivity and supports a large pelagic fishery, again typical of upwelling areas. The California sardine fishery was previously much larger than it is at present, and inspired John Steinbeck's novel *Cannery Row* before it collapsed in the 1940s (Fig. 4.3). Since then, efforts have been made to understand the environmental variability and its effects on fish stocks, aided by a number of large interdisciplinary research projects emanating from marine laboratories in California, Oregon, Washington, Canada, and more recently Mexico. Of these, the California Cooperative Oceanic Fisheries Investigations program (CALCOFI; http://www.calcofi.org/), which has been ongoing since March 1949, is the best known, although it no longer covers the

Fig. 4.3 Hovden's Cannery, now the site of the Monterey Bay Aquarium

whole region but is concentrated around central and southern California (Fig. 4.4). In 2004, the U.S. National Science Foundation funded another major program associated with CALCOFI (California Current Ecosystem Long-Term Ecological Research; CCE-LTER), that has the aims of determining what physical mechanisms lead to changes in ecosystem states and how changing ocean climate can affect community structure and dynamics (Goericke and Ohman 2015).

Since 1997, Mexican researchers have collected data regularly as part of the Investigaciones Mexicanas de la Corrientede California (IMECOCAL). Smaller, shorter-lived projects have concentrated on research in specific regions of the CCS, particularly since 2000. Some of these will be discussed in more detail in this chapter. Since the original overview of the system by Reid et al. (1958), there have been a large number of updates by many authors, while the total number of papers on the system has exploded. Excellent recent reviews of the oceanography of the region are given by Hickey (1998), and Checkley and Barth (2009) for physical aspects, and Mackas (2006) for lower trophic level biology. We have not attempted to discuss all the work done in the region, but we hope we have covered all the major advances.

4.2 History of the Region

Although there is no historical record of large population centres along the west coast of North America before the arrival of European explorers in the16th century, the region is thought to have supported large numbers of people. For example, a recent estimate suggests that one third of the U.S. population in 1492 was located in California (Starr 2005). Mesoamerica, however, certainly had major cultures such

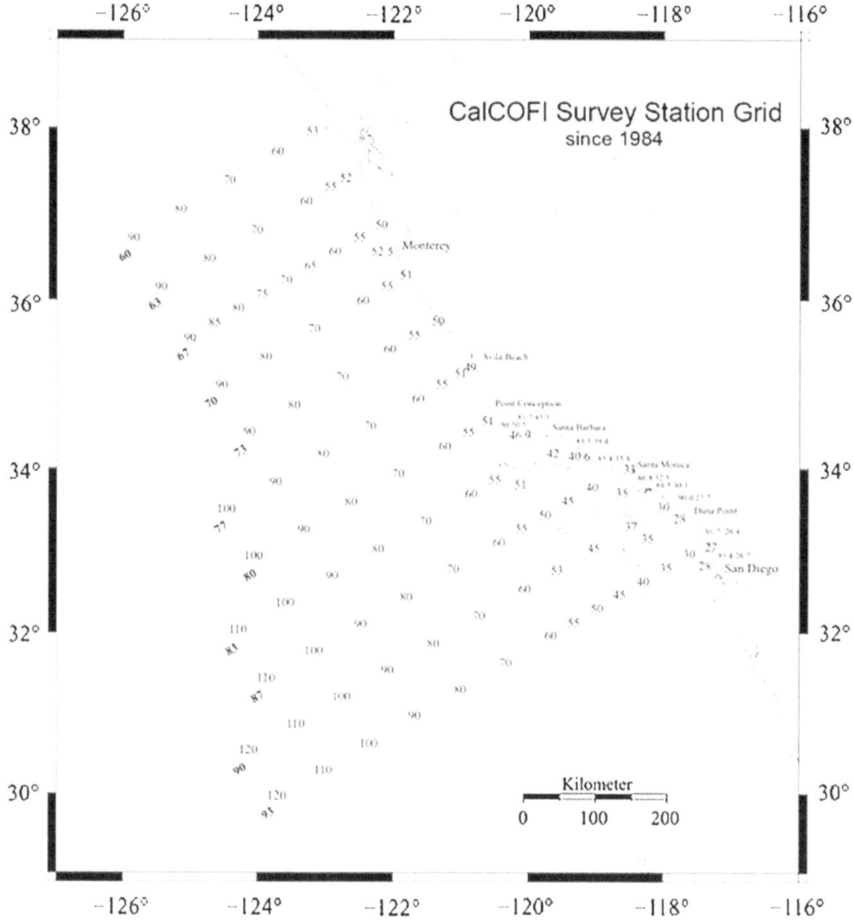

Fig. 4.4 Station map for the CALCOFI grid as occupied since 1984. Biological sampling at stations along lines 60–73 are only surveyed during winter and spring when egg and larval densities are high. The additional stations along line 67 form part of the SECRET monthly time-series, which is collected by researchers at the Monterey Bay Applied Research Institute (MBARI)

as the Capachas, Olmecs, Mixtecs and Mayas after about 1800 BC, whose ancestors are thought to have come down the west coast from the Bering Strait in a succession following the last ice age.

In 1493, the Pope granted Spain exclusive rights to colonize almost all of the Americas apart from eastern Brazil. These rights were confirmed by the Treaties of Tordesillas (1494) and Zaragosa (1529), and the Spanish, under Vasco Nuñez de Balboa, who crossed Panama in 1513, were the first Europeans to reach the west coast of the Americas. Balboa claimed the whole Pacific Ocean for Spain, and it was only in the late 17th century that other European powers began regular visits and attempted to found settlements. Early exploration during the 16th century was based at the Spanish settlement of Acapulco, in southern Mexico, with explorers pushing north in an attempt to find the fabled Strait of Anian, which supposedly connected the Atlantic with the Pacific near 42° N. The Spanish reached what is now San Diego and Monterey in 1542, and may even have reached Oregon at that time. Twenty years later, a Spanish ship sailing northeast from the Philippines made landfall near Cape Mendocino, having used the currents in the North Pacific gyre to aid in the journey, and the Philippine fleets used this route for the next 300 years. In was not until after 1750 that any settlements were founded in California, at San Diego, Monterey, and San Francisco, although the Spanish established a number of missions along the Baja peninsula.

In the meantime, however, Sir Francis Drake had landed north of San Francisco and claimed this part of the country for England, citing as precedent the voyages of John Cabot on the Atlantic coast in 1497. While Drake did not found a settlement, and the English only again became active along this coast in the second half of the 18th century, charters granted after the landings of Drake and Cabot became the basis for the U.S. claim to stretch from coast to coast of the continent after independence from Britain in 1776. Similarly, Russian explorers and trappers hunted sea otters from Alaska to San Diego after 1750 and founded a short-lived trading post at Fort Ross, about 100 km north of San Francisco, in 1812. This was sold to an American investor in the1840s (Thompson 1896).

During the late 18th century, Spanish, Russian and British ships were all busy charting the northern portion of the west coast. While the Russians came from the north, the Spanish and British came from the south. The Spaniard Don Bruno de Heceta was the first western European to see the Columbia River in 1775, eventually reaching 59° N, and later Spanish expeditions reached Valdez and Kodiak Island in Alaska. Meanwhile the Britons James Cook and George Vancouver also charted the coast from California to the Bering Strait. The Spanish gave up their claim to the Pacific northwest after 1794, allowing the British to settle in the region, while the U.S. also gained settlement rights in 1819, but it was not until the California gold rush of 1848 that large-scale immigration to the west coast started, and California even passed the first laws protecting oyster beds in 1851 and salmon runs in 1852.

Rockfish were caught commercially after 1875, although they had been used as a food resource for centuries (Lenarz 1986), and the fishery for sardines began in 1916. However, it was not until 1903 with the establishment of the Scripps Institution for Biological Research, now the world famous Scripps Institution of Oceanography, that marine research began in earnest along the west coast of the U.S., and even then the first discussion of upwelling off California came from German research by Thorade (1909) using merchant ship temperature data. Even after the Scripps Institution was established, expansion was slow until the 1930s (Mills 1991).

4.3 Physical Controls

4.3.1 Large-Scale Physical Controls

The overall circulation of the CCS is driven by the large-scale variability of the North Pacific gyre, which, relative to the coastline, is convergent in the north and divergent in the south. As with all eastern boundary upwelling systems, the local wind field is important, and this varies with latitude, being almost continuously equatorward off California but becoming increasingly intermittent and seasonal to the north, so that off the coast of Washington the periodicity is in the 2–20 day range (Hickey 1998; Barth et al. 2007).

The maximum wind stress occurs in spring in the southern part of the CCS and in summer further north, with the strongest winds occurring offshore of northern California near 38° N, and the wind stress curl diminishes in both directions away from this latitude (Bakun and Nelson 1991). In winter, particularly in the north, the region experiences minimum equatorward stress and even poleward (downwelling) wind stress, which increases northwards. Wind stress along the coast varies by up to about 2 dyn/cm^2 (0.2 Pa) except in the southern California Bight, where it is about 0.2 dyn/cm^2 (0.02 Pa) in summer (Hickey 1992). This means that there is almost no upwelling within the Bight during the summer season, which leads to strong stratification, although offshore in this region the winds are strong and there is replacement via the Santa Barbara Channel from water that has upwelled north of Point Conception. Further south, off Baja California, upwelling begins again, and while the wind stress is generally higher than in the Bight, it is lower than that encountered north of Point Conception.

While there are several wind data sets for calculating wind stress, the most commonly used one is that produced by the NOAA Southwest Fisheries Science Center's Environmental Research Laboratory; these have been used to estimate the upwelling strengths (summarized in Table 4.1) and have been shown generally to compare well with other indices for southern California and Baja California except near 33° S in the Southern California Bight, where the wind field is affected by the bend in the coastline at Point Conception (Perez-Brunius et al. 2007). These calculations ignore the effect of the wind-stress curl, which can be as important as the offshore Ekman transport in terms of supporting biological production, even though the upwelling rate associated with curl is an order of magnitude less (Checkley and Barth 2009; Rykaczewski and Checkley 2008).

Changes in the timing and intensity of seasonal cycles depend on the variability and interaction of the cyclonic Aleutian Low and the anticyclonic North Pacific high pressure systems (see Fig. 2.13 in Chap. 2), which affect the northern and southern portions of the CCS respectively. As might be expected, the Aleutian Low has more effect in winter and along the northern portion of the coastline, while the North Pacific High is more influential in spring and summer, particularly in the southern half of the CCS, with its influence shifting northwards from Baja in spring to central California in summer. The existence of these two contrasting atmospheric

Table 4.1 Zonation of forcing and biological productivity in the California Current system, adapted from U.S. GLOBEC 1994 and PISCO 2004

	Region I	Region II	Region III	Region IV
Storms	Winter storms frequent and strong	Moderate winter storms	Fewer winter storms	Infrequent storms from tropics
Winds	Seasonal wind stress reversal	Winds mostly upwelling favourable	Minimum in longshore wind stress	Modest and persistent longshore wind stress
Upwelling	Moderate upwelling in spring/summer	Strongest upwelling in spring/summer	Weak upwelling	Modest upwelling year-round
Freshwater input	Significant	Minor	Negligible	Negligible
Coastal relief	Relatively smooth coastline	Major coastal promontories	Concave coast, islands, subsurface basins	Several coastal promontories
Shelter/nurseries	Major estuaries/ nursery grounds	A few major bays, estuaries	Sheltered bight, major nursery	Several sheltered bays
Circulation	Moderate advection/ mesoscale activity	Extreme advection, mesoscale activity	Strong local recirculation, weak mesoscale activity, strong stratification, strong mixing	Moderate
Primary productivity	Strongly seasonal	Strongly seasonal	Damped seasonality	Moderate seasonality
Zooplankton biomass	Strongly seasonal (Copepods commonly overwinter at depth)	Seasonal	Modest seasonality	Damped seasonality
Spawning by epipelagic fish	Patchy	Latitudinal minimum	Latitudinal maximum	Patchy

Regions are indicated in Fig. 4.2

systems leads to large differences in precipitation along the coast; south of 35° N the land is almost desert, while the northwestern U.S. is known for its high rainfall, especially in winter.

The interaction of these two systems results in length scales of about 1000 km alongshore and 500 km cross-shelf for both currents and sea level variability (Hickey 1998), but there is additional spatial structure in both directions. This variability has been recognized for over 20 years, and ocean-atmosphere interactions on large-scales, such as the El Niño/Southern Oscillation (ENSO), the Pacific Decadal Oscillation (PDO) and the North Pacific Gyre Oscillation (NPGO) (see Chap. 3) have been highlighted as forcing functions for decadal-scale variability that affects both physical and biological responses (e.g., Mantua et al. 1997; Lynn et al. 1998; McGowan et al. 1998; Lavaniegos and Ohman 2003, 2007; Di Lorenzo et al. 2008).

4.3.2 Basic Description of the CCS

As stated previously, the California Current flows southwards over a generally narrow and steep shelf. In the north the shelf is widest at about 50 km near the Strait of Juan de Fuca at 48°–50° N and narrows to less than 20 km wide northward into the Gulf of Alaska and southward along the coast to Point Conception (Fig. 4.2). In the southern California Bight, with its complex topography of deep basins and islands, the shelf is extremely narrow, but it broadens again off Baja California in the embayments north of Punta Eugenia and Cabo San Lazaro.

These changes in the width of the shelf, together with the changes in alignment of the coastline at Cape Mendocino, Point Conception and San Diego, affect nutrient upwelling and the cross-shelf transport of water, nutrients and particles, and cross-shelf transport within about 100 km of the coast increases where the gradients get steeper. The cross-shelf physical gradients lead to similar gradients in biota and productivity (Table 4.1), with transition points for both physics and biology at Cape Blanco (43° N), Point Conception (35° N) and Punta Baja (30° N). There is also an offshore transition at the edge of the southern California Bight, as seen by the movement of mixing zones and fronts away from the coast to the south (Johnston and Rudnick 2015; Powell and Ohman 2015). This means that off Oregon and Washington, the upwelling is essentially two-dimensional in a narrow band along the coast, while it is more three-dimensional, with offshore filaments and eddies, further south (Hickey and Banas 2008). However, the change in alignment of the coast at Point Conception means that the southern California Bight experiences much less wind stress than elsewhere along the coast, and a cyclonic gyre exists here for much of the year.

The southern limb of this gyre can produce a physical and biological front (the Ensenada front) near 31°–32° N (Lynn and Simpson 1987). Other areas where upwelling is weaker are found in the Strait of Georgia/Puget Sound, San Francisco Bay, Monterey Bay and south of Punta Eugenia (25°–29° N). These all have different salinity regimes because of variations in precipitation, evaporation and topography, so there is no single reason for the changes. All along the coast, offshore of the upwelling front from the Strait of Juan de Fuca to Point Conception, mesoscale eddies, meanders and filaments are frequently seen (Fig. 4.5), with the largest filaments observed between 35°–42° N. Coastal topography is also important; the onshore ridge near Cape Mendicino, for example, affects the local air flow and produces an offshore squirt (Largier et al. 1993). Similar effects are also found at other points where there are changes in the coastal topography.

Although the CCS contains only a few bays and rivers, several submarine canyons cut through the shelf and provide short access routes to deep water. They

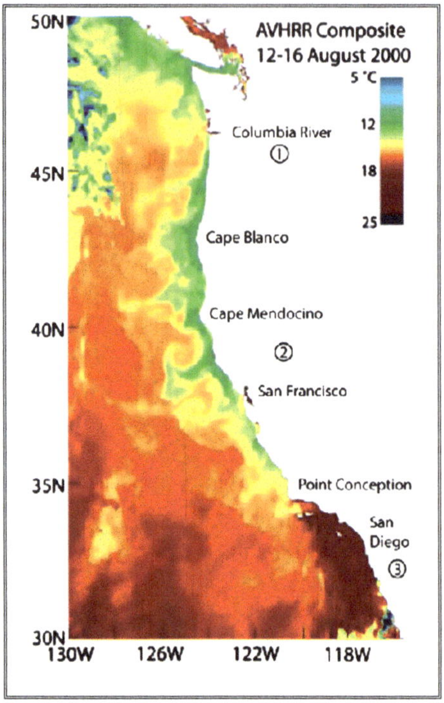

Fig. 4.5 Satellite image of meanders along the CCS, taken from an AVHRR composite image of 12–16 August 2000 (from Checkley and Barth 2009)

can also cause local disruptions of the overall current patterns, e.g., the equatorward flow of the undercurrent seen near 47° N (Hickey 1989). These canyons can result in increased local productivity; for example, seasonal effects have been recorded off the Juan de Fuca canyon near 48.5° N (Freeland and Denman 1982). Other topographic features also affect upwelling in places. Hecata Bank, on the shelf off central Oregon at 44° N, interrupts the southward flow of the coastal current and creates a shelter zone inshore where current velocities are reduced so that phytoplankton production can increase (see Barth et al. 2005).

Off California, the winds are upwelling favourable throughout the year (Chelton et al. 1982), even though upwelling is not permanent. The maximum wind stress is found off northern California and decreases both north and south of 35°–40° N. Along with changes in wind stress, there are also variations in the number of days on which upwelling occurs. As a result, there are frequent current reversals superimposed on the overall equatorward flow, and these increase in both directions away from the region of maximum wind stress near 38° N. Such reversals are particularly prevalent over the inner shelf north of Cape Mendocino, in water less than about 30 m deep, and the currents can reverse within a few hours of local wind reversals (Huyer et al. 1979). Further offshore, current reversals are much less frequent, although the amplitude of the coastal jet varies with wind stress. In winter, upwelling north of the cape is practically non-existent and the main current flow is actually towards the north (the Davidson Current), with speeds of up to about

25 cm/s at the surface (Reid and Schwartzlose 1962). South of Cape Mendocino, filaments, produced as wind-driven jets that separate from the coast, become important in determining surface water structure and movement (Fig. 4.5). Models suggest that the filaments account for much of the offshore variability south of 38° N, and that their controls are complex (Hickey 1998).

In the south, the only long-term current meter records from the shelf off Baja (Barton 1985) show only very weak seasonal or even mean flow but contain barotropic flow reversals with periods varying between 5 and 30 days (Fig. 4.6). Flow during these periods was up to about 50 cm/s at the surface, decreasing to less than 20 cm/s at the bottom in 60 m water depth, and generally parallel to the coastline. Cross-shelf flow showed considerably less variability than alongshelf flow and was

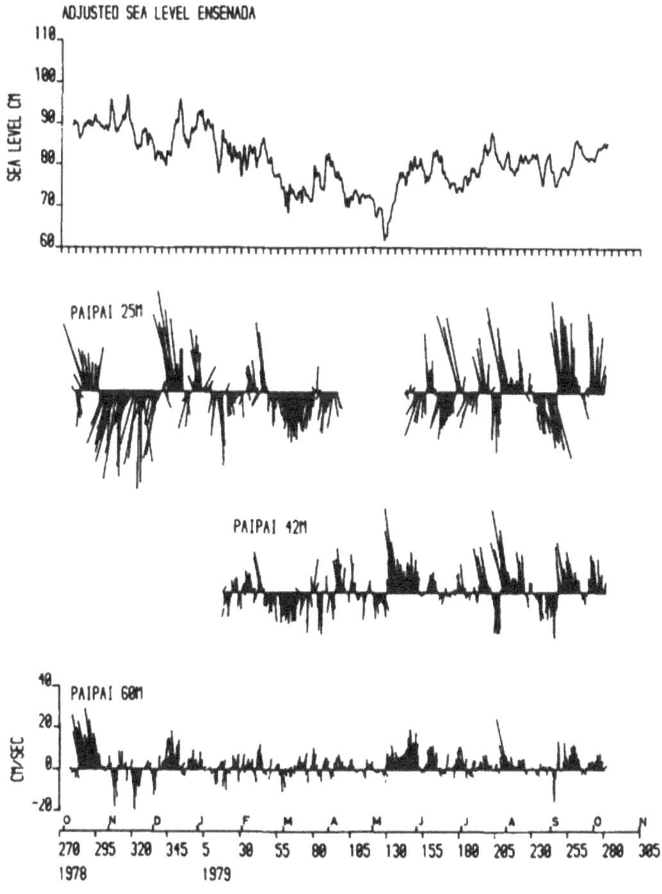

Fig. 4.6 Low-pass filtered sea level at Ensenada adjusted for atmospheric pressure and current vectors at mooring Paipai off Ensenada Bay (31° 50′ N, 116° 50′ W). Vectors are plotted at 24 h intervals, with vertically "up" denoting alongshore and poleward currents towards 356° T. Julian day 270 is 27 September (from Barton 1985)

close to zero at all depths, while scales of flow appear to be about 100 km. This is larger than the scales found in the southern California Bight, but smaller than those seen further north, which can be up to 1,000 km in the longshore direction and 500 km offshore.

As with other eastern boundary currents, the CCS has a persistent poleward undercurrent along the upper slope. Its strength varies throughout the year, and this has been shown off California to be driven by seasonal changes in the longshore pressure gradient (Largier et al. 1993), as originally suggested by Hickey (1989). Interaction with the bottom boundary layer on the shelf leads to the loss of nutrients from the shelf as the undercurrent strengthens, which is the opposite of what happens as upwelling starts and the undercurrent is absent or weak (Hickey and Banas 2008).

The CCS has a large number of coastal tide gauges that stretch from Vancouver Island to Baja California. As a result, the variability in sea level, which shows a distinct seasonal cycle, is well known. The tides are dominated by the semi-diurnal signal, which propagates poleward at $\sim$140 m/s (Rosenfeld and Beardsley 1987), and tidal ellipses are highly elongated and oriented along the topography. The internal tides are also semidiurnal and seen when stratification is high and this seems to be the case from Vancouver Island to California, with baroclinic tides generated at the shelf break and tidal and barotropic currents dominated by Kelvin waves (Hickey 1998). Sea level is generally higher in the north in winter and in the south in summer, but minima occur off northern California in spring and autumn.

Tide gauge records along the California coastline relate well to geostrophic flow (Chelton et al. 1982). When sea level is high, flow is to the north, when it is low, to the south, and the sea level signal leads the transport signal by about three months. So theoretically we can extend transport estimates back the start of sea level measurements in 1900. The sea level record shows El Niño-like conditions in 1902–09, 1914–16, 1940–44, 1951–53, 1957–60, 1969–70, 1972, and 1977–80. However, not all of these were El Niño years, so this analysis does not explain all the anomalies, which presumably are linked to larger scale variability in the North Pacific. However, despite the increasing interest in changes such as the Pacific Decadal Oscillation (PDO) and the North Pacific Gyre Oscillation (NPGO) (see Chap. 3), the larger-scale drivers of sea level variability and the longshore pressure gradient in the CCS are still not known with certainty.

Interannual variability in the CCS is mostly attributed to El Niño/La Niña phases of ENSO as changes in coastal sea level and offshore steric heights are strongly related to sea level variability in the eastern tropical Pacific. Superimposed on this are decadal-scale changes due to changes in the PDO; when this is in the cool, negative phase upwelling tends to increase (Miller et al. 2015). Flow in the CCS decreases during El Niño periods (Chelton et al. 1982) as do nutrient concentrations, phytoplankton and zooplankton biomass and fish catches. During these periods southern species appear further north than usual (Simpson 1992), and the effects of the warmer water are felt far offshore and as deep as 500 m. The largest property anomalies are usually seen in the 50–200 m depth layer, which results in the poleward advection of saltier water from the tropics in the undercurrent (Simpson 1992).

Data from individual years show changes in parameters such as the local wind stress, the timing of the spring transition to the upwelling season, the strength of remote forcing, and observed water properties. While less than 50 % of the variance in these data can be linked to El Niño, poleward flow over the Oregon shelf increases at this time (Huyer and Smith 1985). There are even observable changes at much greater depth, as a 6-year record from a current meter deployment at 350 m depth off Point Sur shows large interannual variability, although this is not necessarily caused by El Niño (Collins et al. 1996). The question arises regarding whether these large-range effects are transmitted via the ocean or atmosphere; the most likely cause is through the ocean via Kelvin waves associated with changes in atmospheric pressure, which then become Rossby waves over the shelf and can travel northwards. This was shown relatively early by models, such as those of Johnson and O'Brien (1990a, b) which showed such a continuous signal from the western Pacific to the U.S. coast and then up to about 50° N.

Off Baja California, it appears that two systems are found, with the break between them occurring at Punta Eugenia (28° N), and the transition appears to be a biological boundary for various fish species (Hewitt 1981). The variability off Baja California from 1998–2008 has been described by Durazo (2009), who showed that temperature and salinity data from this region were indicative of four different states: the 1997–98 El Niño, with warmer and saltier conditions in the near surface layer; La Niña from late 1998–mid 2002 (cooler temperatures but relatively high salinity); a cooling and freshening period between 2002–2006, when subarctic water was present in the upper 200 m (i.e., down to the core of the poleward undercurrent); and a second El Niño-La Niña transition from 2006–2008.

The latitudinal extent of the warm and saline anomalies during the 1997–98 event along the northernmost line agrees with narrow poleward flows seen previously here and along the southern California coast (Durazo and Baumgartner 2002; Lynn and Bograd 2002), while the arrival of water of subarctic origin in October 2002 at the northern line agrees with its southward progress from the northern end of the CCS (Strub and James 2003; Bograd and Lynn 2003). In contrast, the change to renewed El Niño conditions on 2006–2007 was due to the northward advection of warmer and saltier water from the south. Below the surface, a trend towards saltier conditions continued from 2001 to the end of 2008.

4.4 Water Masses

Tibby (1941) showed from a temperature-salinity (T-S) analysis that the water masses off the west coast of North America were essentially a mixture of two extremes, one from the subarctic North Pacific, the other from the equatorial region. He used the data to derive the percentage of each water mass for depths between 100 and 1,000 m, as well as the northward penetration of equatorial water along the

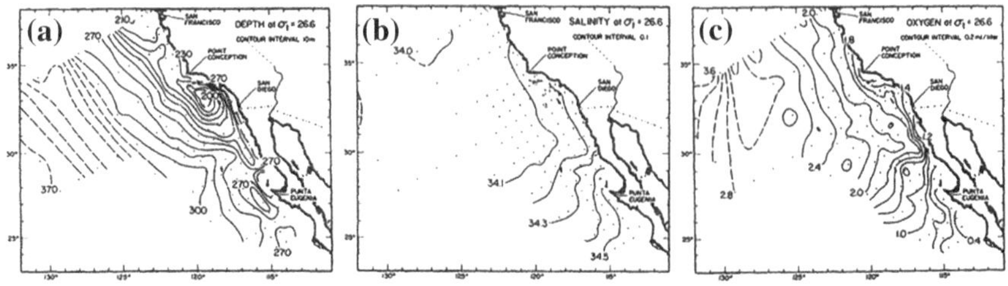

Fig. 4.7 **a** Mean depth (m), **b** mean salinity, and **c** mean dissolved oxygen concentration (mL/L) for the sigma-t = 26.6 density surface in July (means 1950–78) (from Lynn et al. 1982)

California coast. Above 100 m depth, wind mixing affects the T-S relationship, while below about 1,000 m, differences in temperatures and salinities become very small, so the methodology could not be used.

The overall results of this analysis showed that the percentage of equatorial water is higher towards the bottom and towards the shore. In the southern California Bight and off Baja California, the percentage of equatorial water was consistently higher than 50 % below about 200 m near the coast and 250 m offshore, but north of Point Conception northern water predominated throughout the upper 1,000 m of the water column, with 50 % equatorial water being found at 300 m depth only as far north as Monterey (36° N), with cooler, fresher water offshore marking the position of the undercurrent. In general, salinity increases with depth in the CCS but decreases at a given depth as one moves away from the coast. This can be seen by the northward intrusion of high salinity, low oxygen water on the 26.6 sigma-t surface along the coast (Lynn et al. 1982, Fig. 4.7).

More modern analyses (e.g., Lynn and Simpson 1987; Tomczak and Godfrey 1994) suggest that four water masses contribute to upwelling in the California Current System. These are warm, low salinity Tropical Surface Water (T > 24 °C, S < 34) coming from the equatorial regions, North Pacific Central Water (T > 20° in the south, S > 34.4), North Pacific Equatorial Water (8° < T < 15°, S > 33), and North Pacific Subarctic Water (8° < T < 21°, S < 33.6). As they mix, they form the upwelling water mass that typically occupies the region between 8°–18 °C in temperatures and 33.4–34.4 in salinities (Perez-Brunius et al. 2006). Like other eastern boundary current systems, the water that upwells within the California Current comes from midwater, rather than from local sources on the shelf, and much of it originates in the North Pacific subarctic gyre. However, its temperature and salinity are modified over the shelf by interaction with river plumes, especially in the north.

Below the southward flowing surface current field, the California Undercurrent flows in the opposite direction at depths of 150–200 m. The undercurrent transports warm, salty North Pacific Equatorial Water from the subtropics to the north, so that the characteristics of the core of the undercurrent decrease from a temperature of about 9.5 °C and a salinity of 34.6 off southern Baja California to about 9.0 °C and 34.4 off San Diego, and to 7.0 °C and 33.9 off Vancouver Island. Some of this water originates in the Costa Rica Coastal Current which flows northwards along the coastline of Central America and southern Mexico and merges with the California Current just south of Cabo San Lucas at the mouth of the Gulf of California. Northward flow in the Costa Rica Coastal Current is seen during from May-December, with its maximum northward extension occurring after August (Badon-Dangon et al. 1989). The temperature and salinity of the undercurrent at different latitudes vary throughout the year, depending on the intensity of the overall northward flow and mixing along its path, with the warmest and saltiest water at any given latitude (e.g., off Washington) being found in late summer (Hickey 1979).

In addition to these annual changes, the system can be affected by baroclinic (density-related) waves that propagate over large distances. These can lead to seasonal changes in water properties if driven by remote wind events. During El Niño events, this leads to water that is warmer, saltier, and contains fewer nutrients than normal, resulting in reduced phytoplankton production. In contrast, during periods when subarctic water intrudes into the CCS, the water mass on the shelf is colder, fresher and contains more nutrients, leading to enhanced phytoplankton growth. A third type of remote forcing was seen in 2005 (Hickey et al. 2006), when the arrival of upwelling favourable winds in the northern California Current in spring was delayed by about 2 months. This led to temperature-salinity relationships similar to those of the 1983 El Niño year (warmer and more saline), as well as much lower nutrient concentrations in the surface than usual (Fig. 4.8), and affected about 450 km of the shelf, from Vancouver Island to central Oregon. Upwelling of colder, nutrient rich water onto the shelf appeared to be initiated by upwelling favourable winds considerably farther south, off northern California (39° N), in late May, as strong local upwelling started in only mid-July. This allowed the newly upwelled water to reach the inner shelf within a few days, and the system appeared back to “normal” by early August.

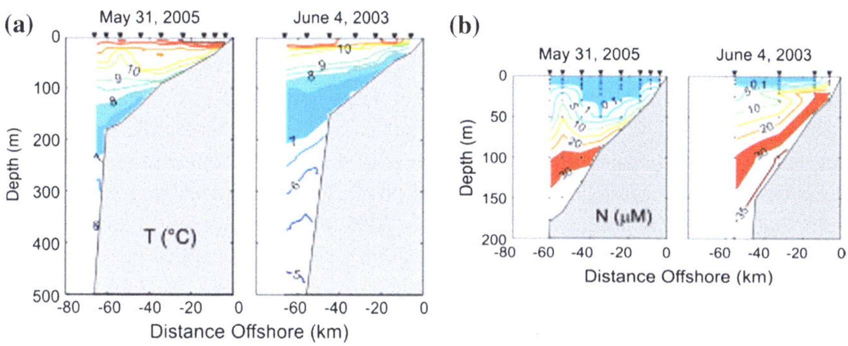

Fig. 4.8 **a** Temperature sections near 47° N in early summer in 2005 and 2003; **b** nitrate concentrations across the Washington shelf and slope on the same dates. *Dots* indicate bottle sampling depths (from Hickey et al. 2006)

4.5 Circulation Patterns and Variability

4.5.1 Overview

Thanks to the California Cooperative Fisheries Investigations (CALCOFI), we have data on the variability of the California Current system from surveys that date back to 1949. Initially, the surveys covered the whole of the California coastline from 42° N, north of Cape Mendocino, to the southern tip of Baja California on a monthly basis at 65 km intervals, and sampled temperature, salinity, oxygen and phosphate concentrations to 500 m depth. In 1961, cruises were reduced to quarterly, but nitrate, nitrite and silicate measurements were added, with chlorophyll added in 1973, ^{14}C primary production in 1984, and CTD profiling in 1993 (www.calcofi.org). However, the cost of the surveys has reduced the size of the station grid and today only 75 stations are occupied between San Diego and Point Conception, i.e., covering the southern California Bight. An additional 38 stations north of Point Conception are added on two cruises each year. From 1969–1982, the regular cruise pattern was disrupted and irregular, with no cruises in 1970 or 1971, only one or two cruises in some years (1973, 1974, 1977, 1980, 1982) and greatly reduced coverage at other times (see the cruise information on the CALCOFI website). In 1997 however, Mexican researchers initiated a new series of cruises off Baja California under the IMECOCAL program. These cruises sample along former CALCOFI lines between 24° and 31° N on a quarterly basis (Durazo 2009).

A number of additional large experiments have augmented the information obtained by the CALCOFI surveys and extended our knowledge of the region's variability farther north. These include, among others, programs such as the Coastal Upwelling Ecosystems Analysis program (CUEA) in 1972–1973, off Oregon and Washington, together with the MESCAL program off Mexico, the Coastal Ocean Dynamics Experiment (CODE) and the Northern California Coastal Study (NCCS), off northern California in the early and late 1980s respectively, and a series of programs in the late 1990s and early 2000s funded through the U.S. NSF's Coastal Ocean Processes (CoOP) program. These last programs (Air-Sea Chemical Exchange, CASCEX; Wind Events and Shelf Transport, WEST; Coastal Advances in Shelf Transport, COAST; and River Influences on Shelf Ecosystems, RISE) were completed off northern California and Oregon, with RISE concentrating on the effects of the Columbia River. All included physics and chemistry and were aimed at elucidating how patchiness and a strongly advective regime could maintain high productivity in the region (Kudela et al. 2008). The CCE-LTER program, that began in 2004 and links physics and chemistry with biology, is a more recent ongoing study (Goericke and Ohman 2015).

The CODE program off northern California examined the dynamics of strong, wind-driven circulation over 100 km of the coastline from 30–500 m depth (Winant et al. 1987). The coastal program was embedded within a larger array that covered the region between 35°–48° N (Denbo and Allen 1987). The experiment included hydrography and meteorology, and was designed to improve modeling of local features such as the frequent current reversals and the effects of coastal trapped waves (Chapman 1987), while the data also allowed computation of mass and heat budgets. Despite the large amount of data collected, the models at this time could not predict the variability in cross-shelf flow to within a factor of five (Hickey 1998), although there have been major improvements since then as a result of increased model resolution and the incorporation of the alongshore pressure gradient (e.g., Combes et al. 2013; McCabe et al. 2015). The work has also led to improvements in knowledge of the importance of the frictional bottom boundary layer.

Later field experiments in the same region in the early 1990s (the Surface Mixed Layer Experiment, SMILE and Sediment Transport Events over Shelves and Slopes, STRESS) studied the bottom boundary layer and sediment transport. The thickness of the bottom mixed layer varies generally between 5 and 15 m, increasing offshore, and depends on several factors of which current direction appears to be the most important (Trowbridge and Lentz 1991). As a result, poleward flows have a thicker bottom boundary layer than equatorward flows. Another interesting finding was that the cross-shelf heat flux is five times greater in summer than winter (Dever and Lentz 1994). As a result of these and other experiments we now have a reasonable idea of the scales of variability between 3 and 60 days. NCCS (Largier et al. 1993) looked at 2.5 years of current meter data from 5 mooring lines along 500 km of the shelf. Most of the variance was related to changes in the along-shelf wind stress and pressure gradients. Referencing the data to 1500 m rather than the more commonly used 500 m increased the transport by the poleward flow over the slope by a factor of two to five times (Bray and Greengrove 1993).

4.5.2 Key Coastal Currents

According to large-scale maps of the flow in the North Pacific (Reid 1997), southward flow along the coastline of North America is confined to the upper 500 m, at speeds of about 10 cm/s, and water below this depth tends to flow northwards. However, even above the 500-m depth level, there is a large northward component to the flow, particularly in winter (November–March) when the Davidson Current develops north of Point Conception and flows north at up to 20–30 cm/s, and in the California Undercurrent which has been measured from Baja to Vancouver Island and can flow continuously for at least 400 km (Collins et al. 1996). The undercurrent may, in fact, be continuous from 32° N–47° N (Pierce et al. 2000), or from even farther south (Badon-Dangon et al. 1989), forming a jet which flows at speeds of up to 30–50 cm/s just offshore and deeper than the shelf break (100–300 m; Hickey 1992). The transport of warm salty water towards the north is between 1 and 3 Sv. The two currents may, in fact, merge during the winter into a continuous northward flow throughout the water column (Lynn and Simpson 1987). Speeds within the undercurrent jet vary considerably depending on latitude and season; off Washington the maximum is usually only about 15 cm/s (Huyer et al. 1989), whereas off Point

Sur it is considerably less (Chelton 1984). At times, the continuous northward flow may be disrupted by "eddy-like features" (Hickey 1998).

The large bend on the coastline at Point Conception disrupts the overall flow regime, leading to a large cyclonic eddy in the southern California Bight that can occasionally round the Point and flow northwards again, when it is known as the *California Countercurrent* (Hickey 1979, 1992; Lynn and Simpson 1987, 1990). The northward flow in the eddy bifurcates near the Santa Barbara Channel with some staying close to the coast and the rest going west south of the offshore islands (Lynn and Simpson 1990). All three currents, the California Current, its undercurrent and the countercurrent, show seasonal minima in spring and maxima in summer/early autumn. As a result of the presence of the eddy, the frontal systems move offshore here (Powell and Ohman 2015).

Over the shelf, currents can flow northwards inshore of the southward flowing upwelling jet. North of Point Conception, near San Francisco, such northward flow appears whenever upwelling winds relax. As a result, equatorward flow over the mid-shelf north of the point is seen only in late spring and summer (Strub et al. 1987), and the southward flow decreases with depth, reversing at about 100 m depth over the outer shelf, 50 m over the mid-shelf, and at 15 m or less over the inner shelf (Winant et al. 1987; Largier et al. 1993). Similar features are seen off Baja California (Barton 1985; Badon-Dangon et al. 1989) and central Oregon (Huyer et al. 1989). Although the flow regime off Baja is still not well described, northward flow appears to be a common feature in that region.

4.5.3 The Onset of the Upwelling Season

An important feature of the CCS is the spring transition that ushers in the start of the upwelling season. This "sudden and dramatic event" (Huyer et al. 1979) is accompanied by a drop in sea level of about 10 cm, an increased upward slope towards the coast of isopycnals, and a current reversal from poleward to equatorward flow within a few days. It is apparently driven by large-scale changes in the wind and pressure fields over the region and results in a very sudden change in conditions (Fig. 4.9). Delays in the spring transition, such as occurred in 2005, can have major effects on productivity, particularly in the northern part of the CCS (Mackas et al. 2006; Barth et al. 2007).

The shelf and slope appear to respond to events with a different time scale to those occurring further offshore (Hickey 1998). Over the shelf and slope, the dominant effect is from wind-forced events with scales of 3–10 days, while offshore regions are dominated by larger-scale changes (e.g., jets and eddies) with scales of 10–40 days. This is particularly true in the northern CCS, off Oregon and Washington, where poleward propagation of coastal-trapped waves is important (e.g., Smith 1974; Halpern 1976). These waves are more common in summer, when the wind decreases towards the north; in winter forcing is driven by the local wind field.

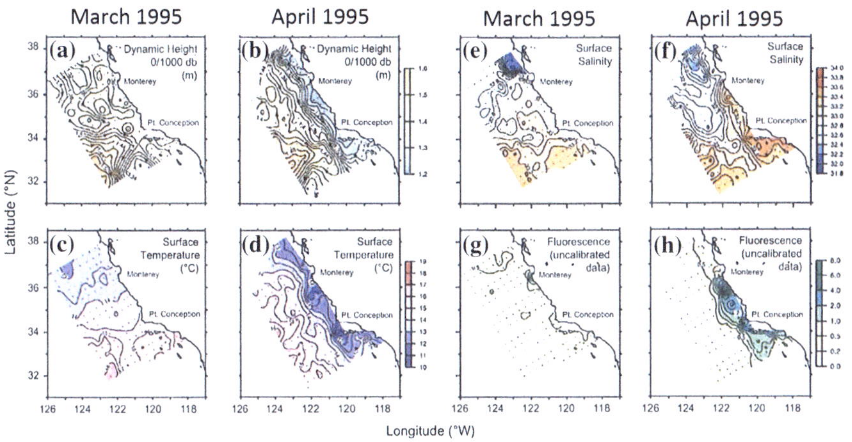

Fig. 4.9 Spring transition in the southern California Current system, March and April 1995, revealed by changes in **a**, **b** dynamic height, **c**, **d** surface temperature, **e**, **f** surface salinity, and **g**, **h** fluorescence (from Lynn et al. 2003)

4.5.4 Circulation in the Southern California Bight

The southern California Bight behaves rather differently from the rest of the CCS. South of Point Conception the scales of the currents decrease from 500 to ~20 km (Winant 1983) and their correlation with wind speed decreases. The shelf here is narrower (2–5 km vs. 40 km) and shallower (60 m vs. 200 m) than in the region to the north but the wind strength is only about one tenth of that north of the point in summer. This leads to greatly reduced localized upwelling, compared with the region north of Point Conception, that occurs in winter and early spring, although upwelled water from north of Point Conception can enter the bight in summer when upwelling to the north is at a maximum. Further south, off Baja California, equatorward flow is reestablished offshore at the surface, accompanied only occasionally by a northward flowing coastal current within about 10 km of the coast, and with the poleward undercurrent flowing in the opposite direction beneath it. Whether the inshore poleward flow is part of the undercurrent or comes from a separate source apparently depends on the local forcing (Barton and Argote 1980; Perez-Brunius et al. 2006).

4.5.5 Eddies and Filaments

Eddies and filaments in the CCS were observed first by Bernstein et al. (1977) and Traganza et al. (1980) and from the tracks of drifters during CODE (Davis 1985a, b). The Coastal Transition Zone experiment (Brink and Cowles 1991) studied these filaments, which were shown to be more than 200 m thick and to separate warm, fresh low-chlorophyll water on the offshore side from cold, salty, chlorophyll-rich upwelled water inshore (Fig. 4.5). The jets have a core speed of ~50 cm/s at the surface, are 50-75 km wide, and transport ~4 Sv of water up to 300 km offshore from Cape Blanco, Cape Mendocino and Point Arena (Huyer et al. 1991; Strub et al. 1991). Satellite imagery shows that they can persist for at least one month in the same position, before reorienting as the wind field changes (Strub et al. 1991). A section through one of these jets is shown in Fig. 4.10, where the strong influence of the nutrient rich water in the jet is shown by the high levels of phytoplankton pigments in the upper 50 m on the inshore side. This increase is maintained not only when the jet is flowing offshore, but also when it returns towards the coast (Fig. 4.10).

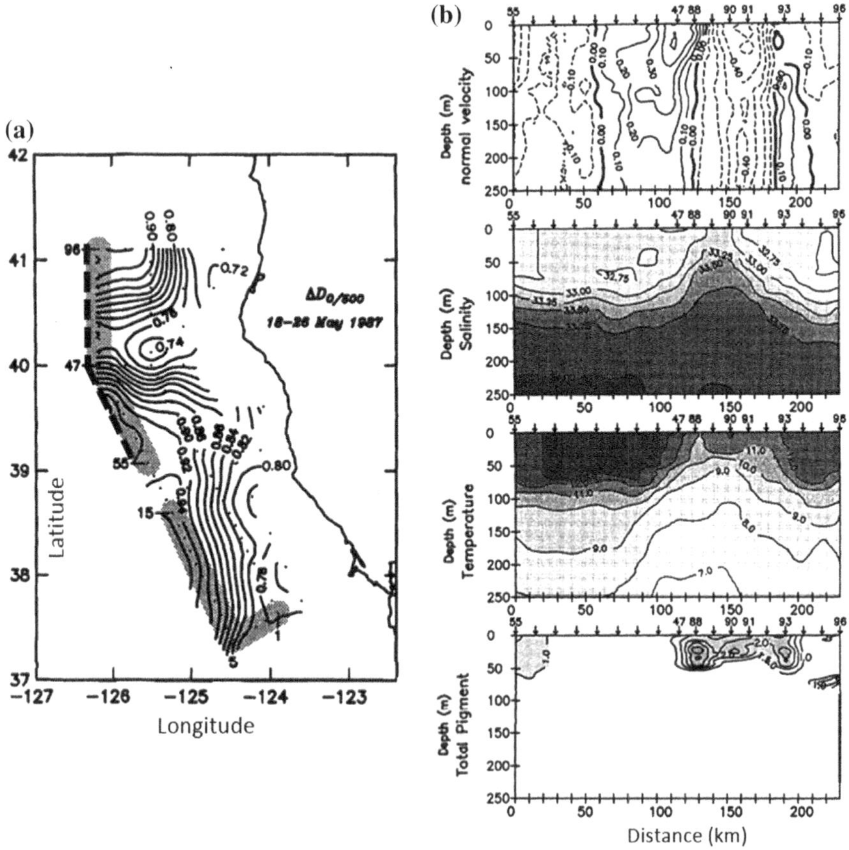

Fig. 4.10 **a** Fields of dynamic height anomaly relative to 500 dbar from 18–26 May 1987 (contour interval is 0.02 dyn m). The *thick dashed line* indicates the location of the transect shown in panel (**b**). **b** Vertical transect along stations 55–96 showing ADCP velocity, salinity, temperature and total pigment. Velocity (contour interval 0.1 m/s) is normal to the transect, with positive values indicating shoreward flow. Salinity contours every 0.25, temperature every 1.0 °C, and pigments 0.5 mg/m^3 (from Strub et al. 1991)

Similar filaments develop at capes south of Point Arena and off Baja California, with a gap in the eastern part of the southern California Bight. The filaments are more common in spring and summer and cause subduction of water as well as lateral advection (Kadko et al. 1991). They appear to form as instabilities when equatorward winds encounter promontories or capes, and therefore are less often found north of about 43° N. Once formed, the filaments then meander equatorward and offshore, but may rejoin the coastal jet or interact with eddies to increase the size of the meanders. Generally they possess lower salinity than the surrounding water, suggesting an upstream origin probably from the Columbia River, particularly for the jets off Cape Blanco and Cape Mendocino (this idea of a northern origin is also shown by zooplankton assemblages, see Mackas et al. 1991). The presence of these filaments leads to a strong offshore front that extends from Cape Blanco to south of Monterey, while similar features are also seen off Baja California, some of which may result from standing waves according to a shelf model for coastal trapped waves by Narimousa and Maxworthy (1989). Biologically, these jets and filaments define a region near the coast that contains the nutrient enriched water and which varies in width from 100–400 km. Off Baja California, similar plumes can develop following Santa Ana winds, which are strong, dry winds blowing offshore through gaps in mountainous terrain, particularly in winter (Trasvina et al. 2003).

Localized experiments show the complexity induced by small-scale features along the CCS coast. For instance, the WEST program of 2001–2003 (Largier et al. 2006) considered the role of the wind in forcing upwelling in the region off Bodega Bay, just north of Point Reyes (38° N), focusing on the physical and biological response of the shelf to wind events and relaxation. Data came from in situ observations and were aided by high-resolution hydrodynamic modeling and ecological modeling. The data showed characteristic periods of both upwelling favourable winds (with alongshore wind speeds >5 m/s) and relaxation periods of 36 h or more during the spring/summer transition (with 36 h low-passed wind speeds of <5 m/s), with the transitions between them being sudden. Northward (relaxation) winds led to a rapid warming of water in the upper 10 m, a slower decrease in salinity, an increase in stratification, and a reversal and weakening of currents on the shelf from southward to northward. Such current reversals bring water from south of Point Reyes into the area (Fig. 4.11). In contrast, upwelling winds led to positive wind stress curl, as well as the eventual export of surface water south of Point Reyes in an offshore filament.

The importance of cross–shelf transport was demonstrated by Dever et al. (2006), who showed that southerly flow was associated with onshore cross-self transport, while relaxation events led to offshore transport. They also calculated divergence estimates from both cross-shelf Ekman transport and surface boundary-layer transport and related them to local winds. Their results implied upwelling rates of about 3×10^{-4} m/s over the shelf between 40 and 90 m depth, and about 1.5×10^{-4} m/s between 90 and 130 m, driven largely by wind-stress curl. This is about half the upwelling rate that occurs along the coastal boundary inshore of the 40 m isobath (about 8×10^{-4} m/s), and agrees with previous data

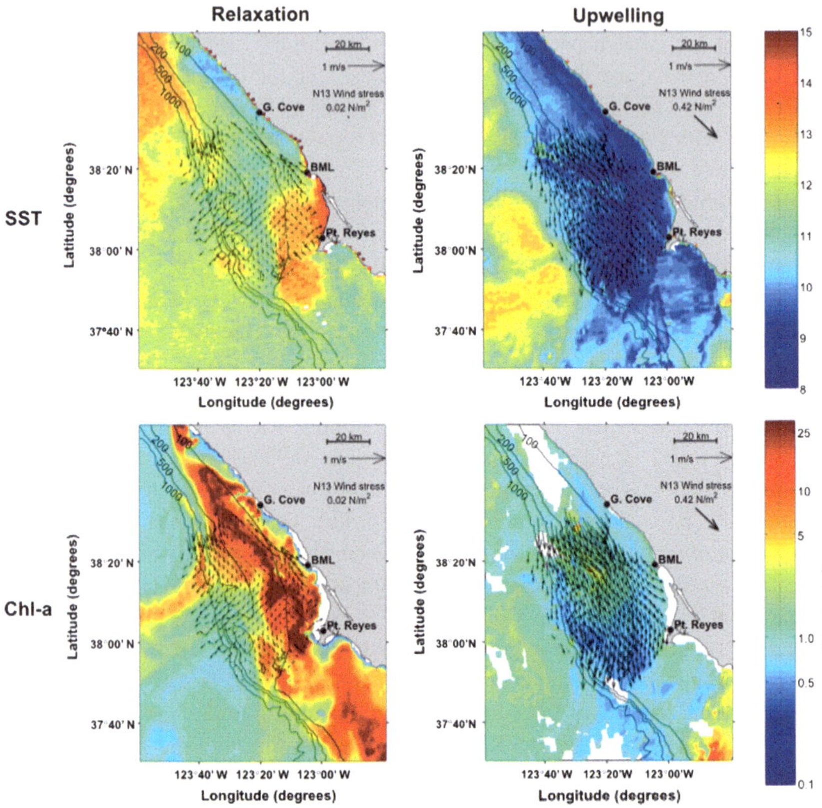

Fig. 4.11 Comparison of upwelling and relaxation periods during the WEST program, showing currents in June 2003 (from HF radar) superimposed on temperature (*upper panels*) and chlorophyll-*a* concentrations (*lower panels*). Relaxation period has current data from 10 June with satellite temperature data from 12–13 days June and satellite chlorophyll data from 11–14 June, while upwelling period has currents from 17 June and satellite data from 19–20 June (from Largier et al. 2006)

(Rudnick and Davis 1988) from further north and simple model results (Enriquez and Friehe 1995; Oke et al. 2002). However, when the areas affected are considered the overall upwelling in the deeper water is much greater than that close to the coast (1.20 m^2/s close to the coast, 2.25 m^2/s between 40–90 m, and 2.36 m^2/s between 90–130 m). Wind-stress-curl-driven upwelling (called *Ekman pumping*) is also found in other upwelling regions such as the Benguela (see Chap. 7).

The overall results suggest that although small-scale changes in wind forcing are not needed to understand how the along-shelf currents vary, they almost certainly control exactly where upwelling occurs, and thus the patchiness observed in phytoplankton distributions and primary productivity. Retention of phytoplankton during and after upwelling for about 3–7 days is also important, as without a stable environment phytoplankton cannot bloom in the aging upwelled water, and such sites are found either side of the Point Reyes promontory (Vander Woude et al. 2006). The effects of such upwelling events are discussed in Sect. 4.7.

4.6 Influence of Continental Discharges

The absence of large rivers along much of the CCS shelf means that buoyancy plumes are only important in the northern portion of the region unless there are major floods. Three large sources of fresh water affect the hydrology of the northern part of the California Current system. The Strait of Juan de Fuca marks the boundary between the U.S. and Canada and provides a route around the southern end of Vancouver Island for freshwater from the Fraser River to reach the Pacific Ocean. The flow out of the Strait of Georgia during spring runoff is estimated to be about 10^5 m^3/s, of which about 1.8×10^4 m^3/s comes from the rivers that empty into it (Masson and Cummins 2004). The Columbia River, on the boundary between Washington and Oregon, is a second major source of fresh water in this region, and influences the coasts of both states. Total tidally-averaged outflow from the Columbia River is about 1.4×10^4 m^3/s, of which about 70 % is freshwater. The Columbia River plume can be found both north and south of the river mouth and can affect both states simultaneously (Hickey et al. 2009; Fig. 4.12), and provides about 75 % of the freshwater inflow between the river and San Francisco Bay. The plume itself could even be traced as far south as San Francisco in 1950 after which the low salinity filament moved offshore to 33.5° N, 127° W (Huyer 1983).

While the Juan de Fuca plume was thought previously to affect only the west coast of Vancouver Island, we now know that in spring and summer a considerable amount of water from the Strait moves southwards along the coast of Washington (MacFayden et al. 2008). This means it can interact with the northward plume from the Columbia River which is known to extend at least 160 km north of the river mouth even during periods when the wind stress is strongly upwelling-favourable (Hickey et al. 2009). Early studies of the Columbia River plume showed it to extend generally northwards in winter, when it can sometimes reach the Strait of Georgia, and southwestwards in summer, but more recent work suggests that the northward plume exists about half the time even in summer (Hickey et al. 2005), when upwelling-favourable winds are the norm and the plume would be expected to move southwards, and can reach the Strait of Juan de Fuca in May or June about one year in two, depending on river flow and wind stress. Depending on when the spring transition occurs, water that has flowed northwards from the Columbia River can then return southwards again in the baroclinic shelf jet (Hickey et al. 2009).

As stated earlier, there can be rapid wind stress reversals along the coast during the upwelling season. When the typical southward upwelling wind reverses during the summer, the southward moving plume contracts and is found close to the coastline. At the same time, the downwelling winds generate a northward moving current within 20–30 km of the coast on the northern side of the river mouth. When

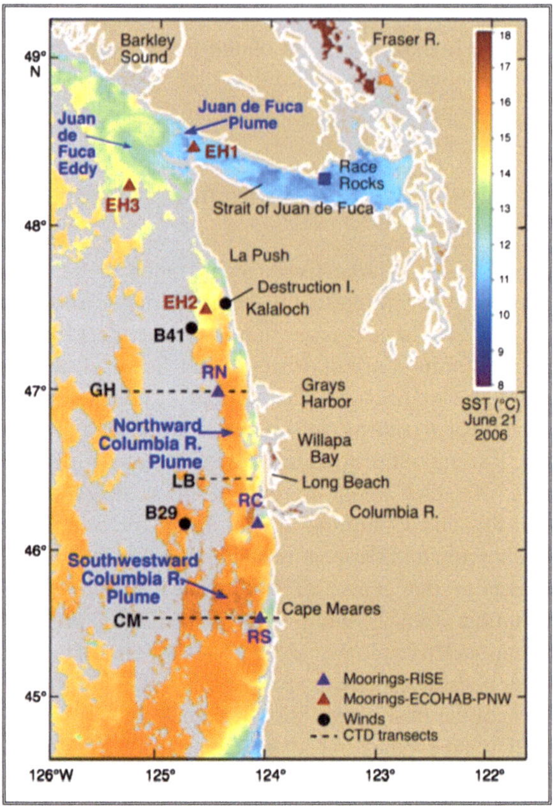

Fig. 4.12 Satellite-derived SST for 21 June 2006, showing three buoyant plumes from the Strait of Juan de Fuca and the northward and southward plumes from the Columbia River (from Hickey et al. 2009)

the winds reverse again and upwelling-favourable conditions are reestablished, the northern plume also reverses its flow and moves offshore, dissipating over the middle and outer shelf (Hickey et al. 2009).

Being much fresher than the surrounding oceanic water, the plumes have a major effect on the local buoyancy. Although the Columbia River plume is only about 10 % of the size of the flux from the Strait of Juan de Fuca, it contains considerably more freshwater and so affects the buoyancy to a much greater extent. The Juan de Fuca outflow has a typical salinity of about 30–31, which is fresher than the coastal surface water, but the Columbia outflow has a salinity of 10–20. This results in the formation of a more intense stratified surface layer in the Columbia River plume, which is typically about 10–20 m thick (Hickey et al. 2009). Both plumes, however, supply terrestrially-derived nutrients to the coastal zone and can entrain upwelling-derived nutrients during mixing events at the base of the plume (how much is entrained depends on the relative concentrations of nutrients in river and ocean water and the strength of the local upwelling). The strongest effect is seen during weak or moderate upwelling, as when upwelling is strong the plume moves offshore. If, however, the upwelling is very weak, then the river plume forms a cap above the deeper, nutrient-rich water, and upwelling stops. This nutrient supply

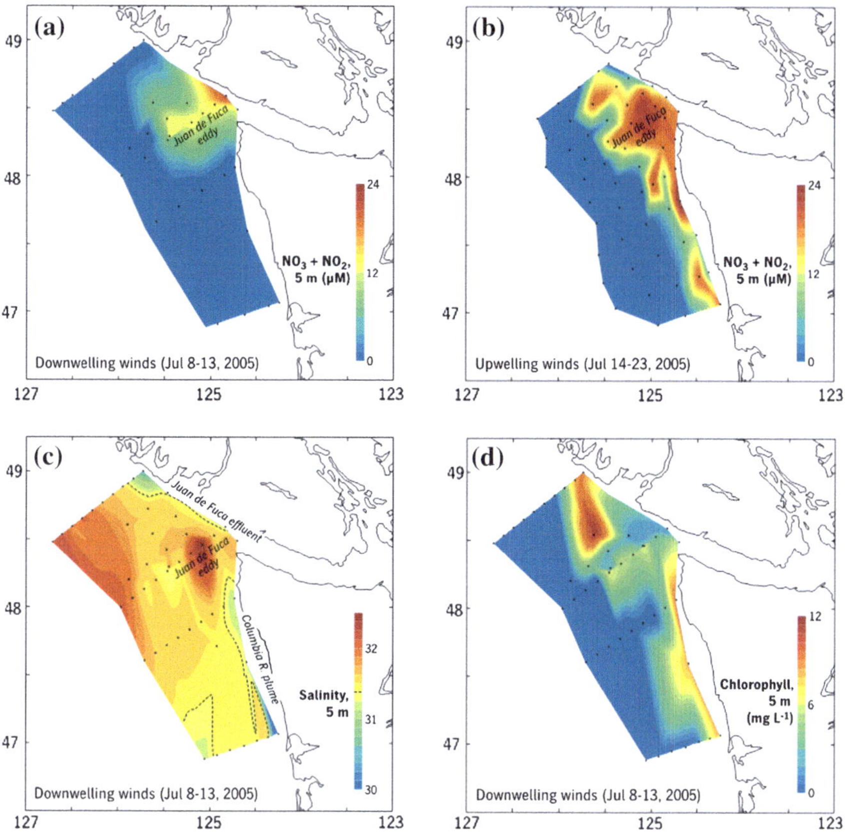

Fig. 4.13 Concentrations of nitrate plus nitrite at 5 m depth in upwelling and downwelling conditions off the Strait of Juan de Fuca, together with salinity and chlorophyll concentrations for downwelling conditions (from Hickey and Banas 2008)

mechanism leads to high chlorophyll production off both Juan de Fuca Strait and in the Columbia River plume even under downwelling conditions (Fig. 4.13).

Much of the water exiting from Juan de Fuca Strait is the same water that upwells further south, and so contains high levels of nutrients. While originating from below 100 m depth offshore, it is drawn upwards via the Juan de Fuca canyon to replace the surface layer leaving the strait. As Hickey and Banas (2008) point out, this mechanism "provides a massive, relatively steady nutrient source to the northern CCS." As the outflow from the strait is much deeper than that from the Columbia River, it fertilizes a much wider zone than the coastal strip influenced by the river plume and the water below the immediate surface layer can be advected southwards along the coast, thus supplying additional nutrients to the outer shelf once the wind changes back to upwelling favourable. This means that even during downwelling conditions there is measurable nitrate offshore of the strait even though concentrations along the coastal strip approach zero (Fig. 4.13).

In contrast, the supply of nutrients from the Columbia River is considerably smaller, as much is used within the estuary and does not make it very far onto the shelf. The exception is following high rainfall events in spring and early summer, when the river plume can affect a larger area. It has been estimated that the total nutrient supply from the Strait of Juan de Fuca during spring is about the same as the amount upwelled along the whole coast between the Strait and the Columbia River. However, the presence of river water means there is no iron limitation anywhere along the northern CCS shelf, unlike the situation off California (see Sects. 4.7.6 and 4.7.7; Chase et al. 2007).

Two other structural features are also important in supplying nutrients to the northern CCS and interact with the plumes from the Juan de Fuca Strait and the Columbia River. The first is the cyclonic Juan de Fuca eddy that forms off the southern end of Vancouver Island after the spring transition and keeps nutrients in the area, thus enhancing local productivity through cross-isopycnal mixing (Hales et al. 2005). The second is the presence of a number of canyons along the Washington and Oregon coasts, which enhance nutrient supply on their south sides because of the cross-shelf pressure gradient associated with the equatorward jet. This forces water to move towards the coast in the bottom boundary layer and it then becomes available for upwelling when local winds are favourable. It is believed that this mechanism can provide approximately the same nutrient flux as local coastal upwelling or the flux from the Strait of Juan de Fuca (Hickey and Banas 2008).

4.7 Chemical and Biological Features

4.7.1 Biological Productivity

The CCS is the region where Dugdale and Goering (1967) and Eppley (1992) developed the concept of new production, driven by nitrate uptake, and regenerated production driven by ammonia. As with all other upwelling systems, the primary productivity in the California Current system is driven by the supply of nutrients from below the surface layer, in this case from deeper than 200 m or so off Washington and Oregon, but from about 150–200 m off northern California (Chavez et al. 2002).

As is typical of upwelling regions, diatoms dominate shelf productivity but smaller organisms, such as cyanobacteria and picoplankton, increase during periods of reduced upwelling and downwelling. However, these smaller cells do not produce enough material to help higher trophic levels (e.g., as occurred in 2005) leading to reduced fish and other catches (Hickey and Banas 2008). Microzooplankton have been shown to consume <66 % of total production (Frame and Lessard 2009) and their grazing pressure seems to be higher on the Oregon shelf (Kudela and Peterson 2009) as part of a general poleward gradient in phytoplankton biomass, but mesozooplankton can also consume large quantities (theoretically >100 %/day during relaxation events), although a more normal range is usually 25–38 % (Slaughter et al. 2006).

The overall productivity of the CCS is closely linked to the physics of the region at several time scales, although the relationship is "fuzzy" in that short and long-term variability (days to years) is a feature of upwelling systems in general. This is particularly important for estimates of phytoplankton production, which show both small-scale patchiness and short-term concentration changes. Mackas (2006) describes well the general productivity of the region, while Rykaczewski and Checkley (2008) show how wind stress curl, with its lower upwelling rates, favours smaller particles (and therefore sardines) in the offshore region, leaving the inshore part of the system to larger particles (and anchovy, which have a coarser filtering mechanism). The advent of ocean colour imagery since the late 1970s has made the task of estimating phytoplankton biomass much easier, and there are now numerous papers on phytoplankton productivity from throughout the CCS, including seasonal and longer-term variability (e.g., Strub et al. 1990; Thomas et al. 2009). These show how the cross-shelf trend varies, with highest concentrations some 10–20 km offshore. The alongshore variability is less marked than the cross-shelf variability, but concentrations are highest in the north and lowest in the southern California Bight.

The Thomas et al. (2009) paper covered the decade from 1998 through 2007, and so included the 1997–98 El Niño event. It compared events in the California Current with those in the Humboldt Current Upwelling System south of the equator. Daily mean biological productivity based on 4 km SeaWiFS imagery was used to calculate monthly means and a decadal climatology. Both systems showed depressed productivity during 1997–98, particularly north of 34° N in the CCS and from 5°–25° S in the Humboldt Current, and this low productivity signal persisted into 1999 in many areas, after which productivity increased again to strong maxima in 2004 south of the equator and through much of 2005 and 2006 off California and Oregon. Changes in chlorophyll and nitrate concentrations along CALCOFI line 67 ($\sim$36° N) during 1997–2001 are shown in Fig. 4.21.

4.7.2 *Seasonality*

The seasonal changes in mean dissolved oxygen, phosphate, satellite-derived chlorophyll and primary productivity are shown in Fig. 4.14 for the upper mixed layer. Clearly shown are the maxima in chlorophyll and production in late summer/autumn in the north, whereas in the south, including Baja California, the maxima occur earlier in spring/early summer. Oxygen and phosphate also show fairly large seasonal variability attaining, as expected, the lowest (oxygen) and highest (phosphate) concentrations in summer months when production is high, while minima are found in all regions in winter.

This is consistent with findings by Thomas et al. (2009) who used EOF analysis and showed how the seasonal variability was strongest and at its widest extent north

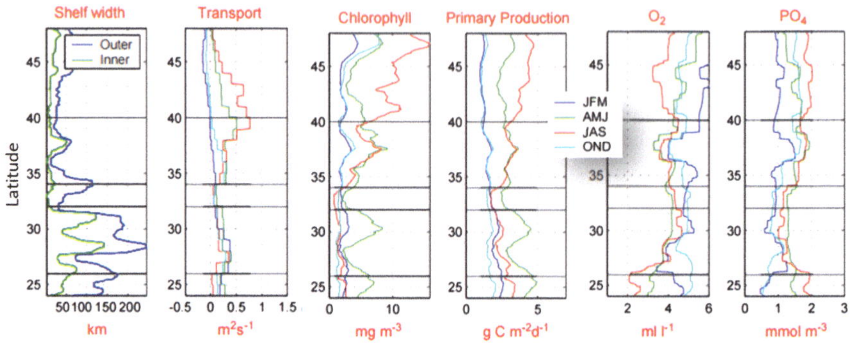

Fig. 4.14 Variation in shelf width and seasonal changes in offshore Ekman transport, chlorophyll concentration, primary production, oxygen and phosphate concentrations with latitude in the California Current system. The *horizontal black lines* divide the system into separate zones, corresponding generally to those shown in Fig. 4.2 (from Carr and Kearns 2003)

of 40° N, weakening and narrowing towards the south (Baja) and close to zero in the southern California Bight. Interestingly, highest productivity anomalies did not appear to be particularly well correlated with the largest upwelling anomalies, either in time or space, supporting the view that larger-scale variability, such as the Pacific Decadal Oscillation, also has to be considered. Correlations between such indices and the local production showed large latitudinal variability.

Individual studies of productivity give somewhat different results to the large-scale means provided by Carr and Kearns (2003). At distances >100 km offshore there is a spring maximum in the northern region, with a weak winter maximum in the south (Mackas 2006). Productivity varies between about 300–600 g C/m^2/year within 100 km of the coast, although it is closer to 150–250 g C/m^2/year in the southern California Bight and further offshore (e.g., Longhurst 1998; Olivieri and Chavez 2000). An interesting feature is that highest production appears on the inshore side of jets (Chavez et al. 1991), but as found in other systems such as the Benguela (see Chap. 7), primary production is generally very patchy and episodic.

4.7.3 Spatial Differences

Spatial differences of nutrient fluxes and ecosystem responses in the CSS are mainly associated with the general shape of the large-scale nutricline, variations in upwelling intensity including retention times, variations in bathymetry (e.g., shelf width, submarine canyons or sheltered retention zones) and river runoff.

Zentara and Kamykowski (1977) produced scatter plots of nitrate against temperature for 10° latitude bands along the west coasts of North and South America and showed that they varied systematically from the poles to the equator, with generally linear relationships for regions having more than 4 μM/L nitrate, which included the CCS. This linear relationship is a consequence of the general shape of the nutricline/thermocline along the eastern boundary of the subtropical gyre (see Chap. 10). Nutrient supply is highest off Vancouver Island due to tidal mixing and upwelling within the deep narrow passages and submarine canyons (Mackas and Harrison 1997). The effect of the canyons increases the nutrient supply—about 58 % of the nutrients off Washington are estimated to come from this source, compared with about 4 % due to tidal mixing, 10 % from surface wind mixing, and 28 % from wind driven upwelling. The oxygen minimum depth is usually below the shelf break (Kamykowski and Zentara 1990), but changes such as the large-scale variability in the North Pacific can have major effects on the biota (Freeland et al. 2003; Grantham et al. 2004; Wheeler et al. 2003).

Kudela et al. (2008) discussed the results of a number of recent research programs within the California Current System, and how these affected phytoplankton productivity. Their conclusions were that the traditional view of upwelling, where equatorward wind stress drives an offshore Ekman flux that is balanced by coastal upwelling, and where phytoplankton (mainly diatoms) and the zooplanktonic species that feed on them are retained in the system through sinking offshore and then returning in the bottom cross-shelf flow, works only for relatively straight coastlines but not those with complex topography. So we should expect differences between the Washington/Oregon coastline and that found farther south.

Retention time is an important control of overall productivity in the CCS. The wind relaxation phase following upwelling determines the optimal conditions for a bloom because the pulsing allows the phytoplankton to grow and reproduce before they are advected offshore. This means that bloom productivity is proportional to the length of the relaxation phase up to the point at which nutrients become exhausted (Wilkerson et al. 2006; Kudela et al. 2006), generally between 3 and 7 days. Too little wind means low production because there are no nutrients, while too much wind means the phytoplankton cannot use them all before they are lost offshore. In addition, the stronger winds increase the depth of the upper mixed layer, reduce light, and mix the phytoplankton cells deeper into the water column. The process has been modeled by Botsford et al. (2006) who showed, perhaps not surprisingly, that narrow shelves lose plankton faster than wider shelves, and from observations by Huyer et al. (2005), who compared two lines north (44.6° N) and south (41.9° N) of Cape Blanco (Oregon), where weak wind stress (in the north) led to nutrient limitation, while stronger winds (in the south) led to light limitation via sediment erosion. The idea that wider shelves result in higher concentrations of phytoplankton is also supported by data from Heceta Bank (44.1° N), where the shelf is at its widest (Barth et al. 2005).

Coastal bathymetry is also important. For example, upwelling shadows can exist downstream of upwelling centres that allow self-seeding as phytoplankton populations age (Graham and Largier 1997). Such sites are known from Monterey Bay and on both sides of Point Reyes (depending on which way the wind is blowing). A similar effect occurs off Oregon where Heceta Bank disrupts southward flow along the coast so that the upwelling jet moves offshore, leaving a shadow zone inshore (Barth and Wheeler 2005). These shadow zones can be strong local sinks for carbon dioxide (Hales et al. 2005).

An interesting feature of the Oregon/Washington portion of the CCS is the interannual variability in nutrient supply and its effect on primary productivity. The importance of the Columbia River and Juan de Fuca plumes and their interaction with deeper upwelling water has been mentioned previously. Water flowing out of the Strait of Juan de Fuca can contain >20 μM/kg dissolved nitrate, which provides a regular nutrient source to the region around the Juan de Fuca eddy and to the southward moving coastal waters. Cross-isopycnal mixing can also occur, particularly in water less than 30 m deep (Hales et al. 2005), which is separate from "normal" upwelling and may supply more than 25 % of the nitrate to the surface mixed layer. The Columbia River plume contains lower nitrate concentrations (2.5–6 μM/kg), which lead to maximum surface levels of about 16–19 μM/kg when it mixes with upwelled water near the "lift-off" region for the plume (Hickey et al. 2009; Bruland et al. 2008); during upwelling quiescent periods, however, nutrients from the river may provide major support for continued production.

This interaction of the plume and the upwelled water means that the high nutrient zone extends further out across the shelf than would normally be expected, thus extending the area of enhanced primary production, while also enhancing cross-shelf flow and perhaps the retention of phytoplankton in the region. While nitrate is generally the limiting nutrient all along this portion of the coast, both phosphorous and even silicate variability can limit production at times (Ruttenberg and Dyhrman 2005; Brzezinski et al. 1997). Another advantage of the river to the Washington/Oregon region is that it supplies large quantities of iron to the coastal zone, thus avoiding the problem of iron limitation further south (discussed in more detail below), while intense internal waves generated at the outer edge of the plume increase mixing, allowing additional nutrient entrainment in the surface layer (Kudela et al. 2008).

As phytoplankton blooms are ~fivefold higher off Oregon than off northern California, even though the upwelling-favourable winds are stronger in the south, there must be better retention methods over this wide straight shelf. Drifter tracks (MacFayden et al. 2008) show little sign of moving offshore from the mouth of the Strait of Juan de Fuca until they reach the Columbia River about 10 days later, although they can be retained for <30 days off the Strait in the Juan de Fuca eddy. The Columbia River plume acts a "porous" barrier to southward flow (Banas et al. 2009; Kudela et al. 2010), pushing water from the north off the shelf in a large "bulge" and creating another shadow zone along the northern Oregon coast. The presence of the river plume moves most of the production offshore into water 100 m deep or deeper and means more inshore organisms can be grazed by zooplankton. Heceta Bank off central Oregon also acts as retention centre (Gan and Allen 2005; Landry et al. 1989; Henson and Thomas 2007), but this region is without a continuous nutrient supply.

4.7.4 Zooplankton

The long-term records from the CALCOFI program show that zooplankton abundance in the upper 250 m of the CCS varies strongly with latitude and season. Minimum changes in biomass during the year (about a factor of two) occur off southern California and increase northward so that the annual variation is about a factor of five off Vancouver Island (Mackas 1992). Maxima in numbers are often found along the shelf break or at the outer edges of upwelling hotspots, as well as in association with eddies and cold core filaments, where, unlike phytoplankton, they are found on the offshore side of the filament (Mackas et al. 1991; Mackas 2006). Like phytoplankton, zooplankton depend on specific physical mechanisms to remain in areas of high production (Barange and Pillar 1992; Peterson 1998), such as the shoreward movement of the upwelling front following relaxation of upwelling-favourable winds, which can bring zooplankton back into the coastal zone from offshore.

Organisms found at higher latitudes tend generally to have longer life spans and show a marked annual cycle in abundance. Community composition in the north is reasonably well understood and has been described by Peterson and Miller (1977) and Mackas (1992). The most common organisms over the shelf are the copepods *Calanus marshallae*, *Pseudocalanus mimus, Acartia longiremis* and the euphausiid *Thysanoessa spinifera*. Further offshore, species from the Alaskan Gyre and some advected from the south also appear. Along California and Baja California different species appear and the composition is more varied, the population being dominated by the copepods *Calanus pacificus, Metridia pacificia,*and *Pleuromamma borealis*, the euphausiid *Euphausia pacifica*, and pelagic tunicates such as *Salpa aspera, S. fusiformis* and *Dolioletta geigenbauri* (Hopcroft et al. 2002; Lavaniegos and Ohman 2003). However, off Washington and Oregon, microzooplankton are thought to have a larger effect on phytoplankton than copepods or mesozooplankton (Banas et al. 2009; Davis et al. 2014).

Because of their longer life cycles than phytoplankton, zooplankton, which include larval fish species, are more dependent on retention zones to avoid being advected out of the upwelling region. Similarly, they need to be able to find sufficient concentrations of food to allow them to compete their development, and larval zooplankton are known to be dependent on thin phytoplankton layers or phytoplankton concentrations at frontal systems, which can be either vertical or horizontal. Convergence zones, such as occur at the shelf break off Vancouver Island, where euphausiids and fish are known to congregate (Mackas et al. 1997; Lu et al. 2003) are also important in this regard. Physical events, such as the shoreward movement of the upwelling front when the upwelling winds drop, or shoreward transport by internal wave trains, are important in terms of returning larval stages, which are generally weak swimmers, to the inner shelf. Adults also make use of such physical events by actively migrating vertically in regions with opposing currents (e.g. Batchelder et al. 2002; Barange and Pillar 1992), while the location of spawning grounds is also tied to the local current field. Since the time and space

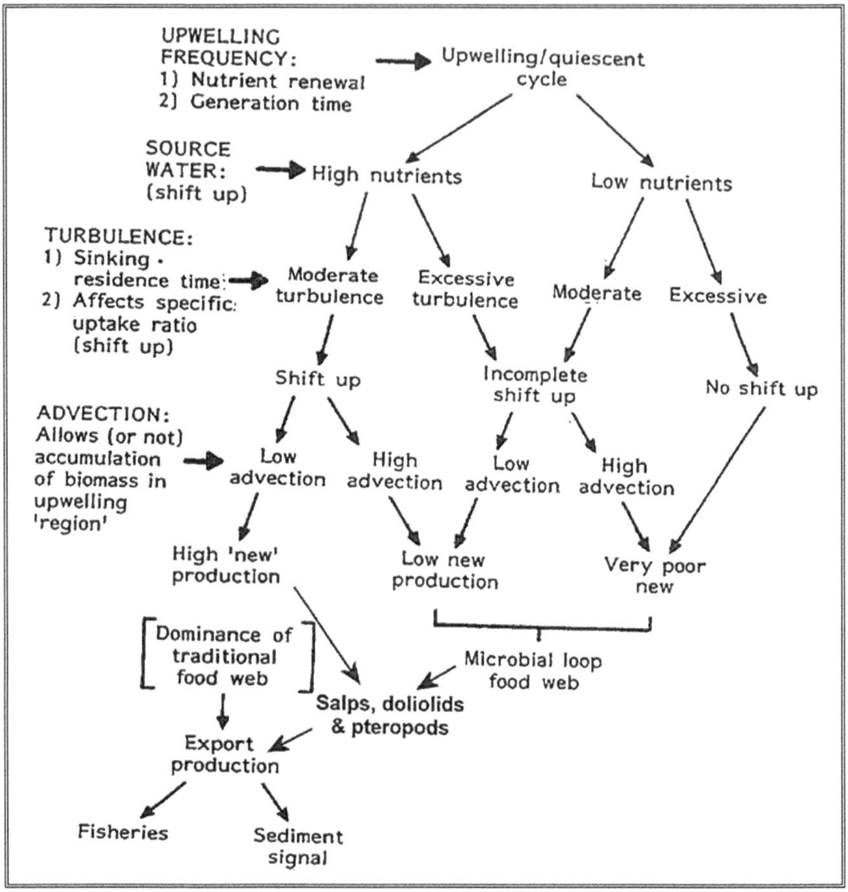

Fig. 4.15 Flow chart showing how nutrient enrichment, advection and turbulence intensity affect food web structure. The CCS food web usually resembles the left hand side of the flow chart, with salps and doliolids returning material from the microbial web back to export production and detritus (Mackas 2006, redrawn from Brink et al. 1995)

scales of the currents are generally small relative to the lifetime of the organism, particle tracking models (e.g., Batchelder et al. 2002) can be helpful in following the movement and retention of organisms in this region, and thus the response of food webs to complex feedback loops such as nutrient supply, advection, or vertical mixing that determine which species profit at any particular time (Fig. 4.15).

Chelton et al. (1982) analyzed 30 years of data (1950–1979) on temperature, salinity, steric height and zooplankton abundance collected by the CALCOFI program and identified several large-scale changes in the biology during this time. They divided the region into four areas (Fig. 4.16), and compared zooplankton abundance with wind stress variability. Area I was dominated by subarctic zooplankton, which could also occur in area II, but didn't dominate the population in

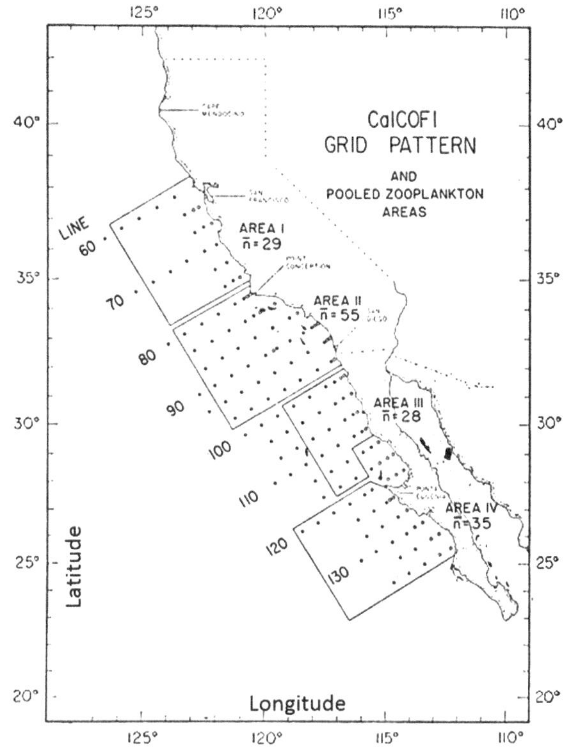

Fig. 4.16 Four geographical areas used by Chelton et al. (1982) to study zooplankton variability. The number n represents the average number of net tows per month for each area (from Chelton et al. 1982)

that region. The fauna of areas III and IV became increasingly dominated by equatorial forms.

The strongest seasonality was found in the north (areas I and II), where 30 % and 23 % of the variance was determined by seasonality respectively (Fig. 4.17), whereas in areas III and IV, the percentages were only 8 and 14 %. In all regions, the maximum biomass occurred in late spring/summer and the minimum in winter, but in area IV (southern Baja) the seasonality was delayed by two months compared to elsewhere. The analysis showed that winds in all areas were upwelling favourable throughout the year, but had their maximum in spring. Area III (northern Baja) showed two wind peaks in May and September, with the biomass maximum coming a month later, but this relationship did not hold in other areas. Chelton et al. (1982) suggested advection as the likely control, in agreement with southward transport estimates.

The anomalies remaining after the seasonal mean was removed, i.e. the non-seasonal component of the signal, showed large year to year changes, with high temperature and salinity corresponding to low zooplankton biomass and vice versa

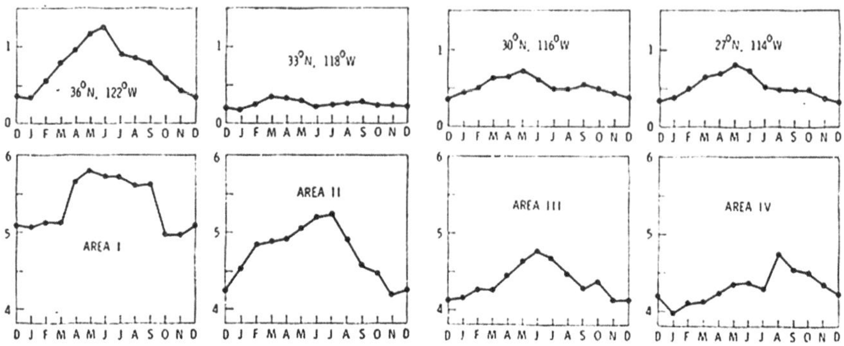

Fig. 4.17 Seasonal components of (*top row*) equatorial wind stress in dynes/cm^2 (1 dyne/cm^2 = 0.1 Pa) and zooplankton volume in units of $\log_e(10^{-3}$ ml/m$^3)$ (from Chelton et al. 1982)

(Chelton et al. 1982; Fig. 4.18). High southward transport meant high zooplankton biomass and the whole system was found to be strongly coherent. A final conclusion was that both horizontal advection and vertical tranport associated with geostrophic tilting of isopycnals from changes in the flow are sources of nutrients for primary productivity.

4.7.5 *Increase in Hypoxia off Oregon and Washington*

An item of particular interest off Oregon and Washington has been the apparent increase in hypoxia (dissolved oxygen concentrations <62 μM/kg) since about 2000. Unlike the situation off Peru (see Chap. 5) or in the northern Benguela (see Chap. 7), anoxia has not generally been a problem in this northern region of the CCS until recently. There has, however, been an overall decrease in oxygen concentration of up to 0.8 μM/kg/year throughout the upper 800 m of the water column both near the coast and offshore, particularly in the poleward undercurrent and at the thermocline. This decline is most obvious since 1998 (Whitney et al. 2007; Pierce et al. 2012; Fig. 4.19).

Historical data from 1950–1986 show that hypoxia occurred regularly across the middle and outer shelf off Washington during the summer (June–September), but more rarely off Oregon (Connolly et al. 2010), presumably at least partly because of the influence of the northward undercurrent, and no anoxia was observed during this time. In 2002, however, severe hypoxia with oxygen concentrations as low as 10 μM/kg, occurred from the shelf break to within 2 km of the coast off Oregon between Stonewall Bank (44.6° N) and Heceta Bank (44.1° N) (Wheeler et al. 2003; Grantham et al. 2004). This is thought to relate to increased upwelling-favourable winds during this summer season and the input of nutrient-rich subarctic water, as nutrient-salinity relationships during 2002 showed

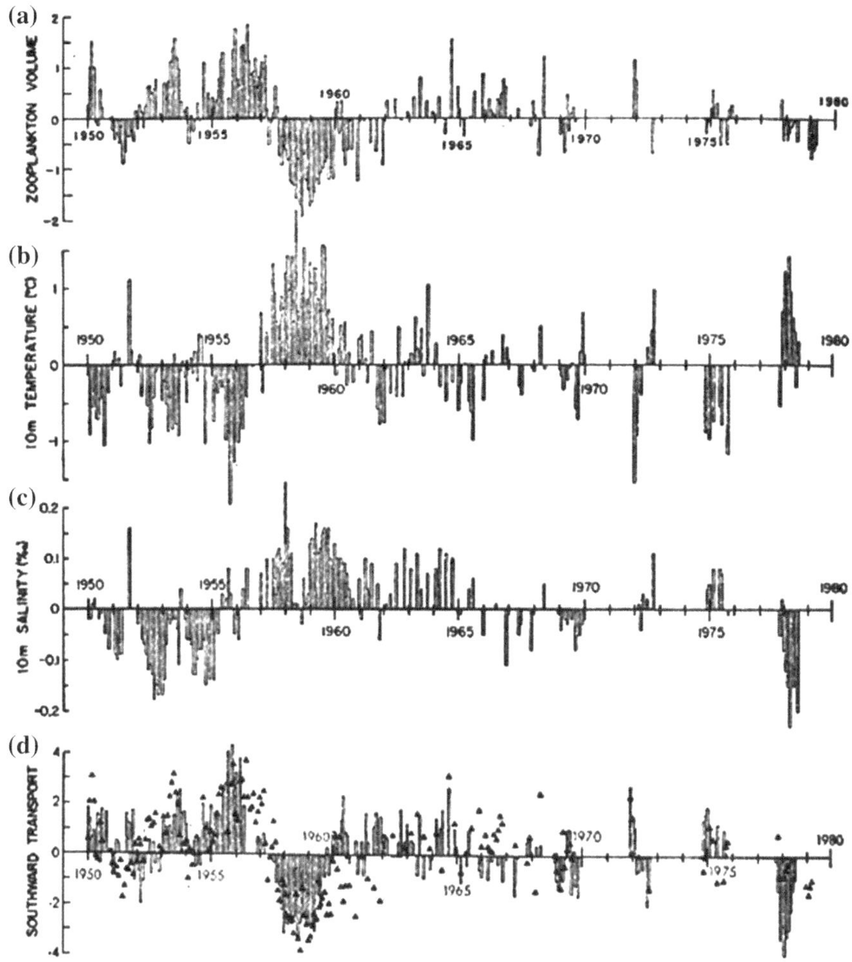

Fig. 4.18 Time series of nonseasonal values of four parameters in the California Current: **a** The average of the individual zooplankton volumes in the four areas $\log_e(10^{-3}\ \text{ml/m}^3)$; **b** the average 10 m temperature over the 150 hydrographic stations in Fig. 4.18; **c** the average 10 m salinity over the 150 hydrographic stations; **d** the amplitude of the principal EOF of steric heights (with reference to 500 dBar), which is proportional to the longshore transport of the California Current (from Chelton et al. 1982)

much higher nutrient concentrations than normal. Since then, summer hypoxia has continued to occur annually off Washington, with lowest dissolved oxygen concentrations near 2 μM/kg or less, particularly in 2006, when oxygen concentrations below 10 μM/kg were found throughout the upper 100 m of the water column. Even anoxic conditions have been observed off the Oregon coast (Chan et al. 2008; Connolly et al. 2010; Pierce et al. 2012; Adams et al. 2013), and there is much concern over whether such conditions will persist.

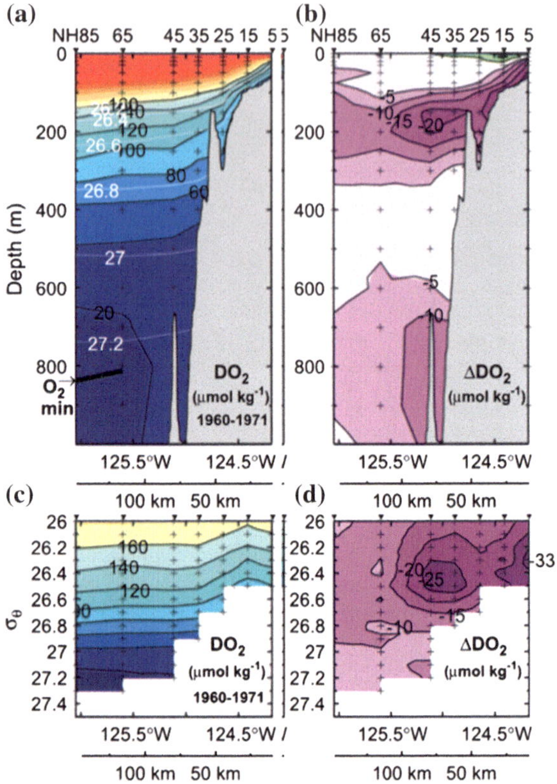

Fig. 4.19 Dissolved oxygen concentrations (μM/kg) for 1960–1971 and changes in dissolved oxygen concentrations along the Newport hydrographic line (~45° N) off the Oregon coast on depth (**a**, **b**) and density (**c**, **d**) surfaces between 1960–1971 and 1998–2009. *White contour lines* in (**a**) show density surfaces (from Pierce et al. 2012)

While some of the variability between years is the result of changes in upwelling strength and water mass variability, it appears that the major influence on bottom water dissolved oxygen depends on changes in water column and sediment biogeochemistry during the upwelling season, with the layer of oxygen-depleted water thickening during the summer over the inner shelf. Lowest absolute oxygen concentrations are seen within 2 weeks of the autumn transition to more ventilated, oxygenated conditions, and model studies (Siedlecki et al. 2014) have suggested that local respiration is responsible for up to about 60 % of the seasonal decline in oxygen off Washington and more off Oregon. Changes are also seen at shorter time intervals within the annual cycle; as the winds switch from downwelling to upwelling, so the currents change direction and this brings cold, low oxygen water onto the shelf (Adams et al. 2013).

4.7.6 Features of Northern California and Iron Limitation

One question of considerable interest to fisheries biologists is why the northern end of the California Current system is so productive, when the upwelling winds are much stronger farther south (by about a factor of eight). Phytoplankton production, for example, is about five times higher off southern British Columbia and Washington than off northern California, which goes against the "classical" paradigm relating upwelling intensity to production (MacIsaac et al. 1985). One possible reason for this is the nutrient-rich freshwater inflow from the Columbia River and Juan de Fuca Strait (Hickey and Banas 2008), which apparently aids juvenile salmon in particular, so that they tend to concentrate between the Columbia River and the Strait.

Northern California, between Point Arena and Point Reyes, was the site for the Wind Events and Shelf Transport (WEST) program from 2000–2003. This work showed that the region has an enhanced nitrogen supply from both Ekman divergence near the coastal boundary amounting to 0.3 μM/L/s and from wind-stress curl further offshore (up to 0.74 μM/L/s, Dever et al. 2006). This leads to maximum nitrate concentrations of about 24 μM/kg, while the effect of the wind stress curl is to double the area over which upwelling occurs. The importance of wind-stress curl in this region was also found in modeling studies by Davis and Di Lorenzo (2015), although it was not as important farther south. While nitrogen has been traditionally considered to be the limiting nutrient in upwelling regions, the iron hypothesis of Martin and Fitzwater (1988) has gained much traction and iron (Fe) is now regarded as being the primary limitation in much of the world's oceans. This appears to be true even in upwelling regions, so that Fe and other trace elements can limit production as much as N, P, or Si (e.g., Hutchins and Bruland 1998). Effects of such limitation are seen in changes in the phytoplankton particulate Si:C and Si:N ratios away from the Redfield ratio, with Fe limitation leading to much greater uptake of silicate. As a result, high particulate Si:N ratios suggest iron stress/limitation, but also lead to more rapid sinking of the silicified phytoplankton, and hence to higher, or at least more efficient, export ratios.

The apparent importance of Fe to primary production in this region has led to the idea of a north-south iron gradient from Oregon into California (Chase et al. 2007) and an "iron curtain" at the shelf edge, inshore of which Fe concentrations are high enough to support enhanced primary production, but beyond which Fe concentrations and production drop off rapidly. This appears to continue at least as far as the southern California Bight. Therefore it seems that productivity in this region depends not only on nitrate, which depends in its turn on upwelling variability as modulated by wind events, but also on Fe supply rates (Dugdale et al. 2006; Wilkerson et al. 2006). Chlorophyll concentrations also depend on the width of the shelf, which varies between 60 km in the north and 10 km off California. Because the wind-driven flow is essentially the same all along the coast at ~0.1 m/s, this means that the mean flow is lower along wider shelves which also affects residence time.

The chemistry off northern California can be summarized from repeated sections along CALCOFI line 67, off Monterey Bay (37° N), even though the time-scales of the region are of order 5–10 days while the sections were generally bimonthly, augmented with biweekly cruises closer to shore (Collins et al. 2003). While temperature minima occur near the coast during April and May, a salinity maximum occurs about a month later, showing the influence of the upwelling water with salinities >33.3–33.4.

Offshore, higher temperature, lower salinity subarctic water is frequently found, although the 1997–98 El Niño displaced this northern water mass. Changes in nutrients follow the changes in temperature/salinity. Nitrate, chlorophyll and primary productivity integrated over the upper 80 m vary in phase and show strong seasonal variability (Fig. 4.20), with highest levels in summer and lowest in winter inshore of the California Current jet. The low values for all three parameters, particularly offshore, during the 1997–98 El Niño are marked. Annual means suggest vertically averaged nitrate maxima at the coast of about 20 μM/kg for the upper 80 m in summer, but <10 μM/kg in winter, with chlorophyll and primary production maxima of about 100 mg/m^2 and 2500 mg C/m^2/day respectively, although a secondary, much lower productivity maximum was found along the edge of the California Current jet at about 175 km offshore. The nutricline and its associated chlorophyll and production maxima were found near 12 m depth inshore, but deepened offshore to 20–30 m in the jet and 40–60 m outside it.

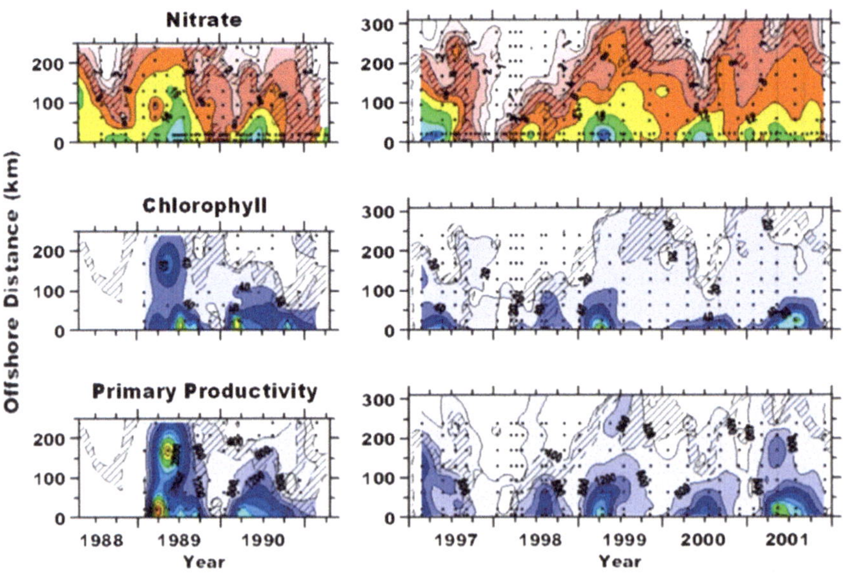

Fig. 4.20 Variations in nitrate concentration, chlorophyll and primary productivity along CALCOFI line 67. Regions where the dynamic height is between 0.45 and 0.49 dynamic metres are hatched. Nitrate values show the mean over the upper 80 dbar, contoured at 1, 2, and 4 μM/kg and at 4 μM/kg intervals above this; chlorophyll concentrations and primary productivity are averaged over the euphotic zone, with contour interval 20 mg/m^2 and 400 mg C/m^2/day respectively (from Collins et al. 2003)

4.7.7 *Features of Southern California*

Oxygen has been declining in the southern California region since the mid-1980s, with annual changes of up to 2.1 μM/kg/year in the upper 500 m of the water column (Bograd et al. 2008, 2015). The decline in the upper 100 m has occurred on the shelf and upper slope, including in the southern California Bight and the California Current inflow from the north. In midwater (200–300 m depth), declines were seen across the whole region sampled, and amounted to up to a 21 % decrease. The declining oxygen levels cover the depths of both the 25.8 and 26.5 sigma-t surfaces, and thus include the whole of the thermocline, and have been accompanied by increasing nutrient levels (Bograd et al. 2015). The rate of change decreased in the 400–500 m depth range, but was still around 15 %, especially in the Santa Barbara Basin, which is presently trending towards anoxia, as it is being accompanied by a decrease in nitrate concentrations in the bottom waters and an order of magnitude increase in nitrite, indicative of denitrification (Goericke et al. 2015). This decline in oxygen concentration has been accompanied by a shoaling of the hypoxic boundary (here defined as oxygen concentrations <60 μM/kg) by <90 m.

What has caused this decrease in dissolved oxygen concentration? It appears to be widespread across the North Pacific and model studies suggest that warming of the surface waters has led to increased stratification, a decrease in the efficiency of the biological pump, and hence reduced downward transport of oxygen from the surface (Bograd and Lynn 2003; Keeling and Garcia 2002; Di Lorenzo et al. 2005). Advection may also contribute, as shown by the decreases in oxygen content in the California Undercurrent and the increasing contribution of high temperature, low oxygen subtropical water (Di Lorenzo et al. 2008).

A recent simple estimate of the carbon cycle budget for the upper 60 m of the water column within the central portion of the CCS between Cape Mendocino (40.5° N) and Point Conception (34.4° N) has been presented by Pennington et al. (2010). The calculation was based on data collected along CALCOFI line 67, which makes landfall near 37° N. At this latitude the offshore jet marking the outer edge of the upwelling system is about 170 km from the coast, but meanders onshore during summer/early autumn. Using an empirical equation for the relationship between temperature and nitrate (Nitrate (μM/L) = $0.6075T^2 - 19.078T + 149.36$), they calculated the upwelling flux of new nitrate. Not all of this is used for production, as there is a significant loss from diffusion and advection before any production occurs (estimated as about 17 %), while particulate carbon is lost through sinking, advection, conversion to dissolved carbon and fish biomass. The calculations lead to total theoretical production of about 44 million tonnes of carbon in the region (345 g C/m^2/year), split almost 50/50 between new and regenerated production, and of which about 50 % is "exported", either literally or because the nitrate is not used. This f-ratio of 0.49 is lower than reported by other workers in the region, who have reported f-ratios up to 0.89 for specific areas in the CCS, but larger than the percentage calculated for the four main upwelling systems by Monteiro (2010). While the upwelling zone close to shore releases carbon dioxide to the atmosphere, this is balanced by input further offshore, so that net CO_2 exchange is about zero. This seems typical of upwelling systems in general, and whether or not a particular region is a source or sink for CO_2 at any given time likely depends on local conditions.

4.7.8 Features of Baja California

There have been fewer studies of nutrients and productivity off the coast of Baja California than farther north, despite regular data collection during the CALCOFI program (1949–mid 1980s), but the Mexican IMECOCAL program (1997 to date) is now leading to more publications. SeaWiFs chlorophyll data show a strong annual cycle along the whole of the Baja peninsula. The largest and most consistent values occur south of Punta Eugenia (27° 30′ N), where the seasonal cycle accounts for about 62 % of the variance, as opposed to only 33 % north of 29° N. Chlorophyll concentrations in this region were apparently unaffected by either the 1997–98 or 2006 El Niños, in contrast to the area north of Punta Eugenia, although higher anomalies were found along the whole of Baja California during the 2002–03 subarctic water intrusion and the 2007–08 La Niña cycle (Gaxiola-Castro et al. 2010).

The Gulf of Ulloa, which covers the Baja coast between Bahia Magdalena (25° N) and Punta Eugenia, is known as a centre of primary production, mainly because of its broad shelf (Fig. 4.15) and its position immediately south of the upwelling centre at Punta Eugenia. Gonzalez-Rodriguez et al. (2012) used satellite altimetry, SST and chlorophyll measurements with a vertical generalized productivity model to estimate changes in local production rates on a seasonal basis between 2003 and 2007. While the wind blows parallel to the coast throughout the year, it is strongest between March and June, with a maximum in April and minimum in July and August. The wind-stress curl also remains positive throughout the year. This means upwelling should be persistent, but productivity is not; a maximum in April leads to decreases through summer and a minimum in winter, as found from in situ sampling (see e.g., Peterson et al. 2006). This is due to the intrusion of warmer, nutrient poor water into the gulf from the south following wind relaxation and the movement offshore of the southward flow around Punta Eugenia. Even if upwelling winds continue during this time, there are too few nutrients to allow phytoplankton production.

A similar situation is found further south in the coastal lagoon of Bahia Magdalena, which receives almost all its nutrient supply from upwelling (Cervantes-Duarte et al. 2013), with nitrate concentrations reaching about 16 μM/L and phosphate up to 3 μM/L. Here, the N:P ratio is much lower than the expected

16:1 Redfield ratio, being about 8.4 during the spring and early summer and less than 2 later in the year, although there is a considerable amount of ammonia which would cause the ratio of dissolved inorganic nitrogen to dissolved inorganic phosphorous to increase considerably. It is also important to note that, even at this latitude, increases and relaxations of the local wind field on time scales of 5–10 days are superimposed on the larger-scale seasonal variability, with the highest energy being found in the north of the peninsula (Zaytsev et al. 2003).

Kamykowski (1972) described the temperature: nitrate relationship at La Jolla as being $[NO_3^-] = 66.25 - 5.60T$, where T is the water temperature, and Ladah (2003) used this relationship with historical data to identify a small number of occasions when water with abnormally high nitrate concentrations was observed off Baja California during El Niño periods. She suggested that such incursions of high temperature, high salinity nutrient enriched water, which may originate within the California Undercurrent because of its low dissolved oxygen concentration, could enhance the productivity of the local kelp forests, which are known to suffer and die back during El Niño events.

4.7.9 Other Features

Of major importance to the inshore benthic biota in the CCS are the kelp forests and rocky reefs that are found throughout coastal zone. The size of the kelp stands and the composition of the species making up these communities are strongly affected by interannual and alongshore changes in flow patterns and nutrient availability (Dayton et al. 1999), which also affect larval supply and settlement. For example, intertidal barnacles and mussels are lowest in areas along central California subject to strong upwelling (Mackas 2006). Similarly, even though productivity is high and there is a large supply of terrigenous material in the north from the Columbia River, the width of the shelf determines the amount of sediment that is deposited, which again affects benthos community structure and density (Jumars 1993). This means that the wider, northern portion of the CCS shelf is a depositional area, whereas farther south, fine sediments are moved down coast and offshore. This offshore transport of sediment can be increased by the presence of many canyons that result in periodic landslides and mean the sediments are deposited into deep water at the base of the continental shelf.

One other point of interest in this region is the occurrence of seeps of hydrocarbons and sulphide-rich water in areas such as Monterey Bay and the deep basins in the southern California Bight. These produce local specialist benthic communities, similar to those found in other regions subject to natural oil leaks or low oxygen concentrations, such as the northern Gulf of Mexico. The communities include *Beggiatoa* mats, as well as crabs, worms and bivalves that are adapted to living under such conditions and which may contain chemosynthetic bacteria, similar to those found at deep sea vents (Kennicutt et al. 1989; Barry et al. 1996).

4.7.10 Harmful Algae Blooms

Species records show that the phytoplankton and zooplankton in the region mirror those found in the North Pacific gyre. There are records from Scripps Pier (33° N) from 1917–1939 and from Port Hueneme (34° N) from 1923–1939. PCA analysis of these shows that there was essentially no change in species during this period (Tont 1987), while later CalCOFI data from the 1990s produced similar species composition inshore (Venrick 1998). During the 1990s offshore records were dominated by the coccolithiophore *Emiliania huxleyi*, which is also abundant in the North Pacific gyre. Records also exist for 1979–89 from southern British Columbia (Forbes and Waters 1993a, b). At all locations, diatoms generally dominate at times of high productivity, as is expected for a coastal upwelling region. However, harmful algal blooms (HABs) of species such as *Alexandrium, Gymnodinium, Pseudo-nitzschia, Chaetoceros*, and the flagellate *Heterosigma akashiwo*, have been noted since the late 18th century in the northern part of the system and for much longer than that off California (Meyer et al. 1928). These species can cause Paralytic Shellfish Poisoning (PSP) or domoic acid poisoning (Amnesiac Shellfish Poisoning, ASP) and occasionally result in human deaths from eating shellfish, despite regular monitoring. Even non-toxic red tides can affect the ecology by reducing the dissolved oxygen concentration as the blooms die and decompose.

A recent paper by Lewitus et al. (2012) summarizes recent HAB occurrences in the CCS, which appear to be increasing in frequency and becoming more widespread, although this may result at least partly from better monitoring efforts. The outer coastlines of Washington and Oregon seem to be less affected than the enclosed bays and fiordic systems in this northern portion of the CCS, but HABs containing relatively low levels of toxins have been found all along the California coast. Levels of toxins seem to be similarly low along Baja California, although PSP toxins are commonly recorded at higher levels further south in Mexico. The blooms are thought generally to originate offshore and be brought onshore during downwelling or quiescent periods, particularly during the autumn, when the surface layer is stratified. Thus, they do not seem to have an anthropogenic source although increased nutrient input to the coastal zone from outfalls, particularly in bays or other sheltered areas, may help keep them going once they start. There is similarly no obvious relationship between their occurrence and large-scale changes such as El Niño, except possibly in the Gulf of California, where blooms may occur more frequently during La Niña conditions, according to records of *G. catenatum* cysts in sediments (Flores-Trujillo et al. 2009).

ASP causing organisms have been identified in the CCS for at least 100 years, but few incidents of human toxicity have been reported. This is not the case for other wildlife, however, and fish (sardines and anchovy), mammals (sea lions, fur seals, sea otters and several species of whales) and birds (pelicans, cormorants, ducks and grebes) have all been affected (see, e.g., Shumway et al. 2003; Bargu et al. 2012). Outbreaks of poisoning by domoic acid, first noticed during a widespread outbreak in 1991, have occurred almost annually since 2000 and have led to the regular closure of shellfish fisheries off Washington, Oregon and California that lead to millions of dollars of lost economic production. The organism responsible, *Pseudo-nitzschia,* is found frequently off coastal Washington and it is thought that the Juan de Fuca eddy is a source region that can infect the coastal zone to the south (MacFadyen et al. 2005). Heceta Bank is thought to be another such source off Oregon, because of the retention of material that occurs there, as are Monterey Bay and the Santa Barbara Channel off California. So far, there have been few reports of domoic acid poisoning from Baja California, except from within the Gulf of California. While anthropogenic nutrient supply does not seem to be a factor in the increased observation of domoic acid producing organisms, the cooler conditions and enhanced upwelling since 1999 off California may be responsible (Lewitus et al. 2012). High Si:N ratios in the upwelling water may also be important in promoting the growth of these organisms, although there is less consistency in the relationship of nutrient concentrations with HAB outbreaks.

4.7.11 Historical Large-Scale Biological Changes

Large-scale biological changes in the CCS were first noticed in the1930s and were followed by the sardine industry crash from an annual catch rate of 550,000 tonnes to 100,000 tonnes between 1945 and 1947. A vital question for the industry was what caused the crash–was it overfishing, biological interactions, or external physical forcing?

Zooplankton biomass in the CCS is known to be inversely related to temperature (Reid 1962), and thus to depend somehow on the southward advection of nutrients. A decrease in copepod numbers by an order of magnitude, most likely due to the 1958/59 El Niño, was seen between October 1955 and October 1958 (Isaacs et al. 1969), and similar events occurred in conjunction with the El Niño event of 1982–83 and the separate regime shift warm event of 1976–77.

Two-hundred-year records from anoxic basins suggest such changes were natural before the 1960s (Soutar and Isaacs 1969, 1974) and that external forcing is the likely cause. While Roemmich and McGowan (1995) showed that the 1957/58 drop in zooplankton essentially reversed in the next two years, they also detailed a continued decline in zooplankton numbers by 80 % over a 43-year period from 1950–1993 along CALCOFI lines 80 (off Point Conception) and 90 (in the southern California Bight). A 5–10 year periodicity was superimposed on the long-term decline in response to the ENSO variability.

This long-term decline corresponded to a warming of surface waters by about 1.2 and 1.6 °C on lines 80 and 90 respectively, which increased the vertical stratification and decreased the nutrient concentrations available in the euphotic layer, even though the wind stress did not change or even increased during this period. While *Calanus pacificus* has always dominated CALCOFI zooplankton assemblages, abrupt abundance changes, both positive and negative, of this species occurred in the late 1950s, late 1960s, mid 1970s, early 1980s, 1990 and 1998–99 (Rebstock 2002).

Similar abrupt changes also occurred for salps and doliolids, although the populations of these organisms have declined continuously since1950 (Lavaniegos and Ohman 2003). However, more recent work (Gaxiola-Castro et al. 2010) suggests that at least off Baja California, copepod and euphausiid numbers have been increasing again since 1998, apart from during the cold-water intrusion in 2002–03.

4.8 Fisheries

We have already mentioned the rise and subsequent collapse of the California sardine fishery in the 1950s. Despite this collapse, nearly all the commercially-caught fisheries resources (with the exception of hake and crustacea) are still pelagics. This includes market squid, anchovy, sardine, mackerel, jack mackerel and herring (see CALCOFI reports). These all have multi-year lifetimes apart from squid, and therefore have to survive in this very variable system. Recent reviews of variability include Smith and Moser (2003), Bakun (2005), and Chavez (2005).

Total landings from California, Oregon and Washington are shown in Fig. 4.21a. During the 1990s, the industry started processing the catch at sea. The discrepancy between reported landings and estimated catch has increased considerably since 1980, presumably because of better monitoring, and is now almost equivalent to the actual landings at between 400–500,000 tonnes (data not shown). Overall, total reported catch rates are relatively steady at around 900 thousand tonnes per annum,

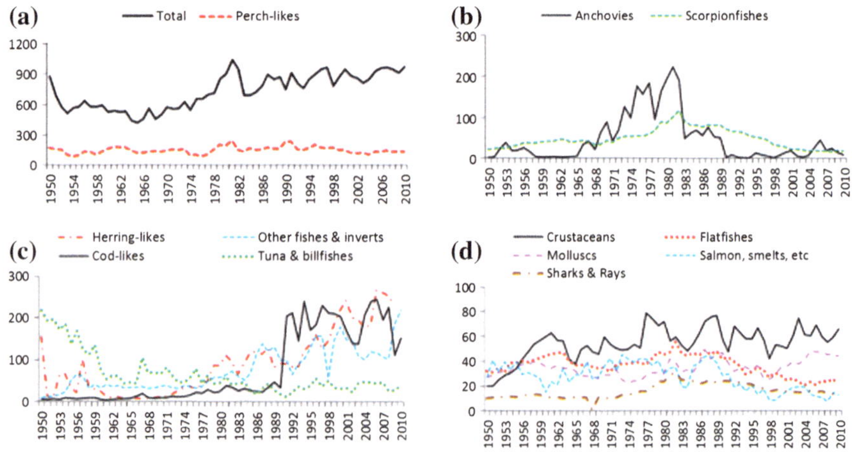

Fig. 4.21 Catches (in units of 1000 tonnes/yr) in the California Current large marine ecosystem for 1950–2010. Data from http://www.seaaroundus.org/. Note that the data in this figure include estimates of fish discarded as bycatch as well as landed weights, and so are higher than officially reported figures which only consider landed catch and catches processed offshore

although that the California Current upwelling region is the least fish-productive region among the big four coastal upwelling systems (see Chap. 10).

Some of the more important individual catches from the California Current system are shown in Fig. 4.21b–d. Note that the figures for herring include a switch from whole fish to herring roe between 1993 and 1994. While sardine and anchovy have traditionally made up the largest component of the catch, large catches of mackerel were also made until the early 1990s, while squid (which dominates the group of "other fishes and inverts" in Fig. 4.21c) catches have been increasing steadily. In 2014, the total value of the California catch alone was about $233 million (see http://www.calcofi.org/publications/calcofireports/v56/Vol56-CalCofi.Journal.2015.pdf).

Apart from the inter-annual variability in all stocks, there appear to be system shifts between sardine (herring) and anchovy catches (see Fig. 4.21) and stocks of jack mackerel (*Trachurus symmetricus*) and chub mackerel (*Scomber japonicus*) (Fig. 4.22). It remains unclear why such alternation of stocks takes place, but this is an effect that has been noticed in other upwelling systems as well (e.g., Chavez et al. 2003; Klyashtorin and Lyubushin 2007), and likely depends on both environmental effects and survival rates of larvae and juveniles. It is certainly not a modern development, as counts of fish scales in the anoxic Santa Barbara Basin show large interdecadal variability in anchovy and sardine numbers dating back to before 500 A.D. (Baumgartner et al. 1992; Skrivanek and Hendy 2015, Fig. 4.23).

Prior to the 1940s, the sardine biomass is estimated as having been over 3 million tonnes. This dropped to less than 10,000 tonnes by 1975 (Smith and Moser 2003) but has since rebounded, helped by strict catch limits, so that the California stock exceeded 1 million tonnes again by the late 1990s, with a total estimate for the CCS of 1.6 million tonnes (see CALCOFI Report 1999). The start of the recovery of the sardine stock coincided with a major regime shift in 1976–77 in the northeast Pacific that was tied to a switch from a cool to a warm phase in the PDO (Venrick et al. 1987), represented by a rise of about 1.5 °C in the 10-m depth temperature in

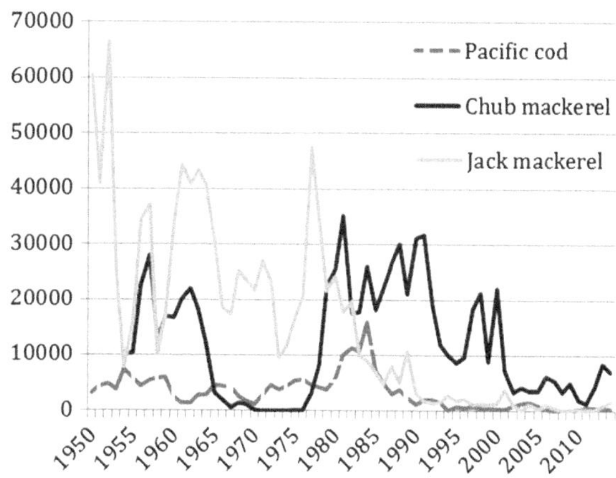

Fig. 4.22 Fish landings of Pacific cod, chub mackerel and jack mackerel along the California Current system from 1950–2014. Data from CALCOFI Reports and NOAA website. All quantities are in tonnes

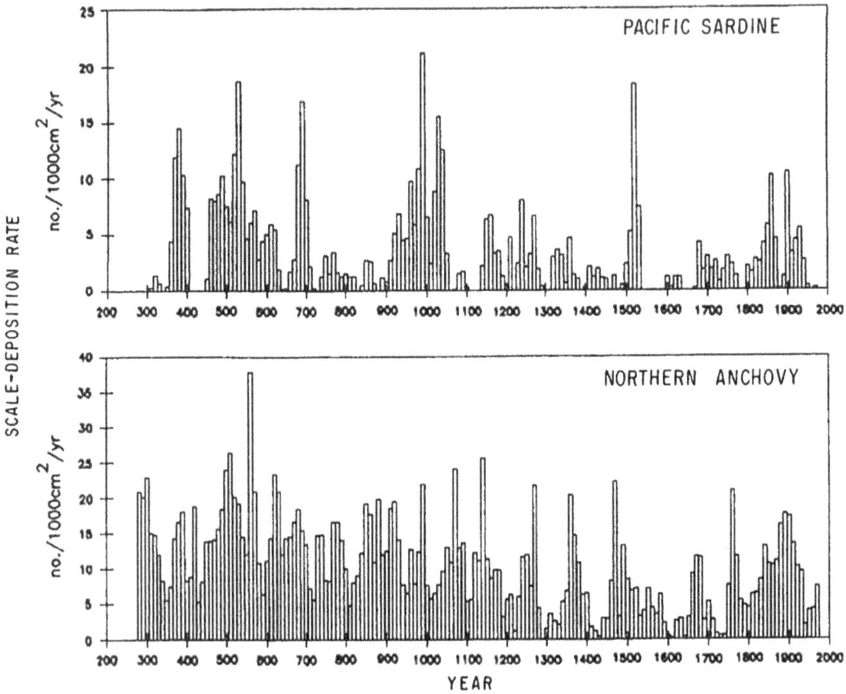

Fig. 4.23 Composite time-series of Pacific sardine and northern anchovy scale deposition rates between 270 and 1970 A.D. based on counts from cores taken in the Santa Barbara basin. For details, see Baumgartner et al. (1992), from which this diagram is taken. http://www.calcofi.org/publications/calcofireports/v33/CalCOFI_Rpt_Vol_33_1992.pdf [accessed on 4 April 2016]

the southern California Bight (the 1982–83 El Niño resulted in another temperature increase of about a degree, and these abnormally warm conditions lasted through the 1998–99 La Niña). This regime change affected many different organisms in several trophic levels (e.g., Roemmich and McGowan 1985; McGowan et al. 1998; Ware 1995; Chavez et al. 2003).

Smith and Moser (2003) give details of changes in the abundance of larvae for 14 fish species in routine surveys of the CALCOFI region from 1950–2000, of which nine showed large increases in egg and larval numbers after the switch, including sardines, jack and chub mackerels and a number of southern offshore species. Other commercial species, however, such as anchovy (*Engraulis mordax*) and hake (*Merluccius productus*), showed either no change or a decline in numbers over this period. Somewhat surprisingly, macrozooplankton did not show any major response, but continued the long-term population decline discussed earlier. Following the La Niña of 1998–99, however, larval numbers of almost all species studied dropped sharply.

As might be expected, the El Niños of 1982–83 and 1997–1998 also affected the fishery with catches of pelagics and squid in particular declining dramatically during both events. The loss of the squid harvest was particularly severe, with an almost complete absence of harvestable squid (see CALCOFI Report 1999), although the Pacific sardine stock continued to increase despite the higher water temperatures, possibly because of the absence of competition for food by other small pelagic fish. However, the PDO also affects populations, with historic highs in both anchovy and sardine stocks coinciding with cool-phase PDO periods (Skrivanek and Hendy 2015). These pelagic species are found generally in the southern part of the CCS, although squid seem to be moving northwards as the temperature warms. Many species use the southern California Bight as a spawning ground, thanks to slower currents that can retain larval and juvenile stages and allow aggregations of prey organisms to develop (Parrish et al. 1981; Bakun and Parrish 1982), although adults will range throughout the region (Checkley and Barth 2009). Examples include sardines, which breed in the southern California Bight in spring, then move north to breed off Washington in autumn, before returning south again, and hake, which spawn somewhat further offshore in the Bight than sardines, but feed all along the coast as far north as the northern tip of Vancouver Island.

Despite the assumed usefulness of fish egg and larval surveys in estimating future stock sizes and the long time-series available from the CCS, however, it is difficult to use this information as a good predictor for fisheries management purposes (Smith and Moser 2003), and the only link to management is the 3-year running temperature mean. When temperature goes up, so too does the allowable catch in the region, but sudden changes such as the 1976–77 shift make prediction difficult.

While salmon are an important component of the CCS biota, most of the catch comes from the northern parts of the system, off Washington, Alaska and British Columbia. Thus it does not figure as a major part of the fish catch in the annual CCS statistics. However, the large-scale environment does affect biomass, with the PDO warm phase favouring high production for Alaskan stocks and low production off Washington, Oregon and California, and the opposite for the cool phase (Hare et al. 1999; see Fig. 4 in Mantua and Hare 2002). The PDO also affects groundfish, e.g., sablefish and halibut, with the halibut reacting in the same way as Alaska salmon.

4.9 Climate Change Impacts in the CCS

4.9.1 Overview

Ecosystems within the California Current are strongly influenced by the natural basin-scale modes of interannual to interdecadal climate variability such as ENSO, the PDO, and the NPGO (Chavez et al. 2011), which modulate the gyre circulation and, hence, the structure and depth of the thermocline and nutricline and wind patterns. These modes of oceanic and atmospheric variability play fundamental roles in structuring the flow of energy and nutrients through food webs across the California Current. Because the timescales related to this climate variability are similar to that of accelerated global warming during the period 1980–2000, it is inherently difficult to identify ecological changes in the California Current that can definitely be attributed to global climate change. For example, the El Niños of 1982–83 and 1997–98 caused severe changes in productivity at all trophic levels, including kelp forests, which seem to be particularly susceptible and take much longer to recover than to succumb (Dayton et al. 1999), as do changes linked to a switch from warm to cold phases of the PDO. Despite this problem, which applies to all ecosystems, not only upwelling systems, multidecadal time series of climate variability and ecosystem responses have provided important insights into the suite of dynamics that will likely govern future changes in this region, again thanks to the long-running CALCOFI program. Whether conclusions derived from this are applicable to other upwelling region is unclear (see Chap. 10 for a comparison of the big four upwelling systems).

Climate change is expected to have several far-reaching consequences for the California Current associated with changes in density stratification, the timing and intensity of upwelling-favourable wind conditions, and source water properties advected into the system (Doney et al. 2012). The system has experienced significant warming (Schwing et al. 2010) and changes in water-column stratification (Palacios et al. 2004) over the past century. As the thermocline deepens and the water-column stability increases, a warming ocean should reduce nutrient fluxes and primary productivity in the system. However, increased surface warming could also increase the temperature difference between land and ocean, thus affecting the pressure gradient and the local wind field. Bakun (1990) was the first to suggest that this effect would increase the local wind stress and hence upwelling. Studies of the CCS by Schwing and Mendelssohn (1997) appear to support this idea, while García-Reyes and Largier (2010) have shown local cooling, or at least not warming, as upwelling-favourable winds increase. Thus, the warming is not uniform across the CCS, but heterogeneous, and the system shows strong spatial gradients in physical properties.

Also, in contrast to expected declines in nutrient concentration as the ocean surface warms, the continuous records of primary productivity or phytoplankton biomass for the central and southern CCS available since the 1980s show how the high-nutrient subsurface waters have shoaled, resulting in increased primary production (Aksnes and Ohman 2009; Chavez et al. 2011). This shoaling has occurred despite increased density stratification. So it appears that the predicted declines in nutrient concentrations caused by warming the water column can be countered by higher wind speeds driving coastal and curl-driven upwelling (Auad et al. 2006), as reported for the southern California Current, for instance, since the 1980s by Rykaczewski and Checkley (2008). Two critical questions for further research are whether the measured changes in wind forcing are secular trends or result from decadal-scale variability in the forcing that affects the CCS, and how continued warming and changes in upwelling will affect future production (Doney et al. 2012). Will there be greater productivity, or will the higher wind speeds merely serve to decrease stability and mix phytoplankton and/or zooplankton deeper into the water column or remove them faster from the coastal zone?

In the southern California current zooplankton biomass has declined dramatically over the past 60 years (see Sect. 4.7.4), as detailed by Roemmich and McGowan (1995), who linked the decline to a warming trend that began in the 1940s (Hsieh et al. 1995). This has since been confirmed by Palacios et al. (2004). Reductions in pelagic tunicates are largely responsible for this decline from the 1950s to early 1990s, possibly as warm-water taxa were favoured by a shift towards greater relative abundance of small prey (picoplankton) (Lavaniegos and Ohman 2003). Non-gelatinous zooplankton in the region have also declined (Hsieh et al. 2009). Such trends, which lead to less abundant, smaller, and lipid-poor subtropical copepods instead of the more normal zooplankton populations, are typical of the transition to warm phases of the PDO (Peterson and Schwing 2003). As a result, continued warming of the California Current is predicted to translate up the food chain to reduce the survival of juvenile salmonids.

Smith and Moser (2003) and Di Lorenzo et al. (2005) showed that the warming trend in the CSS region continued through 1998 after which the 1998/99 La-Niña episode reduced temperatures slightly. After 2000, the temperature trend was essentially neutral (Gaxiola-Castro et al. 2010; CALCOFI Report 2015), accompanied by a decrease in dissolved oxygen concentrations and an increase in nitrate (Fig. 4.24). During late 2013/early 2014, however, a warm water anomaly of ~3 °C relative to the 1950–1979 climatology developed in the Alaskan Basin and affected the whole of the CCS by late 2014 (Bond et al. 2015; CALCOFI Report 2015). The warm anomaly persisted into 2015 and was accompanied by unusually strong

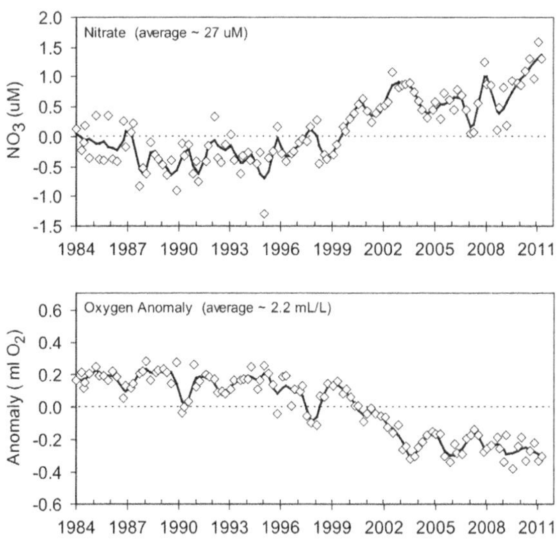

Fig. 4.24 Nitrate and oxygen anomalies on the 26.4 sigma-t isopycnal off southern California during the period 1984–2011, based on CALCOFI data (from Bjorkstedt et al. 2011). http://calcofi.org/publications/calcofireports/v52/Vol_52_CalCOFI_Reports.pdf [accessed 4 April 2016]

upwelling winds and the development of a mild El Niño the same year that affected the equatorial region, and led to the occurrence of many unusual warm water species, such as pelagic red crabs (*Pleuroncodes planipes*) and subtropical krill (*Nyctiphanes simplex* and *Euphausia eximia*) appearing much further north than usual. As might be expected, phytoplankton production and nitrate concentrations in the central CALCOFI region were much lower than usual, as were pilchard and anchovy numbers, despite very large samples of anchovy larvae, and the changed status of the fisheries led to low weights and a large number of strandings of California sea lion pups (*Zalophus californianus*). But despite the warm anomaly throughout much of the region, near-bottom water on the mid-shelf off Oregon was at the same time much colder than usual, showing the pronounced regional variability.

We know also that many biological species, including the commercially important fish stocks, show large population changes with multi-year periodicities that do not always appear to be correlated with these large-scale events (e.g., Bakun 2005; Chavez 2005). These are natural, frequently abrupt, and operate over large distance scales, and the variations in abundance of fish scales in anoxic basins show that such changes have been going on for at least 2000 years (Baumgartner et al. 1992; Fig. 4.23). Examples from California include small pelagic fish stocks (see Sect. 4.8) as well as crabs (Botsford et al. 1994) where the population can change by a factor of 100 over a short time period. Higher trophic levels are likely also affected similarly; bird species richness has declined continuously since routine surveys started in 1987 (e.g., Santora and Syderman 2015; Syderman et al. 2015) and this has been linked to both ocean climate variability and the availability of forage fish food species. Similar results are available from British Columbia and Oregon, where the normal population of zooplankton was replaced during the early 1990s with species from central and southern California. This shift in species composition reversed after 1997–99 and resembled 1980s data again during the 1999–2003 period. However, while the overall trend was similar, annual populations varied by 3–8 times as regards total biomass and <30 times for individual groups of organisms (Mackas 2006).

Strong evidence exists that climate change can affect not only the mean and variance of production at low trophic levels, but higher levels as well (Doney et al. 2012). Both increases in chlorophyll-*a* concentrations and earlier spring blooms have been seen in the southern California Current (Kim et al. 2009). Similar effects have also been seen in the northern California Current, where the timing of the peak biomass of the large copepod, *Neocalanus plumchrus*, has been reduced and brought forward by nearly six weeks over 30 years (Mackas et al. 2007). This leads to potential timing mismatches between predator stocks and prey availability that can affect populations of consumers such as migrating juvenile fish or breeding seabirds that undergo critical life-history transitions over narrow phenological windows. Such mismatches, for example, are blamed for reproductive failures in seabirds, although in some cases, breeding success has merely become more variable with average numbers staying roughly constant (Sydeman et al. 2006, 2009).

Some top predators, such as whales, seals and sea lions, have been increasing recently (Field et al. 2001). It must be remembered, however, that this is related to the recovery of stocks that were decimated by commercial whaling and sealing during the 19th century. As the fish stocks in the CCS that these animals feed on have been fished intensively during the last century while the region has been warming, this makes understanding the true relationship between natural populations and climate variability much more difficult (Field et al. 2006).

A question related to this is why are biological anomalies, such as order of magnitude population changes, very much greater in the CCS than physical/chemical anomalies? This is especially important because such changes are sudden and not gradual. Although the ecosystem has a natural resilience, does the unpredictability in long-period variability overwhelm it during the "crash" phases of a given population? Given the inherent difficulty of modeling ecosystem variations over even short time periods, we have little idea of how climate change will affect the CCS ecosystem except to say that there will be changes in the distribution of many organisms. The 2014–15 warm event, for example, changed the spawning habits of various fish species, with sardines spawning during spring 2015 up to 500 km further north than usual, while jack mackerel spawning was confined to a much smaller area than normal (see CALCOFI Report 2015). Warmer water can also hold less dissolved oxygen, and hypoxic and even anoxic water is now a feature of the Washington/Oregon shelf (Pierce et al. 2012; Fig. 4.19) which can lead not only to the loss of habitat for pelagic or benthic fish and invertebrates but also an expansion of the range of mesopelagics that can withstand lower oxygen concentrations (Bograd et al. 2008). One such is the Humboldt squid (*Dosidicus gigas*), which has expanded its range from Baja California into Alaskan waters (Gilly and Markaida 2007).

4.9.2 Shoaling of Aragonite Saturation Horizon

One area where climate change is already affecting the CCS is in terms of changes in the carbon dioxide system. The continuing dissolution of carbon dioxide from the atmosphere into the ocean has reduced the pH in the surface layer by about 0.1 units since the late 18th century, so that the depths at which calcium carbonate begins to dissolve (the saturation horizon) is shallowing (see Chap. 1). This affects particularly the solubility of the form of calcium carbonate known as aragonite, which is the form that organisms such as oysters, corals and members of the plankton use to produce their carbonate skeletons. In the northeastern Pacific, the saturation horizon has shoaled so much (up to 100 m) that it is now only 100–300 m below the ocean surface (Feely et al. 2008).

A survey in 2007 along the length of the CCS from Queen Charlotte Sound, north of Vancouver Island, to the southern tip of Baja California showed that upwelling enabled water undersaturated in aragonite to rise up onto the continental shelf from its offshore depth of about 150 m (Feely et al. 2008; Fig. 4.25). Almost

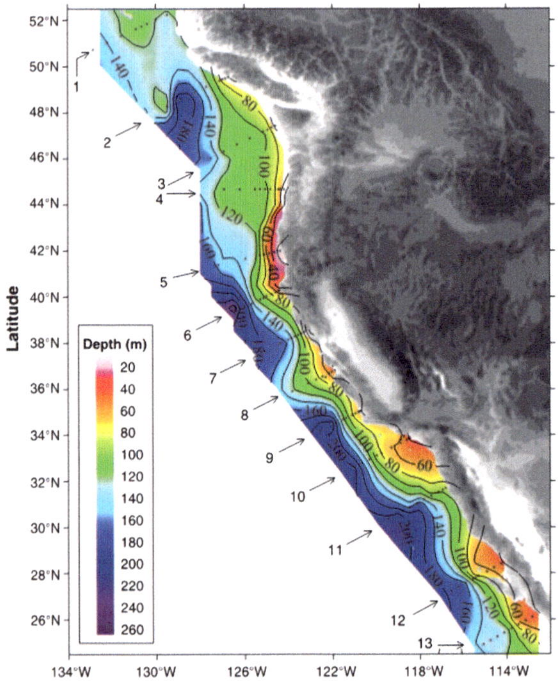

Fig. 4.25 Distribution of the depths of water undersaturated in aragonite (saturation <1.0, pH < 7.75) along the coastline of the California Current system. On line 5, the undersaturated water reached to the surface near the coast. *Black dots* mark station locations (from Feely et al. 2008)

the whole shelf was affected below about 80 m depth, and the undersaturated water even reached the surface some 40 km from the coast between 40° and 44° N. Feely et al. (2008) estimate that the anthropogenic effects of past CO_2 dissolution have brought the aragonite saturation horizon upwards by about 50 m over the past 200 years in this region. Given that these "corrosive" waters are now affecting the shelf during upwelling, the question arises as to how these new conditions will affect, in particular, larval and juvenile organisms that require carbonate for their skeletal material. While climate change impacts on the ecology of the CCS are far from being understood, is it clear that long-term multi-disciplinary monitoring programs, such as CALCOFI, are absolutely essential in the understanding of ecosystem responses in this new epoch that is sometimes referred to as the Anthropocene (e.g., Crutzen and Stoermer 2000).

4.10 Summary

The California Current system can be divided into four component sections. In the north, the straight coasts of Washington and Oregon are aligned roughly north-south, and upwelling is found in a band all along the narrow shelf, with productivity aided by nutrient input from the Strait of Juan de Fuca and the

Columbia River. Strongest winds are found off northern and central California, associated with major changes in coastal orientation. The southern California Bight is a shadow zone that acts as a large retention region for biological organisms, while Baja California again has upwelling concentrated at a number of headlands.

As the CALCOFI research program has been running now for over 60 years, and there have been a number of projects directed at specific aspects of the system, it is not surprising that we probably know more about the California Current system than any other upwelling region. A major driver of the system is the ENSO, which affects the southern portion of the region repeatedly on a timescale of several years, while in the north the seasonal variability of the North Pacific gyre is most important and the PDO causes decadal changes that can result in major incursions of cold water. Recently, it appears that the incidence of hypoxia is increasing along the Washington/Oregon coastline, but the consequences of this are not, at present, well understood.

One of the more obvious features of the California Current, especially in satellite imagery, is the presence of large (several hundred km) filaments that extend from the major capes and provide a route for the offshore transport of organic material. As with other eastern boundary systems, overfishing has had a deleterious effect on biomass, although switches between sardine and anchovy are natural and have been recorded over at least the last thousand years. The system has warmed significantly during the last century, although this is not consistent throughout the region, and the recent (2013–2014) southward advection of a warm water mass from the Alaskan Basin had major effects on both species composition and productivity. Changes in the depth of the aragonite saturation horizon are now affecting the shelf waters, but the long-term future remains unclear.

References

Adams, K.A., J.A. Barth, and F. Chan. 2013. Temporal variability of near-bottom dissolved oxygen during upwelling off central Oregon. *Journal of Geophysical Research* 118: 4839–4854. doi:10.1002/jgrc.20631.

Aksnes, D.L., and M.D. Ohman. 2009. Multi-decadal shoaling of the euphotic zone in the southern sector of the California Current System. *Ecological Research* 54: 1272–1281.

Auad, G., A. Miller, and E. DiLorenzo. 2006. Long-term forecast of oceanic conditions off California and their biological implications. *Journal of Geophysical Research* 111: C9008. doi:10.1029/2005JC003219/.

Badon-Dangon, A., J.M. Robles, and J. Garcia. 1989. Poleward flows off Mexico's Pacific coast. In *Poleward Flows along Eastern Ocean Boundaries*, ed. E.J. Neshyba, C.N.K. Moores, R.L. Smith, and R.T. Barber, 176–201. New York: Springer.

Bakun, A. 1990. Global climate change and intensification of coastal ocean upwelling. *Science* 247: 198–201.

Bakun, A. 2005. Regime shifts. In *The* Sea *(Volume 13, The Global Coastal Ocean: Multiscale Interdisciplinary Processes),* ed. A.R. Robinson and K.H. Brink, 971–1018. Harvard University Press.

Bakun, A., and C.S. Nelson. 1991. The seasonal cycle of wind stress curl in subtropical eastern boundary current regions. *Journal of Physical Oceanography* 21: 1815–1834.

Bakun, A., and R.H. Parrish. 1982. Turbulence, transport and pelagic fish in the California and Peru Current systems. *CALCOFI Report* 23: 99–112.

Banas, N.S., E.J. Lessard, R.M. Kudela, P. MacCready, T.D. Peterson, B.M. Hickey and E. Frame. 2009. Planktonic growth and grazing in the Columbia River plume region: A biophysical model study. *Journal of Geophysical Research* 114: C00B06. doi:10.1029/2008JC004993.

Barange, M., and S. Pillar. 1992. Cross-shelf circulation, zonation and maintenance mechanisms of *Nyctiphanes capensis* and *Euphausia hanseni* (Euphausiacea) in the northern Bengulea upwelling system. *Continental Shelf Research* 12: 1027–1042.

Bargu, S., T. Goldstein, K. Roberts, C. Li, and F. Gulland. 2012. *Pseudo-nitzschia* blooms, domoic acid, and related California sea lion strandings in Monterey Bay, California. *Marine Mammal Science* 28: 237–253.

Barry, J.P., H.G. Greene, D.L. Orange, C.H. Baxter, B.H. Robison, R.E. Kochevar, J.W. Nybakken, D.L. Reed, and C.M. McHugh. 1996. Biologic and geologic characteristics of cold seeps in Monterey Bay, California. *Deep-Sea Research* 43: 1739–1762.

Barth, J.A., and P.A. Wheeler. 2005. Introduction to special section: Coastal advances in shelf transport. *Journal of Geophysical Research* 110: C10S01. doi:10.1029/2005JC003124.

Barth, J.A., S.D. Pierce, and R.M. Castelao. 2005. Time-dependent, wind-driven flow over a shallow mid shelf submarine bank. *Journal of Geophysical Research* 110: C10505. doi:10.1029/2004JC002761.

Barth, J.A., B.A. Menge, J. Lubchenco, F. Chan, J.M. Bane, A.R. Kirincich, M.A. McManus, K.J. Nielsen, S.D. Pierce, and L. Washburn. 2007. Delayed upwelling alters nearshore coastal ocean ecosystem in the northern California Current. *Proceedings of the National Academy of Sciences* 104: 3719–3724.

Barton, E.D. 1985. Low-frequency variability of currents and temperatures on the Pacific continental shelf off northern Baja California, 1978 to 1979. *Continental Shelf Research* 4: 425–443.

Barton, E.D., and M.L. Argote. 1980. Hydrographic variability in an upwelling area off northern Baja California in June 1976. *Journal of Marine Research* 38: 631–649.

Batchelder, H.P., C.A. Edwards, and T.M. Powell. 2002. Individual-based models of copepod populations in coastal upwelling regions: Implications of physiologically and environmentally influenced diel vertical migration on demographic success and nearshore retention. *Progress in Oceanography* 53: 307–333.

Baumgartner, T.R., A. Soutar and V. Ferreira-Bartrina. 1992. Reconstruction of the history of Pacific sardine and northern anchovy populations over the past two millennia from sediments of the Santa Barbara Basin. *CALCOFI Reports* 33: 24–40.

Bernstein, R.L., L. Breaker, and R. Whritner. 1977. California Current eddy formation: Shipboard, air and satellite results. *Science* 195: 353–359.

Bjorkstedt, E.P., R. Goericke, A. McClatchie, et al. 2011. State of the California Current 2010–2011: Regionally variable responses to a strong (but ffieeting) La Niña. *CALCOFI Report* 52: 36–68.

Bograd, S.J., and R.J. Lynn. 2003. Anomalous subarctic influence in the southern California Current during 2002. *Geophysical Research Letters* 30: 8020. doi:10.1029/2003GL017446.

Bograd, S.J., C.G. Castro, E. Di Lorenzo, D.M. Palacios, H. Bailey, W. Gilly, and F.P. Chavez. 2008. Oxygen declines and the shoaling of the hypoxic boundary in the California Current. *Geophysical Research Letters* 35: L12607. doi:10.1029/2008GL034185.

Bograd, S.J., M.P. Buil, E.D. Lorenzo, C.G. Castro, I.D. Schroeder, R. Goericke, C.R. Anderson, C. Benitez-Nelson, and F.A. Whitney. 2015. Changes in source waters to the southern California Bight. *Deep-Sea Research II* 112: 42–52.

Bond, N.A., M.F. Cronin, H. Freeland, and N. Mantua. 2015. Causes and impacts of the 2014 warm anomaly in the NE Pacific. *Geophysical Research Letters* 42: 3414–3420. doi:10.1002/2015GL063306.

Botsford, L.W., C.L. Moloney, A. Hastings, J.L. Largier, T.M. Powell, K. Higgins, and J.F. Quinn. 1994. The influence of spatially and temporally varying oceanographic conditions on microplanktonic metapopulations. *Deep-Sea Research II* 41: 107–145.

Botsford, L.W., C.A. Lawrence, E.P. Dever, A. Hastings, and J. Largier. 2006. Effects of variable winds on biological productivity on continental shelves in coastal upwelling systems. *Deep-Sea Research II* 53: 3116–3140.

Bray, N.H., and C.L. Greengrove. 1993. Circulation over the shelf and slope off northern California. *Journal of Geophysical Research* 98: 18119–18146.

Brink, K.H., and T.J. Cowles. 1991. The Coastal Transition Zone program. *Journal of Geophysical Research* 96: 14637–14647.

Brink, K.H., F.F.G. Abrantes, P.A. Bernal, R.C. Dugdale, M. Estrada, L. Hutchings, R.A. Jahnke, P.J. Muller, and R.L. Smith. 1995. Group Report: How do coastal upwelling systems operate as integrated physical, chemical and biological systems and influence the geological record? The role of physical processes in defining the spatial structures of biological and chemical variables. *Environmental Sciences Research Report ES* 18: 103–124.

Bruland, K.W., M.C. Lohan, A.M. Aguilar-Islas, G.J. Smith, B. Sohst and A. Baptista. 2008. Factors influencing the chemistry of the near-field Columbia River plume: Nitrate, silicic acid, dissolved Fe, and dissolved Mn. *Journal of Geophysical Research* 113: C00B02. doi:10.1029/2007JC004702.

Brzezinski, M.A., D.R. Phillips, F.P. Chavez, G.E. Friederich, and R.C. Dugdale. 1997. Silica production in the Monterey, California, upwelling system. *Limnology and Oceanography* 42: 1694–1705.

Carr, M.-E., and E.J. Kearns. 2003. Production regimes in four eastern boundary current systems. *Deep-Sea Research II* 50: 3199–3221.

Cervantes-Duarte, R., R. Prego, S. Lopez-Lopez, F. Aguirre-Bahena, and N. Ospina-Alvarez. 2013. Annual patterns of nutrients and chlorophyll in a subtropical coastal lagoon under the upwelling influence (SW of Baja-California peninsula). *Estuarine, Coastal and Shelf Science* 120: 54–63.

Chan, F., J. Barth, J. Lubchenco, A. Kirincich, H. Weeks, W.T. Peterson, and B.A. Menge. 2008. Emergence of anoxia in the California Current large marine ecosystem. *Science* 319: 920.

Chapman, D.C. 1987. Application of wind-forced, long, coastal-trapped wave theory along the California coast. *Journal of Geophysical Research* 92: 1798–1816.

Chase, Z, P. Strutton and B. Hales. 2007. Iron links river runoff and shelf width to phytoplankton biomass along the U.S. west coast. *Geophysical Research Letters* 34. doi:10.1029/2006GL028069.

Chavez, F.P. 2005. Biological consequences of interannual to multidecadal variability. In *The* Sea *(Volume 13, The Global Coastal Ocean: Multiscale Interdisciplinary Processes)*, ed. A.R. Robinson and K.H. Brink, 643–677. Harvard University Press.

Chavez, F.P., R.T. Barber, P.M. Kosro, A. Huyer, S.R. Ramp, T.P. Stanton, and B. Rojas. 1991. Horizontal transport and distribution of nutrients in the Coastal Transition Zone off northern California: Effects on primary production, phytoplankton biomass, and species composition. *Journal of Geophysical Research* 96: 14833–14848.

Chavez, F.P., J.T. Pennington, C.G. Castro, J.P. Ryan, R.P. Michisaki, B. Schlining, P. Walz, K.R. Buck, A. MacFadyen, and C.A. Collins. 2002. Biological and chemical consequences of the 1997–1998 El Niño in central California waters. *Progress in Oceanography* 54: 205–232.

Chavez, F.P., J. Ryan, S.E. Lluch-Cota, and M. Ñiquen. 2003. From anchovies to sardines and back: Multidecadal change in the Pacific Ocean. *Science* 299: 217–221.

Chavez, F.P., M. Messié and J. Pennington. 2011. Marine primary production in relation to climate variability and change. *Annual Review of Marine Science* 3: 227–260.

Checkley, D.M., and J.A. Barth. 2009. Patterns and processes in the California Current System. *Progress in Oceanography* 83: 49–64.

Chelton, D.B. 1984. Seasonal variability of along-shelf geostrophic velocity off central California. *Journal of Physical Oceanography* 12: 757–784.

Chelton, D.B., P.A. Bernal, and J.A. McGowan. 1982. Large-scale interannual physical and biological interaction in the California Current. *Journal of Marine Research* 40: 1095–1125.

Collins, C.A., R.G. Paquette, and S.R. Ramp. 1996. Annual variability of ocean currents at 350 m depth over the continental slope off Point Sur, California. *CalCOFI Report* 37: 257–263.

Collins, C.A., J.T. Pennington, C.G. Castro, T.A. Roago, and F.P. Chavez. 2003. The California Current system off Monterey, California: Physical and biological coupling. *Deep-Sea Research II* 50: 2389–2404.

Combes, V., F. Chenillat, E. Di Lorenzo, P. Riviere, M.D. Ohman, and S.J. Bograd. 2013. Cross-shore transport variability in the California Current: Ekman upwelling vs. eddy dynamics. *Progress in Oceanography* 109: 78–89.

Connolly, P., B.M. Hickey, S.L. Geier, and W.P. Cochlan. 2010. Processes influencing seasonal hypoxia in the northern California Current system. *Journal of Geophysical Research* 115. doi:10.1029/2009JC005283.

Crutzen, P., and E. Stoermer. 2000. The "Anthropocene". *Global Change Newsletter* 41: 17–18.

Davis, A., and E. Di Lorenzo. 2015. Interannual forcing mechanisms of California Current transports II: Mesoscale eddies. *Deep-Sea Research II* 112: 31–41.

Davis, K.A., N.S. Banas, S.N. Giddings, S.A. Siedlecki, P. MacCready, E.J. Lessard, R.M. Kudela, and B.M. Hickey. 2014. Estuary-enhanced upwelling of marine nutrients fuels coastal productivity in the U.S. Pacific Northwest. *Journal of Geophysical Research* 119. doi:10.1002/2014JC010248.

Davis, R.E. 1985a. Drifter observations of coastal surface currents during CODE: The method and descriptive view. *Journal of Geophysical Research* 90: 4741–4755.

Davis, R.E. 1985b. Drifter observations of coastal surface currents during CODE: The statistical and dynamical views. *Journal of Geophysical Research* 90: 4756–4772.

Dayton, P.K., M.J. Tegner, P.B. Edwards, and K.L. Riser. 1999. Temporal and spatial scales of kelp demography: The role of oceanographic climate. *Ecological Monographs* 69: 219–250.

Denbo, D.W., and J.S. Allen. 1987. Large-scale response to atmospheric forcing of shelf currents and coastal sea level off California and Oregon: May-July 1981 and 1982. *Journal of Geophysical Research* 92: 1757–1782.

Dever, E.P., and S.J. Lentz. 1994. Heat and salt balances over the northern California shelf in winter and spring. *Journal of Geophysical Research* 99: 16001–16017.

Dever, E.P., C.E. Dorman, and J.L. Largier. 2006. Surface boundary-layer variability off northern California, USA, during upwelling. *Deep-Sea Research II* 53: 2887–2905.

Di Lorenzo, E., A.J. Miller, N. Schneider, and J.C. McWilliams. 2005. The warming of the California Current System: Dynamics and ecosystem implications. *Journal of Physical Oceanography* 35: 336–362.

Di Lorenzo, E., N. Schneider, K.M. Cobb, P.J.S. Franks, K. Chhak, A.J. Miller, J.C. McWilliams, S.J. Bograd, H. Arango, E. Curchitser, T.M. Powell, and P. Riviere. 2008. North Pacific Gyre Oscillation links ocean climate and ecosystem change. *Geophysical Research Letters* 35: L08607. doi:10.1029/2007GL032838.

Doney, S.C., M. Ruckelshaus, J.E. Duffy, J.P. Barry, F. Chan, C.A. English, H.M. Galindo, J.M. Grebmeier, A.B. Hollowed, N. Knowlton, J. Polovina, N.N. Rabalais, W.J. Sydeman, and L.D. Talley. 2012. Climate change impacts on marine ecosystems. *Annual Review of Marine Science* 4: 11–37.

Dugdale, R.C., and J.J. Goering. 1967. Uptake of new and regenerated forms of nitrogen in primary productivity. *Limnology and Oceanography* 12: 196–206.

Dugdale, R., F. Wilkerson, V. Hogue, and A. Marchi. 2006. Nutrient controls on new production in the Bodega Bay, California, coastal upwelling plume. *Deep-Sea Research II* 53: 3049–3062.

Durazo, R. 2009. Climate and upper ocean variability off Baja California, Mexico: 1997–2008. *Progress in Oceanography* 83: 361–368.

Durazo, R., and T. Baumgartner. 2002. Evolution of oceanographic conditions off Baja California: 1997–1999. *Progress in Oceanography* 54: 7–31.

Enriquez, A.G., and C.A. Friehe. 1995. Effects of wind stress and wind stress curl variability on coastal upwelling. *Journal of Physical Oceanography* 25: 1651–1671.

Eppley, R.W. 1992. Chlorophyll, photosynthesis and new production in the southern California Bight. *Progress in Oceanography* 30: 117–150.

Feely, R.A., C.L. Sabine, J.M. Hernandez-Ayon, D. Ianson, and B. Hales. 2008. Evidence for upwelling of corrosive "acidified" water onto the continental shelf. *Science* 320: 1490–1492.

Field, J.C., R.C. Francis, and A. Strom. 2001. Towards a fisheries ecosystem plan for the Northern California Current. *CALCOFI Report* 42: 74–87.
Field, D., D. Cayan, and F. Chavez. 2006. Secular warming in the California Current and North Pacific. *CALCOFI Report* 47: 1–17.
Flores-Trujillo, J.G., J. Helenes, J.C. Herguera, and E. Orellana-Zepeda. 2009. Palynological record (1483–1994) of *Gymnodinium catenatum* in Pescadero Basin, southern Gulf of California, Mexico. *Marine Micropaleontology* 73: 80–89.
Forbes, J.R., and R.E. Waters. 1993a. Phytoplankton species composition and abundance along the Pacific coast of Canada, 1979–1989. Vol. 1: 1979–1984. *Canadian Data Report on Hydrography and Ocean Science* 117(1): 212.
Forbes, J.R., and R.E. Waters. 1993b. Phytoplankton species composition and abundance along the Pacific coast of Canada, 1979–1989. Vol. 2: 1985–1989. *Canadian Data Report on Hydrography and Ocean Science* 117(2): 247.
Frame, E.R., and E.J. Lessard. 2009. Does the Columbia River plume influence phytoplankton community structure along the Washington and Oregon coasts? *Journal of Geophysical Research* 114. doi:10.1029/2008JC004999.
Freeland, H.J., and K.L. Denman. 1982. A topographically controlling upwelling center off southern Vancouver Island. *Journal of Marine Research* 40: 1069–1093.
Freeland, H.J., G. Gatien, A. Huyer, and R.L. Smith. 2003. Cold halocline in the northern California Current: An invasion of subarctic water. *Geophysical Research Letters* 30. doi:10.1029/2002GL016663.
Gan, J., and J.S. Allen. 2005. Modeling upwelling circulation off the Oregon coast. *Journal of Geophysical Research* 110. doi:10.1029/2004JC002692.
García-Reyes, M., and J. Largier. 2010. Observations of increased wind-driven coastal upwelling off central California. *Journal of Geophysical Research* 115: C04011. doi:10.1029/2009JC005576.
Gaxiola-Castro, G., B.E. Laveniegos, A. Martinez, R. Castro and T.L. Espinosa-Carreon. 2010. Pelagic ecosystem response to climate variability in the Pacific Ocean off Baja California. In *Dynamic of the Pelagic Ecosystem off Baja California, 1997–2007*, ed. G. Gaxiola-Castro and R. Durazo. INTECH Open Access publisher http://cdn.intechopen.com/pdfs/11442/InTech-Pelagic_ecosystem_response_to_climate_variability_in_the_pacific_ocean_off_baja_california.pdf. Accessed 4 April 2016.
Gilly, W.F., and U. Markaida. 2007. Perspectives on *Dosidicus gigas* in a changing world. In *The Role of Squid in Open Ocean Ecosystems*, ed. R.J. Olson and J.W. Young, 81–90. Report 24, Global Ocean Ecosystem Dynamics, Honolulu, Hawaii.
Goericke, R., and M.D. Ohman. 2015. Introduction to CCE-LTER: Responses of the California Current Ecosystem to climate forcing. *Deep-Sea Research II* 112: 1–5.
Goericke, R., S.J. Bograd, and D.S. Grundle. 2015. Denitrification and flushing of the Santa Barbara Basin bottom waters. *Deep-Sea Research II* 112: 53–60.
Gonzalez-Rodriguez, E., A. Trasviña-Castro, G. Gaxiola-Castro, L. Zamudio, and R. Cervantes-Duarte. 2012. Net primary productivity, upwelling and coastal currents in the Gulf of Ulloa, Baja California, Mexico. *Ocean Sciences* 8: 703–711.
Graham, W.M., and J.L. Largier. 1997. Upwelling shadows as nearshore retention sites: The example of northern Monterey Bay. *Continental Shelf Research* 17: 509–532.
Grantham, B.A., F. Chan, K.J. Nielsen, D.S. Fox, J.A. Barth, A. Huyer, J. Lubchenco, and B.A. Menge. 2004. Upwelling-driven nearshore hypoxia signals ecosystem and oceanographic changes in the northeast Pacific. *Nature* 429: 749–754.
Hales, B., J. Moum, P. Covert, and A. Perlin. 2005. Irreversible nitrate fluxes due to turbulent mixing in a coastal upwelling system. *Journal of Geophysical Research* 110. doi:10.1029/2004JC002685.
Halpern, D. 1976. Structure of a coastal upwelling event: Observations off Oregon during 1973. *Deep-Sea Research* 23: 495–508.
Hare, S.R., N.J. Mantua, and R.C. Francis. 1999. Inverse production regimes: Alaskan and west coast salmon. *Fisheries* 24: 6–14.

Henson, S.A., and A.C. Thomas. 2007. Phytoplankton scales of variability in the California Current system: 2. Latitudinal variability. *Journal of Geophysical Research* 112. doi:10.1029/2006JC004040.

Hewitt, R. 1981. Eddies and speciation in the California Current. *CalCOFI Reports* 22: 96–98.

Hickey, B.M. 1979. The California Current system: Hypotheses and facts. *Progress in Oceanography* 8: 191–279.

Hickey, B.M. 1992. Circulation over the Santa Monica-San Pedro Basin and shelf. *Progress in Oceanography* 30: 37–115.

Hickey, B.M., and N.S. Banas. 2008. Why is the northern end of the California Current system so productive? *Oceanography* 21(4): 90–107.

Hickey, B.M., R. McCabe, S. Geier, E. Dever, and N. Kachel. 2009. Three interacting freshwater plumes in the northern California Current system. *Journal of Geophysical Research* 114. doi:10.1029/2008JC004907.

Hickey, B.A. 1989. Poleward flow at the northern and southern boundaries of the U.S. west coast. In *Poleward Flows along Eastern Boundaries*, ed. S.J. Neshyba, C.N.K. Mooers, R.L. Smith and R.T. Barber, 160–175. Springer, New York.

Hickey, B.A. 1998. Coastal oceanography of western North America from the tip of Baja California to Vancouver Island. In *The Sea*, ed. A.R. Robinson and K.H. Brink, Vol. 11, 345–393. John Wiley, New York.

Hickey, B., S. Geier, N. Kachel, and A. MacFayden. 2005. A bi-directional river plume: The Columbia in summer. *Continental Shelf Research* 25: 1631–1656.

Hickey, B., A. MacFayden, W. Cochlan, R. Kudela, K. Bruland and C. Trick. 2006. Evolution of chemical, biological and physical water properties in the northern California Current in 2005: Remote or local wind forcing? *Geophysical Research Letters* 33: C00B03. doi:10.1029/2006GL026782.

Hopcroft, R.R., C. Clarke, and F.P. Chavez. 2002. Copepod communities in Monterey Bay during the 1997–1999 El Niño and La Niña. *Progress in Oceanography* 54: 251–264.

Hsieh, W.W., D.M. Ware, and R.E. Thomson. 1995. Wind-induced upwelling along the west coast of North America 1899–1988. *Canadian Journal of Fisheries and Aquatic Sciences* 52: 325–334.

Hsieh, C.-H., H. Kim, W. Watson, E. Di Lorenzo, and G. Sugihara. 2009. Climate-driven changes in abundance and distribution of larvae of oceanic fishes in the southern California region. *Global Change Biology* 15: 2137–2152.

Hutchins, D., and K. Bruland. 1998. Iron-limited diatom growth and Si: N uptake ratios in a coastal upwelling regime. *Nature* 393: 561–564.

Huyer, A. 1983. Upwelling in the California Current system. *Progress in Oceanography* 12: 259–284.

Huyer, A., and R.L. Smith. 1985. The signature of El Niño off Oregon, 1982–1983. *Journal of Geophysical Research* 90: 7133–7142.

Huyer, A., E.J.C. Sobey, and R.L. Smith. 1979. The spring transition in currents over the Oregon continental shelf. *Journal of Geophysical Research* 84: 6995–7011.

Huyer, A., P.M. Kosro, and S.J. Lentz. 1989. Poleward flow in the California Current system. In *Poleward Flows along Eastern Ocean Boundaries*, ed. E.J. Neshyba, C.N.K. Moores, R.L. Smith, and R.T. Barber, 142–156. New York: Springer.

Huyer, A., P.M. Kosro, J. Fleischbein, S.R. Ramo, T. Stanton, L. Washburn, F. Chavez, T. Cowles, S. Pierce, and R.L. Smith. 1991. Currents and water masses of the coastal transition zone off northern California. *Journal of Geophysical Research* 96: 14809–14832.

Huyer, A., J. Fleischbein, J. Keister, P. Kosro, N. Perlin, R. Smith, and P. Wheeler. 2005. Two coastal upwelling domains in the northern California Current system. *Journal of Marine Research* 63: 901–929.

Isaacs, J.D., A. Fleminger and J.K. Miller. 1969. *Distributional atlas of zooplankton biomass in the California Current region: Spring and Fall 1955–1959* (No. 10). Data Collection and Processing Group, Marine Life Research Program, Scripps Institution of Oceanography.

Johnson, M.A., and J.J. O'Brien. 1990a. The northeast Pacific Ocean response to the 1982–1983 El Niño. *Journal of Geophysical Research* 95: 7155–7166.

Johnson, M.A., and J.J. O'Brien. 1990b. The role of coastal Kelvin waves on the northeast Pacific Ocean. *Journal of Marine Systems* 1: 29–38.

Johnston, T., and D.L. Rudnick. 2015. Trapped diurnal internal tides, propagating semidiurnal internal tides, and mixing estimates in the California Current from sustained glider operations, 2006–2012. *Deep-Sea Research II* 112: 61–78.

Jumars, P.A. 1993. *Concepts in Biological Oceanography: An Interdisciplinary Primer*. New York: Oxford University Press. 348 pp.

Kadko, D.C., L. Washburn, and B. Jones. 1991. Evidence of subduction within cold filaments of the northern California transition zone. *Journal of Geophysical Research* 96: 14909–14926.

Kamykowski, D. 1972. *Some Physical and Chemical Aspects of Phytoplankton Ecology of La Jolla Bay*. Ph.D. Thesis, University of California at San Diego.

Kamykowski, D., and S.J. Zentara. 1990. Hypoxia in the world ocean as recorded in the historical data set. *Deep-Sea Research* 37: 1861–1874.

Keeling, R.F., and H.E. Garcia. 2002. The change in oceanic O2 inventory associated with recent global warming. *Proceedings of the National Academy of Sciences* 99: 7848–7853.

Kennicutt, M.C., J.M. Brooks, S.A. Macko, R.R. Bidigare, S.J. McDonald, and D. Adkinson. 1989. An upper slope "cold seep" community in northern California. *Limnology and Oceanography* 34: 635–640.

Kim, H.-J., A. Miller, J. McGowan, and M. Carter. 2009. Coastal phytoplankton blooms in the Southern California Bight. *Progress in Oceanography* 82: 137–147.

Klyashtorin, L., and A. Lyabushin. 2007. *Cyclic Climate Changes and Fish Productivity*. Moscow: VNIRO.

Kudela, R.M., and T.D. Peterson. 2009. Influence of a buoyant river plume on phytoplankton nutrient dynamics: What controls standing stocks and productivity? *Journal of Geophysical Research* 114: C00B11. doi:10.1029/2008JC004913.

Kudela, R.M., N. Garfield, and K.W. Bruland. 2006. Bio-optical signatures and biogeochemistry from intense upwelling and relaxation in coastal California. *Deep-Sea Research II* 53: 2999–3022.

Kudela, R.M., N.S. Banas, J.A. Barth, E.R. Frame, D.A. Jay, J.L. Largier, E.J. Lessard, T.D. Peterson, and A.J. Vander Woude. 2008. New insights into the controls and mechanisms of plankton productivity of the northern California Current system. *Oceanology* 21(4): 46–59.

Kudela, R.M., A.R. Horner-Devine, N.S. Banas, B.M. Hickey, T.D. Peterson, R.M. McCabe, E.J. Lessard, E. Frame, K.W. Bruland, D.A. Jay, J.O. Peterson, W.T. Peterson, P.M. Kosro, S.L. Palacios, M.C. Lohan, and E.P. Dever. 2010. Multiple trophic levels fueled by recirculation in the Columbia River plume. *Geophysical Research Letters* 37: L18607. doi:10.1029/2010GL044342.

Ladah, L.B. 2003. The shoaling of nutrient-enriched subsurface waters as a mechanism to sustain primary productivity off central Baja California during El Niño winters. *Journal of Marine Systems* 42: 145–152.

Landry, M.R., J.R. Postel, W.K. Peterson, and J. Newman. 1989. Broad scale patterns in the distribution of hydrographic variables. In *Coastal Oceanography of Washington and Oregon*, ed. M.R. Landry, and B.A. Hickey, 1–41. Amsterdam: Elsevier Press.

Largier, J.L., B.A. Magnell, and C.D. Winant. 1993. Subtidal circulation over the northern California shelf. *Journal of Geophysical Research* 98: 18147–18180.

Largier, J.L., C.A. Lawrence, M. Roughan, D.M. Kaplan, E.P. Dever, C.E. Dorman, R.M. Kudela, S.M. Bollens, F.P. Wilkerson, R.C. Dugdale, L.W. Botsford, N. Garfield, B. Kuebel Cervantes, and D. Koracin. 2006. WEST: A northern California study of the role of wind-driven transport in the productivity of coastal plankton communities. *Deep-Sea Research II* 53: 2833–2849.

Lavaniegos, B.E., and M.D. Ohman. 2003. Long term changes in pelagc tunicates of the California Current. *Deep-Sea Research II* 50: 2473–2498.

Lavaniegos, B.E., and M.D. Ohman. 2007. Coherence of long-term variations of zooplankton in two sectors of the California Current System. *Progress in Oceanography* 75: 42–69.

Lenarz, W. 1986. The rockfish fishery, a California perspective. In *Rockfish: A focus for research?*, ed. R. Amidei, 3–9. Californian Sea Grant College Program Report, TCSGCP-015, 71 pp.

Lewitus, A.J., R.A. Horner, D.A. Caron, E. Garcia-Mendoza, B.M. Hickey, M. Hunter, D.D. Huppert, R.M. Kudela, G.W. Langlois, J.L. Largier, E.J. Lessard, R. RaLonde, J.E.J. Rensel, P.G. Strutton, V.L. Trainer, and J.F. Tweddie. 2012. Harmful algal blooms along the North American west coast region: History, trends, causes, and impacts. *Harmful Algae* 19: 133–159.

Lu, B.-W., D.L. Mackas, and D.F. Moore. 2003. Cross-shore separation of adult and juvenile euphausiids in a shelf-break alongshore current. *Progress in Oceanography* 57: 381–404.

Lynn, R.J., and S.J. Bograd. 2002. Dynamic evolution of the 1997–1998 El Niño-La Niña cycle in the southern California Current System. *Progress in Oceanography* 54: 59–75.

Lynn, R.S., and J.J. Simpson. 1987. The California Current system: The seasonal variability of its physical characteristics. *Journal of Geophysical Research* 92: 12947–12966.

Lynn, R.S., and J.J. Simpson. 1990. The flow of the undercurrent over the continental borderland off southern California. *Journal of Geophysical Research* 95: 12995–13009.

Lynn, R.J., K.A. Bliss and L.E. Eber. 1982. Vertical and horizontal distributions of seasonal mean temperature, salinity, sigma-t, stability, dynamic height, oxygen, and oxygen saturation in the California Current, 1950–1978. *CALCOFI Atlas* 30: 432.

Lynn, R.J., T. Baumgartner, J. Garcia, et al. 1998. The state of the California Current, 1997–1998: Transition to El Nino conditions. *CALCOFI Report* 39: 25–49.

Lynn, R.J., S.J. Bograd, T.K. Chereskin, and A. Huyer. 2003. Seasonal renewal of the California Current: The spring transition off California. *Journal of Geophysical Research* 108: 3279. doi:10.1029/2003JCoo1787.

MacFadyen, A., B.M. Hickey, and M.G.G. Foreman. 2005. Transport of surface waters from the Juan de Fuca eddy region to the Washington coast. *Continental Shelf Research* 25: 2008–2021.

MacFadyen, A., B.M. Hickey, and W.P. Cochlan. 2008. Influences of the Juan de Fuca eddy on circulation, nutrients and phytoplankton production in the northern California Current system. *Journal of Geophysical Research* 113. doi:10.1029/2007JC004412.

MacIsaac, J.J., R.C. Dugdale, R.T. Barber, D. Blasco, and T.T. Packard. 1985. Primary production cycle in an upwelling center. *Deep-Sea Research* 32: 503–529.

Mackas, D.L. 1992. The seasonal cycle of zooplankton off southwestern British Columbia: 1979–89. *Canadian Journal of Fisheries and Aquatic Sciences* 49: 903–921.

Mackas, D.L. 2006. Interdisciplinary oceanography of the western North American continental margin: Vancouver Island to the tip of Baja California. In *The Sea*, ed. A.R. Robinson and K. Brink, Vol. 14, 441–501. Harvard University Press.

Mackas, D.L., and P.J. Harrison. 1997. Nitrogenous nutrient sources and sinks on the Juan de Fuca Strait/Strait of Georgia/Puget Sound estuarine system: Assessing the potential for eutrophication. *Estuarine, Coastal and Shelf Science* 44: 1–21.

Mackas, D.L., L. Washburn, and S.L. Smith. 1991. Zooplankton community pattern associated with a California Current cold filament. *Journal of Geophysical Research* 96: 14781–14798.

Mackas, D.L., M. Saunders, R. Kieser, R.M. Brown, D.R. Yelland, and D.F. Moore. 1997. Aggregation of euphausiids and hake along the outer continental shelf off Vancouver Island. *Canadian Journal of Fisheries and Aquatic Sciences* 54: 2080–2096.

Mackas, D.L., W.T. Peterson, M.D. Ohman, and B.E. Lavaniegos. 2006. Zooplankton anomalies in the California Current system before and during the warm ocean conditions of 2005. *Geophysical Research Letters* 33: L22507. doi:10.1029/2006GL027930.

Mackas, D., S. Batten, and M. Trudel. 2007. Effects on zooplankton of a warmer ocean: Recent evidence from the Northeast Pacific. *Progress in Oceanography* 75: 223–252.

Mantua, N.J., and S.R. Hare. 2002. The Pacific Decadal Oscillation. *Journal of Oceanography* 58: 35–44.

Mantua, N.J., et al. 1997. A Pacific interdecadal climate oscillation with impacts on salmon production. *Bulletin of the American Meteorology Society* 78: 1069–1079.

Martin, J., and S. Fitzwater. 1988. Iron deficiency limits phytoplankton growth in the northeast Pacific subarctic. *Nature* 331: 341–343.

Masson, D., and P.F. Cummins. 2004. Numerical simulations of a buoyancy-driven coastal countercurrent off Vancouver Island. *Journal of Physical Oceanography* 29: 418–435.

McCabe, R.M., B.M. Hickey, E.P. Dever, and P. MacCready. 2015. Seasonal cross-shelf flow structure, upwelling relaxation, and the alongshelf pressure gradient in the northern California Current system. *Journal of Physical Oceanography* 209–227. doi:10.1175/JPO-D-14-0025.1.

McGowan, J.A., D.R. Cayan, and L.M. Dorman. 1998. Climate-ocean variability and ecosystem response in the northeast Pacific. *Science* 281: 210–217.

Meyer, K.F., H. Sommer, and P. Schoenholz. 1928. Mussel poisoning. *Journal of Preventive Medicine* 2: 365–394.

Miller, A.J., H. Song, and A.C. Subramanian. 2015. The physical oceanographic environment during the CCE-LTER years: Changes in climate and concepts. *Deep-Sea Research II* 112: 6–17.

Mills, E.L. 1991. The oceanography of the Pacific: George F. McEwan, H.U. Sverdrup and the origin of physical oceanography on the west coast of North America. *Annals of Science* 48: 241–266.

Monteiro, P.M.S. 2010. The Benguela Current system. In *Carbon and Nutrient Fluxes in Continental Margins*, ed. K.K. Liu, L. Atkinson, R. Quinones, and L. Talaue-McManus, 65–78. Berlin: Springer.

Narimousa, S., and T. Maxworthy. 1989. Applications of a laboratory model to the interpretation of satellite and field observations of coastal upwelling. *Dynamics of Atmospheres and Oceans* 13: 1–46.

Oke, P.R., J.S. Allen, R.N. Miller, and G.D. Egbert. 2002. A modeling study of the three-dimensional continental shelf off Oregon. Part II: Dynamical analysis. *Journal of Physical Oceanography* 32: 1383–1403.

Olivieri, R.A., and F.P. Chavez. 2000. A model of plankton dynamics for the coastal upwelling system of Monterey Bay, California. *Deep-Sea Research II* 47: 1077–1106.

Palacios, D.P., S.J. Bograd, R. Mendelssohn, and F.B. Schwing. 2004. Long-term and seasonal trends in stratification in the California Current, 1950–1993. *Journal Geophysical Research* 109: C10016.

Parrish, R.H., C.S. Nelson, and A. Bakun. 1981. Transport mechanisms and reproductive success of fishes in the California Current. *Biological Oceanography* 1: 175–203.

Pennington, J.T., G.E. Friederich, C.G. Castro, C.A. Collins, W.W. Evans, and F.P. Chavez. 2010. The northern and central California coastal upwelling system. In *Carbon and Nutrient Fluxes in Continental Margins*, ed. K.K. Liu, L. Atkinson, R. Quinones, and L. Talaue-McManus, 29–44. Berlin: Springer.

Perez-Brunius, P., M. Lopez, and J. Pineda. 2006. Hydrographic conditions near the coast of northwestern Baja California: 1997–2004. *Continental Shelf Research* 26: 885–901.

Perez-Brunius, P., M. Lopez, A. Pares-Sierra and J. Pineda. 2007. Comparison of upwelling indices off Baja California derived from three different wind data sources. *CALCOFI Reports* 48: 204–214.

Peterson, W.T. 1998. Life cycle strategies of copepods in coastal upwelling zones. *Journal of Marine Systems* 15: 313–326.

Peterson, W.T., and C.B. Miller. 1977. The seasonal cycle of zooplankton abundance and species composition along the central Oregon coast. *Fisheries Bulletin of the U.S.* 75: 717–724.

Peterson, W.T., and F.B. Schwing. 2003. A new climate regime in northeast Pacific ecosystems. *Geophysical Research Letters* 30: 1896.

Peterson, W.T., R. Emmett, R. Goericke, et al. 2006. The state of the California Current, 2005–2006: Warm in the north, cold in the south. *CALCOFI Reports* 47: 30–74.

Pierce, S.D., R.L. Smith, P.M. Kosro, J.A. Barth, and C.D. Wilson. 2000. Continuity of the poleward undercurrent along the eastern boundary of the mid-latitude North Pacific. *Deep-Sea Research II* 47: 811–829.

Pierce, S.D., J.A. Barth, R.K. Shearman, and A. Erofeev. 2012. Declining oxygen in the northeast Pacific. *Journal of Physical Oceanography* 42: 495–501.

Powell, J.R., and M.D. Ohman. 2015. Co-variability of zooplankton gradients with glider-detected density fronts in the southern California Current System. *Deep-Sea Research II* 112: 79–90.

Rebstock, G.A. 2002. Climatic regime shifts and decadal-scale variability on calanoid copepod populations off southern California. *Global Change Biology* 8: 71–89.

Reid, J.L. 1962. On circulation, phosphate-phosphorus content, and zooplankton volumes in the upper part of the Pacific Ocean. *Limnology and Oceanography* 7: 287–306.
Reid, J.L. 1997. On the total geostrophic circulation of the Pacific Ocean: Flow patterns, tracers and transports. *Progress in Oceanography* 39: 263–352.
Reid, J.L., and R.A. Schwartzlose. 1962. Direct measurements of the Davidson Current off central California. *Journal of Geophysical Research* 67: 2491–2497.
Reid, J.L., G.I. Roden and J.G. Wyllie. 1958. Studies of the California Current system. *CALCOFI Report* 5: 28–57.
Roemmich, D., and J. McGowan. 1995. Climatic warming and the decline of zooplankton in the California Current. *Science* 267: 1324–1326.
Rosenfeld, L.K., and R.C. Beardsley. 1987. Barotropic semidiurnal tidal currents off northern California during the coastal ocean dynamics experiment (CODE). *Journal of Geophysical Research* 92: 1721–1732.
Rudnick, D.L., and R.E. Davis. 1988. Mass and heat budgets on the Northern California shelf. *Journal of Geophysical Research* 93: 14013–14024.
Ruttenberg, K., and S. Dyhrman. 2005. Temporal and spatial variability of dissolved organic and inorganic phosphorus, and metrics of phosphorus bioavailability in an upwelling-dominated coastal system. *Journal of Geophysical Research* 110. doi:10.1029/2004JC002837.
Rykaczewski, R.R., and D.M. Checkley. 2008. Influence of ocean winds on the pelagic ecosystem in upwelling regimes. *Proceedings of the National Academy of Sciences* 105: 1965–1970.
Santora, J.A., and W.J. Syderman. 2015. Persistence of hotspots and variability of seabird species richness and abundance in the southern California Current. *Ecosphere* 6. doi:10.1890/ES14-00434.1.
Schwing, F.B., and R. Mendelssohn. 1997. Increased coastal upwelling in the California Current System. *Journal of Geophysical Research* 102: 3421–3438.
Schwing, F.B., R. Mendelssohn, S.J. Bograd, J.E. Overland, M. Wang, and S.-I. Ito. 2010. Climate change, teleconnection patterns, and regional processes forcing marine populations in the Pacific. *Journal of Marine Systems* 79: 245–257.
Shumway, S.E., S.M. Allen, and P.D. Boersma. 2003. Marine birds and harmful algal blooms: Sporadic victims or under-reported events. *Harmful Algae* 2: 1–17.
Siedlecki, S.A., N.S. Banas, K.A. Davis, S. Giddings, B.M. Hickey, P. MacCready, T. Connolly, and S. Geler. 2014. Seasonal and interannual oxygen variability on the Washington and Oregon continental shelves. *Journal of Geophysical Research* 120: 608–633. doi:10.1002/2014JC010254.
Simpson, J.J. 1992. Response of the southern California Current system to the mid-latitude North Pacific coastal warming events of 1982–1983 and 1940–1941. *Fisheries Oceanography* 1: 57–77.
Skrivanek, A., and I.L. Hendy. 2015. A 500 year climate catch: Pelagic fish scales and paleoproductivity in the Santa Barbara Basin from the Medieval climate anomaly to the Little Ice Age (AD 1000–1500). *Quaternary International* 387: 36–45.
Slaughter, A.M., S.M. Bollens, and G.R. Bollens. 2006. Grazing impact of mesozooplankton in an upwelling region off northern California, 2000–2003. *Deep-Sea Research II* 53: 3099–3115.
Smith, P.E., and H.G. Moser. 2003. Long-term trends and variability in the larvae of Pacific sardine and associated fish species of the California Current region. *Deep-Sea Research II* 50: 2519–2536.
Smith, R.L. 1974. A description of current, wind and sea level variations during coastal upwelling off the Oregon coast. *Journal of Geophysical Research* 79: 435–443.
Soutar, A., and J.D. Isaacs. 1969. History of fish populations inferred from fish scales in anaerobic sediments off California. *CALCOFI Report* 13: 63–70.
Soutar, A., and J.D. Isaacs. 1974. Abundance of pelagic fish during the 19th and 20th centuries as recorded in anaerobic sediments off the Californias. *Fisheries Bulletin* 72: 257–273.
Starr, K. 2005. *California: A History*, 13. New York: Modern Library.

Strub, P.T., and C. James. 2003. Altimeter estimates of anomalous transports into the northern California Current during 2000–2002. *Geophysical Research Letters* 30: 8025. doi:10.1029/2003GL017513.

Strub, P.T., J.S. Allen, A. Huyer, R.L. Smith, and R.C. Beardsley. 1987. Seasonal cycles of currents, temperatures, winds and sea level over the northeast Pacific continental shelf. *Journal of Geophysical Research* 92: 1507–1526.

Strub, P.T., C. James, A.C. Thomas, and M.R. Abbott. 1990. Seasonal and non-seasonal variability of satellite-derived surface pigment concentration in the California Current. *Journal of Geophysical Research* 95: 11501–11530.

Strub, P.T., P.M. Kosro, and A. Huyer. 1991. The nature of the cold filaments in the California Current system. *Journal of Geophysical Research* 96: 14743–14768.

Sydeman,W., R. Bradley, P. Warzybok, C. Abraham, J. Jahncke, et al. 2006. Planktivorous auklet *Ptychoramphus aleuticus* responses to ocean climate, 2005: Unusual atmospheric blocking? *Geophysical Research Letters* 33: L22S09.

Sydeman,W., K. Mills, J. Santora and S. Thompson. 2009. Seabirds and climate in the California Current: a synthesis of change. *CalCOFI Report* 50: 82–104.

Syderman, W.J., S.A. Thompson, J.A. Santora, J.A. Koslow, R. Goericke, and M.D. Ohman. 2015. Climate-ecosystem change off southern California: Time-dependent seabird predator-prey numerical responses. *Deep-Sea Research II* 62: 158–170.

Thomas, A.C., P. Brickley, and R. Weatherbee. 2009. Interannual variability in chlorophyll concentrations in the Humboldt and California Current systems. *Progress in Oceanography* 83: 386–392.

Thompson, R.A. 1896. *The Russian settlement in California known as Fort Ross, founded 1812, abandoned 1841: Why the Russians came and why they left*. Western Americana, frontier history of the trans-Mississippi West, 1550–1900. Sonoma Democrat Publishing Company. Santa Rose, CA, p. 5.

Thorade, H. 1909. Uber die kalifornische Meeresstromungen, Oberflachen-temperaturen und Stromungen, an der westkuste Nordamerikas. *Annalen der Hydrographie und Maritimen Meteorologie* 37(17–34): 63–76.

Tibby, R.B. 1941. The water masses off the west coast of North America. *Journal of Marine Research* 4: 112–121.

Tomczak, M., and S.J. Godfrey. 1994. *Regional Oceanography: An Introduction*. London: Elsevier. 422 pp.

Tont, S. 1987. Variability of diatom species populations from days to years. *Journal of Marine Research* 45: 985–1006.

Traganza, E.D., D.A. Nestor, and A.K. McDonald. 1980. Satellite observations of a nutrient upwelling off the coast of California. *Journal of Geophysical Research* 85: 4104–4106.

Trasvina, A., M. Ortiz-Figueroa, H. Herrera, M.A. Cosio, and E. Gonzalez. 2003. "Santa Ana" winds and upwelling filaments off northern Baja California. *Dynamics of Atmospheres and Oceans* 37: 113–129.

Trowbridge, J.H., and S.J.K. Lentz. 1991. Asymmetric behavior of an oceanic boundary layer above a sloping bottom. *Journal of Physical Oceanography* 21: 1171–1185.

U.S. GLOBEC. 1994. Eastern Boundary Current Program: A Science Plan for the California Current. *U.S. Global Ocean Ecosystems Dynamics Report* 11: 134.

Vander Woude, A., J.L. Largier, and R.M. Kudela. 2006. Nearshore retention of upwelled waters north and south of Point Reyes (northern California)—patterns of surface temperature and chlorophyll observed in CoOP WEST. *Deep-Sea Research II* 53(25–26): 2985–2998.

Venrick, E.L. 1998. Spring in the California Current: The distribution of phytoplankton species. *Marine Ecology Progress Series* 167: 73–88.

Venrick, E.L., J.A. McGowan, D.R. Cayan, and T.L. Hayward. 1987. Climate and chlorophyll a: Long-term trends in the central North Pacific Ocean. *Science* 238: 70–72.

Ware, D.M. 1995. A century and a half of change in the climate of the NE Pacific. *Fisheries Oceanography* 4: 267–277.

Wheeler, P.A., A. Huyer, and J. Fleischbein. 2003. Cool halocline, increased nutrients and higher productivity off Oregon in 2002. *Geophysical Research Letters* 30. doi:10.1029/2003GL017395.

Whitney, F.A., H.J. Freeland, and M. Robert. 2007. Persistently declining oxygen levels in the interior waters of the eastern subarctic Pacific. *Progress in Oceanography* 75: 179–199.

Wilkerson, F.P., A.M. Lassiter, R.C. Dugdale, A. Marchi, and V.E. Hogue. 2006. The phytoplankton bloom response to wind events and upwelled nutrients during the CoOP WEST study. *Deep-Sea Research II* 53: 3023–3048.

Winant, C.D. 1983. Longshore coherence of currents on the southern California shelf during the summer. *Journal of Physical Oceanography* 11: 71–86.

Winant, C.D., R.C. Beardsley, and R.E. Davis. 1987. Moored wind, temperature and current observations made during the Coastal Ocean Dynamics Experiments 1 and 2 over the northern California continental shelf and upper slope. *Journal of Geophysical Research* 92: 1569–1604.

Zaytsev, O., R. Cervantes-Dyarte, O. Montante, and A. Gallegos-Garcia. 2003. Coastal upwelling activity on the Pacific shelf of the Baja California peninsula. *Journal of Oceanography* 59: 489–502.

Zentara, S.-J., and D. Kamykowski. 1977. Latitudinal relationships among temperature and selected plant nutrients along the west coast of North and South America. *Journal of Marine Research* 25(2): 312–337.

Chapter 5
The Peruvian-Chilean Coastal Upwelling System

Abstract This chapter describes the Peruvian-Chilean coastal upwelling system which is by far the most productive system in terms of pelagic fish among the Big Four coastal upwelling systems—a feature that is known as the *Peruvian Puzzle*. Also discussed are fluctuations of fish and bird abundance in the context of the El Niño Southern Oscillation, overfishing and influences of nutrient limitation and interactions with the oxygen minimum zone.

Keywords Upwelling · Oceanography · Humboldt current · Peruvian puzzle · El Niño southern oscillation · Marine biogeochemistry · Marine ecology · Ecological regime shifts · Fisheries · Fish stock variations · Climate-change impacts

We all have a sea inside us. Can you hear it?

Can you hear the ocean roaring?

Dianna Hardy (Taken from *Cry of the Wolf—Eye of the Storm*, 2013, reproduced with permission)

5.1 Introduction

The Peru-Chile upwelling system (Fig. 5.1) is one of the most productive ecosystems on Earth. The upwelling system is classified as a highly productive Large Marine Ecosystem with a productivity of >300 g C/m^2/yr (Class 1). The scientific understanding of the physics, ecosystem dynamics and carbon cycle of this particular region has been pivotal in the progress of the climate sciences. For instance, the irregular disappearance of upwelling-enhanced productivity in the region, originally referred to by fishermen as El Niño, has led to the discovery of the El Niño Southern Oscillation (ENSO) which is one of the leading modes of global climate variability. Pauly and Tsukayama (1987) compiled one of the first comprehensive interdisciplinary data bases on the Peruvian upwelling ecosystem covering three decades of

© Springer International Publishing Switzerland 2016
J. Kämpf and P. Chapman, *Upwelling Systems of the World*,
DOI 10.1007/978-3-319-42524-5_5

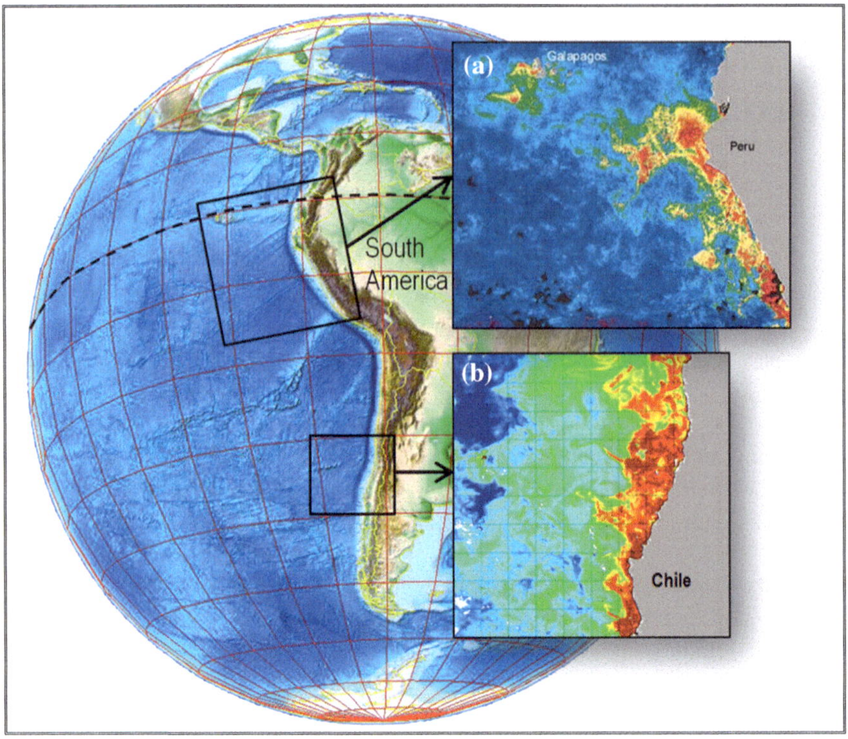

Fig. 5.1 The Peru-Chile coastal upwelling system. Image source: Google Earth with ETOPO 1 bathymetry layer. **a** Image credit: NASA, **b** *Source* http://www.esa.int/spaceinimages/Images/2004/12/Upwelling_zone_off_Chile [accessed on 5 April 2016]

change. Most of the current knowledge on physical-ecological interactions in this upwelling system followed on from this extraordinary work. Another excellent source of information is the comprehensive review of more recent work by Thiel et al. (2007).

The *Humboldt Current* system, also called the *Peru-Chile Current* system, is located in the eastern South Pacific. It extends along the west coast of South America from southern-central Chile (around 42° S) up to Ecuador and the Galapagos Islands near the equator (Thiel et al. 2007; Montecino and Lange 2009; Quinones et al. 2010). Together with the California, Canary and Benguela current systems, the Peru-Chile current system belongs to the four major eastern boundary upwelling systems.

Along the coasts of Peru and northern and central Chile, upwelling is localized and its occurrence changes from being mostly continuous (aseasonal) in Peru and northern Chile to a more seasonal pattern in southern-central Chile. Several important upwelling centres along the Chilean coast are interspersed with long stretches of coast without or with sporadic and less intense upwelling. Large-scale climatic phenomena, such as ENSO, are superimposed onto this regional pattern, which results in a high spatio-temporal heterogeneity, complicating the prediction of ecological processes along the Chilean coast.

From 5° S to ~40° S, the isotherms and isohalines of the upper ~100 m are tilted permanently upward towards the coast and associated with quasi-continuous upwelling. Off Peru, the seasonally maximum winter winds and the small Coriolis term combine to create strong offshore Ekman transport with upwelling effects evident at least as far as 400 km offshore. The continental shelf near Pisco (13.7° S), in particular, is a region of enhanced upwelling. Along the Chilean coast, four regions are recognized for especially strong upwelling, most likely due to topographic enhancement by headlands, as listed by e.g. Figueroa and Moffat (2000) and Mesías et al. (2003). Regional upwelling centres are found at:

- Antofagasta and Mejillones Peninsula (23° S)
- Coquimbo Bay (30° S)
- Valparaíso (33° S), and
- Bay of Concepción (37° S)

Chilean pelagic fisheries, typically concentrated near main upwelling centres around 20°–22° S, 32°–34° S, and 36°–38° S, take an important share of the fish production, thereby affecting trophic interactions in the Humboldt Current system. Interestingly, El Niño events in northern Chile do not appear to cause a dramatic decline in primary or zooplankton production but rather a shift in species composition, which affects trophic efficiency of and interactions among higher-level consumers (Thiel et al. 2007).

Northern Chile, with the Atacama Desert, is an extremely arid zone with no major fluvial inputs to the ocean. The maximum annual precipitation varies between only ~0.5–78 mm/year, which is significantly lower than in central and southern Chile (300 to >1000 mm/year). Off southern-central Chile (36°–40° S), coastal stratification imposed by freshwater runoff becomes important even during summer upwelling conditions (Atkinson et al. 2002). In general, rivers play an important role in the fluxes to coastal waters of trace metals, nutrients and particulate matter, all components that are important factors determining primary production in the water column. For example, iron has been proposed as a factor limiting primary production in waters enriched with other macronutrients but low in pigment concentrations (Martin and Gordon 1988; Martin et al. 1993; Coale et al. 1998).

5.2 Cultural, Social and Economic Relevance

The socioeconomic and cultural impacts of coastal upwelling along the west coast of South America are best described by the consequences that arise from the periodic slackening of the trade winds and the consequent reduction or even

Fig. 5.2 Peruvian anchoveta (Engraulis ringens). Image source: http://perumaritimo.pe/uploads/noticia105.jpg [accessed on 5 April 2016]

cessation of upwelling during El Niño events. In terms of their cultural context, El Niños are believed to have affected pre-Columbian Incas and led to the demise of the Moche and other pre-Columbian Peruvian cultures (Fagan 1999). Peru's fisheries contribution to the GDP varies annually, depending to a large extent on oceanic conditions. Fishing, which is very dependent on the availability of anchoveta (*Engraulis ringens*, Fig. 5.2) "normally" accounts for roughly 3.5 % of the country's GDP and employs approximately 80,000 people. In El Niño years, such as 1998, the sector's contribution falls to below 1 % of GDP. Beyond their immediate influence on fish landings, El Niños have other widespread socioeconomic impacts associated with severe flooding and extensive weather-related damages throughout Latin America, and flood-related damages to infrastructure also result in substantial economic losses in El Niño years. For instance, the 1997/98 El Niño caused damages of about USD 3.5 billion, or about 4.5 % of GDP. Hence, the economies of Ecuador, Peru and Chile are particularly vulnerable to El Niño events.

Environmental disturbances associated with El Niño have triggered socioeconomic and political reactions in Peru altering aspects of society. For instance, a strong El Niño event caused the anchovy fishery to collapse in 1993. This, coupled with political change in Peru, led to a nationalization of the Peruvian fishing industry, resulting in massive layoffs and a restructuring of the industry (Glantz 1979). During the 1997/1998 El Niño, the fishermen's labor union was virtually powerless, and unable to secure governmental aid beyond some provision of foodstuffs (Broad 1999).

Chile's fisheries account for about 2 % of GDP and 11 % of its global exports. In contrast to Peru, Chile's seafood export market is strongly founded on aquaculture products (e.g., salmon, trout, oysters, abalone), which are externally fed and are thus largely independent of oceanic nutrient sources. Chile's dependence on wild fisheries is relatively low, and El Niño events have little impacts on the country's catch rates of anchoveta (0.3 Million tonnes per year), which are only 10 % of Peru's decimated fish landings during El Niño (<3 Million tonnes).

5.3 History of Discovery

Alexander von Humboldt (1769–1859) and Aimé Bonplant (1773–1858) were the first to scientifically explore the flora, fauna, and topography of Latin America. While on the west coast of South America, von Humboldt discovered and scientifically documented the Peru Current in 1802, which, over the objections of von Humboldt himself, has been referred to as the "Humboldt Current" (Berghaus 1837–1847). Humboldt (1846), in his book *Cosmos*, wrote on page 301:

> in the Southern Pacific Ocean, ... a current the effect of whose low temperature on the climate of the adjacent coast was first brought into notice by myself in the autumn of 1802. This current brings the cold water of high southern latitudes to the coast of Chile, and follows its shores and those of Peru northward.

At the congress of the Geographical Society in Lima in 1892, Peruvian Navy Captain Camilo Carrillo's account is the first reference to El Niño, although it was certainly known about long before this by the local fishermen. He made the following statement (Carrillo 1892):

> Peruvian sailors from the port of Paita in northern Peru, who frequently navigate along the coast in small crafts, either to the north or to the south of Paita, named this current "El Niño" without doubt because it is most noticeable and felt after Christmas.

Between the first two World Wars, two important expeditions covered the region: the *Carnegie* expedition from the USA in 1928–1931 (Sverdrup 1930), and the *William Scoresby* expedition from Britain in 1931 (Gunther 1936). These expeditions represented a major early contribution to the knowledge of the Chilean coast sector of the Humboldt Current. The *Scoresby* expedition (May–September 1931) covered the region between 3°–48° S from the coast to 300 miles off Peru and, in more detail, off Chile to 50 miles offshore, collecting data on temperature and salinity, the effects of wind on water movement, and the consequent effect on the phosphate content and, hence, marine life. Whereas *Carnegie*'s survey referred to the *Oceanic Peru Current*, Gunther's report dealt with the *Coastal Peru Current*, although the upwelling system in this region had long before been described by von Humboldt. Later, Wooster (1970) proposed using the name *Humboldt Current* to describe the whole of the Chile-Peru Current system.

Since the late 1960s numerous research cruises have focussed on the Peru upwelling regions. Noticeable is the *Pisco* cruise (e.g. Smith et al. 1971) from 20 March–3 May 1969 which collected almost daily surface maps of temperature, salinity, nitrate, silicate, phosphate, and fluorescence. As part of the multi-national Surface Ocean-Lower Atmosphere Study (SOLAS), which was established in 2004, observational studies started to focus on air-sea gas fluxes, oxygen minimum zones and the biogeochemistry in the Peru and other eastern boundary upwelling regions.

Studies of the ocean and weather conditions during the 1960s emphasized the anomalous warm years, and referred to those episodes as El Niños. Bjerknes (1966, 1969) documented coherent warm equatorial SST anomalies from the dateline to the coast of Ecuador, and related this feature to both the warming at the Peru coast and to planetary-scale changes in the tropical atmosphere, the "Southern Oscillation". In subsequent years, the term "El Niño" has been used in the literature to describe basin-scale equatorial Pacific warmings, and this has blurred the distinction with the coastal phenomenon, which, while related to it, does not exhibit a one-to-one correspondence with the basin-scale SST variability (e.g., Trenberth and Stepaniak 2001). This usage has led to confusion and contradictions in the use of terminology (Aceituno 1992). Trenberth (1977) explored possible definitions of El Niño and concluded that the definition is still evolving and alternative criteria might be used. In 1983, Working Group 55 of the Scientific Committee for Ocean Research (SCOR) defined El Niño as "the appearance of anomalously warm water along the coast of Ecuador and Peru as far south as Lima (12° S)". This implied a normalized SST anomaly exceeding one standard deviation for at least four consecutive months at three of five Peruvian coastal stations. Nowadays, scientists frequently use SST anomalies in the region 5° N to 5° S, 170°–120° W (referred to as "Niño 3.4") as an operational definition for monitoring and prediction purposes. This index is based on SST anomalies (3-month average SST anomaly >=0.5°C with respect to the 1971–2000 average) in the Niño 3.4 region. A brief history of El Niño definitions is given by Stewart (2008).

5.4 Bathymetry and Atmospheric Forcing

Figure 5.1 shows the general bathymetry and topography for western South America (see also Fig. 5.6). The continental region is complex near Panama (9° N), where some wider shelf regions (width > 100 km) are found. The shelf becomes narrow off Colombia and Ecuador, wider (up to 100 km) off Peru, very narrow off northern Chile, and wider again off southern Chile. The shelf is virtually missing entirely off parts of northern and central Chile, e.g., at 30° S it rapidly drops to 800 m by about 8 km offshore. Near Concepción (36°–37° S), the shelf expands to widths of 20–60 km, similar to other upwelling regions. South of 42° S, the shelf is several hundred kilometres wide and it is covered with scattered islands offshore of fjords.

The Peruvian coastline is relatively straight except north of about 5° N, where the coastline orientation changes at it bends into Ecuador's Gulf of Guayaquil. The continental shelf near Pisco (13.7° S) is particularly productive and forms a distinct upwelling centre. Here, upwelling is enhanced via the influence of the irregular coastline of the Paracas Pensinula (located south of Pisco) and larger islands such as

Isla De Saugayan. Overall, the coastline of northern and central Chile (18° S to ~40° S) is also relatively straight, but in the nearshore region small-scale geographic features produce a high spatial heterogeneity, which also influences oceanographic conditions in this area. Several bay systems having complex internal circulation are found along the coast of northern-central Chile. Headlands favour the generation of powerful coastal flow structures (squirts) that transport surface waters up to 100 km offshore (Marín et al. 2003b). In contrast, long stretches of exposed outer coast without headlands or bays, as found for example in northern Chile between 20° S and 22° S, favour alongshore currents leading to relatively homogeneous conditions and downstream transport (Palma et al. 2006).

On land, the Andes Mountain Range extends along most of the length of Peru and Chile at a distance of 100–200 km from the coast. The marine climate of the region results from interactions of larger scale atmospheric systems and the mountain range. The most influential atmospheric pattern is the South Pacific High (see Fig. 2.13), also known as the southeastern Pacific Subtropical Anticyclone, which drives equatorward winds along the coasts of Chile and Peru. The equatorward winds are the principle agent of coastal upwelling. The South Pacific High is bounded in the north by the Inter-Tropical Convergence Zone (ITCZ) and to the south by the polar front and its disturbances. The ITCZ moves from ~10° N in austral winter (June–September) to 2°–5° N in austral summer (December to March). When the ITCZ is farthest north, the southeast trade winds turn eastward off northern Ecuador and Colombia, creating downwelling-favourable (poleward) coastal winds (Strub et al. 1998). When the ITCZ is in its northernmost position in austral winter, upwelling favourable winds exist north of 4° N.

South of 27° S in winter, the effects of polar front disturbances become more pronounced, with time scales of 7–10 days. These synoptic-scale disturbances trigger coastal atmospheric lows between 27° and 32° S that propagate poleward as atmospheric coastal trapped waves and cause upwelling enhancement-relaxation cycles (Rutllant 1993). Between 35° and 45° S, upwelling favourable winds display a high degree of synoptic scale of variability (Djurfeldt 1989). South of 45° S, polar-front synoptic disturbances create mean downwelling-favourable winds all year round.

5.5 Physical Oceanography

The coastal upwelling off Peru is embedded with several more or less independent currents interacting in a rather complicated manner (Wyrtki 1966). In the shallow upper ocean, the Peru-Chile upwelling system is dominated by an equatorward flow of fresh, relatively cool subantarctic surface water, originating from the northern branch of the Antarctic Circumpolar Current. This northward-flowing, cool surface current is the *Humboldt Current* (or *Peru Current*). In the north (~4° S), the Humboldt Current feeds the westward flowing *South Equatorial Current*. Upwelling-favourable coastal winds lead to the formation of a coastal upwelling jet,

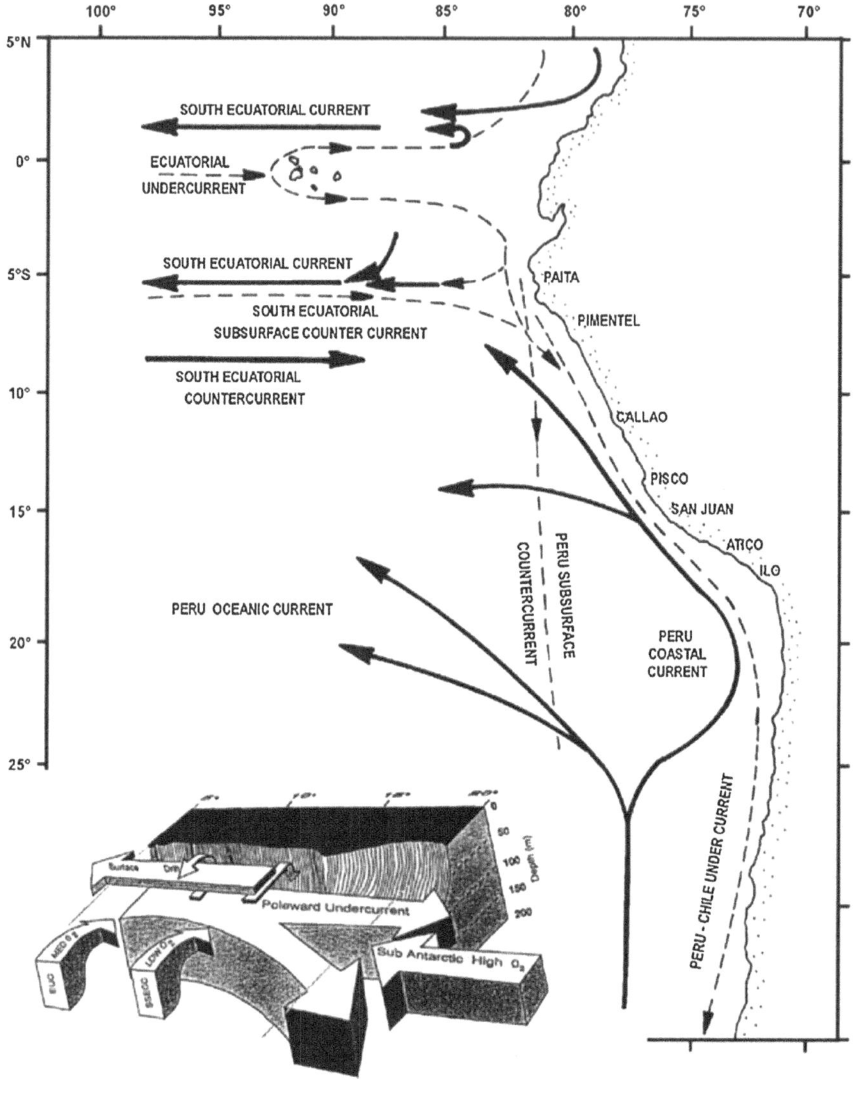

Fig. 5.3 Key currents of the Peru-Chile current system (from Codispoti et al. 1989)

known as *Peru Coastal Current* (Fig. 5.3). Generally, the Peru Coastal Current is strongest from April to September. Gunther (1936) first distinguished a poleward countercurrent situated between these equatorward currents. This intermediate current, called the *Peru Countercurrent* or *Gunther Current*, is a weak and irregular southward flow along 80° W and is usually observed only as a subsurface current. It is strongest near 100 m depth, but reaches to about 500 m.

Beneath the equatorward surface currents, a poleward flowing undercurrent dominates the subsurface and the shelf (Brink et al. 1983; Penven et al. 2005;

Montecino and Lange 2009). This so-called *Peru-Chile Undercurrent* is the dominant undercurrent within the first 180 km from the coast (Huyer 1980), and serves as a major source of upwelled water (Penven et al. 2005; Kessler 2006; Pennington et al. 2006). It originates from the eastward flowing *Equatorial Undercurrent*, which splits at the Galapagos Islands into two branches, one propagating to the south to form the undercurrent, the other flowing southeastward reaching the coast at around 5° N (Lukas 1986; Penven et al. 2005; Kessler 2006). In addition to coastal upwelling, the trade winds cause oceanic divergence or surface Ekman transport away from the equator. This divergence forces local upwelling along the equator, which produces a region of shallow thermoclines and above-normal productivity that extends westward along the equator from the coast (see also Chap. 9).

The vertical displacement of the upwelling surface is typically less than 50 m between 15° S and 34° S. Source water is primarily Equatorial Sub-Surface Water (ESSW) north of 18° S and Sub-Antarctic Water (SAW) south of that; beyond 35° S there is an influence of Antarctic Intermediate water (AIW) (Blanco et al. 2001). The two water masses are clearly distinct in their oxygen levels. ESSW has an oxygen concentration of <2 ml/l, whereas SAW has a concentration of 4 ml/l.

Due to its close connection to the eastern limit of the equatorial currents, the Peru-Chile current system is strongly affected by large-scale, basin-wide interannual variability caused by ENSO events (Brainard and McLain 1987; Carr et al. 2002). During stronger El Niño events, wind anomalies reduce or eliminate the east-west downward tilt of the equatorial thermocline. In conjunction with eastward propagating equatorial Kelvin waves comes the increase in sea surface height and the deepening of the thermocline, oxycline, and nutricline in the eastern equatorial Pacific Ocean (Pennington et al. 2006). As a result, the equatorward water flow in the Humboldt Current decreases or even reverses, and warm and nutrient-depleted, but oxygen-richer equatorial water appears on the Peruvian shelf (Thiel et al. 2007). The deepening of the thermocline results in a substantial decrease in upwelling intensity and, consequently, in reduced nutrient supply and primary production.

The other phase of ENSO La Niña, leads to a shallower thermocline and a strengthened upwelling, associated with colder sea surface temperatures on the Peruvian-Chilean shelf (Pennington et al. 2006; Thiel et al. 2007; Quinones et al. 2010). On the more local scale, seasonality and anomalies in local wind systems, as well as other local factors, such as coastal topography, the width of the continental shelf, and the source of the upwelled waters, produce strong seasonal and latitudinal variability in primary production within the Peru-Chile current system (Thomas et al. 2004). Coastally trapped waves appear especially energetic during El Niño periods and weaker during La Niña periods and austral winter (Shaffer et al. 1999).

Coastally trapped waves frequently propagate poleward along the entire Peru and Chile coastlines, traceable to wind fluctuations in equatorial regions (Hormazábal et al. 2001). The waves raise and lower the pycnocline/nutricline, influencing the effectiveness of upwelling, with dominant frequencies of days to weeks off Peru (Enfield et al. 1987) and ~50-day periods off northern and central Chile (e.g., Rutllant et al. 2004b). Ramos et al. (2006) show that variability in equatorial source waters at both annual and semi-annual periods strongly modulates isotherm depth along the coast. At the shorter time-/space scales, diurnal cycles in wind stress are important contributors to forcing along the arid northern Chilean coast, especially in summer (Rutllant et al. 1998), but become less important with increasing latitude, where storm-mediated variability on 3- to 7-day cycles increases (Strub et al. 1998), with a maximum in austral winter.

5.6 Regional Aspects

While the continental shelf near Pisco (Peru) is a region of enhanced upwelling, there are a number of individual upwelling centres along the Chilean coastline. South of 26° S, within the area where coastal upwelling is more seasonal, Coquimbo Bay (30° S) and the Bay of Concepción (37° S) are important regional coastal upwelling centres.

The Coquimbo Bay system is located between two capes (Punta Lengua de Vaca and Punta Pájaros) and is the site of intensive fisheries and eco-tourism. Filaments, including bifurcated upwelling filaments (Moraga et al. 2001), are known to generate at Punta Lengua de Vaca, contributing cold, nutrient-high waters to the coastal system (e.g., Montecino et al. 2006). Marín et al. (2003b) have shown that cold-water *squirts* (i.e. filaments leading to counter-rotating eddy pair) generate at the northern end of the bay system (near Punta Pájaros).

Satellite observations (Fig. 5.4) and Lagrangian drifter data (Marín and Delgado 2007) have shown that this squirt is a recurrent feature in the area, reaching distances on the order of 140 km offshore. Squirt speeds, estimated both through feature-tracking analysis (Marín et al. 2003b) and Lagrangian drifters (Marín and Delgado 2007), range between 0.2 and 0.3 m/s. Thus, considering that the lifetime

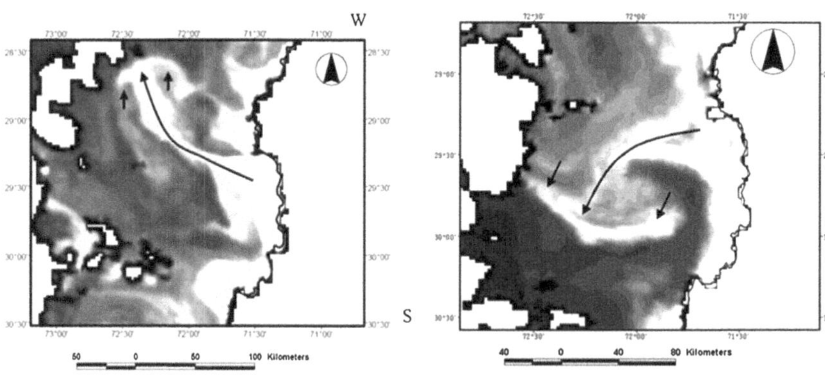

Fig. 5.4 SeaWiFS image (chlorophyll-*a*) from 20 January 1999 (*left panel*) and 19 January 2002 (*right panel*). *Long arrows* show the main axis of the squirt; *short arrows* correspond to the "tip of the hammer." Taken from Marín et al. (2003b)

of a single squirt is related to the active period of equatorward wind events, which for the area range between 3 and 7 days (Rutllant et al. 2004a), coastal organisms trapped within the squirt are likely to reach 100–200 km offshore in a period of less than a week.

Another important upwelling focus in the northern zone is the Mejillones Peninsula (23° S). Observational (e.g., Marín et al. 2003a) and modelling studies (Escribano et al. 2004) have shown that the dynamics of the coastal ecosystems in that area largely depend on the generation of upwelling filaments here. The generation of filaments in the northern tip of the peninsula (Punta Angamos) has been identified as the main mechanism of nutrient enrichment in the surface layers (Marín and Olivares 1999). Furthermore, upwelling shadows within Mejillones Bay, an equator-facing bay located in the northern end of the peninsula, have been dynamically linked to the generation of bifurcated filaments at Punta Angamos (Marín et al. 2003a). This shadow is an important physical structure within the bay, affecting primary production (Marín et al. 2003a) and the retention of planktonic organisms (Olivares 2001).

An alternative mechanism, described as an *upwelling trap* by Castilla et al. (2002) and also related to the coastal upwelling dynamics, generates higher temperatures inside Antofagasta Bay, a pole-facing bay at the southern end of the Mejillones Peninsula. In this case also the physically generated structure contributes to the retention of planktonic organisms. Thus, mesoscale flow features (upwelling shadows and upwelling traps), associated with cold-water filaments, seem to play an important role not only in relation to the biological productivity of coastal upwelling regions but also as mechanisms for the retention of coastal planktonic species.

5.7 Seasonality

5.7.1 Ekman Transport

From the standpoint of coastal wind forcing, the Peru-Chile upwelling system can be divided into two latitudinal areas with a transition near 26° S in northern Chile (Figueroa 2002). From 26° S to the north, upwelling-favourable winds exist year-round; south of this latitude greater seasonality is observed (Thiel et al. 2007). Time series of the upwelling index (see Sect. 2.1) provides information on seasonal and interdecadal variations of the alongshore wind conditions. To illustrate spatial variations of wind conditions along the west coast of South America from 15° S to 45° S, we use monthly averaged upwelling indices for the period 1981–2002, provided by NOAA's Pacific Fisheries Environmental Laboratory (PFEL) (Fig. 5.5). Note that this data set excludes the highly productive Peruvian upwelling region located at 5°–15° S. Mendo and Castillo (1987) present upwelling-index data for the period 1953–1985 for the Peruvian coast north of 15° S.

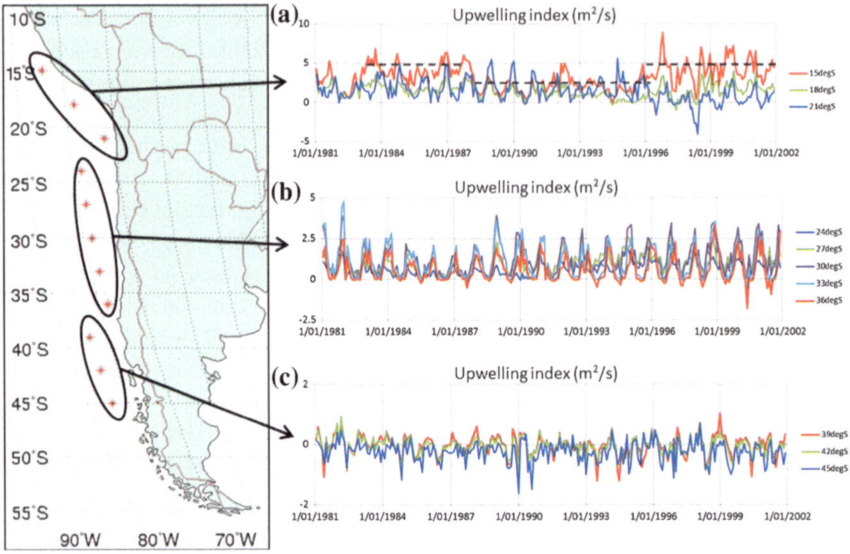

Fig. 5.5 Time series of monthly mean upwelling index at various locations along the coasts of Peru and Chile for the period 1 January 1981 to 31/12/2001. Data source: NOAA, http://www.pfeg.noaa.gov/products/PFEL/modeled/indices/upwelling/SA/upwell_menu_SA.html [accessed 5 April 2016]

From this time series, it is evident that the overall magnitude of upwelling favourable wind stress strongly decreases southward along the coast. At the northernmost Peruvian location (15° S), the upwelling index peaks at values >5 m^2/s (corresponding to an offshore mass transport of >600 tonnes per 100 m of coastline). The magnitude of the upwelling index reduces to ~2.5 m^2/s in the northern portion of Chile (24°–36° S) and drops to relatively low values of <1 m^2/s farther to the south.

At 15° S, upwelling-favourable alongshore winds are a largely continuous phenomenon (Fig. 5.5a). Interestingly, the time series indicates *climate shifts* of surface wind systems and the associated magnitude of the upwelling index. That is, the periods 1983–1988 and 1996–2002 experienced stronger upwelling winds with a upwelling index of ~5 m^2/s on average, but these periods were interrupted by a period (1989–1995) during which the upwelling index was markedly reduced by half to ~2.5 m^2/s.

At 21° S, weak but continuous upwelling-favourable winds occurred until 1996, followed by a transition to a regime of markedly reduced alongshore winds. In this northern region (15°–21° S), we can also clearly identify ENSO-related wind disturbances. Substantial weakening of the upwelling index occurred during the particularly strong 1982/83 and 1998/99 El Niño events. In contrast, alongshore coastal winds farther away from the equator are not much influenced by such ENSO events.

The upwelling dynamics off the northern portion of Chile (24° S–36° S) have a pronounced seasonal cycle (Fig. 5.5b). Upwelling-favourable winds develop regularly during austral summer, but tend to vanish during austral winter. This behaviour is characteristic of seasonal coastal upwelling systems. The wintertime disappearance of offshore Ekman drift generally supports the survival of anchoveta eggs off Chile. While the continental shelf south of 36° S still occasionally experiences upwelling-favourable wind events, the magnitude of such events is much smaller. At times, this region is exposed to extended periods of reversed, downwelling favourable wind conditions, which explains the relatively low productivity of this region.

5.7.2 *Primary Production and Influences of Sub-Surface Currents*

The continental shelf of the Peru-Chile coastal upwelling system is narrower compared with the other eastern boundary currents (compare with Figs. 4.14, 6.16 and 7.14). The widths of both the inner and outer shelves are far less than 50 km between 15° and 35° S (Fig. 5.6, first panel), and the continental slope is very steep, especially between 20° S and 34° S. Between 7° and 10° S, both inner and outer shelves broaden and extend beyond 100 km. The shelf width can be used as a proxy for the availability of benthic iron, which for narrow shelves can limit primary production (Carr and Kearns 2003).

Ekman transport is offshore year-round throughout most of this system. The regions north of 15° S and between 22° and 40° S display the largest Ekman transports of 0.5–1 m^2/s in July–September and >0.5 m^2/s in October–December, respectively (Fig. 5.6, second panel). Both transport and seasonal range are minimum between 16° and 22° S (compare with Fig. 5.5). Maximum sea-surface-temperature

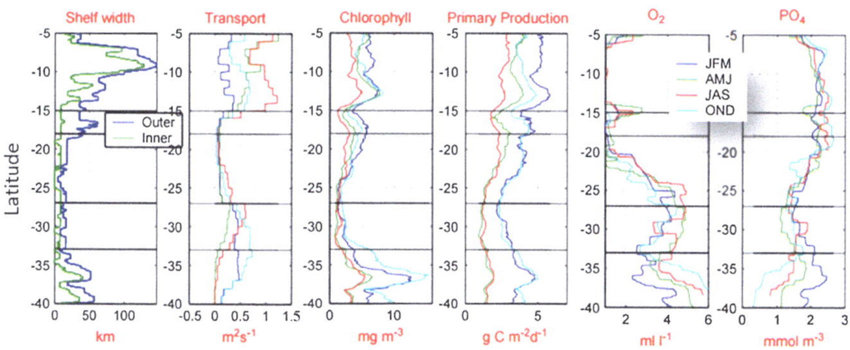

Fig. 5.6 Variation in shelf width and seasonal changes in offshore Ekman transport, chlorophyll concentration, primary production, oxygen and phosphate concentrations with latitude in the Peru-Chile upwelling system. The *horizontal black lines* divide the system into separate zones (from Carr and Kearns 2003)

anomalies (not shown) occur in April to June in most of the region with maximum values >5 °C north of 15° S.

Biomass peaks in January–March north of 22° S and in October–December to the south. The maximum chlorophyll-*a* concentration exceeds 5 mg/m^3 north of 20° S and south of 35° S; seasonal ranges are at maximum in these regions as well (Fig. 5.6, third panel). Primary production has the same seasonal cycle as chlorophyll concentration and is at maximum (>4 g C/m^2/day) north of 18° S in January–March and south of 34° S in October–December) (Fig. 5.6, fourth panel). Primary production attains a minimum at 27° S (~2 g C/m^2/day).

As stated above, upwelling source water is primarily Equatorial Sub-Surface Water (ESSW) north of 18° S and Sub-Antarctic Water (SAW) south of that. Beyond 35° S there is an influence of Antarctic Intermediate Water (AIW) (Blanco et al. 2001). The two water masses are clearly distinct in latitudinal variations in oxygen concentration (Fig. 5.6, fifth panel). The phosphate content of the source water is the highest of all eastern boundary currents (Fig. 5.6, last panel; compare with Figs. 4.14, 6.16 and 7.14) and exceeds 2 mM/m^3 both north of 20° S and south of 34° S. The seasonal variation in source water characteristics reflects the poleward transport of ESSW in the undercurrent (maximum in April–June) and the equatorward transport of SAW (maximum in July–September) as seen in seasonal variations in oxygen. The minimum in nutrient concentration of the source water near 27° S is reflected in the chlorophyll concentrations.

Friedrich and Codispoti (1981) located the front between SAW and ESSW at ~15° S and found maximum influence of low nutrient-high oxygen SAW in July–September (which is also the strongest upwelling period). Carr and Kearns (2003) identified a similar maximum in SAW influences in the same months, but maximum variability was observed at 18° S rather than 15° S. Carr and Kearns (2003) point out that this difference is due to different methodologies rather than physical processes.

5.7.3 Phytoplankton Blooms and Anchoveta Spawning off Peru

Winds along the coastline of Peru are favourable for upwelling throughout most of the year. Generally, the alongshore wind magnitude intensifies during austral winter months (June–September) (Fig. 5.7). Due to upwelling and incursions of warmer tropical water, the coastal waters off Peru undergo a seasonal temperature cycle in which the warmest (coldest) periods occur at the end of austral summer (winter). The reader should note that seasonal atmospheric heat-flux variations, which are less pronounced near the equator than at higher latitudes, are not the cause of this pronounced seasonal temperature cycle. Instead, SST anomalies are rather controlled by two opposing processes:

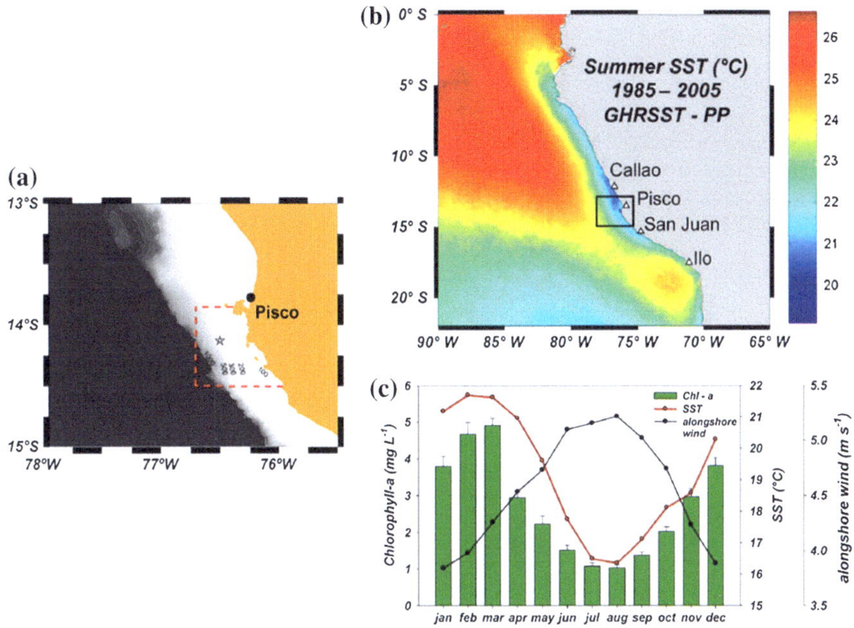

Fig. 5.7 Seasonal variations of chlorophyll-*a* concentration, sea surface temperature and alongshore wind speed in the upwelling centre offshore from Pisco, Peru. Image source: Gutiérrez et al. (2011)

(1) Coastal upwelling events lowering SST, and
(2) Inflow of tropical waters increasing SST.

Hence, the influence of coastal upwelling over inflow of tropical water clearly intensifies during austral winter months—in coincidence with the peak spawning period of anchoveta.

Despite this intensification of upwelling influences in austral winter, the yearly cycle in total phytoplankton biovolume on the Peruvian shelf including Ancón Bay (11° 12′ S) shows the highest values during summer–autumn months and the lowest values during winter months (Fig. 5.8; Ochoa et al. 2010). This seasonality has also been estimated through remote sensing (SeaWiFS) of chlorophyll off Peru (e.g., Echevin et al. 2008; Gutiérrez et al. 2011). Lower chlorophyll-*a* levels during winter months in the larger Peruvian upwelling system are also observable in situ (Montecino et al. 2006; Pennington et al. 2006; Echevin et al. 2008). Generally, increased chlorophyll levels in the Humboldt Current system begin earlier in spring with a slightly later summer–autumn peak than the observed biovolume peak in Ancón Bay (Ochoa et al. 2010).

Modelling investigations of the larger Peruvian system attribute these seasonal peaks in chlorophyll-*a* to the dynamics of upwelling-related nutrient availability in combination with destratification–restratification dynamics resulting from changes

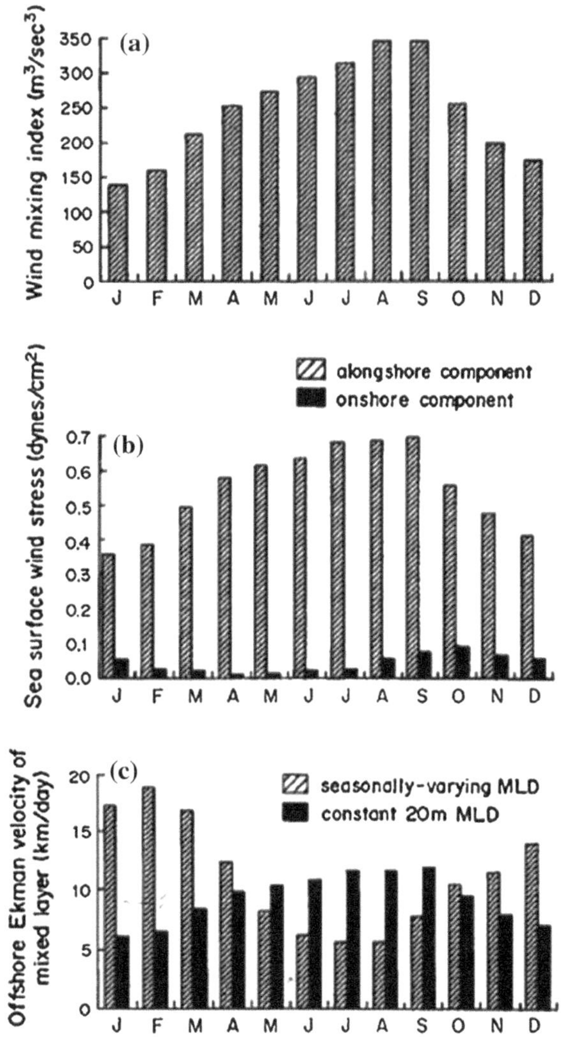

Fig. 5.8 Typical seasonal cycles off Peru of **a** wind mixing index (wind speed cubed), **b** wind stress components and **c** offshore Ekman velocity for a seasonally varying or constant mixed layer depth. Taken from Bakun (1987)

in the depth of the mixed layer. Specifically, increased surface winds during winter deepen the mixed layer and reduce the surface chlorophyll accumulation due to a dilution effect and light limitation (Echevin et al. 2008). In north-central Chile, highest chlorophyll levels are also observed during more neutral rather than cold upwelling periods, suggesting that transitional periods with less turbulent mixing can result in higher phytoplankton biomass (Montecino et al. 2006).

Anchoveta eggs float passively with the currents. Hence, the risk of egg losses is greatest during periods of strong offshore surface flows. The seasonal variation in mixed layer depth off Peru proceeds in phase with that of transport, in response to the seasonality in turbulent mixing (Fig. 5.8a), but has a greater relative amplitude.

At first sight, it is puzzling that the anchoveta's spawning period in austral winter coincides with the time of strongest upwelling-favourable winds (Fig. 5.9b) corresponding to maximal offshore volume transports in the surface Ekman layer. However, when accounting for the fact that the volume transport is proportional to the product of speed and mixed layer depth, it turns out that the offshore speed of Ekman flows is much smaller in the spawning season than it is during austral summer (Parrish et al. 1983). The result is that drifting organisms (i.e., anchoveta eggs) distributed though the surface mixed layer would experience a faster offshore drift in the thinner surface mixed layer of austral summer than in the deeper mixed layer of winter, even though the winter volume transport is much larger. Hence, the timing of anchoveta spawning matches the period of weakest offshore drift speeds (Fig. 5.8c).

5.7.4 Phytoplankton Blooms Off Chile

Chile's northern ocean margin (~ 18°–30° S) presents distinctive characteristics in topography, climatic and oceanographic conditions, which modulate primary production and water column chemistry. It features a comparatively low primary production for the Humboldt Current system, and despite semi-permanent wind-driven upwelling some areas are considered as high-nutrient low-chlorophyll environments (e.g., Torres 1995). These observations are in contrast to the localized prominent upwelling cells with relatively high primary production of 0.5–9.3 g C/m^2/day, such as off Iquique (21° S), Antofagasta (23° S) and Coquimbo (30° S) (González et al. 1998; Daneri et al. 2000; Thomas et al. 2001).

In northern Chile (off Iquique and Antofagasta), a permanent upwelling sustains regionally high primary production throughout the entire year (e.g., Iriarte and González 2004). There is only weak seasonality in chlorophyll-*a* concentration, which shows slightly higher values during the winter and early spring. The highest chlorophyll-*a* concentrations (15–20 mg/m^3) were mostly restricted to a narrow inshore zone (<37 km wide) and were associated mostly with microphytoplankton larger than 20 μm. High microphytoplankton abundances in the sediment record correlate positively with intense/more frequent upwelling events (higher primary production) in waters of Mejillones Bay (Ortlieb et al. 2000).

Off central Chile (33°–37° S) upwelling-favourable south-southwest winds predominate during austral spring and summer months when chlorophyll-*a* concentrations range between 3.8 and 26 mg/m^3 compared with 1.0–2.5 mg/m^3 recorded during winter (González et al. 1989; Montecino et al. 2004). In Concepción Bay (37° S), few numerically dominant diatom species represented >80 % of the total diatom abundance during upwelling events (González et al. 1987), and in correspondence with this, the highest biomass was concentrated in the microphytoplankton fraction (>20 μm) from winter through spring (González et al. 1989).

In the southern area (37°–45° S) the upwelling events are mainly concentrated in the spring–summer period (Strub et al. 1998; Thomas et al. 2001). Pico- and nano-phytoplankton (pico: 0.7–2.0 μm; nano: 2.0–20 μm) predominate in the offshore, mostly oligotrophic zone, associated with the intrusion of nutrient-depleted, warmer subtropical waters (Morales et al. 1996; Iriarte and González 2004). In such cases, small dinoflagellates (*Gymnodinium* spp., 5–25 μm) and autotrophic flagellates dominated the phytoplankton assemblages (Iriarte et al. 2000).

5.8 The Peruvian Puzzle

The Peruvian upwelling system produces 3–10 times more fish than other upwelling ecosystems (Cury et al. 1998; Fréon et al. 2009), which is known as the *Peruvian puzzle* to fish biologists. One key to this puzzle is that the location near the equator allows for strong upwelling under relatively moderate wind forcing. This is because Ekman transport is inversely proportional to the sine of geographical latitude (see Eq. 2.3 in Sect. 2.1). The second key is the absence of strong wind-mixing events and associated dilution effects on phytoplankton growth. The turbulent mixing energy off Peru (Fig. 5.8a) is low compared to other anchovy reproductive habitat, even at its seasonal maximum (Bakun 1987). This absence of wind-mixing events creates a stable, shallow thermocline that, in conjunction with nutrient upwelling, is an optimal environment for the development of plankton communities (Bakun 1996; Bakun and Weeks 2008).

The third key to the Peruvian puzzle is the relatively high concentration of the micronutrient iron that is released from shelf sediments. Observations in the Peru upwelling regime during August and September 2000 (Bruland et al. 2005) demonstrated that the supply of iron, relative to that of the macronutrients nitrate, phosphate and silicic acid, plays a critical role in allowing extensive diatom blooms to develop in the region. Iron is a key limiting factor for photosynthesis in high-nitrate–low-chlorophyll regions, such as the Peru/Humboldt Current system and the adjacent eastern equatorial Pacific (Moore et al. 2013). Within the Peruvian upwelling area, bioavailable iron is released from reducing continental margin sediments (Bruland et al. 2005) with broader sections of the continental shelf releasing more iron than narrower sections. Thus, massive diatom blooms with maximal photochemical efficiencies occur in the iron-rich upwelling region observed over the broad continental shelf off northern and central Peru, leading to the extremely high chlorophyll concentrations within Peru's "brown waters" (where chlorophyll-*a* concentrations are between 20 and 45 μg/L).

The source of the upwelled water in this region is the nutrient-rich subsurface countercurrent in contact with the organic rich shelf sediments. The shelf system off Callao (12° S) has a relatively gradual slope to 150 m depth. The offshore Ekman transport takes place within the upper 20 m or so of the water column, with onshore transport down to about 80 m depth that supplies the upwelling water (Gutiérrez et al. 2008). This subsurface shelf water is suboxic and has extremely high concentrations of dissolved Fe (>50 nM) in the near-bottom waters. In marked contrast, relatively low-chlorophyll "blue waters" (with chlorophyll-*a* concentrations of 2 μg/L) containing low concentrations of dissolved Fe (~0.1 nM) and high unutilized macronutrient concentrations are observed in the coastal upwelled waters along the southern coast of Peru and in the offshore regions of the Humboldt Current. Hence, the micronutrient iron appears to play a critical role in the "brown waters" of Peru.

Biogenic opal and organic carbon vertical rain rates in sediment cores reveal a strong cyclicity in the productivity of the upwelling system off presently arid northern Chile during the last 100,000 years (Dezileau et al. 2004). The data indicate that changes in productivity are in phase with the precessional cycle (~20,000 years) and with iron inputs from the continent. During austral summer insolation maxima, increased precipitation and river runoff in the region appear to have brought high inputs of iron, mainly from the Andes, to the coastal ocean enhancing primary productivity there. Dezileau et al. (2004) interpret their results as evidence for iron control of past productivity in this upwelling system and for a tight link between productivity and orbital forcing at midlatitudes.

Saito et al. (2004) demonstrated that, similar to iron, cobalt is also a limiting micronutrient for phytoplankton. Its abundance in the environment is low and it can become limiting to certain phytoplankton groups, thereby influencing community composition. The Peru surface water cobalt concentrations presented by Saito et al. (2004) likely originate as a flux from anoxic sediments, and this flux of cobalt may constitute a major source of cobalt to the coastal Peru ecosystem and perhaps to the surface waters of the South Pacific.

5.9 Impacts of El Niño-Southern Oscillation

The Peru-Chile coastal upwelling system is closely connected to the equatorial upwelling of the Pacific Ocean (see Sect. 9.3) via the equatorial undercurrent passing on its water mass characteristics to the poleward undercurrent. Due to its large extent, the equatorial Pacific upwelling region is a major source of CO_2 to the atmosphere (Tans et al. 1990) and it contributes to >18 % of global new biological production (Chavez and Toggweiler 1995). ENSO events strongly control CO_2 exchanges with the atmosphere over the entire equatorial Pacific region. The CO_2 release to the atmosphere amounts to ~0.8–1.0 Pg C/yr during non-El Niño years, whereas El Niño events transiently reduce the CO_2 flux to the atmosphere substantially to 0.2–0.4 Pg C/yr (Feely et al. 2002). Hence, ENSO events (extending to the Chile-Peru upwelling system) play a dominant role in the climate variability of the global carbon cycle.

While it is obvious that El Niño events have a profound effect on the marine food web and anchoveta stocks in the Peruvian upwelling region, it is interesting to note that this effect does not extend to the Chilean upwelling region. Instead, Ulloa et al. (2001) reported during the 1997–98 El Niño event that both temperature and oxygen contents of the coastal waters off northern Chile were noticeably higher than during non-El Niño conditions, as expected, but phytoplankton and zooplankton biomasses were not dramatically altered. In contrast, the herbivorous copepod *Calanus chilensis*, which shows a limited tolerance to low-oxygen conditions, exhibited greater abundances, higher growth rates, and a significant reduction in adult body size. Ulloa et al. (2001) concluded that there can be a positive effect from the changes in oceanographic conditions due to El Niño on zooplankton growth and production in this region.

5.10 Longer-Term Variability and Trends

In conjunction with a slight but gradual cooling of upwelling waters off Peru on multidecadal timescales, Gutiérrez et al. (2011) detected an overall increasing positive trend in SeaWiFs chlorophyll-*a* concentrations for the period 1965–2004 (Fig. 5.9). They attributed these trends to an intensification of alongshore winds and hence upwelling, particularly during austral spring.

Many zooplankton eat phytoplankton, and are in turn preyed upon by fish larvae and many adult planktivorous fish. Conversely, certain zooplankton groups (e.g. medusae) also prey on fish eggs and larvae (Ayón et al. 2008). Thus, due to their intermediate position in the food web between primary producers and predators, the zooplankton serve as a link between bottom-up climate-related control of phytoplankton and fish. In contrast to the positive phytoplankton trend seen in Fig. 5.9, estimated zooplankton volumes were initially high at >0.6 ml/m^3 until 1972, before dropping to values as low as <0.2 ml/m^3 in the early 1970s and slowly increasing to levels of ~0.4 ml/m^2 over the following 20 years (Fig. 5.10). Therefore, no direct relationship between the primary production level and the zooplankton biomass has been observed. This observational evidence suggests that consumption by forage fish (anchoveta and mackerel) largely controls zooplankton abundance in the Peruvian upwelling system, which is characteristic of a wasp-waist control of the marine food web.

During the period 1963–2002, the anchoveta biomass experienced dramatic interannual variability with relatively large stocks occurring in the 1960s and the 1990s, and almost depleted stocks during the period 1975–1983. The magnitude of the population collapse in the late 60s and early 70s is similar for both anchovy and zooplankton, but the subsequent recovery was much better for anchovies.

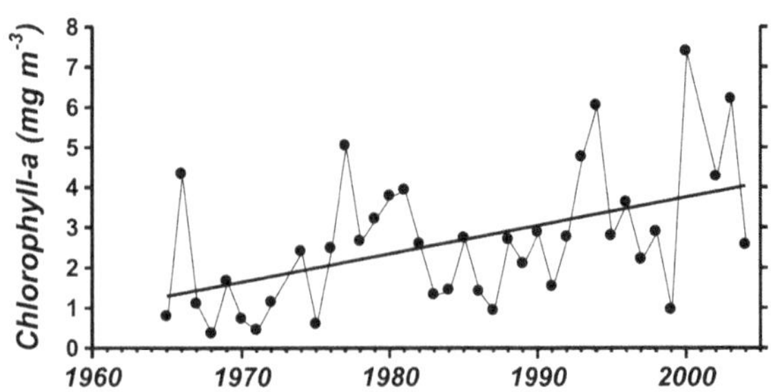

Fig. 5.9 Annual average chlorophyll-*a* concentration for the Peruvian upwelling region. Taken from Gutiérrez et al. (2011). Note that minima correspond to stronger El Niño events (see Fig. 3.9)

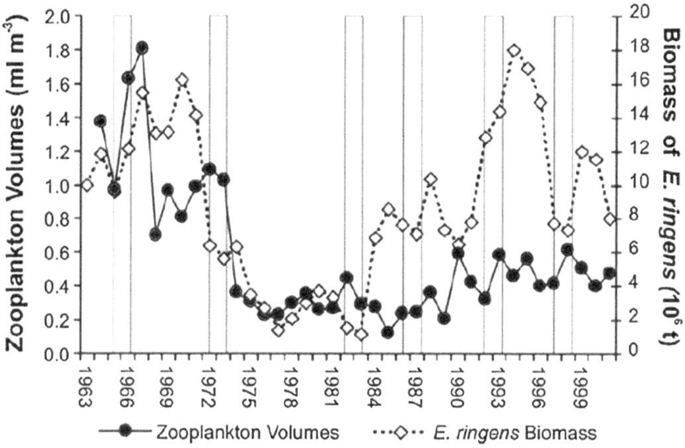

Fig. 5.10 Estimated zooplankton volumes and biomass of Peruvian anchoveta (*Engraulis ringens*) for the Peruvian upwelling zone for the period 1963–2002. Taken from Ayón et al. (2008)

In other coastal upwelling systems, such as the Benguela system, zooplankton abundance and pelagic fish abundance/catches are negatively related to each other, which is characteristic of a top-down control (Cury et al. 2000). The apparent long-term decline in biomass of large zooplankton in the Humboldt Current system off Peru during the early 1970s, following the well-documented 1972 El Niño and the collapse of the local anchoveta population (Alheit and Bernal 1993), appears to be an exception. From a food-chain perspective, the enormous reduction in anchoveta biomass should have reduced predation on zooplankton populations and also reduced competition for phytoplankton between zooplankton and anchoveta. Accordingly, zooplankton standing stocks might have been expected to increase much more rapidly than has been observed. The opposite effect suggests a basic change in ecosystem functioning, concomitant with a switch in the diet of pelagic predators, such as horse mackerel (*Trachurus symmetricus murphyi*) and chub mackerel, from anchoveta to zooplankton (Carrasco and Lozano 1989) (Fig. 5.11).

Corredor-Acosta et al. (2015) studied satellite data of sea-surface temperature, chlorophyll-*a* concentrations and wind stresses off central-southern Chile for the period 2002–2012. In contrast to the Peru region, which has experienced a positive trend in chlorophyll-*a* concentrations (Fig. 5.9), Corredor-Acosta et al. (2015) determined a slight but significant trend for central-southern Chile, negative for chlorophyll-*a* concentrations and positive for offshore Ekman transport. Their finding was unexpected since three La Niña related conditions were identified during the 2007–2012 period.

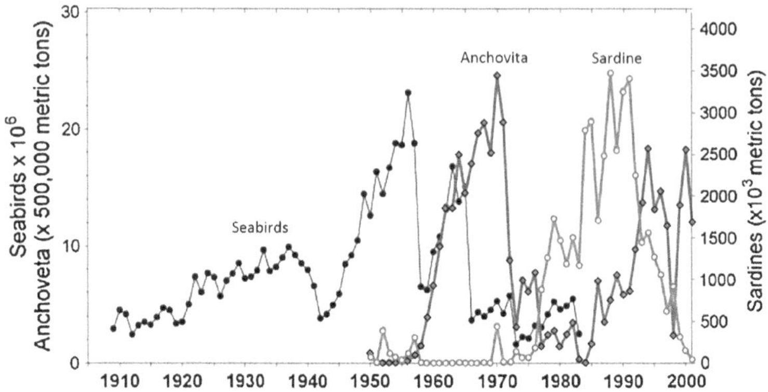

Fig. 5.11 Time series of the abundance of seabirds, anchoveta and sardines in the Peruvian upwelling region. Redrawn after Chavez et al. (2003)

5.11 Fisheries and the "Rivalry" Between Anchoveta and Sardines

The Peruvian anchoveta (*Engraulis ringens*, Fig. 5.2) is a member of the family of *Engraulidae* and plays a central role in regional ecosystem dynamics, supporting one of the world's most economically important fisheries (Penven et al. 2005; Pennington et al. 2006; Montecino and Lange 2009; Quinones et al. 2010). It occurs exclusively along the eastern coast of South America, from 4° 30′ S off Peru to 42° 30′ S off Chile. The heaviest concentration occurs along the coast of northern and central Peru, downstream of the strongest upwelling area. At the northern end of their range anchoveta biomass drops off rapidly, whereas towards the south, the biomass only tapers off gradually (Fig. 5.12). Peruvian anchoveta spawn primarily during the period from August to November each year (Pauly and Soriano 1987), but do not migrate far seasonally and do not expand far enough poleward during warm regimes to avoid the increased temperatures; they tend to recover during cooler periods (Lluch-Belda et al. 1992).

Anchovy and sardine are the dominant species among the small-size fish consumers in the Humboldt Current region. Anchovy feed on large zooplankton, whereas Pacific sardine (*Sardinops sagax*) feed on phytoplankton and small zooplankton and can expand their range poleward in warmer periods. At a gross level, the anchoveta off northern/central Peru are distinct from those from southern Peru/Chile, as the latter have less coarse grill rakers (Tsukayama 1966) and shorter guts (Rojas 1971), both features suggesting that these fish rely on zooplankton more than their northern counterparts. Large predators in the system include the jack mackerel (*Trachurus murphyi*), hake (*Merluccius gayi*) and cephalopods (Neira and Arancibia 2004), while the top predators are large pelagic fish such as tuna (*Thunnus orientalis*) and swordfish (*Xiphias gladius*), southern sea lions (*Otaria flavescens*) and seabirds.

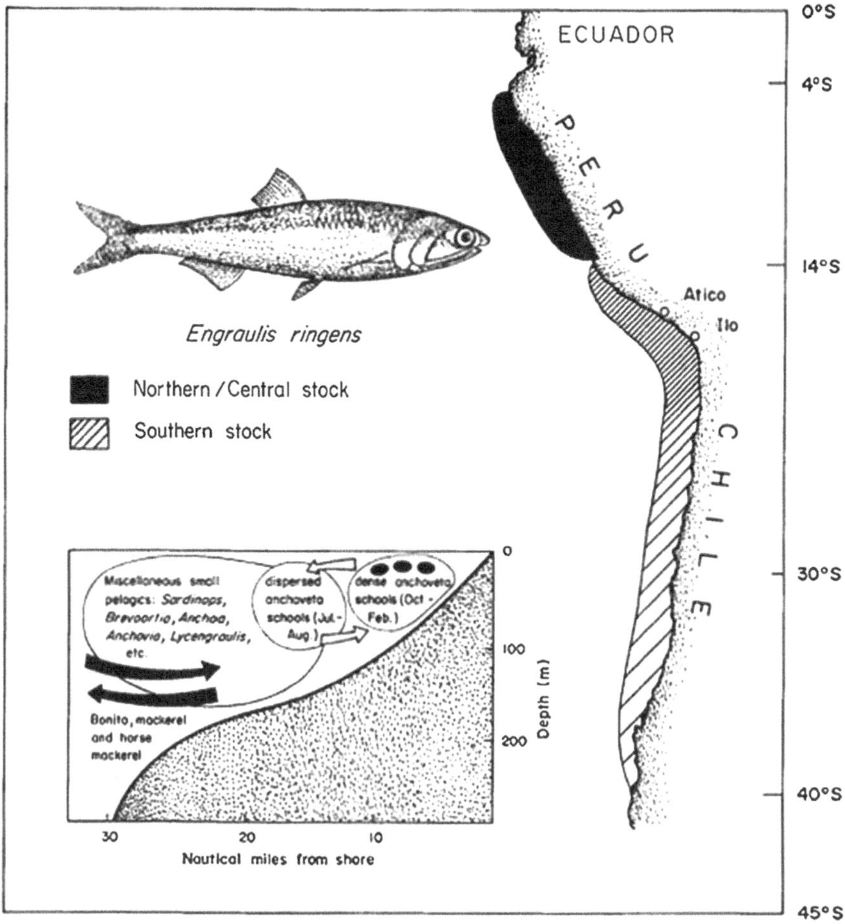

Fig. 5.12 Distribution of anchoveta stocks along the west coast of South America (from Pauly and Tsukayama 1987)

Total catches in the Peru/Chile upwelling system exceed those in the other eastern boundary upwelling systems by far (see Sect. 10.2), mostly because of the extensive anchovy stocks (Fig. 5.13a). While anchovy catches declined dramatically in the early 1970s, the catch rates of herring-like and perch-like fish increased significantly from the mid-1970s to the late 1990s, reaching total weights of ~10 million tonnes in 1990 (Fig. 5.13b), almost as much as the anchovy catch was in the late 1960s. Note that the regime shift in the early 1970s that led to the sudden loss of anchovy stocks was mirrored by a similar sharp decrease in the catch rates of tuna and billfishes, two of the top predators. The influence of strong El Niños, such as the 1997/1998 event, are also apparent in variations in anchovy catch rates. Overall, catch rates of different species in this upwelling system show very large

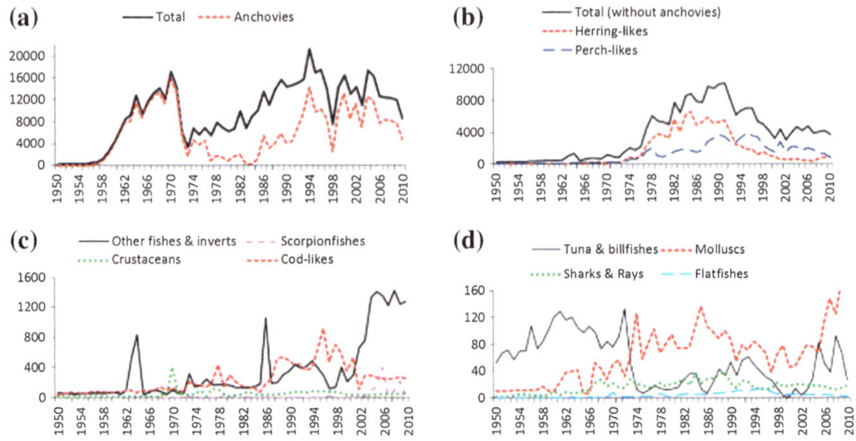

Fig. 5.13 Catches (in units of 1000 tonnes/yr) in the Humboldt Current large marine ecosystem for 1950–2010. Data from http://www.seaaroundus.org/. Note that the data in this figure include estimates of fish discarded as bycatch as well as landed weights, and so are higher than officially reported figures which only consider the landed catch

variability, but the total catch has stayed consistently high and even the landed weights are above 4 million tonnes per year (Fig. 5.14).

Sardines and anchovies coexist and have been extensively fished in five regions in the world ocean; that is, off Japan and in the big four coastal upwelling systems. Populations of sardines and anchovies have undergone large changes in abundance in each of the five regions. Often, anchovies were abundant when sardines were relatively scarce and vice versa (Fig. 5.11; Lluch-Belda et al. 1992; see also Fig. 3.15). Sardines in the different regions of the Pacific Ocean (and in the Canary Current) tend to be abundant at almost the same time (e.g., Kawasaki 1983).

The observed regime cycles of the abundance of sardines and anchovies follow two distinct types (Lluch-Belda et al. 1992). Type 1 species, notably sardines, acquire smaller body sizes at the same age and the capability to swim longer distances in the high-abundance phase. In the low-abundance phase, they are more or less restricted to limited areas with a larger body size and are non-migratory. In particular, sardine populations expand rapidly poleward during warming periods, greatly shifting their main spawning and feeding areas. Type 2 species, such as anchovies, are distributed more evenly along their normal geographical range regardless of conditions, although population size can change rapidly by large amounts.

Probably one of the most well-known examples of ecological regime shifts in the ocean in the context of overfishing is that between anchoveta and sardines in the Peruvian upwelling region (Fig. 5.11). Here, we can identify either anchoveta-dominated periods of low sardine stocks (1960s, 1990s) or vice versa (1980s). Similar ecological regime shifts have been identified in other major upwelling regions (Chavez et al. 2003; Klyashtorin and Lyubushin 2007). The key

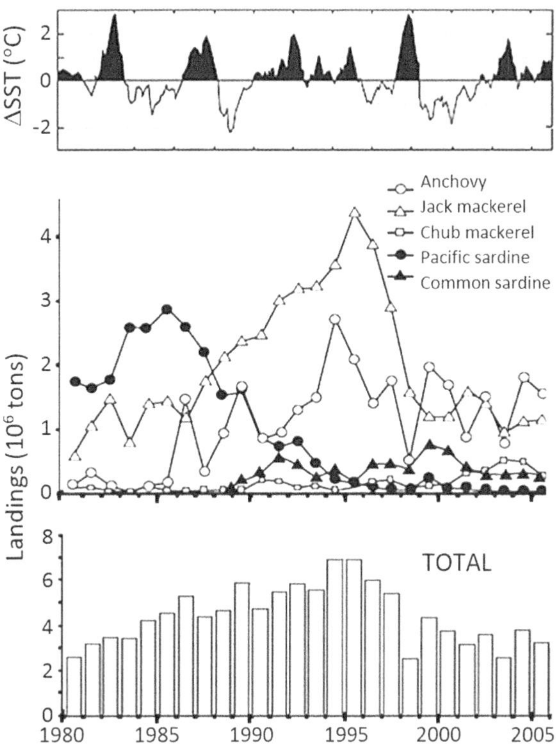

Fig. 5.14 Total annual landings for the five most important pelagic species caught by the Chilean purse seine fleet during the period 1980–2005. *Top panel* shows SST anomaly, the *middle panel* total landings and the *bottom panel* landings of the five species. Reproduced from data from Thiel et al. (2007)

question here is whether such dramatic changes are caused by climate variability such as El Niño events or the Pacific Decadal Oscillation (see Chap. 3), by *overfishing*, or by a combination thereof. As the reader may imagine, debates on this issue can be fairly heated.

It has been proposed that the regime shift from a sardine-dominated system to an anchovy-dominated system (or vice versa) may ultimately be mediated by trophic feedbacks. The consequences of these regime shifts, which have been analysed for Peru and northern Chile by Alheit and Niquen (2004), also extend to central Chile. During warm periods, the preferred prey items of anchovy (large copepods and euphausids) become less available, while predation pressure on adult anchovies increases, due to invasions of jack mackerel into coastal waters (Alheit and Niquen 2004). Simultaneously, sardines, which are also important predators on anchovy eggs (Alheit 1987), may be favoured because they have a wider prey spectrum that includes phytoplankton.

While large temporal fluctuations of individual fish stocks and their associated multidecadal regime shifts can be attributed initially to natural fluctuations within marine ecosystems (Klyashtorin and Lyubushin 2007), the amount of fish caught

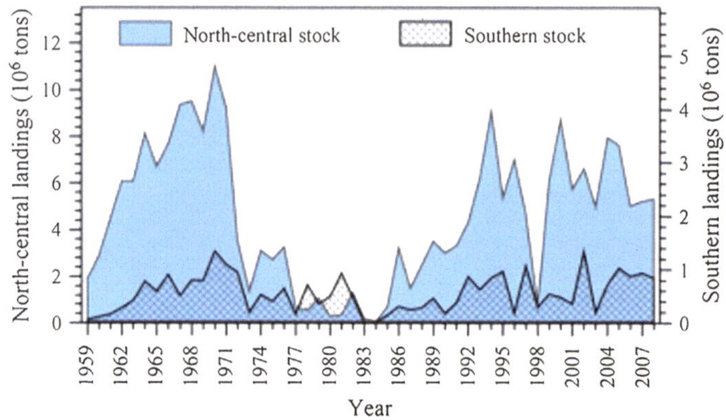

Fig. 5.15 North-central (5°–14° S) and southern (15°–18° S) stock Peruvian anchovy landings between 1959 and 2008 (*Source* IMARPE). Taken from Schreiber et al. (2011)

relative to the total stock has played a central role in the collapse of some commercial fish stocks. For instance, the collapse of Peruvian anchoveta in the early 1970s can be primarily attributed to unsustainable high catch rates during a sequence of stronger El Niño events (Pauly and Tsukayama 1987). In contrast, Peruvian anchoveta stocks remained relatively high from the late 1980s onwards, during a period dominated by La Niña events, despite the short-term and transient dramatic decline during the 1997/1998 El Niño (Fig. 5.15).

In the coastal waters off Chile, comparison with the SST anomalies shows that landings of anchovy correlate negatively with SST anomalies (Yáñez et al. 2001) while chub mackerel landings may also correlate negatively with SST (Fig. 5.14). In contrast, interannual variations in the landings of the other three key species (jack mackerel, Pacific sardine and common sardine) seem to be largely independent of variations in SST. The stabilisation of maximum anchovy landings between 1.5 and 2.7 million tonnes/yr during the 1990s and the parallel decline of the landings of the Pacific sardine in the entire southeastern Pacific reflects a regime shift from the warm 'sardine regime' to a cool 'anchovy regime' (Chavez et al. 2003; Alheit and Niquen 2004; Halpin et al. 2004), also seen in Chilean coastal waters. Overall, however, Chilean fish landings seem less vulnerable to El Niño events, mainly because they rely less on anchovy catches.

For completeness it should be noted that sharp decreases in the number of Peruvian seabirds (Fig. 5.11), based on *guano* production, in the mid-1940s, the late 1950s and during 1965 were the results of El Niño events making their main prey, the anchoveta, scarce or unavailable (Duffy and Siegfried 1987). The avian population never recovered fully from the 1965 El Niño, probably because heavy commercial fishing for the anchoveta reduced the birds' food supply leading to a large non-breeding population and reduced rates of increase.

5.12 Effects of the Oxygen Minimum Zone

The Peruvian-Chilean upwelling system is characterized by high photoautotrophic primary production, driven by the upwelling of nutrient-rich waters, and the biomass produced supports large fish populations. A significant proportion of the produced biomass, however, also sinks through the water column and is remineralized below the surface, contributing to oxygen depletion in intermediate water depths. This contributes to the formation and expansion of the eastern Pacific oxygen minimum zone, which is a prominent feature in Peru and northern Chile closely related to the Equatorial Subsurface Water. The Equatorial Subsurface Water, which normally occupies the intermediate (200–500 m) layer (Blanco et al. 2001), may ascend to much shallower depths (<50 m) near the coast due to upwelling (Morales et al. 1999). The influence of this low-oxygen water on pelagic communities in coastal waters is not well understood.

The oxygen minimum zone off Peru, Chile and Ecuador in the South Pacific Ocean is the largest oceanic area where oxygen concentrations are reported to fall below the detection limit of the most sensitive oxygen sensors (10–100 nM) (Fig. 5.16; e.g., Canfield et al. 2010). In the absence of oxygen, organic carbon degradation has been historically attributed to heterotrophic denitrification, the reduction of nitrate (NO_3^-) via nitrite (NO_2^-) to dinitrogen gas (N_2) (e.g., van de Vossenberg et al. 2008) as shown in Fig. 1.11. While some in situ experiments have confirmed active heterotrophic denitrification in oxygen minimum waters, however, numerous studies have demonstrated that *anammox*, the anaerobic oxidation of ammonium (NH_4^+) with nitrite (NO_2^-) to N_2, is responsible for the major loss of fixed nitrogen from the oxygen minimum zones off Peru (Van Mooy et al. 2002; Lam et al. 2009), Chile (Deutsch et al. 2007) and other upwelling regions (see Sect. 2.2). Although oxygen minimum zones constitute only about 0.1 % of the ocean volume worldwide, it is estimated that anywhere between 20 and 40 % of the total loss of oceanic nitrogen occurs in these zones (Gruber 2004).

Quinones et al. (2010) have attempted a box model of the nutrient budgets for the whole of the Humboldt Current system, rather than the smaller areas covered by earlier researchers. They used the methodology of Gordon et al. (1996), which relies on developing coupled water and salt budgets to estimate material exchanges and fluxes, and using local Redfield ratios to specify individual chemical species in five regions along the coasts of Peru and Chile. Offshore limits were determined from the local coast-ocean chlorophyll gradient. The results suggest that northern (18°–27° S) and central Chile (27°–32° S) have the highest net inputs of both nitrogen and phosphorous, and that these inputs are about an order of magnitude higher than for the other three regions. While this suggests these regions should be highly productive, central Chile is not, possibly because of a high rate of export of water from the region so that phytoplankton "seed" cells cannot become established in time to make use of the available nutrients. The model also suggested that the Humboldt Current system has a severe shortage of nitrogen, because of

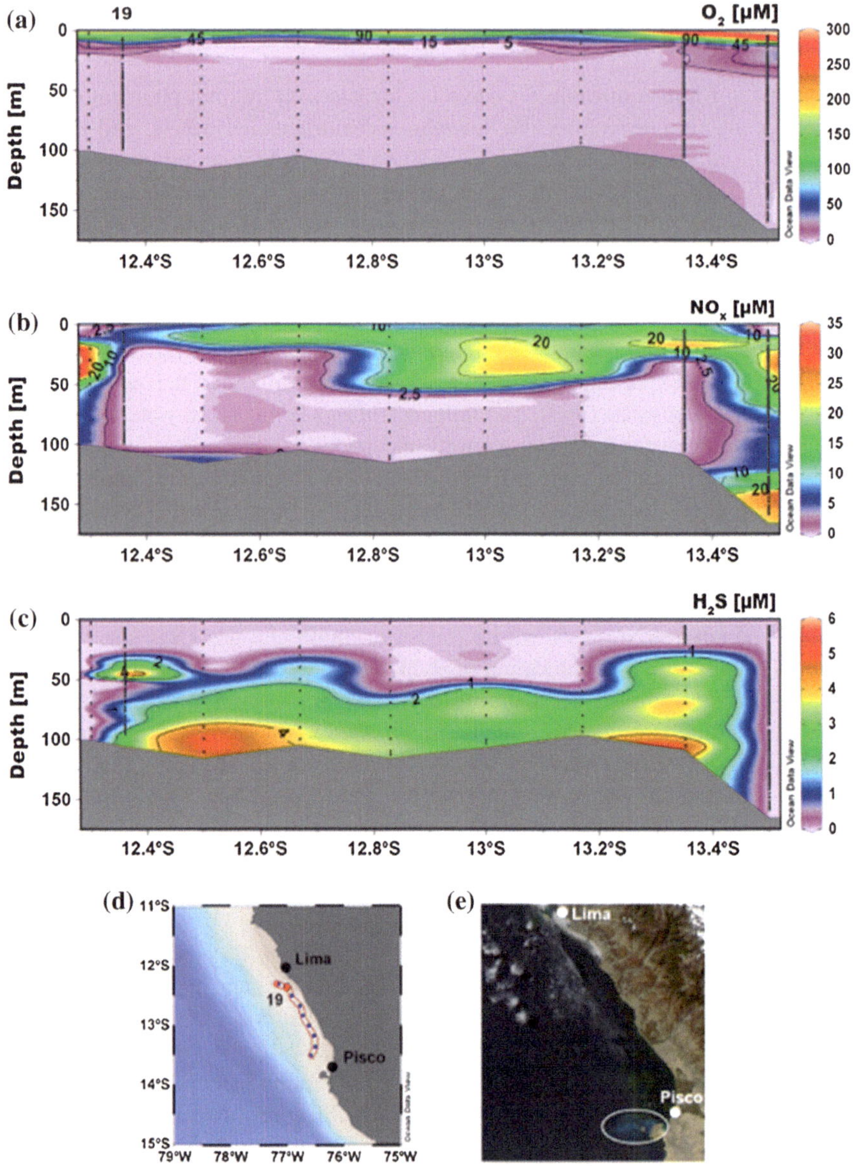

Fig. 5.16 Extent of the sulfidic plume off the Peruvian coast. **a** Vertical distribution of oxygen concentrations. **b** Vertical distribution of NOx (the sum of nitrate and nitrite) concentrations. **c** Vertical distribution of hydrogen sulphide concentrations. **d** Stations sampled along the Peruvian coast between Lima and Pisco. **e** Satellite image (MODIS) showing a colloidal S° plume (*white circle*) on May 8th, 2009, which is turquoise discoloured surface-water that forms upon hydrogen sulphide oxidation. Taken from Schunck et al. (2013)

denitrification, amounting to between 22.5 and 55.9 × 10^{12} g N/yr, depending on how the calculation is made.

The shelf from southern Peru to northern Chile is extremely narrow at 10–15 km width in comparison with central Peru and southern-central Chile, where it is 40–60 km wide (Strub et al. 1998), and where the oxygen minimum zone extends over a wide area of the shelf, promoting distinct biogeochemical processes (Gutiérrez 2000; Gutiérrez et al. 2008; Neira et al. 2001). This characteristic of northern areas could thus affect pathways (i.e., aerobic or anaerobic) associated with organic matter degradation in the sediments, which is an important source of regenerated nutrients to the water column. Off Mejillones (23° S), for instance, a high percentage (86 %) of photosynthetically produced particulated protein is degraded within the upper 30 m of the water column (Pantoja et al. 2004), coinciding with oxygenated waters. In consequence, organic matter reaching the sediments at greater depths is depleted of proteins. Over the shallower shelf sediments, where preserved fish debris and bones are also found (Milessi et al. 2005), high pigment concentrations have been reported (e.g., Muñoz et al. 2005), suggesting much greater burial of phytoplankton detritus.

Thus, there is a narrow band of inshore sediments that are enriched in fresh organic matter coming from the water column. The remineralization of this material can generate an important flux of nutrients contributing to fertilization of the water column. Similar predictions can be made for other areas of high primary production along the coast of northern Chile, where upwelled waters containing preformed nutrients are enriched with recycled nutrients derived from the degradation of organic matter in shelf sediments. However, the relevance of the sea floor as a whole in the system to water column fertilization and biological productivity as well as its relevance in the global carbon cycle has not been well examined. Walsh (1981) made the first estimate of a carbon budget for the region, including the effects of the sea floor, but since then information is available only for the role of sediments near the main upwelling centres, and almost nothing is known about the biogeochemical processes along the large extent of the margin between them (Thiel et al. 2007).

The low oxygen concentrations in subsurface waters of the Humboldt Current system influence predator-prey interactions in the plankton by preventing some species from migrating to deeper waters. Dominant zooplankton, which usually aggregate near the upwelling centres (Escribano and Hidalgo 2000), must cope with such low-oxygen conditions. The options are either to avoid the oxygen minimum zone, which restricts the population to the upper oxygenated layer, as Escribano (1998) has reported for some dominant copepods, or to evolve some metabolic adaptations to withstand poor oxygen conditions, such as those described in González and Quiñones (2002). It has been observed that several abundant epipelagic species do concentrate in the upper 50 m without exhibiting diel vertical migration (e.g., Escribano 1998), although some euphausiids, such as *Euphausia mucronata*, may temporarily enter the oxygen minimum zone (Antezana 2002) or even reside in it, like the copepod *Eucalanus inermis* (Hidalgo et al. 2005). Thus, the oxygen minimum zone cannot be considered only as a constraint to an organism's vertical distribution because several species may use it as their habitat, either temporarily or permanently.

Considering the general effect of the oxygen minimum zone on benthic communities, and based on the limited amount of biological sampling available at that time, Gallardo (1963) proposed the existence of three main benthic zones for the local eukaryotic communities:

(1) an upper sublittoral zone, up to 50 m deep, with favourable conditions for the development of 'normal' benthic communities,
(2) a lower sublittoral zone, from 50 to 300–400 m (varying with latitude and coinciding with the extent of the oxygen minimum zone), in which only those organisms highly adapted to cope with oxygen deficiency and high organic loadings are able to thrive (basically this means small polychaetes, oligochaetes, nematodes and a few molluscs), and
(3) a bathyal area, associated mainly with oxygen-richer Antarctic Intermediate Water, with a diverse and rich fauna (dominated by annelids, crustaceans, molluscs and echinoderms) that benefits from enhanced oxygen and good quality and quantity of sediment organic matter.

The oxygen minimum zone has a strong effect on the bathymetric distribution of sublittoral soft-bottom communities along the coast in the system. One of the most distinguishing features of benthic shelf communities within sediments affected by the oxygen minimum zone off Chile is the presence of extensive mats of the filamentous, sulphide-oxidising bacteria *Thioploca* and *Beggiatoa* (e.g., Gallardo 1963; Arntz et al. 2006) that can dominate the biomass. These mats are also found on the central and southern Peruvian shelf (Rosenberg et al. 1983). However, the contribution of these microbial communities to the total primary production of the system and their function in structuring communities of the oxygen-minimum zone is still scarcely known.

Macrofaunal assemblages of the oxygen minimum zone have low diversity and are typically composed of organisms with morphological and metabolic adaptations and feeding strategies suited to these conditions. Gutiérrez et al. (2008) examined 12 years of monthly data from the shelf near Callao (12° S) and showed how the passage of coastal trapped waves caused an increase in bottom oxygen concentrations one month later. They identified three successions within the benthos: in conditions of severe hypoxia or anoxia the only benthos (both macro- and meiofaunal) consisted of nematodes, with very few macrofaunal species. When oxygen concentrations were in the 10–20 μM/kg range, the dominant species was mats of the sulphide bacterium *Thioploca*. Oxygen concentrations exceeding 40 μM/kg were high enough for the third level to develop, with macrofauna beginning to colonize the region and eventually developing sustainable populations. Changes between the three regimes were found to be rapid, over a period of months as oxygen levels increased.

Oxygen minimum zones can sporadically accumulate hydrogen sulphide (H_2S), which is toxic to most multicellular organisms and has been implicated in massive fish kills. During a cruise to the oxygen minimum zone off Peru in January 2009, Schunck et al. (2013) observed such a sulphidic plume in continental shelf waters, covering an area of 5500 km^2 and which contained $\sim$2.2 $\times$ 10^4 tonnes of H_2S (Fig. 5.16). This was the first time that H_2S was measured in the Peruvian region and with a volume of $\sim$440 km^3 was the largest sulphidic plume ever reported for oceanic waters. Similar features are a regular occurrence in the Benguela upwelling system (see Chap. 7).

5.13 Carbon Fluxes

Owing to the relatively high productivity and physical exchanges with deeper water and the adjacent ocean, continental shelves play a particularly important role in the ocean's carbon cycle. The solubility and biological pumps, which also operate on the much smaller shelf scale, are known as the *continental shelf pump* (Tsunogai et al. 1999). Carbon fluxes of the continental shelf pump in upwelling systems are complex. In principal, atmospheric cooling of upwelled CO_2-rich subsurface water leads to supersaturation, facilitating the efflux of CO_2 back into the atmosphere. On the other hand, enhanced primary production can significantly reduce the CO_2 concentration in surface water, reducing or even reversing the air-sea gas exchange.

High primary production in the Humboldt Current system (3.7–5.2 g C/m^2/day off Peru and 3.0–7.0 g C/m^2/day off Chile; Montecino et al. 2006) constitutes an important way of sequestering CO_2 and supports a high rate of particulate organic matter export to deeper waters (González et al. 2000; Pantoja et al. 2004). This material, which is partly remineralised in the water column, strengthens the oxygen minimum zone and promotes biogeochemical anaerobic processes. In this sense, a sequence of mechanisms that are determined by the oceanographic conditions is regulating the chemistry of the water column and the seafloor. These estimates of primary production along the Chilean coast are similar to those of the Peru and about double those of the California upwelling system (1000–2500 mg C/m^2/day; Olivieri and Chavez 2000).

While Walsh (1981) produced the first carbon budget for the region, the first studies that considered several trophic levels of the pelagic system (Peterson et al. 1988) and trophic models of carbon flux (Bernal et al. 1989) were conducted in the coastal area off Concepción during the late 1980s. These initial studies supported the classical view that upwelling areas are characterized by short food chains dominated by large chain-forming diatoms and few small clupeiform fish species or the 'traditional food chain' (Ryther 1969). More recently, this view has been challenged, highlighting the relevance of the microbial loop (e.g., Troncoso et al. 2003) and the gelatinous food web (González et al. 2004). These trophic flows are important throughout the whole year in oceanic areas and are highly relevant during the non-productive periods (including El Niño events) in coastal upwelling areas.

The carbon budget of the photosynthetically generated organic matter in the coastal areas of the Humboldt Current system has been under debate for many years (e.g., Bernal et al. 1989). It is accepted nowadays that the fraction of the primary production is that removed from the photic zone, which is highly variable on an annual basis (Hebbeln et al. 2000), strongly depends on the various biological (internal metabolism), physical (stratification/mixing), and chemical (nutrient rich/poor waters) processes involved as well as the time of year. However, the sources of this variability, both in space and time, have been poorly analysed until recently (Morales and Lange 2004), mainly because of the lack of long-term time-series studies.

Although there are no direct measurements of bacterial secondary production off Peru, recent studies report very high bacterial secondary production (<5 g C/m^2/day) in the coastal area of northern-central Chile (e.g., Troncoso et al. 2003). This suggests there is a tight coupling between primary production and bacterial secondary production, and measurements from this region suggest that anywhere between 10 and 96 % of primary production can be used by bacteria (Eissler and Quiñones 1999; Troncoso et al. 2003; Cuevas et al. 2004), although percentages in the range of 19–50 % seem more realistic, and are well within the range (3–55 % of primary production) of those described for other upwelling systems in the world oceans (Ducklow 2000). In addition, both zooplankton grazing and export production (e.g., Grünewald et al. 2002) gave values of 2–10 % of primary production, while particle fluxes can be up to about 20 % at 65 m depth and 3–8 % at 300 m depth (González et al. 2000, 2004), and are composed primarily of zooplankton faecal material. These carbon flows are more representative of the coastal upwelling systems of Antofagasta and Concepción because they are the most studied areas (from an oceanographic point of view) along the Chilean coast, although particle fluxes off Peru are similar (Quinones et al. 2010) and at present there is no good estimate of the carbon budget for the whole system. In fact, calculations have even suggested that phytoplankton production cannot support the known catch rates of the fishery in central and southern Chile (Cubillos et al. (1998)! So there is certainly a need for more advanced biogeochemical models that incorporate the full complexity of existing trophic webs for this region.

As regards carbon dioxide, even though the Peru-Chile coastal upwelling system is among the most productive oceanic areas in the world ocean, huge oversaturation of CO_2 with respect to the atmosphere has been reported. Partial gas pressures (pCO_2) of up to 1200 μatm have been measured, compared with a partial pressure of CO_2 in the atmosphere of about 400 μatm. Conversely, very low values down to 140 μatm and the associated gas undersaturation have also been observed as a result of inorganic carbon fixation by phytoplankton (e.g., Torres et al. 1999). Such dramatic changes in the CO_2 concentrations can occur on timescales of a week.

The export of carbon from upwelling areas occurs in the form of sedimentation of particulate organic matter and offshore dispersion by oceanic eddies in the form of sub-mesoscale filaments. The subsequent warming of the CO_2-rich water contained in such filaments induces carbon outgassing in the ambient ocean. Given the short-term variability of carbon sequestration and the complexity of ocean dynamics in coastal upwelling regions, it is extremely difficult to derive accurate carbon budgets for such systems. Nevertheless, observational evidence of CO_2 gas

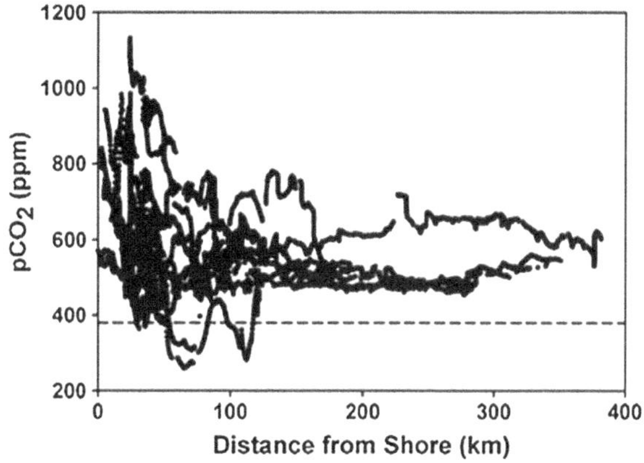

Fig. 5.17 Sea surface CO_2 gas pressure (pCO_2) versus distance from the shore off central Peru at 12° S. The *dashed line* indicates the mean atmospheric pCO_2 of 378 ppm at the time of cruises. Taken from Friedrich et al. (2008)

pressures in the coastal waters off Peru reveal that the upwelling region operates overall as a carbon source for the atmosphere (Friederich et al. 2008) (Fig. 5.17).

The Peru-Chile upwelling system is strongly influenced by oxygen-depleted water inherent with the Oxygen Minimum Zone (OMZ) (Fig. 5.7). The comparison with other upwelling regions indicates that coastal upwelling areas associated with OMZs, such as the Arabian Sea (Goyet et al. 1998) and the Peruvian and Chilean coasts (Friederich et al. 2008; Paulmier et al. 2008), are sources of CO_2 to the atmosphere because denitrification leads to lower concentrations of nitrate and excess of dissolved inorganic carbon (DIC) relative to nitrogen.

5.14 Summary

This overview of the Peru-Chile upwelling system illustrates the complex physical-ecological interplay between nutrient upwelling, oxygen minimum zones, different trophic levels of the food web, climate variability and fisheries on seasonal, interannual and interdecadal timescales. In particular, this upwelling system is exposed to anoxic subsurface water of the oxygen minimum zone of the eastern tropical Pacific Ocean. Complex microbial biogeochemical processes, which are still not fully understood, take place in this anoxic zone in close interaction with anoxic sediments and bacterial secondary production is an important component of the coastal carbon cycle. On one hand, the anoxic environment facilitates the seabed release of the otherwise limiting micronutrients iron and cobalt. On the other hand, we have seen the creation of a gigantic toxic sulphidic plume on the continental shelf. Given the high productivity of this upwelling region and the associated carbon fluxes, continued research effort is required to monitor and further explore the ecosystem response of the Peru-Chile upwelling region to global warming.

References

Aceituno, P. 1992. El Niño, the Southern Oscillation, and ENSO: Confusing names for a complex ocean-atmosphere interaction. *Bulletin of the American Meteorological Society* 73: 483–485.

Alheit, J. 1987. Egg cannibalism versus egg predation: Their significance in anchovies. *South African Journal of Marine Science* 5: 467–470.

Alheit, J., and P. Bernal. 1993. Effects of physical and biological changes on the biomass yield of the Humboldt Current ecosystem. In *Large Marine Ecosystems: Stress, Mitigation, and Sustainability*, ed. K. Sherman, L.M. Alexander, and B.D. Gold, 53–68. Washington, D.C: American Association for the Advancement of Science.

Alheit, J., and M. Niquen. 2004. Regime shifts in the Humboldt Current ecosystem. *Progress in Oceanography* 60: 201–222.

Antezana, T. 2002. Vertical distribution and diel migration of *Euphausia mucronata* in the oxygen minimum layer of the Humboldt Current. In *Oceanography of the Eastern Pacific II*, ed. J. Farber, 13–28. Ensenada, Mexico: CICESE.

Arntz, W.E., V.A. Gallardo, D. Gutiérrez, E. Isla, L.A. Levin, J. Mendo, C. Neira, G.T. Rowe, J. Tarazona, and M. Wolff. 2006. El Niño and similar perturbation effects on the benthos of the Humboldt, California, and Benguela Current upwelling ecosystems. *Advances in Geosciences* 6: 243–265.

Atkinson, L.P., A. Valle-Levinson, D. Figueroa, R. De Pol-Holz, V.A. Gallardo, W. Schneider, J. L. Blanco, and M. Schmidt. 2002. Oceanographic observations in Chilean coastal waters between Valdivia and Concepción. *Journal of Geophysical Research* 107: C7. doi:10.1029/2001JC000991.

Ayón, P., M.I. Criales-Hernandez, R. Schwamborn, and H.J. Hirche. 2008. Zooplankton research off Peru: a review. *Progress in Oceanography* 79: 238–255.

Bakun, A. 1987. Monthly variability in the ocean habitat off Peru as deduced from maritime observations from 1953 to 1984. In Pauly, D., and I. Tsukayama (see below), p. 46–74.

Bakun, A. 1996. *Patterns in the ocean: Ocean processes and marine population dynamics*. University of California Sea Grant, California, USA, in cooperation with Centro de Investigaciones Biologicas de Noroeste, La Paz, Baja California Sur, Mexico, 323 pp.

Bakun, A., and S.J. Weeks. 2008. The marine ecosystem off Peru: what are the secrets of its fishery productivity and what might its future hold? *Progress in Oceanography* 79: 290–299.

Berghaus, H.K.W. 1837–1847. *Berghaus' Physikalischer Atlas*. Gotha.

Bernal, P.A., R. Ahumada, H. González, S. Pantoja, and A. Troncoso. 1989. Flujo de energía en un modelo trófico en la Bahía de Concepción. *Biología Pesquera* 18: 5–14.

Bjerknes, J. 1966. A possible response of the atmospheric Hadley circulation to equatorial anomalies of ocean temperature. *Tellus* 18: 820–829.

Bjerknes, J. 1969. Atmospheric teleconnections from the equatorial Pacific. *Monthly Weather Review* 97: 163–172.

Blanco, J.L., A. Thomas, M.-E. Carr, and P.T. Strub. 2001. Seasonal climatology of hydrographic conditions in the upwelling region off northern Chile. *Journal of Geophysical Research* 106: 11451–11467.

Brainard, R., and D. McLain. 1987. Seasonal and interannual subsurface temperature variability off Peru, 1952 to 1984. In Pauly, D., and I. Tsukayama (see below), pp. 14–45.

Brink, K., D. Halpern, A. Huyer, and R. Smith. 1983. The physical-environment of the Peruvian upwelling system. *Progress in Oceanography* 12(3): 285–305.

Broad, K. 1999. *Climate, Culture and Peruvian Fisheries*. Ph.D. Thesis, Lamont-Doherty Earth Observatory, Palisades, New York.

Bruland, K.W., E.L. Ruea, G.J. Smitha, and G.R. DiTullio. 2005. Iron, macronutrients and diatom blooms in the Peru upwelling regime: Brown waters of Peru versus blue waters. *Marine Chemistry* 93: 81–103.

Canfield, D.E., F.J. Stewart, B. Thamdrup, L. De Brabandere, T. Dalsgaard, et al. 2010. A cryptic sulfur cycle in oxygen-minimum-zone waters off the Chilean coast. *Science* 330(6009): 1375–1378.

Carr, M.-E., and E.J. Kearns. 2003. Production regimes in four eastern boundary current systems. *Deep-Sea Research II* 50: 3199–3221.

Carr, M.E., P.T. Strub, A.C. Thomas, and J.L. Blanco. 2002. Evolution of 1996-1999 La Niña and El Niño conditions off the western coast of South America: A remote sensing perspective. *Journal of Geophysical Research* 107(C12): 3236. doi:10.1029/2001/JC001183.

Carrasco, S., and O. Lozano. 1989. Seasonal and long-term variations of zooplankton volumes in the Peruvian Sea, 1964–1987. In *The Peruvian Upwelling Ecosystem: Dynamicsand Interactions*, ed. D. Pauly, P. Muck, J. Mendo and I. Tsukyama, 82–85. ICLARM Conference Proceedings 18. IMARPE, GTZ, and ICLARM, Manila.

Carrillo, C. 1892. Hidrografía oceánica: Las corrientes oceánicas y estudios de la Corriente Peruana ó de Humboldt. *Boletín de la Sociedad Geográfica de Lima* 2: 72–110.

Castilla, J.C., N.A. Lagos, R. Guiñez and J. Largier. 2002. Embayments and nearshore retention of plankton: The Antofagasta Bay and other examples. In *The Oceanography and Ecology of the Nearshore and Bays in Chile*, ed. J.C. Castilla and J.L. Largier, 179–203. Santiago, Chile: Ediciones Universidad Católica de Chile.

Chavez, F.P., and J.R. Toggweiler. 1995. Physical estimates of global new production: the upwelling contribution. In *Upwelling in the Ocean: Modern Processes and Ancient Records*, ed. C.P. Summerhayes, K.-C. Emeis, M.V. Angel, R.L. Smith and B. Zeitschel, 313–320. Chichester: Wiley.

Chavez, F.P., J. Ryan, S.E. Lluch-Cotaand, and M. Ñiquen. 2003. From anchovies to sardines and back: Multidecadal change in the Pacific Ocean. *Science* 299: 217–221.

Coale, K.H., K.S. Johnson, S.E. Fitzwater, S.P. Blain, T.P. Stanton, and T.L. Coley. 1998. Ironex-I, an *in situ* iron enrichment experiment: experimental design, implementation and results. *Deep-Sea Research II* 45: 919–945.

Codispoti, L.A., R.T. Barber and G.E. Friederich. 1989. Do nitrogen transformations in the poleward undercurrent off Peru and Chile have a globally significant influence? In *Poleward Flows Along Eastern Ocean Boundaries*, ed. S.J. Neshyba, C.H.K. Mooeres, R.L. Smith and R.T. Barber. New York: Springer. doi:10.1029/CE034p0281.

Corredor-Acosta, A., C.E. Morales, S. Hormazabal, I. Andrade, and M.A. Correa-Ramirez. 2015. Phytoplankton phenology in the coastal upwelling region off central-southern Chile (35° S–38° S): Time-space variability, coupling to environmental factors, and sources of uncertainty in the estimates. *Journal of Geophysical Research* 120(2): 813–831. doi:10.1002/2014JC010330.

Cubillos, L.A., E. Nunez, and R. Arcos. 1998. Produccion primaria requerida para sustenar el desembarque de peces pelagicos en Chile. *Investigaciones marinas* 26: 83–96.

Cuevas, L.A., G. Danieri, B. Jacob, and P. Monteiro. 2004. Microbial abundance and activity in the seasonal upwelling area off Concepción ($\sim$36° S), central Chile: a comparison of upwelling and non-upwelling conditions. *Deep-Sea Research II* 51: 2427–2440.

Cury, P., C. Roy, and V. Faure. 1998. Environmental constraints and pelagic fisheries in upwelling areas: the Peruvian puzzle. *South African Journal of Marine Science* 19: 159–167.

Cury, P., A. Bakun, R.J.M. Crawford, A. Jarre, R.A. Quiñones, L.J. Shannon, and H.M. Verheye. 2000. Small pelagics in upwelling systems: patterns of interaction and structural changes in "wasp-waist" ecosystems. *ICES Journal of Marine Science* 57: 603–618.

Daneri, G., V. Dellarossa, R. Quiñones, B. Jacob, P. Montero, and O. Ulloa. 2000. Primary production and community respiration in the Humboldt Current System off Chile and associated oceanic areas. *Marine Ecology Progress Series* 197: 41–49.

Deutsch, C., J.L. Sarmiento, D.M. Sigman, N. Gruber, and J.P. Dunne. 2007. Spatial coupling of nitrogen inputs and losses in the ocean. *Nature* 445: 163–167.

Dezileau, L., O. Ulloa, D. Hebbeln, F. Lamy, J.L. Reyss and M. Fontugne. 2004. Iron control of past productivity in the coastal upwelling system off the Atacama Desert, Chile. *Paleoceanography* 19: PA3012. doi:10.1029/2004PA001006.

Djurfeldt, L. 1989. Circulation and mixing in a coastal upwelling embayment, Gulf of Arauco. Chile. *Continental Shelf Research* 9(11): 1003–1016.

Ducklow, H.W. 2000. Bacterial production and biomass in the oceans. In *Microbial Ecology of the Oceans,* ed. D. Kirchman, pp 85–120. New York: Wiley-Liss.

Duffy, D.C., and W.R. Siegfried. 1987. Historical variations in food consumption by breeding seabirds of the Humboldt and Benguela upwelling regions. In *Seabirds: Feeding Biology and Role in Marine Ecosystems*, ed. J.P. Croxall, 327–346. Cambridge University Press.

Echevin, V., O. Aumont, J. Ledesmaand, and G. Flores. 2008. The seasonal cycle of surface chlorophyll in the Peruvian upwelling system: A modeling study. *Progress in Oceanography* 79: 167–176.

Eissler, Y., and R.A. Quiñones. 1999. Microplanktonic respiration off northern Chile during El Niño 1997–1998. *Journal of Plankton Research* 21: 2263–2283.

Enfield, D.B., M.P. Cornejo-Rodriguez, R.L. Smith and P.A. Newberger. 1987. The equatorial source of propagating variability along the Peru coast during the 1982–1983 El Niño. *Journal of Geophysical Research* 92: 14,335–14,346. doi:10.1029/JC092iC13p14335.

Escribano, R. 1998. Population dynamics of *Calanus chilensis* in the Chilean Eastern Boundary Humboldt Current. *Fisheries Oceanography* 7: 245–251.

Escribano, R., and P. Hidalgo. 2000. Influence of El Niño and La Niña on the population dynamics of *Calanus chilensis* in the Humboldt Current ecosystem of northern Chile. *ICES Journal of Marine Science* 57: 1867–1874.

Escribano, R., S.A. Rosales, and J.L. Blanco. 2004. Understanding upwelling circulation off Antofagasta (northern Chile): a 3-dimensional numerical-modeling approach. *Continental Shelf Research* 24: 37–53.

Fagan, B. 1999. *Floods, Famines and Emperors: El Niño and the Fate of Civilizations*, 119–138. New York: Basic Books.

Feely, R.A., J. Boutin, C.E. Cosca, Y. Dandonneau, J. Etcheto, H.Y. Inoue, M. Ishii, C. Le Quéré, D.J. Mackey, M. McPhaden, N. Metzl, A. Poisson, and R. Wanninkhof. 2002. Seasonal and interannual variability of CO_2 in the equatorial Pacific. *Deep Sea Research II* 49(13–14): 2443–2469. doi:10.1016/S0967-0645(02)00044-9.

Figueroa, D. 2002. Forcing of physical exchanges in the nearshore Chilean ocean. In *The Oceanography and Ecology of the Nearshore and Bays in Chile*, ed. J.C. Castilla, and J.L. Largier, pp 31–43. Santiago, Chile: Ediciones Universidad Católica de Chile.

Figueroa, D., and C. Moffat. 2000. On the influence of topography in the induction of coastal upwelling along the Chilean coast. *Geophysical Research Letters* 27: 3905–3908.

Fréon, P., M. Barange, and J. Arístegui. 2009. Eastern boundary upwelling ecosystems: Integrative and comparative approaches. *Progress in Oceanography* 83: 1–14.

Friedrich, G.E., and L. Codispoti. 1981. The effects of mixing and regeneration on the nutrient content of upwelling waters off Peru. In *Coastal Upwelling*, ed. F.A. Richards, 221–227. Washington, D.C., USA: American Geophysical Union.

Friederich, G.E., J. Ledesma, O. Ulloa, and F.P. Chavez. 2008. Air–sea carbon dioxide fluxes in the coastal southeastern tropical Pacific. *Progress in Oceanography* 79(2–4): 156–166.

Gallardo, V.A. 1963. Notas sobre la densidad de la fauna bentónica en el sublitoral del norte de Chile. *Gayana Oceanológica* 10: 3–15.

Glantz, M.H. 1979. Science, politics, and economics of the Peruvian anchovy fishery. *Marine Policy*, July, 201–210.

González, H.E., P. Bernal, and R. Ahumada. 1987. Desarrollo de dominancia local en la taxocenosis de fitoplancton de Bahía de Concepción, Chile, durante un evento de surgencia. *Revista Chilena de Historia Natural* 60: 19–35.

González, H.E., S. Pantoja, J. Iriarte, and P.A. Bernal. 1989. Winter-spring variability of size fractioned autotrophic biomass in Concepción Bay, Chile. *Journal of Plankton Research* 11: 1157–1167.

González, H.E., G. Daneri, D. Figueroa, J.L. Iriarte, N. Lefevre, G. Pizarro, R. Quiñones, M. Sobarzo, and A. Troncoso. 1998. Primary production and its fate in the pelagic food web and deep-sea, and ocean-atmosphere CO_2 exchange in the northern Humboldt Current (23°S): possible effects of the 1997–1998 El Niño in Chile. *Revista Chilena de Historia Natural* 71: 429–458.

González, H.E., V.C. Ortíz, and M. Sobarzo. 2000. The role of faecal material in the particulate organic carbon flux in the northern Humboldt Current, Chile (23°S), before and during the 1997–1998 El Niño. *Journal of Plankton Research* 22: 499–529.

González, H.E., R. Giesecke, C.A. Vargas, A. Pávez, J. Iriarte, P. Santibáñez, L. Castro, R. Escribano, and F. Pages. 2004. Carbon cycling through the pelagic foodweb in the northern Humboldt Current off Chile (23° S). *ICES Journal of Marine Science* 61: 572–584.

González, R.R., and R.A. Quiñones. 2002. Ldh activity in *Euphausia mucronata* and *Calanus chilensis*: implications for vertical migration behaviour. *Journal of Plankton Research* 24: 1349–1356.

Gordon, D.C., P.R. Boudreau, K.H. Mann, J.-E. Ong, W.L. Silvert, S.V. Smith, G. Wattayakorn, F. Wulff and T. Yanagi. 1996. LOICZ Biogeochemical Modelling Guidelines. *LOICZ Reports and Studies*, 5, LOICZ Core Project Office, Den Burg-Texel, Netherlands, 104 pp.

Goyet, C., F.J. Millero, D.W. O'Sullivan, G. Eischeid, S.J. McCue and R.G.J. Bellerby. 1998. Temporal variations of pCO_2 in surface seawater of the Arabian sea in 1995. *Deep Sea Res I* 45 (4–5): 609–623.

Gruber, N. 2004. The dynamics of the marine nitrogen cycle and its influence on atmospheric CO2 variations. In *The Ocean Carbon Cycle and Climate*, ed. M. Follows and T. Oguz, 97–148. Dordrecht: NATO ASI Series, Kluwer Academic.

Grünewald, A., C. Morales, H.E. González, C. Sylvester, and L. Castro. 2002. Grazing impact of copepod assemblages and gravitational flux in coastal and oceanic waters off central Chile during two contrasting seasons. *Journal of Plankton Research* 24: 55–68.

Gunther, E.R. 1936. A report on oceanographic investigations in the Peru Coastal Current. *Discovery Reports* 13: 107–276.

Gutiérrez, D. 2000. *Bioperturbación y macrofauna en fondos sublitorales de un área de surgencias frente a Chile Central (36°30'S): variación espacial y temporal en el período 1997–1999.* Ph.D. thesis, Universidad de Concepción, Concepción, Chile.

Gutiérrez, D., E. Enriquez, S. Purca, L. Quipúzcoa, R. Marquina, G. Flores and M. Graco. 2008. Oxygenation episodes on the continental shelf of central Peru: Remote forcing and benthic ecosystem response. *Progress in Oceanography* 79: 177–189. doi:10.1016/j.pocean.2008.10.025.

Gutiérrez, D., et al. 2011. Coastal cooling and increased productivity in the main upwelling zone off Peru since the mid twentieth century. *Journal of Geophysical Research* 38: L07603. doi:10.1029/2010GL046324.

Halpin, P.M., P.T. Strub, W.T. Peterson, and T.R. Baumgartner. 2004. An overview of interactions among oceanography, marine ecosystems, climatic and human disruptions along the eastern margins of the Pacific Ocean. *Revista Chilena de Historia Natural* 77: 371–409.

Hebbeln, D., M. Marchant, T. Freudenthal, and G. Wefer. 2000. Surface sediment distribution along the Chilean continental slope related to upwelling and productivity. *Marine Geology* 164: 119–137.

Hidalgo, P., R. Escribano, and C.E. Morales. 2005. Ontogenetic vertical distribution and diel migration of the copepod *Eucalanus inermis* in the oxygen minimum zone off northern Chile (20–21°S). *Journal of Plankton Research* 27: 519–529.

Hormazábal, S., G. Shaffer, L. Letelier, and O. Ulloa. 2001. Local and remote forcing of sea surface temperature in the coastal upwelling system off Chile. *Journal of Geophysical Research* 106: 16657–16671.

Humboldt, A. 1846. *Cosmos: A Sketch of a Physical Description of the Universe*. London: Green and Longmans.

Huyer, A. 1980. The offshore structure and subsurface expression of sea-level variations off Peru, 1976-1977. *Journal of Physical Oceanography* 10(11): 1755–1768.

Iriarte, J.L., and H.E. González. 2004. Phytoplankton size structure during and after the 1997/1998 El Niño in a coastal upwelling area of the northern Humboldt Current System. *Marine Ecology Progress Series* 269: 83–90.

Iriarte, J.L., G. Pizarro, V.A. Troncoso, and M. Sobarzo. 2000. Primary production and biomass of size-fractionated phytoplankton off Antofagasta, Chile (23–24°S) during pre-El Niño and El Niño 1997. *Journal of Marine Systems* 26: 37–51.

Kawasaki, T. 1983. Why do some pelagic fishes have wide fluctuation in their numbers? Biological basis of fluctuation from the viewpoint of evolutionary ecology. *FAO Fishery Report* 291(3): 1065–1080.

Kessler, W. 2006. The circulation of the eastern tropical Pacific: A review. *Progress in Oceanography* 69(2–4): 181–217.

Klyashtorin, L., and A. Lyubushin. 2007. *Cyclic Climate Changes and Fish Productivity*. Moscow: VNIRO. 223 pp.

Lam, P., G. Lavik, M.M. Jensen, J. van de Vossenberg, M. Schmid, et al. 2009. Revising the nitrogen cycle in the Peruvian oxygen minimum zone. *Proceedings of the National Academy of Sciences USA* 106: 4752–4757.

Lluch-Belda, D., R.A. Schwartzlose, R. Serra, R.H. Parrish, T. Kawasaki, D. Hedgecock, and R.J. M. Crawford. 1992. Sardine and anchovy regime fluctuations of abundance in four regions of the world oceans: a workshop report. *Fisheries Oceanography* 1(4): 339–347.

Lukas, R. 1986. The termination of the Equatorial Undercurrent in the Eastern Pacific. *Progress in Oceanography* 16(2): 63–90.

Marín, V.H., and G.R. Olivares. 1999. Seasonality of primary productivity in Mejillones del Sur Bay (Chile): a process-functional approach. *Revista Chilena de Historia Natural* 72: 629–641.

Marín, V.H., and L.E. Delgado. 2007. Lagrangian observations of surface coastal flows north of 30°S in the Humboldt Current System. *Continental Shelf Research* 27: 731–743.

Marín, V.H., L.E. Delgado, and R. Escribano. 2003a. Upwelling shadows at Mejillones Bay (northern Chilean coast): a remote sensing *in situ* analysis. *Investigaciones Marinas, Valparaíso* 31: 47–55.

Marín, V.H., L.E. Delgado, and G. Luna-Jorquera. 2003b. S-Chlorophyll squirts at 30°S off the Chilean coast (eastern South Pacific): feature-tracking analysis. *Journal of Geophysical Research* 108(C12): 3378. doi:10.1029/2003JC001935.

Martin, J.H., and R.M. Gordon. 1988. Northeast Pacific iron distributions in relation to phytoplankton productivity. *Deep-Sea Research A* 35: 177–196.

Martin, J.H., S.E. Fitzwater, R.M. Gordon, C.N. Hunter, and S.J. Tanner. 1993. Iron, primary production and carbon nitrogen flux studies during the JGOFS North-Atlantic bloom experiment. *Deep-Sea Research II* 40: 115–134.

Mendo, J., L. Pizarro, and S. Castillo. 1987. Monthly upwelling and turbulence indices off the Peruvian coast at Callao and Trujillo during the period 1953–1983. In Pauly, D., and I. Tsukayama (see below), pp. 75–88.

Mesías, J.M., R.P. Matano, and P.T. Strub. 2003. Dynamical analysis of the upwelling circulation off central Chile. *Journal of Geophysical Research* 108(C3): 3085. doi:10.1029/2001JC001135.

Milessi, A., J. Sellanes, V. Gallardo, and C.B. Lange. 2005. Osseous skeletal material and fish scales in marine sediments under the oxygen minimum zone off northern and central Chile. *Estuarine, Coastal and Shelf Science* 64: 185–190.

Montecino, V., and C.B. Lange. 2009. The Humboldt Current System: Ecosystem components and processes, fisheries, and sediment studies. *Progress in Oceanography* 83(1–4, Sp. Iss. SI): 65–79.

Montecino, V., R. Astoreca, G. Alarcón, L. Retamal, and G. Pizarro. 2004. Bio-optical characteristics and primary productivity during upwelling and non-upwelling conditions in a highly productive coastal ecosystem off central Chile (~ 36°S). *Deep-Sea Research II* 51: 2413–2426.

Montecino, V., T.P. Strub, F. Chavez, A. Thomas, J. Tarazona and T. Baumgartner. 2006. Bio-physical interactions off western south-America. In *The Global Coastal Ocean. Interdisciplinary Regional Studies and Syntheses*, ed. A.R. Robinson and K. Brink, 329–390. Cambridge, MA: Harvard University Press.

Moore, C.M., et al. 2013. Processes and patterns of oceanic nutrient limitation. *Nature Geoscience* 6: 701–710.

Moraga, J., E. Valdebenito, and J. Rutllant. 2001. Condiciones oceanográficas durante la fase de relajación de un evento de surgencia invernal frente a Punta Lengua de Vaca, Coquimbo. *Investigaciones Marinas, Valparaíso* 29: 59–71.

Morales, C.E., and C.B. Lange. 2004. Oceanographic studies in the Humboldt Current System off Chile: an introduction. *Deep-Sea Research II* 51: 2345–2348.

Morales, C.E., J.L. Blanco, M. Braun, H. Reyes, and N. Silva. 1996. Chlorophyll-*a* distribution and associated oceanographic conditions in the upwelling region off northern Chile during the winter and spring 1993. *Deep-Sea Research I* 43: 267–289.

Morales, C., S. Hormazabal, and J.L. Blanco. 1999. Interannual variability in the mesoscale distribution of the depth of the upper boundary of the oxygen minimum layer off northern Chile (18°–24° S): Implications for the pelagic system and biogeochemical cycling. *Journal of Marine Research* 57: 909–932.

Muñoz, P., J. Sellanes, C. Neira, M. Palma, L. Levin, S. Pantoja, C.B. Lange and S.A. Salamanca. 2005. Sedimentación v/s mezcla biológica en sedimentos del norte de Chile (Iquique, 20°S) asociados a la zona de mínimo oxígeno (ZMO). *Libro de Resúmenes, Congreso de Ciencias del Mar, Sociedad de Ciencias del Mar de Chile, Chile*, May, 172.

Neira, S., and H. Arancibia. 2004. Trophic interactions and community structure in the upwelling system off Central Chile (33–39° S). *Journal of Experimental Marine Biology and Ecology* 312: 349–366.

Neira, C., J. Sellanes, A. Soto, D. Gutiérrez, and V.A. Gallardo. 2001. Meiofauna and sedimentary organic matter off Central Chile: response to changes caused by the 1997–1998 El Niño. *Oceanologica Acta* 24: 313–328.

Ochoa, N., et al. 2010. Intra- and interannual variability of nearshore phytoplankton biovolume and community changes in the northern Humboldt Current system. *Journal of Plankton Research* 32: 843–855.

Olivares, G. 2001. *Mecanismos de interacción físico-químicas en una zona de surgencia costera: retención de larvas y cierre del ciclo de vida de* Euphausia mucronata. M.Sc. thesis, Universidad de Chile, Santiago, Chile.

Olivieri, R.A., and F.P. Chavez. 2000. A model of plankton dynamics for the coastal upwelling system of Monterey Bay, California. *Deep-Sea Research II* 47: 1077–1106.

Ortlieb, L., R. Escribano, R. Follegati, O. Zuniga, I. Kong, L. Rodriguez, J. Valdes, N. Guzmán, and P. Iratchet. 2000. Recording of ocean-climate changes during the last 2,000 years in a hypoxic marine environment off northern Chile (23° S). *Revista Chilena de Historia Natural* 73: 221–242.

Palma, W., R. Escribano, and S.A. Rosales. 2006. Modeling study of seasonal and inter-annual variability of circulation in the coastal upwelling site of the El Loa River off northern Chile. *Estuarine, Coastal and Shelf Science* 67: 93–107.

Pantoja, S., J. Sepulveda, and H.E. González. 2004. Decomposition of sinking proteinaceous material during fall in the oxygen minimum zone off northern Chile. *Deep-Sea Research I* 51: 55–70.

Parrish, R.H., A. Bakun, D.M. Husby and C.S. Nelson. 1983. Comparative climatology of selected environmental processes in relation to eastern boundary current pelagic fish reproduction. In *Proceedings of the expert consultation to examine changes in abundance and species composition of neritic fish resources*, ed. G.D. Sharp and J. Csirke. *FAO Fishery Reports* 291: 731–778.

Paulmier, A., D. Ruiz-Pino, and V. Garçon. 2008. The oxygen minimum zone (OMZ) off Chile as intense source of CO_2 and N_2O. *Continental Shelf Research* 28: 2746–2756.

Pauly, D., and M. Soriano. 1987. Monthly spawning stock and egg production of Peruvian anchoveta (*Engraulis ringens*), 1953 to 1982. In Pauly, D., and I. Tsukayama (see below), pp. 167–178.

Pauly, D., and I. Tsukayama. 1987. The Peruvian anchoveta and its upwelling ecosystem: three decades of change. Instituto del Mar del Perú (IMARPE), Callao, Perú; Deutsche Gesellschaft fur Technische Zusammenarbeit (GTZ), GmbH, Eschborn, Federal Republic of Germany; and International Center for living Aquatic Resources Management, Manila, Philippines. *ICLARM Studies and Reviews* 15: 351. http://pdf.usaid.gov/pdf_docs/PNAAY567.pdf. Accessed 5 April 2016.

Pennington, J.T., K.L. Mahoney, V.S. Kuwahara, D.D. Kolber, R. Calienes, and F.P. Chavez. 2006. Primary production in the eastern tropical Pacific: A review. *Progress in Oceanography* 69(2–4): 285–317.

Penven, P., V. Echevin, J. Pasapera, F. Colas, and J. Tam. 2005. Average circulation, seasonal cycle, and mesoscale dynamics of the Peru Current System: A modeling approach. *Journal of Geophysical Research* 110(C10021). doi:10.1029/2005JC002945.

Peterson, W.T., D.F. Arcos, G.B. McManus, H. Dam, D. Bellantoni, T. Johnson, and P. Tiselius. 1988. The nearshore zone during coastal upwelling: Daily variability and coupling between primary and secondary production off central Chile. *Progress in Oceanography* 20: 1–40.

Quinones, R.A., M.H. Gutierrez, G. Daneri, D.G. Aguilar, H.E. Gonzalez, and F.P. Chavez. 2010. The Humboldt Current System. In *Liu, Carbon and Nutrient Fluxes in Continental Margins, Global Change—The IGBP Series*, ed. K.-K.L. Atkinson, R. Quinones, and L. Talaue-McManus, pp 44–64. Berlin-Heidelberg: Springer.

Ramos, M., O. Pizarro, L. Bravo, and B. Dewitte. 2006. Seasonal variability of the permanent thermocline off northern Chile. *Geophysical Research Letters* 33: L09608. doi:10.1029/2006GL025882.

Rojas, B.R. 1971. Some observations on the feeding of the Peruvian anchoveta *Engraulis ringens* J. in two regions of the Peruvian coast. In *Fertility of the Sea. Volume 2*, ed. J.D. Costlow, 417–440. New York: Gordon and Breach Scientific Publishing.

Rosenberg, R., W. Arntz, E. Chumán de Flores, L.A. Flores, G. Carbajal, I. Finger, and J. Tarazona. 1983. Benthos biomass and oxygen deficiency in the Peruvian upwelling system. *Journal of Marine Research* 41: 263–279.

Rutllant, J.A. 1993. Coastal lows and associated southerly wind events in north-central Chile. *Reprints 4th International Conference on Southern Hemisphere Meteorology and Oceanography*, 268–269. Boston: American Meteorological Society.

Rutllant, J.A., H. Fuenzalida, R. Torres, and D. Figueroa. 1998. Land-air-sea interaction at the Antofagasta region (Chile, 23°S). The DICLIMA experiment. *Revista Chilena de Historia Natural* 71: 405–427.

Rutllant, J.A., B. Rosenbluth, and S. Hormazábal. 2004a. Intraseasonal variability of wind-forced coastal upwelling off central Chile (30°S). *Continental Shelf Research* 24: 789–804.

Rutllant, J.A., I. Masotti, J. Calderon, and S.A. Vega. 2004b. A comparison of spring coastal upwelling off central Chile at the extremes of the 1996–1997 ENSO cycle. *Continental Shelf Research* 24: 773–787.

Ryther, J.H. 1969. Photosynthesis and fish production in the sea. *Science* 166: 72–76.

Saito, M.A., J.W. Moffett and G.R. DiTullio. 2004. Cobalt and nickel in the Peru upwelling region: A major flux of labile cobalt utilized as a micronutrient. *Global Biogeochemical Cycles* 18: GB4030. doi:10.1029/2003GB002216.

Schunck, H., G. Lavik, D.K. Desai, T. Großkopf, T. Kalvelage, et al. 2013. Giant hydrogen sulfide plume in the oxygen minimum zone off Peru supports chemolithoautotrophy. *PLoS ONE* 8(8): e68661. doi:10.1371/journal.pone.0068661.

Schreiber, M.A., M. Ñiquen and M. Bouchon. 2011. Coping strategies to deal with environmental variability and extreme climatic events in the Peruvian anchovy fishery. *Sustainability* 3: 823–846. doi:10.3390/su3060823.

Shaffer, G., S. Hormázabal, O. Pizarro, and S. Salinas. 1999. Seasonal and interannual variability of currents and temperature off central Chile. *Journal of Geophysical Research* 104: 29951–29961.

Smith, R.L., D.B. Enfield, T.S. Hopkins and R.D. Pillsbury. 1971. The circulation in an upwelling ecosystem: The *Pisco* Cruise. *Investigation Pesquera* 35(1): 9–24.

Stewart, R.H. 2008. Introduction to Physical Oceanography. http://oceanworld.tamu.edu/resources/ocng_textbook/PDF_files/book_pdf_files.html. Accessed 31 March 2016.

Strub, P.T., J.M. Mesías, V. Montecino and J. Ruttlant. 1998. Coastal ocean circulation off western South America. In *The Sea*, ed. A.R. Robinson and K.H. Brink, Vol. 11, Wiley, 273–313.

Sverdrup, H.U. 1930. Some oceanographic results of the *CARNEGIE*'s Work in the Pacific—The Peruvian Current. *Transactions of the American Geophysical Union* 257–264.

Tans, P.P., I.Y. Fung, and T. Takahashi. 1990. Observational constraints on the global atmospheric CO_2 budget. *Science* 247: 1431–1438.

Thiel, M., E.C. Macaya, E. Acuna, W.E. Arntz, H. Bastias, et al. 2007. The Humboldt Current System of northern and central Chile. *Oceanography and Marine Biology: An Annual Review* 45: 195–344.

Thomas, A.C., M.E. Carr, and P.T. Strub. 2001. Chlorophyll variability in Eastern Boundary Currents. *Geophysical Research Letters* 28: 3421–3424.

Thomas, A., P. Strub, M. Carr, and R. Weatherbee. 2004. Comparisons of chlorophyll variability between the four major global eastern boundary currents. *International Journal of Remote Sensing* 25(7–8): 1443–1447.

Torres, R. 1995. *Condiciones oceanográficas y baja concentración de clorofila frente a Coquimbo, Chile (Lat. 30° S) durante 1992–1994*. M.Sc. thesis, Universidad de Concepción, Concepción, Chile.

Torres, R., D.R. Turner, N. Silva and J. Rutllant. 1999. High short-term variability of CO2 fluxes during an upwelling event off the Chilean coast at 30° S. *Deep Sea Research I* 46: 1161–1179.

Trenberth, K.E. 1977. The definition of El Niño. *Bulletinof the American Meteorological Society* 78: 2771–2777.

Trenberth, K.E., and D.P. Stepaniak. 2001. Indices of El Niño evolution. *Journal of Climate* 14: 1697–1701.

Troncoso, V.A., G. Daneri, L.A. Cuevas, B. Jacob, and P. Montero. 2003. Bacterial carbon flow in the Humboldt Current System off Chile. *Marine Ecology Progress Series* 250: 1–12.

Tsukayama, I. 1966. El número de branquispinas como carácter diferencial de sub-poblaciones de anchoveta (*Engraulis ringens* Jenyns) en la costa del Perú. In *Memorias 1er. Seminario Latino-Americano sobre el Océano Pacífico Oriental*, 139–145. Universidad Nacional Mayor de San Marcos, Lima, Perú.

Tsunogai, S., S. Watanabe, and T. Sato. 1999. Is there a "continental shelf pump" for the absorption of atmospheric CO2? *Tellus Series B* 51: 701–712.

Ulloa, O., R. Escribano, S. Hormazábal, R. Quiñones, R. Gonzalez, and M. Ramos. 2001. Evolution and biological effects of the 1997–98 El Niño in the upwelling ecosystem off northern Chile. *Geophysical Research Letters* 28: 591–1594.

van de Vossenberg, J., J.E. Rattray, W. Geerts, B. Kartal, L. van Niftrik L, E.G. van Donselaar J.S. Sinninghe Damsté, M. Strous, and M.S. Jetten. 2008. Enrichment and characterization of marine anammox bacteria associated with global nitrogen gas production. *Environmental Microbiology* 10: 3120–3129.

Van Mooy, B.A.S., R.G. Keil, and A.H. Devol. 2002. Impact of suboxia on sinking particulate organic carbon: Enhanced carbon flux and preferential degradation of amino acids via denitrification. *Geochimica et Cosmochimica Acta* 66: 457–465.

Walsh, J.J. 1981. A carbon budget for overfishing off Perú. *Nature* 290: 300–304.

Wooster, W.S. 1970. Eastern boundary currents in the South Pacific. In *Scientific Exploration of the South Pacific*, ed. W.S. Wooster, 60–68. Washington, D.C.: National Academy of Sciences.

Wyrtki, K. 1966. Oceanography of the eastern equatorial Pacific Ocean. *Oceanography & Marine Biology—An Annual Review* 4: 33–68.

Yáñez, E., M.A. Barbieri, C. Silva, K. Nieto, and F. Espíndola. 2001. Climate variability and pelagic fisheries in northern Chile. *Progress in Oceanography* 49: 581–596.

Chapter 6
The Canary/Iberia Current Upwelling System

Abstract This chapter describes the major coastal upwelling system on the continental shelves off Iberia (Portugal) and northwest Africa, its physical origin, ecological, cultural, social and economic relevance, and how it has been impacted under the pressures of fishing, climate variability and climate change.

Keywords Upwelling · Oceanography · Canary current · Iberia current · Marine biogeochemistry · Marine ecology · Canary eddy corridor · Fisheries · Fish stock variations · Climate-change impacts

I am longing to be with you, and by the sea,

where we can talk together freely & build our castles in the air

Bram Stoker (1847–1912) (taken from *Dracula*, 1897)

6.1 Introduction

The Canary Current Upwelling System is named after the *Canary Current*, which is an eastern branch of the subtropical gyre in the North Atlantic. Despite its name, the system also comprises the coastal upwelling region off the Iberian coast (Portugal), which can be affiliated with the offshore *Portugal Current*. In its broadest sense the large marine ecosystem of this upwelling system covers the latitudinal range 12°–43° N (Fig. 6.1), but both the northern and southern limits shift seasonally. The entire area is separated into two distinct upwelling regions with apparently little continuity of flows between them. This separation is caused by the interruption of the coastline at the Strait of Gibraltar, which allows the exchange of water between the Eurafrican Mediterranean Sea and the Atlantic Ocean.

As one of the four eastern boundary upwelling systems, the ecosystem is highly productive and supports a variety of fisheries in the countries of the region, which are northwestern Spain (Galicia), Portugal, Morocco, Western Sahara (which is

© Springer International Publishing Switzerland 2016

J. Kämpf and P. Chapman, *Upwelling Systems of the World*,
DOI 10.1007/978-3-319-42524-5_6

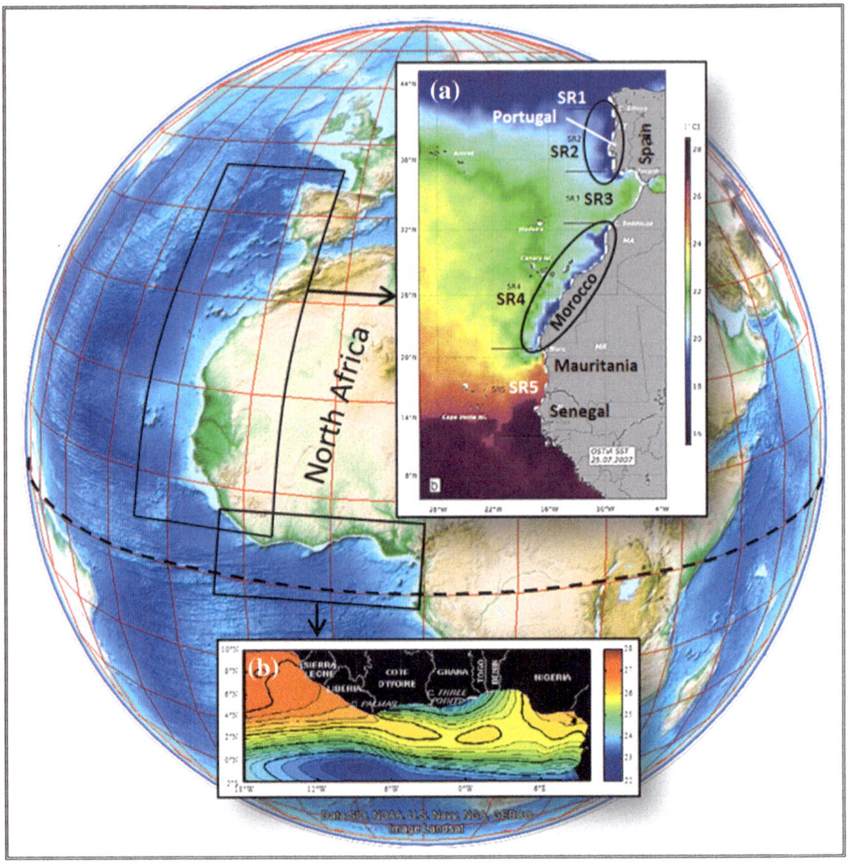

Fig. 6.1 Key upwelling regions in the eastern North Atlantic. The Canary current upwelling system comprises upwelling off the Iberian coast and off northwest Africa (panel **a**; displayed are SSTs). Five subregions are marked (*SR1* Galician; *SR2* Portuguese; *SR3* Gulf of Cadiz; *SR4* Moroccan; *SR5* Mauritanian). Panel **b**displays an upwelling situation along the equatorward facing coastline of central Africa. This specific upwelling, which is unrelated to winds, is further described in Chap. 9. *Image sources* Google earth with ETOPO 1 bathymetry layer (*background image*), Arístegui et al. (2009) (panel **a**), Toualy et al. (2012) (panel **b**)

administered by the Kingdom of Morocco), Mauritania and Senegal. Fisheries agreements (the latest being the 2006 EU-Morocco Fisheries Partnership Agreement, FPA) have been signed periodically between Morocco and the EU since the 1980s, allowing European vessels (especially Spanish and Portuguese) to fish in Moroccan waters in exchange for a monetary contribution. With little opportunity to develop their own infrastructure, sub-Saharan West African coastal states of the Canary Current Upwelling System have also signed agreements with the European Union which permits foreign countries to fish in their territories for financial compensation. Kaczynski and Fluharty (2002) make the case that this type of relationship is against both the long-term interests of the coastal states and the sustainability of the coastal resources and that the countries concerned will face severe overexploitation of their resources and subsequent drop in license revenues. Fluctuations in the abundance of small pelagic fish species in this ecosystem have a serious impact on the socio-economy of the countries of the region.

6.2 Historical and Cultural Context

The wide and slow moving Canary Current is thought to have been exploited by the early Phoenicians, whose main base was in present day Lebanon, during their navigation and settlement along the coast of western Morocco between 800 and 700 BC. The ancient Phoenicians not only exploited numerous fisheries within this current zone, but also established a factory at Iles Purpuraires (the site of present day Essaouira) for extracting a Tyrian purple dye from a marine gastropod *Murex* species (Clark et al. 1975). The Phoenicians also sailed much farther south; the

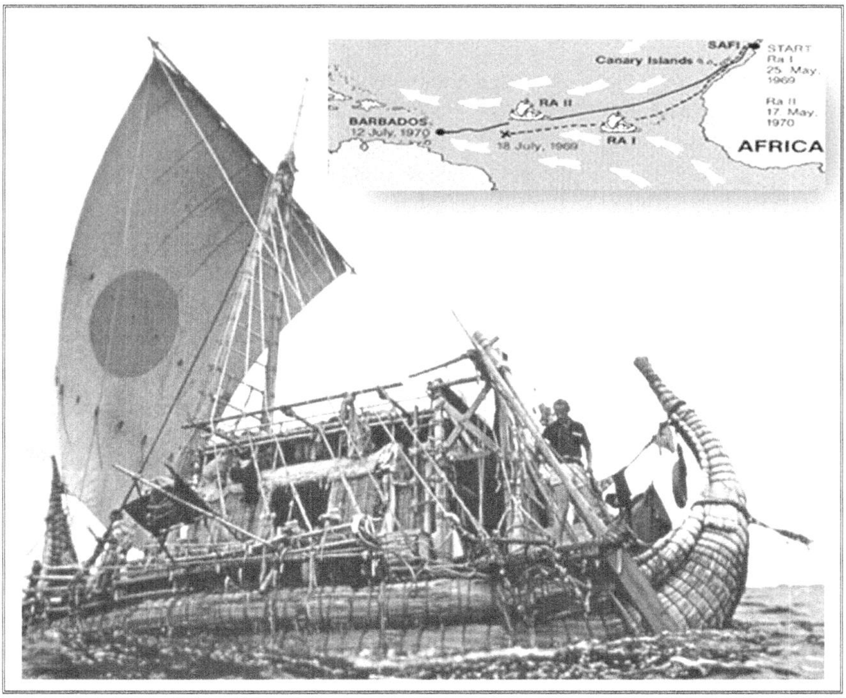

Fig. 6.2 Thor Heyerdahl on the Ra II vessel and approximate routes of his two trans-Atlantic voyages. *Picture source* http://www.heyerdahl-institute.no [accessed on 5 April 2016]

Greek historian Herodotus describes how they set sail from the Red Sea and sailed south "with the sun rising on their left hand side" until they rounded Cape Agulhas in South Africa (see Chap. 7), whereupon the sun rose on the right hand side of the ship, and eventually reached their North African settlements again after a three-year voyage. This is believed to have occurred around 590 BC.

The Canary Current was pivotal for civilization of the Americas. Christopher Columbus took advantage of the Canary Current, setting out from the Canary Islands, on his first crossing of the Atlantic in 1492 AD, when the Italian-Spanish navigator was searching for a route to Asia but ended up in the Americas instead. The resulting Spanish treasure fleets continued to use the Canary Current for the next 300 years until 1790. Based on archaeological evidence, world-renowned Norwegian explorer Thor Heyerdahl concluded that ancient sailors undertook cross-Atlantic voyages using the Canary Current at much earlier times. To prove this point, Heyerdahl built ancient vessels designed like Egyptian reed boats and managed to cross the entire Atlantic in his second attempt on the Ra II in 57 days (Fig. 6.2).

6.3 History of Scientific Discovery

Research effort has been unevenly distributed through the whole region. The upwelling region of northwest Africa was first intensively studied in the 1970s, during the International Decade of Ocean Exploration (IDOE), under the international programme of Cooperative Investigation of the Northern Part of Eastern Central Atlantic (CINECA; Hempel 1982), which was followed by the U.S. funded JOINT-I program, part of the Coastal Ecosystems Upwelling Analysis program (CUEA, see also Chap. 5) in 1974 (King 1981). First oceanographic findings were published by Schemainda et al. (1975), Wooster et al. (1976) and Peters (1976). The region was almost ignored during the 1980s. In the following decade, interdisciplinary studies focused on the cross-shelf export of organic matter on the North-West African shelf, such as the French EUMELI (Eutrophic, Mesotrophic, Oligotrophic) project (Morel et al. 1996), and the European Canaries-Coastal Transition Zone (Barton et al. 1998) and CANIGO (Canary Islands, Azores, Gibraltar Observations; Parrilla et al. 2002) projects.

The western and northern coasts of the Iberian Peninsula have been extensively studied from the hydrographic and dynamic point of view (Barton et al. 1998). However, most of the field studies on carbon and nutrient biogeochemistry in the Iberian margin have concentrated on the Galician and Cantabrian coast (42°–44° N, 5°–10° W). Intensive research at this boundary between the temperate and subpolar regimes of the North Atlantic has been supported by the European scientific community in the frame of several research projects, like "The Control of Phytoplankton Dominance" (Figueiras et al. 1994; Moncoiffé et al. 2000),

MORENA (Multidisciplinary Oceanographic Research in the Eastern Boundary of the North Atlantic; Fiúza et al. 1998; Pérez et al. 1999, 2001), or OMEX II (Ocean Margin Exchange; Joint and Wassmann 2001; Huthnance et al. 2002; Joint et al. 2002; Van Weering and McCave 2002). Recent projects focus on both management and establishing fishery and ecosystem governance frameworks such as the Canary Current Large Marine Ecosystem (CCLME) project. To facilitate future research, the UNESCO has recently compiled a directory of atmospheric, hydrographic and biological datasets for the Canary Current Upwelling System (Déniz-González et al. 2014).

6.4 Ecosystem Subregions

Arístegui et al. (2009) separated the upwelling system into five different subregions (Fig. 6.3), each defined in terms of factors such as whether its coastline orientation makes it more or less susceptible to a given coastal wind event, the presence/absence of river input, and the existence of large bays. The Iberian sector, for instance, can be separated into the Galician and Portuguese subregions, both of which are strongly influenced by continental runoff. The Gulf of Cadiz subregion represents a major interruption in the continuity of the system, given its coastline orientation, which is unfavourable for coastal upwelling, but facilitates exchanges between the Atlantic and the Mediterranean Sea, with low salinity surface flow into the Mediterranean and a deep, saltier outflow that governs the salinity of much of the subtropical North Atlantic.

Farther south, the Moroccan subregion between Cape Sim (31° 59′ N) and Cape Blanc (20° 46′ N, now known as Ras Nouadhibou) benefits from year-round upwelling and exhibits a high level of mesoscale oceanographic variability arising from its geographical heterogeneity. Variations in shelf width, the presence of major capes and dynamical perturbations caused by the Canary Islands produce extended upwelling filaments and island-induced eddies (see Fig. 6.3).

The fifth subregion defines the Mauritanian–Senegalese coastal waters south of Cape Blanc. The northward limit of this subregion is defined by the separation of the Canary Current from the coast. The southward limit is defined by the reach of wintertime upwelling-favourable coastal winds. This highly-productive subregion is the only one of the five where hypoxic conditions are found in the oxygen minimum zone (Karstensen et al. 2008). Its location adjacent to the vast Sahara desert exposes it to one of the highest rates of deposition of airborne dust, which has a major impact on the biogeochemistry of the region. Freshwater runoff from rivers again becomes important only at the southernmost limit of the upwelling region (Fig. 6.4).

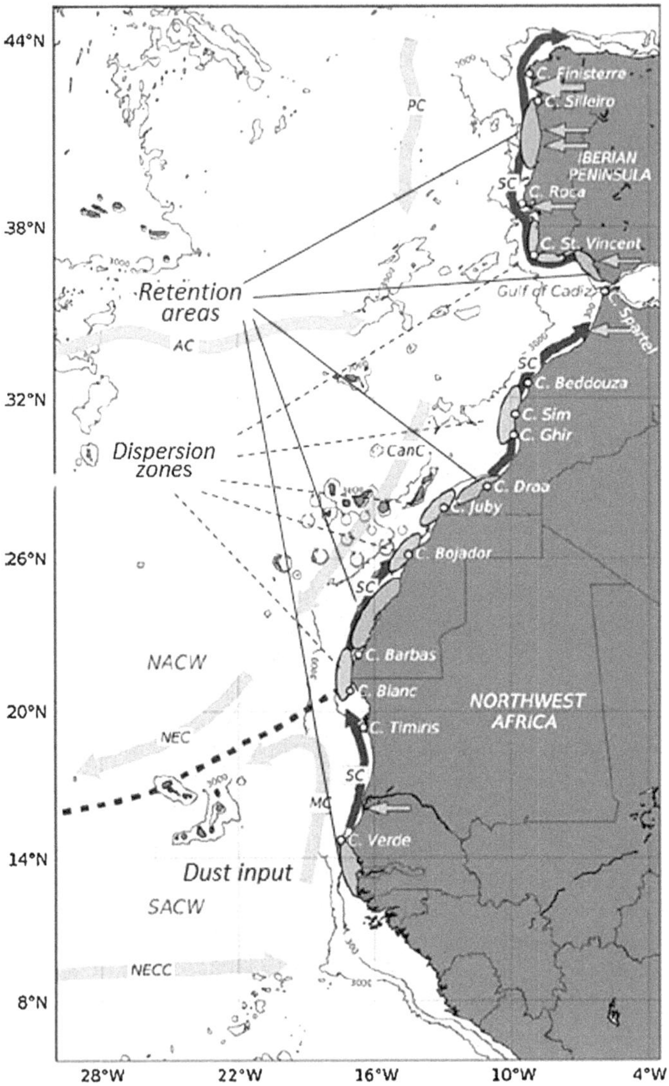

Fig. 6.3 Schematic map of the Canary current upwelling system showing the main surface and slope (SC) currents, major capes, freshwater and dust inputs (>10 g C/m^2/yr, large shaded area), retention and dispersion zones on the shelf, frontal zone between water masses (*dashed lines*) and mesoscale eddies south of the Canary Islands. *NACW* North Atlantic Central Water; *SACW* South Atlantic Central Water; *AC* Azores Current; *CanC* Canary Current; *MC* Mauritanian Current; *NEC* North Equatorial Current; *NECC* North Equatorial Countercurrent; *PC* Portuguese Current; *SC* Slope Current. Modified from Arístegui et al. (2009)

Fig. 6.4 River discharges seen as discoloured turbid coastal waters in south Senegal, Gambia and Guinea Bissau. *Image source* ESA (European Space Agency)

6.5 Bathymetry, Climate and Atmospheric Forcing

6.5.1 Bathymetry

The northwest African/southern Iberian margin is bordered by a continental shelf that is generally 40–60 km wide, although maximum shelf widths of >100 km occur off the western Saharan coast. Minimum shelf widths (<20 km) are found off the western Portuguese margin and also off Cape Verde and Cape Ghir on the northwest African continental margin. The shelf has a seaward slope of <1° (Weaver and Canals 2003) with the shelf break occurring generally at a water depth of 100–200 m. Beyond the shelf break, the width of the continental slope and rise varies from over 1000 km off Mauritania to about 50 km off southwest Portugal. The continental slope displays slope angles of 1°– 6°. In some regions, such as adjacent to the Western Sahara Canyon System, slope angles may reach 40°. The coastal orientation is predominantly meridional or has a strong meridional orientation, barring the northern and southern Iberian coasts which are zonal. Several submarine canyons, which are associated with eddy generation and are also sites of coastal sediment deposition, are found along both the Iberian shelf (e.g., Fiúza et al. 1998) and the northwest African shelf (e.g., Barton et al. 2004). A further feature of major topographical importance is the Canaries archipelago at 28° N, which lies at the transition zone between coastal and oceanic waters.

The northwest Iberian shelf is relatively narrow, ranging in width from 30 km in the north to 50 km off the Douro Rover mouth in the south. The shelf break lies at a depth of 160–180 m. The mid-shelf region, at depths of 100–140 m, hosts the northwest Iberian mud belt—the major depocentre for fine-grained sediment from the late Holocene to present (e.g., Oberle et al. 2014). The mudbelt is located between plutonic and metamorphic outcrops on the inner shelf and transgressive to modern well-sorted fine sands on the outer shelf.

The eastern Gulf of Cadiz continental margin is characterised by a crescent-shaped continental shelf up to 45 km wide, with the shelf break at 130–140 m depth. The continental slope displays an irregular bathymetry and is defined by rough ridge and valley morphology, bordered by two wide terraces between 400 and 800 m water depth (Baraza et al. 2003). The Gulf of Cadiz is an area where the Atlantic and Mediterranean waters meet, resulting in a complex current pattern.

The continental shelf of northwest Africa is generally 40–60 km wide, but ranges from a minimum of 13 km width at Cape Ghir (30.5° N) to 120 km width between Cape Bojador and just to the south of Cape Blanc (Seibold 1982). The shelf edge shoals from about 150 m offshore of Morocco to an average of about 110 m seaward of the Sahara, and ranges between 130 and 150 m near Senegal (Tooms et al. 1971). Compared to other eastern boundary current upwelling regions, the northwest African shelf is relatively shallow with average depths of only 60–80 m. Numerous canyons and gullies cut the slope, except for the section between 25° S and Cape Blanc.

6.5.2 Climate and Atmospheric Forcing

The Canary Current Coastal Upwelling System covers almost 30° of latitude, from the Senegalese coast at about 15° N to the Cantabrian coast at 44° N (see Fig. 6.3). The area of interest therefore encompasses a range of climatic conditions, from tropical to the temperate mid-latitudes. As described by Ekman (1905) and later by Bakun (1973), coastal upwelling of the region is the consequence of the alongshore component of the trade winds; that is, coastal winds associated with the eastern flanks of either the Bermuda High in boreal summer or the Azores High in boreal winter (see Fig. 2.13). On longer decadal to century time scales, however, it appears from core studies that upwelling-driven SSTs off Morocco vary out of phase with northern hemisphere temperature anomalies (McGregor et al. 2007), so that warm conditions existed during the Little Ice Age and cold conditions during the Medieval Warm Period. In general, there has been consistent cooling and increased upwelling throughout the 20th century, in agreement with increasing wind speeds at Cape Ghir.

Fig. 6.5 Long-term average seasonal cycle of the offshore Ekman transport ($-Q_x$, in m^2/s), calculated at the different subregions of the Canary current upwelling system (from Arístegui et al. 2009)

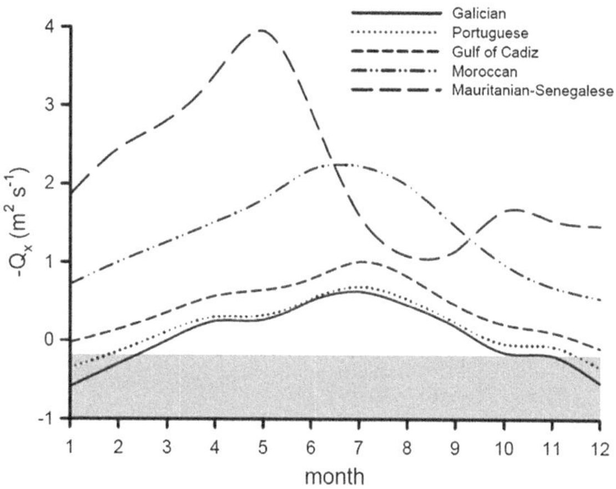

The wind-driven offshore transport in the surface Ekman transport varies by one order of magnitude between the southern reaches of the upwelling region (2.16 m^2/s, 12-month average at 17°30′ N) and the north (0.38 m^2/s, 6-month average at 41° 30′ N) (Fig. 6.5). There is a pronounced seasonal variability in offshore Ekman transports in the upwelling region associated with seasonal variations of the coastal winds. In most of the region, the long-term mean upwelling index (offshore Ekman transport) peaks in boreal summer (July–August) and attains minimum values in winter (December). The Mauritanian–Senegalese region behaves differently and shows an earlier peak in May–June and a minimum in offshore Ekman transports in August–September. Note that in contrast to this long-term average, seasonal patterns of alongshore winds in the Mauritanian–Senegalese region vary significantly from year to year (Fig. 6.6). Hence, the average seasonal pattern shown in Fig. 6.5 can be misleading.

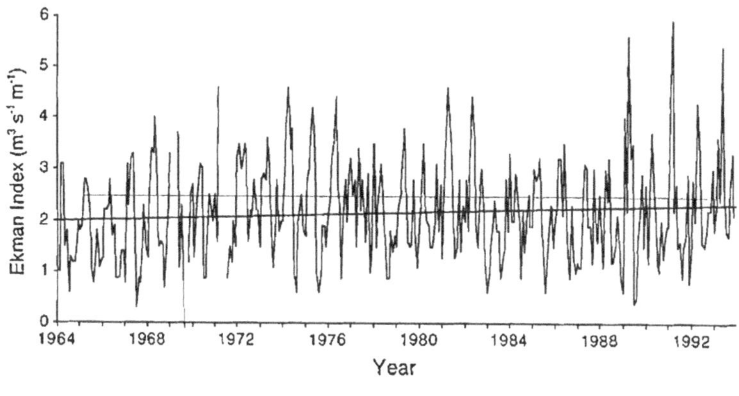

Fig. 6.6 Upwelling index calculated from ship data in the Mauritania region, from 1964 to 1992 (from Demarcq 1998)

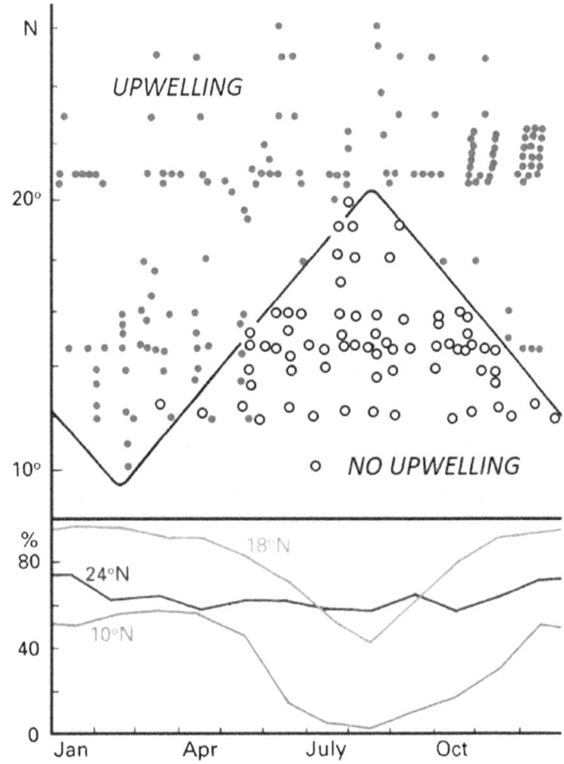

Fig. 6.7 **a** Seasonal variability of the southern limit of the Canary current upwelling system. *Full dots* indicate observed upwelling, *circles* indicate observed absence of upwelling. **b** Frequency of occurrence of winds favourable for upwelling (wind direction is in the quarter between alongshore towards the south and exactly offshore). Taken from Tomczak and Godfrey (2003), adapted from Schemainda et al. (1975)

The coastal region between 20° and 30° N is characterized by upwelling-favourable wind stress all year-round (Bakun and Nelson 1991). Maximum magnitudes are found north of 21° N during boreal summer and south of 21° N during boreal winter (Mittelstaedt 1991). South of 15° N the Inter-Tropical Convergence Zone (ITCZ) becomes the main driver of the upwelling seasonality. As a consequence, the southern limit of the upwelling system shifts northward and southward seasonally by more than 1000 km (Fig. 6.7). The Canary Current upwelling reaches its southernmost extent in winter when the trade wind is strongest. It then extends well past Cape Blanc—the separation point of the Canary Current from the African coast.

Over the Iberian Peninsula (37°–43° N), upwelling-favourable equatorward winds dominate during spring and summer months, whereas downwelling-favourable mean winds prevail in the other months. It is remarkable that due to the different orientation of the western and northern Galician coasts, northerly winds produce upwelling off the western coast whereas easterly winds do it off the northern coast. The orientation of the coast changes abruptly north of Cape Finisterre, in such a way that both northerly and easterly winds are upwelling favourable there (McClain et al. 1986; Torres et al. 2003). Similar considerations apply to Cape St Vincent (Fiúza 1983; Relvas and Barton 2002).

6.5.3 Atmospheric Nutrient Inputs

Atmospheric inputs are relatively minor along the coastal upwelling systems of the west coast of the Americas and southern Africa, although they do occur in the Benguela system (see Chap. 7). In the Canary Current upwelling system, however, atmospheric inputs represent a major source of nutrients, particularly iron (Neuer et al. 2004). The Sahara desert is the major supplier of dust to the subtropical North Atlantic and Eurafrican Mediterranean with an estimated annual deposition of 220 × 10^9 kg per annum (Duce and Tindale 1991). The dust plume can be traced year-round by satellite and its area of maximum deposition shifts from around 5° N in winter to around 20° N in summer, driven by the shift of the Inter-Tropical Convergence Zone (Moulin et al. 1997). The Canary Island region (28°–29° N) is located on the northern margin of the main dust cloud and thus receives episodic pulses with peaks reaching the area mainly in winter and in summer/autumn (Torres-Padrón et al. 2002).

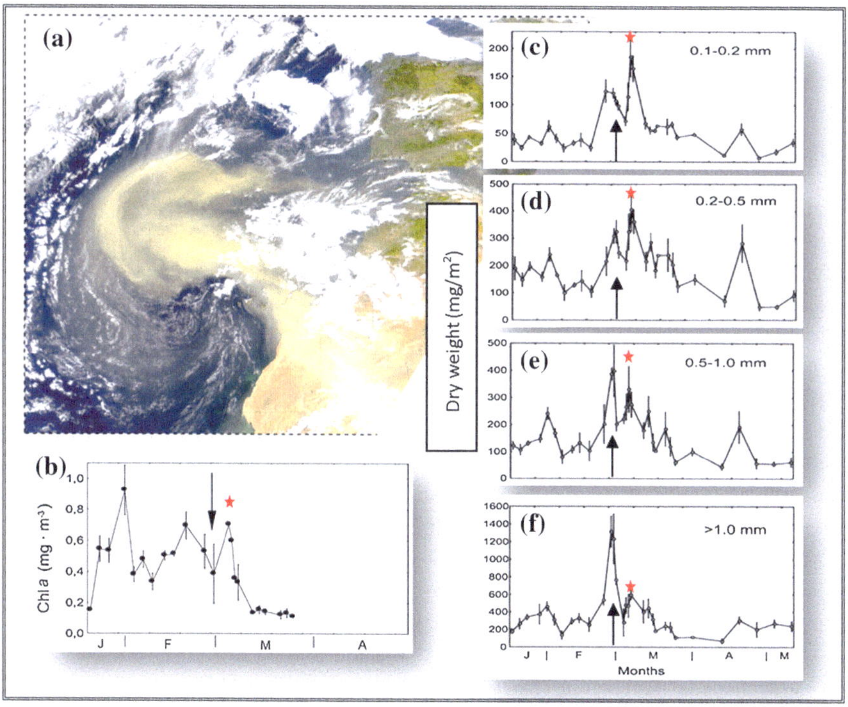

Fig. 6.8 Phytoplankton and zooplankton response north of Gran Canaria (Canary Islands) after a major sand storm from the Sahara. Panel (**a**) SeaWiFs satellite image of the dust storm from 26 February 2000 [http://visibleearth.nasa.gov/cgibin/viewrecord?22352, (accessed 5 April 2016)]. Panel (**b**) Chlorophyll-*a* (mg/m^3) at 5 m depth. Panel (**c–f**) Average zooplankton biomass (mg dry weight/m^2) and standard error separately represented for four zooplankton size fractions. The date of the dust storm in late February is marked with *arrows*. *Stars* indicate chlorophyll and zooplankton responses to the dust storm one week later. Taken from Hernández-León et al. (2004)

Atmospheric dust has two different impacts on the ocean's biogeochemistry. First it can provide nutrients for phytoplankton growth, both macronutrients and trace metals (e.g., Gao et al. 2003). Second, it can accelerate or induce carbon sedimentation by adsorption, ballasting and possibly aggregation of marine particles such as detritus or faecal pellets (e.g. Fischer et al. 2007). Neuer et al. (2004) found a close correlation between the atmospheric dust concentration in winter and the magnitude of the downward particle flux at a station near Gran Canaria (Canary Islands). At the end of February 2000, an impressive cloud of dust originating in the Sahara desert probably contributed the additional nutrients that led to an increase in chlorophyll and small zooplankton in the Canary Island region a week later (Fig. 6.8; Hernández-León et al. 2004).

6.6 Physical Oceanography

6.6.1 Circulation

The main large-scale currents associated with the eastern part of the anticyclonic North Atlantic subtropical gyre are the *North Atlantic Current*, the *Azores Current* and the *Canary Current* (Fig. 6.9). The North Atlantic Current branches into three southward-flowing streams as it moves eastwards across the basin, forming the broad, slow and generally southward-flowing *Portugal Current* and the Azores Current on the eastern boundary. The Portugal Current connects the northern branch of the North Atlantic Current with the Canary Current (Pérez et al. 2001), flowing southwards year round from 45°–50° N to 10°–20° W (Krauss 1986), although its contribution is small (Barton 2001). The remainder of the North Atlantic Current heads northeastward, becoming the *North Atlantic Drift* located between Iceland and the British Isles. The Canary Current is supplied also by an eastward branch of the Azores Current, which passes north of and around Madeira. The total transport of the Canary Current is seasonally constant, although its geostrophic circulation varies; it is stronger in summer near the African coast, while in winter it is stronger west of the Canary Islands (Stramma and Siedler 1988).

Three papers by Pelíz and Fiúza (1999) and Pelíz et al. (2002, 2005) characterize the surface circulation off Iberia. The circulation pattern near the Iberian shelf is complex (Fig. 6.10). A pronounced year-round feature is a poleward coastal flow associated with a low-salinity water lens, which extends all along the coast as a narrow band—*the Western Iberia Buoyant Plume* (WIBP). This plume is related to buoyancy input from the many regional rivers (the Douro, Minho and Mondego Rivers, other smaller rivers, and the Rías Baixas) (Pelíz et al. 2002). The WIBP influences upwelling by increasing stratification over the shelf and by the creation of an inshore frontal region that promotes northward baroclinic transport. Apart from this year-round feature, currents on the west Iberian shelf experience a marked

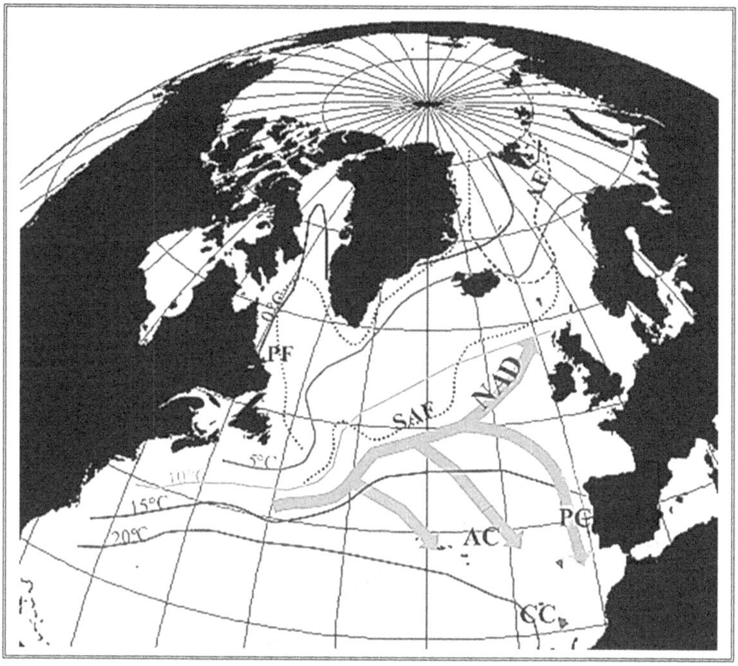

Fig. 6.9 General structure of the recirculation branches of the North Atlantic Current. *NAD* North Atlantic Drift; *PC* Portugal Current; *AC* Azores Current; *CC* Canary Current; *SAF* Sub-Artic Front; *PF* Polar Front; *AF* Arctic Front. *Lines* are SST contours. Taken from Eynaud et al. (2007)

seasonal variability defined by the coastal wind regime along the western and northern coast.

During spring and summer (from March/April to September/October) northeasterly winds predominate in the Iberian basin (Wooster et al. 1976; Bakun and Nelson 1991), producing the southward flowing *Portugal Coastal Current* at the surface (<100 m) (Fig. 6.10a) and the northward flowing *Portugal Coastal Undercurrent* on the slope. The Portugal Coastal Current is associated with seasonal coastal upwelling at the Iberian margin, and the export of coastal surface waters to the open ocean, especially at the recurrent upwelling centres and filaments along the western Iberian coast (Fig. 6.11; e.g., Haynes et al. 1993; Pelíz and Fiúza 1999).

southwesterly winds are predominant during the autumn and winter months, provoking a reversal of the surface circulation to form the *Iberia Polar Current* (also called *Portugal Coastal Counter Current*) (Fig. 6.10b). The existence of this poleward surface flow was described first by Wooster et al. (1976) and subsequently by Frouin et al. (1990) and Haynes and Barton (1990) along the western Iberian coast and by Pingree and Le Cann (1990) along the Cantabrian coast. At about 39°–40° N, winter imagery of sea surface temperatures reveals a recurrent meandering frontal system, named the *Western Iberia Winter Front*

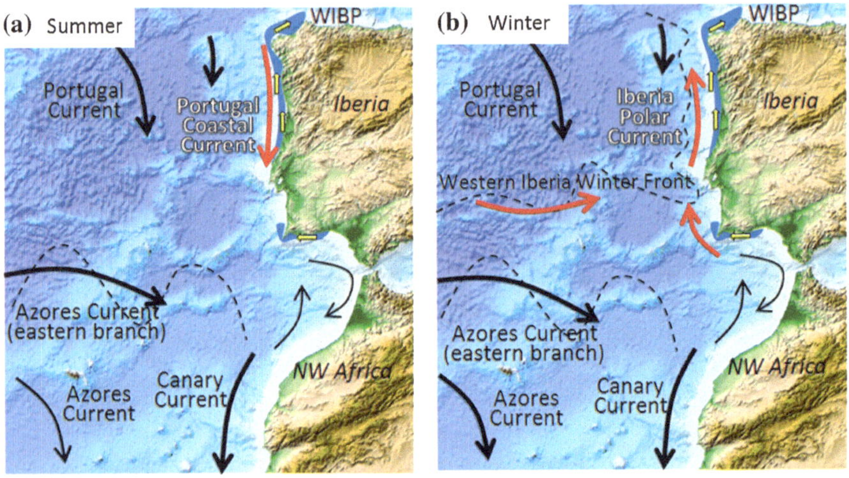

Fig. 6.10 Schematic of the surface circulation (*arrows*) with a focus on the west Iberian shelf. *Dashed lines* indicate frontal regions. *WIBP* Western Iberia Buoyant Plume. *Small arrows* near the coast indicate the flow direction of river plumes. **a** Summer. **b** Winter

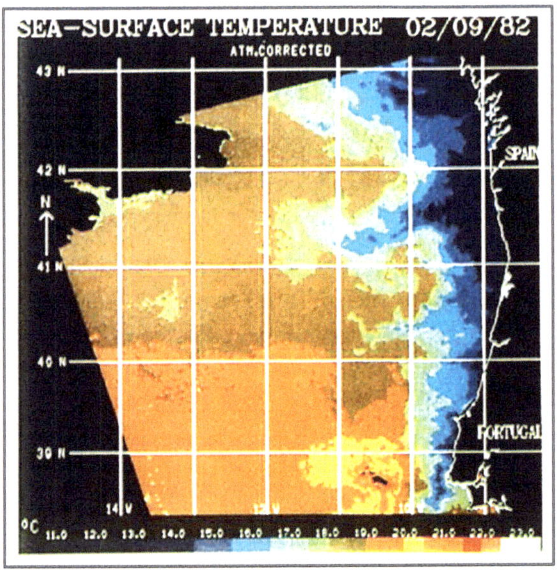

Fig. 6.11 Early satellite-based evidence of filaments forming along the western Iberian coast for September 2, 1982. Taken from Haynes et al. (1993)

(see Pelíz et al. 2005). This front represents the transition to the southern area of the Iberian Basin, where the Portugal Current is less influential, and supports the eastward advection of relatively warm and salty water that feeds into the Iberia Polar Current (Pelíz et al. 2005).

The Iberia Poleward Current is a narrow (25–40 km) slope-trapped tongue-like structure that flows northward for more than 1500 km off the coasts of the Iberian Peninsula and southeast France. It normally arrives in the Cantabrian Sea at the beginning of every winter (this is well documented, see e.g., Gil 2003). The January warm-water extension of the Iberia Poleward Current along the Cantabrian coast has been referred to as *Navidad* because it starts to be evident around Christmas and New Year. Due to the inability of the poleward flow to follow abrupt changes of topography, such as around Cape Ortegal, just east of Cape Finisterre, and the Cape Ferret Canyon in the southeast Bay of Biscay, the Iberian Poleward Current exhibits a turbulent nature that results in local instabilities and gives rise to eddies that separate from the current and move offshore into the Bay of Biscay region (e.g. Pingree and Le Cann 1990; García-Soto et al. 2002).

The coastal upwelling region from Gibraltar to Cape Blanc is maintained by the presence of favourable northeasterly winds throughout the year, although winds and upwelling are more intense during the summer months (see Fig. 6.5). In contrast with the Iberian coast, the northwest African coast is largely influenced by the general circulation of the North Atlantic subtropical gyre, particularly by its eastern branch—the Canary Current. This current flows equatorward while interacting with the coastal upwelling waters. It detaches from the coast near Cape Blanc (21° N), flowing westward at the latitude of Cape Vert (15° N). South of Cape Blanc, a large permanent cyclonic recirculation—the *Mauritanian Gyre*—develops as a consequence of the offshore displacement of the Canary Current. In winter, a near shore, narrow equatorward flow develops over the shelf in response to coastal upwelling forced by the southward migration of the trade wind band (Hughes and Barton 1974). Several studies (e.g., Stramma and Siedler 1988; Siedler and Onken 1996) have described the seasonal variability of the Canary Current, confirming the existence of the inflow of water from the open ocean into the coastal upwelling region north of the Canary Islands.

The *poleward undercurrent* (Barton 1989) is a persistent feature off the northwest African coast (Fig. 6.12). It appears to be strongest at about 100–200 m depth, but extends down to about 1000 m. The undercurrent has a width of <100 km and a mean speed of 10 cm/s at the level of maximum flow. Generally the poleward undercurrent follows the shelf break. South of Cape Blanc, the poleward undercurrent is indistinguishable from the surface Mauritanian Current. North of Cape Blanc the upwelling front is found far offshore and creates a near-slope northward path to the undercurrent (Peña-Izquierdo et al. 2012). Trajectories of subsurface drifters indicate that a substantial volume fraction of the undercurrent recirculates offshore just south of Cape Blanc.

A number of authors have provided estimates of the flow in the Canary Current, but these are very variable depending on how they were obtained. Early work, summarized by Stramma (1984), suggested that the eastward flow of the Azores Current is up to ~14 Sv. As shown above, however, the flow splits into a number of separate streams between 35° W and the African coast, all of which are generally weak (2–8 cm/s), and it is not always easy to determine which flow stream is being discussed. Stramma (1984) suggested that the Portugal Current transports about

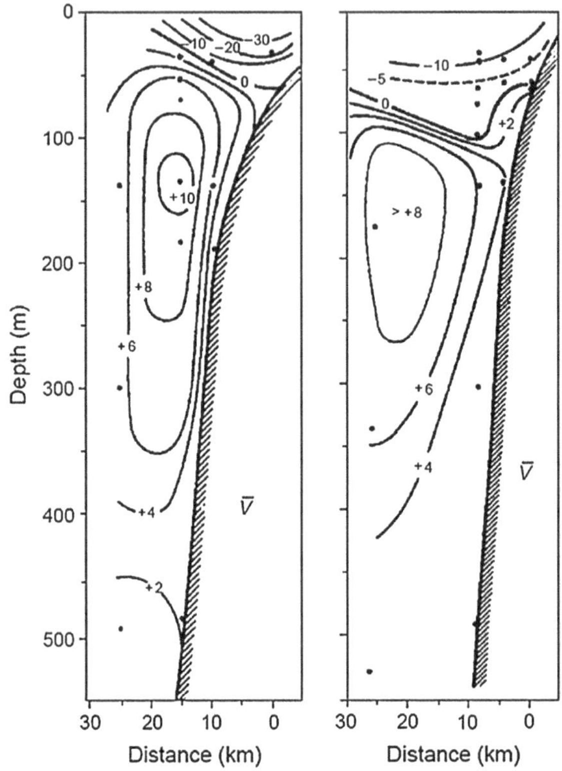

Fig. 6.12 Two sections of alongshore flow measured by current meters (*solid dots*) near 20° N off northwest Africa, showing the structure of the poleward undercurrent. Speeds are given in cm/s, northward positive. Taken from Barton (2001)

3 Sv southward, but only about 1 Sv of this actually feeds the Canary Current along the coast of Morocco. Later work (e.g., Navarro-Pérez and Barton 2001; Knoll et al. 2002; Hernandez-Guerra et al. 2005; Mason et al. 2011) using both hydrography, current meter measurements and fine-scale models, suggests that the Azores Current transports only about 10 Sv, and that the mean southward flow along the African coastline is less than 2 Sv, possibly as low as 0.2 Sv (Knoll et al. 2002). This flow appears to be seasonal, and while offshore there may be net southward flow of up to about 4.5 Sv, along the coast the flow can be northward, particularly in winter and spring.

6.6.2 Bathymetric Features and Frontal Zones

The complex bathymetry of both the Iberian Peninsula and the North African coast leads to persistent hydrological features in the Canary Current system. The Rías Baixas, a series of large coastal inlets at the northern end of the Iberian coast, forms an isolated ecosystem that amplifies upwelling-induced biogeochemical signals (Álvarez et al. 2005). Nutrient supply to the rias is enhanced during the upwelling season, while the local topography gives rise to cool filaments during this time and slope water eddies during the downwelling season (see Fig. 6.11). Both filaments and eddies enhance the exchange of water, dissolved species, and suspended material between the coastal region and the open ocean (Álvarez-Salgado et al. 2010). Further south, capes and promontories such as Cape Roca and Cape St. Vincent in Portugal, and Cape Ghir and Cape Blanc in Africa, produce additional filaments of upwelled coastal water that can stretch up to several hundred kilometres into the open ocean, transporting waters rich in organic matter into the impoverished oligotrophic waters of the subtropical gyre.

Another singularity of this eastern boundary system is the presence of the Canary Archipelago, close to the northwest African coast, which interrupts the main flow of the Canary Current and introduces large mesoscale variability, mainly in the form of vortex streets downstream of the islands that form a consistent pathway that the eddies follow (Arístegui et al. 1994). The interaction between the island eddies and upwelling filaments in this *Canary-Coastal Transition Zone* (also called the *Canary eddy corridor*; see Sect. 6.7.3) is another mechanism for the exchange of water properties, and thus acts as an efficient route for transporting organic matter to the open ocean (e.g., Pelegrí et al. 2005). Research cruises have shown how the water recirculates southward along the continental slope, where quasi-permanent filaments stretch offshore and exchange water properties with island eddies (Arístegui et al. 1997; Barton et al. 1998). Reversals in the main flow have been observed close to the Canary-Coastal Transition Zone during late autumn and winter (e.g. Navarro-Pérez and Barton 2001). These flow diversions, probably caused by a weakening of the trade winds south of Cape Ghir (Pelegrí et al. 2005), allow a northward flow to develop between Cape Blanc and Cape Juby, and consequently an offshore spread of organic matter produced in upwelling waters near the Canary Islands region (Arístegui et al. 1997).

Pelegrí et al. (2005) suggested that the Canary Current region is characterised by the presence of two cells transporting upwelled waters into the open ocean. The first one would be the standard vertical cell, present in all upwelling systems, with Ekman offshore transport responding to the wind variations. The second one would be the horizontal circulation cell that starts with open ocean water reaching the coast north of Cape Ghir and which is closed by the offshore export of water through several upwelling filaments and the flow diversion at Cape Ghir. The joint action of both cells would cause this upwelling region to be a key region for export of organic matter and nutrients to the open ocean.

The *Cape Verde Frontal Zone*, located at around 15° N offshore, separates water of tropical (southern) origin and water of subtropical (northern) origin. Water carried southward by the Canary Current and water carried northward by the *Mauritania Current* converge in this zone and feed into the North Equatorial Current (Fig. 6.3). This frontal zone is found to intersect the slope north of Cape Blanc, with its position varying between 21° and 23° N, although tropical influences can sometimes predominate over the slope as far north as 24° N (Peña-Izquierdo et al. 2012).

6.6.3 Water Masses and Nutrient Concentrations

Most of the region, from Cape Finisterre to Cape Blanc, is dominated by ENACW, which is responsible for the fertility of coastal waters during upwelling events. Two branches of ENACW of either subtropical or subpolar origin occupy the depth range from 50–100 to 450–750 m, depending on latitude. The subtropical portion sits above subpolar waters off the western Iberian coast and north of the Canary region, so nutrient-poorer (0–6 μM of nitrate) subtropical water upwells first, followed by nutrient-richer (6–10 μM of nitrate) subpolar water that enters the shelf only during strong upwelling events. Bay of Biscay Central Water, characterized by a temperature of 12 °C and 6 μM of nitrate, dominates the central waters off the Cantabrian coast (e.g., Botas et al. 1989).

The nutrient regime in coastal waters of northwest Africa is conditioned by the presence of two different water masses. A marked front at about 21° N (Cape Blanc), which is the coastward extension of the *Cape Verde Frontal Zone*, separates North Atlantic Central Water (NACW) from the slightly cooler, less saline, and nutrient-richer South Atlantic Central Water (SACW) . SACW crosses the equator in the North Brazil Current and then is carried eastwards in the North Equatorial Countercurrent to feed into the northward flow off the coast of West Africa. The boundary between these two water masses is convoluted, variable in location and characterized by intense mixing and interleaving processes (e.g., Hagen and Schemainda 1987). Since the SACW is richer in nutrients than the NACW, a meridionally decreasing nutrient gradient is apparent in the northward flowing waters. Compared to other eastern boundary upwelling regions, the source waters off northwest Africa are somewhat poorer in nutrients, but richer in oxygen, largely as a result of the basin-scale circulation and the fact that the upwelling ENACW comes from the ventilated thermocline, unlike in other regions (Codispoti et al. 1982; Castro et al. 2000; Álvarez-Salgado et al. 2010).

6.6.4 Spatial Differences in Upwelling Dynamics

Arístegui et al. (2009) discuss differences in the upwelling dynamics in different subregions of the Canary Current upwelling region in comparison to the classical mechanism. The classical upwelling circulation for eastern boundary systems (Fig. 6.13a) consists of an equatorward wind stress provoking offshore transport in the surface Ekman layer which is replaced by onshore flow in and below the upwelled pycnocline. The offshore surface Ekman drift creates an equatorward geostrophic coastal upwelling jet, and a poleward undercurrent is found trapped on the continental slope (see Chap. 2).

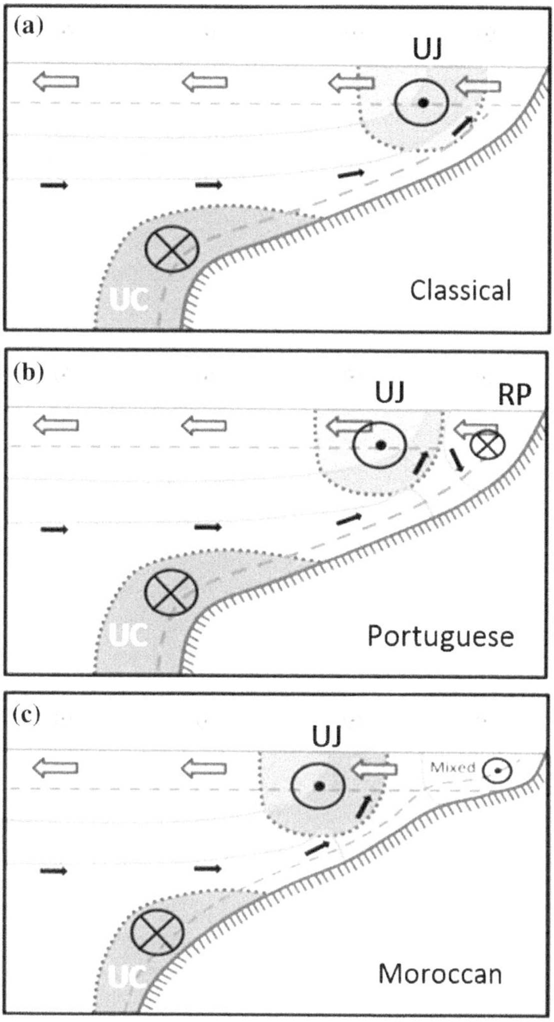

Fig. 6.13 Schematic of different types of upwelling patterns found in the Canary current upwelling system. See text for a discussion. Shown are wind stress (*small circles*), offshore transport in the surface Ekman layer (*top arrows*) replaced by onshore flow (*black arrows*) in and below the upwelled pycnocline (*solid line*), the equatorward upwelling jet (UJ), northward spreading river plumes (RP), and the poleward undercurrent (UC) trapped on the continental slope. Redrawn after Arístegui et al. (2009)

The Galician and Portuguese subregions are distinguished by riverine influences. These lead to the formation of a low-salinity wedge spreading northward along the coast as the West Iberian Buoyant Plume (Fig. 6.13b). During the upwelling season (May–October), the classic mode of upwelling develops offshore from this plume, and the biogeochemical response strongly depends on complex interactions between the upwelling jet, the buoyant plume and orographic influences. However, even though the rias are sheltered from the prevailing winds, they act as an extension of the shelf and bottom water continues to move onshore during upwelling events into the outer portions of the rias, with lower salinity surface water moving offshore. This is driven by a drop in coastal sea level during upwelling events. Riverine influences are thus limited to the inner reaches of the coastal inlets.

Conversely, downwelling-favourable southerly winds reverse this flow pattern and create an outflow of bottom water from the rias (Álvarez-Salgado et al. 2010).

The wind variability on periods of 10–15 days (Nogueira et al. 1997) provokes repeated spin up and relaxation of upwelling. During upwelling, the coastal sea level is lowered, the thermocline is raised and a baroclinic coastal jet develops, flowing at speeds of 15–20 cm/s along the temperature front between upwelled and oceanic waters (Castro et al. 1994). With strong upwelling winds, the plume is spread offshore in a thin layer by Ekman transport and eventually disperses. However, observations (Torres and Barton 2007) demonstrate that both upwelling jets and riverine counterflows frequently co-exist. Upwelling jets often originate at sites of more intense upwelling off specific capes, like Cape Roca (38.6° N), forced topographically or by locally enhanced wind (Relvas and Barton 2005; Oliveira et al. 2008). The interplay of these equatorward and poleward flows is important to the fate of ichthyoplankton (Santos et al. 2004) and harmful algal blooms (Sordo et al. 2001).

Along most of the Moroccan subregion upwelling is year round and strongest in late summer (Wooster et al. 1976). Filament formation is especially notable off Cape Ghir (30° N) and Cape Juby (28° N). Pelegrí et al. (2005, 2006) suggest the Cape Ghir filament in autumn represents a major separation of the Canary Current from the coast, potentially exporting large amounts of organic material into the open ocean. The event-scale variability follows the classical pattern in areas where the shelf is narrow (e.g., Cape Bojador), but exhibits shelf-edge upwelling (Fig. 6.13c) in areas where the shelf is broad (Barton et al. 1977). In the latter case a combination of weak stratification, a wide shelf and persistent upwelling-favourable wind events produces a progressive separation of the main upwelling cell from the coast (Marchesiello and Estrade 2007). This separation is associated with the extremely shallow shelf, only 60–80 m deep. In such shallow depths, the surface and bottom Ekman layers merge and a classical upwelling cell can only form in deeper portions of the shelf. Mittelstaedt (1991) cites reports of nearshore countercurrents in that area, but generally there is little information available on nearshore circulation. River influences in the Mauritanian–Senegalese subregions may lead to the scenario displayed in Fig. 6.13b.

6.7 Primary Production

6.7.1 General Features and Seasonality

A pronounced feature of the Canary Current upwelling system is the transition from well oxygenated but phosphate (and nitrate) poor NACW north of 25° N to hypoxic but phosphate-rich SACW south of 20° N across the Cape Verde Frontal Zone (Fig. 6.14). This large-scale water-mass distribution indicates that productivity in the northern region is limited by low nutrient availability while greater nutrient

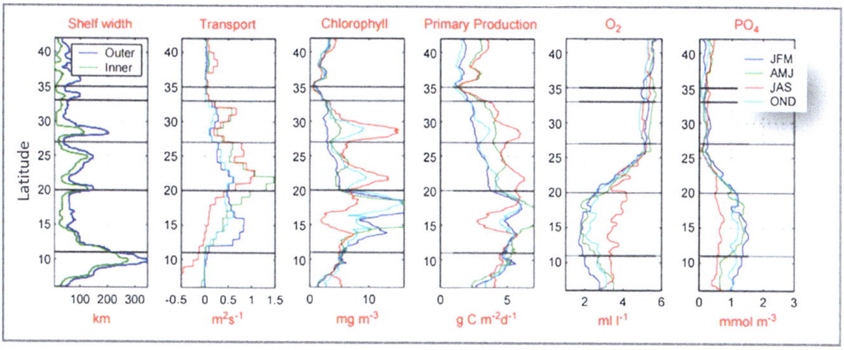

Fig. 6.14 Variation in shelf width and seasonal changes in offshore Ekman transport, chlorophyll concentration, primary production, oxygen and phosphate concentrations with latitude in the Canary-Iberia upwelling system. The horizontal black lines divide the system into separate zones (from Carr and Kearns 2003)

availability and aeolian iron input is deemed to support higher productivity in the southern region (see Carr and Kearns 2003 and references therein). Exceptions are the highly productive Canary Eddy Corridor at 27°–30° N (see Sect. 6.7.3).

The region north of 20° N experiences maximal chlorophyll levels and peak primary production in boreal summer, which coincides with enhanced offshore Ekman transport and peak photosynthetically active radiation (PAR). In contrast to the northern region, the southern region between 15° and 18° N experiences maximal chlorophyll levels in off-summer seasons and the highest primary production in spring. The decrease in summer production is consistent with less upwelling-favourable wind conditions during this season (see Fig. 6.5). The strongest seasonal variations are seen in the Cape Verde Frontal Zone around 15° N which experiences pronounced summer minima in chlorophyll levels and primary production.

6.7.2 Features of Iberian Coastal Waters

Arístegui et al. (2006) discuss the seasonal cycles of upwelling index and biochemical parameters for the Iberian upwelling subregion, based on the study by Nogueira et al. (1997) in the Ría de Vigo—a large (2.5 km^3) coastal embayment that, as stated above, behaves largely as an extension of the shelf (Figueiras et al. 2002) (Fig. 6.15). The extension and intensity of seasonal upwelling and downwelling favourable periods varies considerably from year to year as a consequence of NAO influences (Fig. 6.15a). Whereas the onset of the upwelling season can occur anytime during the two months from the beginning of April to the end of May, its cessation is more marked, falling within a period of just one month, from the middle of September to the middle of October. However, this seasonal cycle

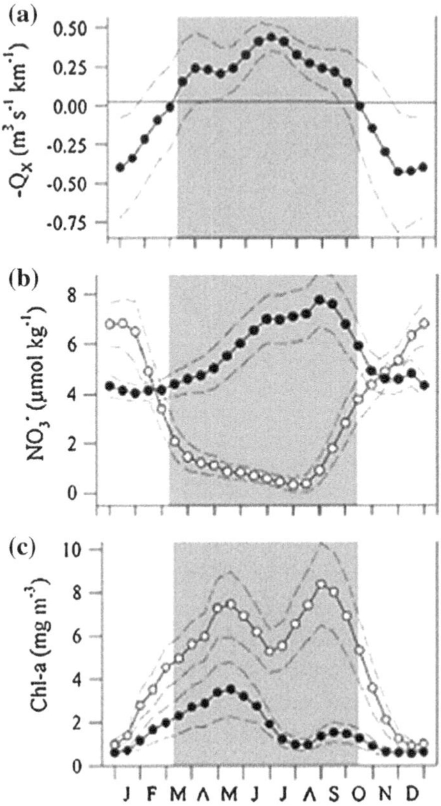

Fig. 6.15 Seasonal cycles of **a** offshore Ekman transport ($-Q_X$, m^3/s/km) in a 2° × 2° geostrophic cell centred at 43° N 11° W; **b** surface and bottom nitrate (NO_3^-, μM/kg); and **c** Chlorophyll (mg/m^3) in the coastal upwelling ecosystem of the Ría de Vigo. *Shaded area* spring-summer upwelling season. *Black circles* in panel **a** 1987–96 fortnightly average; *black circles* in panels **b**–**c** 1987–96 fortnightly bottom average; *white circles* in panels **b**–**c** 1987–96 fortnightly surface average; *dashed lines* average ± standard deviation (redrawn from Arístegui et al. 2006)

only explains about 10 % of the variability of the wind regime off northwest Spain, whereas more than 70 % of the variability arises from periods <30 days (Álvarez-Salgado et al. 2002, 2003). In fact, the upwelling season in this region is described best as a succession of upwelling/relaxation events of periods between 1 and 3 weeks (e.g., Da Silva 1992; Álvarez–Salgado et al. 1993), similar to what occurs in the southern Benguela Current (see Chap. 7).

The seasonal evolution of inorganic nitrogen (Fig. 6.15b) allows clear definition of the spring–summer upwelling period, when nitrate levels are the highest in the cold bottom layer, because of intermittent coastal upwelling, and the lowest in the warm surface layer, because of efficient utilisation of upwelled nitrate by coastal phytoplankton populations. At the short timescale of an upwelling episode (1–3 weeks), the sequence consists of upwelling of nitrate rich ENACW followed by nutrient consumption during the subsequent upwelling relaxation (Pérez et al. 2000).

In contrast, the autumn–winter downwelling period is characterised by higher nitrate levels throughout the water column, which reach their highest levels in the surface layer in association with continental runoff and intense regeneration processes. The importance of regeneration during the autumn–winter period is indicated by the succession of ammonium, nitrite and nitrate maxima in the surface layer, each separated by ~ 1 month, which is characteristic of nitrification processes (Wada and Hattori 1991). The succession of ammonium, nitrite and nitrate maxima also occurs in the bottom layer during the spring–summer period, indicating that regeneration processes are also important here. It should be noted that episodic upwelling events also exist during winter months. For example, Santos et al. (2004) demonstrated a retention mechanism for sardine larvae produced by the interaction of a strong winter upwelling event, the Iberia Poleward Current and the buoyant river plume off western Iberia in February 2000.

The effect of spring–summer upwelling and autumn–winter downwelling periods in the productivity of the Iberian margin is illustrated by the 1987–1996 average seasonal cycle of chlorophyll–*a* in the Ría de Vigo (Fig. 6.15c; Nogueira et al. 1997). Apart from the spring and autumn chlorophyll maxima—characteristic of many temperate ecosystems—chlorophyll levels remain relatively high throughout the summer because of nutrient fertilisation by coastal upwelling. In addition, the spring and autumn chlorophyll maxima occur within the transitional periods of the onset and cessation of the upwelling season (Fig. 6.15a), respectively. These changes therefore determine whether the accumulated organic matter is exported off the shelf or remineralized in situ (Álvarez-Salgado et al. 2003).

6.7.3 The Canary Eddy Corridor

As stated previously, current instabilities and mesoscale eddies in the *Canary Coastal Transition Zone* play an important part in the offshore export of nutrients and carbon. The temporal and spatial variability in plankton biomass, community structure and metabolic activity in this region largely results from mesoscale variability created by island induced perturbations of the main flow (Arístegui et al. 2006). Figure 6.16 summarizes the functioning of the biogeochemistry of the region, as discussed in the following.

On average, the far-field nitracline lies below the compensation depth for phytoplankton growth, and so production is low, carried out by pico- or nanoplankton, and likely sustained by recycled ammonium. However, both cyclonic and anticyclonic eddies disturb this ambient setting. Cyclonic eddies operate to lift isopycnals and the nitracline relative to the compensation depth, so locally stimulating new production and creating a sub-surface (or "deep") chlorophyll-*a* maximum (Fig. 6.17). Anticyclonic eddies, on the other hand, depress isopycnals and the nitracline, deepening the deep chlorophyll maximum below the compensation depth. Counter-paired cyclonic and anticyclonic eddies act as a two-way biological pump enhancing the formation and vertical transport of organic matter in the water column. The deep chlorophyll maximum intensifies towards the coast and higher

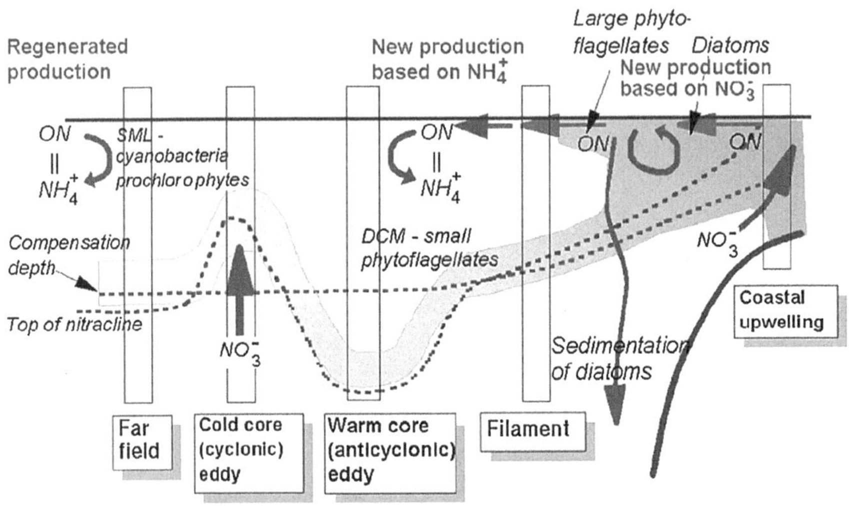

Fig. 6.16 Schematic of vertical processes along the Canary-coastal transition zone. *Vertical rectangles* indicate typical situations. Taken from Arístegui et al. (2006), modified from Barton et al. (1998)

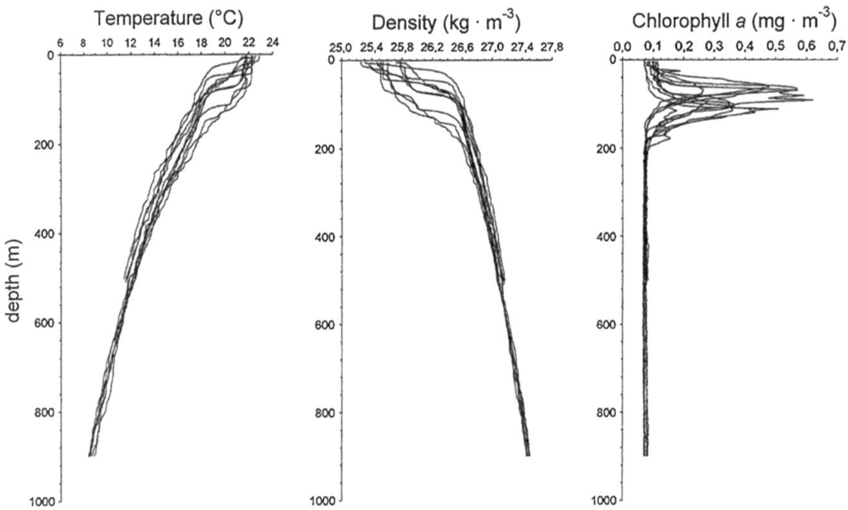

Fig. 6.17 Vertical profiles of temperature, salinity and fluorescence observed between 5 and 26 August 1993 in waters near the Canary Islands (from Hernández-León et al. 2001)

nitrate concentrations are exposed above the compensation depth. In the coastal upwelling region the nitracline reaches the surface mixed layer, resulting in the high production of diatoms that convert nitrate into organic nitrogen. These diatoms sink or are consumed as upwelled water moves offshore in filaments, leaving large phytoflagellates as the dominant producers, supported by recycled ammonia from the organic matter transported by the filament. At the outer edge of the filaments, productivity depends again on nano- and picoplankton.

6.8 Zooplankton

Between 1970 and 1977, a total of eight German research expeditions, known as the "Rostock zooplankton studies", explored the zooplankton dynamics off West Africa in the Canary and Benguela upwelling regions as part of the CINECA program (Cooperative Investigation on the Northern Part of the Eastern Central Atlantic; carried out under the umbrella of ICES), which included 14 countries and about 100 expeditions. Field observations covered scales ranging from minutes to several years in time and from hundreds of meters to thousands of kilometers in space. Postel et al. (1995) summarized the research findings of this intensive field program. The northwest Africa expeditions covered the latitudes between Bahía de Garnet (25° N) and Cabo Roxo (10° N), from the near coastal area to the 21° W meridian. Zooplankton samples were taken from depths of up to 200 m. During this period and as part of the Coastal Upwelling Ecosystems Analysis (CUEA) program, the JOINT-1 investigation was undertaken from March to May 1974 off the southern part of Spanish Sahara, now Mauritania (see Blackburn 1979).

Between 1994 and 2006 systematic zooplankton sampling was carried out during Russian expeditions aboard *AtlantNIRO* vessels along the Moroccan coast from 21° to 36° N as part of an intergovernmental agreement between the Russian Federation, the Kingdom of Morocco and the Islamic Republic of Mauritania (Lidvanov 2011). These data provided insights into interannual variations of the abundance and biomass of different zooplankton communities.

Epipelagic mesozooplankton consisted mainly of copepods, especially *Calanoids* which have development times of 20–23 days. Findings indicate that zooplankton are able to double their biomass after an upwelling event, independent of coastal distance and depth. This upwelling response lasted about 3 weeks in near-surface waters and 6–8 weeks in waters deeper than 75 m. Additionally there was a relationship between cumulative increases in zooplankton biomass and the number of individual upwelling events. The resultant net growth rate of zooplankton biomass was most pronounced at the shelf break (see Fig. 6.18), coinciding with the area of highest fish biomass, and in the upper 25 m of the water column.

The CINECA measurements also documented the strong influence of mesoscale phenomena on zooplankton variability including coastally trapped waves and eddies. Moreover, the CINECA findings indicate that different water masses can be distinguished by indicator species, species combinations or the significant absence

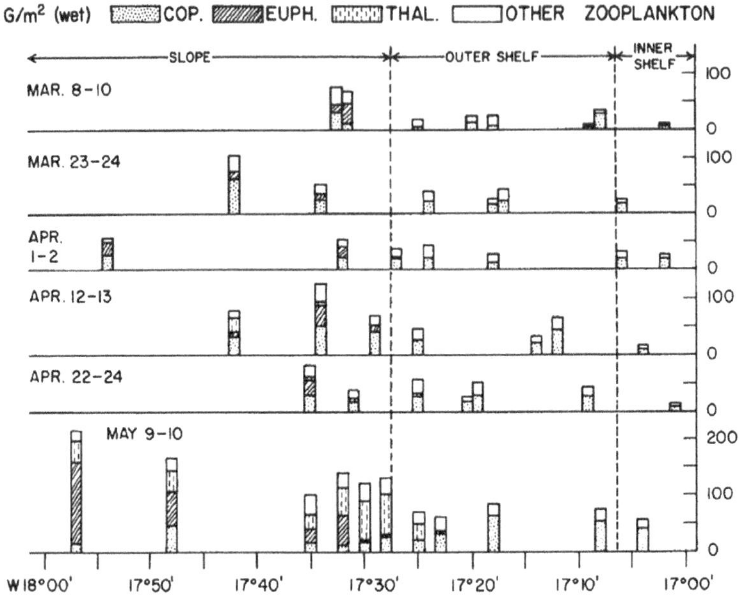

Fig. 6.18 Wet weight biomass (g/m^2) of total zooplankton for 46 hauls on the continental margin of northwest Africa between 21°35′ and 21°45′ from March to May 1974, arranged by area and period. Biomass is shown divided by principal taxa. Shown are Copepoda, Euphausiscea, Thaliacea and other zooplankton species (from Blackburn 1979)

of species. For instance, the clinoid *Calanus helgolandicus* indicates North Atlantic Central Water, whereas *Calanoides carinatus* is an indicator of South Atlantic Central Water.

The Russian *AtlantNIRO* surveys revealed dramatic interannual variations of abundance and biomass of dominant zooplankton communities off the Moroccan coast (Fig. 6.19). This included a significant upward trend in both quantities; i.e. the total biomass of neritic zooplankton doubled from 300 to ~600 mg/m^3 from 1994 to 2006. There was, however, a pronounced mismatch in the evolutions of biomass and abundance, particularly in the period from 1999 to 2004. While the total biomass of neritic zooplankton remained largely the same, total abundance increased by a factor of four. This indicates a regime shift towards smaller zooplankton species, which is consistent with intensification of upwelling and a corresponding cooling of the Canary current waters after 1999 (Lidvanov 2011). In addition, there was a significant transient drop in total biomass in 1998, which almost wiped out the neritic-oceanic zooplankton community. Reasons for this phenomenon are not known, although 1998 was an El Niño year.

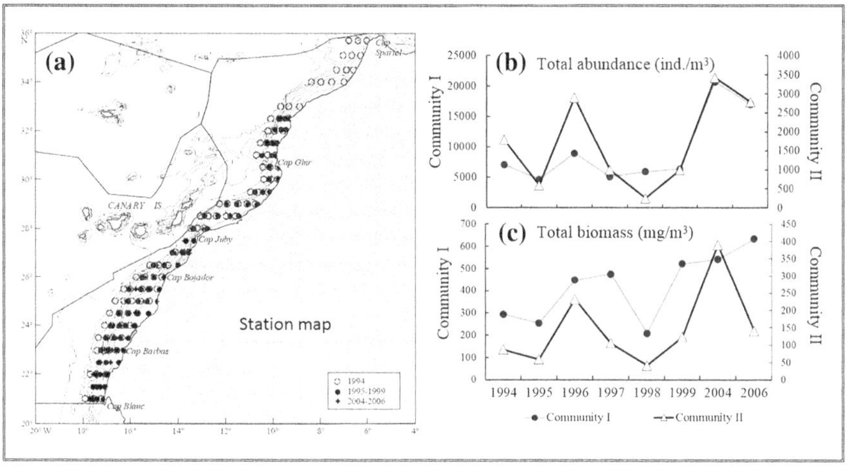

Fig. 6.19 **a** Sampling stations of Russian *AtlantNIRO* surveys for the period from 1994 to 2006. Interannual variations of total abundance (ind./m^3) and total biomass (mg/m^3) of two different zooplankton communities. Community I is formed mostly by near-coastal neritic species; community II by neritic-oceanic species (taken from Lidvanov 2011). Note that the time-axis in panels **b** and **c** is not linear (there are no data for 2000–2003 and 2005)

6.9 Fisheries

6.9.1 Overview

Fisheries in the Canary Current Upwelling System developed several centuries ago in northern Iberia, and continued later off northwest Africa as European fishermen sought new fishing grounds during the 16th century (Chauveau 1989). European fishing fleets initially targeted sardine and some abundant large demersal fish (e.g., grunts, croakers and sharks), using passive gear and purse seines (e.g., Ribeiro et al. 2004). The purse-seiner fishery was conducted mainly along the Iberian and northern Moroccan coasts, whereas ground fish were exploited particularly in the Moroccan subregion (Chauveau 1989). The Iberian fisheries landed the greatest catches in the mid-1960s. Since then, landings have dropped to less than 20 % of the total catches after 1970 (Hill and Coelho 2001).

The upwelling system supports a variety of fish assemblages ranging from boreal and temperate groups in its northern part to subtropical and tropical groups in the south. The Galician and Portuguese subregions contain more boreal species but lower species richness than southern areas (Sousa et al. 2006), while the Gulf of Cadiz and Moroccan subregions constitute a transition zone between temperate and subtropical provinces (Belvèze and Bravo de Laguna 1980). The Mauritanian–Senegalese subregion, which has a large annual temperature range due to the seasonal shifts of the southern extent of the upwelling region (see below), contains a complex assemblage that is dominated by subtropical and tropical species, although some small pelagic temperate species extend their distributional range to this area.

During and after the World Wars, the introduction of an artisanal bottom-trawl fishery in the Moroccan and Gulf of Cadiz subregions increased the contribution of many other demersal species (smaller sparids, hake and shrimps) to the catches. Industrial-scale exploitation developed after 1958. The main targets were demersal fish and other ground fish caught with bottom-trawls (Guénette et al. 2001; Ribeiro et al. 2004). Subsequently, pelagic and midwater species, such as sardinellas, horse mackerel and mackerel were taken in the early 1960s, and the fishery changed from semi-pelagic to midwater-trawling. Finally, with another switch in targeted catch to sardines, the industry became increasingly based on purse seiners (Fréon et al. 1978).

Indirect evidence indicates that removal of the larger-sized demersal species may have resulted in a weakening of the top-down predatory control of short-lived demersal species, at least in some fishing grounds that were subject to uncontrolled heavy fishing (Gulland and Garcia 1984; Caddy and Rodhouse 1998; Kifani et al. 2008). Gascuel et al. (2007) report that increasing fishing pressure has led to a reduction by 75 % of demersal resources (in particular of top predators) in the Mauritanian–Senegalese subregion over the past 25 years. Similarly, seal harvesting almost led to led to extinction of the monk seal (*Monachus monachus*), which was formerly abundant in the Mediterranean Sea and in coastal waters off northwest Africa. Monk seals show no signs of recovery despite increasing protection of the few hundred individuals that survive in the southern part of the upwelling region (CMS 2005).

6.9.2 Food Web Structure and Dominant Forage Fish

As found in other coastal upwelling systems, the ecology of the Canary Current Upwelling System exhibits a distinct wasp-waist trophic structure. Overall, the European sardine (*Sardina pilchardus*) (Fig. 6.20) currently dominates the trophic level of small pelagics in most of the upwelling system, except for its southern reaches off Mauritania and Senegal where sardinella (*Sardinella aurita* and *S. maderensis*) and Atlantic horse mackerel (*T. trachurus*) play a dominant role. While sardines use a filter-feeding strategy with a phytoplankton diet, sardinella and horse mackerel large largely feed on zooplankton (Arístegui et al. 2009).

Unlike other major coastal upwelling systems during the historical period, anchovy (*E. encrasicolus*) constitutes one of the less abundant small pelagic species in the Canary Current upwelling system and there is no evidence of oppositely phased population swings between sardine and anchovy (Alheit et al. 2008). The small-pelagic fish assemblages off Galicia and Portugal and in the Gulf of Cadiz fish assemblages are thus dominated by sardine, with anchovy as a secondary resource. The demersal assemblage is dominated by snipefishes (*Macroramphosus spp.*), blue whiting (*Micromesistius poutassou*), boardfish (*Capros aper*), horse

Fig. 6.20 School of European sardine (*Sardina pilchardus*). *Picture source* http://www.britannica.com/EBchecked/topic/524126/sardine [accessed on 5 April 2016]

mackerel and hake (*Merluccius merluccius*) (Sousa et al. 2006). Megrim (*Lepidorhombus boscii*) in Galicia, sparids in Portugal and Gulf of Cadiz, and deep-water shrimps and small cephalopods in the Gulf of Cadiz (Gomes et al. 2001; Sousa et al. 2005) are also important components of the demersal community.

The Moroccan subregion is characterized by higher overall fish abundance than neighbouring subregions (Belvèze and Bravo de Laguna 1980), and by pronounced long-term fluctuations in fish assemblages. Sardines currently dominate the small-pelagic fish assemblage, although there is evidence that a mix of sardinellas, horse mackerels and chub mackerel could have composed the coastal pelagic assemblage before the sardine outburst between Cape Bojador and Cape Blanc in the 1970s (Boëly and Fréon 1979; Gulland and Garcia 1984; see Sect. 6.10). The demersal assemblage on the shelf is represented by sparids and cephalopods (Gulland and Garcia 1984; Caddy and Rodhouse 1998). Abundant species on the outer shelf include hairtails (*Trichiurus lepturus*) and hakes (*Merluccius senegalensis* and *M. polli*). The coastal pelagic fish assemblage in the Mauritanian–Senegalese subregion is dominated by sardinellas and horse mackerels (Boëly and Fréon 1979; Josse 1989).

6.9.3 Seasonal Migration

Interactions between upwelling-induced currents, complex bathymetry and an irregular coastline create distinct *dispersal zones*, corresponding to upwelling centres and strong local upwelling jets, and *retention zones*, associated with weaker circulation or recirculation in the lee of capes and headlands (see Fig. 6.3). The location of these zones (which underlie seasonal and inter-annual variability) influences the passive drift behaviour of fish larvae as well as the distribution patterns of adult pelagic fish.

The general pattern that emerges from studies of pelagic fish distributions in the Canary Current Upwelling System is that species undertake seasonal latitudinal migrations depending on their geographical and bathymetric positioning in relation to dispersal and retention zones. Arístegui et al. (2009) point out that, in the northern inter-tropical transition area of the Canary Coastal Transition Zone, north–south migrations prevail for almost all small-pelagic fish, many large pelagic fish (Fonteneau and Marcille 1993), epipelagic large sharks (Zeeberg et al. 2006) and demersal fish (Champagnat and Domain 1978). Based on seasonal distribution patterns in Arístegui et al. (2009, their Fig. 5) we conclude that sardines, round sardinella and European and African horse mackerel all exhibit their southernmost distributions during winter, but the two sub-tropical species (round sardinella and African horse mackerel) also display pronounced northward migrations during the summer from the region around Cape Verde (14° N) to Cape Blanc (20° N).

In the northern part of the Moroccan subregion, only sardines show latitudinal migration (Furnestin and Furnestin 1970). In the Galician and Portuguese subregions sardines, as they mature, migrate from their recruitment grounds in the north of the Portuguese subregion either towards the Bay of Biscay or the Gulf of Cadiz, but with a preferred direction over different time periods (Silva et al. 2009). These movements seem to be rather limited and do not indicate any large-scale migration but rather connectivity between sub-populations (Carrera and Porteiro 2003; Silva et al. 2009).

6.9.4 Catch Statistics

Total catches in the Canary Current Upwelling System experienced a huge and rapid increase at the beginning of the 1970s (Fig. 6.21). It was during this period that the national industrial northwest African fishing fleets emerged (Belvèze and Bravo de Laguna 1980; Troadec and Garcia 1979). Improvements in the Senegalese artisanal fishery, especially by adding motors to the pirogue fleet and adopting the use of purse seines (Boëly 1982; Bonfil et al. 1998), and the improvement of semi-artisanal purse seiners in the Moroccan fleet (Belvèze and Bravo de Laguna 1980) contributed to the increased catch. The modification of the Law of the Sea, which led to the introduction of a 200-mile exclusive economic zone in most of the

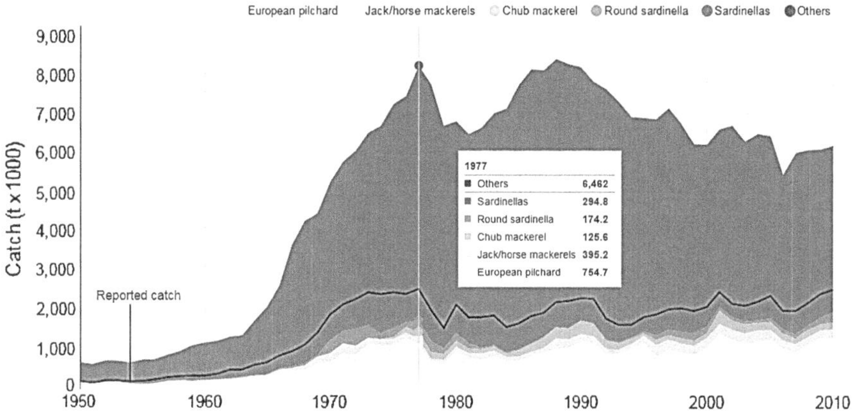

Fig. 6.21 Catches of selected taxa in the Canary Current upwelling system. Data from http://www.seaaroundus.org/. Note the large proportion of unreported catch

world's oceans, facilitated the increasing contribution of different African coastal countries to the total catch and fishing capacity during the 1980s.

During the 1970s and 1980s foreign nations were particularly active in the Canary Current with a total of 25 long-range fishing fleets operating mainly in the central part of the region (FAO 1997; Maus 1997). This pattern changed drastically in the 1990s following the political transformation in the former Soviet bloc countries of Eastern Europe. By targeting mostly pelagic species, these fleets contributed up to 50 % of the total regional marine catches, but their contribution has been declining rapidly since the early 1990s, with the partial withdrawal of eastern European and former USSR fleets from West Africa (FAO 1997).

The overall reported marine catch during the early 2000s was around 2 million tonnes. Small-pelagic fish (mainly European sardine) represent 32 % of the 1950–2004 average catch, followed by medium-size pelagic fish (sardinellas, horse mackerel and chub mackerel) and cephalopods (mainly octopuses), contributing 29 and 5 %, respectively. However, in many production zones, fisheries have experienced a drop in landings (Chavance et al. 2004). The FAO estimates that 40 % of the northwest African fisheries are currently in a senescent, biologically ageing phase, and the rest, which are providing about 90 % of the catches (mainly small and medium pelagics), are in a mature phase (i.e., "plateauing" at a high exploitation level) (Heilemen and Tandstad 2008).

In contrast to the Peru-Chile upwelling system , which is prone to major changes in catch rates because of El Niño events, reconstructed catch rates in the Canary Current have been relatively steady during the period 1975–2010 (Fig. 6.22), ranging between 6 and 8 million tonnes per annum. There seems to be a gradual downward trend in the total catch from the late 1980s onward, but it should be noted that unreported catches inferred from other metrics such as fish consumption are extremely high (up to 70 %) in this region. It is interesting to note that anchovy

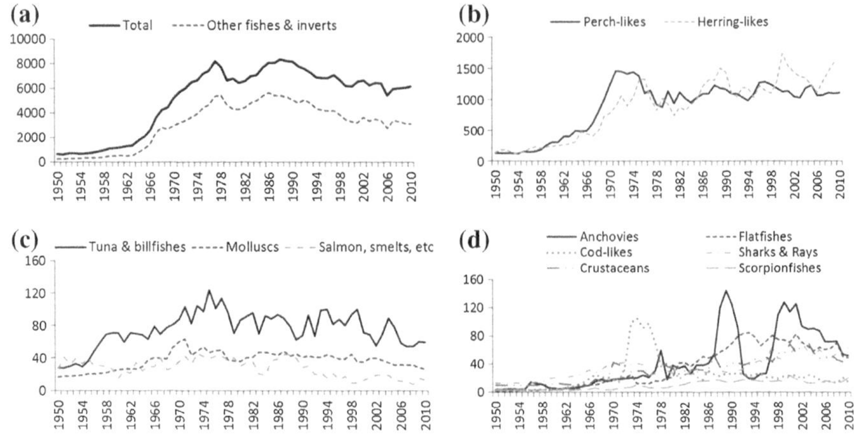

Fig. 6.22 Catches (in units of 1000 tonnes/yr) in the Canary Current large marine ecosystem for 1950–2010. Data from http://www.seaaroundus.org/. Note that the data in this figure include estimates of fish discarded as by catch as well as landed weights, and so are higher than officially reported figures which only consider landed catch. Unreported data are extremely high (up to 70 % of total catch) for the region

catches are generally low, except for two pronounced peaks around 1998 and from 1999 to 2002, when catches exceeded 100,000 tonnes/yr.

6.9.5 *Social and Economic Relevance*

Fisheries play an important role in the socio-economy of coastal states of the Canary Current Upwelling System. Based on 2006 data, Morocco's fisheries, for instance, contributed 3 % to the country's GDP and were associated with 100,000 direct and 400,000 indirect jobs. Portugal's fisheries were worth 4 % of the GDP with 27,000 direct jobs. In contrast, despite being a major deep water fishing nation, Spain's fisheries contributed only <0.5 % to the GDP. Senegal's fisheries contributed 11 % to the country's GDP and were associated with 100,000 direct and 600,000 indirect jobs, the latter amounting to 17 % of the country's workforce. The fishing sector plays a dominant in Mauritania's economy accounting for around half of the country's GDP. Given the relatively high poverty levels (i.e., 29.6 % of the Senegalese population were below the international poverty line of US$1.25 per day in 2007–2011; UNICEF data), regional fisheries, in particular artisanal fisheries, are extremely important for the socio-economic functioning of those coastal states.

Artisanal fisheries in West Africa have been in operation for centuries. Since the early 1960s, they have considerably improved their efficiency by integrating new technologies such as outboard engines and new fishing gear (Chauveau 1991). The regional importance of the artisanal sector is best illustrated by the Senegalese

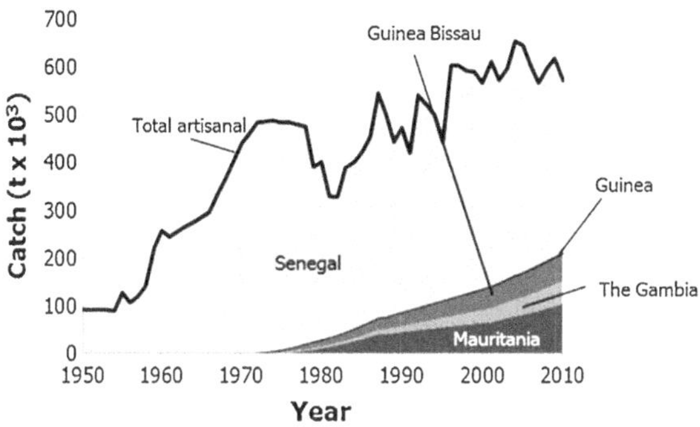

Fig. 6.23 Total reconstructed artisanal catches landed in Senegal, with the origin of the migrant fishers' catch 1950–2010 (from Belhabib et al. 2014)

small-scale fisheries which drastically "mutated" over 30 years to become as socially and economically efficient as the industrial fishery sector (Kébé 1994). With a fleet of about 5000 motorized canoes, this fleet provides employment to

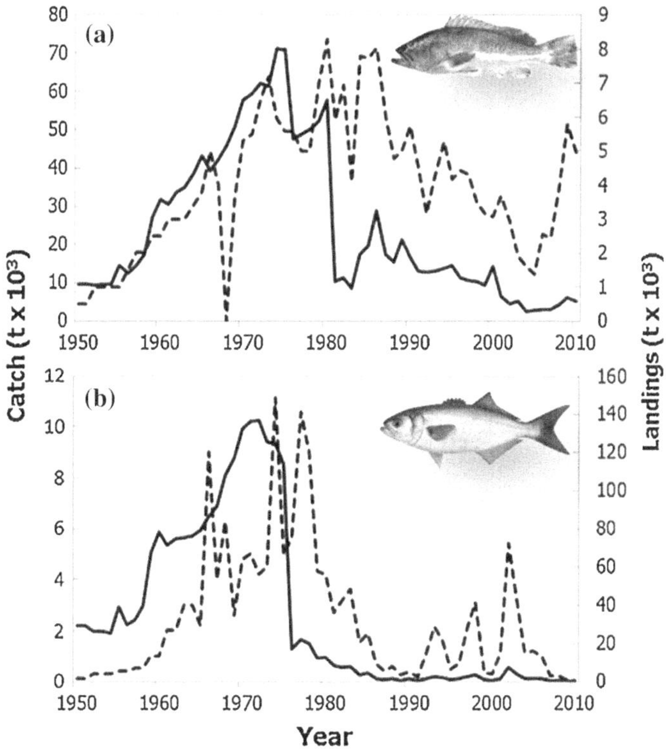

Fig. 6.24 Total reconstructed catches of **a** groupers (*Epinephelus spp.*) (*dashed line*) compared to official landing data (*solid line*); **b** bluefish (*Pomatomus saltatrix*) (*dashed line*) catch compared to official landing data (*solid line*) (from Belhabib et al. 2014)

about 35,000 fishermen. Total catches were estimated at 270,000 metric tonnes of pelagic fish and 46,000 tonnes of demersal fish in the early 1990s (Ferraris et al. 1998), which was about 15 % of the total catch in the region at that time. This is consistent with reconstructed artisanal catches, which due to migrant fisheries have increased from 500,000 metric tonnes in 1970 to around 600,000 metric tonnes in 2010 (Fig. 6.23). The increase of the unregulated Senegalese fleet in the 1970s has been linked to the collapse of white grouper stocks, *Epinephelus aeneus*, also known as 'false cod' (Thiao et al. 2012), and those of other large pelagic fish in the region (Fig. 6.24).

6.10 Interannual Variability, Trends and Regime Shifts

Several fish population outbursts have been observed in the Canary Current during the last 60 years. Species that were known to be rare suddenly developed huge biomass over several years which decreased again equally quickly over a relatively short time span. Off Morocco, for instance, an outburst of blue whiting (*M. poutassou*) was recorded in the 1960s, which later disappeared. In the 1970s snipefishes (*Macrorhamphosus scolopax* and *M. gracilis*) were common all along the Moroccan coast with an estimated biomass of 1 tonne. In the 1980s, the abundance was reduced drastically and today the two species are rarely encountered (Arístegui et al. 2006). In the late 1960s, trigger fish (*Balistes carolinensis*) was scarce in the central Atlantic, but between 1972 and 1980 the biomass of this species increased dramatically in the Gulf of Guinea, exceeding 1 tonne. This increase in abundance was associated with a wide geographical spreading that reached the southern part of the Canary Current, south of 20° N (Gulland and Garcia 1984). Off Senegal, the peak in abundance was recorded in the early 1980s. A sharp decline in abundance followed the outburst of the population at the end of the 1980s, such that this species almost disappeared.

Similarly, the octopus (*Octopus vulgaris*) population increased significantly in the Canary Current in the mid-1960s (Caddy and Rodhouse 1998). Three stocks of octopus are presently exploited by both industrial and artisanal fisheries off southern Morocco, Mauritania and Senegal. The rapid increase of the octopus stock in Mauritania in the late 1960s led to historical yields of 52,900 tonnes and 45,600 tonnes recorded in 1976 and 1987, respectively. During recent years, production has decreased to about 20,000 tonnes (Faure et al. 2000). A rapid expansion of the stock was observed off Morocco with catches reaching 100,000 tonnes at the beginning of the 1980s, while in the mid-1980s octopus abundance also started to increase off Senegal.

On longer time scales, decadal fluctuations in fisheries landings—particularly north of 20° N—have been related to environmental changes due to the North Atlantic Oscillation (NAO) (Borges et al. 2003), that induces changes in sea surface temperature (Ottersen et al. 2003; Visbeck et al. 2003), variations in upwelling patterns and associated nutrient availability (Borges et al. 2003; Santos et al. 2005), and changes in various marine trophic levels of the eastern North Atlantic (Drinkwater et al. 2003). In the southern part of the region, the influence of the El Niño Southern Oscillation (ENSO) may be also responsible of some of the recorded inter-annual variability in fisheries landings (Roy and Reason 2001), although evidence of the teleconnections that drive such changes remains slight. There is also evidence of distinct longer-term ecosystem variations and shifts of fish stocks, some of which could be linked to wind-induced temporal changes of circulation patterns in the ambient ocean (see below).

Cumulative total catches of pelagic and demersal resources in the entire upwelling system are predominantly determined by fish landings in northwest Africa in Moroccan waters which, since the 1970s tend to exceed Iberian fish landings by far (Fig. 6.25 shows sardine catches). Interestingly, the patterns of variability of sardine landings differ substantially between the regions. For instance, sardine landings in Iberian waters dropped substantially in the 1970s which coincided with peak landings in northwest African waters. Similarly, northwest Africa

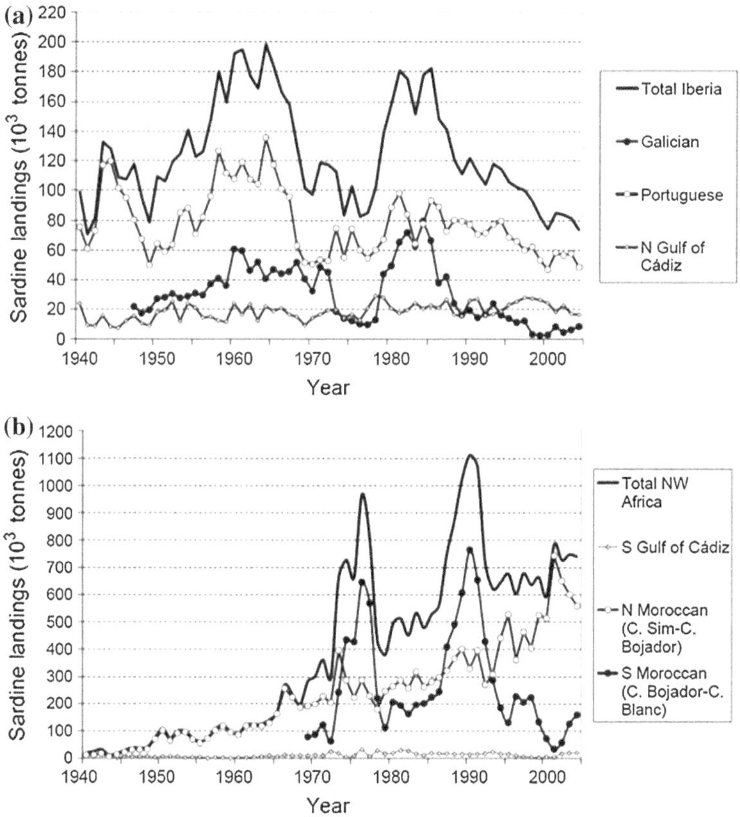

Fig. 6.25 Sub-regional long-term variability in sardine catches off **a** Iberia and **b** northwest Africa (data from ICES 2006) (from Arístegui et al. 2009)

sardine landings declines substantially in the early 1980s during a time when fish landings off Iberia substantially increased to maximal catches which lasted about five years. Reasons for the strongest declines of sardine landings in the 1970s and 1980s in Portuguese and Galician waters are not clearly understood.

There is a somewhat better understanding of sardine variability in northwest Africa. For instance, the outburst of sardines off the Sahara in the late 1960s could be attributed to a spatial shift of sardine and sardinella distributions (e.g., Gulland and Garcia 1984). Belvèze and Erzini (1983) ascribed the decline of sardine abundance in their foraging ground north of Cape Sim-Cape Ghir to a weakening of upwelling intensity in the vicinity of Cape Ghir during the 1970s. Conversely, the southward extension of sardine was linked to strengthening of the trade wind intensity and upwelling activity off the Sahara during the same time period (e.g., Binet 1988). Several studies agreed that this shift benefitted the sardine population (filter-feeding on phytoplankton) south of Cape Blanc and led to a retreat of sardinella (mainly feeding on zooplankton) from the region. Coinciding with the Saharan sardine outburst, some species that were considered rare in the upwelling system transiently developed high biomass for a few years. Such outbursts include snipefish (*Macrorhamphosus app.*) in the Gulf of Cadiz and triggerfish (*B. carolinensis*) in Mauritanian–Senegalese waters, but there remains no good explanation for particular population booms.

Arístegui et al. (2009) report that the abundance of sardines off southern Morocco decreased dramatically from more than 5 million tonnes in 1996 to less than 1 million tonnes in 1997 without any known change of fishing pressure, presumably because of a transient northward expansion of hypoxic South Atlantic Central Water that led to the population moving to the north. While the sardine abundance recovered steadily afterwards, sardinella have gradually increased their presence north of Cape Blanc since the collapse of sardine, and have been observed north of Cape Juby (Arístegui et al. 2009). More than 50 % of the total regional biomass of sardinella was located off southern Morocco after the mid-1990s, while the bulk of biomass was found in the Mauritanian–Senegalese sub-region in the 1980s (Saetersdal et al. 1999). Arístegui et al. (2009) speculate that the 1996–1997 warming event may have been a nested episode in a longer-term shift of the system to a warmer regime, as seems to be indicated by the increase in abundance of some tropical species, like croakers and the Atlantic bumper (*Chloroscombrus chrysurus*) in the Mauritanian–Senegalese subregion during the two last decades (Lobry et al. 2003).

Despite Bakun's (1990) hypothesis that, because the continents warm more than the ocean, global warming will enhance the cross-shore atmospheric pressure gradient, enhance upwelling, and hence trigger the appearance of cooler water in the upwelling regions, most analyses of sea-surface temperatures and coastal winds run counter to this. For instance, observations in the Portuguese subsection for the period between 1941 and 2000 showed a progressive weakening of upwelling and slight warming of near-shore water (Lemos and Pires 2004). In the Galician subregion, Álvarez-Salgado et al. (2008) reported a 30 % decrease in the duration of the upwelling season based on data from between 1968 and 2008. Carson and

Harrison (2008) showed, in a global analysis of sub-surface temperatures from the World Ocean Database 2005, that the layers of the Canary Current above 300 m experienced a general warming over the last 50 years. Available satellite-derived chlorophyll-*a* records do not show strong trends in either sense, but there seems to be a slight overall decreasing trend in the Mauritania–Senegalese subregion (Arístegui et al. 2009). However, as mentioned previously (Sect. 6.5.2), century-scale records suggest that temperatures have cooled and upwelling has increased off Morocco during the 20th century, in line with Bakun (1990) (McGregor et al. 2007).

6.11 Air-Sea Carbon Fluxes

The behaviour of coastal upwelling systems as sources or sinks of CO_2 to the atmosphere depends on the balance of two opposing factors. On one side, the higher the nutrient concentration of the source water, the higher is its partial CO_2 pressure (pCO_2). On the other, the higher the production rates (enhanced by the nutrient input), the higher must be the reduction of pCO_2 in the surface layer (e.g., Watson 1995; Borges and Frankignoulle 2002a). Given these general considerations, the water masses and the dynamics of the Iberian margin would favour its behaviour as an overall CO_2 sink (Arístegui et al. 2006). First, nutrients and, consequently, pCO_2 levels of upwelled Eastern North Atlantic Central Water are relatively low (400–500 μatm) compared with the aged central waters of the South Atlantic, the Indian and the Pacific Ocean. Second, the intermittency of coastal upwelling in the Iberian margin allows efficient utilisation of upwelled nutrients, leading to production rates comparable with other coastal upwelling systems, where the nutrient loads of upwelled waters are much higher.

Surface pCO_2 measurements in the western Iberian margin confirm this general view. Despite seasonal upwelling, surface pCO_2 undersaturation occurs throughout the spring and summer upwelling period, except at the Cape Finisterre upwelling centre (Borges and Frankignoulle 2002b) and along the northern Portuguese coast during strong upwelling events (Pérez et al. 1999). During the autumn and winter period, pCO_2 undersaturation is associated with low-salinity continental runoff, which allows a sequence of stratification, chlorophyll accumulation and pCO_2 decrease. On the other hand, pCO_2 supersaturation occurs anywhere continental runoff is reduced, because the aged shelf bottom waters are in contact with the atmosphere as a result of irradiative loss and strong vertical mixing (Fiúza et al. 1998; Vitorino et al. 2002). Equilibrium with the atmosphere or slight undersaturation is usually found in the high-salinity subtropical waters occupying the Iberian-Coastal Transition Zone and the surrounding ocean year-round (Pérez et al. 1999; Borges and Frankignoulle 2002a).

Borges and Frankignoulle (2002b) calculated air–sea exchange fluxes of CO_2 on the western Iberian shelf from 42° to 44° N yielding a net influx in the range of −2.3 to −4.7 mM C/m^2/day (CO_2 uptake by the ocean) during the upwelling season and of −3.5 to −7.0 mM C/m^2/day on an annual basis, using different formulations of the CO_2 exchange coefficient. CO_2 uptake is maximal during the spring and autumn blooms, when influxes of up to −4.3 mM C/m^2/day were recorded in the middle shelf. In contrast, the area acted as a CO_2 source to the atmosphere during the winter, with maximum fluxes of 2.3 mM C/m^2/day again on the middle shelf.

Recent studies in the northwest Africa upwelling near the Canary Islands region, have identified the coastal upwelling as a weak CO_2 source, with average carbon fluxes of 0.5 mM C/m^2/day (Santana-Casiano et al. 2001; Pelegrí et al. 2005). However, most of these studies were performed during autumn and winter, when winds are low to moderate and upwelling is weaker than during the rest of the year. It is therefore plausible that during strong upwelling events and higher productivity, the system behaves as a carbon sink, as off the north Iberian coast. A recent study of CO_2 parameters at the ESTOC Station (European Station of Time Series in the Ocean, Canary Islands) for the period 1995–2004 (Santana-Casiano et al. 2007) confirms that this region behaves as a minor sink of CO_2 with an annually averaged flux of around -0.14 ± 0.1 mM C/m^2/day.

Upwelling filaments can lead to the export of excess inorganic carbon from the coastal upwelling region to the open ocean, providing carbon uptake by phytoplankton is not large enough to decrease significantly the pCO_2 along the filament extension. Pelegrí et al. (2005) observed a net offshore surface flux of CO_2 from the coast to the open ocean through the Cape Ghir filament during October 1999. The calculated biological consumption of CO_2 along the filament was low enough to allow supersaturation of CO_2 in the warmer open ocean waters, increasing the net flux of CO_2 to the atmosphere.

Revisiting previously published data, Álvarez-Salgado et al. (2007) showed that upwelling filaments off Iberia and northwest Africa export between 35 and 58 % of the net community production generated in the coastal upwelling system to the offshore region, largely as dissolved organic matter. Transport by filaments accounts for 2.5–4.5 times the offshore carbon export driven by Ekman transport. The fate of this carbon is unknown, although conservative mass balance analysis in the subtropical northeast Atlantic region suggests that 16 % of the exported carbon may be respired in the Canary Coastal Transition Zone. The remainder of the exported organic matter is transported to and accumulated in the subtropical gyre (Hansell 2002), where it contributes to local oxygen reduction during respiration.

6.12 Summary

The Galician and Portuguese upwelling regions reflect the influence of boreal and temperate affinities in their fish assemblages, whereas the Moroccan and Mauritanian–Senegalese upwelling regions are characterized by subtropical and tropical assemblages. Sardine is the main pelagic resource, except in the Mauritanian–Senegalese regions, where sardinella dominates. In contrast to other eastern boundary upwelling systems, anchovy constitutes a less abundant small pelagic fish species.

The Moroccan upwelling region supports the highest fish abundance, presumably because of high year-round productivity and favourable environmental conditions for larval survival and recruitment. Pelagic and demersal resources in this region have been exposed to marked shifts in the past 50 years in their ranges of distribution and abundances. Seasonal shifts in the upwelling centres along the northwest African coast have produced regional migratory movements, at least in a few pelagic fish species such as sardine and sardinella, which take advantage of plankton seasonal variability. However, the distributional ranges of sardine and sardinella seem to be more controlled by thermal than productivity gradients. Thus, long-term changes in sardine abundances and distribution would be susceptible to environmental forcing under a global climate change scenario.

Most of the evidence suggests that the Canary Current Upwelling Region as a whole has been experiencing a progressive warming and a decrease in productivity over the past 20 years. This overall trend seems not to be directly reflected in the fisheries of the ecosystem. Despite some regional research efforts, the overall understanding of the ecosystem functioning of this Upwelling System remains incomplete. In particular, most of the Moroccan subregion, which is of great significance to fisheries, has been poorly studied in terms of oceanography and biogeochemistry since the first research programs of the 1970s.

References

Alheit, J., S. Kifani, and C. Roy. 2008. Interdecadal variability in populations of small pelagic fish. In *Predicted effects of climate change on SPACC systems*, ed. Checkley D., Roy C. and Alheit J. Cambridge University Press.

Álvarez, I., M. deCastro, M. Gomez-Gesteira, and R. Prego. 2005. Inter- and intra-annual analysis of the salinity and temperature evolution in the Galician Rias Baixas-ocean boundary (northwest Spain). *Journal of Geophysical Research* 110: C04008. doi:10.1029/2004JC002504.

Álvarez-Salgado, X.A., G. Rosón, F.F. Pérez, and Y. Pazos. 1993. Hydrographic variability off the Rías Baixas (NW Spain) during the upwelling season. *Journal of Geophysical Research* 98: 14447–14455.

Álvarez-Salgado, X.A., S. Beloso, I. Joint, E. Nogueira, L. Chou, F.F. Pérez, S. Groom, J.M. Cabanas, A.P. Rees, and M. Elskens. 2002. New production of the NW Iberian shelf during the upwelling season over the period 1982–1999. *Deep Sea Research I* 49: 1725–1739.

Álvarez-Salgado, X.A., F.G. Figueiras, F.F. Pérez, S. Groom, E. Nogueira, A. Borges, L. Chou, C. G. Castro, G. Moncoiffé, A.F. Ríos, A.E.J. Miller, M. Frankignoulle, G. Savidge, and R. Wollast. 2003. The Portugal coastal counter current off NW Spain: new insight on its biogeochemical variability. *Progress in Oceanography* 56: 281–321.

Álvarez-Salgado, X.A., J. Arístegui, E.D. Barton, and D.A. Hansell. 2007. Contribution of upwelling filaments to offshore carbon export in the subtropical Northeast Atlantic Ocean. *Limnology and Oceanography* 52: 1287–1292.

Álvarez-Salgado, X.A., U. Labarta, M.J. Fernández-Reiriz, F.G. Figueiras, G. Rosón, S. Piedracoba, R. Filgueira, and J.M. Cabanas. 2008. Renewal time and the impact of harmful algal blooms on the extensive mussel raft culture of the Iberian coastal upwelling system (SW Europe). *Harmful Algae* 7: 849–855.

Álvarez-Salgado, X.A., A.V. Borges, F.G. Figueras, and L. Chou. 2010. Iberian Margin: the Rias. In *Carbon and nutrient fluxes in continental margins*, ed. K.K. Liu, et al., 103–120. Berlin: CMTT volume, Springer.

Arístegui, J., P. Sangrà, S. Hernández-León, M. Cantón, A. Hernández_Guerra, and J.L. Kerling. 1994. Island-induced eddies in the Canary Islands. *Deep-Sea Research I* 41: 1509–1525.

Arístegui, J., P. Tett, A. Hernández-Guerra, G. Basterretxea, M.F. Montero, K. Wild, P. Sangrà, S. Hernández-León, M. Cantón, J.A. García-Braun, M. Pacheco, and E.D. Barton. 1997. The influence of island-generated eddies on chlorophyll distribution: a study of mesoscale variation around Gran Canaria. *Deep-Sea Research I* 44: 71–96.

Arístegui, J., X.A. Álvarez-Salgado, E.D. Barton, F.G. Figueiras, S. Hernández-León, C. Roy, and A.M.P. Santos. 2006. Oceanography and fisheries of the Canary Current Iberian region of the Eastern North Atlantic. In *The Sea,* vol. 14, ed. A. Robinson and K.H. Brink, 877–931. Harvard University Press.

Arístegui, J., E.D. Barton, X.A. Álvarez-Salgado, A.M.P. Santos, F.G. Figueiras, S. Kifani, S. Hernández-León, E. Mason, and E. Machú. 2009. Sub-regional ecosystem variability in the Canary Current upwelling. *Progress in Oceanography* 83(1–4): 33–48.

Bakun, A. 1973. *Coastal upwelling indices, west coast of North America, 1946–71.* U.S. Department of Commerce, NOAA Technical Report, NMFS SSRF-671, 103 p.

Bakun, A. 1990. Global climate change and intensification of coastal upwelling. *Science* 247: 198–201.

Bakun, A., and C.S. Nelson. 1991. The seasonal cycle of wind-stress curl in subtropical eastern boundary current regions. *Journal of Physical Oceanography* 21: 1815–1834.

Baraza, J., G. Ercilla, B. Alonso, and D. Casas. 2003. Sediment dynamic features on the eastern Gulf of Cadiz (SW Spain). In *European margin sediment dynamics, side-scan sonar and seismic images,* ed. J. Mienert and P. Weaver, 261–266. Berlin: Springer.

Barton, E. 1989. The poleward undercurrent of the eastern boundary of the subtropical North Atlantic. In *Poleward flows along eastern ocean boundaries*, vol. 34, ed. S. Neshyba, et al., 82–95. New York: Springer.

Barton, E. 2001. Canary and Portugal currents. In *Encyclopedia of ocean sciences,* vol. 1, ed. J. Steele, K. Turekian and S. Thorpe, 380–389. Academic Press.

Barton, E.D., A. Huyer, and R.L. Smith. 1977. Temporal variation observed in the hydrographic regime near Cabo Corveiro in the NW African upwelling region, February to April 1974. *Deep-Sea Research* 24: 7–24.

Barton, E.D., J. Arístegui, P. Tett, M. Cantón, J. García-Braun, S. Hernández-León, L. Nykjaer, C. Almeida, J. Almunia, S. Ballesteros, G. Basterretxea, J. Escánez, L. García-Weil, A. Hernández-Guerra, F. Ópez-Laatzen, R. Molina, M.F. Montero, E. Navarro-Pérez, J.M. Rodríguez, K. van Lenning, H. Vélez, and K. Wild. 1998. The transition zone of the Canary Current upwelling region. *Progress in Oceanography* 41: 455–504.

Barton, E., J. Arístegui, P. Tett, and E. Navarro-Pérez. 2004. Variability in the Canary Islands area of filament-eddy exchanges. *Progress in Oceanography* 62: 71–94.

Belhabib, D., V. Koutob, A. Sall, V.W.Y. Lam, and D. Pauly. 2014. Fisheries catch misreporting and its implications: the case of Senegal. *Fisheries Research* 151: 1–11. doi:10.1016/j.fishres. 2013.12.006.

Belvèze, H., and J. Bravo de Laguna. 1980. Les ressources halieutiques de l'Atlantique Centre-Est. Deuxième partie. Les ressources de la côte ouest-africaine entre 24°N et le détroit de Gibraltar. *FAO Documents Techniques sur les Pêches* (186.2).

Belvèze, H., and K. Erzini. 1983. The influence of hydro-climatic factors on the availability of the sardine (*Sardina pilchardus* Walbaum) in the Moroccan Atlantic fishery. *FAO Fisheries Report* 291: 285–328.

Binet, D. 1988. Rôle possible d'une intensification des alizés sur le changement de répartition des sardines et sardinelles de long de la côte ouest-africaine. *Aquatic Living Resources* 1: 115–132.

Blackburn, M. 1979. Zooplankton in an upwelling area off northwest Africa: composition, distribution and ecology. *Deep Sea Research* 26(1): 41–56. doi:10.1016/0198-0149(79)90084-0.

Boëly, T. 1982. Les resources en poissons pélagiques des côtes ouest-africaines entre la Mauritanie et le fleuve Congo. *Rapport et Procès-Verbaux des Réunions, Conseil International pour l'Exploration de la Mer* 180: 423–431.

Boëly, T., and P. Fréon. 1979. Les ressources pélagiques cotières. In Les Ressources Halieutiques de l'Atlantique Centre-Est. 1. Les Ressources du Golfe de Guinée, de l' Angola à la Mauritanie. *FAO Fishery Technical Documents* 186: 13–78.

Bonfil, R., G. Munro, U.R. Sumaila, H. Valtysson, M. Wright, T. Pitcher, D. Preikshot, N. Haggan, and D. Pauly. 1998. Distant water fleets: an ecological, economic and social assessment. *Fisheries Centre Research Reports* 6(6).

Borges, A.V., and M. Frankignoulle. 2002a. Aspects of dissolved inorganic carbon dynamics in the upwelling system off the Galician coast. *Journal of Marine Systems* 32: 181–198.

Borges, A.V., and M. Frankignoulle. 2002b. Distribution of surface carbon dioxide and air-sea exchange in the upwelling system off the Galician coast. *Global Biogeochemical Cycles* 16. doi:10.1029/2000GB001385.

Borges, M.F., A.M.P. Santos, N. Crato, H. Mendes, and B. Mota. 2003. Sardine regime shifts off Portugal: a time series analysis of catches and wind conditions. *Scientia Marina* 67: 235–244.

Botas, J.A., E. Fernández, A. Bode, and R. Anadón. 1989. Water masses off the central Cantabrian coast. *Scientia Marina* 53: 755–761.

Caddy, J.F., and P.G. Rodhouse. 1998. Cephalopod and groundfish catches: evidence for ecological change in global fisheries? *Review of Fish Biology and Fisheries* 8: 431–444.

Carr, M.-E., and E.J. Kearns. 2003. Production regimes in four eastern boundary current systems. *Deep-Sea Research II* 50: 3199–3221.

Carrera, P., and C. Porteiro. 2003. Stock dynamics of the Iberian sardine (*Sardina pilchardus, W.*) and its implication on the fishery off Galicia (NW Spain). *Scientia Marina* 67: 245–258.

Carson, M., and D.E. Harrison. 2008. Is the upper ocean warming? Comparisons of 50-year trends from different analyses. *Journal of Climate* 21: 2259–2268.

Castro, C.G., F. Pérez, X. Álvarez-Salgado, G. Rosón, and A.F. Ríos. 1994. Hydrographic conditions associated with the relaxation of an upwelling event off the Galician coast (NW Spain). *Journal of Geophysical Research* 99: 5135–5147.

Castro, C.G., F.F. Pérez, X. Álvarez-Salgado, et al. 2000. Coupling between thermohaline and chemical fields during two contrasting upwelling events off the NW Iberian Peninsula. *Continental Shelf Research* 20: 189–210.

Champagnat, C., and F. Domain. 1978. Migrations des poissons démersaux le long des cotes ouest africaines de 10 à 24° de latitude nord. *Cahiers ORSTOM, Série Océanographie SVI* 5-4: 239–261.

Chauveau, J.-P. 1989. Histoire de la pêche industrielle au Sénégal et politiques d'industrialisation. 1ère partie: Cinq siècles de pêche européenne (du XVe siècle au milieu des années 1950). *Cahiers des Sciences Humaines* 25(1–2): 237–288.

Chauveau, J.-P. 1991. Les variations spatiales et temporelles de l'environnement soci-économique et l'évolution de la pêche maritime artisanale sur les côtes ouest-africaines. Essai d'analyse en longue période: XVè-XXè siècle. In *Pêcheries ouest-africaines: variabilité, instabilité et changement*, ed. P. Cury and C. Roy, 14–25. Paris: ORSTOM Editions.

Chavance, P., M. Bâ, D. Gascuel, J.M. Vakily, and D. Pauly, eds. 2004. *Marine fisheries, ecosystems and societies in West Africa: half a century of change*. Actes du symposium international, Dakar (Sénégal), 24–28 juin 2002. Collection des rapports de recherche halieutique A.C.P.-U.E.No. 15.

Clark, D.J., R.A. Oliver, J.D. Fage, and A.D. Roberts. 1975. *The Cambridge history of Africa*. Cambridge University Press.

CMS. 2005. Action plan for the recovery of the Mediterranean monk seal in the Eastern Atlantic. Thirteenth meeting of the CMS Scientific Council. Nairobi, Kenya, 16–18 November 2005. CMS/Scientific Council,13/Inf.3.

Codispoti, L.A., R.C. Dugdale, and H.J. Minas. 1982. A comparison of the nutrient regimes off northwest Africa, Peru and Baja California. *Rapport et Procès-Verbaux des Réunions, Conseil International pour l'Exploration de la Mer* 180: 177–194.

Da Silva, A.T. 1992. *Dependence of upwelling related circulation on wind forcing and stratification over the Portuguese northern shelf, ICES C.M.* 1992/C, 17, 12. Copenhagen: International Council for the Exploration of the Sea.

Demarcq, H. 1998. Spatial and temporal dynamic of the upwelling off Senegal and Mauritania: local change and trend. In *Global versus local changes in upwelling systems,* ed. M.-H. Durand, P. Cury, R. Mendelssohn, C. Roy, A. Bakun and D. Pauly, 149–166. Paris: Orstom ed.

Déniz-González, I., P.J. Pascual-Alayón, J. Chioua, M.T. García-Santamaría, and J.L. Valdés. 2014. *Directory of atmospheric, hydrographic and biological datasets for the Canary Current large marine ecosystem,* vol. 110. IOC Technical Series, 214. Paris: IOC-UNESCO.

Drinkwater, K.F., A. Belgrano, A. Borja, A. Conversi, M. Edwards, C.H. Greene, G. Ottersen, A. J. Pershing, and H. Walker. 2003. The response of marine ecosystems to climate variability associated with the North Atlantic Oscillation. In *The North Atlantic Oscillation: climatic significance and environmental impact,* ed. J. Hurrell, Y. Kushnir, G. Ottersen and M. Visbeck, 211–234. Washington DC: AGU Geophysical Monograph 134.

Duce, R.A., and N.W. Tindale. 1991. Atmospheric transport of iron and its deposition in the ocean. *Limnology and Oceanography* 36: 1715–1726.

Ekman, V.W. 1905. On the influence of the earth's rotation on ocean currents. *Arkiv för Matematik, Astronomi och Fysik*, 2 (11): 1–52.

Eynaud, F., S. Zaragosi, J.D. Scourse, M. Mojtahid, J.F. Bourillet, I.R. Hall, A. Penaud, M. Locascio, and A. Reijonen. 2007. Deglacial laminated facies on the NW European continental margin: the hydrographic significance of British-Irish ice sheet deglaciation and Fleuve Manche paleoriver discharges. *Geochemistry, Geophysics, Geosystems*, 8, Q06019. doi:10.1029/2006GC001496.

FAO. 1997. *Review of the state of world fishery resources: marine fisheries*. FIRM/C920, Fisheries Circular No. 920.

Faure, V., C.A. Inejih, H. Demarcq, and P. Cury. 2000. The importance of retention processes in upwelling areas for recruitment of *Octopus vulgaris:* the example of the Arguin Bank (Mauritania). *Fisheries Oceanography* 9: 343–355. doi:10.1046/j.1365-2419.2000.00149.x.

Ferraris, J., K. Koranteng, and A. Samba. 1998. Comparative study of the dynamics of small-scale marine fisheries in Senegal and Ghana. In *Global versus local changes in upwelling systems*, ed. M.H. Durand, P. Cury, R. Mendelssohn, C. Roy, A. Bakun, and D. Pauly, 447–464. Paris: Editions ORSTOM.

Figueiras, F.G., K. Jones, A.M. Mosquera, X.A. Álvarez-Salgado, A. Edwards, and N. MacDougall. 1994. Red tide assemblage formation in an estuarine upwelling ecosystem: Ría de Vigo. *Journal of Plankton Research* 16: 857–878.

Figueiras, F.G., U. Labarta, and M.J. Fernnández Reiriz. 2002. Coastal upwelling, primary production and mussel growth in the Rías Baixas of Galicia. *Hydrobiologia* 484: 121–131.

Fischer, G., S. Neuer, R. Davenport, O. Romero, V. Ratmeyer, B. Donner, T. Freudenthal, H. Meggers, and G. Wefer. 2007. Control of ballast minerals on organic carbon export in the eastern boundary current system (EBCs) off NW Africa. In *Carbon and nutrient fluxes in continental margins,* ed. K.K. Liu, et al. Berlin: CMTT volume, Springer.

Fiúza, A.F.G. 1983. Upwelling patterns off Portugal. In *Coastal upwelling: its sediment record,* ed. E. Suess and J. Thiede, 85–98. New York: A Plenum Press.

Fiúza, A.F.G., M. Hamann, I. Ambar, G. Díaz del Río, N. González, and J.M. Cabanas. 1998. Water masses and their circulation off western Iberia during May 1993. *Deep Sea Research* 45: 1127–1160.

Fonteneau, A., and J. Marcille 1993. *Resources, fishing and biology of the tropical tunas of the eastern central Atlantic*, vol. 292. FAO Fisheries Document Paper, 354. Rome: FAO.
Fréon, P., B. Stéquert, and T. Boely. 1978. La pêche des poissons pélagiques côtiers en Afrique de l'Ouest des îles Bissagos au nord de la Mauritanie: description des types d'exploitation. *Cahiers ORSTOM Série Océanographie* 18: 209–228.
Frouin, R., A.F.G. Fiúza, I. Ambar, and T.J. Boyd. 1990. Observations of a poleward surface current off the coasts of Portugal and Spain during winter. *Journal of Geophysical Research* 95: 679–691.
Furnestin, J., and M.L. Furnestin. 1970. La sardine marocaine et sa pêche. Migrations trophique et génétique en relation avec l'hydrologie et le plankton. *Rapport et Procès-Verbaux des Réunions, Conseil International pour l'Exploration de la Mer* 159: 165–175.
Gao, Y., F. Song-Miao, and J.L. Sarmiento. 2003. Aeolian iron input to the ocean through precipitation scavenging: a modeling perspective and its implication for natural iron fertilization in the ocean. *Journal of Geophysical Research* 108(D7): 4221. doi:10.1029/2002JD002420.
García-Soto, C., R.D. Pingree, and L. Valdés. 2002. Navidad development in the southern Bay of Biscay: climate change and swoddy structure from remote sensing and in situ measurements. *Journal of Geophysical Research* 107: C8. doi:10.1029/2001JC001012.
Gascuel, D., P. Labrosse, B. Meissa, M.O. Taleb Sidi, and S. Guénette. 2007. Decline of demersal resources in northwest Africa: an analysis of Mauritanian trawl-survey data over the past 25 years. *African Journal of Marine Science* 29: 331–345.
Gil, J. 2003. Changes in the pattern of water masses resulting from a poleward slope current in the Cantabrian Sea (Bay of Biscay). *Estuarine, Coastal and Shelf Science* 57: 1139–1149.
Gomes, M.C., E. Serrão, and M.F. Borges. 2001. Spatial patterns of groundfish assemblages on the continental shelf of Portugal. *ICES Journal of Marine Science* 58: 633–647.
Guénette, S., E. Balguerías, and M.T. García Santamaría. 2001. Spanish fishing activities along the Saharan and Moroccan coasts. In *Fisheries impacts on North Atlantic ecosystems: catch, effort and national/regional data sets*, vol. 9(3), ed. D. Zeller, R. Watson and D. Pauly, 206–213. Part III: South-eastern North Atlantic: Fisheries Centre Research Reports.
Gulland, J.A., and S. Garcia. 1984. Observed patterns in multispecies fisheries. In *Life sciences research report*, vol. 32, ed. May, R.M., 155–190. Exploitation of Marine Communities, Report of the Dahlem Workshop on exploitation of marine communities. Berlin, 1–6 April 1984. Berlin: Springer-Verlag.
Hagen, E., and R. Schemainda. 1987. On the zonal distribution of South Atlantic Central Water (SACW) along a section off Cape Blanc, Northwest Africa. *Oceanologica Acta* 6: 61–70.
Hansell, D.A. 2002. DOC in the global ocean carbon cycle. In *Biogeochemistry of marine dissolved organic matter*, ed. D.A. Hansell and C.A. Carlson, 685–715. Academic Press.
Haynes, R., and E.D. Barton. 1990. A poleward flow along the Atlantic coast of the Iberian Peninsula. *Journal of Geophysical Research* 95: 11425–11441.
Haynes, R., E.D. Barton, and I. Pilling. 1993. Development, persistence and variability of upwelling filaments off the Atlantic coast of the Iberian Peninsula. *Journal of Geophysical Research* 98: 22681–22692.
Heilemen, S., and M. Tandstad. 2008. Canary current LME# 27. http://www.lme.noaa.gov/Portal/LME_Report/lme_27.pdf. Accessed 1 April 2016.
Hempel, G. 1982. The Canary Current: studies of an upwelling system. Introduction. *Rapports et Proces-verbaux des Réunions/Conseil International pour l'Éxploration de la Mer* 180: 7–8.
Hernandez-Guerra, A., E. Fraile-Nuez, F. Lopez-Laatzen, A. Martinez, G. Parrilla, and V. Perez-Belchi. 2005. Canary Current and North Equatorial Current from an inverse box model. *Journal of Geophysical Research* 110. doi:10.1029/2005JC003032.
Hernández-León, S., M. Gómez, M. Pagazaurtundua, A. Portillo-Hahnefeld, I. Montero, et al. 2001. Vertical distribution of zooplankton in Canary Island waters: implications for export flux. *Deep-Sea Research I* 48: 1071–1092. doi:10.1016/s0967-0637(00)00074-1.

Hernández-León, S., C. Almeida, P. Bécognée, L. Yebra, and J. Arístegui. 2004. Zooplankton biomass and indices of grazing and metabolism during a late winter bloom in subtropical waters. *Marine Biology* 145: 1191–1200. doi:10.1007/s00227-004-1396-5.

Hill, L., and M.L. Coelho. 2001. Portuguese fisheries in Portugal for the period 1950–1999. Comparison with ICES data. Fisheries impact on North Atlantic ecosytems: catch, effort and national/regional data set. In *Fisheries impacts on north atlantic ecosystems: catch, effort and national/regional data sets* vol. 9(3), ed. D. Zeller, R. Watson and D. Pauly, 187–190. Part III: Fisheries Centre Research Reports.

Hughes, P., and E.D. Barton. 1974. Stratification and water mass structure in the upwelling area off northwest Africa. *Memoires de la Societé Royale des Sciences de Liège,* 6 sér., X: 31–42.

Huthnance, J.M., H.M.V. Aken, M. White, E.D. Barton, B. Le Cann, E.F. Coelho, E.A. Fanjul, P. Miller, and J. Vitorino. 2002. Ocean margin exchange—water flux estimates. *Journal of Marine Systems* 32: 107–137.

Joint, I., and P. Wassmann. 2001. Lagrangian studies of the Iberian upwelling system—an introduction. A study of the temporal evolution of surface production and fate of organic matter during upwelling off the NW Spanish continental margin. *Progress in Oceanography* 51: 217–221.

Joint, I., S.B. Groom, R. Wollast, L. Chou, G.H. Tilstone, F.G. Figueiras, M. Loijens, and T. J. Smyth. 2002. The response of phytoplankton production to periodic upwelling and relaxation events at the Iberian shelf break: estimates by the 14C method and by satellite remote sensing. *Journal of Marine Systems* 32: 219–238.

Josse, E. 1989. Les ressources halieutiques de la ZEE mauritanienne description, évaluation et aménagement. Rapport du deuxième *Groupe de travail CNROP/FAO/ORSTOM*. COPACE/PACE Series 89/49.

Kaczynski, V.M., and D.L. Fluharty. 2002. European policies in West Africa: who benefits from fisheries agreements? *Marine Policy Journal*, 26: 75–93.

Karstensen, J., L. Stramma, and M. Visbeck. 2008. Oxygen minimum zones in the eastern tropical Atlantic and Pacific oceans. *Progress in Oceanography* 77: 331–350.

Kébé, M. 1994. Principales mutations de la pêche artisanale maritime sénégalaise. In *L'évaluation des resources exploitables par la pieche artisanale sénégalaise,* ed. M.B. Barry-Gérard, T. Diouf and A. Fonteneau, 43–58. Paris: ORSTOM Editions.

Kifani, S., H. Masski, and A. Faraj. 2008. The need of an ecosystem approach to fisheries: the Moroccan upwelling-related resources case. *Fisheries Research* 94: 36–42.

King, L.R. 1981. The coastal upwelling ecosystem analysis program as an experience in international cooperation. *Ocean Development and Regional Law* 9: 269–288.

Knoll, M., A. Hernandez-Guerra, B. Lenz, F. Lopez-Laatzen, F. Machin, T.J. Muller, and G. Siedler. 2002. The eastern boundary current system between the Canary Islands and the African coast. *Deep-Sea Research II* 49: 3427–3440.

Krauss, W. 1986. The North Atlantic Current. *Journal of Geophysical Research* 91(C4): 5061–5074.

Lemos, R.T., and H.O. Pires. 2004. The upwelling regime off the West Portuguese coast, 1941–2000. *International Journal of Climatology* 24: 511–524.

Lidvanov, V.V. 2011. Interannual dynamics of zooplankton in the Canary Current ecosystem. Extended abstract. ICES annual science conference 19–23 September 2011, Gdańsk Music and Congress Center, Gdańsk, Poland. Available at http://www.ices.dk/sites/pub/CM%20Doccuments/CM-2011/M/M1811.pdf. Accessed 28 April 2016.

Lobry, J., D. Gascuel, and F. Domain. 2003. La biodiversité spécifique des ressources démersales du plateau continental guinéen: utilisation d'indices classiques pour un diagnostic sur l'évolution de l'écosystème. *Aquatic Living Resources* 16: 59–68.

Marchesiello, P., and P. Estrade. 2007. Eddy activity and mixing in upwelling systems: a comparative study of Northwest Africa and California regions. *International Journal of Earth Sciences* 98(2): 299–308. doi:10.1007/s00531-007-0235-6.

Mason, E., F. Colas, J. Molemaker, A.F. Shchepetkin, C. Troupin, J.C. McWilliams and P. Sangra. 2011. Seasonal variability of the Canary Current: a numerical study. *Journal of Geophysical Research* 116. doi:10.1029/2010JC006665.

Maus, J. 1997. *Sustainable fisheries information management in Mauritania*. PhD thesis. Ecosystem analysis and Management Group, 269. Coventry: University of Warwick.

McClain, C.R., S.-Y. Chao, L.P. Atkinson, J.O. Blanton, and F.F. de Castillejo. 1986. Wind-driven upwelling in the vicinity of Cape Finisterre, Spain. *Journal of Geophysical Research* 91: 8470–8486.

McGregor, H.V., M. Dima, H.W. Fischer, and S. Mulitza. 2007. Rapid 20th-century increase in coastal upwelling off Northwest Africa. *Science* 315: 637–639.

Mittelstaedt, E. 1991. The ocean boundary along the northwest African coast: circulation and oceanographic properties at the sea surface. *Progress in Oceanography* 26: 307–355.

Moncoiffé, G., X.A. Álvarez-Salgado, F.G. Figueiras, and G. Savidge. 2000. Seasonal and short timescale dynamics of microplankton community production and respiration in an inshore upwelling system. *Marine Ecology Progress Series* 196: 111–126.

Morel, A., D. Antoine, M. Babin, and Y. Dandonneau. 1996. Measured and modeled primary production in the northeast Atlantic (EUMELI JGOFS program): the impact of natural variations in photosynthetic parameters on model predictive skills. *Deep-Sea Research* 43: 1273–1304.

Moulin, C., F. Guillard, F. Dulac, and C.E. Lambert. 1997. Long-term daily monitoring of Saharan dust load over ocean using Meteosat ISCCP-B2 data: 1. Methodology and preliminary results for 1983–1994 in the Mediterranean. *Journal of Geophysical Research* 102. doi:10.1029/96JD02620.

Navarro-Pérez, E., and E.D. Barton. 2001. Seasonal and interannual variability of the Canary Current. *Scientia Marina* 65(S1): 205–213.

Neuer, S., M.E. Torres-Padrón, M.D. Gelado-Caballero, M.J. Rueda, J. Hernández-Brito, R. Davenport, and G. Wefer. 2004. Dust deposition pulses to the eastern subtropical North Atlantic gyre: does ocean's biogeochemistry respond? *Global Biogeochemical Cycles* 18: GB4020. doi:10.1029/2004GB002228.

Nogueira, E., F.F. Perez, and A.F. Rios. 1997. Seasonal patterns and long-term trends in an estuarine upwelling ecosystem (Ria de Vigo, NW Spain). *Estuarine, Coastal and Shelf Science* 44: 285–300.

Oberle, F.K.J., T.J.J. Hanebuth, B. Baasch, and T. Schwenk. 2014. Volumetric budget calculation of sediment and carbon storage and export for a late Holocene mid-shelf mudbelt system (NW Iberia). *Continental Shelf Research* 76: 12–24. doi:10.1016/j.csr.2013.12.012.

Oliveira, P.B., R. Nolasco, J. Dubert, T. Moita, and A. Peliz. 2008. Surface temperature, chlorophyll and advection patterns during a summer upwelling event off central Portugal. *Continental Shelf Research* 29: 759–774.

Ottersen, G., H. Loeng, B. Ådlandsvik, and R. Ingvaldsen. 2003. Temperature variability in the northeast Atlantic. *ICES Marine Science Symposium* 219: 86–94.

Parrilla, G., S. Neuer, P.-Y. Le Traon, and E. Fernández-Suárez. 2002. Topical studies in oceanography: Canary Islands Azores Gibraltar Observations (CANIGO). Volume 1: studies in the northern Canary Islands basin. *Deep Sea Research II* 49: 3409–3413.

Pelegrí, J.L., J. Arístegui, L. Cana, M. González-Dávila, A. Hernández-Guerra, S. Hernández-León, A. Marrero-Díaz, M.F. Montero, P. Sangrà, and M. Santana-Casiano. 2005. Coupling between the open ocean and the coastal upwelling region off Northwest Africa: water recirculation and offshore pumping of organic matter. *Journal of Marine Systems* 54: 3–37.

Pelegrí, J.L., A. Marrero-Diaz, and A.W. Ratsimandresy. 2006. Nutrient irrigation of the North Atlantic. *Progress in Oceanography* 70: 366–406.

Pelíz, Á.J., and A.F.G. Fiúza. 1999. Temporal and spatial variability of CZCS-derived phytoplankton pigment concentrations off the western Iberian Peninsula. *International Journal of Remote Sensing* 20: 1363–1403.

Pelíz, Á.J., T.L. Rosa, A.M.P. Santos, and J.L. Pissarra. 2002. Fronts, jets, and counter-flows in the Western Iberian upwelling system. *Journal of Marine Systems* 35: 61–77.

Pelíz, Á.J., J. Dubert, A.M.P. Santos, P.B. Oliveira, and B. Le Cann. 2005. Winter upper ocean circulation in the Western Iberian basin—fronts, eddies and poleward flows: an overview. *Deep-Sea Research* 52: 621–646.

Peña-Izquierdo, J., J.L. Pelegrí, M.V. Pastor, P. Castellanos, M. Emelianov, J. Salvador, and E. Vázquez-Domínguez. 2012. The continental slope current system between Cape Verde and the Canary Islands, *Scientia Marina* 76S1: 65–78.

Pérez, F.F., A.F. Ríos, and G. Rosón. 1999. Sea surface carbon dioxide off the Iberian Peninsula (NE Atlantic Ocean). *Journal of Marine Systems* 19: 27–46.

Pérez, F.F., X.A. Álvarez-Salgado, and G. Rosón. 2000. Stochiometry of nutrients (C, N, P and Si) consumption and organic matter production in a coastal inlet affected by upwelling. *Marine Chemistry* 69: 217–236.

Pérez, F.F., C.G. Castro, X.A. Álvarez-Salgado, and A.F. Ríos. 2001. Coupling between the Iberian basin—scale circulation and the Portugal boundary current system. A chemical study. *Deep Sea Research* 48: 1519–1533.

Peters, H. 1976. The spreading of water masses of the Banc d'Arguin in the upwelling area off the northern Mauritanian coast. *'Meteor' Forschungsergebnisse* A, 18: 78–100.

Pingree, R.D., and B. Le Cann. 1990. Structure, strength and seasonality of the slope currents in the Bay of Biscay region. *Journal of the Marine Biology Association of the United Kingdom* 70: 857–885.

Postel, L., E.A. Arndt, and U. Brenning. 1995. Rostock zooplankton studies off West Africa. *Helgoländer Meeresuntersuchungen* 49(1): 829–847.

Relvas, P., and E.D. Barton. 2002. Mesoscale patterns in the Cape São Vicente (Iberian Peninsula) upwelling region. *Journal of Geophysical Research* 107(C10): 28/1-28/23. doi:10.1029/2000JC000456.

Relvas, P., and E.D. Barton. 2005. A separated jet and coastal counterflow during an upwelling relaxation off Cape Sao Vicente (Iberian Peninsula). *Continental Shelf Research* 25: 29–49.

Ribeiro, C., P.J. Gonçalves, A. Moreira, and K.A. Stobberup. 2004. The Portuguese industrial fisheries in Northwest Africa during the 20th century. In *Marine fisheries, ecosystems and societies in West Africa: half a century of change,* ed. P. Chavance, M. Bâ, D. Gascuel, J.M. Vakily and D. Pauly, 79–98. Actes du Symposium International, Dakar (Sénégal), 24–28 Juin 2002. Collection des rapports de recherche halieutique A.C.P.-U.E., No. 15.

Roy, C., and C. Reason. 2001. ENSO related modulation of coastal upwelling in the eastern Atlantic. *Progress in Oceanography* 49: 245–255.

Saetersdal, G., G. Bianchi, T. Stroemme, and S.C. Venema. 1999. The Dr. Fridtjof Nansen Programme 1975–1993. Investigations of fishery resources in developing regions: history of the programme and review of results. *FAO Fisheries Technical Paper*, 391.

Santana-Casiano, J.M., M. González-Dávila, L.M. Laglera-Baquer, and M.J. Rodriguez-Somoza. 2001. Carbon dioxide system in the Canary region during October 1995. *Scientia Marina* 65: 41–49.

Santana-Casiano, J.M., M. González-Dávila, M.-J. Rueda, O. Llinás, and E.-F. González-Dávila. 2007. The interannual variability of oceanic CO_2 parameters in the northeast Atlantic subtropical gyre at the ESTOC site. *Global Biogeochemical Cycles*, 21: GB1015. doi:10.1029/2006GB002788.

Santos, A.M.P., A. Peliz, J. Dubert, P.B. Oliveira, M.M. Angelico, and P. Ré. 2004. Impact of a winter upwelling event on the distribution and transport of sardine eggs and larvae off western Iberia: a retention mechanism. *Continental Shelf Research* 24: 149–165.

Santos, A.M.P., A.S. Kazmin, and A. Peliz. 2005. Decadal changes in the Canary upwelling system as revealed by satellite observations: their impact on productivity. *Journal of Marine Research* 63: 359–379.

Schemainda, R., D. Nehring, and S. Schulz. 1975. Ozeanologische Untersuchungen zum Produktionspotential der nordwestafrikanischen Wasserauftriebsregionen 1970–1973. *Geodätische und Geophysikalische Veröffentlichungen*, Reihe IV: 1–88.

Seibold, E. 1982. Sediments in upwelling areas, particularly off Northwest Africa. *Rapports et Proces-verbaux des Réunions/Conseil International pour l'Éxploration de la Mer* 180: 315–322.

Siedler, G., and R. Onken. 1996. Eastern recirculation. In *The warm-water sphere of the North Atlantic Ocean*, ed. W. Krauss, 339–364. Berlin: Gebrüder Borntraeger.

Silva, A., D.W. Skagen, A. Uriarte, J. Massé, M.B. Santos, V. Marques, P. Carrera, P. Beillois, G. Pestana, C. Porteiro, and Y. Stratoudakis. 2009. Geographic variability of sardine dynamics in the Iberian Biscay region. *ICES Journal of Marine Science* 66: 495–508.

Sordo, I., E.D. Barton, J.M. Cotos, and Y. Pazos. 2001. An inshore poleward current in the NW of the Iberian Peninsula detected from satellite images, and its relation with *G. catenatum and D. acuminata* blooms in the Galician rías. *Estuarine, Coastal and Shelf Science* 53: 787–799.

Sousa, P., M. Azevedo, and M.C. Gomes. 2005. Demersal assemblages off Portugal: mapping, seasonal, and temporal patterns. *Fisheries Research* 75: 120–137.

Sousa, P., M. Azevedo, and M.C. Gomes. 2006. Species-richness patterns in space, depth, and time (1989–1999) of the Portuguese fauna sampled by bottom trawl. *Aquatic Living Resources* 19: 93–103.

Stramma, L. 1984. Geostrophic transport in the warm water sphere of the eastern subtropical North Atlantic. *Journal of Marine Research* 42: 537–558.

Stramma, L., and G. Siedler. 1988. Seasonal changes in the North Atlantic subtropical gyre. *Journal of Geophysical Research* 93: 8111–8118.

Thiao, D., C. Chaboud, A. Samba, F. Laloë, and P.M. Cury. 2012. Economic dimension of the collapse of the 'false cod' *Epinephelus aeneus* in a context of ineffective management of the small-scale fisheries in Senegal. *African Journal of Marine Science* 34(3): 305–311. doi:10.2989/1814232x.2012.725278.

Tomczak, M., and J.S. Godfrey. 2003. *Regional oceanography: an introduction*. 2nd Edn., Daya Publishing House, 390.

Tooms, J.S., C.P. Summerhayes, and R.L. McMaster. 1971. Marine geological studies on the northwest African margin. In *The geology of the East African continental margin*, vol. 14, ed. F.M. Delany, 31, 9–26. Cambridge: Institute of Geological Sciences Report 70/16 ICSU: SCOR Working Party.

Torres, R., and E.D. Barton. 2007. Onset of the Iberian upwelling along the Galician coast. *Continental Shelf Research* 27(13): 1759–1778. doi:10.1016/j.csr.2007.02.005.

Torres, R., E.D. Barton, P. Miller, and R. Fanjul. 2003. Spatial patterns of wind and sea surface temperature in the Galician upwelling region. *Journal of Geophysical Research* 108(C4): 3130. doi:10.1029/2002JC001361.

Torres-Padrón, M.E., M.D. Gelado-Caballero, C. Collado-Sánchez, V.F. Siruela-Matos, P.J. Cardona-Castellano, and J.J. Hernández-Brito. 2002. Variability of dust inputs to the CANIGO zone. *Deep Sea Research II* 49: 3455–3464.

Toualy, E., G. Stanojevic, K.Y. Kouadio, and A. Aman. 2012. Multi-decadal variability of Sea surface temperature in the northern coast of Gulf of Guinea. *Asian Journal of Applied Sciences* 5: 552–562. doi:10.3923/ajaps.2012.552.562.

Troadec, J.P., and S. Garcia. 1979. Les ressources halieutiques de l'Atlantique Centre-Est. Première partie: les ressources du Golfe de Guinée de l'Angola à la Mauritanie. *FAO Documents Techniques sur les Pêches* (1986.1), 167.

Van Weering, T.C., and I.N. McCave. 2002. Benthic processes and dynamics at the NW Iberian margin: an introduction. *Progress in Oceanography* 52: 123–128.

Visbeck, M., E. Chassignet, R. Curry, T. Delworth, B. Dickson, and G. Krahmann 2003. The ocean's response to North Atlantic Oscillation variability. In *The North Atlantic oscillation*, ed. J.W. Hurrell, Y. Kushnir, G. Ottersen and M. Visbeck, 113–146. AGU.

Vitorino, J.O., J.M. Jouanneau, and T. Drago. 2002. Winter dynamics on the northern Portuguese shelf. Part I: physical processes. *Progress in Oceanography* 52: 129–153.

Wada, E., and A. Hattori. 1991. *Nitrogen in the Sea: forms, abundances and rate processes*. Boca Raton: CRC Press.

Watson, A.J. 1995. Are upwelling zones sources or sinks of CO2? In *Upwelling in the ocean. modern processes and ancient records,* ed. C.P. Summerhayes, K.-C. Emeis, M.V. Angel, R.L. Smith, and B. Zeitzschel, 321–336. Wiley and Sons.
Weaver, P.P.E., and M. Canals. 2003. The Iberian and Canaries margin including NW Africa. In *European margin sediment dynamics,* ed. J. Mienert and P.P.E. Weaver, 251–260. Springer-Verlag.
Wooster, W.S., A. Bakun, and D. McLain. 1976. The seasonal upwelling cycle along the eastern boundary of the North Atlantic. *Journal of Marine Research* 34(2): 131–141.
Zeeberg, J.J., A. Corten, and E. de Graaf. 2006. By-catch and release of pelagic megafauna in industrial trawler fisheries off Northwest Africa. *Fisheries Research* 78: 186–195.

Chapter 7
The Benguela Current Upwelling System

Abstract This chapter describes the Benguela upwelling system along the western coast of southern Africa and its interactions with the tropical ocean to the north and the Agulhas Current to the south. It covers the physics, chemistry and biology in the region, as well as the fishery potential under the increasing pressure of climate change.

Keywords Upwelling · Oceanography · Benguela current · Marine ecology · Marine biogeochemistry · Fisheries · Fisk stock variations · Climate-change impacts

The sea is the vast reservoir of nature

The globe began with sea, so to speak and who knows if it will not end with it?

Jules Verne (1828–1905)

(Taken from *Twenty Thousand Leagues Under the Sea*, 1870)

7.1 Introduction

The Benguela Current upwelling system stretches along the southwest coast of Africa, essentially from the region north of Cape Frio, at about 15° S in Angola, to Cape Agulhas, the southern tip of the continent near 35° S. There is also intermittent upwelling at capes further east towards Port Elizabeth (Figs. 7.2 and 7.3), but these are not considered part of the Benguela system proper. At its southern end the system is bounded by the warm Agulhas Current, which sweeps down from Mozambique along the south coast of South Africa, while at the northern end, just north of the southern border of Angola, lies the Angola Front, a second boundary with warm water. Having warm water at both ends of the system makes the Benguela unique when compared with the other major upwelling areas. As with the

© Springer International Publishing Switzerland 2016

J. Kämpf and P. Chapman, *Upwelling Systems of the World*,
DOI 10.1007/978-3-319-42524-5_7

other three major eastern boundary current systems, the region supports important fisheries, and this has in the past led to over-exploitation, particularly in the period before 1982 when the concept of a 200 nautical mile exclusive economic zone was adopted as part of the UN Convention on the Law of the Sea (Fig. 7.1).

Much of the west coast of South Africa and Namibia is either desert or semi-desert, with the only large riverine inputs being the Cunene, at the northern end of the region, and the Orange, at 29° N (Fig. 7.3). Two smaller rivers, the Berg and the Olifants, reach the sea just north of Cape Columbine and about 100 km north of this cape respectively. These rivers have strongly seasonal flows, which reach a maximum in winter (May–October) and are greatly reduced in summer. Maximum flows in the Orange and Cunene rivers, in contrast, are in summer (October–April and January–April respectively), but all four rivers tend to dam up

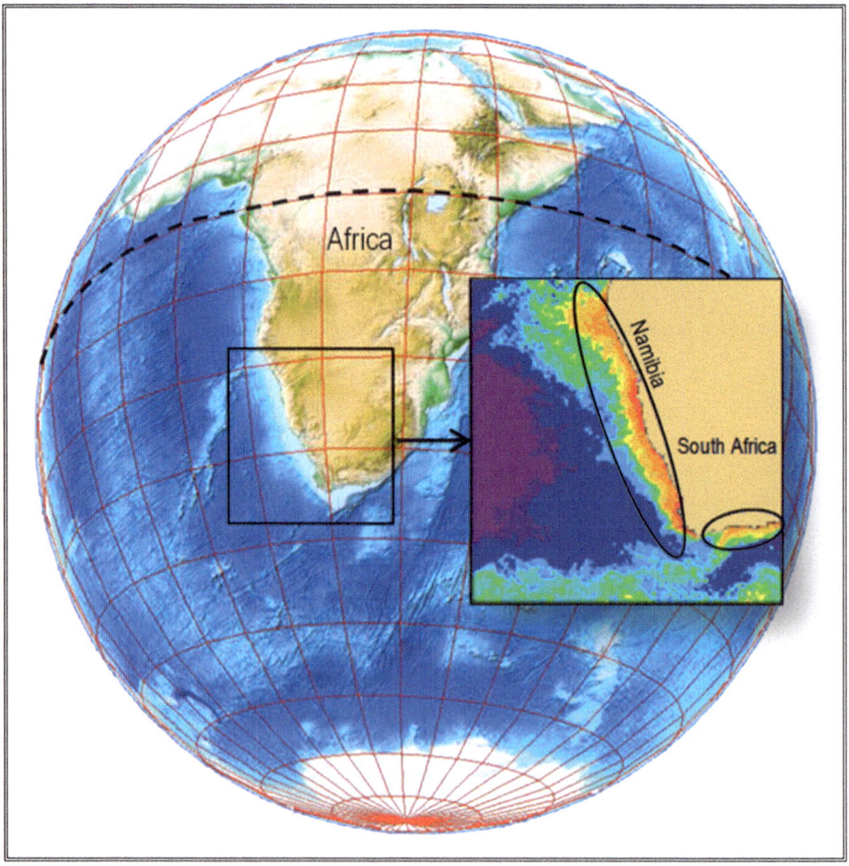

Fig. 7.1 The Benguela current upwelling region in the eastern South Atlantic. The *inset* shows ocean-colour-derived chlorophyll-*a* concentrations. Upwelling on the Agulhas Bank is also evident (see Sect. 9.5 for more details). *Background image* Google Earth with ETOPO 1 bathymetry layer. *Insert* MODIS-Aqua data from NASA's Giovanni interface

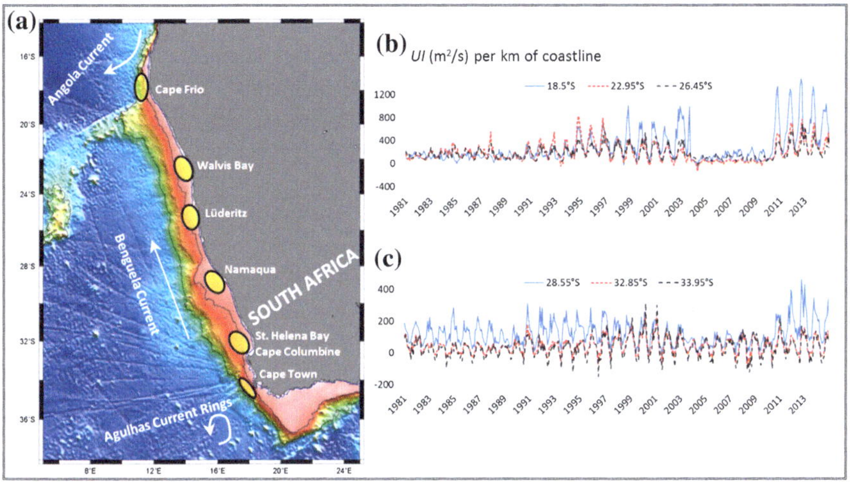

Fig. 7.2 The Benguela Current region showing the shelf area, upwelling centres, and upwelling indices derived from wind stress at Cape Frio (18.5° S), Walvis Bay (22.95° S), Lüderitz (26.45° S), Namaqualand (28.55° S), Cape Columbine (32.85° S) and Cape Town (33.95° S). Map from Liu and Dittert (2010), upwelling indices from NOAA (http://www.pfeg.noaa.gov/products/PFEL/modeled/indices/upwelling/SA/upwell_menu_SA.html)

with natural sand bars during low flow periods. Other small inflows in Namibia come from the Swakop and Ugab rivers, which flow very intermittently and deliver only minor volumes of fresh water to the coastal region.

While the volume of water from the Cunene has remained relatively consistent, the volumes of fresh water introduced by other South African rivers (i.e. the Orange, Olifants and Berg rivers) have decreased considerably by 30–50 % during the past 30 years when compared to data from prior to 1983 (see Chapman and Shannon 1985). It is not clear if this is due to reduced rainfall in their catchments or to greater abstraction for farming, mining and other uses, although decreasing trends in the flow of a number of southern African rivers have been noted by Alemaw and Chaoka (2002), who attributed this at least partly to the relatively high incidence of ENSO events between 1950 and 1998 (Alemaw and Chaoka 2006). These tend to lead to reduced rainfall in the eastern part of South Africa and in central Namibia (Landman and Klopper 1998; Jury 2003; Alemaw and Chaoka 2006). As the Orange River and its major tributaries rise in the eastern part of South Africa, and it also receives water from the Fish River, which flows southwards down the centre of Namibia, this may account for at least some of the decrease in flow.

A series of review papers on the Benguela was published in *Oceanography and Marine Biology Annual Reviews* during the 1980s and 1990s (Shannon 1985a; Chapman and Shannon 1985; Shannon and Pillar 1986; Crawford et al. 1987; Branch and Griffiths 1988; Berruti et al. 1989; Rogers and Bremner 1991). These papers summarize the work done in the region until this date, and we have drawn

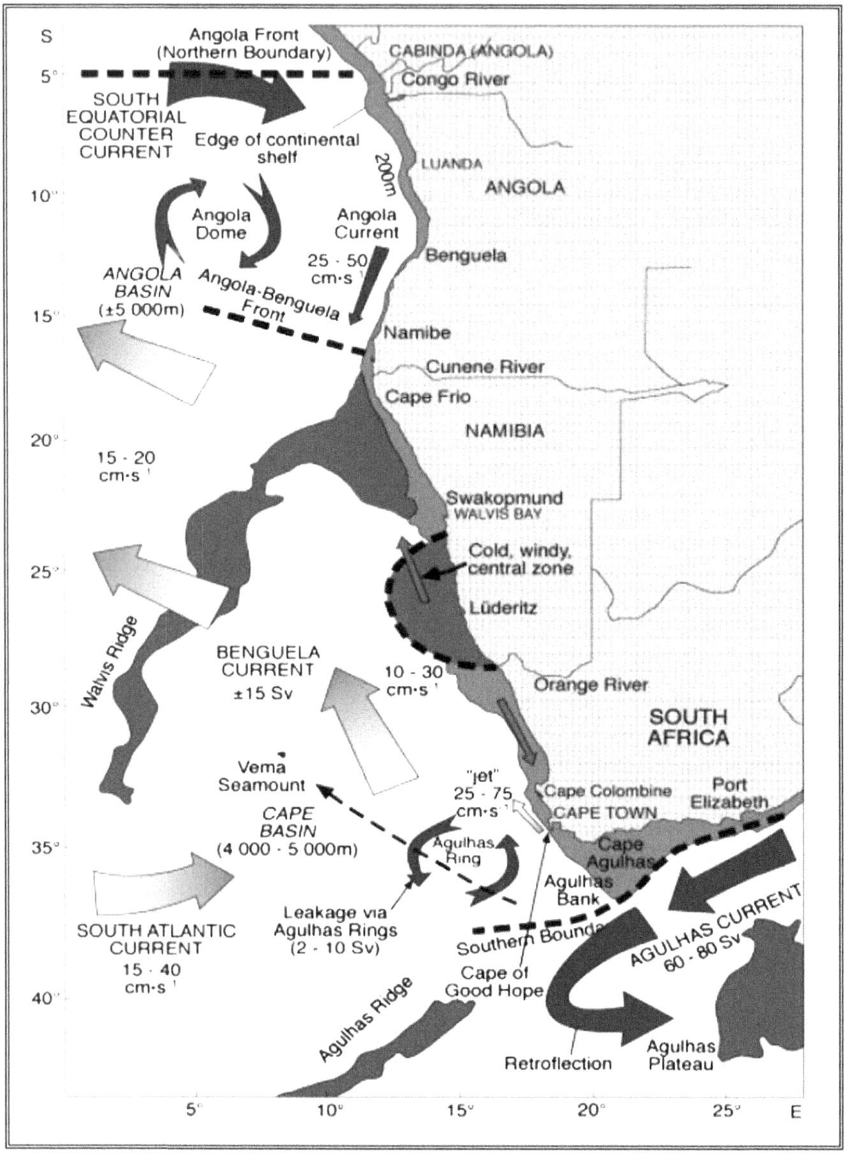

Fig. 7.3 Main features of the Benguela region, showing prevailing currents and interaction with the warmer waters off Angola and in the Agulhas Current (from Shannon 2006)

extensively on them for the general description of the system. Since then, additional reviews have appeared in a number of outlets, including those by Griffiths et al. (2004), Field and Shillington (2005) and Shannon et al. (2006).

7.2 History of Exploration in the Benguela

The San Bushman population was the first to take advantage of the bounty provided by the Benguela upwelling, and shell middens, consisting of limpets, whelks, black mussels and oysters, are known all along the western and southern coasts of southern Africa, dating back to at least 6000 years BP (Jerardino 2010). Some of these middens were enormous, occupying several thousand cubic meters and containing many tonnes of shell material (Parkington et al. 1988; Jerardino and Yates 1997). A sudden decline in the occurrence of middens after about 2000 BP is thought to be due to a change from a hunter-gatherer society to stock breeding rather than from over-utilization of these resources, (e.g., Sealy and Yates 1994; Smith 2006), although the gathering of sea food undoubtedly continued after this time.

The Portuguese were the first "modern" Europeans to explore the west coast of southern Africa, and Diego Cao is thought to have sailed as far south as 22° 09′ S during his voyage of 1484–86. Two years later, Bartholomeu Dias landed in Walvis Bay and, after being blown off course by a storm, eventually reached Mossel Bay, east of Cape Town, in February 1488. He carried on eastwards but turned back a month later and erected three stone crosses, one in what is now the eastern Cape, one just west of Cape Point (which he saw for the first time on his return voyage), and one in Namibia. However, according to Herodotus, the Portuguese were preceded by the Phoenicians, who sailed round Africa the other way, from the Red Sea to the Straits of Gibraltar, in about 600 BCE. Herodotus doubted this story, as the Phoenicians reported that at the southern end of the continent the sun rose on their right instead of on their left as it would have done in the northern hemisphere when sailing south, but this is likely the best confirmation of their voyage and that they had rounded Cape Point and were sailing north. There is no mention, however, of whether the sailors took advantage of the available fish stocks in the Benguela, although this seems likely, as they would have tried to stay in sight of land during the voyage.

Ten years after Dias, Vasco da Gama pioneered the route around southern Africa to the "spice islands" in Indonesia. The Portuguese also established the first European settlement in the region in 1587, a fort at Benguela in Angola that became the city of the same name in 1617, and from whence they exported slaves to Brazil and Cuba. Cape Town itself was only settled by the Dutch under Jan van Riebeek in 1652 as a supply station for the ships of the Dutch East India Company, providing freshwater, meat and vegetables, as well as a place to repair ships and tend to the sick among their crews. The lack of a well-protected anchorage at Cape Town was a drawback, but the much better harbour at Saldanha Bay, near Cape Columbine, had no good freshwater supply. The early settlers also took advantage of the sea's bounty, using nets and handlines for both personal and commercial purposes, although commercial fishing in the modern sense began only in the late 1800s, with the start of the trawl fishery for hake. The pelagic industry began in 1935, with the main target being horse mackerel for canning. Only in the 1960s did the commercial fishery for anchovy and pilchard start (Crawford et al. 1987). This is discussed further in Sect. 7.7.

While Rennell (1832) was the first to describe a cold current flowing up the western coast of southern Africa, this work was generally forgotten, and the era of large-scale oceanographic exploration only began in earnest after the voyage of the Challenger from 1872–1876. Earlier workers, such as Muhry (1862) and Petermann (1865) thought that the flow of water along the southwest coast of Africa was merely a northward extension of the West Wind Drift, although Ross (1847) had suggested otherwise, and his temperature data seemed to confirm this. Meyer (1923) was the first to show that the Benguela Current was separated from the Southern Ocean by the subtropical convergence, and the results of expeditions such as the German cruises of the *Valdivia* (1898–1899) and *Meteor* (1925–1927), and the British cruises on the *Discovery* and *William Scoresby* (1925–1951) all provided a large amount of data on the physics, chemistry and biology of the region (Schott 1902; Hentschel 1928; Hentschel and Wattenberg 1930; Clowes 1938; Hart and Currie 1960).

The rich fishing harvest along the Benguela coast was exploited by the local settlers for many years without being well understood. At the beginning of the 20th century, however, the South African government began a systematic research effort into the fish stocks, led by Dr. J. Gilchrist (Shannon and Pillar 1986). The results of this ongoing research are discussed more fully in Sect. 7.7. Apart from the fishery for both pelagic and demersal fish, there was also a whaling industry. The Dutch tried whaling in the 17th century, but the industry did not take off for another century, with American whalers becoming active off Saldanha Bay after 1788 and continuing until about 1840, after which catches declined until the advent of modern harpoon guns in the early 20th century (Best, 1987, 1994). As a result, the catches of southern right whales and humpback whales had declined precipitously by 1920. Whaling continued sporadically until 1967, when whales became scarce and the government closed the industry. Sealing similarly went through a period of boom and bust, starting with the first Dutch seal harvest in 1610, although seal numbers are now increasing again (Hutchings et al. 2009a).

7.3 History of Marine Mining and Other Extractive Industries

Apart from the fishery, there is also a thriving and extremely lucrative diamond extraction industry along the Benguela coastline (Heyes 2006). All natural diamonds are believed to form at least 150 km deep in the crust under intense heat and pressure and reach the surface following eruptions. Southern African diamonds result from eruptions about 125–115 million years BP or 90–80 million years BP during the late Cretaceous period in modern-day Botswana, South Africa and Zimbabwe. Since that time, the riverine drainage system has been towards the west,

Fig. 7.4 An alluvial diamond mine on an old beach terrace. The diamonds are trapped in grooves in the bedrock. Image courtesy of Diamond Fields International (http://www.diamondfields.com/s/Namibia.asp?reportID=88890, [accessed on 5 April 2016])

into the Atlantic Ocean, and erosional processes have brought diamonds from their original formation sites to the coast, particularly in the drainage basin of the Orange River. Redistribution along the northern coast of South Africa and Namibia has resulted from wave and tidal action, with a general movement towards the north. As diamonds are heavier than most minerals in sand and gravel, they tend to move downwards into diamond traps, grooves and depressions on the bedrock (Fig. 7.4). These serve to concentrate the diamonds in specific pockets along the coast. As sea level has changed over the past several million years, so the layers containing the diamonds have migrated onshore and offshore and these processes continue today. This migration process also destroyed diamonds containing imperfections in their lattice structure, so that sea diamonds are generally higher quality than stones from inland sites (Heyes 2006).

Although the South African diamond mining industry began in 1860 in Kimberley, the first sea diamonds were discovered only in 1908 in an old beach terrace near Lüderitz. The resulting diamond rush meant that by 1913, Namibia was producing 20 % of the world's diamonds from these alluvial deposits. Initially, all mining was done by removing the surface sediment down to bedrock, at first with machines, but with the final sorting being done by hand. As the industry followed the bedrock layers into the ocean, starting in the 1960s, berms were built to prevent the workers from drowning. This is not possible in deeper water, however, where the industry now uses shallow, diver-operated suction pumps in depths less than about 30 m, and more sophisticated, ship-based vertical or horizontal vacuum pump systems that suck the sediments onto the drill ship for areas as deep as about 300 m (Heyes 2006).

There is also salt production at Walvis Bay, Swakopmund, and further north at Cape Cross (Barnes and Alberts 2008). The companies involved have developed salt pans along the coast, which produce about 800,000 tonnes of coarse salt a year, used mainly in chemical manufacturing. While natural gas and/or oil production has not so far been started in Namibia, the offshore Kudu gas field is targeted for development to fire an electricity generation facility, which is likely to be located within the coastal zone.

Commercial guano production began in the early 20th century, based on the large populations of penguins, gannets and cormorants on the islands along the South African and Namibian coasts (Cooper et al. 1982). Platforms were erected at a number of places in Namibia during the 1930s to encourage nesting and make it easier for guano collectors to operate. Up to 3,600 metric tonnes of guano was collected each year from these platforms between 1931 and 1978 (Cooper et al. 1982), with higher quantities (<5,000 tonnes/yr) thereafter (Schwartzlose et al. 1999). Additionally, up to about 10,000 tonnes of guano per year was collected from the natural islands (Schwartzlose et al. 1999). Although initial collections at natural sites during 1895–1903 were much larger, totaling over 200,000 short tons (1 short ton = 907.19 kg) (Rand 1952), this quantity certainly included both old and recent deposits, so should not be used to show that the quantity of guano has decreased since then. Data do show, however, that there is considerable interannual variability with about an 8-year periodicity, presumably depending on fish production in the area (see Sect. 7.7.8). The recent data suggest that a reasonable current estimate for sustained annual production in Namibia is about 2,000–3,000 tonnes (Barnes and Alberts 2008; Crawford et al. 2007).

7.4 Physical Controls and Subsystems

7.4.1 Large-Scale Atmospheric Controls

As stated earlier, the Benguela is unusual in that it has warm water at both ends of the system. Thus, its variability is controlled not only by what happens in the South Atlantic, but also by what happens in the tropical Indian Ocean (Shannon and Nelson 1996; Reason et al. 2006). South Atlantic meteorology is driven largely by the presence of the quasi-permanent high-pressure anticyclone, centered near 28° S, 8° W, which moves north in winter and south in summer on a regular basis (Shannon 1985a; Preston-Whyte and Tyson 1988). Over the continent, the pressure changes from a summer low to a weak high during winter as the temperature decreases. The strong pressure gradient that forms between this maritime high-pressure cell and the low pressure along the west coast of the continent forces the equatorward upwelling winds in summer. These winds are also affected by the changing topography as one moves inland from the coast, being accelerated by the increased pressure gradient as the land rises to about 1,000 m (Shannon 1985a, b; Shillington 1998). As the high-pressure system moves north during the austral winter, so the southern part of the Benguela becomes subject to westerly winds associated with the regular cyclonic low pressure systems that move from west to east south of the continent (Shillington 1998). This change in wind stress induces a divergence zone north of 28° S, originally suggested by Stander (1964), maximum wind stress offshore, and changes in seasonal variability north and south of Lüderitz.

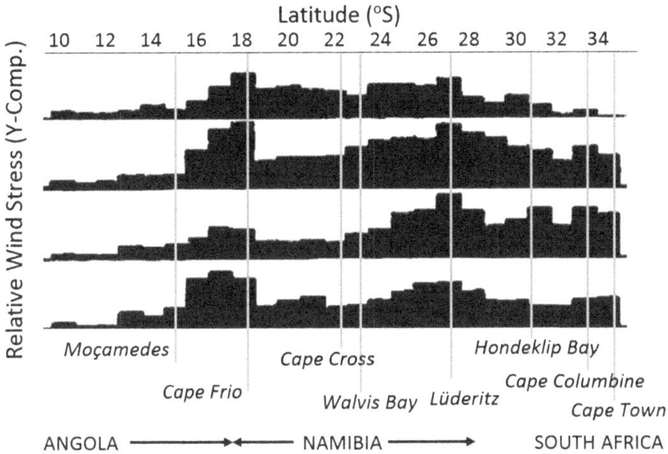

Fig. 7.5 Northward (Y) component of wind stress calculated for 1° boxes along the coast. Redrawn after Shannon (1985a)

Bakun and Nelson (1991) also showed that the strongest wind-stress curl (initiating additional flow divergence and upwelling aside from boundary effects) was found just south of 25° S. The variability in the Y-component of the wind stress (Fig. 7.5) clearly coincides with the specific upwelling centres system is therefore unlike that off the in the Benguela. The system is therefore unlike that off the northwest United States, where the upwelling forms more of a zonal band, and resembles that off the coast of Chile. The upwelling centres occur at Cape Frio, Walvis Bay and Lüderitz, where upwelling occurs throughout the year but with a minimum in summer (as noted by Boyd 1987), and at Cape Cross (22° S), with a peak in autumn, Hondeklip Bay (31° S), with a peak in summer, and in the south near Cape Point and Cape Columbine where maximum upwelling is found in the spring and summer seasons. Upwelling can also occur at the various capes between Cape Point and Cape Agulhas in summer, depending on wind direction, but these are not major upwelling centres. Similar distribution of upwelling is found in model output, such as that of Veitch et al. (2009) (Fig. 7.6), confirming that the southern region has the strongest seasonal signal, decreasing to the north, and that the two northernmost upwelling centres show strongest upwelling in winter, while the remainder have the greatest upwelling in summer.

More recently, Risien et al. (2004) have reexamined wind fields over the Benguela from 32.5° S to 10° S, using wavelet analysis of 16 months of QuikSCAT scatterometer data between August 1999 and November 2000. We should note that this period coincided with a Pacific La Niña event, when stronger southeasterly winds are expected in the Benguela region, and so may not be indicative of general conditions. North of 24° S, maximum wind stresses developed during winter months with the strongest stresses, which lasted almost throughout the period of observations, being found near Lüderitz between 24°–28.5° S, as expected from

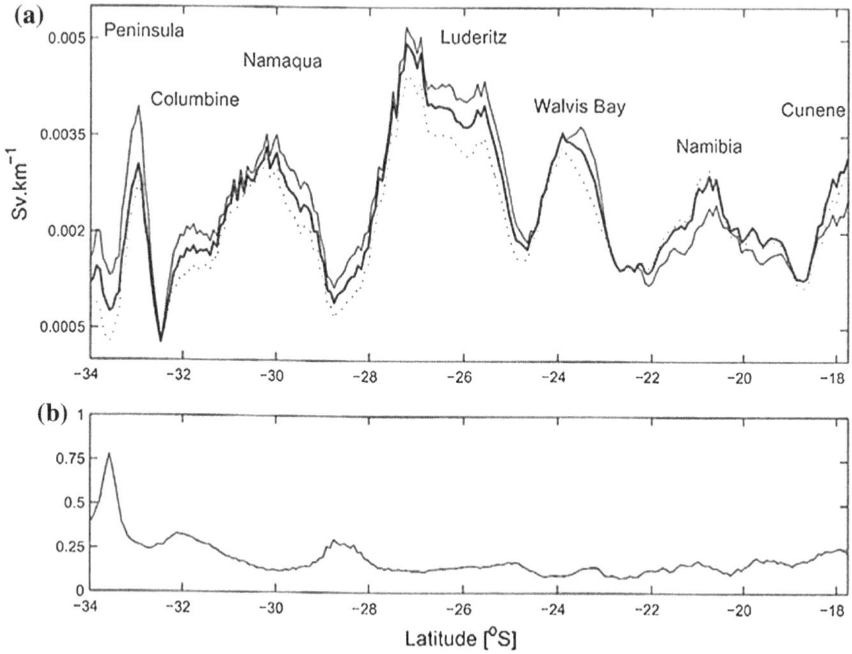

Fig. 7.6 **a** Model-derived offshore upwelling transport per km of coastline (Sv/km) from 34° S to 17.75° S, showing the annual mean (*bold*), summer (*solid*) and winter (*dotted*) fluxes. **b** Normalized mean standard deviation (standard deviation divided by the mean) of the output (from Veitch et al. 2009)

earlier work. South of Lüderitz strongest winds occurred during the summer, and synoptic variability increased towards the southern end of the continent, as the effects of the anticyclonic circulation increased. Variability in wind stress was dominated by 4–16 day periodicity north of 19° S, while south of 24° S, there was a bimodal distribution, with peaks in the 4–12 and 25–50 day bands, concentrating around 40 days. The region from 19° S to 23.5° S was shown to be a transition region between zones controlled by tropical and subtropical influences respectively.

Apart from large-scale seasonal variability of winds patterns, there is also shorter-term variability, particularly in the southern part of the region, on timescales of about one or two weeks (Nelson and Hutchings 1983; Jury et al. 1990). When the South Atlantic high-pressure system is well developed in boreal summer (see Fig. 2.13a), a coastal low develops frequently near Lüderitz. The high-pressure system can ridge round to the south of the continent, bringing gale-force upwelling winds to Cape Town. The regular passage of cyclones south of the continent induces the high-pressure system to move northwards, while the coastal low travels south along the coast as a trapped wave at speeds of up to 750 km/day (Nelson and Hutchings 1983), temporarily suppressing the upwelling winds. Once the low-pressure system has passed, the South Atlantic high strengthens again and upwelling resumes.

While summer conditions generally result merely in a slackening of upwelling-favourable winds, in winter the passage of the cyclones can bring strong northwesterly gales to the region between Cape Town and Cape Columbine. These storms have been responsible for numerous shipwrecks along this part of the coast. In autumn and winter, when a high-pressure system sets up over the southeastern part of southern Africa, strong katabatic, offshore easterly or northeasterly wind events, known as Berg winds, can occur (Jackson 1947; Hart and Currie 1960). These winds can reach velocities of up to 15 m/s, and deliver large quantities of sand and dust to the coastal zone and up to at least 150 km offshore (Shannon and Anderson 1982). Unlike the situation in northwest Africa (see Chap. 6), the dust plumes are concentrated at gaps in the topography, such as near Walvis Bay or in river valleys such as near the Orange River mouth. This likely intensifies the wind and increases the offshore transport of the dust. Berg winds causing dust plumes are generally short-term events, lasting for only 1 or 2 days, but can still deliver large quantities of material, amounting to several million tonnes per event.

The other short-term effect along this coast is the land-sea breeze, particularly north of Cape Columbine. Because the coast is so arid, the land heats up rapidly during the day, leading to a switch from weak easterly or even northerly (offshore) winds during the morning to stronger southerly or southwesterly (onshore) winds during the afternoon (Hart and Currie 1960; Stander 1958, 1964). The weak northerly winds are found most commonly north of Lüderitz, especially in summer. The strong temperature contrast between the cold surface water in the upwelling region and the much warmer land also leads to the frequent occurrence of fog all along the west coasts of South Africa and Namibia, especially in the area near Walvis Bay (Shannon 1985a, b). In this, the Benguela resembles other eastern boundary regions such as the California Current and Peru. Fog, onshore winds, and the few significant landmarks along the coast led to many shipwrecks in the region, particularly off Namibia, which was named the Skeleton Coast by early sailors.

There are only a limited number of sites along the west coast of Africa where regular wind measurements are taken, and no offshore moorings such as make up the PIRATA array in the equatorial and western South Atlantic. As a result, knowledge of the in situ interannual variability in upwelling intensity in the Benguela is incomplete. A time-series of upwelling-favourable winds from Cape Point between 1961 and 1983 integrated over the summer upwelling period (Nelson and Hutchings 1983; Shannon 1985a) indicated that 1966 was a year of weaker upwelling-favourable winds, while stronger upwelling-favourable winds occurred in 1975. Reanalysis using monthly means (Hutchings et al. 1984) showed more frequent variability within years and both 1965–66 and 1982–83 showed consistently lower upwelling winds in this second analysis. Off Cape Columbine, however, Johnson and Nelson (1999) showed large interannual changes in an upwelling index based on wind records from the Cape Columbine lighthouse (Fig. 7.7), with a particularly low index for the period between 1983 and 1989, after which it appeared to recover. While the wind strengths appeared to regain their pre-1979 strength after that time, the latest data came from a different site and cannot be considered an extension of the earlier records (Johnson and Nelson 1999). Cape

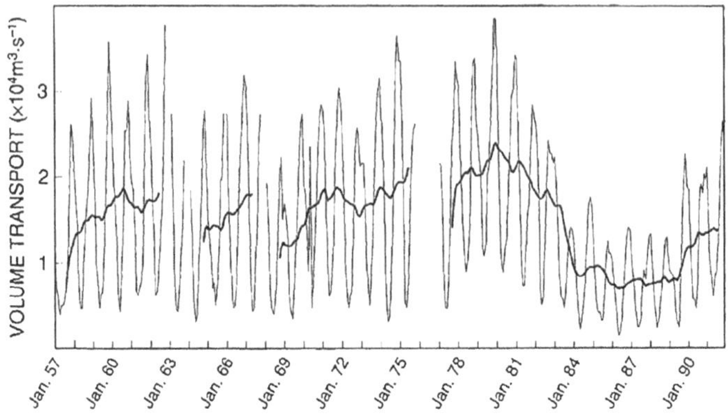

Fig. 7.7 Three-monthly filtered offshore Ekman total transport at Cape Columbine and the seasonally filtered plot (from Johnson and Nelson 1999)

Point, however, is not particularly representative of the rest of the region (Shannon 1985a) and Hutchings and Taunton-Clark (1990) demonstrated that there was little spatial coherence between 30-year time series of winds from Cape Columbine and Cape Point. The importance of local variability in the southern Benguela (south of 28° S) was shown also by Penven et al. (2001), whose model produced significant mesoscale dynamic variability without any synoptic or interannual atmospheric variability (Blanke et al. 2002). Another such drop in offshore Ekman transport in the early 2000s is seen throughout the system in the upwelling indices in Fig. 7.2.

Long-term data, although unfortunately few, show that there can be teleconnections between the Benguela and other regions, of which the most important seems to be with El Niño (Shannon and Nelson 1996; Reason et al. 2006), which lead to a series of warm and cold events. The El Niño phase of ENSO leads to a negative pressure anomaly between Marion (46° 52′ S, 37° 51′ E) and Gough (40° 19′ S, 9° 55′ W) Islands, which tends to lead to weaker winds in the southern Benguela. During La Niña these winds strengthen as a result of changes in the Rossby wave structure in the westerlies south of the continent. The El Niño phase also often induces droughts in southeast Africa (Alemaw and Chaoka 2006; Landman and Klopper 1998; Reason et al. 2006).

Major incursions of warm water from the north, dubbed "Benguela Niños" by Shannon et al. (1986), although there is no specific definition for these, occurred in 1963, 1972/74, 1984, and 1994/95, with possible additional events in 1934, 1950, and 1998/99 (see Shannon et al. 1986; Gammelsrod et al. 1998; Veitch et al. 2006; Rouault 2012). The events lasted for at least 6 months, and the 1984 event penetrated at least as far south as 25° S. These events are thought to result from modulation of the trade winds, leading to reduced zonal wind stress in the western tropical Atlantic and the formation of an equatorial Kelvin wave that propagates along the equatorial wave guide and then poleward along the African coast (Lass et al. 2000; Florenchie et al. 2004), thus affecting the atmospheric circulation in the northern Benguela (Rouault et al. 2003). However, Richter et al. (2010) have put forward an alternative hypothesis, based on coupled ocean-atmosphere model output and observations, that meridional anomalies in the wind field along the southwest African coast also affect Benguela Niño occurrences. As the wind anomalies cause a weakening of the South Atlantic high pressure system, this allows tropical water to push further south some 3 months later.

Benguela Niños can affect the biology. Pollock et al. (1997), for example, suggest that the 1994/95 event had a major effect on west coast rock lobster (*Jasus lalandii*), causing a drop in growth rate and subsequent large reduction in catch rates, although reduced upwelling, as measured by wind velocities, actually started rather earlier in 1990. Whether the observed decreases in guano production along the Namibian coast (Schwartzlose et al. 1999) are also related to Benguela Niños is unclear, but the approximate 10-year periodicity of such events makes it a possibility (see Sect. 7.7.8).

Veitch et al. (2006) also showed that there has been a general warming of the northern Benguela system since the early 1980s, amounting to about 3 °C off Angola north of the Angola-Benguela front and 1 °C south of it. Whether these warm events were directly associated with ENSO events in the Pacific Ocean is unclear, as there was no such event in 1963 or 1994/95, although there were strong ENSO events in 1982/83 and 1998/99 that may have forced the 1984 and 1998/99 events in the Benguela. For the period from 1982 to 2002, Rouault (2012) has shown that incursions of tropical water occur biannually across the Angola–Benguela frontal zone. The coldest events, which suggest increased upwelling intensity, occurred in 1982/83, 1987, 1990, 1992 and 1997 (Fig. 7.8). Further south, mesoscale activity is more important, and only the largest events (1982–83, 1984, 1995, 1997 and 2001) seem to have major effects on northern Namibia.

At the other end of the continent, there may be some effects from the Antarctic Oscillation (AAO), which exists as a pressure anomaly of one sign over the Antarctic continent and one of the opposite sign over the region near 40°–50° S (Reason et al. 2006).The AAO modulates atmospheric circulation and heat fluxes over the mid-latitude South Atlantic, with additional effects on winds, so may be expected to affect upwelling in the southern Benguela region. The local variability however, makes it difficult to see any obvious effects on the upwelling system, although Rouault et al. (2005) have shown that the AAO can affect rainfall, temperature and local circulation near Marion Island further east in the southern Indian Ocean.

7.4.2 Water Masses in the Benguela

Water in the Benguela is a mixture of warm, nutrient-poor water from the equatorial Atlantic, together with cooler waters richer in nutrients coming from the South Atlantic and South Indian Oceans. Thus, there is a considerable gradient in properties between water in the southern part of the region and that in the northern

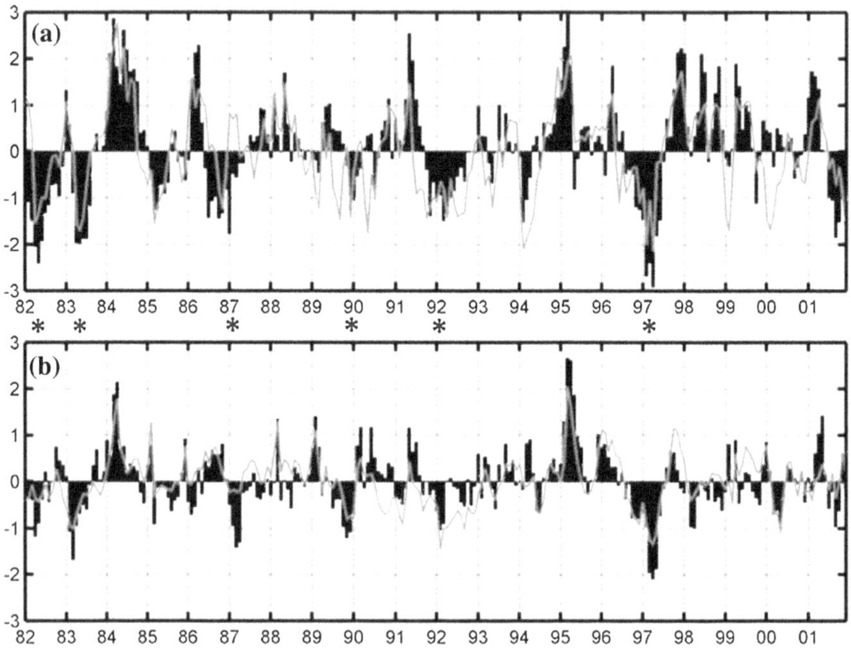

Fig. 7.8 Detrended normalized monthly anomalies of SST in **a** Southern Angola averaged from 10° S to 15° S and from the coast to 1° offshore and **b** Northern Namibia averaged from 19° S to 24° S and from the coast to 1° offshore from 1982 to 2002 from observations (*bar*) and model (*solid line*). Taken from Rouault (2012). *Asterisks* indicate periods of concurrent cooling events in both regions

part. Deep waters are constrained by the Walvis Ridge, which is a barrier to northward flow in the Cape Basin, so that Antarctic Bottom Water (ABW) is confined almost entirely south of the ridge, while North Atlantic Deep Water (NADW) can pass over the ridge into the Angola Basin. While the characteristics of deep and bottom waters vary slightly with latitude, they are not important to the upwelling process and will not be discussed further here.

Antarctic Intermediate Water (AIW) is found throughout the region, with its core at about 1100 m in the southern Indian Ocean and the Agulhas retroflection and between 700–800 m in the southeast Atlantic. The Benguela also receives some water at this level from the South Atlantic Current to the southwest of the continent (Stramma and Peterson 1989), but because of their similarity in temperature and salinity it is difficult to distinguish between the water from these three sources without using tracers such as CFCs (Gordon et al. 1992). Salinity at the core of the AIW increases from about 34.3 in the Cape Basin and southeast Atlantic to >34.5 in the Angola Basin north of 20° S, while its oxygen concentration decreases from about 4.5 mL/L (200 μM) to less than 2.0 mL/L (90 μM) over the same latitude range (Shannon and Hunter 1988).

South Atlantic Central Water (SACW), which occupies the linear portion of the theta-S diagram above the AIW, between about 6°, 34.4 and 16°, 35.5, is the source for upwelling water throughout the Benguela region (see Shannon 1985a; Shillington et al. 2006, and references cited therein), but the northern and southern components have different origins. Lower salinity SACW in the Cape Basin includes Indian Ocean CW, whereas that found in the Angola Basin is derived from water originally from the Brazil Current region on the western side of the South Atlantic (Shillington et al. 2006). Again, it is hard to distinguish individual sources of central water. The overall flow of SACW is to the northwest as far north as about 20° S, but north of this the general movement seems to be to the south around the Angola Dome (Shannon 1985a; Boyd et al. 1987; see also below). As this water upwells, it is warmed with little change in salinity. Despite the surface warming, there is a clear difference between this Modified Upwelled Water (MUW) and the regular surface water within the Angola Basin, the latter being generally considerably more saline (>35.5) and warmer (>20 °C). In general, colder, less saline waters are found in the south, warmer, more saline waters in the north, for all water masses within the intermediate, central or upper layers (Duncombe Rae 2005).

7.4.3 The Northern and Southern Frontal Zones

As stated previously, the Benguela is unique among eastern boundary upwelling systems in having warm water at both the northern and southern extremity. To the south, at Cape Agulhas, the cooler upwelling waters along the southwestern boundary of the continent meet the warm waters that cover the shallow (~150 m deep) Agulhas Bank, which are derived from the nutrient-poor Agulhas Current (for a comprehensive review of the Agulhas Current, see Lutjeharms 2006). The shoreward edge of the Agulhas Current affects the water over the bank as it meanders onshore and offshore, while shear-edge eddies and plumes of warm, surface water also interact with the cooler water closer to the coast (Lutjeharms et al. 1989; Schumann and van Heerden 1988). This leads to complex interactions of various water masses to the southeast of Cape Point and over the western Agulhas Bank. Further offshore, filaments of warm surface water, usually less than 50 m thick, and deep eddies (~1500 m) released at the Agulhas Retroflection move slowly northwards into the South Atlantic (Gordon 1985, 1986; Lutjeharms and Cooper 1996). According to Shannon (1985a), these filaments rarely extend north of 33° S, but the rings are more persistent, and many have been observed to move into the subtropical Atlantic (e.g., McCartney and Woodgate-Jones 1991). Occasionally there are major incursions of Agulhas water into the Benguela region. Shannon et al. (1990) described such an incursion, 240 km wide, in 1985–86 that eventually split off the Agulhas Current as a ring. It is thought that such incursions could be caused by slackening of the wind field in the southwest Indian Ocean, leading to a reduced flow in the Agulhas Current that allows the warm water to round the Cape into the southeast Atlantic rather than retroflecting.

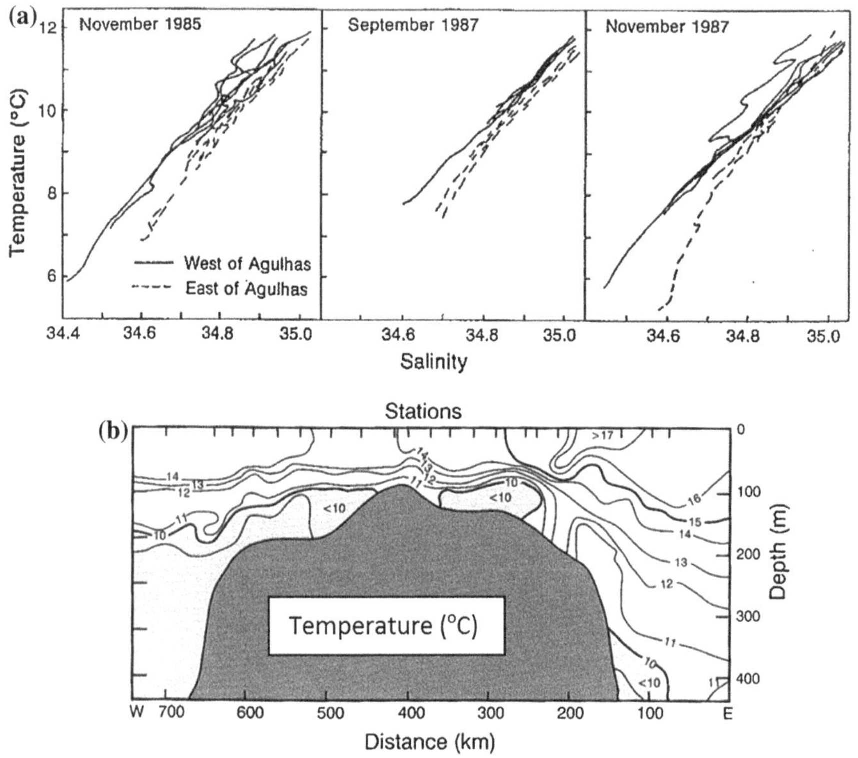

Fig. 7.9 **a** Temperature-salinity plots of stations from west and east of 20° 30′ E showing lower salinity water on the western Agulhas Bank west of Cape Agulhas (from Chapman and Largier 1989). **b** Zonal section across the western Agulhas Bank along 35° 30′ S showing the upwelling of cold (<10 °C) water in the west, which is not connected with the two cold water masses east of the 400 km marker. The latter originate at the eastern edge of the Agulhas Bank (from Lutjeharms 2006)

Studies of water masses over the western Agulhas Bank by Chapman and Largier (1989) and Largier et al. (1992) have shown that the central water found in this region comes primarily from the southeast Atlantic, not the Agulhas, and confirmed that this region is the southernmost extension of the Benguela Current system (Fig. 7.9). They also showed that this area has three sub-regions: an inshore upwelling zone dominated by wind forcing, a region of strong vertical stratification over the mid-shelf, and an offshore region dominated by oceanic forcing. However, upwelling north and south of Cape Point occurs under different wind regimes because of the different orientation of the coastline (Lutjeharms and Stockton 1991), and upwelling can be absent east of Cape Point even when strong upwelling occurs to the north. Despite this, there is a strong seasonal cycle in both wind stress and temperature structure over the offshore portion of the western Agulhas Bank, with strong stratification setting in during the summer months (Boyd et al. 1985; Lutjeharms et al. 1997; Demarcq et al. 2003). This also affects chlorophyll variability, which increases in this area during the upwelling summer season (Demarcq et al. 2003).

Water from the western Agulhas Bank certainly also intrudes into the Benguela west and north of Cape Point, particularly in summer. Shannon (1985a) describes historic studies of ships' drift data (Harris 1978), isopycnal and isentropic analysis (Shannon 1966), and drift cards (Duncan and Nell 1969; Shannon et al. 1973) that all point to a westward movement of water from the western Agulhas Bank. Biological material, such as fish eggs and larvae (Shelton and Hutchings 1982; Shelton 1984) and *Physalia* (Shannon and Chapman 1983) also support this interpretation, as do tar balls (Shannon et al. 1983). Larger intrusions of Agulhas water were also reported in 1986, 1992, and 2000 (Hardman-Mountford et al. 2003; Roy et al. 2001). When the wind is westerly, however, there is certainly flow in the other direction as well.

The northern boundary of the Benguela has been less well studied than the southern boundary, but the consensus is that it is centered near 16° S, while its extremes can vary between about 14° and 17° S (Shannon 1985a; Parrish et al. 1983; Meeuwis and Lutjeharms 1990; Veitch et al. 2006), or even further south if a Benguela Niño is occurring. Certainly there is upwelling north of Cape Frio at 18° N, and it appears that Benguela water can penetrate on occasions as far as 12°–13° S (Moroshkin et al. 1970). The associated frontal zone can be very sharp, as was shown by Hart and Currie (1960), who reported a decrease in sea surface temperature from 27° to 20.5 °C during a one-hour passage southwards in March 1960, but can also be seen as a series of multiple, smaller fronts. It can extend at least 250 km out from the coast (Shannon and Nelson 1996). The front coincides with an abrupt change in wind stress vectors (Shannon et al. 1987), and divides regions of differing biological characteristics (Shannon and Pillar 1986; Crawford et al. 1987). The front is generally a shallow feature, with high stratification in the north, particularly in summer, but its limited zonal extent is clear (Fig. 7.10). While often limited to the upper 50 m of the water column, it may be observed as deep as 200 m (Shannon and Nelson 1996). As well as a strong surface temperature gradient, the front also shows a surface salinity gradient, with salinities less than 35.4 to the south and salinities above 35.6 to the north, although lower salinity water can be found close to the Angolan coast north of Cape Frio (Shannon et al. 1987), and the front is least well-defined during July–September. Veitch et al. (2006) showed that this is because the northern and southern boundaries of the front move in opposition to each other, with the northern boundary moving northwards as the southern boundary moves south, and vice versa.

The water circulation in this region appears to be controlled partly by the Angola Dome (see Chap. 9), a cyclonic feature centered near 10° S, 10° E, whose northern edge is formed by the eastward flowing South Equatorial Counter Current. This induces a poleward current, the Angola Current in the upper 200 m along the Angolan continental shelf, which is found throughout the year, weakening during late winter and spring (Dias 1983). The southern edge of the Angola Dome is bounded by westward flowing water at the northern edge of the Benguela as it moves out into the South Atlantic (see Fig. 7.3). Warm, saline Angolan water

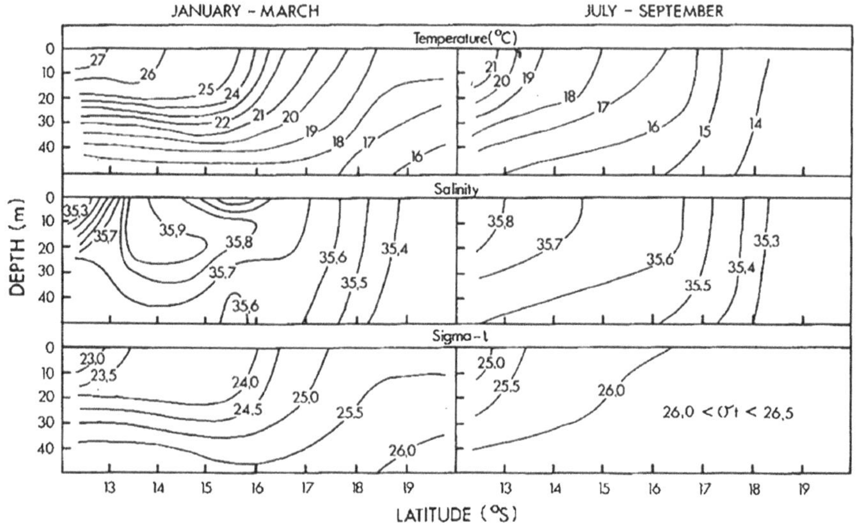

Fig. 7.10 Temperature, salinity and sigma-t in the upper 50 m across the Angola-Benguela between 12° and 20° S for two seasons based on data from one-degree squares (from Shannon et al. 1987)

pushes south into northern Namibia in late summer/early autumn, in contrast to the northward and offshore (westward) movement of the Benguela water, which is driven by the southeasterly trade winds. This leads to a wedge of Angolan water along the coast that can penetrate as far south as 22° S (Boyd et al. 1987), but upwelling can occur on the coastal side of this warm tongue.

Because of the convergence of the two counter-moving currents—the Benguela and Angola currents—there must be offshore movement along the outer edge of the frontal zone, shown near 20° S in Fig. 7.11, although onshore flow can occur nearer the coast. Mesoscale activity, as shown by changes in daily SST, can be large in this region (Shannon et al. 1987; Shannon and Nelson 1996). More recently, Rouault (2012) has suggested, based on a combination of multiple sources of SST, altimetric sea level anomalies, and QuickScat wind stress data combined with a 0.25° horizontal resolution model, that the southward movement of Angolan water occurs twice a year and is tied to the annual cycle of sea level anomaly in the region, which has maxima in October and February and minima in June and December. Occasionally, extreme incursions of warm, saline surface water can occur. In May/June 1963, for instance, warm water was found to penetrate as far as 24° S causing deleterious effects on the pelagic fishery that year (Stander and De Decker 1969).

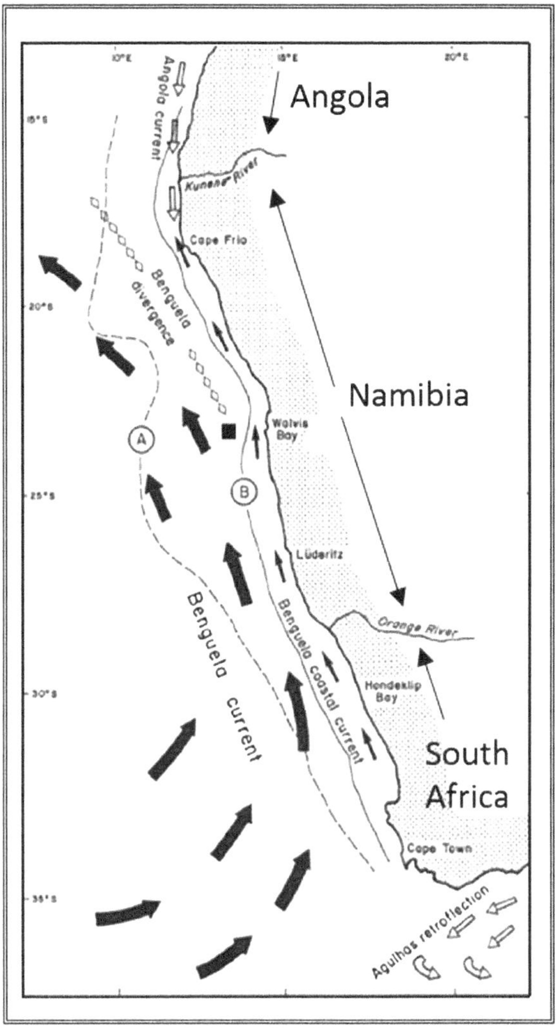

Fig. 7.11 Circulation patterns of the Benguela system after Shannon (1985a) and Lutjeharms and Stockton (1987). *A* donates the offshore limit of the filamentous mixing domain, *B* the offshore limit of the active upwelling regime. Taken from Giraudeau et al. (2000)

7.5 Large-Scale and Coastal Circulation Patterns

7.5.1 General Circulation

Defant (1936) was to first to describe the general circulation in the Benguela. Using data from 1° rectangles, he identified a broad current, some 200–300 km wide, extending northwards off the west coast of southern Africa, diverging from the coast as it moved north, and bending to the west north of 20° S (Fig. 7.11). This general movement was confirmed later from the tracks of drifters (Harris and Shannon 1979; Nelson and Hutchings 1983) and drift cards (Stander et al. 1969;

Shannon et al. 1973). Typical surface current speeds are in the 10–30 cm/s range, although the speed of individual jets can be higher than this, up to about 50 cm/s. This work, and that of other authors, was synthesized by Shannon (1985a), who estimated the overall transport to be about 15 Sv, and has been supported by model results (e.g., Fennel 1999).

More recently, Stramma and Peterson (1989) discussed the general circulation in the area south of 10° S and east of 0° longitude, and suggested a geostrophic net flow of surface water towards the north of 21 Sv at 32° S and 18 Sv at 28° S, relative to a level of no motion corresponding to the 27.75 sigma-theta surface. Added to this is a flow of 4–5 Sv of AIW, also towards the north. North of 28° S the northward flow is much reduced and the main flow path of the surface water is towards the west, all but about 3 Sv passing south of Valdivia Bank (near 25° S, 5° E) while warm, saline water from Angola moves southwards nearer the coast. This coastal water originates within the Angola Basin as described above. The turn to the west of the northward flowing Benguela Current, which occurs largely because of the influence of the Walvis Ridge, is also reproduced in models (e.g., Veitch et al. 2009) and float tracks (Richardson and Garzoli 2003). While most of the water in the Benguela is from the South Atlantic, up to about one quarter of the total flow of $\sim$25 Sv is believed to come from the southwest Indian Ocean (van Ballegooyen et al. 1994). This occurs mainly by transport within Agulhas rings, which add both heat and salt to the southern Benguela and move through the region at speeds of 5–8 cm/s (Gordon 1986; Gordon et al. 1992; Olson and Evans 1986; Richardson and Garzoli 2003). They can often be identified by their reduced oxygen concentrations (Chapman 1988). The schematic view of the overall surface circulation in the region, that incorporates the above, is shown in Figs. 7.3 and 7.11.

Below the surface layer, flow changes considerably. Instead of the overall northward flow seen at the surface, southward flow is prevalent within about 3° of the coast from Angola to about 27° S off Lüderitz. This southward flow is strongest along the shelf edge and plays a major role in the transport of low oxygen water into the northern Benguela region (Chapman and Shannon 1985). While the Walvis Ridge does affect flow at this level, it is too deep to prevent the continued offshore northwestward movement across it. Southward flow along the shelf edge is also found further south, reaching Cape Point, but this flow is separate from that found further north (Chapman and Shannon 1985; see below). Nelson (1989), based on data from numerous current meter deployments along the coast, suggested that the overall movement of subsurface water on the shelf and over the continental slope is actually southward, and that northward flow, as expected from classical Ekman dynamics, occurs only in the uppermost 20 m of so of the water column that is affected by the local wind field and by the offshore jet. The meridional poleward component over the shelf is about 1.6 Sv, while the equatorward component is about 0.7 Sv (Shannon and Nelson 1996). Barotropic southward flow has been observed inshore of the offshore jet over the steeper topography near Cape Town (Shannon and Nelson 1996), suggesting the superimposition of a coastal trapped shelf wave on the general flow field.

One of the most interesting features of the Benguela general circulation is the effect that the Lüderitz upwelling region has in dividing the upwelling regime into separate northern and southern parts. This was first noticed by Copenhagen (1953) and Hart and Currie (1960). Duncombe Rae (2005) has summarized the research in this region, pointing out that the shelf narrows abruptly just south of Lüderitz, in contrast to the wide shallow shelf to the south that incorporates the sediment cone from the Orange River. Observations suggest that the presence of this strong upwelling cell forces central and intermediate water masses, from both the north and south, to move offshore in this region (Duncombe Rae 2005), but there are not enough data to determine if this is a permanent effect or if it varies seasonally. Despite this uncertainty, the physical discontinuity resulting from the Lüderitz upwelling centre leads to differences in chemistry and biology north and south of the upwelling cell (Monteiro 1996; Shannon and Pillar 1986; Agenbag and Shannon 1988; Barange and Pillar 1992), possibly through increased turbulence that prevents organisms passing through the region.

7.5.2 Inter-annual and Seasonal Variability

There is considerable variability over long time scales and over latitudinal regions within the Benguela system, from the year-round upwelling north of about 30° S to the more intense summer seasonal upwelling in the south (Field and Shillington 2005). However, there is also an overall seasonality in large-scale structures, such as temperature and salinity patterns, in response to local climatic changes (Christensen 1980; Demarcq et al. 2003), which leads to corresponding changes in chlorophyll abundance (Carr 2001; Demarcq et al. 2003; Fig. 7.12).

While the mean temperature along the coast and the annual mean chlorophyll concentration are relatively constant, the standard variation of the data changes considerably with latitude (Figs. 7.12b, d). In winter and spring, temperatures cooler than 16 °C are found all along the western coastline of southern Africa from about 18° S to Cape Agulhas, extending up to 300 km offshore (Boyd and Agenbag 1984a; Cole and Villacastin 2000). In summer, the area covered by this cool water contracts considerably, both zonally and meridionally, and is not found north of Walvis Bay (23° S). Simultaneously, the offshore temperature gradient increases south of this latitude (see the monthly images in Cole and Villacastin 2000; Demarcq et al. 2003), and the warmest waters off Namibia and the southern Cape coast are found in late summer/autumn. Changes in salinity in the upper 50 m off central Namibia (21°–24° S) range typically between about 35.1 in spring to 35.4 in autumn (Boyd and Agenbag 1984b), with both positive and negative excursions from this range. Salinity changes typically follow temperature changes with a lag of 1–2 months.

Farther south, off Cape Columbine (32° 40′ S), the temperature and salinity of the upper 50 m range between about 11°–17° and 34.75–35.35 respectively, with decreasing surface temperatures and salinities in summer reflecting the seasonality

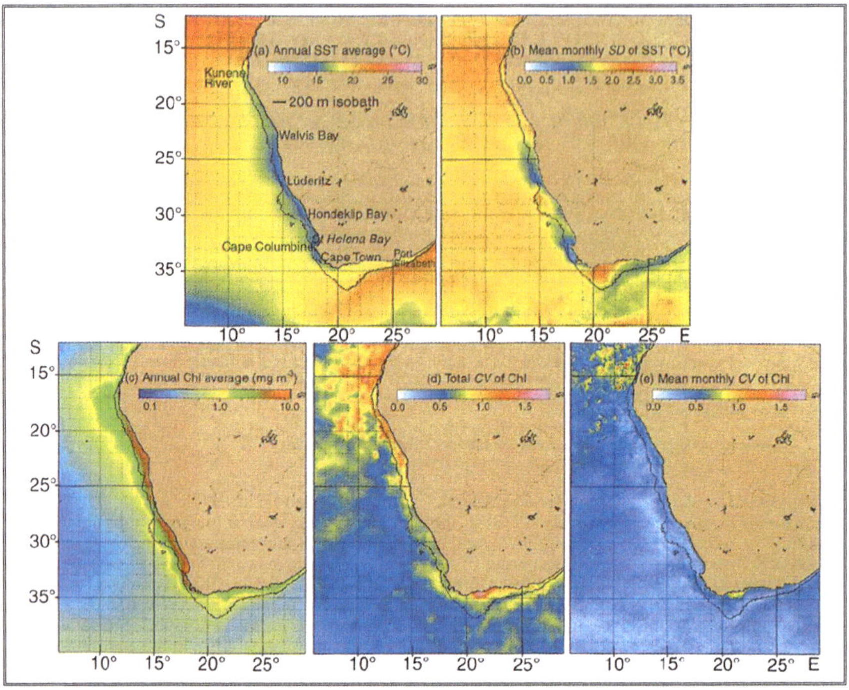

Fig. 7.12 **a** Annual average of SST and **b** mean monthly standard deviation (*SD*) for SST computed from monthly composites for the period 1982–1999; **c** annual average chlorophyll (Chl) concentration and **d** total coefficient of variation (*CV* = *SD* divided by average) of Chl computed from 5-day weighted composites from September 1997 to April 2002, and **e** mean monthly *CV* of Chl computed from monthly composites. Taken from Demarcq et al. (2003)

of upwelling at this latitude (Buys 1959). Lower salinities are typically found at 50 m rather than at the surface, unless upwelling is occurring, and as expected, the maximum change occurs as one goes zonally from inshore to offshore waters outside the upwelling front. While upwelling is generally continuous off northern Namibia, pulsing at scales of about a week in the south leads to a series of upwelling events during the summer season (e.g., Andrews and Hutchings 1980; Bailey and Chapman 1991). The situation is more complex at both northern and southern ends of the system because of intraseasonal eddy variability (Chavez and Messié 2009) and the passage of coastally-trapped waves (Rouault et al. 2007).

Although there have been a number of studies of multiyear variability in the Benguela (see Shannon (1985a) for a description of early work), these have usually only looked at a few years of data. The longest record is that of sea level. Brundrit et al. (1984) and Brundrit (1995) showed strong coherency between monthly sea level records from nine sites between Walvis Bay and Cape Agulhas, with a delay of about 2 days as the sea level signal travelled from north to south. The agreement between sites was particularly close during the warm anomaly in the southern

Benguela in 1982, with all sites showing a downward trend in sea level from 1979 onwards. Over the longer term, records beginning in 1959 showed high sea levels in 1962–63, 1968–69, 1980–81, and 1986–87 with levels lower by 10 cm in 1965–66 and 1971. The data agree well with changes in atmospheric pressure and presumably relate to large-scale changes in the pressure field over the tropical Atlantic, which result in Kelvin wave activity along the African west coast.

More recent comparison studies of model output with observed (satellite) winds and SST measurements (Blanke et al. 2002) show spectral peaks at about 14, 7, and 4–5 days for the region south of Lüderitz, and a mean upwelling rate of about 0.49 Sv, which is about 20 % less than the theoretical Ekman transport along this boundary. Tim et al. (2015) applied a high-resolution model (0.1° grid, 80 levels) driven by the 60-year NCEP meteorological reanalysis product, but instead of using an upwelling index based on SST, they used the estimated vertical mass transport at 52 m depth (close to the modeled mixed layer depth). The model showed differences in the magnitude and seasonality of upwelling between the regions north and south of Lüderitz, in good agreement with observations. Interannual standard deviations relative to the long-term mean were smaller in the northern region than in the south, and were also much more coherent across seasons in the north. The model suggested that decadal changes in upwelling intensity had occurred in the northern region but these were not significant at the 95 % level. There were, however, significant correlations in different seasons with large-scale drivers such as ENSO and the Antarctic Oscillation (in both regions), and the Atlantic Meridional Mode (in summer in the south).

These seasonal and longer term changes in the physics also affect the biology (see Sect. 7.7). In brief, pelagic fish spawn mainly in summer, using batch spawning to take advantage of the short-term upwelling events (Shelton 1987), while hake spawn in spring (October-December) in the northern Benguela and also in February–March in the southern Benguela (Botha 1986; Sundby et al. 2001).

7.5.3 Mesoscale Variability and Coastal Circulation

As in other eastern boundary upwelling systems, the Benguela shows typical inshore features including an inshore upwelling front, upwelling jets, an offshore divergence belt, and a poleward undercurrent (Bang 1971; Nelson and Hutchings 1983). Given the importance of individual upwelling cells along the coast, this means that rather than the continuous belt of upwelling referred to in early texts on oceanography, upwelling in the Benguela is determined more by local conditions, which vary on several time scales and are not always in phase along the almost 2,000 km of the system. The longer time-scales have been discussed earlier, and this section concentrates more on shorter time and space scales.

Around Cape Point and Cape Columbine, Andrews and Cram (1969), Bang and Andrews (1984), Andrews and Hutchings (1980), and Nelson (1985), among others, have all shown the existence of northward flowing jets that are now considered to be vital to the continuing maintenance of the local pelagic and demersal fish stocks (e.g., Shelton and Hutchings 1982; Parada et al. 2003). These jets, which can attain speeds of up to 150 cm/s, are accompanied by poleward subsurface flows that meander along the shelf edge. At times there may be two jets and two undercurrents; these typically occur where the shelf is widest and has distinct steps in its topography. Hart and Currie (1960) and De Decker (1970) suggested that the undercurrent is responsible for bringing water from the Angola Dome with low oxygen concentrations into the northern Benguela, which is discussed more fully by Chapman and Shannon (1985) and Monteiro (1996).

Frontal jets and undercurrents are important features of both the northern and southern Benguela regions, but it is unclear if they are continuous along the whole of the Benguela system, particularly near Lüderitz, where a cyclonic gyre may exist (Nelson and Hutchings 1983). While surface currents are predominantly alongshore, substantial cross-shelf flows can develop below the pycnocline. For example, hydrographic data from repeated sections across the shelf near 20° S revealed the presence of substantial transient cross-shelf water exchanges (Hagen 1985). Within 100 km of the coast, the response to changes in the wind field occurred within a day, and such responses had a mean period of 5.6 days. Shannon (1985a) discussed early work in the region, listing a number of reports of rapid changes in surface water temperature within 1 or 2 days of wind changes, in both the northern and southern Benguela (e.g., Bang 1971; Andrews and Hutchings 1980). Rapid wind-induced changes in water structure in the northern Benguela were also observed by Bailey (1979) (Fig. 7.13). Note that the origin of the upwelled central water is generally between 200 and 300 m in the north, although off the Cunene River (bordering Angola and Namibia) it may be as shallow as 100 m or even less. Dynamic topography primes the system (Stander 1964) so that Central Water is found close to the surface and can be uplifted quickly into the euphotic zone once the wind becomes upwelling-favourable.

The areal extent of the upwelling plumes that develop at the various sites along the coast varies considerably, again in response to the local wind stress. At both Cape Point and Cape Columbine, the mountainous topography affects the wind field, and the currents can change rapidly as the wind stress varies. Andrews and Hutchings (1980), Taunton-Clark (1985), Jury (1985) and Nelson (1985) discuss details of the upwelling tongue off Cape Point, while Shannon et al. (1984), Holden (1985), and Kamstra (1985) discuss the development of the upwelling tongue off Cape Columbine. Their studies consistently demonstrate that both upwelling tongues can grow over a period of several days under consistent winds, but collapse very quickly once the wind regime changes. As they grow, the cold tongues tend to bend cyclonicly towards the coast, creating a mechanism that traps water in Table Bay, north of Cape Town and in St. Helena Bay, north of Cape Columbine, respectively. In addition, shelf-break canyons (the Cape Point Canyon southwest of Cape Point and the Cape Canyon about 30 km immediately west of Cape Columbine) facilitate enhanced cross-shelf exchanges and rapid upwelling with upwelling rates of about 20 m/day in both areas (Shannon 1985a).

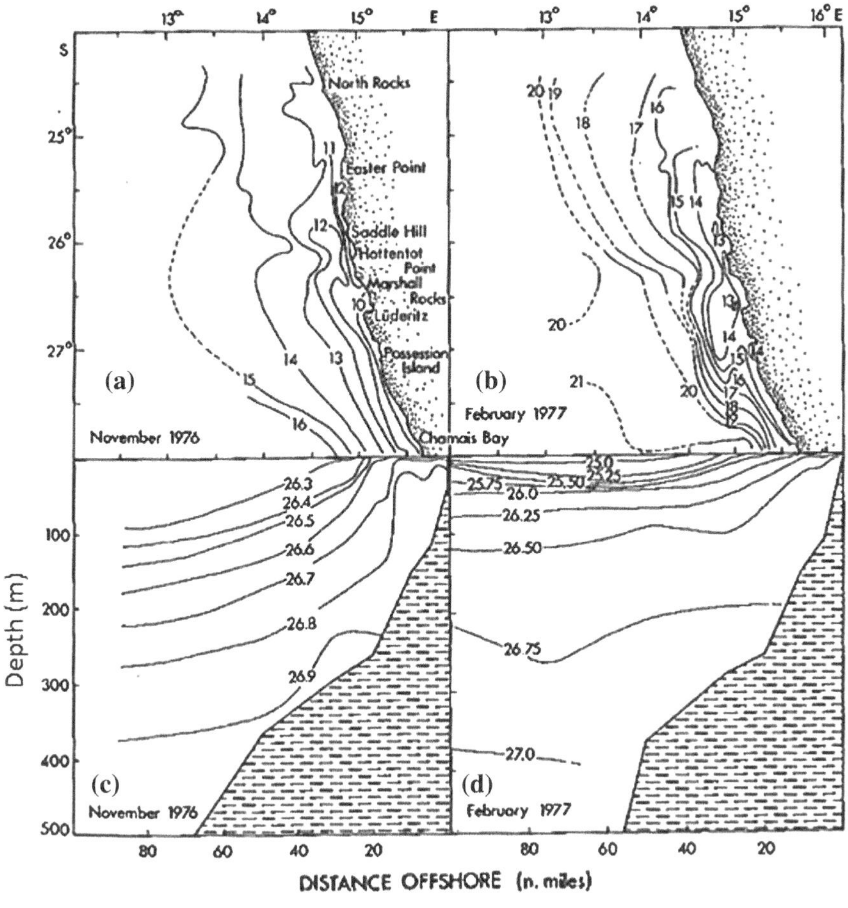

Fig. 7.13 Surface temperature (°C) near Lüderitz and sigma-t profile along a line of stations at 26° 21′ S during upwelling (**a**, **c**) and quiescent (**b**, **d**) periods (from Bailey 1979)

While advection is central to cross-shelf exchange processes, mixing by internal waves can play an important role in vertical fluxes of nutrients and other water-mass properties on the shelf. Chapman and Shannon (1985), using data from Schultz et al. (1979), revealed the existence of internal waves with a frequency of about a day, and amplitudes of 10–20 m, over the shelf north of Walvis Bay. The shelf in this region is about 80 m deep, and their suggestion was that such waves could be strong enough to mix the water layer below the pycnocline and cause periodic disruption of the water-sediment interface. This interpretation is borne out by current meter records that show upward and downward phase propagation of inertial wave motions with a period of 30.7 h (Lass and Mohrholz 2005). The energy transfer is found predominantly in the upper 20 m of the water column, but

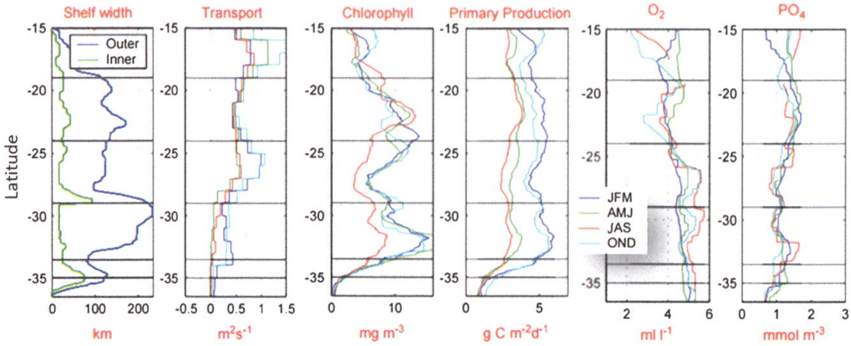

Fig. 7.14 Variation in shelf width and seasonal changes in offshore Ekman transport, chlorophyll concentration, primary production, oxygen and phosphate concentrations with latitude in the Benguela. The *horizontal black lines* divide the region into 6 separate zones (from Carr and Kearns 2003)

extends down to 60–70 m on occasion. Interestingly, onshore flow occurred at depths between about 15–70 m, with offshore flow above and below this.

An additional feature of the Benguela, typical of many other upwelling zones, is the presence of large filaments of upwelled water that can extend up to 1,000 km into the open ocean (Lutjeharms et al. 1991). These upwelling filaments are thought to result from either interaction with Agulhas rings, or from the influence of the katabatic (Berg) winds that blow off the Namibian desert. Hydrographic data show that these filaments are typically only about 50–100 m thick (Shillington et al. 1990), and that they consist of aged upwelled water. They can occur all along the Benguela coast; for instance, Nelson et al. (1998) reported on such filaments off Cape Town. While these filaments are ephemeral, their effect on the short term can be important to the biology of the region. Duncombe Rae et al. (1992) suggested that the interaction of these filaments with Agulhas rings can lead to the loss from the coastal zone of large volumes of surface water containing recruiting fish larvae (Fig. 7.14).

7.6 Chemistry and Related Processes

7.6.1 Overview

Although the German *Valdivia*, *Mowe* and *Meteor* cruises occupied a number of stations in the region between 1898 and the 1920s, the first planned large-scale chemical surveys of the Benguela Current system were made only in 1950 as part of the *Discovery* expeditions (Hart and Currie 1960). Even then, almost all stations sampled were north of the Orange River off what is now Namibia. Much of the information available since this date comes from routine measurements made by the

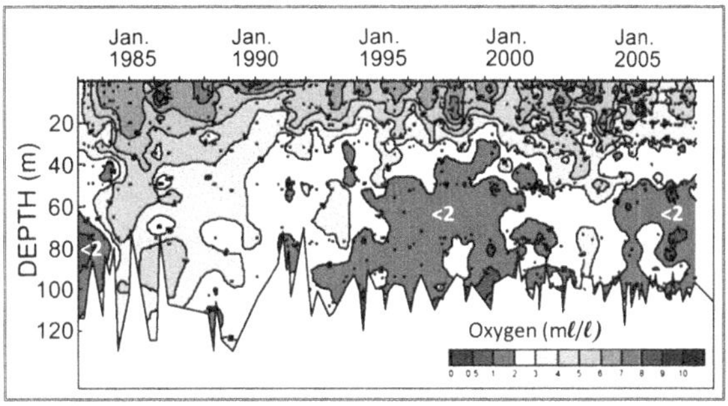

Fig. 7.15 Variability of dissolved oxygen at a station off Cape Columbine for the period 1983–2007. Note the difference between the 1980s and 1990s as regards low oxygen occurrences (from Hutchings et al. 2009b)

Sea Fisheries Research Institute in Cape Town, the government agency responsible for advising the relevant Minister on fisheries stock assessments, or by researchers at the Council for Scientific and Industrial Research in Stellenbosch. More recently, research in Namibian waters has been carried out by scientists at the National Marine Information and Research Centre in Swakopmund, north of Walvis Bay. Chapman and Shannon (1985) reviewed the early work on the chemistry of the system. Carr and Kearns (2003) released a seasonal mean plot of the variability of a number of physical and chemical parameters for all four major western boundary currents; the changes in shelf width, together with the seasonal variability of chlorophyll and primary productivity within the upper 50 m of the water column, and oxygen and phosphate concentrations for the densest water in the upper 50 m of the Benguela near the coast, are shown in Fig. 7.14. The two maxima in chlorophyll correspond to the highly productive regions near Walvis Bay (23° S) and Cape Columbine (33° S) respectively, and it is clear that summer chlorophyll and productivity are considerably higher than winter values. Conversely, there is little seasonal change in either oxygen or phosphate concentrations except in the northernmost sector.

7.6.2 Upwelling Chemistry: Oxygen and Nutrients

Wattenberg (1938, 1939), using data collected by the *Meteor* expedition, was the first to report the existence of a large mass of oxygen-depleted water in the southeast Atlantic. The core depth of this layer was at 300–400 m, and the lowest concentrations were below 20 μM between the 0° meridian and the Angolan coastline and from 8°–15° S. The oxygen-depleted layer stretched across the whole of the tropical Atlantic to the Brazilian coast, where values below 100 μM were measured. Later studies (e.g., Copenhagen 1953; Hart and Currie 1960; De Decker 1970) showed that the low oxygen water mass, accompanied by higher nutrient concentrations than normal for offshore South Atlantic Central Water, extended along almost the whole of the Namibian and South African shelves. Bubnov (1972) suggested that there are two sources for oxygen-depleted water in the northern part of the system, with the main source being in the region of the Angola Dome while additional depletion occurs over the shelf because of the high phytoplankton production induced by upwelling. Voituriez and Herbland (1982), however, suggested that it was equatorial upwelling, rather than Angola Dome production, that created the main oxygen minimum. In either case, water from offshore is transported southwards in a poleward undercurrent near the shelf break, and this brings water from the main formation region onto the shelf, from where it can be upwelled.

De Decker's (1970) suggestion that the poleward undercurrent was continuous from northern Namibia to Cape Town was contradicted by findings of Chapman and Shannon (1985), who showed that the temperature, salinity and density ranges of the oxygen-depleted water decreased from Namibia towards the south, the opposite of what would be expected if the water mass was continuous. For example, the range in sigma-*t* off Namibia is between 26.4 and 27.0 (Stander 1964; De Decker 1970), while off Cape Columbine and Cape Point it is more constrained at 26.8–26.9 (Andrews and Hutchings 1980; Bailey and Chapman 1985). Chapman and Shannon (1985) also noted a discontinuity in the low oxygen undercurrent near Lüderitz, and suggested that south of 27° S local respiration over the shelf in retention areas such as the Orange Bight and St. Helena Bay contributed to the low dissolved oxygen concentrations seen along the coast.

Monteiro (1996) introduced the "Gate hypothesis", that there are three discrete circulation regimes along the coast between Cape Frio and Cape Point, with "gates" at the northern end of each where SACW can upwell onto the shelf. Thus, instead of the traditional Ekman model of upwelling, where the overall flow of water is onshore to offshore, from south to north, the general flow over the shelf is actually southward, particularly in the bottom boundary layer. This research added carbon dioxide chemistry in support of the physical oceanographic data discussed in Sect. 7.5.1. Monteiro put the gates just south of Cape Frio, south of Lüderitz, and near the Olifants River at 33° S. Immediately north of each gate there is a corresponding offshore flow that compensates for the onshore advection at the gate. The theory means that along with an offshore surface upwelling plume, there is a more general southward flow of aged upwelled water along the inner shelf that transports phytoplanktonic particulate carbon, which is eventually deposited offshore at the southern end of each region, i.e., near 26°–28° S, 33°–34° S, and near 36°–37° at the bottom of the Cape Point Valley.

Forcing factors in the three regions are believed to differ (Monteiro and van der Plas 2006). In the northern region, it is believed that forcing is the result of variability in the eastern tropical Atlantic, particularly changes in the easterly winds over the Angolan system that drive the cyclonic circulation in this area. Stronger winds mean a stronger pycnocline, which increases the southward flow of oxygen-depleted water. Off central Namibia, however changes result from a combination of the poleward undercurrent from the north (which affects the overall oxygen content of water upwelled onto the shelf) and local production over the shelf. In the southernmost region, oxygen depletion depends on the strength of the equatorward wind stress (Johnson and Nelson 1999) and local biochemical changes on the shelf (Bailey and Chapman 1985; 1991), with more upwelling leading to lower oxygen concentrations. Seasonal changes in the local wind field mean that the gates can move meridionally or even disappear at times, and there are seasonal variations in the oxygen minimum concentrations. This accounts for the observed seasonal movement off Namibia of the oxygen-depleted water noted by Chapman and Shannon (1985), who showed that maximum southward penetration occurred in late summer, when minimum oxygen concentrations are also found in St. Helena Bay.

There is also some decadal variability in both oxygen and nutrients, as the 1980s were characterized by generally weak winds in the southern region, in contrast to the 1990s (see Fig. 7.7). This led to major differences in the bottom oxygen concentrations in St. Helena Bay, and other multi-year changes have been seen since (Monteiro and van der Plas 2006; Hutchings et al. 2009b, 2012; Fig. 7.15). Nutrient concentrations also changed (Hutchings et al. 2012). While there was no change in dissolved silicate, both nitrate and phosphate concentrations increased by about 40–50 % below the thermocline in St Helena Bay between 1978 and 1998, after which they declined again to their original concentrations by 2007. This is thought to be due to a change in the chemistry of the upwelling South Atlantic Central Water rather than to any local change in biogeochemistry.

The contrast in the wind fields between the northern and southern regions and the fact that the water that upwells north of Lüderitz is already considerably depleted in oxygen compared to that upwelling in the south means that there are major differences between the nutrient distributions in the two regions. There have been three sets of time series that sampled repeatedly over two to four weeks, unlike the multi-year or decadal time series shown in e.g., Monteiro and van der Plas (2006), which were only monthly or quarterly at best. Chapman and Shannon (1985) showed data from a 15-day time-series station off Walvis Bay in November 1976 at 22° 16′ S, 14° 05′ E (Schultz et al. 1979), while Bailey and Chapman (1991) presented data from a similar time period in June/July 1978 at 22° 00′ S, 14° 01′ E (Boyd 1983), and from a 1 month station at 32° 33′ S, 18° 05′ E in St. Helena Bay during March/April 1987. Winds during all three time-series were quiescent (<5 m/s).

The two Walvis Bay data sets, which were sampled every three hours, showed relatively constant conditions, with very low dissolved oxygen at the bottom and continuous nitrate reduction below 20 m depth from a maximum of 25–30 μM to less than 10 μM at the bottom. In contrast, the St. Helena Bay time series (Fig. 7.16), sampled on a four-hourly repeat, showed freshly upwelled water aging and developing a bloom of large-celled phytoplankton that slowly sank during the course of the time-series. Oxygen concentrations in the bottom layer slowly decreased in St. Helena Bay, but only dropped below 22 μM occasionally, in contrast to the two northern time series where they remained below this level throughout the period of the experiment.

This difference in bottom water oxygen characteristics affects the benthos; in St. Helena Bay the oxygen concentrations are almost always high enough to support a thriving bottom faunal community, including juvenile hake recruiting to the fishery, whereas off Walvis Bay only anaerobic bacteria are found and there is a thick,

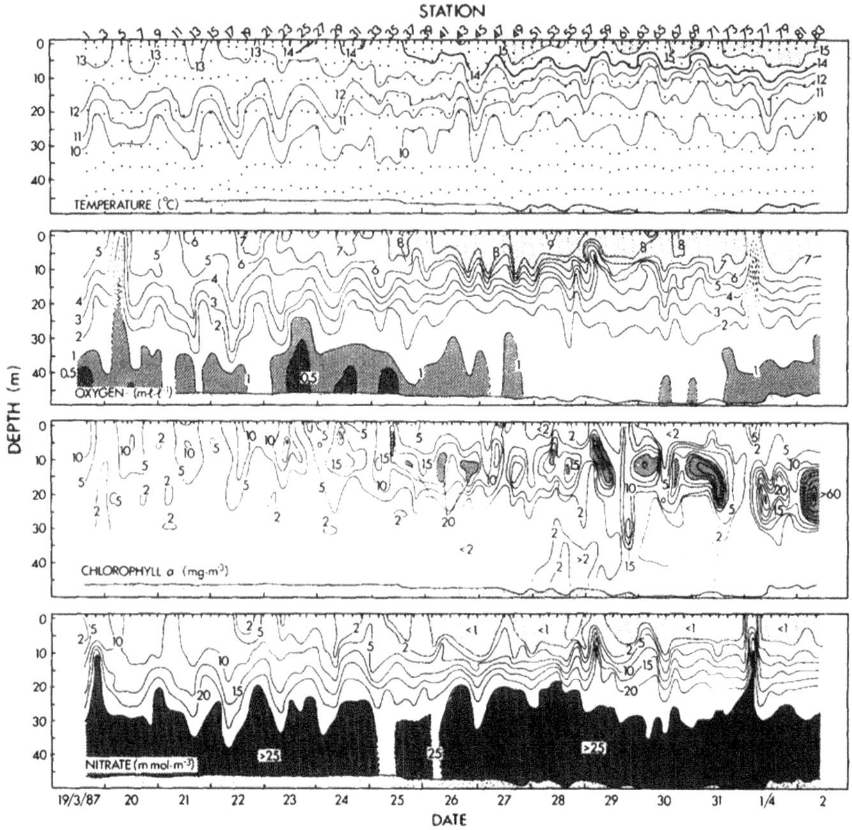

Fig. 7.16 Time series of temperature, oxygen, chlorophyll-*a* and nitrate during a time series at 32° 33′ S, 18° 05′ E in March 1987 (from Bailey and Chapman 1991)

anaerobic mud belt (Bremner 1983). Bottom temperatures at the two sites also differed, being >12.5 °C off Walvis Bay, but <10 °C in St. Helena Bay. This supports the ideas of Monteiro and van der Plas (2006) regarding the ages of upwelled water on the shelf in the two regions, and also means that the pycnocline in St. Helena Bay is stronger than that further north.

Other studies have confirmed the relatively short time-scales seen for upwelling events in the southern Benguela. A number of studies of upwelling plumes (e.g., Olivieri 1983; Brown and Hutchings 1987) have shown that blooms can develop rapidly once the wind relaxes, with chlorophyll-*a* concentrations increasing from <1 to >15 μg/L in about 3 or 4 days. This leads to depletion of nitrate from >20 μM to less than 1 μM over the same time period, while oxygen saturation levels in the surface euphotic layer can increase simultaneously to more than 150 %. Typical sections across the narrow, steep shelf off Cape Columbine are shown in Chap. 2 (Fig. 2.15).

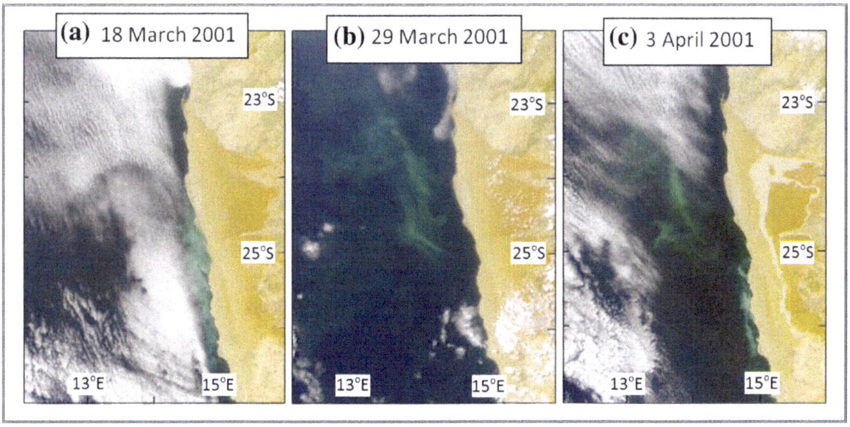

Fig. 7.17 Quasi-true colour images for the region 12°–16° E, 22°–27° S for **a** 18 March, **b** 29 March and **c** 3 April 2001. Images generated from Orbview-2 SeaWifs satellite. The milky turquoise colouration represents high concentrations of suspended sulphur granules (from Weeks et al. 2002)

The large anaerobic region in the northern Benguela leads to much greater reduction than occurs south of Lüderitz. In particular, this leads to nitrogen reduction in the water column, which, as discussed in Chap. 2, is largely accomplished by anammox bacteria (using anaerobic oxidation of ammonia by nitrite). There is also bacterial generation of free hydrogen sulphide and methane in the sediments (Emeis et al. 2004), as well as the irregular appearance of temporary "mud islands". Such eruptions have been known for a long time; Copenhagen (1953) and Hart and Currie (1960) reported that hydrogen sulphide could be smelled 60 km inland, while lead-based white paint and metalwork turned black from reacting with airborne sulphide. More importantly, these eruptions result in fish kills and walkouts by rock lobster. More recently, satellite imagery has led to the discovery of large areas offshore of up to 20,000 km^2 where sulphur crystals reach the surface (Weeks et al. 2002, 2004; Fig. 7.17). Once thought to be rare, they are now known to be much more commonplace, especially during summer, with 14 eruptions covering more than 2400 km^2 and another of 400 km^2 observed between March 2001 and March 2002 (Weeks et al. 2004).

7.6.3 Primary Productivity and Nutrient Cycling

There have been numerous estimates of the productivity of the Benguela upwelling region. Shannon and Field (1985) summarized the previous work, which was based on discrete sampling cruises, and concluded that in the southern region between Cape Point and Cape Columbine, annual production averaged 2.8 g C/m^2/day, while off Namibia it was about 2.0 g C/m^2/day. This compared with estimates of about 1.9 g C/m^2/day for Peru and northwest Africa by Barber and Smith (1981). They calculated total production in the southern portion of the system (Cape Columbine to Cape Point) as being 40 $\times$ 10^6 tonnes/year. Brown et al. (1991) extended this work and calculated production values of 77.4 $\times$ 10^6, 76.4 $\times$ 10^6, and 79.0 $\times$ 10^6 tonnes/year for the Namibian, Cape, and south coasts (as far as 27° E) respectively, for a total of 232.8 $\times$ 10^6 tonnes/year, which compares well with an earlier estimate of 274 $\times$ 10^6 tonnes/year by Cushing (1969), who used an abnormally large area for his estimate.

Most of this work was completed before the advent of ocean colour satellites, although South Africa had a strong interest in the Coastal Zone Color Scanner on the Nimbus-7 satellite and carried out a large-scale environmental survey and calibration exercise during 1979 (see the papers in Shannon 1985b). Since the advent of the SeaWifs satellite, in 1997, global coverage of ocean colour and its relation to primary production has become more routine. Carr (2001) analyzed the first 2 years of data from SeaWifs, which was extended to a third year by Carr and Kearns (2003; Fig. 7.14). Silio-Calzada et al. (2008) have also used satellite imagery to estimate productivity based on the relationship between SST and nitrate for the whole Benguela region, but do not give any annual estimate because of the problems inherent in assuming a suitable C:N ratio during growth.

Carr and Kearns (2003) data suggest that the overall productivity of the Benguela is about 3 g C/m^2/day in winter, and 5 g C/m^2/day in summer, with annual production being about 0.37 Gt/year (Carr 2001). This would make the Benguela the most productive of the four eastern boundary upwelling systems, although her figures for all systems were revised downwards considerably by Monteiro (2010), based on new estimates of their respective *f*-ratios (fraction of total primary production fueled by nitrate). For the southern Benguela, Monteiro used an f-ratio of 0.2–0.3, based on work by Probyn (1992). This is less than half of what is normally expected for a diatom dominated region, but Probyn claimed that the southern Benguela is, in fact dominated by small phytoplankton that use ammonium in preference to nitrate. Similar figures for the *f*-ratio in the northern Benguela have been published by Benavides et al. (2014).

The discrepancy in the different sets of estimates shows how hard it is, even with global satellite coverage, to account for the high local variability in production, given that (i) the satellite sensors only reflect surface values, while production may occupy at least 50 m of the water column, and (ii) the ratio of chlorophyll observed to carbon produced can vary considerably over short time intervals. The Benguela, however, does appear to be at the top end of the productivity range with the Canary Current system. Ismail et al. (2015) have shown that there is only a poor relationship between time series of upwelling intensity and SST and salinity values in St. Helena Bay, casting further doubts on the use of remote sensing techniques as indicators of upwelling in this region, but this may be merely a question of the timescale of interest, as high-frequency data suggest a better relationship in the southern Benguela (Weeks et al. 2006). Even if the high productivity suggested by satellite data is correct, this does not seem to be translated directly into higher trophic levels, and fish catches in the Benguela are lower than expected given the relatively high observed production rates (Hutchings 1992).

Hutchings (1992) pointed out that it is not enough to assume that upwelling will always produce high concentrations of higher trophic level organisms, and listed a number of mechanisms that may limit fish production in the southern Benguela. These included (i) how long eggs survive on the spawning grounds, (ii) food availability there, (iii) success during transport from the spawning area over the western Agulhas Bank to the recruiting area in St Helena Bay, including the ability of the larvae to migrate vertically and transfer across the shelf from the jet current to the quieter coastal region, and (iv) feeding success and predation of juveniles before they return to the spawning area. It seems likely that mismatches between the timing of larval transport and food availability are a major constraint on recruitment success, and is one reason that pelagic forage fish have developed serial spawning, which helps ensure that at least some of the larvae survive to recruit into the fishery.

When conditions are right—during quiescent periods following upwelling events—red or brown "tides" can form along the coast in wind shadow regions inshore of upwelling centres (Pitcher and Weeks 2006; Jury 2014), as found also in other regions (e.g., Graham and Largier 1997). These can be transported onshore across the shelf and then southwards during wind relaxation or reversal. Production rates within these blooms, which are most common during late summer and autumn and are pulsed typically on 3–5 day periodicity, depending on stratification, can be in excess of 500 mg C/m^3/hr. These red tides have been noted all along the coast since early in the 19th century (Shannon and Pillar 1986), and whether or not they are toxic depends on the species involved. Both toxic and non-toxic species can cause fish mortalities, either from the toxins released or from simple clogging of their gills. Crustacean and shellfish mortalities also occur, usually from the formation of low oxygen water close to the coast as the bloom decays. This has caused numerous "walkouts" of the commercially valuable rock lobster *Jasus lalandii*, particularly along the northern shores of St. Helena Bay. Once formed, the blooms can be advected away from their source regions to affect other parts of the coastline. Dinoflagellates are usually the culprits in toxic blooms, while ciliates may also be involved. This occurs because once upwelling ceases, diatoms strip out the nutrients and dinoflagellates, which prefer calm, stratified conditions, are then able to outcompete them. Common non-toxic organisms causing these blooms include *Noctiluca scintilans*, *Mesodinium rubrum*, and various species of *Gymnodinium*, among others, while the toxic blooms, which can cause human fatalities if they affect mussels that are later harvested and eaten, seem to be caused mainly by *Alexandrium (*formerly *Gonyaulax*) *catenella* and *Protoceratium reticulatum* (formerly *G. grindleyi*), although other toxic species, such as *Dinophysis acuminata* and *D. fortii* have since been identified (Pitcher and Weeks 2006).

Following the time-series station data shown by Bailey and Chapman (1991; Fig. 7.16), Cochrane et al. (1991) modelled the chemical and some of the biological components (phytoplankton and zooplankton), based on nitrogen as the major chemical component. While the agreement with the chemistry was reasonable, the model failed to predict the abrupt changes seen in the biological components, probably because of limitations in the parameterization of such effects as grazing, light limitation, vertical migration, and patchiness. Touratier et al. (2003) reexamined the same data set with a new ecosystem model that incorporated nitrogen, silica and carbon. The model output agreed well with the observations during the time-series and with other observations made in the southern Benguela, although patchiness was still a problem. Overall, the model suggested that the local sediments played a large role in supplying the nutrients, particularly nitrate, necessary to support the observed productivity. Additionally, the authors suggest that the carbon produced is split, with about 25 % being used by grazers, 28 % dying and entering the microbial loop, 41 % exuded as dissolved organic matter, and only 8 % ending up in the sediments.

7.6.4 Zooplankton

A recent time series, based on satellite imagery, of the variability in chlorophyll concentration is given in Fig. 7.18, while monthly and long-term changes in zooplankton biomass are shown in Fig. 7.19. Primary production in the northern and southern parts of the system can be completely out of phase and adjacent regions

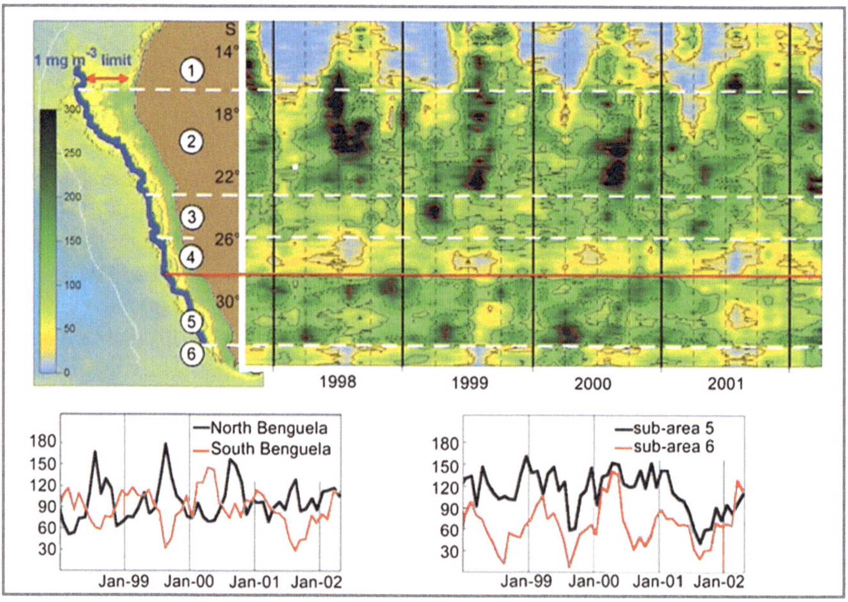

Fig. 7.18 Patterns and indices of phytoplankton enrichment on the southern African west coast from Angola to Cape Agulhas, based on satellite colour imagery and divided into six zones. Lower panels show enrichment indices based on the sum of pixels between the coast and the 1 mg/m^3 chlorophyll-*a* isoline showing monthly and interannual variability. From Hutchings et al. (2009b)

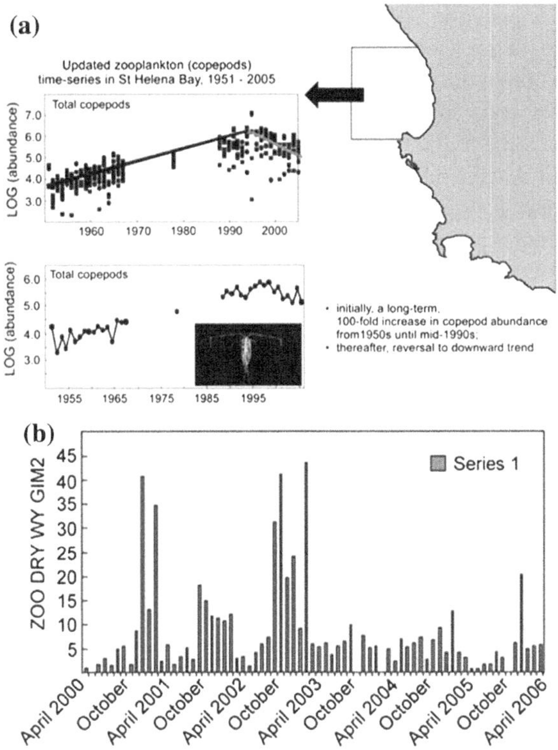

Fig. 7.19 **a** Long-term changes in planktonic copepods in St. Helena Bay during April–June, 1951–2005, showing a 100-fold increase until 1995; **b** monthly variations in total zooplankton biomass (g dry weight/m^2) at five stations across the southern part of the bay. From Hutchings et al. (2009b)

can also vary considerably (Fig. 7.18), which is not obvious from the means presented by Carr and Kearns (2003; Fig. 7.14). Figure 7.19 demonstrates both the large monthly variability in zooplankton biomass in St. Helena Bay and a long-term increase, by two orders of magnitude, followed by an ongoing decrease in copepod abundance by a factor of 10 in the same region. The increase from 1985–1995 seems to have been related to a parallel increase in the concentrations of both surface nutrients and chlorophyll-*a* during this period (Verheye 2000), while Hutchings et al. (2012) suggest that changes in predation by small pelagic fish could cause the observed changes in biomass. Overall, production in the region seems to be controlled largely by nitrate availability (Chapman and Shannon 1985; Wasmund et al. 2014), even for diatoms, which are often limited by silica, and phytoplankton are able to deplete surface nutrients rapidly as they move downstream in upwelling plumes and the system stabilizes after upwelling winds relax (Olivieri 1983). This leads to rapid turnover of nutrients in the region and the eventual export of carbon to the sediments and the open ocean.

A number of recent investigations focused on the zooplankton distributions in the northern Benguela upwelling system on the Namibian shelf (Bode et al. 2014; Martin et al. 2014), where copepods usually dominate zooplankton communities by 70–85 % (Hansen et al. 2005). From these studies it can be concluded that zooplankton abundance and copepod species composition is highly variable. For instance, Martin et al. (2014) analyzed zooplankton data for four transects at 23° S (Walvis Bay), 20° S (Walvis Ridge), 19° S (Rocky Point) and 17.3° S (Kunene River) in March 2008, December 2009, September 2010 and February 2011. These authors found that both the upwelling index and chlorophyll concentrations peaked in August/September, but that zooplankton abundance peaked in December 2010 and February 2011 with up to 6 months delay. Bode et al. (2014) analyzed approximately monthly zooplankton data for a vertical transect at 20° S perpendicular to the coast for the period from 2005 to 2011. Martin et al. (2014), they found the largest copepod abundance between 30 and 60 nautical miles of the coast, coinciding with the shelf break and upper continental slope.

The data used by Bode et al. (2014) did not show obvious seasonal cycles of copepod abundance however. Instead seasonal patterns in abundance and species composition varied significantly from year to year and total annual abundance responded to variations in upwelling intensity and intrusions of warm and nutrient-depleted waters from the Angola Current associated with changes in the location of the Angola-Benguela frontal zone (usually located between 14° S and 17° S). Apart from pronounced interannual variability associated with such warm-water intrusions, the time series is too short for the identification of longer-term trends.

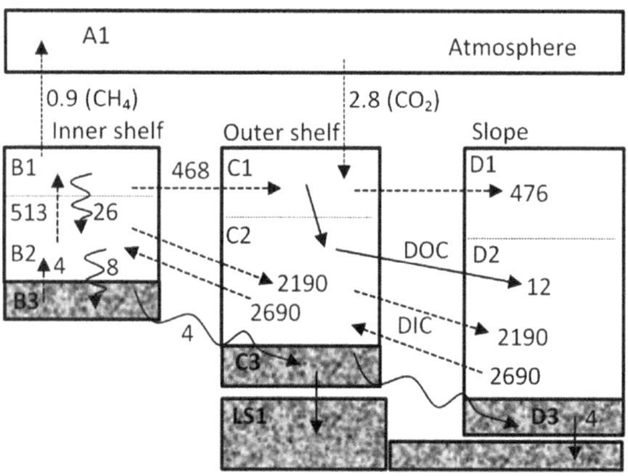

Fig. 7.20 Graphical representation of system fluxes of carbon in a box model using typology of Liu et al. (2000). *A* is atmosphere, *B*, *C* and *D* the *inner shelf*, *outer shelf* and *slope*, divided into surface and deep compartments, while the shaded compartments are the sediments in each zone. Dashed lines denote advection, solid lines diffusion and/or sedimentation. All units are in tonnes C/year. Reproduced from Monteiro (2010)

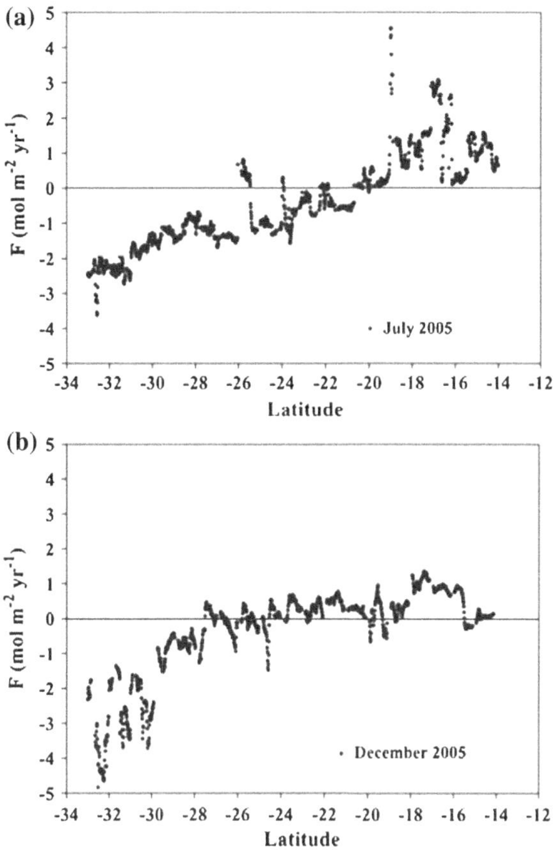

Fig. 7.21 Fluxes of CO_2 ($M/m^2/yr$) computed using weekly wind speed maps at 1° resolution following the Wanninkhof (1992) parameterization for the Benguela for **a** July and **b** December 2005, based on continuous measurement of CO_2 (from Santana Casiano et al. 2009)

7.6.5 Carbon Fluxes

Monteiro (2010) has produced a box model of carbon fluxes within and between the inner shelf, outer shelf, and slope for the whole Benguela region (Fig. 7.20). When split into northern and southern components, it suggests that the northern Benguela is a net sink for carbon, which contributes to the low oxygen found over the wide shelf in that region, but that in the southern region production and consumption are fairly well matched. When we consider exchange with the atmosphere, these figures, together with those of Santana-Casiano et al. (2009) suggest that the region as a whole is a small net source for methane and a small net sink for carbon dioxide on an annual basis (Fig. 7.21).

While the data show considerable variability, with highest levels being related to the positions of active upwelling sites, overall the area north of 20° S appears to be a slight source of carbon dioxide (0.1 to 1.5 $M/m^2/yr$, depending on season), while these results and those of Gregor and Monteiro (2013) suggest that the southern region is a net sink (−0.71 to −2.43 $M/m^2/yr$; Fig. 7.21), although the boundary

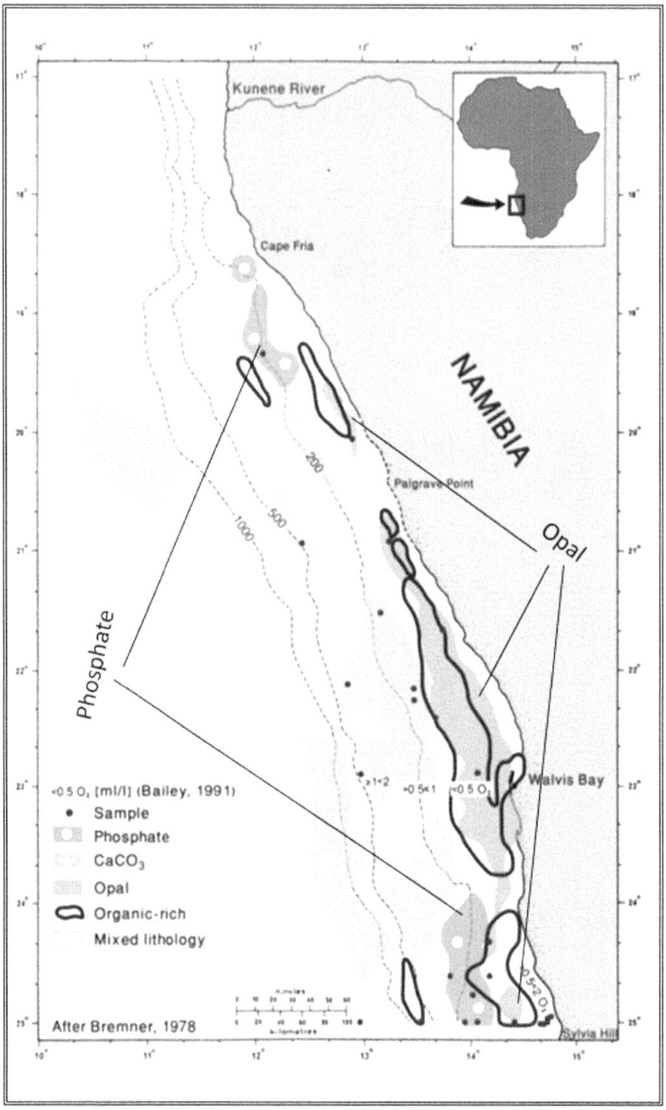

Fig. 7.22 Distribution of sediments with highest weight percentages of opal, organic-matter, $CaCO_3$ and phosphate (50–88 %, 19–24 %, >75 % and >20 %, respectively) on Walvis Shelf between 17° and 25° S based on Bremner (1978). Seafloor oxygen levels measured by Bailey (1991) are superimposed on the key lithologies (from Edelman-Furstenberg 2014)

between source and sink does vary seasonally. It should be noted that physical transport processes play a large role in the air-sea exchange of CO_2, as deep water that upwells contains more dissolved CO_2 than the surface layer, so causes slight outgassing, while surface new production reduces the surface concentration to less than that of the atmosphere within a few days. Similarly, upwelling filaments, which also contain CO_2 concentrations above atmospheric equilibrium, transport both dissolved and particulate carbon across the upwelling front into deeper water.

The idea of the northern region being a net sink for carbon is borne out by sediment composition. Bremner (1978, 1983) and Rogers and Bremner (1991) have reviewed much of the geological work in the region with regard to the distribution and composition of surface sediments. Their maps of surface composition confirm the very high deposition of biological material along the shelf and slope (Fig. 7.22). The outer shelf and upper slope are dominated by carbonate sediments, with a zone of higher benthic faecal pellet concentration along the outer shelf off Namibia north of Lüderitz (see Chapman and Shannon 1985). The highest organic matter concentrations are seen on the inner shelf off Walvis Bay, accompanied by diatomaceous opal but very few benthic faecal pellets, because of the extremely low oxygen environment and free sulphide in the sediment pore water. The presence of sulphides is likely the reason for the high concentrations of copper, nickel, lead and zinc found associated with the organic matter on the inner shelf (Calvert and Price 1970).

Farther south, a lower concentration of organic matter, diluted by inorganic material from the Orange River, is found along the inner and middle shelf between the river mouth and Cape Columbine, and faecal pellets are considerably more abundant since the region supports a rich benthic community. Estimates of depositional fluxes from sediment traps off Namibia (e.g., Giraudeau et al. 2000; Monteiro et al. 2006) and in St. Helena Bay (Bailey 1991), agree with the overall sediment composition data, with a tenfold decrease off Namibia over the outer shelf compared to the inner shelf. In St. Helena Bay, the upwelling plume and the pulsed wind stress current pattern lead to a retention zone that concentrates organic matter here.

7.7 Fisheries

7.7.1 General Description

As with the other upwelling areas, the Benguela has long been a major fishery. The major commercial species include two species of hake, the shallow water *Merluccius capensis* and the deepwater *M. paradoxus* (the most valuable fishery), sole (*Austroglossus* species), horse mackerel (*Trachurus capensis*), and the small pelagic species pilchard (*Sardinops sagax*), anchovy (*Engraulis encrasicolus*) and round herring or red eye (*Etrumeus whiteheadi*). There is also an important fishery for crayfish (*Jasus lalandii*) within the coastal kelp beds, while squid and tuna are also caught offshore. Fishing methods include demersal and midwater trawling, purse seining and long-lining, plus line fishing for a number of other species (Crawford et al. 1987, Griffiths et al. 2004; DAFF 2014). Purse seining is responsible for the largest percentage of the catch We are not considering either

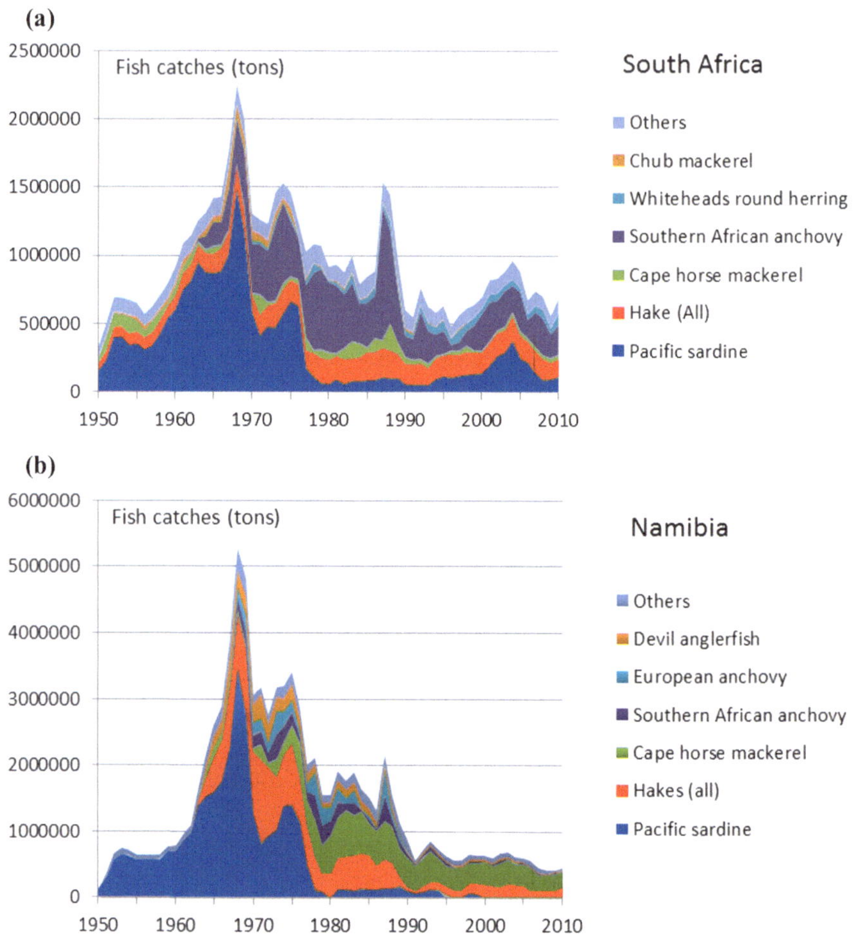

Fig. 7.23 Fisheries catches (tonnes/yr) for **a** South Africa and **b** Namibia for the period 1950–2010 (data from www.seaaroundus.org). Note that the data in this figure include estimates of fish discarded as bycatch as well as landed weights, and so are higher than official figures which only consider landed catch; the angler fish catch statistics in particular seem very high

handlining or the inshore fishery although these are important socio-economically, especially along the west coast of South Africa (Griffiths et al., 2004).

Before November 1977, when South Africa declared a 200-mile exclusive economic zone, and the independence of Namibia in 1990, there was massive uncontrolled fishing by foreign fleets outside the 12-mile territorial waters of both countries. Peak landings in 1968 are estimated at over 3 million tonnes for the region as a whole, after which catches dropped to about 2 million tonnes in the mid-1980s as the pilchard and anchovy catches crashed (total catch estimates are even higher, see Fig. 7.23). Similarly, stocks of hake and horse mackerel off Namibia were targeted by deep-water fleets, particularly from eastern Europe.

Numerous management changes, such as permitted net sizes, quotas (single or mixed species), fleet size and fishing season did little to help the stocks recover, and despite continuing management controls and catch limits the stocks have not yet recovered to anything like their former levels.

As stated above, the Lüderitz region is an environmental barrier to biological transfer on the shelf, with most epipelagic stocks spawning either north or south of the barrier. While genetic studies show no obvious differences in muscle proteins between northern and southern stocks of pilchard (Grant 1985), tagging studies have shown almost no movement between the two regions (Newman 1970). Similarly, the Angola-Benguela front forms a northern barrier, so that Namibian and South African fish stocks are mostly isolated (Sumaila et al. 2004a).

Early figures for catch statistics likely suffer from underreporting of catches, particularly of bycatch, dumping of unwanted catches, or from misidentification of species during times when single species quotas were imposed (Crawford et al. 1987; Boyer and Hampton 2001). This accounts for the discrepancies between the numbers given below in the text and those shown in Fig. 7.23, which tries to account for all fish actually caught during a particular year. Much of the discussion that follows is taken from Crawford et al. (1987), Griffiths et al. (2004) and the most recent annual report of the South African Department of Agriculture, Fisheries and Forestry (DAFF 2014) Griffiths et al. (2004) is a particularly good source on the history of the different fisheries. Data on Namibian fisheries come from various sources, including Boyer and Hampton (2001), Sumaila et al. (2004b) and references therein. The overall trend is a change from long-lived, piscivorous fishes to smaller, shorter-lived planktivorous fish, as well as a decrease in trophic level of fish taken (Willemse and Pauly 2004). The more important species are discussed briefly below.

7.7.2 Hake

Trawling for hake in the Benguela only started off South Africa during the First World War, and annual catches were about 1,000 tonnes by 1918. The fishery expanded slowly until the Second World War and reached 50,000 tonnes by the early 1950s, when the introduction of freezer trawlers led to a surge in catches to 115,000 tonnes by 1955. In the 1960s, deepwater fleets from Bulgaria, East Germany, Japan, Poland, Romania, Russia and Spain moved into deep waters off both South Africa and Namibia, leading to large increases in catches, which included many small sub-adult fish, and a decline in catch rate. The catch peaked in 1972 at 1.1 million tonnes, of which the South African catch was about 300,000 tonnes (today it is about 130,000 tonnes), and then declined until 1977 when foreign fleets were banned. Since then, South Africa has established annual quotas, and the catch remained stable at about 150,000 tonnes from 1977 until 2002, after which it declined, probably because of a drop in biomass of the deepwater species after poor recruitment. Off Namibia, catches stayed at about 500–600,000 tonnes until independence, after which stricter quotas were introduced and the fishery has since been divided between trawling, longline and handline operators, with the aim of encouraging non-white operators. Catches are reasonably stable at about 200,000 tonnes, with much of it being made by Spanish trawling companies.

It is thought that there are four main hake stocks, off Namibia, the Orange River, the western Cape, and the south coast. A third species, the Benguela hake (*M. polli*), is found off Angola and overlaps the ranges of the other two species in northern Namibia. Hake diet consists of euphausiids and gobies for *M. capensis*, with the amount of other fish increasing as the hake get bigger, while *M. paradoxus* also eats decapods and cephalopods. Both species are cannibalistic. They are preyed on by fur seals and other fish (kingklip, monkfish). Hake spawn throughout the year along the shelf edge from the western Agulhas Bank to the Cunene River, apart from near the Orange River-Lüderitz upwelling zone, and recruit inshore all along the coast.

7.7.3 Sole

A. microlepis is found north of Port Nolloth as far as the Cunene River, but *A. pectoralis* occurs along the South African south coast. The fishery started in 1930s off Namibia and off the Orange River in the 1950s. While the south coast stock has been reasonably stable with catches varying generally between 600–1000 tonnes from 1940–2000, the catch dropped after this largely because of a reduction in effort. The catch of the Namibian, west coast stock, however, has been very variable at between 300–3000 tonnes. Sole need muddy bottoms and so are found only in localized regions. They spawn throughout the year but mainly in October/November. An interesting change seen off Namibia is that the adult stock seems to have moved offshore during the 1970s so that while before then they were caught generally in shallow water of less than 120 m depth, now all are caught in deep water (>275 m).

7.7.4 Horse Mackerel

The horse mackerel, *T. capensis*, is a major component of the midwater trawl fishery and is also caught by purse seiners. There are three main stocks, off Namibia, off the western Cape and east of Cape Agulhas, although the large stocks of adults that were found formerly off the west coast of South Africa during the 1950s had disappeared by the early 1970s. Much of the horse mackerel catch in the early years was based on a particularly strong set of year classes in 1946–1948 (Geldenhuys 1973). Off Namibia, catches climbed from 46,000 tonnes in 1969 to 400–500,000 tonnes in the 1970s, and 600,000 tonnes from 1982, but have been maintained at half this amount since the country declared independence. Adult horse mackerel eat copepods, euphausiids and smaller fish (lanternfish and lightfish) and in turn are eaten by hake, fur seals and seabirds. Off Namibia, they spawn during the summer with the smallest fish being found in the north of the country and these then move south towards Walvis Bay. There is also a recruiting area in St. Helena Bay, following a spawning maximum in spring. Juveniles are found inshore from False Bay to the Orange River and from Walvis Bay to the Cunene River, but there is substantial movement both alongshore and cross-shelf as the fish mature.

7.7.5 Tuna

Several species of tuna were caught by sport fishermen during the 1940s/50s but there was no commercial exploitation until 1960 when longlining started off the western Cape. The industry collapsed in 1964 because of the poor quality of the fish landed, but started again in the late 1970s—the 1979 catch was >6,000 tonnes after which it declined again. However, Japanese, Taiwanese and Korean boats were catching albacore (*Thunnus alalunga*) and yellowfin (*T. albacares*) under bilateral agreements from the 1980s until the 2000s (the annual catch from Benguela waters was ~4,000–5,000 tonnes/yr). There is also a pole-line industry for the same species. Southern Bluefin tuna (*T. maccoyii*) were abundant during the 1960s, but heavy exploitation throughout their range has led to very low quotas managed by the International Commission for the Conservation of Atlantic Tuna (ICCAT). Stock assessments for all tuna species in the Benguela suggest they are over-exploited, with the possible exception of bigeye tuna (*T. obesus*).

7.7.6 Small Pelagic Species

The three small pelagics, sardine, anchovy, and round herring are discussed together because of their tendency to form mixed shoals that are exploited by the purse-seine fleets. Although pilchard were formerly the mainstay of both the Namibian and South African fisheries, the stocks in both regions had collapsed by the late 1970s because of over fishing and environmental changes. Such changes are known to affect the relative abundance of sardine and anchovy stocks both locally and worldwide over decadal timescales (see Chap. 3).

Sardines were first canned in the western Cape in 1935, but commercial operations only started in 1943 during the Second World War when canned food was needed, with pilchard and horse mackerel being targeted in the area between Lambert's Bay (north of St Helena Bay) and the eastern edge of False Bay. This is still the main area for the South African fishery. These species dominated catches until the 1960s, with anchovy becoming the dominant species after 1966, with a peak catch of 350,000 tonnes in 1974. The landed sardine catch peaked at 400,000 tonnes off South Africa in the early 1960s, based on strong year classes during the late 1950s. Off Namibia, the highest catch was 1.4 million tonnes in 1968, which collapsed to 11,000 tonnes by 1980 as fisheries quotas were relaxed during the 1970s. Warm water intrusions in 1963 and 1984 moved the Namibian purse seine fishery south from Walvis Bay towards Lüderitz, but yields declined. While the South African fishery for both sardine and anchovy recovered from lows during the mid-1990s, with booms in anchovy catches in particular in 1987–88 (>500,000 tonnes) and for sardine between 2001 and 2006 (<300,000 tonnes; Hutchings et al. 2012) this was not the case in Namibia, and catches of both species have been very low for many years. The decrease in South African sardine catches since 2004 is blamed on several years of low recruitment, while the poor recruitment in 1989 is blamed on "predation" of the recruits by an Agulhas ring that passed close to the shelf edge and removed a large volume of shelf water in a filament (Duncombe Rae et al. 1992).

Both sardine and anchovy have their spawning grounds on the western Agulhas Bank, and after hatching, the larvae are transported north around Cape Point and Cape Columbine in the offshore jet described in Sect. 7.5. The juveniles recruit along the west coast of South Africa from March onwards, with the major recruitment area being St. Helena Bay, while adults migrate south again to spawn during the spring. The reduction in size of small pelagics caught off St Helena Bay since the 1950s is likely due to improved fishing apparatus and increased fishing pressure (Hutchings et al. 2012). During the 1950s, the industry was catching 5–6 year old fish, but now they catch predominantly juveniles or 1–2 year olds.

In Namibia, recruitment occurs off Walvis Bay between March and August for sardine, after which they move north, sometimes to southern Angola, but the largest fish occur at the southern end of their range, where they go to spawn, as well as off Dune Point (19°–21° S). Anchovy in this region tend to spawn slightly later, between May and August. Both sardine and anchovy are serial spawners, but egg production varies over the year because of an overall reduction of the fish's energy reserves as the season progresses, changes in internal energy partitioning, and seasonal changes in stock age structure. Juveniles of both species eat calanoid copepods, but adults eat mostly phytoplankton. James (1987), however, states that their diet also includes zooplankton. Their predators are other fish, fur seals and seabirds.

While juvenile round herring can be caught with sardine and anchovy, mainly in St. Helena Bay, adult round herring are mainly taken in directed fishing operations between January and May. All three species seem to have moved slowly south and east since the mid-1990s (van der Lingen et al. 2005), but this trend seems to have reversed for round herring and pilchard at least. What causes such movements is presently unknown.

7.7.7 Rock Lobster

Fishing for rock lobster (*Jasus lelandii*), with hoop nets in water depths of up to about 25 m and with traps in deeper water of up to about 100 m, began off the west coast of South Africa in the late 1800s and had reached an annual catch of about 8,000 tonnes by 1918. The peak catch in the early 1950s was about 18,000 tonnes/yr. The catch dropped below 10,000 tonnes/yr during the 1960s and has decreased steadily since then, so that it is now only around 2,000 tonnes/yr (DAFF 2014). While they occupy inshore waters from 25° S all the way to Cape Agulhas, their distribution north of 31° 50′ S is determined by the presence of low oxygen water, so they occur close inshore in the northern areas (Pollock and Shannon 1987). Recently, the population has been moving east and the catches along the west coast have reduced considerably. The causes of this movement and of the drop in catches are uncertain but may be related to more frequent occurrences of low oxygen water (a number of mass strandings of lobsters occurred between 1990 and 2005), reduced growth rates, incursions of warm water, and to over-exploitation of the resource, including poaching (Pollock et al. 1997; Blamey et al. 2012, 2015).

Size limits on carapace length (89 mm) were first introduced as early as 1933. While this is still the limit in the southern Cape, it has been reduced to 75 mm in South Africa and 65 mm in Namibia, because of changes in growth rate in the north. Rock lobster eat mainly ribbed mussels (*Aulocomya ater*), but can eat echinoderms and gastropods as well, and high *Jasus* growth rates are encouraged in areas where small mussels, with their high productivity, are found. Mature females moult May-June, carry berry until spring and spawn in October/November. As the larvae travel around the South Atlantic gyre before settling, it is thought that spawning is synchronized with the start of strong upwelling that allows the larvae to move northwards and offshore (Shannon et al. 1989; Pollock 1990).

7.7.8 Fish Stock Variability and Regime Shifts

While it might be thought that over-exploitation of the fish stocks has had the largest effect on fisheries in the Benguela (it has certainly meant that the ability for stocks to recover has been reduced), it is clear from old records such as guano collections that the stocks were highly variable even before commercial fishing started (Fig. 7.24), as seen in other systems (e.g., Blaxter and Hunter 1982; Schwartzlose et al. 1999). Similarly, changes in population structure that lead to the dominance of different species at different times are also well known. This means that despite recent advances in the statistics of fish stock assessment, predicting population growth and availability is difficult, and while each year's estimate is tuned based on the most recent data, this does not seem to improve future predictions (Field and Shillington 2005). This is true both for single species stocks and for the more recent multi-species models in all upwelling systems (e.g., Cury et al. 2000; Shannon et al. 2000, 2008), because of the many variables that can affect egg survival and juvenile recruitment (Hutchings 1992).

We do not know, for instance, why the large increases in sardine and anchovy stocks in the southern part of the Benguela system around 2000 were not repeated in the north (Field and Shillington 2005), although apparently there is a positive

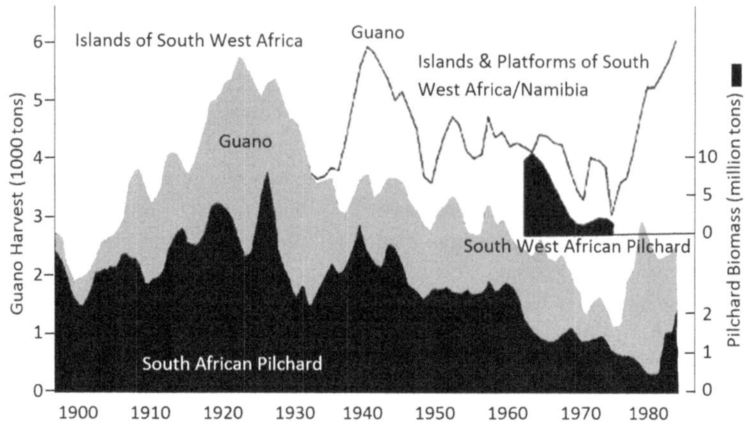

Fig. 7.24 Trends in guano production at islands and platforms off South Africa and Namibia and in the biomass of the two sardine stocks. Redrawn after Crawford et al. (1987)

relationship between Benguela Niños and hake recruitment off Namibia (Voges et al. 2002). One can argue whether a decadal variation can be extracted from a 14 year time-series with a sensible correlation, but the Voges model, which uses two upwelling indices and assumes the fish need reduced upwelling for spawning and increased upwelling a year later when the juveniles are old enough to use vertical migration to prevent being swept offshore, did predict three years of poor recruitment well.

Cumulative catch rates, as shown in Fig. 7.23, can be misleading since the graph reveals only the variations of the dominant species caught and of the total catch. Variations in other fish species remain uncertain from this graphical display method. If we instead produce time series of individual commercial groups (Fig. 7.25), we can see the transient dominance of catches of herring-like fish (sardines) until 1975 and a transient regime shift towards higher catches of anchovies and cod-like fish from the early 1960s till the late 1990s.

The increasing availability of time-series data relating to biological organisms has led to much interest in the idea of regime shifts, when a stable ecosystem state changes in a short time frame (essentially a step function) into a different state that may be as stable or more variable. These are often driven by physical forcing functions, such as changes in wind strength or temperature (Polovina 2005), although human induced effects, such as over fishing, can also cause regime shifts (Scheffer and Carpenter 2003). Blamey et al. (2012, 2015) discuss the incidence of regime shifts in the Benguela and have identified a number of changes since the 1950s. The eastward movement of rock lobsters (*J. lelandii*) during the early 1990s from the Cape west coast to the coastline along the western Agulhas Bank has been mentioned previously (Sect. 7.7.7); apart from the possibilities suggested earlier, Blamey et al. (2012) have suggested additional reasons such as an increase in upwelling variability along the west coast at this time and an increase in upwelling

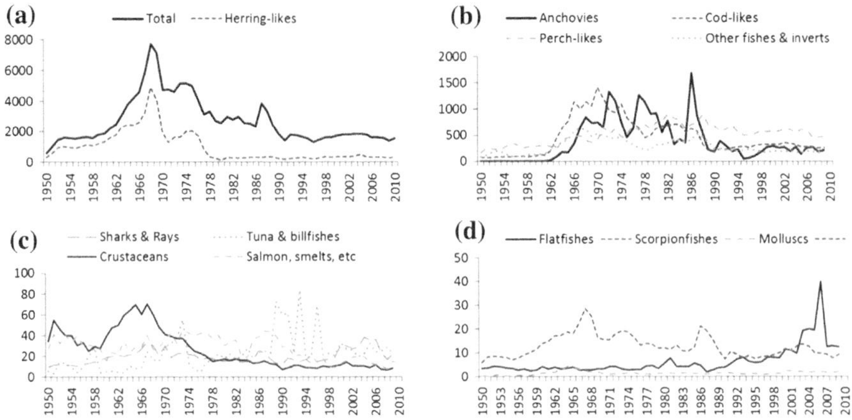

Fig. 7.25 Catches (in units of 1000 tonnes per annum) in the Benguela Current large marine ecosystem for 1950–2010. Data from http://www.seaaroundus.org/. Note that the data in this figure include estimates of fish discarded as bycatch as well as landed weights, and so are higher than officially reported figures which only consider landed catch

strength in about 1995. While the change in location of rock lobsters was an obvious signal, this has led to adverse socio-economic effects in the west coast lobster fishery, as well as to other changes in the inshore ecosystem east of False Bay, such as a decline in sea urchins (which are eaten by *Jasus*) and the almost complete disappearance of abalone (*Haliotis midae*) as juvenile abalone use sea urchins for protection from predators. As abalone also previously supported a lucrative fishing industry, their disappearance is another "knock on" effect of the regime change.

Other examples of changes that can be classified as regime shifts in the southern Benguela include an increase in harmful algal blooms since the 1960s (Stephen and Hockey 2007), an eastward movement of both sardine and anchovy after 1991 (van der Lingen et al. 2005; Blamey et al. 2012), changes in the numbers of seabirds, which tend to follow fish stocks (see Sect. 7.7.9), and an eastward movement of kelp beds (*Ecklonia maxima*). Apart from such changes in species abundance, other indicators, such as the size of organisms taken during stock assessment cruises, also show significant changes that suggest the ecosystem has deteriorated since the 1980s (Blamey et al. 2015).

While monitoring such shifts is clearly possible from changes in populations or organism size ranges, it is harder to identify the drivers of these changes. Overfishing is certainly one potential cause, but was there also a physical one? Wind records show a decrease in upwelling-favourable winds during the 1980s, followed by an increase in upwelling strength during the 1990s, and the eastward movement of cold water species, as well as smaller declines in warm water species, suggest that temperature is a major driver. Satellite SST measurements agree with this scenario, although, somewhat surprisingly, it appears to apply only to the region south of Cape Columbine, north of which warming has occurred, especially off Namibia and Angola (Blamey et al. 2015). Hutchings et al. (2012) reported on 50 years of sampling in St Helena Bay, and while they could show there have been decadal shifts in a number of parameters during this time, such as the local water chemistry, zooplankton biomass, and fish abundance, they could not identify any one change that had lasting effects on the biology.

It seems quite likely that both physics and overfishing can force observed changes, with the balance between them varying from year to year, but the present time-series data for both physical and biological parameters suffer from data gaps and changes in measurement methodology and scale that mean it is difficult to draw really firm conclusions, especially given the inherent small-scale variability in the system in both time and space. This applies equally in other systems and not just the Benguela.

7.7.9 Marine Birds and Mammals

The Benguela supports 12 breeding seabirds, which is fewer than the other three major eastern boundary upwelling systems, and 36 regular non-breeding visitors (Berruti et al. 1989). The three most abundant species are the Cape gannet (*Morus capensis*), African penguin (*Spheniscus demersus*), also known as the jackass or black-footed penguin and Cape cormorant (*Phalacrocorax capensis*). All three species rely on the sardine and anchovy populations for food, and their breeding success is largely dependent on changes in the biomass of these two species. However, all three also eat juvenile horse mackerel, while the gannets eat saury (*Scomberesox saurus*) and the penguins and cormorants eat the pelagic goby (*Sufflogobius bibarbatus*). Between them, it is estimated that these three species account for up to about 100,000 tonnes of commercial fish each year, almost all of which is caught near the surface during the day (Berruti et al. 1989).

As shown by the guano collection data (Fig. 7.22), the seabird populations were highly variable even before commercial fishing started, but populations have decreased considerably since the 1950s. The gannet populations at the three breeding colonies off Namibia (Mercury Island, Ichaboe Island, and Possession Island) decreased from about 55,000 to about 6,000 between 1956 and 1985, while the two populations in the southern Benguela saw modest increases from about 11,000 to 16,000 birds over the same time period. This seems to be related directly to sardine availability, as the gannet colony in Algoa Bay, in the central Agulhas Bank, increased considerably over the same time period as the fish stocks moved eastwards (Batchelor and Ross 1984), but may also indicate birds moving from one colony to another (Crawford et al. 1983; Berruti 1985). West coast gannets continued to decrease after 2003, although colonies on the south coast continued to increase, with almost 100,000 pairs in 2005/6 (Crawford et al. 2008; 2011). Population counts, however, can be very variable, and have not been made every year so apparent trends may be biased, although penguin counts of breeding pairs have been made every year since 1991.

According to the records, Cape cormorants have actually increased over the thirty years from 1956 from about 65,000 to 190,000 in the southern Benguela and from 25,000 to 780,000 off Namibia (Berruti et al. 1989). However, the accuracy of counts, particularly on guano platforms off Namibia is unknown. The large sardine classes in the late 1960s could certainly have affected the population; Berruti et al. (1989) state:

> Because Cape cormorants are mobile and have a high rate of potential increase, they can take advantage of short-term regional increases in prey abundance.... Conversely, the species appears to be vulnerable to temporary food shortages, leading to mass desertion of nests and death of adults.

Such a mass desertion occurred in 1985–86 in all west coast colonies. More recently, numbers at the northern colonies along the South African west cast have decreased, but those at the southern sites have increased (Crawford et al. 2008).

During the 1956–1986 time period, penguins off Namibia decreased from 150,000 to 80,000, and off South Africa from 160,000 to about 30,000. Some of the reduction was caused by increases in the numbers of fur seals (*Arctocephalus pusillus*), as well as egg harvesting (since banned) and competition from the fisheries, but the most likely reason was the decline in sardine stocks (Crawford et al. 1987). Since then, the population has continued to decrease, with South African breeding pairs declining from about 56,000 in 2001 to 21,000 pairs in 2009, together with 5,000 pairs in Namibia, and the species is now classified as endangered (Crawford et al. 2011). The more recent decline is attributed to the general southeastward movement of their prey after 2000, leading to a mismatch between prey and the west coast breeding grounds. Because of their limited foraging range during the breeding season, consideration is being given to closing areas around their breeding grounds to fishing. There is also pressure from fur seals, which both compete for breeding space with penguins and other seabirds and prey on adults and chicks (Crawford et al. 2001, 2011).

Changes in the trophic structure of the Benguela have likely affected how energy flows through the bird populations (Crawford et al. 1987). The collapse of the sardine stocks off Namibia in 1970 led to an expansion of the pelagic goby stocks, which were exploited by penguins and cormorants at the expense of Cape gannets. Birds are also known to be affected by large-scale or global events such as ENSO (Crawford et al. 2011). The warm event of 1982/83 in the Benguela similarly had a major effect on breeding success, with both penguins and Cape cormorants producing far fewer young than usual, although gannets apparently switched their diet to fishing industry offal and were much less affected (Duffy et al. 1984). More recently, there seems to have been a change in trophic levels, particularly in the northern Benguela, with jellyfish becoming more prevalent (Shannon et al. 2008). The effect of the increased coelenterate population, which feed on larvae and juvenile fish, on the populations of higher trophic levels, is unknown. Otherwise, the main problem for seabirds in the Benguela, especially penguins, comes from oil spills, with 20,000 being oiled during a spill in 2000 (Berruti et al. 1989; Crawford et al. 2011).

Cape fur seals (*Arctocephalus pusillus pusillus*) have been hunted along the Benguela coastline since before the founding of Cape Town in 1642, with the first harvest recorded in 1610. Indiscriminate killing during the17th–19th centuries reduced the population very considerably until the species was given legal protection in 1893, and at least 23 colonies died out during this period (David et al. 2003). The industry persisted until 1990, under a controlled license system, when it was banned in South Africa after more than another 2.5 million seals had been killed. The controlled harvesting, however, did allow the population to recover, and by 2000 it was estimated at between 1.5 and 2 million animals (Hutchings et al. 2009a). Seals eat fish and squid, mostly the five main commercial species, and are therefore direct competitors with birds and humans for the food resource. Thus, just as bird populations wax and wane with the biomass of the commercial fishery, so do seal populations. Voges et al. (2002), for instance, looked at the effect of Benguela Niños on seal colonies in Namibia and found a large intermittent drop following the 1994 event. A similar mass mortality occurred in 2001 (Hutchings et al. 2009a).

While the seals feed primarily on fish, they have been shown to attack and kill seabirds, including gannets, Cape cormorants and penguins, all of which face conservation pressure (David et al. 2003). Such attacks have likely led to the elimination of a number of smaller bird colonies, and expanding seal numbers can also lead to loss of nesting sites. For example, counts showed that seals killed 6,000 gannets around Malgas Island (near Cape Columbine) during the 2000/2001 breeding season (Makhado et al. 2006). The main culprits are juvenile or sub-adult males, and the South African Government initiated a selective culling program in 1993 aimed at taking out those individual seals responsible. Additional details on birds and mammals in the Benguela, with emphasis on historical changes, are given in Griffiths et al. (2004).

7.8 Climate Change and the Benguela

Given the generally highly variable nature of upwelling systems, trying to sort out the likely effects of climate change on systems that are already perturbed by humans is not easy. In a discussion on the fish stocks of northern Namibia, where annual upwelling is more consistent than in the southern region, Shannon et al. (1987) stated:

> In the area of Northern Namibia, conditions are favourable for spawning during the first quarter when wind speeds are lowest (reduced turbulence), when temperatures are optimum, when stratification is strongest, and when there is still sufficient upwelling of nutrients to support a high plankton biomass.

Thus environmental changes, such as a slackening of the southerly wind regime, might be expected to have less effect north of Lüderitz, and might even increase fish recruitment there. The southern Benguela is more seasonal, so the effects of changes to the wind field on upwelling and production will depend on whether these occur in summer or winter. Strengthening of the pycnocline, which makes the system more stable, and which can be induced either by less wind or by higher temperatures, could similarly increase production, although if the pycnocline strengthens to the extent that newly-upwelled water can no longer reach the surface, then productivity will drop.

While global SST temperature trends have increased since 1985, this has not been the case for the Benguela for the period 1998–2007 (Demarq 2009). While SST in the southern Benguela did increase by about 0.1 °C, in the north it decreased by the same amount. Chlorophyll concentrations for the same decade estimated from SeaWifs data show a general decrease in the stratified ocean, but a general increase in upwelling systems, apart from the Benguela where the trend has been the reverse of the SST, an increase in the north and a decrease in the south, but both changes are less than 0.5 mg chlorophyll-*a*/m^3, so the significance is unclear. In the southern Benguela there was an increase along the inner shelf north of 32° S, but a decrease offshore beyond the shelf break. Changes in the influx of warm water from either the equator or the Agulhas Currents are also possible, depending on how the global circulation and the equatorial wind systems vary. Increases in the warm water flux could reduce the latitudinal extent of the upwelling system, but at present the likelihood of long-term changes in upwelling in the Benguela from this source remains unclear.

From his study of all four systems, Demarq (2009) concluded that changes in SST *per se* had only minor effects on productivity, and that the equatorward wind strength was more important. This, and similar work by Santos et al. (2012), support Bakun's (1990) suggestion that global warming will increase the land-sea temperature difference and increase the alongshore wind strength, although a modeling study by Tim et al. (2015) disputes this as they did not find the expected relationship between increased wind strength and changes in the offshore surface pressure gradient over a 60-year period. However, other modeling studies (e.g., Wang et al. 2015) suggest a trend towards increasing length of the upwelling season by up to 2 days per decade between 1950 and 2099 for the region south of 31° S, as well as an increase in upwelling intensity at all latitudes, leading to less heterogeneity between the northern and southern regions. Despite the advent of satellite-derived winds from Quikscat, the required finer-scale data are not presently available and our ability to model potential changes in upwelling strength is limited by the poor resolution of the large-scale wind field near the coast, which is known to drop compared to wind scales offshore and be affected by orography (Desbiolles et al. 2016). Global models cannot resolve the winds to better than 1°–2° of longitude, and even regional shelf models, which can handle much finer scales ($\sim$a few km), contain a "blind zone" at the coastal boundary. As an example, Desbiolles et al. (2016) compared the effects of 0.25° and 0.5° winds in a 1/12th degree ROMS model of the Benguela and showed that the wind stress and Ekman transport were stronger in the 0.5° experiment, producing a stronger southward-flowing jet, and that differences in SST between the two model runs were >2 °C after 2 months. As SST exerts a feedback mechanism on local wind speed, there is a need for better wind fields at the coast if we are to predict the effects of climate variability on upwelling.

7.9 Summary

As the only upwelling system with warm water at both the equatorward and poleward ends, the Benguela has links to the equatorial regions in both the Atlantic and Indian Oceans. In the north, south of the Angola-Benguela frontal region, it receives water from the large oxygen minimum zone near the cyclonic Angola Dome which contributes to the oxygen minimum found along the Namibian shelf. In the south, large anticyclonic rings from the Agulhas Current push past the narrow shelf near Cape Town, providing heat and salt to the southeastern Cape Basin as well as removing shelf water and any organisms contained in it.

Although the whole system is driven by the variability of the South Atlantic high pressure system, it is divided into two by the massive upwelling centre at Lüderitz (27° S), which forms a biological barrier for many species and leads to differences in chemistry and biology to north and south. Walvis Bay, which at 23° S is downstream of Lüderitz, is the most productive centre in the Benguela, and is associated with a major zone of anoxia. Upwelling at Lüderitz and Cape Frio is continuous throughout the year, but in the south there are several smaller upwelling centres separated by shadow zones where biological recruitment can take place. Upwelling here is seasonal, peaking in spring and summer with multiple upwelling events during the season and becoming downwelling favourable during the winter months.

Like other eastern boundary upwelling systems, the Benguela fisheries have alternated between favouring sardines or anchovy on decadal scales, but this does not seem to be related to specific changes in physical or chemical forcing, and overfishing has had more of an impact on the biomass. Despite this, the past 20 years have seen a general eastward movement of pelagic fish from west of Cape Point to the Cape Agulhas region, possibly associated with a cooling of water south of Cape Columbine. While this seems to be a clear case of a "regime shift," it is unclear if it is permanent or if the system will tilt back again within a few years. Similarly, there is little agreement on how the system as a whole will respond to continued global warming.

References

Agenbag, J.J., and L.V. Shannon. 1988. A suggested physical explanation for the existence of abiological boundary at 24° 30′ S in the Benguela system. *South African Journal of Marine Science* 6: 119–132.

Alemaw, B.F., and T.R. Chaoka. 2002. Trends in the flow regime of the southern African rivers as visualized from rescaled adjusted partial sums (RAPS). *African Journal of Science and Technology* 3: 70–79. doi:10.4314/ajst.v3i1.15288.

Alemaw, B.F., and T.R. Chaoka. 2006. The 1950-1998 warm ENSO events and regional implications to river flow variability in southern Africa. *Water SA* 32: 459–463.

Andrews, W.R.H., and D.L. Cram. 1969. Combined aerial and shipboard upwelling study in the Benguela Current. *Nature* 224: 902–904.

Andrews, W.R.H., and L. Hutchings. 1980. Upwelling in the southern Benguela Current. *Progress in Oceanography* 9: 1–81.

Bailey, G.W. 1979. *Physical and chemical aspects of the Benguela current in the Lüderitz region, M.Sc. Thesis*. Cape Town: University of Cape Town. 225 pp.

Bailey, G.W., and P. Chapman. 1985. Nutrient status in the St. Helena Bay region in February 1979. In *South African Ocean Colour and Upwelling Experiment*, ed. L.V. Shannon, 125–145. South Africa: Sea Fisheries Research Institute.

Bailey, G.W. 1991. Organic carbon flux and development of oxygen deficiency on the modern Benguela continental shelf south of 22° S: Spatial and temporal variability. In *Modern and ancient continental shelf anoxia*, ed. R.V. Tyson, and T.H. Pearson, 171–183. London: Geological Society.

Bailey, G.W., and P. Chapman. 1991. Short-term variability during an anchor station study in the southern Benguela upwelling system: Chemical and physical oceanography. *Progress in Oceanography* 28: 9–37.

Bakun, A. 1990. Global climate change and intensification of coastal ocean upwelling. *Science* 247: 198–201.

Bakun, A., and C.S. Nelson. 1991. The seasonal cycle of wind stress curl in subtropical eastern boundary current regions. *Journal of Physical Oceanography* 21: 1815–1834.

Bang, N.D. 1971. The southern Benguela current region in February, 1966: Part II. Bathythermography and air-sea interactions. *Deep-Sea Research* 18: 209–224.

Bang, N.D., and W.R.H. Andrews. 1984. Direct current measurements of a shelf-edge frontal jet in the southern Benguela system. *Journal of Marine Research* 32: 405–417.

Barange, M., and S.C. Pillar. 1992. Cross-shelf circulation, zonation and maintenance mechanisms of the euphausiids *Nyctiphanes capensis* and *Euphausia hanseni* (Euphausiacea) in the northern Benguela upwelling system. *Continental Shelf Research* 12: 1027–1042.

Barber, R.T., and R.L. Smith. 1981. Coastal upwelling ecosystems. In *Analysis of marine ecosystems*, ed. A.R. Longhurst, 31–68. London: Academic Press.

Barnes, J.I., and M. Alberts. 2008. Sustainable natural resource use on the coast of Namibia. Directorate of Environmental Affairs Discussion paper, #78, 37 pp. http://www.dlist-benguela.org/sites/default/files/doclib/nam_sus%20coast%20resource%20use.pdf. Accessed 1 Apr 2016.

Batchelor, A.L., and G.J.B. Ross. 1984. The diet and implications of dietary change of Cape gannets on Bird Island, Algoa Bay. *Ostrich* 55: 45–63.

Benavides, M., Y. Santana-Falcon, N. Wasmund and J. Aristegui. 2014. Microbial uptake and regeneration of inorganic nitrogen off the coastal Namibian upwelling system. *Journal of Marine Systems*, 140 B: 123–129.

Berruti, A. 1985. Cape Gannets *Sula capensis* ashore at Dyer and Robben Islands, southwestern Cape, South Africa. *Cormorant* 13: 71–72.

Berruti, A., N.J. Adams, and S. Jackson. 1989. The Benguela ecosystem. Part VI. Seabirds. *Oceanography and Marine Biology Annual Reviews* 27: 273–335.

Best, P.B. 1987. Estimates of the landed catch of right (and other whalebone) whales in the American fishery, 1805-1909. Fishery Bulletin, 85: 403–418.

Best, P.B. 1994. A review of the catch statistics for modern whaling in southern Africa, 1908-1930. Report of the International Whaling Commission, 44: 467–485.

Blamey, L.K., J.A.E. Howard, J. Agenbag, and A. Jarre. 2012. Regime-shifts in the southern Benguela shelf and inshore region. *Progress in Oceanography* 106: 80–95.

Blamey, L.K., L.J. Shannon, J.J. Bolton, R.J.M. Crawford, F. Dufois, H. Evers-King, C.L. Griffiths, L. Hutchings, A. Jarre, M. Rouault, K.E. Watermeyer, and H. Winker. 2015. Ecosystem changes in the southern Benguela and the underlying processes. *Journal of Marine Systems* 144: 9–29.

Blanke, B., C. Roy, P. Penven, S. Speich, J. McWilliams and G. Nelson. 2002. Linking wind and interannual upwelling variability in a regional model of the southern Benguela. *Geophysical Research Letters*, 29. doi:10.1029/2002GL015718.

Blaxter, J.H.S., and J.R. Hunter. 1982. Biology of the clupeoid fishes. *Advances in Marine Biology* 20: 1–223.

Bode, M., A. Kreiner, A.K. van der Plas, D.C. Louw, R. Horaeb, et al. 2014. Spatio-temporal variability of copepod abundance along the 20° S monitoring transect in the northern Benguela upwelling system from 2005 to 2011. *PLoS ONE* 9(5): e97738. doi:10.1371/journal.pone.0097738.

Botha, L. 1986. Reproduction, sex ratio and rate of natural mortality of Cape hakes *Merluccius capensis* Cast. and *M. paradoxus* Franca in the Cape of Good Hope area. *South African Journal of Marine Science* 4: 23–35.

Boyd, A.J. 1983. Intensive study of the currents, winds and hydrology at a coastal site off central South West Africa, June/July 1978. *Investigational Report, Sea Fisheries Research Institute, South Africa*, 126, 47 pp.

Boyd, A.J. 1987. *The Oceanography of the Namibian Shelf.* Ph.D. thesis, University of Cape Town, 190 pp.

Boyd, A.J., and J.J. Agenbag. 1984a. Seasonal trends in the longshore distribution of surface temperatures off southwestern Africa 18–34° S, and their relation to subsurface conditions and currents in the area 21–24° S. *International Symposium on the Most Important Upwelling Areas off Western Africa (Cape Blanco and Benguela),* vol. 1. 1985.

Boyd, A.J., and J.J. Agenbag. 1984b. Seasonal temperature and salinity trends off central Namibia between 1978 and 1983 with particular reference to the cool winter of 1982. *South African Journal of Science* 80: 77–79.

Boyd, A.J., B.B.S. Tromp, and D.A. Horstman. 1985. The hydrology of the South African south-western coast between Cape Point and Danger Point on 1975. *South African Journal of Marine Science* 3: 145–168.

Boyd, A.J., J. Salat and M. Maso. 1987. The seasonal intrusion of relatively saline water on the shelf off northern and central Namibia. In *The Benguela and Comparable Ecosystems*, ed. A.I. L. Payne, J.A. Gulland, and K.H. Brink, *South African Journal of Marine Science*, 5: 107–120.

Boyer, D.C., and I. Hampton. 2001. An overview of the living marine resources of Namibia. *South African Journal of Marine Science* 23: 5–35.

Branch, G.M., and C.L. Griffiths. 1988. The Benguela ecosystem. Part V. The Coastal Zone. *Oceanography and Marine Biology Annual Reviews* 26: 395–486.

Bremner, J.M. 1978. Sediments on the continental margin off South West Africa between latitudes 17° and 25° S. (PhD Thesis) Geology Department University of Cape Town (300 pp.).

Bremner, J.M. 1983. Biogenic sediments on the South West African (Namibian) continental margin. In *Coastal Upwelling: Its Sedimentary Record. Part B. Sedimentary Records of Ancient Coastal Upwelling*, ed. J. Thiede and E. Suess, 73–103. New York: Plenum Press.

Brown, P.C., and L. Hutchings. 1987. The development and decline of phytoplankton blooms in the southern Benguela upwelling system. 1. Drogue movements, hydrography and bloom development. *South African Journal of Marine Science* 5: 357–391.

Brown, P.C., S.J. Painting, and K.L. Cochrane. 1991. Estimates of phytoplankton and bacterial biomass and production in the northern and southern Benguela ecosystems. *South African Journal of Marine Science* 11: 537–564.

Brundrit, G.B. 1995. Trends of southern African sea level: Statistical analysis and interpretation. *South African Journal of Marine Science* 16: 9–17.

Brundrit, G.B., B. de Cuevas and A.M. Shipley. 1984. Significant sea-level variations along the west coast of Southern Africa 1979–83. *South African Journal of Science*, 80: 80–82.

Bubnov, V.A. 1972. Structure and characteristics of oxygen minimum layer in southeastern Atlantic. *Oceanology* 12: 193–201.

Buys, M.E.L. 1959. Hydrographic environment and the commercial catches, 1950–57. *Division of Fisheries, South Africa, Investigational Report,* 37, 176 pp.

Calvert, S.E., and N.B. Price. 1970. Minor metal contents of recent organic-rich sediments off South West Africa. *Nature* 227: 593–595.

Carr, M.-E. 2001. Estimation of potential productivity in eastern boundary currents using remote sensing. *Deep-Sea Research II* 49: 59–80.

Carr, M.-E., and E.J. Kearns. 2003. Production regimes in four eastern boundary current systems. *Deep-Sea Research II* 50: 3199–3221.

Chapman, P. 1988. On the occurrence of oxygen-depleted water south of Africa and its implications for Agulhas-Atlantic mixing. *South African Journal of Marine Science* 7: 267–294.

Chapman, P., and J.L. Largier. 1989. On the origin of Agulhas Bank bottom water. *South African Journal of Science* 85: 515–529.

Chapman, P., and L.V. Shannon. 1985. The Benguela ecosystem. Part II. Chemistry and related processes. *Oceanography and Marine Biology Annual Reviews* 23: 183–251.

Chavez, F.P., and M. Messié. 2009. A comparison of eastern boundary upwelling ecosystems. *Progress in Oceanography* 83: 80–96.

Christensen, M.S. 1980. Sea surface temperature charts for southern Africa, south of 26° S. *South African Journal of Science* 76: 541–546.

Clowes, A.J. 1938. Phosphate and silicate in the Southern Ocean. *Discovery Reports*, 19: 1–120, plates 1–25.

Cochrane, K.L., A.G. James, B.A. Mitchell-Innes, G.C. Pitcher, H.M. Verheye, and D.R. Walker. 1991. Short-term variability during an anchor station study in the southern Benguela upwelling system: A simulation model. *Progress in Oceanography* 28: 121–152.

Cole, J., and C. Villacastin. 2000. Sea surface temperature variability in the northern Benguela upwelling system, and implications for fisheries research. *International Journal of Remote Sensing* 21: 1597–1617.

Cooper, J., R.I. Brooke, P.A. Shelton, and R.J.M. Crawford. 1982. Distribution, population size, and conservation of the Cape cormorant (*Phalacrocorax capensis*). *Fisheries Bulletin South Africa* 16: 121–143.

Copenhagen, W.J. 1953. The periodic mortality of fish in the Walvis Bay region—a phenomenon within the Benguela Current. *Investigational Report, Division of Sea Fisheries, South Africa*, 14, 32 pp.

Crawford, R.J.M., P.A. Shelton, J. Cooper, and R.K. Brooke. 1983. Distribution, population size and conservation of the Cape gannet *Morus capensis*. *South African Journal of Marine Science* 1: 153–174.

Crawford, R.J.M., L.V. Shannon, and D.E. Pollock. 1987. The Benguela ecosystem. Part IV. The major fish and invertebrate resources. *Oceanography and Marine Biology Annual Review* 25: 353–505.

Crawford, R.J.M., J.H.M. David, L.J. Shannon, J. Kemper, N.T.W. Klages, J.P. Roux, L.G. Underhill, V.L. Ward, A.J. Williams, and A.C. Wolfaardt. 2001. African penguins as predators and prey–coping (or not) with change. *South African Journal of Marine Science* 23: 435–447.

Crawford, R.J.M., B.M. Dyer, J. Kemper, R.E. Simmons, and L. Upfold. 2007. Trends in numbers of Cape Cormorants (*Phalacrocorax capensis*) over a 50-year period, 1956–57 to 2006–07. *Emu* 107: 253–261.

Crawford, R.J.M., A.J. Tree, P.A. Whittington, J. Visagie, L. Upfold, K.J. Roxburg, A.P. Martin, and B.M. Dyer. 2008. Recent distributional changes of seabirds in South Africa: Is climate having an impact? *African Journal of Marine Science* 30: 189–193.

Crawford, R.J.M., R. Altwegg, B.J. Barham, P.J. Barham, J.M. Durant, B.M. Dyer, D. Geldenhuys, A.B. Makhado, L. Pichegru, P.G. Ryan, L.G. Underhill, J. Visagie, L.J. Waller, and P.A. Whittington. 2011. Collapse of South Africa's penguins in the early 21st century. *African Journal of Marine Science* 33: 139–156.

Cury, P., A. Bakun, R.J.M. Crawford, A. Quinones, L.J. Shannon, and H.M. Verheye. 2000. Small pelagics in upwelling systems: Patterns of interaction and structural changes in "wasp-waist" ecosystems. *ICES Journal of Marine Science* 57: 603–618.

Cushing, D.H. 1969. Upwelling and fish production. *FAO Fisheries Technical Paper* 84: 1–40.

DAFF. 2014. *Status of the South African marine fishery resources 2014*. Cape Town: Department of Agriculture, Forestry and Fisheries. 74 pp.

David, J.H.M., P. Cury, R.J.M. Crawford, R.M. Randall, L.G. Underhill, and M.A. Meyer. 2003. Assessing conservation priorities in the Benguela ecosystem, South Africa: Analyzing predation by seals on threatened seabirds. *Biological Conservation* 114: 289–292.

De Decker, A.H.B. 1970. Notes on an oxygen-depleted subsurface current off the west coast of South Africa. *Investigational Report, Division of Sea Fisheries, South Africa*, 84, 24 pp.

Defant, A. 1936. *Das Kaltwasserauftriebsgebiet vor der Kusteudwestafrikas*, 52–66. Stuttgart: Landerkdl. Forsch. Festchrift N. Krebs.

Demarcq, H., R.G. Barlow, and F.A. Shillington. 2003. Climatology and variability of sea surface temperature and surface chlorophyll in the Benguela and Agulhas ecosystems as observed by satellite imagery. *African Journal of Marine Science* 25: 363–372.

Demarq, H. 2009. Trends in primary production, sea surface temperature and wind in upwelling systems. *Progress in Oceanography* 83: 376–385.

Desbiolles, F., B. Blanke, A. Bentamy, and C. Roy. 2016. Response of the southern Benguela upwelling system to fine-scale modifications of the coastal wind. *Journal of Marine Systems* 156: 46–55.

Dias, C.A. 1983. Notes on the evidence of a permanent southward flow of upper oceanic tropospheric waters off Angola at 12° S. *Collection of scientific papers of the International Commission for South East Atlantic Fisheries* 10: 99–102.

Duffy, D.C., A. Berruti, R.M. Randall, and J. Cooper. 1984. Effects of the 1982–3 warm water event on the breeding of South African seabirds. *South African Journal of Science* 80: 65–69.

Duncan, C.P., and J.H. Nell. 1969. Surface currents off the Cape coast. *Investigational Report*, Division of Sea Fisheries, South Africa, 76, 19 pp.

Duncombe Rae, C.M. 2005. A demonstration of the hydrographic partition of the Benguela upwelling ecosystem at 26° 40′ S. *African Journal of Marine Science* 27: 617–628.

Duncombe Rae, C.M., A.J. Boyd, and R.J.M. Crawford. 1992. "Predation" of anchovy by an Agulhas ring: A possible contributory cause for the very poor year class of 1989. *South African Journal of Marine Science* 12: 167–173.

Edelman-Furstenberg, Y. 2014. Distribution and paleoecology of molluscan skeletal remains along an upwelling tract: Benguela system, Namibian shelf. *Marine Geology* 353: 153–162. doi:10.1016/j.margeo.2014.04.011.

Emeis, K.-C., V. Bruchert, B. Currie, R. Endler, T. Ferdelman, A. Kiessling, T. Leipe, K. Noli-Peard, U. Struck, and T. Vogt. 2004. Shallow gas in shelf sediments of the Namibian coastal upwelling ecosystem. *Continental Shelf Research* 24: 627–642.

Fennel, W. 1999. Theory of the Benguela upwelling system. *Journal of Physical Oceanography* 29: 177–190.

Field, J.G., and F.A. Shillington. 2005. Variability of the Benguela Current system (16° E). In *The Sea, Volume 14*, ed. A.R. Robinson and K.H. Brink, 835–863.

Florenchie, P., C.J.C. Reason, J.R.E. Lutjeharms, M. Rouault, C. Roy, and S. Masson. 2004. Evolution of interannual warm and cold events in the southeast Atlantic Ocean. *Geophysical Research Letters* 30: 1505. doi:10.1029/2003GL017172.

Gammelsrod, T., C.H. Bartholomae, D.C. Boyer, V.I.L. Filipe, and M.J. O'Toole. 1998. Intrusion of warm surface water along the Angolan-Namibian coast in February-March 1995: The 1995 Benguela Niño. *South African Journal of Marine Science* 19: 51–56.

Geldenhuys, N.D. 1973. Growth of the South African maasbanker *Trachurus trachurus* Linnaeus and age composition of the catches, 1950–1971. *Investigational Report, Division of Sea Fisheries, South Africa*, 101, 24 pp.

Giraudeau, J., G.W. Bailey and C. Pujol. 2000. A high-resolution time-series analysis of particle fluxes in the northern Benguela coastal upwelling system: Carbonate record of changes in biogenic production and particle transfer processes. *Deep-Sea Research* II, 1999–2028.

Gordon, A.L. 1985. Indian-Atlantic transfer of thermocline water at the Agulhas retroflection. *Science* 227: 1030–1033.

Gordon, A.L. 1986. Inter-ocean exchange of thermocline water. *Journal of Geophysical Research* 91: 5037–5046.

Gordon, A.L., R.F. Weiss, W.M. Smethie, and M. Warner. 1992. Thermocline and intermediate water communication between the South Atlantic and Indian Oceans. *Journal of Geophysical Research* 97: 7223–7240.

Graham, W.M., and J.L. Largier. 1997. Upwelling shadows as nearshore retention sites: The example of northern Monterey Bay. *Continental Shelf Research* 17: 509–532.

Grant, W.S. 1985. Biochemical genetic stock structure of the southern African anchovy, *Engraulis capensis* Gilchrist. *Journal of Fish Biology* 27: 23–29.

Gregor, L. and P.M.S. Monteiro. 2013. Is the southern Benguela a significant regional sink of CO2? *South African Journal of Science*, 109. doi:10.1590/sajs.2013/20120094.

Griffiths, C.L., L. van Sittert, P.B. Best, A.C. Brown, B.M. Clark, P.A. Cook, R.J.M. Crawford, J. H.M. David, B.R. Davies, M.H. Griffiths, L. Hutchings, A. Jerardino, N. Kruger, S. Lamberth, R.W. Leslie, R. Melville-Smith, R. Tarr, and C.D. van Lingen. 2004. Impacts of human activities on marine animal life in the Benguela: A historical overview. *Oceanography and Marine Biology Annual Reviews* 42: 303–392.

Hagen, E. 1985. Meso-scale upwellings off Namibian coast. In *Proceedings of the International Symposium on the Most Important Upwelling Areas off Western Africa (Cape Blanco and Benguela)*, Barcelona, Spain, 161–179.

Hansen, F.C., R.R. Cloete, and H.M.V. Verheye. 2005. Seasonal and spatial variability of dominant copepods along a transect off Walvis Bay (23° S) Namibia. *African Journal of Marine Science* 27(1): 55–63.

Hardman-Mountford, N.J., A.J. Richardson, J.J. Agenbag, E. Hagen, L. Nykjaer, F.A. Shillington, and C. Villacastin. 2003. Climate variability of the South East Atlantic ocean observed from satellite. *Progress in Oceanography* 59: 181–221.

Harris, T.F.W. 1978. Review of coastal currents in southern African waters. *South African National Scientific Programmes Report* 30, CSIR, Pretoria, 103 pp.

Harris, T.F.W., and L.V. Shannon. 1979. Satellite-tracked drifter in the Benguela Current system. *South African Journal of Science* 75: 316–317.

Hart, T.J., and R.I. Currie. 1960. *The Benguela Current. Discovery Reports* 31: 123–298.

Hentschel, E. 1928. Die Grundzüge der Planktonverteilung im Südatlantischen Ozean. *International Review of Hydrobiology* 21: 1–16.

Hentschel, E., and H. Wattenberg. 1930. Plankton und Phosphat in der Oberflachenschicht des Südatlantischen Ozeans. *Annalen der Hydrographischer, Berlin* 58: 273–277.

Heyes, C. 2006. Offshore diamond mining. http://www.marcon.com/library/articles/2006/diamondmining/diamondmining.pdf. Accessed 1 Apr 2016.

Holden, C. 1985. Currents in St Helena Bay inferred from radio-tracked drifters. In *South African Ocean Colour and Upwelling Experiment*, ed. L.V. Shannon, 97–109. South Africa: Sea Fisheries Research Institute.

Hutchings, L. 1992. Fish harvesting in a variable, productive environment–searching for rules or searching for exceptions? *South African Journal of Marine Science* 12: 297–318.

Hutchings, L., and J. Taunton-Clark. 1990. The monitoring of gradual change in areas of high mesoscale variability. *South African Journal of Science* 90: 109–113.

Hutchings, L., C.J. Holden, and B. Mitchell-Innes. 1984. Hydrological and shipboard monitoring of upwelling off the Cape Peninsula. *South African Journal of Science* 80: 83–89.

Hutchings, L., C.J. Augustyn, A. Cockcroft, C. van der Lingen, J. Coetzee, R.W. Leslie, R.J. Tarr, H. Oosthuizen, M.R. Lipinski, M.R. Roberts, C. Wilke, R. Crawford, L.J. Shannon, and M. Mayekiso. 2009a. Marine fisheries monitoring programmes in South Africa. *South African Journal of Science* 105: 182–192.

Hutchings, L., M.R. Roberts, and H.M. Verheye. 2009b. Marine environmental monitoring programmes in South Africa: A review. *South African Journal of Science* 105: 94–102.

Hutchings, L., A. Jarre, T. Lamont, M. van den Berg, and S.P. Kirkman. 2012. St Helena Bay (southern Benguela) then and now: Muted climate signals, large human impact. *African Journal of Marine Science* 34: 559–583.

Ismail, H.E., J.J. Agenbag, S. de Villiers, and B.J. Ximba. 2015. Relation between upwelling intensity and the variability of physical and chemical parameters in the southern Benguela upwelling system. *International Journal of Oceanography* 2015: 510713. doi:10.1155/2015/510713.

Jackson, S.P. 1947. Air masses and circulation over the plateau and coasts of South Africa. *South African Geographical Journal* 29: 1–15.

James, A.G. 1987. Feeding ecology, diet and field-based studies on feeding selectivity of the Cape anchovy Engraulis capensis Gilchrist. *South African Journal of Marine Science* 5: 673–692.

Jerardino, A. 2010. Large shell middens in Lamberts Bay, South Africa: A case of hunter-gatherer resource intensification. *Journal of Archaeological Science* 37: 2291–2302.

Jerardino, A., and R. Yates. 1997. Excavations at Mike Taylor's Midden: A summary report and implications for a re-characterisation of megamiddens. *South African Archaeological Bulletin* 52: 43–51.

Johnson, A.S., and G. Nelson. 1999. Ekman estimates of upwelling at Cape Columbine based on measurements of longshore wind from a 35 year time series. *South African Journal of Marine Science* 21: 433–436.

Jury, M.R. 1985. Case studies of alongshore variations in wind-driven upwelling in the southern Benguela region. In *South African Ocean Colour and Upwelling Experiment*, ed. L.V. Shannon, 29–46. South Africa: Sea Fisheries Research Institute.

Jury, M.R. 2003. The coherent variability of African river flows: Composite climate structure and the Atlantic circulation. *Water SA* 29: 1–10.

Jury, M.R. 2014. Environmental forcing of red tides in the southern Benguela. *International Journal of Oceanography*, doi:10.1155/2014/325321.

Jury, M.R., C.I. McArthur, and G.B. Brundrit. 1990. Pulsing of the Benguela upwelling region: Large-scale atmospheric controls. *South African Journal of Marine Science* 9: 27–41.

Kamstra, F. 1985. Environmental features of the southern Benguela with special reference to the wind stress. In *South African Ocean Colour and Upwelling Experiment*, ed. L.V. Shannon, 13–27. South Africa: Sea Fisheries Research Institute.

Landman, W.A., and E. Klopper. 1998. 15-year simulation of December to March rainfall season of the 1980s and 1990s using canonical correlation analysis (CCA). *Water SA* 24: 281–285.

Largier, J.L., P. Chapman, W.T. Peterson, and V.P. Swart. 1992. The western Agulhas Bank: Circulation, stratification, and ecology. *South African Journal of Marine Science* 12: 319–339.

Lass, H.U., and V. Mohrholz. 2005. On the fluctuations and vertical structure of the shelf circulation off Walvis Bay, Namibia. *Continental Shelf Research* 25: 1473–1497.

Lass, H.U., M. Schmidt, V. Mohrholz, and G. Nausch. 2000. Hydrographic and current measurements in the area of the Angola-Benguela front. *Journal of Physical Oceanography* 30: 2589–2609.

Liu, K.-K., and N. Dittert. 2010 Web-based electronic supplements, Appendix C. In *Carbon and nutrient fluxes in continental margins*, ed. K.-K. Liu, L. Atkinson, R.A. Quiñones and L. Talaue-McManus. Springer, Berlin, http://cmtt.pangaea.de/. Accessed 11 June 2015.

Liu, K.-K., L. Atkinson, and C.-T.A. Chen. 2000. Exploring continental margin carbon fluxes in the global context. *Eos* 81: 641, 642, 644.

Lutjeharms, J.R.E. 2006. *The Agulhas current*. Berlin: Springer. 329 pp.

Lutjeharms, J.R.E., and J. Cooper. 1996. Interbasin leakage through Agulhas Current filaments. *Deep-Sea Research I* 43: 213–238.

Lutjeharms, J.R.E., and P.L. Stockton. 1991. Aspects of the upwelling regime between Cape Point and Cape Agulhas. *South African Journal of Marine Science* 10: 91–102.

Lutjeharms, J.R.E., R. Catzel, and H.R. Valentine. 1989. Eddies and other border phenomena of the Agulhas Current. *Continental Shelf Research* 9: 597–616.

Lutjeharms, J.R.E., F.A. Shillington, and C.M. Duncombe Rae. 1991. Observations of extreme upwelling filaments in the southeast Atlantic Ocean. *Science* 253: 774–776.

Lutjeharms, J.R.E., A.A. Meyer, I.J. Ansorge, G.A. Eagle, and M.J. Orren. 1997. The nutrient characteristics of the Agulhas Bank. *South African Journal of Marine Science* 17: 253–274.

Makhado, A.B., R.J.M. Crawford, and L.G. Underhill. 2006. Impact of predation by Cape fur seals *Arctocephalus pusillus pusillus* on Cape gannets *Morus capensis* at Malgas Island, Western Cape, South Africa. *South African Journal of Marine Science* 28: 681–687.

Martin, B., A. Eggert, V. Mohrholz, M. Schmidt, R. Diekmann, and R. Koppelmann. 2014. Spatio-temporal variability of mesozooplankton biomass and environmental control in the northern Benguela Upwelling System: Field investigations and model simulation. *Marine Ecology* 36: 637–658. doi:10.1111/maec.12173.

McCartney, M.S., and M.E. Woodgate-Jones. 1991. A deep-reaching anticyclonic gyre in the subtropical gyre of the eastern South Atlantic. *Deep-Sea Research* 38(S1): S411–S443.

Meeuwis, J.M., and J.R.E. Lutjeharms. 1990. Surface thermal characteristics of the Angola-Benguela front. *South African Journal of Marine Science* 9: 261–279.

Meyer, H.H.F. 1923. Die Oberflächenströmungen des Atlantischen Ozeans im Februar. *Veröffentlichungen des Instituts für Meereskunde,* Berlin N.F., Reihe A, 11: 1–35.

Monteiro, P.M.S. 1996. *The oceanography, the biogeochemistry and the fluxes of carbon dioxide in the Benguela upwelling system*. Ph.D. thesis, University of Cape Town, 354 pp.

Monteiro, P.M.S., A. van der Plas, V. Mohrholz, E. Mabille, A. Pascall and W. Joubert. 2006. Variability of natural hypoxia and methane in a coastal upwelling system: Oceanic physics or shelf biology? *Geophysical Research Letters*, L16614. doi:10.1029/2006GL026234.

Monteiro, P.M.S. 2010. The Benguela current system. In *Carbon and nutrient fluxes in continental margins*, ed. K.-K. Liu, L. Atkinson, R.A. Quiñones, and L. Talaue-McManus, 65–78. Berlin: Springer.

Monteiro, P.M.S., and A.K. van der Plas. 2006. Low oxygen water (LOW) variability in the Benguela system: Key processes and forcing scales relevant to forecasting. In *Benguela: Predicting a large marine ecosystem*, ed. V. Shannon, G. Hempel, P. Malanotte-Rizzoli, C. Moloney, and J. Woods, 71–90. Amsterdam: Elsevier.

Moroshkin, K.V., V.A. Bubnov, and R.P. Bulatov. 1970. Water circulation in the eastern South Atlantic Ocean. *Oceanology* 10: 27–34.

Muhry, A. 1862. *Klimatologische Ubersicht der Erde*. Leipzig.

Nelson, G. 1985. Notes on the physical oceanography of the Cape Peninsula upwelling system. In *South African ocean colour and upwelling experiment*, ed. L.V. Shannon, 63–95. South Africa: Sea Fisheries Research Institute.

Nelson, G. 1989. Poleward motion in the Benguela area. In *Poleward flows along eastern ocean boundaries*, ed. S.J. Neshyba, C.N.K. Mooers, R.L. Smith, and R.T. Barber, 110–130. New York: Springer.

Nelson, G., and L. Hutchings. 1983. The Benguela upwelling area. *Progress in Oceanography* 12: 333–356.

Nelson, G., A.J. Boyd, J.J. Agenbag, and C.M. Duncombe Rae. 1998. An upwelling filament north-west of Cape Town, South Africa. *South African Journal of Marine Science* 19: 75–88.

Newman, G.G. 1970. Stock assessment of the pilchard *Sardinops ocellata* at Walvis Bay, South West Africa. *Investigational Report, Division of Sea Fisheries, South Africa*, 85, 13 pp.

Olivieri, E.T. 1983. Colonization, adaptations and temporal changes in diversity and biomass of a phytoplankton community in upwelled water off the Cape Peninsula, South Africa, in December 1979. *South African Journal of Marine Science* 1: 77–109.

Olson, D.B., and R.H. Evans. 1986. Rings of the Agulhas current. *Deep-Sea Research* 33: 27–42.

Parada, C., C.D. van der Lingen, C. Mullon, and P. Penven. 2003. Modeling the effect of buoyancy on the transport of anchovy (*Engraulis capensis*) eggs from spawning to nursery grounds in the southern Benguela: An IBM approach. *Fisheries Oceanography* 12: 170–184.

Parkington, J.E., C. Poggenpoel, W. Buchanan, T. Robey, A. Manhire, and J. Sealy. 1988. Holocene coastal settlement patterns in the Western Cape. In *The archaeology of prehistoric coastlines*, ed. G. Bailey, and J.E. Parkington, 22–41. Cambridge: Cambridge University Press.

Parrish, R.H., A. Bakun, D.M. Husby and C.S. Nelson. 1983. Comparative climatology of selected environmental processes in relation to eastern boundary current fish production. In *Proceedings of the expert consultation to examine changes in abundance and species of neritic fish resources, San Jose, Costa Rica, 1983*, ed. G.D. Sharp and J. Csirke, *FAO Fisheries Report*, 291(3): 731–777.

Penven, P., C. Roy, G.B. Brundrit, A. Colin de Verdiere, P. Freon, A.S. Johnson, J.R.E. Lutjeharms, and F.A. Shillington. 2001. A regional hydrodynamic model of the southern Benguela upwelling. *South African Journal of Science* 97: 472–475.

Petermann, A. 1865. Der Nordpol und Südpol, die Wichtigkeit ihrer Erforschung in geographischer und kulturhistorischer Beziehung. *Petermann's Geographische Mitteilungen* 11: 146–160.
Pitcher, G.C., and S.J. Weeks. 2006. The variability and potential for prediction of harmful algal blooms in the southern Benguela ecosystem. In *Benguela: Predicting a large marine ecosystem*, ed. V. Shannon, G. Hempel, P. Malanotte-Rizzoli, C. Moloney, and J. Woods, 125–146. Amsterdam: Elsevier.
Pollock, D.E. 1990. Palaeoceanography and speciation in the spiny lobster genus *Jasus*. *Bulletin of Marine Science* 46: 387–405.
Pollock, D.E., and L.V. Shannon. 1987. Response of rock lobster populations in the Benguela ecosystem to environmental change—a hypothesis. *South African Journal of Marine Science* 5: 887–899.
Pollock, D.E., A.C. Cockcroft, and P.C. Goosen. 1997. A note on reduced rock lobster growth rates and related environmental anomalies in the southern Benguela, 1988–1995. *South African Journal of Marine Science* 18: 287–293.
Polovina, J.J. 2005. Climate variation, regime shifts, and implications for sustainable fisheries. *Bulletin of Marine Science* 76: 233–244.
Preston-Whyte, R.A., and P.D. Tyson. 1988. *The atmosphere and weather of Southern Africa*. Cape Town: Oxford University Press. 375 pp.
Probyn, T.A. 1992. The inorganic nitrogen nutrition of phytoplankton in the southern Benguela: New production, phytoplankton size and implications for pelagic foodwebs. *South African Journal of Marine Science* 12: 411–420.
Rand, R.W. 1952. Guano enterprise in SW Africa. *Ostrich* 23: 169–185.
Reason, C.J.C., P. Florenchie, M. Rouault, and J. Veitch. 2006. Influences of large scale climate modes and Agulhas system variability on the BCLME region. In *Benguela: Predicting a large marine ecosystem*, ed. V. Shannon, G. Hempel, P. Malanotte-Rizzoli, C. Moloney, and J. Woods, 223–238. Amsterdam: Elsevier.
Rennell, J. 1832. *An Investigation of the Currents of the Atlantic Ocean, and of those which Prevail between the Indian Ocean and the Atlantic*. London: J.G. & F. Rivington. 359 pp.
Richardson, P.L., and S.L. Garzoli. 2003. Characteristics of intermediate water flow in the Benguela Current as measured with RAFOS floats. *Deep-Sea Research II* 50: 87–118.
Richter, I., S.K. Behera, Y. Masumoto, Z.B. Taguchi, N. Komori, and T. Yamagata. 2010. On the triggering of Benguela Niños: Remote equatorial versus local influences. *Geophysical Research Letters* 37: L20604. doi:10.1029/2010GL044461.
Risien, C.M., C.J.C. Reason, F.A. Shillington, and D.B. Chelton. 2004. Variability in satellite winds over the Benguela upwelling system during 1999–2000. *Journal of Geophysical Research* 109: C03010. doi:10.1029/2003JC001880.
Rogers, J., and J.M. Bremner. 1991. The Benguela ecosystem. Part VII. Marine geological aspects. *Oceanography and Marine Biology Annual Review* 29: 1–85.
Ross, J.C. 1847. *Voyage of discovery and research in the southern and Antarctic regions during the years 1839–1843, 2 volumes*. London: John Murray.
Rouault, M. 2012. Bi-annual intrusion of tropical water in the northern Benguela upwelling. *Geophysical Research Letters*, 39. doi:10.1029/2012GL052099.
Rouault, M., P. Florenchie, N. Faucherau and C.J.C. Reason. 2003. South east Atlantic warm events and southern African rainfall. *Geophysical Research Letters*, 30, 8099. doi:10.1029/2002GL014840.
Rouault, M., J-L. Melice, C.R.C. Reason and J.R.E. Lutjeharms. 2005. Climate variability at Marion Island, Southern Ocean, since 1960. *Journal of Geophysical Research* 110, C05008. doi:10.1029/2004JC002492.
Rouault, M., S. Illig, C. Bartholomae, C.J.C. Reason, and A. Bentamy. 2007. Propagation and origin of warm anomalies in the Angola Benguela upwelling system in 2001. *Journal of Marine Systems* 68: 473–488.
Roy, C., S. Weeks, M. Rouault, G. Nelson, R. Barlow, and C. van der Lingen. 2001. Extreme oceanographic events recorded in the southern Benguela during the 1999–2000 summer season. *South African Journal of Science* 97: 465–471.

Santana-Casiano, J.M., M. Gonzalez-Davila, and I.R. Ucha. 2009. Carbon dioxide fluxes in the Benguela upwelling system during winter and spring: A comparison between 2005 and 2006. *Deep-Sea Research II* 56: 533–541.

Santos, F., M. Gomes-Gesteira, M. deCastro, and L. Alvarez. 2012. Differences in coastal and oceanic SST trends due to the strengthening of coastal upwelling along the Benguela Current system. *Continental Shelf Research* 34: 79–86.

Scheffer, M., and S.R. Carpenter. 2003. Catastrophic regime shifts in ecosystems: Linking theory to observations. *Trends in Ecology & Evolution* 18: 648–656.

Schott, G. 1902. Die Auftriebszone an der Küste von südwest Afrika. *Wissenschaftliche Ergebnisse des Deutsche Tief-see-Expedition der Valdivia* 1: 1–403.

Schultz, S., R. Schemainda and D. Nehring. 1979. Beitrage der DDR zur Erforschung der kustennahen Wasserauftriebsprozesse, Teil VIII: Das ozeanologische Beobachtungsmaterial der Messfahrt vom 21.9 bis 20.12.1976 nach Sudwestafrika. *Geodatische und geophysikalische Veroffentlichungen*, Reihe IV, 28, 50 pp.

Schumann, E.H., and I.L. van Heerden. 1988. Observations of Agulhas Current frontal features south of Africa, October 1983. *Deep-Sea Research* 35: 1355–1362.

Schwartzlose, R.A., J. Alheit, A. Bakun, T.R. Baumgartner, R. Cloete, R.J.M. Crawford, W. J. Fletcher, Y. Gree-Ruiz, E. Hagen, T. Kawasaki, D. Lluch-Belda, S.E. Lluch-Cota, A.D. MacCall, Y. Matsuura, M.O. Nevarez-Martinez, R.H. Parrish, C. Roy, R. Serra, K.V. Shust, M. N. Ward, and J.Z. Zununaga. 1999. Worldwide large-scale fluctuations of sardine and anchovy populations. *South African Journal of Marine Science* 21: 289–347.

Sealy, J., and R. Yates. 1994. The chronology of the introduction of pastoralism in the Cape, South Africa. *Antiquity* 68: 58–67.

Shannon, L.J., P. Cury, and A. Jarre. 2000. Modeling effects of fishing in the southern Benguela ecosystem. *ICES Journal of Marine Science* 57: 720–722.

Shannon, L.J., S. Neira, and M. Taylor. 2008. Comparing internal and external drivers in the southern Benguela and the southern and northern Humboldt upwelling ecosystems. *African Journal of Marine Science* 30: 63–84.

Shannon, L.V. 1966. Hydrology of the south and west coasts of South Africa. *Investigational Report, Division of Sea Fisheries, South Africa*, 58, 52 pp.

Shannon, L.V., G.H. Stander and J.A. Campbell. 1973. Oceanic circulation deduced from plastic drift cards. *Investigational Report, Sea Fisheries Branch, South Africa*, 108, 31 pp.

Shannon, L.V. 1985a. The Benguela Ecosystem. Part I. Evolution of the Benguela, physical features and processes. *Oceanography and Marine Biology: An Annual Review*, 23: 105–182.

Shannon, L.V. 1985b. *South African ocean colour and upwelling experiment*. South Africa: Sea Fisheries Research Institute. 270 pp.

Shannon, V. 2006. A plan comes together. In *Benguela: Predicting a large marine ecosystem*, ed. V. Shannon, G. Hempel, P. Malanotte-Rizzoli, C. Moloney, and J. Woods, 3–10. Amsterdam: Elsevier.

Shannon, L.V., and F.P. Anderson. 1982. Applications of satellite ocean colour imagery in the study of the Benguela Current. *South African Journal of Photogrammetry, Remote Sensing and Cartography* 13: 153–169.

Shannon, L.V., and P. Chapman. 1983. Incidence of *Physalia* on beaches in the south western Cape Province during January 1983. *South African Journal of Science* 79: 454–458.

Shannon, L.V., and J.G. Field. 1985. Are fish stocks limited in the southern Benguela pelagic ecosystem? *Marine Ecology Progress Series* 22: 7–19.

Shannon, L.V., and D. Hunter. 1988. Notes on Antarctic intermediate water around southern Africa. *South African Journal of Marine Science* 6: 107–117.

Shannon, L.V., and G. Nelson. 1996. The Benguela: Large scale features and processes and system variability. In *The South Atlantic past and present circulation*, ed. G. Wefer, W.H. Berger, G. Siedler, and D.J. Webb, 163–210. Berlin: Springer.

Shannon, L.V., and S.C. Pillar. 1986. The Benguela ecosystem. Part III. Plankton. *Oceanography and Marine Biology Annual Reviews* 24: 65–170.

Shannon, L.V., J.J. Agenbag and M.E.L. Buys. 1987. Large- and mesoscale features of the Angola-Benguela front. In *The Benguela and Comparable Ecosystems*, ed. by A.I.L. Payne, J. A. Gulland and K.H. Brink, *South African Journal of Marine Science* 5: 11–34.

Shannon, L.V., P. Chapman, G.A. Eagle, and T.P. McClurg. 1983. A comparative study of tar ball distributions and movement in two boundary current regimes. *Oil and Petrochemical Pollution* 1: 243–259.

Shannon, L.V., P. Schlittenhardt, and S.A. Mostert. 1984. The NIMBUS 7 CZCS experiment in the Benguela Current region off southern Africa, February 1980. 2. Interpretation of imagery and oceanographic implications. *Journal of Geophysical Research* 89: 4968–4976.

Shannon, L.V., A.J. Boyd, G.B. Brundrit, and J. Taunton-Clark. 1986. On the existence of an El Niño-type phenomenon in the Benguela system. *Journal of Marine Research* 44: 495–520.

Shannon, L.V., D.E. Pollock, P. Chapman, and A.A. Robertson. 1989. South-east Atlantic expedition of *R.S. Africana*, April 1989—some preliminary results. *South African Journal of Science* 85: 665–669.

Shannon, L.V., J.J. Agenbag, N.D. Walker, and J.R.E. Lutjeharms. 1990. A major perturbation in the Agulhas retroflection area in 1986. *Deep-Sea Research* 37: 493–512.

Shannon, V., G. Hempel, P. Malanotte-Rizzoli, C. Moloney, and J. Woods (eds.). 2006. *Benguela: Predicting a large marine ecosystem*. Amsterdam: Elsevier. 410 pp.

Shelton, P.A. 1984. Notes on the spawning of anchovy during the summer of 1982–3. *South African Journal of Science* 80: 69–71.

Shelton, P.A. 1987. Life history traits displayed by neritic fish in the Benguela Current ecosystem. *South African Journal of Marine Science* 5: 235–242.

Shelton, P.A., and L. Hutchings. 1982. Transport of anchovy, *Engraulis capensis* Gilchrist, eggs and early larvae by a frontal jet current. *Journal du Conseil* 40: 185–198.

Shillington, F.A. 1998. The Benguela upwelling system off southwestern Africa. In *The Sea*, ed. A.R. Robinson and K.H. Brink, vol. 11, 583–604. Chichester: Wiley.

Shillington, F.A., W.T. Peterson, L. Hutchings, T.A. Probyn, H.N. Waldron, and J.J. Agenbag. 1990. A cool upwelling filament off Namibia, southwest Africa: Preliminary measurements of physical and biological features. *Deep-Sea Research* 37: 1753–1772.

Shillington, F.A., C.J.C. Reason, C.M. Duncombe Rae, P. Florencie, and P. Penven. 2006. Large scale physical variability of the Benguela Current Large Marine Ecosystem (BCLME). In *Benguela: Predicting a large marine ecosystem*, ed. V. Shannon, G. Hempel, P. Malanotte-Rizzoli, C. Moloney, and J. Woods, 49–70. Amsterdam: Elsevier.

Silio-Calzada, A., A. Bricaud, J. Uitz, and B. Gentili. 2008. Estimation of new primary production in the Benguela upwelling area, using ENVISAT satellite data and a model dependent on the phytoplankton community size structure. *Journal of Geophysical Research* 113: C11023. doi:10.1029/2007JC004588.

Smith, A.B. 2006. Excavations at Kasteelberg and the origins of the Khoekhoen in the Western Cape, South Africa. Cambridge Monographs in African Archaeology 66. In *British archaeological reports international series*, vol. 1537.

Stander, G.H. 1958. The variations of temperature in the surface layer of the sea near Walvis Bay during 1954–57 with an analysis of some wind data from Pelican Point. *Division of Fisheries, South Africa, Investigational Report,* 35, 40 pp.

Stander, G.H. 1964. The Benguela Current off South West Africa. *Marine Research Laboratory, South West Africa, Investigational Report* 12, 120 pp.

Stander, G.H., and A.H.B. De Decker. 1969. Some physical and biological aspects of an oceanographic anomaly off South West Africa in 1963. *Division of Fisheries, South Africa, Investigational Report,* 81, 46 pp.

Stander, G.H., L.V. Shannon, and J.A. Campbell. 1969. Average velocities of some ocean currents as deduced from the recovery of plastic drift cards. *Journal of Marine Research* 27: 293–300.

Stephen, V.C., and P.A.R. Hockey. 2007. Evidence for an increasing incidence and severity of harmful algal blooms in the southern Benguela region. *South African Journal of Science* 103: 223–231.

Stramma, L., and R.G. Peterson. 1989. Geostrophic transport in the Benguela Current region. *Journal of Physical Oceanography* 19: 1440–1448.

Sumaila, U.R., D. Boyer, M.D. Skogen, and S.I. Steinshamn. 2004a. Namibia's fisheries: Introduction and overview. In *Namibia's fisheries*, ed. U.R. Sumaila, D. Boyer, M.D. Skogen, and S.I. Steinshamn, 1–9. Delft, The Netherlands: Eburon.

Sumaila, U.R., D. Boyer, M.D. Skogen, and S.I. Steinshamn (eds.). 2004b. *Namibia's fisheries*. Delft, The Netherlands: Eburon. 363 pp.

Sundby, S., A.J. Boyd, L. Hutchings, M.J. O'Toole, K. Thorisson, and A. Thorsen. 2001. Interaction between Cape hake spawning and the circulation in the northern Benguela upwelling ecosystem. *South African Journal of Marine Science* 23: 317–336.

Taunton-Clark, J. 1985. The formation, growth and decay of upwelling tongues in response to the mesoscale wind field during summer. In *South African ocean colour and upwelling experiment*, ed. L.V. Shannon, 47–61. South Africa: Sea Fisheries Research Institute.

Tim, N., E. Zorita, and B. Hunicke. 2015. Decadal variability and trends of the Benguela upwelling system as simulated in a high-resolution ocean simulation. *Ocean Science* 11: 483–502.

Touratier, F., J.G. Field, and C.L. Moloney. 2003. Simulated carbon and nitrogen flows of the planktonic food web during an upwelling relaxation period in St. Helena Bay (southern Benguela ecosystem). *Progress in Oceanography* 58: 1–41.

Van Ballegooyen, C., M.L. Grundlingh, and J.R.E. Lutjeharms. 1994. Eddy fluxes of heat and salt from the south west Indian ocean into the southeast Atlantic Ocean: A case study. *Journal of Geophysical Research* 99: 14053–14070.

Van der Lingen, C.D., J.C. Coetzee, H. Demarcq, L. Drapeau, T.P. Fairweather, and L. Hutchings. 2005. An eastward shift in the distribution of southern Benguela sardine. *GLOBEC Newsletter* 11: 17–22.

Veitch, J.A., P. Florenchie, and F.A. Shillington. 2006. Seasonal and interannual fluctuations of the Angola-Benguela frontal zone (ABFZ) using 4.5 km resolution satellite imagery from 1982 to 1999. *International Journal of Remote Sensing* 27: 987–998.

Veitch, J., P. Penven, and F. Shillington. 2009. The Benguela: A laboratory for comparative modeling studies. *Progress in Oceanography* 83: 296–302.

Verheye, H. 2000. Decadal-scale trends across several marine trophic levels in the southern Benguela upwelling system off South Africa. *Ambio* 29: 30–34.

Voges, E., A. Gordon, C.H. Bartholomae, and J.G. Field. 2002. Estimating the probability of different levels of recruitment for Cape hakes *Merluccius capensis* off Namibia, using environmental indices. *Fisheries Research* 1350: 1–8.

Voituriez, B., and A. Herbland. 1982. A comparative study of the productive systems of the tropical east Atlantic: Thermal domes, coastal upwellings and equatorial upwelling. *Rapports et Process-Verbaux des Reunions du Conseil Permanent pour l'International Exploration de la Mer* 180: 114–130.

Wang, D., T.C. Gouhier, B.A. Menge, and A.R. Ganguly. 2015. Intensification and spatial homogenization of coastal upwelling under climate change. *Nature* 518: 390–394.

Wanninkhof, R. 1992. Relationship between wind speed and gas exchange over the ocean. *Journal of Geophysical Research* 97: 7373–7382.

Wasmund, N., G. Nausch and A. Hansen. 2014. Phytoplankton succession in an isolated upwelled Benguela water body in relation to different initial nutrient conditions. *Journal of Marine Systems* 140 B: 163–174.

Wattenberg, H. 1938. Die Verteilung des Sauerstoffs im Atlantischen Ozean. *Wissenschaftliche Ergebnisse der Deutschen Atlantischen Expedition der Meteor* 1925–1927(9): 1–132.

Wattenberg, H. 1939. Atlas zu: Die Verteilung des Sauerstoffs im Atlantischen Ozean. *Wissenschaftliche Ergebnisse der Deutschen Atlantischen Expedition der Meteor, 1925–1927*. 9 (Atlas), 72 plates.

Weeks, S.J., B. Currie, and A. Bakun. 2002. Massive emissions of toxic gas in the Atlantic. *Nature* 415: 493–494.

Weeks, S.J., B. Currie, A. Bakun, and K.R. Peard. 2004. Hydrogen sulphide eruptions in the Atlantic ocean off southern Africa: Implications of a new view based on SeaWifs satellite imagery. *Deep-Sea Research I* 51: 153–172.

Weeks, S.J., R. Barlow, C. Roy, and F.A. Shillington. 2006. Remotely sensed variability of temperature and chlorophyll in the southern Benguela: Upwelling frequency and phytoplankton response. *African Journal of Marine Science* 28: 493–509.

Willemse, N.E., and D. Pauly. 2004. Reconstruction and interpretation of marine fisheries catches from Namibian waters, 1950 to 2000. In *Namibia's fisheries*, ed. U.R. Sumaila, D. Boyer, M. D. Skogen, and S.I. Steinshamn, 99–112. Delft, The Netherlands: Eburon.

Chapter 8
Seasonal Wind-Driven Coastal Upwelling Systems

Abstract This chapter describes some significant seasonal wind-driven coastal upwelling systems, most of which are associated with Large Marine Ecosystems. Starting in the west Pacific Ocean with the productive ecosystems in the East and South China Seas, the scientific journey continues through the Indonesian Seas, where the Arafura Sea stands out as a highly productive region, to coastal upwelling regions on the southern shelf of Australia and around New Zealand. Descriptions of upwelling regions in the Arabian Sea of the northern Indian Ocean, where the world's swiftest western boundary current—the Somali Current—forms, are followed by a journey to seasonal upwelling regions of the Atlantic Ocean: the Gulf of Mexico, the southern Caribbean Sea, the Brazilian shelf, and the Eurafrican Mediterranean Sea. These upwelling systems are discussed in the context of regional fisheries and specific regional characteristics, such as oxygen minimum zones and air-sea CO_2 fluxes.

Keywords Seasonal upwelling systems · Oceanography · East & South China Seas · Somali Current · Miscellaneous upwelling systems · Regional fisheries

There is, one knows not what sweet mystery about this sea,

whose gently awful stirrings seems to speak of some hidden soul beneath.

Herman Melville (1819–1891)

(Taken from *Moby Dick*, 1851)

8.1 Introduction

8.1.1 Overview

In contrast to the four eastern boundary current systems described in Chaps. 4–7 where upwelling occurs for much or all of the year, seasonal wind-driven coastal upwelling systems experience upwelling only seasonally. Monsoonal wind forcing

© Springer International Publishing Switzerland 2016

J. Kämpf and P. Chapman, *Upwelling Systems of the World*,
DOI 10.1007/978-3-319-42524-5_8

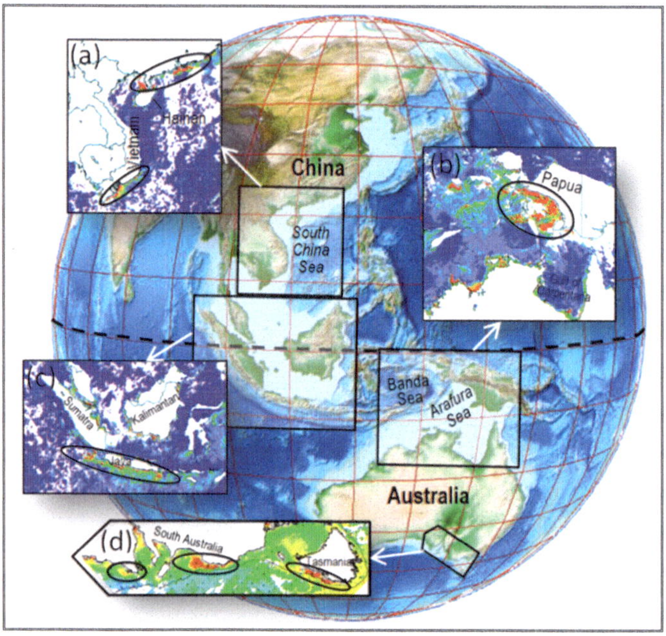

Fig. 8.1 Selected seasonal coastal upwelling regions in the West Pacific and Oceania. Shown are satellite-derived elevated chlorophyll-*a* concentrations of upwelling regions **a** in the South China Sea, **b** in the Arafura Sea, **c** off Sumatra and Java, and **d** on the southern shelf of Australia. Data source: Giovanni (NASA). Background image source: Google Earth with ETOPO 1 bathymetry layer

creates significant seasonal coastal upwelling systems in the northern Indian Ocean and in southeast Asian waters. Despite their seasonal nature, these upwelling systems are regionally important nutrient sources for marine food webs. The significance of upwelling in the northern Indian Ocean and southeast Asian waters is supported by the fact that China and India are the two largest producers of wild-caught seafood (if the fishery statistics can be trusted), followed by the southeast Asian countries of Indonesia, Thailand, the Philippines, and Vietnam. Other regionally important seasonal coastal upwelling systems of the world ocean are found on Australia's southern shelf, around New Zealand, in the Gulf of Mexico and the southern Caribbean Sea, on the Brazilian shelf and in the Eurafrican Mediterranean Sea (Fig. 8.1).

8.1.2 Southeast Asia: A Centre of Global Seafood Production

Prior to the early 1900s, the population of the whole of southeast Asia was about 40 million with the majority of that population living in villages (Morgan and Staples 2006). The countries of the area were not yet urbanized and therefore fishing, both marine and inland, was directed mainly at the need to supply food for these village communities. There was some limited trade, often between coastal and inland villages, and only simple preservation and processing methods such as salting, drying and the manufacture of fish sauce were developed and used to facilitate this commerce. The supply of fresh, preserved or processed fish for large centralized markets, either domestic or international, was largely unknown. Restricted by simple fishing gear and vessels, and with abundant coastal fish resources, most fishing was undertaken in near-shore waters using artisanal methods.

As fishing in southeast Asian coastal waters moved away from a subsistence-level activity and became increasingly oriented towards supplying more remote markets in the late 1800s and early 1900s, a spectacular growth in catches occurred (Sugiyama et al. 2004). This was much earlier than most other global fisheries, apart from the cod and herring fisheries in the North Atlantic. This expanding demand for marine products was created by urbanization and the growth in population but also by the development of transport and marketing systems and changes in the techniques in preserving fish that prompted people to produce for the market. The expansion of marine landings in southeast Asia has been so spectacular that two nations in the region (Thailand and Indonesia) are now among the world's top fish producing nations and marine fisheries production accounts for more than 1 % of GDP in almost all countries of the region (Sugiyama et al. 2004). The enormous fishing pressure now being exerted on stocks, coupled with the lack of governmental management of the fisheries and other marine resources and increasing coastal pollution as coastal development continues, is becoming an increasing problem in southeast Asia.

8.2 West Pacific and Eastern Indian Ocean

8.2.1 South China Sea

The South China Sea is bordered by China, Taiwan, the Philippines, Malaysia, Brunei, Indonesia, Singapore and Vietnam. The multitude of countries involved makes management of the region difficult, particularly because of competing claims of sovereignty, most of which result from China claiming almost the whole area. Recent efforts by China to establish local sovereignty by building naval and air facilities on contested islands and reefs is only exacerbating the situation (see www.cfr.org/asia-and-pacific/chinas-maritime-disputes/p31345#1/p31345).

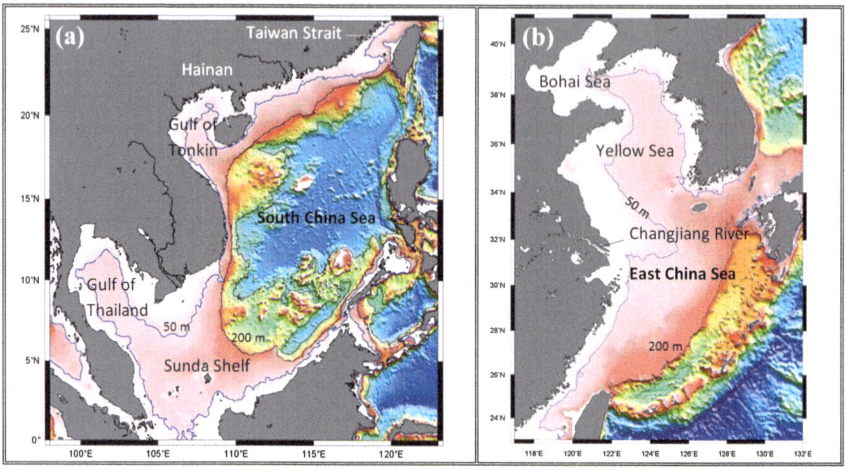

Fig. 8.2 Bathymetry of **a** the South China Sea and **b** the East China Sea and the Yellow and Bohai Sea. Image source: Liu and Dittert (2010). The shelf regions <200 m deep are shown in *pink shades*

The region includes a tremendous diversity of natural ecosystems, ranging from coral reefs, rocky and sandy shores, mangroves, estuaries and mudflats to open waters and abyssal habitats. Although the South China Sea contains a deep basin with maximum depths greater than 5000 m, large areas are relatively shallow (<200 m), particularly along the western boundary (Fig. 8.2a). These are influenced by both marine processes and riverine and terrestrial inputs, with about 120 rivers supplying nutrients and sediment. The two main continental shelf areas are the Gulf of Thailand and the Sunda Shelf in the southwest, and the northern shelf off the coasts of southeast China and northern Vietnam. Between these two shallow and wide shelves is the narrow slope east of Vietnam. Given its limited connection to the ambient open ocean, the South China Sea is a semi-enclosed Mediterranean Sea.

Most of the South China Sea lies within the subtropical and equatorial zones and its climate is controlled by the East Asian monsoon system (Wyrtki 1961). In boreal winter (December–February), the region is dominated by strong northeasterly monsoon winds, whereas in boreal summer (June–August) the winds reverse their direction to southwesterly and are upwelling-favourable over the northern inner continental shelf of the South China Sea. The southwesterly monsoon first appears in the central basin in May and then spreads over the entire basin in July and August (Shaw and Chao 1994).

The summer monsoon supports classical Ekman-driven coastal upwelling along the southern Vietnamese coast (Xie et al. 2003), downstream of the large island of Hainan (Jing et al. 2008), and along the northern coast of the Guangdong Province, which reaches from Hainan to the entrance of the Strait of Taiwan (Fig. 8.2). A coastal upwelling jet forms along most of the coastline from the southern

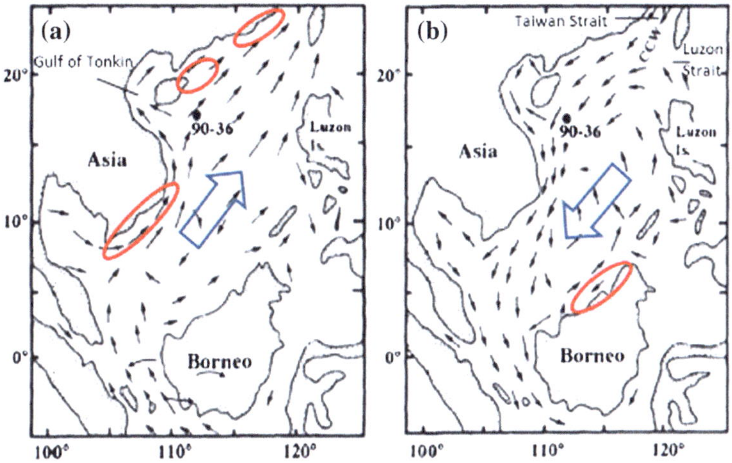

Fig. 8.3 Seasonal variations of mean monsoonal wind direction (*large arrows*), ocean currents (*small arrows*) and the location of coastal upwelling centres (*ellipses*) in the South China Sea. Modified after Chen et al. (1985)

Vietnamese coast into the Taiwan Strait, possibly influencing the nutrient budget of the Gulf of Tonkin (Fig. 8.3). The monsoon reversal in winter triggers localized coastal upwelling off northern Borneo (Yunwei et al. 2015).

Summer upwelling in the northern South China Sea was first described by Wyrtki (1961) off the coast of Hongkong and by Niino and Emery (1961) to the east of the Hainan Island and along the southern coast of Shantou (Fig. 8.4). Later, Guan and Chen (1964) systematically investigated the upwelling based on the First Nationwide Survey of China Seas during 1959–1960, and their research indicated that the summer East Asian (southwesterly) monsoon is the dominant dynamical driver of coastal upwelling.

Since then, intensive studies have been carried out to understand the local upwelling in the inshore areas off the coast of Shantou to Nanri island (referred to as *Yuedong Upwelling*) and the upwelling to the east of Hainan Island, the east of the Leizhou Peninsula and the southeast of Zhanjiang Bay (referred to as the *Qiongdong Upwelling*) using various field observations and modelling (Zeng 1986; Yu 1987; Han and Ma 1988; Han et al. 1990; Gan et al. 2009; Jing et al. 2009; Fig. 8.5). This work has led to the idea that the coastal winds, the Kuroshio Current passing through the Luzon Strait in the deeper layer (e.g., Su and Wang 1990), and complex topography (Morton and Blackmore 2000) all combine to produce a series of local upwelling cells rather than larger areas of continuous upwelling. For example, Hu et al. (2001) analysed AVHRR sea surface temperatures during 1996–1999, compared them with the survey data collected during 1997–1999, and found the new evidences of four regional upwelling regions in the Taiwan Strait. Interestingly, the water that upwells in the southern part of the Taiwan Strait contains only low concentrations of nutrients as it originates from a depth of 50–100 m at about 115° E

Fig. 8.4 Major seasonal coastal upwelling systems in the South China Sea

and is brought onto the shelf some 150–300 km to the south of its eventual upwelling site (Hu et al. 2015).

The South China Sea is classified as a Class II (moderate productivity) Large Marine Ecosystem. Regional marine fisheries are important to the food security and economy of the bordering countries and targeted pelagic species include flying fish, tuna, billfish, mackerels and sharks. Due to illegal fishing activities, total catch rates and trends are subject to substantial uncertainty. The biodiversity of the Chinese cold-water marine biota is not as high as that of the neighbouring Japanese waters, and the biodiversity of the warm-water species is distinctly lower than that of the Philippines–Indonesia–New Guinea triangle—the world's centre of tropical marine biota. Among Chinese shelf seas, however, the South China Sea has a greater biodiversity (~247 species per 100 km of coastline) than the East China Sea (~149 species) or the Yellow and Bohai Seas (~60 species) (Liu 2013). Nevertheless, Chinese marine fisheries production—from both capture fisheries and mariculture—has surpassed that of all other countries since the end of the last century. The serious overexploitation of living resources has led to declines in natural fish and shrimp stocks, and inhibited the sustainable development of marine fisheries.

Coastal development, destructive fishing practises, pollution and siltation have resulted in severe marine habitat and community modification in the South China Sea. Significant fractions of coral reefs have been degraded or are severe threat. The original area of mangroves has decreased by ~70 % during the last 70 years, and

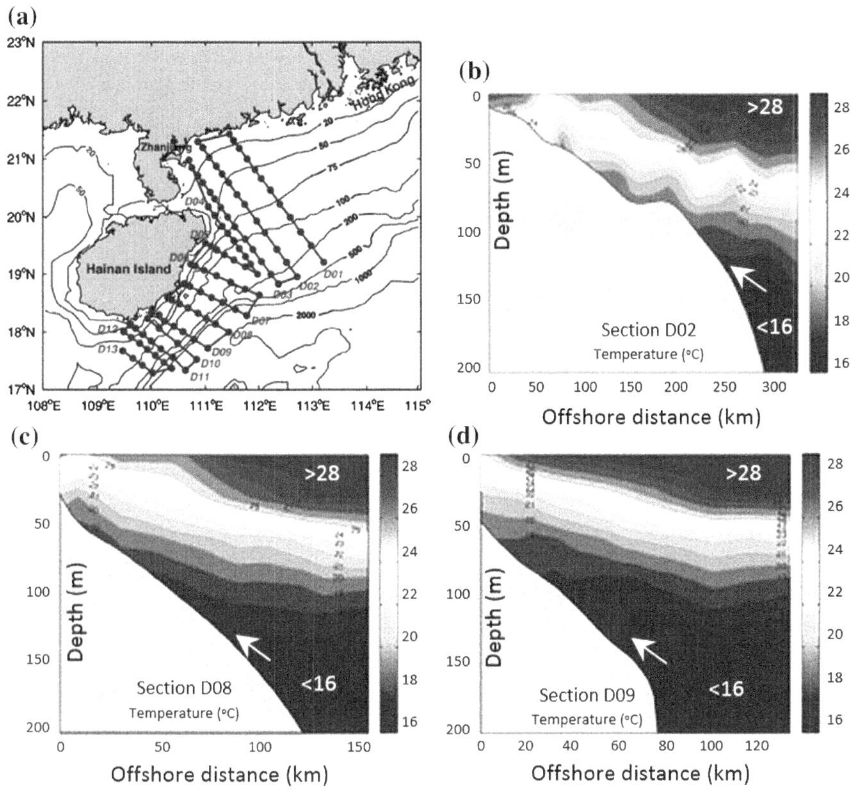

Fig. 8.5 **a** Station map of field survey during 19 July–06 August 2006 in the South China Sea. Vertical distributions of the observed temperature for transects **b** D02, **c** D08 and **d** D09 (from Jing et al. 2009)

there is evidence of widespread modification of seagrass habitats throughout the region, with 20–50 % of severe damage of seagrass beds (Talaue-McManus 2000).

8.2.2 *East China Sea*

The East China Sea together with the Yellow and Bohai Seas is a semi-enclosed sea surrounded by Japan, China and Korea with a wide and shallow continental shelf area (Fig. 8.2b). The continental shelf has a combined total area of about 0.9×10^6 km^2 and is one of the largest in the world. It is also one of the most productive areas of the world's oceans thanks to the high nutrient input by two of the largest rivers in the world, the Yangtze River (Changjiang) and the Yellow River (Huanghe), although flow in the latter has declined drastically since the 1960s. The

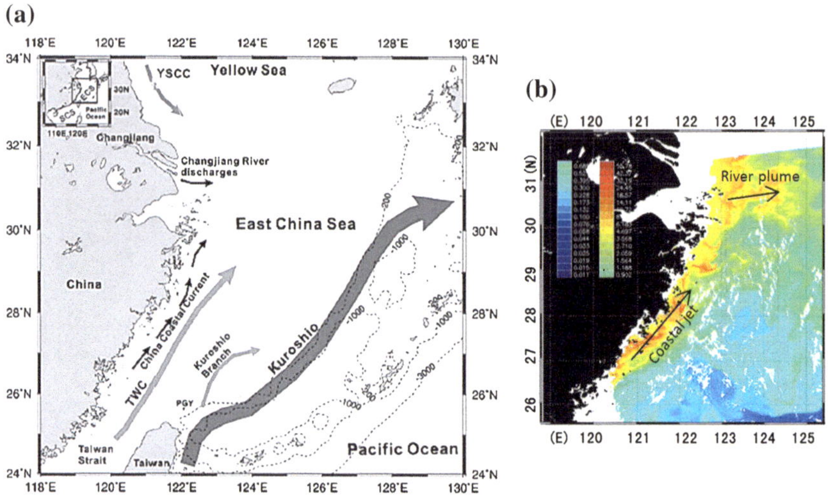

Fig. 8.6 **a** Summer currents in the East China Sea. *TWC* Taiwan Warm Current *YSCC* Yellow Sea Coastal Current. Taken from Chen et al. (2004). **b** Example of coastal phytoplankton blooms as seen in satellite chlorophyll-*a* concentrations. *Arrows* indicate upwelling and river-plume influences. Modified from Chang et al. (2012)

Changjiang River is the fifth largest river in the world in terms of volume discharge and the discharge peaks at about 40,000 m^3/s during the period from May to September (Chen et al. 2009). In contrast, the Yellow River has an annual discharge of about 1,800 m^3/s, although this can increase by an order of magnitude during summer floods. Apart from the rivers, the shelf also receives nutrients from the intermediate water of the Kuroshio Current (Chen 1996) which flows northeastward along the shelf break of the region (Fig. 8.6a). Overall, the East China Sea is classified as a highly productive (Class I) Large Marine Ecosystem.

The region comes under the influence of the southwest monsoon. While the coastal wind forcing of the summer monsoon should support classical coastal upwelling, the ocean dynamics of the region are far more complex as they involve interactions with river plumes, frontal instabilities, dynamical interactions with the Kuroshio Current, including associated exchanges across the shelf break, and tidal mixing. Mao et al. (1964) were to first to report the existence of summertime upwelling at 29° N. In summer, southwesterly winds create a coastal upwelling jet (Fig. 8.6a) known as the *China Coastal Current* (Chen et al. 2004) or the *Fujian-Zhejiang Current* (Yang et al. 2011). This interferes with the river plume of the Changjiang River, which otherwise tends to flow to the south along the coast (see Li et al. 2012). A detailed analysis of in situ hydrographic data for July and August 2006 (Du et al. 2011) shows that the wind-driven Fujian-Zhejiang Current operates to prevent the southward migration of the river plume in summer, such that river influences are confined to the northern part of the Zhejiang coast and only in nearshore waters (<20 m deep) elsewhere. The upwelling, coupled with the discharge from the Changjiang River and increasing stratification as the surface water warms, leads to the development of a large zone of hypoxia ($<\sim 20{,}000\ km^2$), largely to the north of the river mouth in summer within the 50 m isobath (e.g., see Rabouille et al. 2008; Wang et al. 2012 and references therein). This low oxygen water mass forms during late May/early June and may persist until November, depending on how the local wind field changes with the passage of frontal systems.

Because of this interaction between the river plume and the coastal current, typical summer phytoplankton patterns seen in satellite images (Fig. 8.6b) are presumably the result of two distinct processes:

(i) Classical coastal upwelling and/or tidal mixing along most of the Zhejiang coast, and
(ii) Eastward deflection of the *Changjiang River Plume.*

The analysis by Du et al. (2011) shows also that the bottom water of the Taiwan Warm Current is derived from the nutrient-rich intermediate water of the Kuroshio Current Hence, the Kuroshio Current is the main source of nutrients for coastal upwelling off the coast of Zhejiang Province. Nutrient budgets by Chen and Wang (1999) suggest that this contribution, especially for phosphorous, is far more than the inputs from the phosphorous-deficient Yangtze and Yellow Rivers. In addition, mass balance calculations reveal that the net denitrification rate is more than the total riverine supply of nitrogen. Tseng et al. (2014) have suggested that the East China Sea behaves as an overall sink of atmospheric CO_2 with a flux of 1–4 M $C/m^2/yr$ and that although the East China Sea contains only $\sim 0.2\ \%$ of global ocean area, it could yield an annual carbon uptake of 0.01–0.03 Gt C, representing 0.5–2.0 % of the global uptake.

Rapid development of coastal agriculture and the growth of coastal cities in China has caused an increase in the organic nitrogen and phosphate content of coastal waters. For instance, the nutrient input from the Changjiang River (Yangtze River) has been increasing dramatically since the 1960s. According to Zhou et al. (2008), at the mouth of the Changjiang River, the nitrate concentration has increased about three-fold in 40 years, from 20.5 μM/L in the 1960s to 59.1 μM/L in the 1980s and to 80.6 μM/L in 1990–2004. Phosphate concentration similarly increased by a factor of 30 %, from 0.59 μM/L in the 1980s to 0.77 μM/L in 1990–2004. As a result, the shelf of the East China Sea is highly productive year round, but phytoplankton abundance tends to increase in boreal summer from July to September, and attains a secondary peak in April (e.g., Wang and Wang 2007). However, the nutrient inputs have also led to eutrophication in coastal waters and embayments leading to harmful algal blooms of diatoms and dinoflagellates, so that the sea area adjacent to the Changjiang River estuary is one of the most notorious regions in China for such blooms. About one quarter of the recorded red-tide events have occurred in this region (Zhou 2010), although no formal records were recorded until the beginning of the 1980s. Red tides of *Noctiluca scintillans* were recorded in June and August 1981, while in 1982, red tides of *Trichodesmium sp.*, *Noctiluca scintillans* and *Ceratium sp.* were also reported. Since then, many more red tide events have been recorded, with 12 events from 1981 to 1989, and an additional 39 events from 1990 to 2001 (Jiang et al. 1992). The red tide hazards in eastern and northern China have resulted in the mass mortality of natural and cultured fish or/and shellfish populations and great economic damage (Qi et al. 2003; Zhou et al. 2008; Zhou 2012).

The launch of the Geostationary Ocean Color Imager (GOCI) by the Korea Ocean Satellite Center on 27 June 2010 made it possible to monitor red tides in northeast Asia at both high temporal (eight measurements per day) and spatial resolution. A bloom of *Prorocentrum donghaiense* along the coast of Zhejiang province from mid-May to early June 2011 provided one of the first opportunities to test GOCI's potential for studying red tides in a turbid-water environment. The 23-day bloom near Wenzhou and Taizhou (Fig. 8.7) caused fish mortality and had adverse impacts on the local economy as well as on the marine environment (see Lou and Hu (2014) for more details). While the development of red tides along the coast of the Zhejiang Province coincides with the season of upwelling-favourable winds and the onset of the China Coastal Current, details of these interactions remain unclear.

The East China Sea has experienced a dramatic 2 °C warming since 1982 (Belkin 2009), following more than two decades (1957–1981) of relatively stable temperatures. Data reanalysis indicates that climate warming has caused a weakening of both the winter and the summer monsoon winds over the East China and Yellow Seas since 1976, leading to the observed increase in temperatures (Cai et al. 2006). In summer, except for the northern region, warming occurred in most parts of the sea, including the Kuroshio and the Taiwan Warm Current. In winter, the fastest warming rate occurred in the Taiwan Warm Current, suggesting rapid warming of its source of its source, the Kuroshio Current.

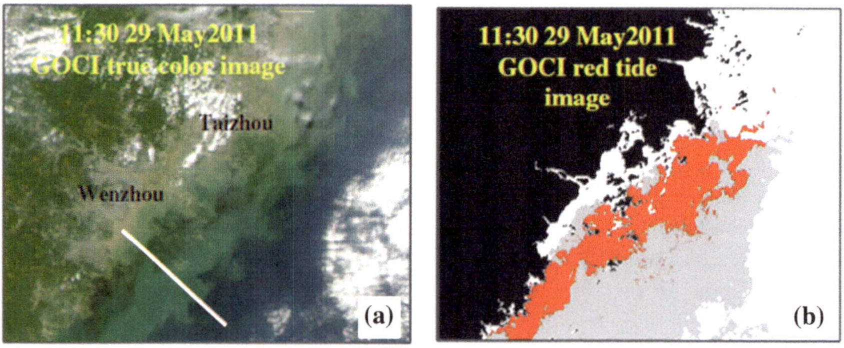

Fig. 8.7 True-colour and *red-tide* images recorded by GOCI on 29 May 2011 off the coast of Zhejiang Province (from Lou and Hu 2014)

Fish and other living resources are heavily exploited in the East China Sea and the adjacent Yellow Sea, and the coastal upwelling of the Zhejiang Province lies within the *Zhoushan Fishery Ground*, which is the largest fishing ground of China. Significant changes in fish biomass and catch composition in the region are attributed to overexploitation and pollution (Chen and Shen 1999). Chinese fisheries are recognizing the problems of severe overexploitation and have reduced fishing effort and intensity, particularly during the summer. A three-month suspension of fishing was initiated in 1995 (Tang 2006).

8.2.3 Indonesian Seas (Excluding South China Sea)

The bathymetry of the *Indonesian Seas*, also called *Australasian Mediterranean Sea* or *Indonesian Throughflow*, is complex as it involves numerous islands, narrow straits, ridges and deep ocean basins (Fig. 8.8). Indonesia is the world's largest archipelagic nation with a coastline exceeding 84,000 km and an estimated 17,805 islands. The region includes the Philippines-Indonesia-New Guinea triangle which is part of the *Coral Triangle*—one of the world's foremost biodiversity hotspots.

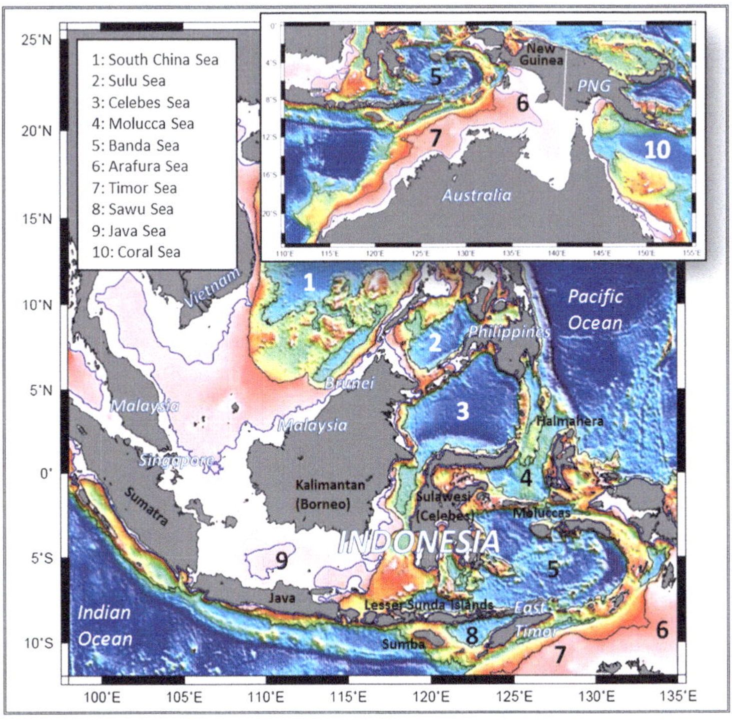

Fig. 8.8 Bathymetry of the Indonesian Seas. Image source: Liu and Dittert (2010)

The Indonesian Seas constitute a Class I Large Marine Ecosystem (>300 g C/m^2/yr), although many areas beyond the continental shelf are oligotrophic. Northern Australian waters are dominated by picoplankton, especially cyanobacteria (Burford and Rothlisberg 1999). Waters are relatively clear offshore and the euphotic zone can extend down to 100 m across the shelf. Primary productivity is limited by nutrient availability and the influence of winds and tides in mobilising sediments (Burford and Rothlisberg 1999). Nutrient discharges from rivers are restricted to the summer wet season and are highly variable within and between years. Tidal mixing is believed to be a major contributor to the nutrient dynamics of this generally shallow region. Monsoonal winds and tropical cyclones also contribute to nutrient enrichment in shelf waters (Condie and Dunn 2006). Major features of the sheltered, semi-enclosed waters of the region include mangroves and freshwater and estuarine wetlands, and between them Indonesia (3,112,989 ha) and Australia (977,975 ha) account for nearly 30 % of the global area of mangrove forest (Giri et al. 2011). Extensive coral reefs and seagrass beds are also symptomatic of the region.

The deep-water trenches, passages and current systems of the Indonesian Seas are increasingly being recognised as a global 'hot spot' and migratory pathway for many species of megafauna that frequently traverse the region. These include globally threatened cetaceans, such as blue, sperm, fin and humpback whales, turtles, elasmobranchii and other pelagic fish species (Dethmers et al. 2009). The Sawu Sea, for example, is recognised as one of the world's largest nurseries for six whale species, including humpback whales, pilot whales and the highly endangered blue whales. The waters around the island of Timor, which include the 3 km deep passages of the Ombai Strait and Timor Passage, adjacent to the north and south coast of Timor respectively, are particularly important as major and globally significant corridors for migratory marine species, especially megafauna (Dethmers et al. 2009). This region is also the only known nursery ground of the Southern Bluefin Tuna (*Thunnus maccoyii*) (Kennedy 1999).

The Indonesian Seas contain the warmest waters of the oceans, commonly known as the *Warm Pool* that plays a central role in the development of El Niño events, and the climate of this region is dominated by the tropical monsoon system with a dry winter (November–April) and a rainy summer (May–October). The wet tropical islands of the Indonesia Seas receive between 5–10 m of annual rainfall, the runoff from which supplies the coastal seas with a disproportionately large contribution of organic matter, nutrients and sediments. Precipitation and the resulting continental runoff create strong density stratification in the upper ocean (e.g., Chu et al. 2002), which isolates deep nutrient-rich (and oxygen-depleted) waters and prevents them from reaching the surface mixed layer from the thermocline or deeper. Hence, wind-driven coastal upwelling and other forms of upwelling play an important role in supplying nutrients in this region.

Owing to the great variety of coastline orientations, coastal upwelling occurs on many different shelves within the Indonesian Seas and through several different forcing mechanisms. For instance, the southeast monsoon (June–August) triggers seasonal coastal upwelling along the southern coastlines of Java, Sumatra, the

Lesser Sunda Islands and Timor Leste (East Timor) (see Fig. 8.1; Susanto et al. 2001; Ningsih et al. 2013). In contrast, year-round, tidal mixing brings nutrient-enriched water to the surface over the Sulu Ridge (the ridge between the Sulu Sea and the Celebes Sea) (Jing et al. 2012).

Probably the most significant area of marine primary productivity of the Indonesian Seas is the shallow Arafura Sea, which is only ~40–50 m deep (Fig. 8.9a). Freshwater input to the Arafura Sea comes mostly from more than thirty southwest-flowing rivers in Papua and, secondarily, from some rivers flowing into the Gulf of Carpentaria and onto the northern Australian shelf. Satellite data indicate large areas of chlorophyll-*a* concentrations in excess of 10 mg/m^3, which peaks during the southeast monsoon. While riverine nutrient discharges are a possible candidate for the creation of phytoplankton blooms, they cannot explain the concurrent temperature decrease in shelf water by up to 2 °C (Fig. 8.9b), which points to the existence of upwelling of colder water to the surface.

Wyrtki (1958, 1961) and Rochford (1966) were the first to examine the oceanography of the Banda and Arafura Seas, followed by the more detailed observations of the Indonesian-Dutch Snellius-II Expedition (1984/1985), which focussed on temperature and salinity distributions (Zijlstra et al. 1990), nutrient concentrations (Wetsteyn et al. 1990), chlorophyll distributions (Gieskes et al. 1988), and zooplankton abundance (Baars et al. 1990). Data from an early field expedition in August 1964 revealed the existence a high-nutrient water mass in the Arafura Sea that was derived from Banda Sea slope water from depths of 100–150 m (Rochford 1966). The data indicated that the resulting high surface oxygen saturation (139 %) and the accumulation of organic phosphorous in near-bottom waters of the Arafura Sea were the result of an uplift of Banda slope waters. A later study by Moore et al. (2003) confirmed that upwelling of Banda slope waters during the southeast monsoon is a regular phenomenon. While Gordon and Susanto (2001) showed that larger-scale Ekman pumping induced by the wind-stress curl over the Banda Sea initiates upwelling in the eastern Banda Sea from April through

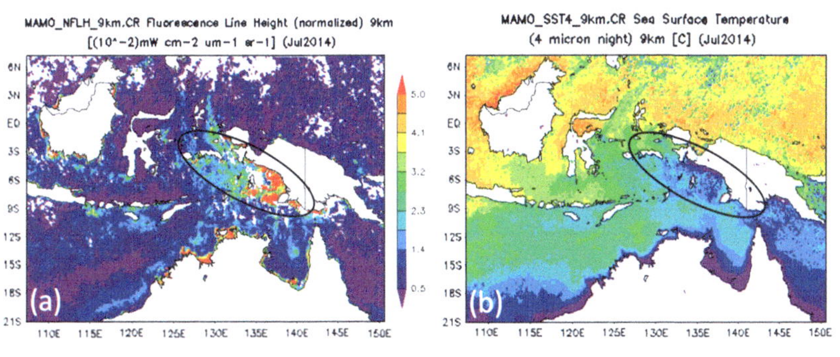

Fig. 8.9 Example of a massive phytoplankton bloom in the Arafura Sea in July 2014. Panel **a** shows fluorescence line height, which can be taken as a measure of chlorophyll-*a* concentrations. Panel **b** shows sea surface temperatures. Data source: NASA (Giovanni)

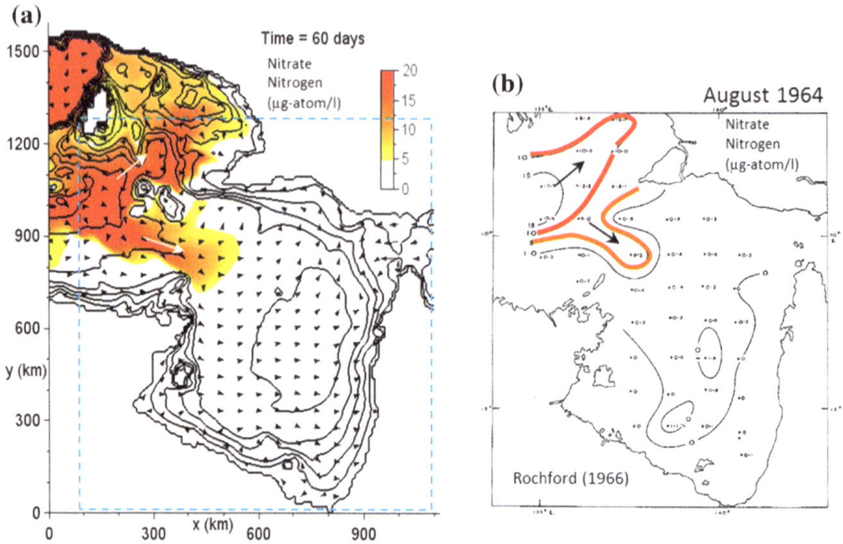

Fig. 8.10 **a** Numerical simulation and **b** observational evidence (Rochford 1966) of the pathway of nitrate nitrogen in the Arafura Sea during the southeasterly monsoon. Taken from Kämpf (2016)

December, this mechanism does not explain how upwelled water is moved horizontally over vast distances (~300 km) into the northwestern Arafura Sea.

Only recently, Kämpf (2015, 2016) demonstrated that the southeasterly monsoon creates widespread undercurrents in the Arafura Sea that are the principal source of import and upwelling of Banda Sea slope water into the northwestern Arafura Sea (Fig. 8.10). The undercurrents are a signature of the classical lee effect ("Leewirkung" in German) (Hela 1976), which is frequently observed in lakes (e.g., Monismith 1986) and coastal oceans (Svannsson 1975). The lee effect develops also in the Arafura Sea for three reasons: (i) the bay-like configuration of the Arafura Sea, (ii) its shallow-water depth, and (iii) its close vicinity to the equator. The bay-like configuration creates the required eastward pressure-gradient forcing; the other two features lead to partially or fully overlapping surface and bottom Ekman layers, which facilitates the creation of undercurrents.

The fisheries of the region are complex and diverse, reflecting the region's extraordinarily heterogeneous geography and species richness. While most of fish catch is artisanal, industrial fisheries contribute considerably more in terms of value since they target high-value shrimp and tuna stocks. Major species caught include tuna, sardines, anchovy, mackerel, as well as a range of reef fishes. The coastal and marine biodiversity and resources of the Coral Triangle are, however, under serious threat, with increased exploitation fuelled by exploding population growth and poor management (Burke et al. 2011), so that many fish stocks have been exploited well beyond biological limits and their continued viability is in question. This is especially the case in the coastal zone, which is exploited by 85 % of Indonesian fishermen and supports approximately 60 % of its 212 million people (Alongi et al. 2011). Arguably the best empirical evidence for overfishing comes from the Aru Sea and the Arafura Sea, where high intensity fishing has caused severe overexploitation of both demersal fish and shrimp stocks. The Ministry of Marine Affairs and Fisheries reported that catch per unit effort in the Arafura Sea declined from 95 kg/haul in 1974 to 38 kg/haul in 1996. While shrimp harvests are being taken beyond the maximum sustainable yield, pelagic fish stocks are still in a moderate state of exploitation (Alongi et al. 2011).

8.2.4 Australia's Southern Shelf

The southern shelf of Australia (Fig. 8.11a) is located in an arid climate region largely devoid of continental runoff. The shelf, which exceeds 3000 km in length, includes the Great Australian Bight, two large inverse estuaries, Spencer Gulf and Gulf St. Vincent, and the shallow Bass Strait between Tasmania and the mainland. The surface circulation on the continental shelf is dominated by an eastward flow that follows the shelf break either as the extension of the Leeuwin Current (Ridgway and Condie 2004) or as the South Australian Current forming during austral summer in the western Great Australian Bight (Herzfeld 1997). On the western shelf of Tasmania, this flow continues southward as the Zeehan Current. Below shelf-break depth, the Flinders Current flows westward on the upper continental slope (Middleton and Bye 2007).

The upwelling season on the southern shelf of Australia occurs during austral summer months (December–April; see Fig. 2.7). Coastal upwelling is associated with synoptic anticyclonic weather systems that transiently induce upwelling-favourable southeasterly coastal winds (Fig. 8.11b). The existence of coastal upwelling along the Bonney Coast, known as the *Bonney Upwelling* (Fig. 8.12), was first described by Rochford (1977) and Lewis (1981). Interestingly, the discovery by Rochford (1977)

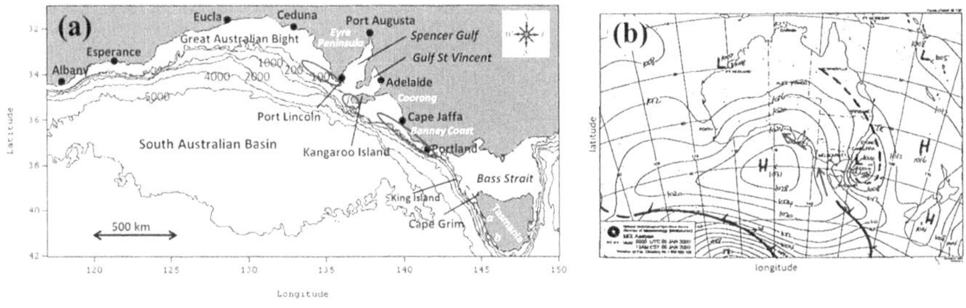

Fig. 8.11 **a** The geography of Australia's southern shelves. Isobath depths are in metres. Ellipses display known locations of coastal upwelling centres. **b** Mean sea-level pressure field for 6 January 2000, courtesy of the Bureau of Meteorology (Australia) http://www.bom.gov.au/cgi-bin/charts/. *Arrows* indicate upwelling-favourable coastal winds. Taken from Kämpf (2015)

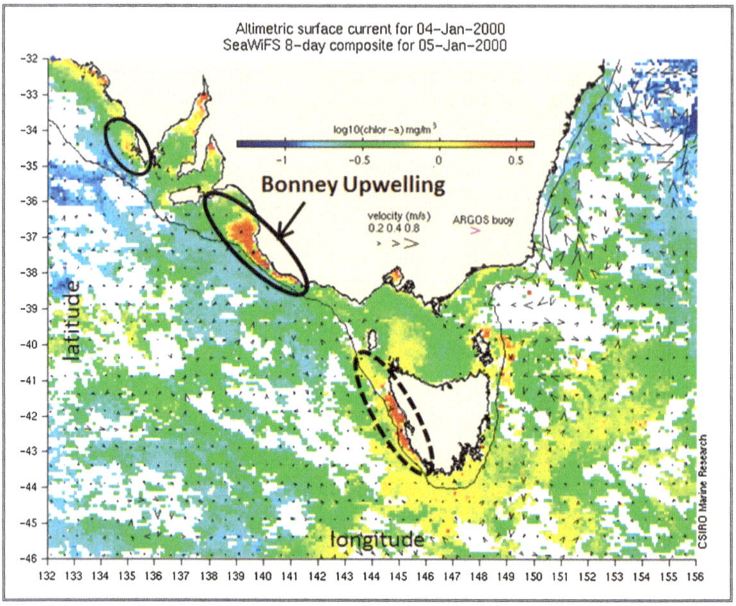

Fig. 8.12 Example of establishment of upwelling centres (highlighted by *ellipses*), seen as elevated chlorophyll-*a* concentrations, on Australia's southern shelf and the west coast of Tasmania. Taken from Kämpf (2015)

was based on an upwelling event in austral winter, which is a rare occurrence (see Kämpf 2015). Based on in situ observations and satellite data, Kämpf et al. (2004) identified other upwelling centres in the eastern Great Australian Bight (GAB), namely off the southern tip of the Eyre Peninsula and off the southwestern coast of Kangaroo Island at the southern end of Gulf St. Vincent. The existence of upwelling in the eastern GAB supports a large biomass of sardine which is vital as food for juvenile Southern Bluefin Tuna (*Thunnus maccoyii*) (Kennedy 1999). This so-called *Great South Australian Coastal Upwelling System* is also an important food source for Blue Whales (*Balaenoptera musculus*) (Gill et al. 2011). Kämpf (2015) has recently identified another upwelling centre off the west coast of Tasmania (Fig. 8.12).

While upwelling on the narrow continental shelf along the Bonney Coast and western Tasmania follows the classical Ekman theory, the upwelling dynamics on the wider shelf of the eastern Great Australian Bight is more complex (Kämpf 2010). It involves a preconditioning process during which a subsurface pool of nutrient-rich water is formed to the west of Kangaroo Island, known as the *Kangaroo Island Pool* (McClatchie et al. 2006), followed by a sequence of coastal wind events that induce the upwelling of nutrient-rich pool water to the surface. Water mass analysis (Kämpf 2010; see Fig. 3.4) indicates that upwelling water comes from an average depth of ~310 m, which is 110 m below shelf-break

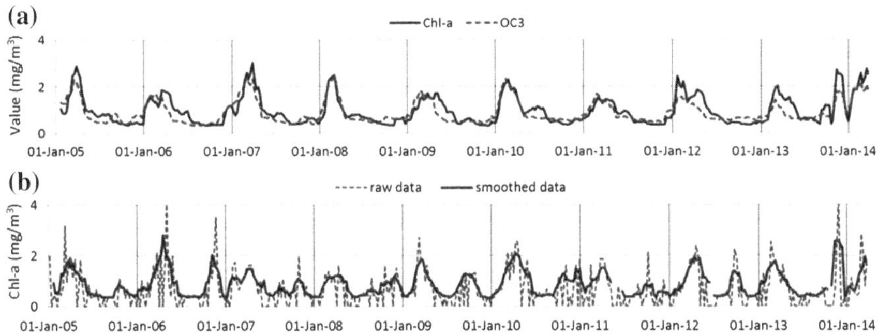

Fig. 8.13 Reconstructed time series of chlorophyll-*a* concentrations for upwelling centres off the Bonney coast and the west Tasmanian coast (from Kämpf 2015)

depth, and that localized subsurface upwelling in submarine shelf-break canyons (located south of Kangaroo Island) plays a fundamental role in the formation of the Kangaroo Island Pool.

There are generally only two or three larger, episodic upwelling events during each year's upwelling season associated with synoptic weather patterns (Kämpf et al. 2004), and maximum chlorophyll-*a* concentrations stay below 3 mg/m^3 (Fig. 8.13). This is due largely to the fact that nutrient concentrations in the surface mixed layer are consistently low, particularly for silica, but the clarity of the water means that the euphotic zone can extend below the mixed layer to where the nutrient concentrations are somewhat higher. Productivity therefore depends on how much water from the high nutrient pool makes it into the euphotic zone, and the overall productivity of the region is relatively low at less than 800 mg C/m^2/day apart from hotspots close to the upwelling centres where it can reach up to 3.9 g C/m^2/day (van Ruth et al. 2010a, b). This low productivity explains why the region does not classify as a Large Marine Ecosystem, but because of its importance as a key regional feeding zone for juvenile southern Bluefin tuna, blue whales and other large pelagic species (i.e., sharks, seals and sea lions), it is significant.

The onset of industrial fishing in the 1950s, in conjunction with improving technologies and the knowledge of migration routes, has led to the exploitation of southern bluefin tuna across its entire range. After peak catches in the late 1970s the total population of southern bluefin tuna has since declined by 85 % over 36 years from 1973 to 2009 so that it is now critically endangered and there is no sign of recovery yet (Collette et al. 2011), while despite continual monitoring of individual country catch quotas, fishing pressure is still a major concern. After the collapse of Australia's longline tuna fisheries in the early 1980s, Australia's catches started to focus on juvenile tuna in the eastern Great Australian Bight and developed a lucrative fish farming economy at the southern end of the Eyre Peninsula.

8.2.5 Upwelling Around New Zealand

In New Zealand, upwelling occurs along the eastern shores of Northland, north-eastern New Zealand, and in places along the South Island's west coast (Fig. 8.14a). Upwelling of cold water to the sea surface off Farewell Spit (Fig. 8.14b) was first observed by Garner (1954) and further documented by Stanton (1976). The upwelling centre is a source of nutrient enrichment of the generally oligotrophic eastern boundary of the Tasman Sea and results in locally elevated zooplankton abundance (Bradford and Roberts 1978; Bradford 1983). Bowman et al. (1983) observed the upwelling as an elongated cold-core plume extending from Kahurangi Point northwards into the South Taranaki Bight. Upwelling is generally shallow; i.e., the upwelled water comes from a depth of ~ 100 m (Shlrtcliffe et al. 1990).

The existence of coastal upwelling on South Island's west coast is interlinked with the existence of the northward-flowing Westland Current first identified by Brodie (1960), which itself is a response to the southwesterly winds, prevalent along the west coast, as suggested by Stanton (1976). A study by Shlrtcliffe et al. (1990) indicates that upwelling near Cape Farewell depends on the existence of the intermittent Westland Current and that it is intensified by transient onshore winds. Various field measurements, including hydrography and float deployments, indicate that the formation of the Westland Current and its associated upwelling events are intermittent phenomena (e.g., Shlrtcliffe et al. 1990), presumably due to the

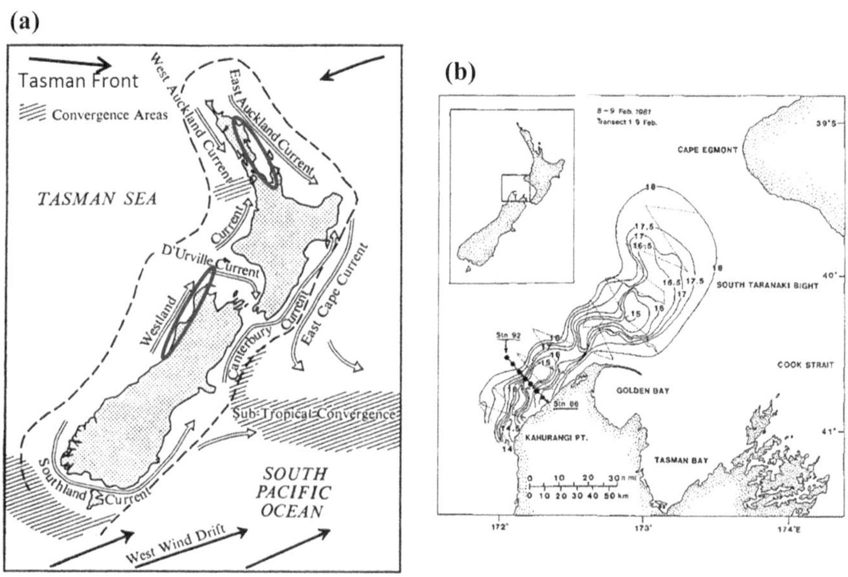

Fig. 8.14 **a** Schematic of currents around New Zeeland. **b** Temperature observations of the Cape Farewell upwelling plume (from Foster and Battaerd 1985)

intermittency of the coastal wind forcing, and that the current flows for no longer than about 4–6 weeks at a time, and generally only for up to about 15 days.

Bowman et al. (1983) discussed the physics of the upwelling system and concluded that on that occasion it resulted from wind forcing of the Westland Current around the coastal bend at Kahurangi Point and over the Kahurangi Shoals and Paturau Bank just to the north (Fig. 8.14b), where anticyclonic eddies and meanders are formed. Southerly winds modulate the intensity of the Westland Current and the shape of the upwelling plume. In the Bight, the plume can be advected eastwards in the D'Urville Current before being mixed by turbulence or by viscous dissipation. Cyclogenesis involved in the formation of the plume may help to keep colder eddy centres more or winds, known as the about cross-frontal mixing of waters of inshore and offshore origins.

Intermittent wind-driven coastal upwelling from a depth of ~ 100 m occurs also along the eastern shores of Northland and has been previously observed in the outer Hauraki Gulf (e.g., Zeldis et al. 2004) and the Bay of Plenty (Longdill et al. 2008).

8.3 Northern Indian Ocean

8.3.1 Overview

Upwelling in the northern Indian Ocean (Fig. 8.15) occurs annually as a seasonal phenomenon during the southwest monsoon (June–September) off the coasts of Somalia and eastern Arabia (Yemen and Oman), the southwest coast of India, and the south coast of Sri Lanka, driven by the wind stress. There is also upwelling offshore, driven by wind stress curl inshore of the strongest winds, known as the *Findlater Jet* (e.g., Smith and Bottero 1977). Three major international research programmes have previously been undertaken in the Arabian Sea of the Indian Ocean:

(1) The first International Indian Ocean Expedition (IIOE) in the 1960s;
(2) The Indian Ocean programme of Joint Global Ocean Flux Study (JGOFS) during 1992–1997; and
(3) The Indian Ocean expedition of the World Ocean Circulation Experiment (WOCE) from late 1994 to early 1996.

There has also been a decade-long deployment of a cabled observatory off the coast of Oman from 2003–2013 (du Vall et al. 2011; Wang et al. 2013). This ceased operation in 2013, but proved the concept as a way of monitoring the ocean in this region.

During the IIOE in the 1960s, a fairly large number of physical, chemical, and biological observations were made off the southern seaboard of Oman, most of which extended from the sea surface to the seabed. Since the upwelling is a seasonal phenomenon associated with the southwest monsoon during the summer months from May to October, the observations made by R.R.S *Discovery* in 1963

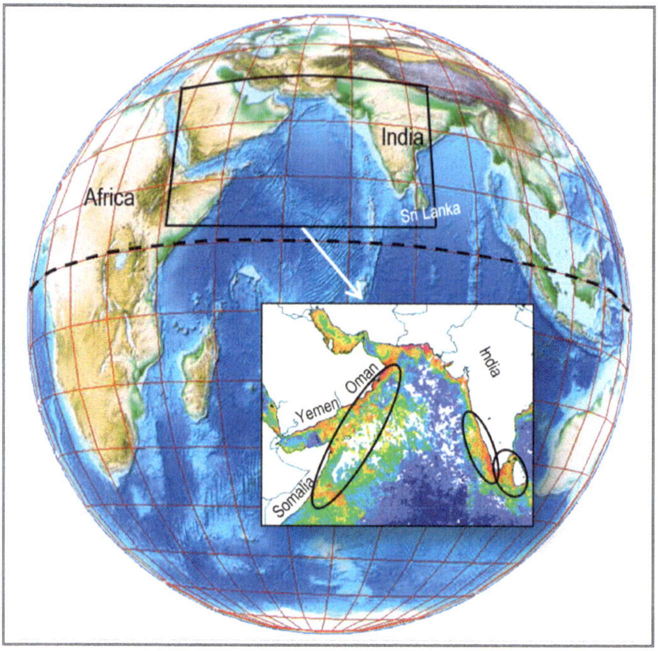

Fig. 8.15 Seasonal coastal upwelling regions in the northern Indian Ocean as seen in satellite-derived elevated chlorophyll-*a* concentrations. Data source: Giovanni (NASA). Background image source: Google Earth with ETOPO 1 bathymetry layer

are of particular interest as they formed the first integrated coverage of the area during the season of active upwelling and highest productivity. Wyrtki (1973), Currie et al. (1973), Schott (1983) and Currie (1992) have summarized the early research findings. The JGOFS program also carried out an extensive series of cruises, led by Dutch, German, Indian, Pakistani, U.K. and U.S. researchers during the early 1990s. The U.S. program off Oman investigated the physics, chemistry and biology of the region from September 1994–December 1995, while simultaneous WOCE cruises provided information on the larger scale variability of the Arabian Sea region. An overview of the hydrographic data resulting from the U.S. JGOFS program is given by Morrison et al. (1998), and numerous other papers on WOCE and JGOFS studies have been published in special issues of Deep-Sea Research II.

8.3.2 Somali Current

The southwesterly monsoon, commencing in May–June, blowing more or less parallel to the Somali and Arabian coasts. It reaches its peak in July–August and

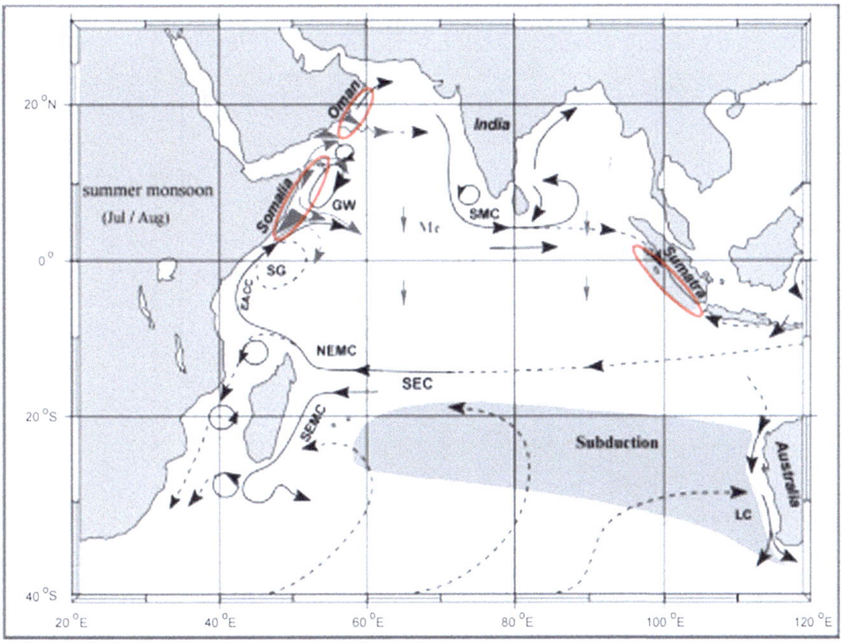

Fig. 8.16 A schematic representation of identified current branches in the Indian Ocean during the Southwest Monsoon. Marked are the winter subduction area in the southern subtropics, upwelling areas (*ellipses*) and southward Ekman transport on both sides of the equator (*small arrows*). Current branches indicated are the South Equatorial Current (SEC) , Northeast and Southeast Madagascar Current (NEMC and SEMC), East African Coast Current (EACC), Somali Current (along coast of Somalia), Southern Gyre (SG) and Great Whirl (GW), Southwest Monsoon Current (SMC), and Leeuwin Current (LC). Modified from Schott et al. (2002)

fades away in September–October. Figure 8.16 displays the surface circulation that establishes during the southwesterly monsoon. The southwesterly monsoon coincides with the existence of strong, northward flowing western boundary current, known as the *Somali Current*. Likewise a pronounced flow to the northeast is evident off the Arabian coast (Royal Society 1963), called the *East Arabian Current*. The Somali Current is the swiftest boundary current of the global oceans reaching speeds exceeding 3 m/s (Swallow and Bruce 1966; Schott 1983; Fischer et al. 1996). Associated with the creation of these boundary currents is classical wind-driven Ekman upwelling of subsurface water from a depth of about 100–150 m (Wyrtki 1973) (Fig. 8.17; Currie et al. 1973). Coastal upwelling also occurs along the coast of Oman (Fig. 8.18), the water being sourced from the same depth level as further south, and corresponding in density to the core of the highly saline Arabian Sea Water, which is formed locally by evaporation and cooling during the northeast monsoon but includes a small fraction of denser water coming from the Persian Gulf outflow (Morrison et al. 1998).

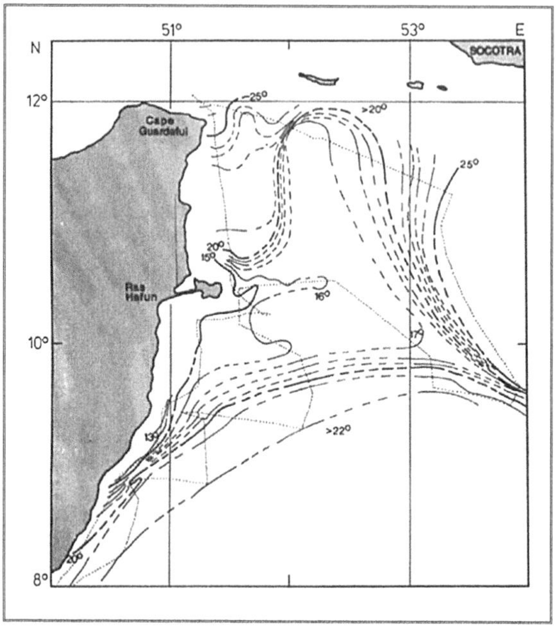

Fig. 8.17 Coastal upwelling along the coast of Somalia seen in distributions of surface temperature (°C) in the summer of 1964. ("*Discovery*" and "*Argo*" observations) (from Currie et al. 1973)

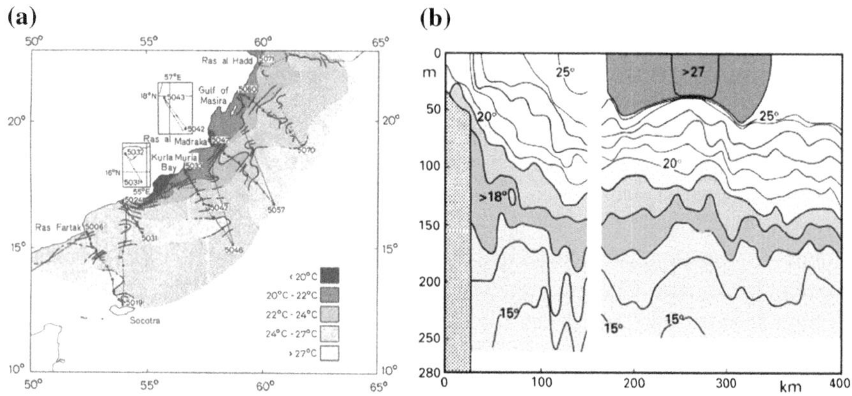

Fig. 8.18 Coastal upwelling along the coast of Oman. **a** Horizontal surface temperature distribution off the SE Arabian coast during the SW monsoon of 1963 from observations made by *RRS Discovery*. **b** A temperature section across the Arabian upwelling region during July 1983, from the Kuria Muria Islands (18° N) towards the southeast. The break in the isotherm slope about 150 km from the coast separates the coastal upwelling from the Ekman suction region associated with a warm anticyclonic eddy of temperatures >27 °C. After Currie et al. (1973)

The extent of the southeastern Arabian upwelling region has been recorded previously (Puff 1890; Bobzin 1922; Royal Society 1963; Currie 1964; Bottero 1969; Currie et al. 1973; Bruce 1974; Smith and Bottero 1977). The region occupies a length of nearly 1000 km of the coast of Oman extending from Ras Fartak in the West, to Ras al Hadd at the mouth of the Sea of Oman in the East, but the areas of most pronounced negative sea-surface temperature anomalies are found in the vicinity of the Kuria Muria islands (17° 30′ N, 56° 0′ E) and around Ra's al Madrakah (19° 0′ N, 57° 50′ E).

In response to the onset of the southwesterly monsoon two large anticyclonic eddies develop annually off the east coast of Africa. These are the *Southern Gyre* near the equator and the *Great Whirl* at about 10° N. Both eddies are about 500 km in diameter (Figs. 8.16 and 8.19; Swallow and Fieux 1982; Schott et al. 1997; Beal and Donohue 2013). The Great Whirl contributes substantially to the upwelling along the northern Somali coast and to the eastward transport of cold water into the interior Arabian Sea. As such, the Great Whirl is part of the upward branch of the shallow tropical–subtropical circulation cell in the Indian Ocean and strongly affects the heat balance of the Arabian Sea (Schott et al. 2002). The Great Whirl extends throughout the upper 1500 m of the water column, but current speeds above 1 m/s are only observed in the top 200 m.

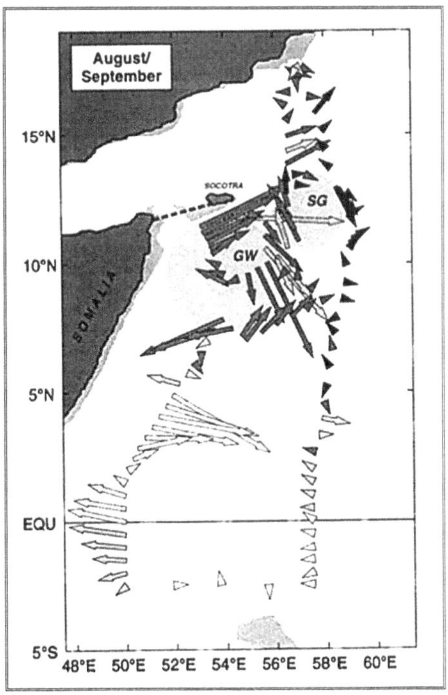

Fig. 8.19 Near-surface currents at 25 m from shipboard ADCP measurements during R/V *Meteor* cruises of August/September 1995. Water masses are characterized by salinity classes shown by shading grades; approximate locations of the Great Whirl (GW) and Socotra Gyre (SG) are shown as shaded circles (from Schott et al. 1997)

The strength of the Somali Current and the variability of the annual wind stress mean that downstream propagation of physical and chemical features is highly variable. This variability is found throughout the upper 1,000 m of the water column, and can be seen in satellite imagery, temperature/salinity relationships, and current velocities (Morrison et al. 1998). Somewhat surprisingly, the U.S. JGOFS program found that during the 1994–1995 period, deepest mixed-layer depths occurred during the weaker northwest monsoon rather than during the southwest monsoon, as had been expected (Morrison et al. 1998).

The Somali Current upwelling system is classified as a highly productive ecosystem (>300 g C/m^2/yr). The seasonal upwelling off Somalia is one of the most intense coastal upwelling systems of the world (Baars et al. 1998; Bakun et al. 1998). However, the mean phytoplankton density and productivity are lower than expected, presumably because of dilution effects due to the strong winds and the high speed of the Somali Current (Baars et al. 1998). In some years, the highest productivity has been noted 200 km offshore, likely due to the dynamics of mesocale eddies enhancing productivity, rather than as the consequence of coastal upwelling (Hitchcock and Olson 1992). Euphausiids make up about 25 % of the total zooplankton biomass, while copepods make up most of the remainder (Okemwa 1998). Fisheries in the Somali Current zone largely consist of artisanal fisheries, which are confined to its inshore areas, due to the lack of technology to fish in offshore waters, and there appears to be low diversity with considerable endemism, especially in the Red Sea (van der Elst et al. 2005). Following several years of research, a fishery concentrating on mesopelagic organisms, particularly *Benthosema pterotum*, began in 1996 (FAO 1997). The fish stock statistics for the region are unreliable given the lack of both information and management practices, and even landings are not well determined, but an FAO estimate for 1999 was 2.2 million tonnes for the entire Arabian Sea. Much of this was caught along the Indian coast and consisted of herrings, sardines, anchovies and crustaceans, while offshore foreign fleets have concentrated on tuna (FAO 2003a, b).

Farther north, in the Arabian Sea, the presence of the oxygen minimum zone (see Sect. 8.3.5) has a major effect on the presence of pelagic organisms. This was particularly marked during the U.S. JGOFS study, and there was an overall inverse relationship between organisms and oxygen concentration. Large numbers of diel migrators, mainly copepods, euphausiids, salps and myctophid fish, however, were found to move into surface waters during the night but utilized the suboxic waters of the oxygen minimum layer during the day (Smith et al. 1998; Wishner et al. 1998; Morrison et al. 1999). Because of the oxygen minimum zone, it is expected that a large percentage of sinking biological material passes through this layer relatively untouched, providing more food resources to organisms in the deeper waters than would normally be the case, and an increase in organisms was seen also at the lower boundary of the oxygen minimum zone.

8.3.3 Southwest Indian Shelf

Winds of the summer monsoon attain a slight southward component along the southwest coast of India and also initiate wind-driven upwelling there (Fig. 8.15; Jayaram et al. 2010). Although this upwelling is less intense compared with other upwelling regimes of the Arabian Sea, it has profound impacts on the coastal fisheries of India. While the west coast of India accounts for 70 % of the total Arabian Sea fish production (Luis and Kawamura 2004), the southwest coast alone accounts for 53 % (Sanjeevan et al. 2009). As a result of the upwelling in the region, the sediments along the western coast of India are organic-rich and anoxic (Naqvi 1991; Naqvi et al. 2000), similar to the situation in other eastern boundary upwelling areas. The anoxia arises because of the low oxygen concentration in the upwelling water (~ 22 μM) coupled with local production, and anoxic conditions can cover the whole of the shelf zone along the Indian west coast (Naqvi et al. 2000). This low oxygen zone is separate from the main zone of low oxygen in the northern Indian Ocean, and is characterized by high rates of denitrification and production of nitrous oxide.

8.3.4 Sri Lanka

Upwelling off Sri Lanka also occurs during the southwesterly monsoon (Fig. 8.15) and chlorophyll concentrations (>5 mg/m^3) have been recorded, particularly along the country's southern coast, which appears to be a major upwelling region (Vinayachandran et al. 2004). These elevated chlorophyll concentrations persist for more than 4 months and have been attributed to coastal upwelling, advection by the South Monsoon Current and open-ocean Ekman pumping (Vinayachandran et al. 2004). During the northeasterly monsoon, on the other hand, chlorophyll concentrations appear to be low, but there is evidence of high productivity through the documented feeding aggregations of blue whales (*Balaenoptera musculus*) along the southern coast of Sri Lanka (de Vos et al. 2014a). Recent modelling studies (de Vos et al. 2014b) indicate that the upwelling is generated by different mechanisms during the two monsoon periods.

8.3.5 Chemistry and Productivity

Using in situ data during collected 1992–1997, under the JGOFS Indian programme, Kumar et al. (2000) show that lateral advection of nutrients by mesoscale eddies supports productivity in central parts of the Arabian Sea during the summer monsoon. They also showed that relatively high productivity persists in the northern Arabian Sea during the winter monsoon, supported by nutrients entrained by convective mixed-layer deepening (this was also found by U.S. researchers; see Marra and Barber 2005). Findings of an eddy-resolving bio-physical model indicate that:

(i) eddy-induced vertical advection provides the major input of nutrients (up to 60–90 %) supplied to upwelling regions during the early stage of the summer bloom period (Resplandy et al. 2011), and
(ii) mesoscale filaments are the main process during the summer bloom period exporting 50–70 % of the nutrients from coastal upwelling regions into the central Arabian Sea, which is in general agreement with observational evidence (Kumar et al. 2000).

Phytoplankton biomass in the Arabian Sea varies both temporally on diurnal to seasonal scales and spatially at fine and large scales (Marra and Barber 2005), although it appears to be limited by neither light nor nutrient availability. This is somewhat surprising given that nitrate is always more limiting than phosphate in this region. Phytoplankton productivity, as measured by satellite-derived chlorophyll-*a* concentrations, is at a maximum in February and August–September, i.e., at the end of both monsoon periods, after mixing of the upper water column brings additional nutrients to into the euphotic zone, and the suggestion is that it is the rate of mixing within the water column that exerts the most control on productivity, along with grazing by zooplankton. Typical productivity values are in the range of 1.7–2.5 g $C/m^2/day$, but these vary enormously and are considerably reduced during the inter-monsoon periods (Marra and Barber 2005), so that mean rates during the SW and NE monsoons are more in the range of 1.0–1.7 g $C/m^2/day$ (Smith 2001).

Observations suggest that the Indian Ocean region north of 18° S is a net source of atmospheric CO_2, while the region south of 18° S is a net sink (Fig. 8.20; Sarma et al. 2013). The large source of CO_2 to the atmosphere is driven by coastal upwelling in the northwestern Indian Ocean, i.e., off the coasts of Somalia and Oman coast, where pCO_2 levels as high as >600 μatm have been reported during peak southwesterly monsoon periods (e.g., Kortzinger and Duinker 1997). While the most intense changes in pCO_2 occur during the southwest monsoon, the area that is affected is less at that time than during the northeast monsoon, because convective mixing occurs over a larger area. Offshore, however, changes in pCO_2 are less than 10 μatm between monsoonal and inter-monsoonal periods (Smith 2001). The South Equatorial Current normally situated at 15°–18° S (Schott and McCreary 2001) is the boundary between surface waters of high pCO_2 with CO_2 fluxes to the atmosphere, and waters of relatively low pCO_2 and CO_2 fluxes into the sea (Sabine et al. 2000). This transition zone is more easily observed in subsurface features than at the surface, marking the boundary between the low oxygen, high-nutrient waters of the northern Indian Ocean and the high oxygen, low nutrient values of the subtropical gyre (Wyrtki 1973). While the reversal in sign across the South Equatorial Current is clear, trends in total CO_2 fluxes in the Indian Ocean are uncertain (Sarma et al. 2013).

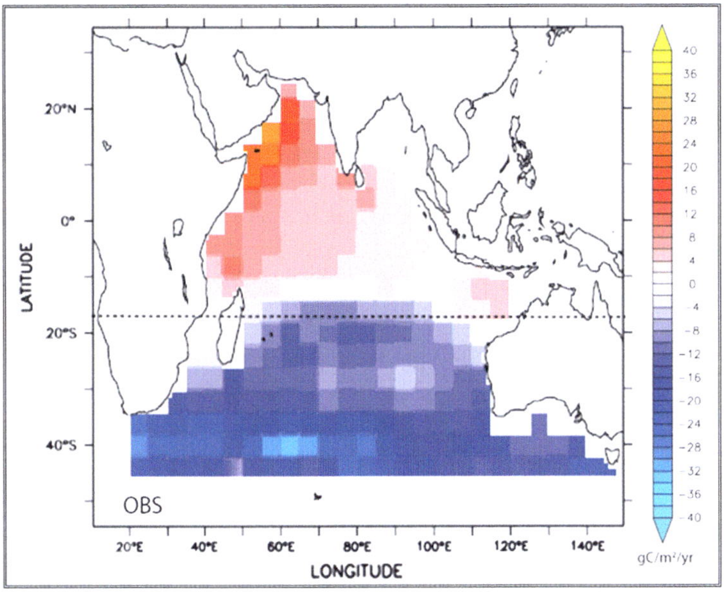

Fig. 8.20 CO_2 air-sea fluxes in the Indian Ocean based on available observations. Positive fluxes are directed to the atmosphere (from Sarma et al. 2013)

Despite the clear latitudinal distinction between regions that are sources and sinks of CO_2 shown in Fig. 8.20, however, there is certainly a large export of carbon to the deep ocean (Smith 2001). This is borne out by the extensive area of sediments with high organic carbon content of up 7.5 % in the upwelling areas off Oman and Yemen, which is as high as found anywhere (Shimmield et al. 1990), and much of it appears to be from diatoms (Honjo et al. 1999).

The oxygen minimum zone is a pronounced feature in both basins of the northern Indian Ocean, the Arabian Sea and the Bay of Bengal. Here, oxygen concentrations of <5 μM (≈0.448 mL/L) are found between 100 and 1000 m, with zones containing no measurable oxygen several hundred metres thick (e.g., Paulmier and Ruiz-Pino 2009; Talley 2013). Unlike the Atlantic and Pacific Oceans, the most intense oxygen minimum zone in the Indian Ocean is not located along its eastern boundary, but in the north. This anomaly, like other distinguishing features of the Indian Ocean, arises from its unusual geography, i.e. the presence of the Asian landmass that places its northern boundary within the tropics, not allowing adequate ventilation of the thermocline from the north. Coupled with the high productivity that occurs following the coastal and offshore upwelling and a relatively long residence time in the northern part of the basin, the result is that local respiration keeps oxygen concentrations at very low levels, and there is little seasonality even during the height of the southwest monsoon, when replenishment of oxygen concentrations might be expected from the south (Morrison et al. 1999).

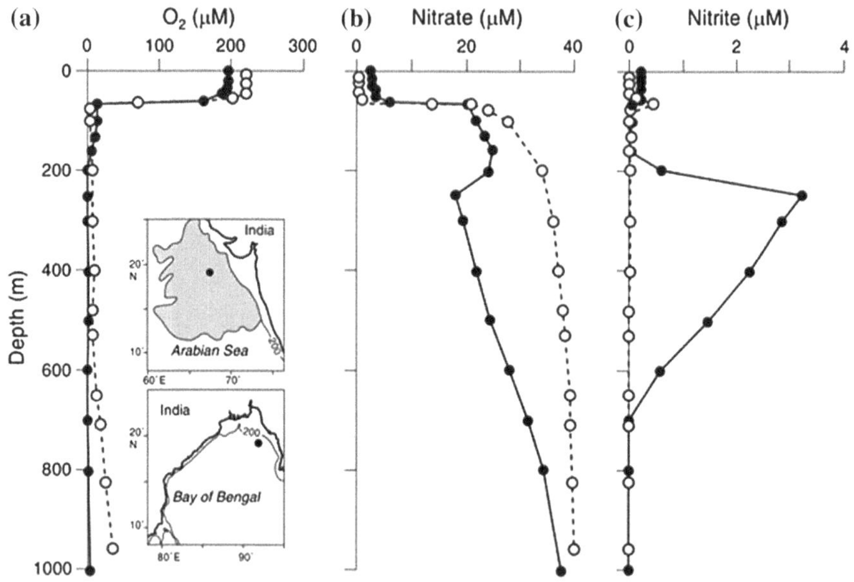

Fig. 8.21 Comparison of vertical profiles of **a** O_2, **b** NO_3^- and **c** NO_2^- in the Arabian Sea (*filled circles*) and the Bay of Bengal (*open circles*). The *insets* show the station locations and the geographical extent of the perennial open-ocean anoxic zone (shaded area) (from Naqvi 2006)

The size of the oxygen minimum zone in the Arabian Sea is considerably larger than in the Bay of Bengal, while the minimum concentration is also lower. While the lowest concentrations in the Arabian Sea often reach zero, this is rare in the Bay of Bengal, but this minute difference results in starkly contrasting biogeochemical cycling in the two basins. Oxygen concentrations within the oxygen minimum zone in the Arabian Sea often fall below the threshold (<1 μM) required for denitrification (microbial reduction of nitrate to nitrogen gas) and reduction of other polyvalent elements, and consequently one finds bacterial populations that use nitrate as an electron acceptor rather than oxygen (see Chap. 2). Nitrogen concentration data thus typically show nitrate deficits of up to 10–20 μM and elevated nitrite concentrations of up to about 5 μM in the 100–1000 m depth range in the Arabian Sea but not in the Bay of Bengal (Fig. 8.21; Naqvi 1991, Naqvi 2006). The nitrite maximum coincides generally with the core of the Persian Gulf Water of density (sigma-t) ~26.6, while the base of the oxygen minimum zone is roughly coincident with the density of Red Sea Water (27.2) (Morrison et al. 1999). Calculated denitrification rates in the region lie somewhere between 10 and 30 Tg/year (Mantoura et al. 1993; Naqvi et al. 1992). During the 1994–95 U.S. program, N:P ratios were found to be considerably lower than the standard Redfield ratio of 16:1 within the upper 1,000 m showing the importance of denitrification in this region, and nitrate is apparently always more limiting than phosphate. Despite this, the water that upwelled along the coast of Oman during the southwest monsoon had N:Si ratios that were considerably higher, by a factor of about two, than the normally accepted 1:1 ratio, and strong diatom growth was only seen during the late summer monsoon period, even though iron was present in abundance (Morrison et al. 1998).

The perennial suboxic zone of the Arabian Sea (shaded area in Fig. 8.21) differs from the other open-ocean suboxic zones in that it is located well away from the sites of the most intense upwelling in the western Arabian Sea and is most developed in the least productive central Arabian Sea. Suboxic (even anoxic) conditions do develop over the western Indian shelf, but this happens only seasonally (during late summer and autumn) and the two zones are not contiguous as they are separated by a slightly oxygenated poleward undercurrent (e.g., Naqvi et al. 2000). Resplandy et al. (2011) show that the oxygen drawdown in the region of highest productivity in the western Arabian Sea is counterbalanced by the supply of oxygenated waters originating from the south and advected horizontally by the western boundary current.

8.4 Atlantic Ocean

8.4.1 Gulf of Mexico

The Gulf of Mexico is a semi-enclosed sea having maximum depths exceeding 3,800 m as well as extensive shelf regions comprising the West Florida shelf, the Texas-Louisiana shelf, the Tamaulipas-Veracruz shelf, and the Campeche shelf (Fig. 8.22a). Together with the neighbouring Caribbean Sea it forms the *American Mediterranean Sea*. Productivity in the Gulf of Mexico Large Marine Ecosystem is moderate (<300 g C/m^2/yr). On the seasonal scale, satellite data indicate that the

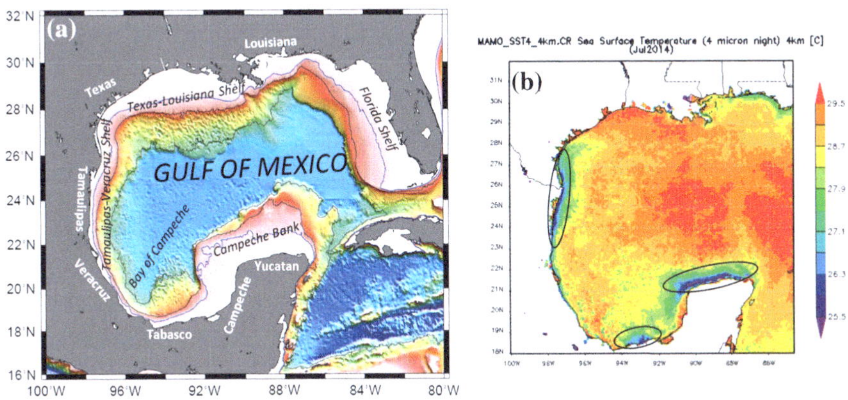

Fig. 8.22 **a** Bathymetry of the Gulf of Mexico. Image source: Liu and Dittert (2010). **b** Lowered SSTs (highlighted by *ellipses*) during coastal upwelling events in July 2014. Data source: Giovanni (NASA)

shelves off Tamaulipas and Veracruz experience upwelling favourable winds from April to August, when southeasterly winds are dominant (Fig. 8.22b). However, during summer, values of chlorophyll-*a* concentration are lower than those in autumn and winter, as a result of currents bringing biological material into the region (Zavala-Hidalgo et al. 2006). During winter, there is a cold front on the Tamaulipas shelf produced by advection of cold water from the Texas–Louisiana shelf and not due to upwelling. On the eastern Campeche Bank, persistent upwelling is observed due to favourable winds throughout the year. The strongest upwelling occurs along the inner shelf from May to September (Zavala-Hidalgo et al. 2006).

While winds on the northern coast of the Gulf of Mexico are easterly (i.e., downwelling-favourable) for most of the year, coastal upwelling develops west of the Mississippi delta during summer months (June–August) on the northern Gulf of Mexico shelf when the wind switches to westerly. This region is subject to large freshwater and nutrient inputs from the Mississippi-Atchafalaya River system, for which the annual mean flow is about 14,000 m^3/s, and which feed massive phytoplankton blooms. The stratification produced by the freshwater inflow, combined with the local respiration of the decaying phytoplankton, results in the annual formation of a hypoxic zone west of the Mississippi delta that covers on average about 16,000 km^2 of the Texas-Louisiana shelf (the maximum was 22,000 km^2 in 2002), and can extend as far west as Matagorda Bay (29° 25′ N, 91° 30′ W; Rabalais et al. 2007; Obenour et al. 2013). Hypoxia formation can begin as early as February, but is usually seen between April/May and September, after which storm fronts pass across the shelf and break down the stratification. If the summer westerlies blow consistently for long periods, they decrease the area affected by hypoxia by changing the position of the low salinity river plume over the shelf, moving it offshore over deep water and preventing it from spreading as far west, and wind duration is more important than wind strength (Feng et al. 2012, 2014).

8.4.2 Caribbean Sea

The southern Caribbean Sea (Fig. 8.23a) experiences strong wind-driven coastal upwelling from approximately January to May along the continental margin between about 61° W to 74° W and 10°–13° N, along the coasts of Trinidad, Venezuela and Colombia (Richards 1960; Gordon 1967; Herrera and Febres-Ortega 1975; Muller-Karger and Aparicio 1994; Muller-Karger et al. 2004; Ruiz-Ochoa et al. 2012). This upwelling process maintains a highly productive ecosystem in the region—called the *southern Caribbean upwelling system*. The Caribbean Sea as a whole is classified as a Large Marine Ecosystem of moderate productivity (150–300 g $C/m^2/yr$).

Corredor (1979) commented on the effects of upwelling on phytoplankton growth in the eastern Colombian Basin (near 72° W). By following a drogue placed in the upwelling tongue and set at 10 m depth, he was able to follow a parcel of

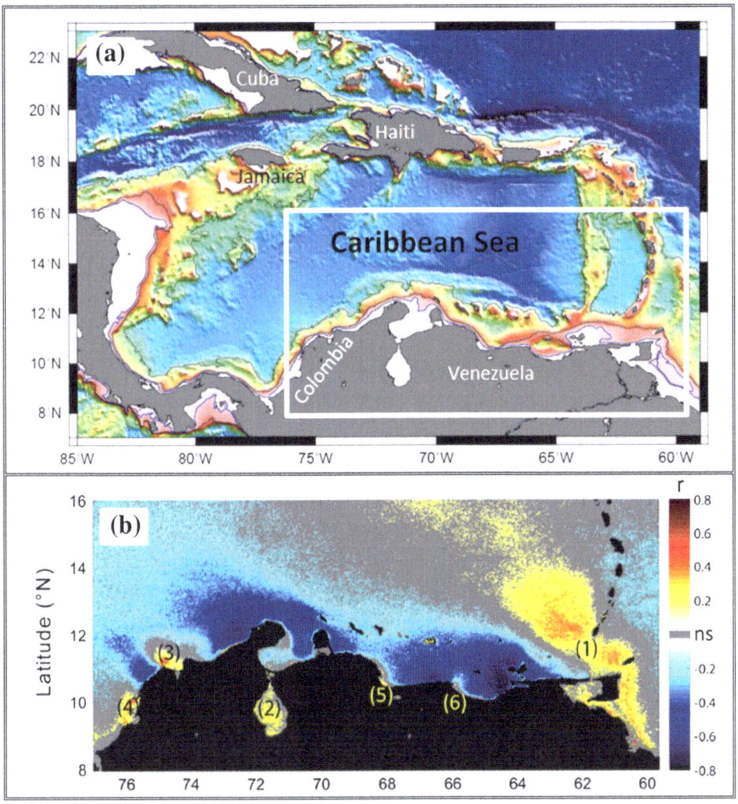

Fig. 8.23 **a** Bathymetry of the Caribbean Sea. Image source: Liu and Dittert (2010). **b** Correlation coefficient (r) between weekly time series of satellite SST and the logarithm of chlorophyll-*a* concentrations for the period 1998–2009. Upwelling areas show strong inverse correlations with phytoplankton biomass, while areas with freshwater influence show a direct correlation caused by their high content in colored dissolved organic matter (CDOM). The principal freshwater sources that influence the correlation are: (*1*) Orinoco River, (*2*) Maracaibo Lake, (*3*) the Magdalena River and Ciénaga Grande, and (*4*) the Sinú river and Canal del Dique. Other areas showing contamination of the chlorophyll product very close to the coast due to smaller local freshwater discharges are (*5*) the Tocuyo River and (*6*) the Tuy River (from Rueda-Roa and Muller-Karger 2013)

upwelling water to the west for 33 h. As expected, the surface layer warmed, chlorophyll-*a* concentrations increased after an initial lag phase and the depth of the chlorophyll maximum shallowed, and nitrate concentrations decreased. Productivity measurements suggested that the instantaneous phytoplankton productivity was about 200 mg C/m^2/hour, which is similar to other upwelling regions. However, the low concentrations of nitrate ($\sim$2.4 μM) in the upwelling high salinity Subtropical Underwater, which is found here at about 50–100 m depth, mean that such rates of production cannot be sustained for long periods. Thus, the overall productivity of this region, and that of the Venezuelan upwelling farther east, which is also supplied by Subtropical Underwater, is considerably lower than in the major eastern boundary currents.

Four acoustic fisheries surveys were conducted during 1988 (covering all seasons) aboard the *R/V Dr. Fridtjof Nansen* on the shelf region from Suriname to the western border of the Colombian Caribbean (Stromme and Saetersdal 1989). The total biomass of small pelagics in the southern Caribbean upwelling system was estimated at 1,580,000 tonnes, which includes clupeids, anchovies, carangids, scombrids and barracudas. Nearly all (95 %) of the biomass was concentrated in two areas, an eastern (63°–65° W) and a western (70°–73° W) upwelling area (Fig. 8.23b). However, the eastern area had a much higher biomass of small pelagics (78 %).

A recent study (Rueda-Roa and Muller-Karger 2013) compared these two regions in terms of wind conditions and responses in surface temperatures and chlorophyll-*a* concentrations. Based on 6 years of data (1994–2009), the eastern area featured the lowest SST (25.24 °C) and the highest chlorophyll-*a* concentrations (1.65 mg/m^3); it also had moderate wind intensity (6.12 m/s) and shallower isotherms. In contrast, the western area had stronger winds (8.23 m/s), but deeper isotherms, slightly higher SST (25.53 °C) and only moderate chlorophyll-*a* concentrations (1.15 mg/m^3). Upwelling in the eastern area was more prolonged than in the western area during this time period. According to the 'optimal environmental window' theory (see Sect. 2.4), small clupeoid recruitment is a dome-shaped function of the upwelling intensity, turbulence and SST, with an optimum wind speed around 5–6 m/s. The eastern upwelling area wind speed was close to this optimum value. The western upwelling area showed much higher wind speed that caused high levels of turbulence and strong offshore transport that could hinder the recruitment of small pelagics in that area (Rueda-Roa and Muller-Karger 2013).

8.4.3 Brazil

Coastal upwelling occurs on the Brazilian shelf in three different regions (Fig. 8.24):

(i) off southern Bahia (15° S–19° S)
(ii) off Cabo Frio (20° S–24° S), and
(iii) off Cabo de Santa Marta.

The upwelling in these regions is largely wind-driven (Mazzini and Barth 2013). Persistent upwelling-favourable, northeasterly winds are present throughout the year north of Rio de Janeiro (22.9° S) to approximately 19.5° S and 17.5° S during autumn and winter, respectively, and extend during spring and summer to Ilha de Santa Catarina (27.6° S). The large-scale, quasi-steady feature in the atmosphere responsible for this scenario is the South Atlantic high pressure system (see Fig. 2.13). Significant

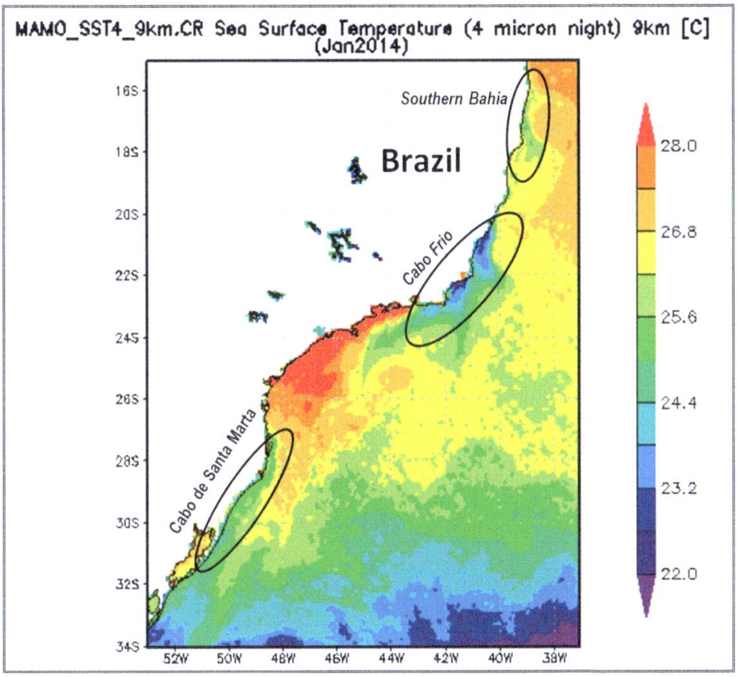

Fig. 8.24 January mean satellite SST data illustrating coastal upwelling in the regions of (1) southern Bahia, (2) Cabo Frio and (3) Cabo de Santa Marta. Data source: Giovanni (NASA)

perturbations of the wind field over the region commonly occur due to the passage of synoptic weather systems, which travel from the southwest towards the northeast, and become weaker as they reach lower latitudes. Perturbations are more energetic during autumn and particularly in the wintertime, when the wind field can reverse to southerly causing the upwelling to shut down (Stech and Lorenzzetti 1992).

Using scatterometer wind stress data for the upwelling system around Cabo Frio, Castelão and Barth (2006) showed the wind-stress-curl driven Ekman suction mechanism is as important as the coastal upwelling for the total vertical transport of the region. However, using a similar method, Campos et al. (2013) concluded that the Ekman suction mechanism is insignificant in the Cabo de Santa Marta region (28°–32° S), where upwelling follows mainly from episodic upwelling-favourable synoptic wind events. Using a hydrodynamic model, Mazzini and Barth (2013) demonstrated that topographically-driven upwelling plays a significant role in the vicinity of capes and other coastal irregularities.

Both the East Brazil Shelf (from the equator to 23° S) and the South Brazil Shelf (27°–32° S) classify as Large Marine Ecosystems of moderate productivity, not only because of the productivity associated with coastal upwelling, but also because of substantial nutrient input by numerous rivers. The East Brazil Shelf is a typical oligotrophic system, poor in nutrients and phytoplankton biomass, except in areas of upwelling where primary production is enhanced (Gaeta et al. 1999). The fisheries in this region are mainly artisanal.

The South Brazil Shelf contributes about half of Brazil's commercial fisheries yield. In 2002, artisanal fisheries accounted for about 22 % of the total commercial catch in this region with sardines representing the most important group in shelf catches (FAO 2003a, b). Historical catches have been dominated by Brazilian sardinella (*Sardinella brasiliensis*). A combination of overexploitation and oceanic anomalies in upwelling intensity are believed to have accounted for the fluctuations of the sardine and anchovy fisheries in this region (Bakun and Parrish 1991; Paiva 1997; Matsuura 1998). Eutrophication in lagoons, estuaries and bays along the Brazil coast situated downstream from densely occupied urban areas, industrial activities and agriculture runoff is currently a serious environmental issue (UNEP 2004).

8.4.4 Eurafrican Mediterranean Sea

The Eurafrican Mediterranean Sea is a complex system of relatively low productivity; that is, it is a generally oligothrophic sea and classified as a low productivity Large Marine Ecosystem. Nutrient-poor, oxygenated Atlantic surface water through the Strait of Gibraltar is the major inflow into the Mediterranean, while the outflow is at depth, spilling high salinity water over the sill to form the Mediterranean Outflow Water that pervades the North Atlantic below the thermocline. Deep convection in winter results in generally well-oxygenated bottom waters. Sources of nutrients are atmospheric dust, river discharges, coastal upwelling and mesoscale cyclonic eddies (see Siokou-Frangou et al. (2010) and references therein), but nutrients are generally low in concentration, particularly phosphorous with N:P ratios of up to 60. Analyses of wind-driven surface Ekman transport by Agostini and Bakun (2002) suggest that intermittent coastal upwelling events are likely to take place in regions such as the Alboran Sea, the Balearic Sea, the Straits of Sicily, the eastern Adriatic Sea and the northeastern Aegean Sea. These regions are relatively unproductive in terms of phytoplankton production, however, because the upwelling water itself contains low nutrient concentrations (considerably lower than in most other upwelling regions), most likely because upwelling events are short-lived and because of the inverse estuary circulation in the Mediterranean (Siokou-Frangou et al. 2010).

However, relative to its low primary productivity and vast surface area, fish landings in the Mediterranean Sea are remarkably high at around 2 million tonnes per annum (see http://www.seaaroundus.org/), which is similar to total catches in the Sea of Japan or the Benguela upwelling region. The Mediterranean hosts several important summer spawning grounds. For instance, anchovy abundance and catches are highest in the Catalonian and Adriatic Seas; and the Aegean Sea supports various fish species, especially small pelagic fish (Stergiou and Christou 1996).

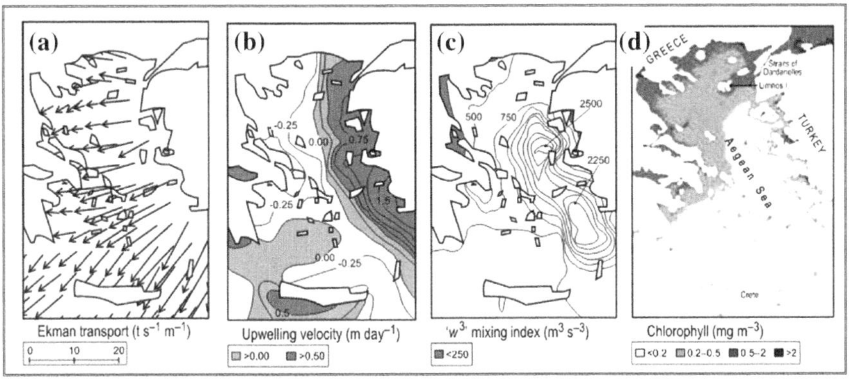

Fig. 8.25 Typical summer distributions of **a** Ekman transport, **b** upwelling velocity (calculated from the horizontal divergence of Ekman transports), **c** mixing index (calculated from the cube of wind speed), and **d** surface chlorophyll in the Aegean Sea (from Agostini and Bakun 2002)

The analysis of wind fields by Bakun and Agostini (2001) indicates the existence of several wind-driven coastal upwelling systems. Two of these systems in the Aegean and Catalan Seas are briefly discussed in the following. Summer coastal upwelling in the Aegean Sea (Fig. 8.25) is driven by persistent northerly cold and dry "Etesian" winds. This creates strong westward Ekman transports (Fig. 8.25a) and upwelling velocities (Fig. 8.25b) in the eastern part of the sea, where strong winds also trigger intensive mixing (Fig. 8.25c). Nevertheless, there is almost no phytoplankton response in the upwelling regions, presumably due to low nutrient fluxes. Instead, the highest chlorophyll levels are found in the northwestern portion of the sea and in the nutrient-enriched outflow from the Black Sea (Fig. 8.25d), which are regions of low wind mixing.

Agostini and Bakun (2002) propose that phytoplankton blooms in the northwestern Aegean Sea are triggered to some extent by the Ekman drift of nutrient-enriched water from the upwelling region into regions of low wind energy. However, a more likely explanation is that discharges from major rivers draining southeastern Europe and associated density fronts are the foremost nutrient source for the region. Similarly, strong wind jets (i.e., the "Tramontagne" and the "Mistral") create summer upwelling in the Gulf of Lions (Fig. 8.26), but, again, regions of high primary productivity coincide with major discharges by the Rhone and Ebro rivers. Nevertheless, Agostini and Bakun (2002) make the point of the importance of processes favouring retention within or the drift towards appropriate habitat. In the examples of the Aegean and Catalan Seas, the Ekman drift facilitates the westward movement of fish larvae into more sheltered regions of low wind energy. Furthermore, the horizontal convergence of Ekman transports near coasts (in the Aegean Sea) or in the open sea (in the Catalan Sea) facilitates the concentration of fish larvae in regions which are known spawning grounds of various fish species. Altogether, the process chain of (i) enrichment, (ii) concentration and

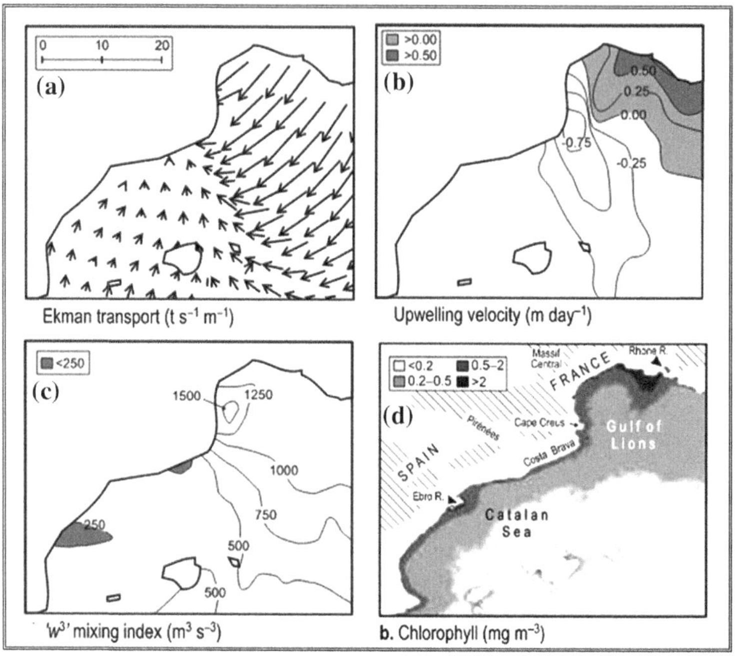

Fig. 8.26 Same as Fig. 8.25, but for the Catalan Sea and the Gulf of Lions (from Agostini and Bakun 2002)

(iii) retention found in the Mediterranean is the well-known and fundamental "ocean triad" (Bakun 1996) which has substantial explanatory power.

Overall the western part of the Mediterranean Sea operates as a sink of atmospheric CO_2, while the eastern part is a source (D'Ortenzio et al. 2008). This is likely because of the higher productivity in the western Mediterranean ($\sim$150 g C/m^2/yr) compared to the eastern basins ($\sim$60 g C/m^2/year; Siokou-Frangou et al. 2010). While a comprehensive review of the oceanography, biogeochemistry and marine ecology of the Eurafrican Mediterranean Sea is far beyond the scope of this work, interested readers are referred to the recent collection of scientific papers in a book edited by Goffredo and Dubinsky (2014).

8.5 Summary

The seasonal coastal upwelling regions of the northern Indian Ocean and southeast Asian coastal waters stand out among other seasonal upwelling systems given their extremely high fishery yield and economic value, and, in the cases of the East China and South China Seas, indications of severe overexploitation. Seasonal coastal upwelling in the Arafura Sea, which stands out in satellite chlorophyll-*a* data as a region of extremely high productivity (see Fig. 8.8), is clearly understudied and more research efforts are required to acquire a basic understanding of the functioning of this system.

While the monsoonal wind pattern guarantees a consistent forcing in support of coastal upwelling in the northern Indian Ocean and the Indonesian Seas, coastal upwelling events in other regions, such as along the southern shelf of Australia or the western shelf of New Zealand's South Island, develop rather episodically in response to synoptic weather events. Such regions are generally associated with relatively low productivity (based on annual averages) and therefore do not classify as Large Marine Ecosystems. Nonetheless, some of these seasonal upwelling systems play an extremely important role in marine life cycles, particularly for migrating large pelagic fish and megafauna, due to the specific timing of upwelling events and the specific location of the upwelling region. For instance, the Great South Australian Coastal Upwelling System, despite its relatively low overall productivity, is a key feeding area for migrating Southern Bluefin Tuna and Blue Whales. Hence, the concept of Large Marine Ecosystems based on annual-mean primary productivity should be interpreted with caution, as it underestimates the significance of upwelling systems that, despite low productivity, play an important role for migrating marine species.

References

Introduction

Morgan, G.R., and D.J. Staples. 2006. *The history of industrial marine fisheries in Southeast Asia*, 28. Bangkok: FAO, Regional Office for Asia and the Pacific.

Sugiyama, S., D. Staples and S.J. Funge-Smith. 2004. *Status and potential of fisheries and aquaculture in Asia and the Pacific*. FAO Regional Office for Asia and the Pacific, RAP Publication 2004/25, 53 pp.

South China Sea

Chen, S., T. Chen, X. Xu, Z. Chen and S. Sul, eds. 1985. *The Vast South China Sea*, (in Chinese) 218 pp. Beijing: China Science Press.

Gan, J., L. Li, D. Wang, and X. Guo. 2009. Interaction of a river plume with coastal upwelling in the northeastern South China Sea. *Continental Shelf Research* 29: 728–740.

Guan, B.X., and S.J. Chen. 1964. *Ocean Current system in East China Sea and South China Sea*. Qingdao: Institute of Oceanology, Chinese Academy of Sciences.

Han, W.Y., and K.M. Ma. 1988. Study of Yuedong coastal upwelling. *Acta Oceanologica Sinica* 10: 52–59.

Han, W.Y., M.B. Wang, and K.B. Ma. 1990. The lowest surface water temperature area of China sea in summer—the upwelling along the east coast of Hainan Island. *Oceanologia Et Limnologia Sinica* 21: 167–275.

Hu, J.Y., H. Kawamura, H.S. Hong, M. Suetsugu, and M.S. Lin. 2001. Hydrographic and satellite observations of summertime upwelling in the Taiwan Strait: A preliminary description. *Terrestrial, Atmospheric and Oceanic Sciences* 12(2): 415–430.

Hu, J., W. Lan, B. Huang, K.-P. Chiang, and H. Hong. 2015. Low nutrient and high chlorophyll a coastal upwelling system—a case study in the southern Taiwan Strait. *Estuarine, Coastal and Shelf Science* 166: 170–177.

Jing, Z., Y. Qi, Z. Hua and H. Zhang. 2009. Numerical study on the summer upwelling system in the northern continental shelf of the South China Sea. *Continental Shelf Research*. doi:10.1016/j.csr.2008.11.008.

Liu, K.-K., and N. Dittert. 2010. Web-based electronic supplements, Appendix C. In *Carbon and Nutrient Fluxes in Continental Margins*, ed. K.-K. Liu, L. Atkinson, R.A. Quiñones and L. Talaue-McManus. Springer, Berlin. http://cmtt.pangaea.de/. Accessed 11 June 2015.

Liu, J.Y. 2013. Status of marine biodiversity of the China Seas. *PLoS ONE* 8(1): e50719. doi:10.1371/journal.pone.0050719.

Morton, B., and G. Blackmore. 2000. South China Sea. *Marine Pollution Bulletin* 42(12): 1236–1263. doi:10.1016/S0025-326X(01)00240-5.

Niino, H., and O. Emery. 1961. Sediment of shallow portions of East China Sea and South China Sea. *Geological Society of American Bulletin* 72: 731–761.

Shaw, P.T., and S.Y. Chao. 1994. Surface circulation in the South China Sea. *Deep-Sea Research I* 40(11/12): 1663–1683.

Su, J.L., and W. Wang. 1990. On the sources of Taiwan Warm current from South China Sea. *Chinese Journal of Donghai Marine Science* 8(3): 1–9. (in Chinese).

Talaue-McManus, L. 2000. *Trans-boundary diagnostic analysis for the South China Sea*. EAS/RCU Technical Report Series 14. United Nations Environment Programme, Bangkok, Thailand.

Wyrtki, K. 1961. Physical oceanography of the Southeast Asia waters. *NAGA Report* 2: 1–195.

Xie, S.-P., Q. Xie, D. Wang, and W.T. Liu. 2003. Summer upwelling in the South China Sea and its role in regional climate variations. *Journal of Geophysical Research* 108(C8): 3261. doi:10.1029/2003JC001867.

Yu, W.Q. 1987. A preliminary approach of the upwelling for the northern South China Sea. *Marine Science* 6: 7–10.

Yunwei, Y., L. Zheng, and C. Changlin. 2015. Winter coastal upwelling off northwest Borneo in the South China Sea. *Acta Oceanologica Sinica* 34(1): 3–10. doi:10.1007/s13131-015-0590-2.

Zeng, L.M. 1986. A preliminary analysis of indicators of offshore upwelling off eastern Guangdong. *Tropic Oceanology* 5: 68–73.

East China Sea

Belkin, I.M. 2009. Rapid warming of Large Marine ecosystems. *Progress in Oceanography* 81: 207–213.

Cai, R., J. Chen, and R. Huang. 2006. The response of marine environment in the offshore area of China and its adjacent ocean to recent global climate change. *Chinese Journal of Atmospheric Science (in Chinese)* 30: 1019–1033.

Chang, J., W. Huang, X. Lou, and A. Shi. 2012. Upwelling region wind speed correction method for wind retrieval from synthetic aperture radar imagery in the East China Sea. *Journal of Applied Remote Sensing* 6(1): 063540. doi:10.1117/1.JRS.6.063540.

Chen, C.-T.A. 1996. The Kuroshio intermediate water is the major source of nutrients on the East China Sea continental shelf. *Oceanologica Acta* 19(5): 523–527.

Chen, C.-T.A., and S.-L. Wang. 1999. Carbon, alkalinity and nutrient budgets on the East China Sea continental shelf. *Journal of Geophysical Research* 104(C9): 20675–20686.

Chen, C.-C., F.-K. Shiah, K.-P. Chiang, G.-C. Gong, and W.M. Kemp. 2009. Effects of the Changjiang (Yangtze) River discharge on planktonic community respiration in the East China Sea. *Journal of Geophysical Research* 114: C03005. doi:10.1029/2008JC004891.

Chen, Y.Q., and X.Q. Shen. 1999. Changes in the biomass of the East China Sea Ecosystem. In *Large Marine Ecosystems of the Pacific Rim—Assessment, Sustainability and Management*, ed. K. Sherman, and Q. Tang. U.S: Blackwell Science, Malden.

Chen, Y.-L., H.-Y. Chen, G.-C. Gong, Y.-H. Lin, S. Jand, and M. Takahashi. 2004. Phytoplankton production during a summer coastal upwelling in the East China Sea. *Continental Shelf Research* 24: 1321–1338.

Du, P., H. Zhang, W. Xiao, X. Kang, and Q. Guan. 2011. Analysis of thermohaline and current distribution characteristics of Zhejiang and Fujian waters in summer. *Acta Oceanologica Sinica* 30(5): 71–83. doi:10.1007/s13131-011-0149-9.

Jiang, X.-S., J.-C. Hong, G.-L. Wang, and X.-Q. Huang. 1992. Analysis of *Noctiluca scintillans* red tide occurred in red tide frequent area of Changjiang Estuary. *Journal of Jinan University* 13: 134–139. (in Chinese).

Li, X., T.S. Bianchi, M.A. Allison, P. Chapman, S. Mitra, Z. Zhang, G. Yang and Z. Yu. 2012. Composition, abundance and age of total organic carbon in surface sediments from the inner shelf of the East China Sea. *Marine Chemistry*, 145–147: 37–52.

Lou, X., and C. Hu. 2014. Diurnal changes of a harmful algal bloom in the East China Sea: Observations from GOCI. *Remote Sensing of Environment* 140: 562–572.

Mao, H.L., Y.W. Ren, and G.D. Sun. 1964. Preliminary analysis of summer hydrographic characteristics and water masses in the southern Huanghai Sea and northern East China Sea. *Studia Marina Sinica* 1: 23–77. (in Chinese with English abstract).

Qi, Y., J. Zou, and S. Liang (eds.). 2003. *Red tides along Chinese coast*. Beijing: Science Press.

Rabouille, C., D.J. Conley, M.H. Dai, W.-J. Cai, C.T.A. Chen, B. Lansard, R. Green, K. Yin, P. J. Harrison, M. Dagg, and B. McKee. 2008. Comparison of hypoxia among four river-dominated ocean margins: The Changjiang (Yangtze), Mississippi, Pearl, and Rhone Rivers. *Continental Shelf Research* 28: 1527–1537.

Tang, Q. ed. 2006. *Living Marine resources and inhabiting environment in the Chinese EEZ*, 1238 pp. Science Press, Beijing (In Chinese).

Tseng, C.-M., P.-Y. Shen, and K.-K. Liu. 2014. Synthesis of observed air–sea CO_2 exchange fluxes in the river-dominated East China Sea and improved estimates of annual and seasonal net mean fluxes. *Biogeosciences* 11: 3855–3870.

Wang, B., and X. Wang. 2007. Chemical hydrography of coastal upwelling in the East China Sea. *Chinese Journal of Oceanology and Limnology* 25(1): 16–26.

Wang, B., Q. Wei, J. Chen, and L. Xie. 2012. Annual cycle of hypoxia off the Changjiang (Yangtze) River Estuary. *Marine Environmental Research* 77: 1–5.

Yang, D., B. Yin, Z. Liu, and X. Feng. 2011. Numerical study of the ocean circulation on the East China Sea shelf and a Kuroshio bottom branch northeast of Taiwan in summer. *Journal of Geophysical Research* 116: C05015. doi:10.1029/2010JC006777.

Zhou, M. 2010. Environmental settings and harmful algal blooms in the sea area adjacent to the Changjiang River Estuary. In *Coastal environmental and ecosystem issues of the East China Sea*, ed. A. Ishimatsu, and H.-J. Lie, 133–149. TERRAPUB and Nagasaki: University.

Zhou, M. 2012. Marine disasters. In *Oceanography of China*, ed. Y.Wang, J. Su and R. Liu, Chapter 17. Beijing: Science Press.

Zhou, M.J., Z.L. Shen, and R.C. Yu. 2008. Responses of a coastal phytoplankton community to increased nutrient input from the Changjiang (Yangtze) River. *Continental Shelf Research* 28: 1483–1489.

Indonesian Seas (excluding South China Sea)

Alongi, D.M., K. Edyvane, M.O. do Ceu Guterres, W.S. Pranowo, S. Wirasantosa and R. Wasson. 2011. *Biophysical Profile of the Arafura and Timor Seas*. Report prepared for the Arafura Timor Seas Ecosystem Action (ATSEA) Program. 32 pp.

Baars, M.A., A.B. Sutumo, S.S. Oosterhuis, and O.H. Arinardi. 1990. Zooplankton abundance in the eastern Banda Sea and northern Arafura Sea during and after the upwelling season, August 1984 and February 1985. *Netherland Journal of Sea Research* 25(4): 527–543.

Burford, M.A., and P.C. Rothlisberg. 1999. Factors limiting phytoplankton production in a tropical continental shelf ecosystem. *Estuarine, Coastal and Shelf Science* 48(5): 541–549.

Burke, L., K. Reytar, M. Spalding, and A. Perry. 2011. *Reefs at risk revisited*. Washington, DC: World Resources Institute. 114 pp.

Chu, P.C., Q. Liu, Y. Jia, and C. Fan. 2002. Evidence of a barrier layer in the Sulu and Celebes Seas. *Journal of Physical Oceanography* 32: 3299–3309.

Condie, A., and J. Dunn. 2006. Seasonal characteristics of the surface mixed layer in the Australasian region: implications for primary production regimes and biogeography. *Marine & Freshwater Research* 57: 569–590.

Dethmers, K.E.M., R. Chatto, M. Meekan, A. Amaral, C. de Cunha, N. de Carvalho, and K. Edyvane. 2009. *Marine megafauna surveys in Timor Leste: identifying opportunities for potential ecotourism—Final Report*. Government of Timor Leste: Ministry of Agriculture & Fisheries.

Gieskes, W.W.C., G.W. Kraay, A. Nontji, D. Setiapermana and Sutomo. 1988. Monsoonal alternation of a mixed and a layered structure in the phytoplankton of the euphotic zone of the Banda Sea (Indonesia): a mathematical analysis of algal pigment fingerprints. *Netherland Journal of Sea Research*, 22: 123–137.

Giri, C., E. Ochieng, L. Tieszen, Z. Zhu, A. Singh, T. Loveland, J. Masek, and N. Duke. 2011. Status and distribution of mangrove forests of the world using earth observation satellite data. *Global Ecology Biogeography* 20(1): 154–159.

Gordon, A.L., and R.D. Susanto. 2001. Banda Sea surface layer divergence. *Ocean Dynamics* 52 (1): 2–10. doi:10.1007/s10236-001-8172-6.

Hela, I. 1976. Vertical velocity of the upwelling in the sea. *Societas Scientiarum Fennica—Commentationes Physico-Mathematicae* 46: 9–24.

Jing, Z., Y. Qi, and Y. Du. 2012. Persistent upwelling and front over the Sulu Ridge and their variations. *Journal of Geophysical Research* 117: C11011. doi:10.1029/2012JC008355.

Kämpf, J. 2015. Undercurrent-driven upwelling in the northwestern Arafura Sea. *Geophysical Research Letters* 42: 9362–9368. doi:10.1002/2015GL066163.

Kämpf, J. 2016. On the majestic seasonal upwelling system of the Arafura Sea. *Journal of Geophysical Research* 121(1218–1228): 2015J. doi:10.1002/C011197.

Kennedy, J. 1999. A dynamic model of cooperative and non-cooperative harvesting of Southern Bluefin Tuna with an open access fringe, paper presented at the *World Conference on Natural Resource Modeling*, Saint Mary's University, Halifax, Canada, 23–25 June.

Liu, K.-K., and N. Dittert. 2010. Web-based electronic supplements, Appendix C. In *Carbon and nutrient fluxes in continental margins*, ed. K.-K. Liu, L. Atkinson, R.A. Quiñones and L. Talaue-McManus. Springer, Berlin. http://cmtt.pangaea.de/. Accessed 11 June 2015.

Monismith, S.G. 1986. An experimental study of the upwelling response of stratified reservoirs to surface shear stress. *Journal of Fluid Mechanics* 171: 407–439.

Moore, T.S., J. Marra, and A. Alkatiri. 2003. Response of the Banda Sea to the southeast monsoon. *Marine Ecology Progress Series* 261: 41–49.

Ningsih, N.S., N. Rakhmaputeri, and A.B. Harto. 2013. Upwelling variability along the southern coast of Bali and in Nusa Tenggara waters. *Ocean Science Journal* 48: 49–57. doi:10.1007/s12601-013-0004-3.

Rochford, D.J. 1966. Some hydrological features of the eastern Arafura Sea and the Gulf of Carpentaria in August 1964. *Australian Journal of Marine and Freshwater Research* 17: 31–60.

Susanto, R.D., A.L. Gordon, and Q. Zheng. 2001. Upwelling along the coasts of Java and Sumatra and its relation to ENSO. *Geophysical Research Letters* 28: 1559–1602.

Svannsson, A. 1975. Interaction between the coastal zone and the open sea. *Finnish Marine Research* 239: 11–28.

Wetsteyn, F.J., A.G. Ilahude, and M.A. Baars. 1990. Nutrient distribution in the upper 300 m of the eastern Banda Sea and northern Arafura Sea during and after the upwelling season, August 1984 and February 1985. *Proceedings of the Snellius-II Symposium, Netherland Journal of Sea Research* 25: 449–464.

Wyrtki, K. 1958. The water exchange between the Pacific and the Indian Oceans in relation to upwelling processes. *Proceedings of the Ninth Pacific Science Congress* 16: 61–66.

Wyrtki, K. 1961. Physical oceanography of the Southeast Asia waters. *NAGA Report* 2: 1–195.

Zijlstra, J.J., M.A. Baars, S.B. Tijssen, F.J. Wetsteyn, J.J. Witte, A.G. Ilahude and Hadikusumah. 1990. Monsoonal effects on the hydrography of the upper waters (<300 m) of the eastern Banda Sea and northern Arafura Sea, with special reference to vertical transport processes. *Proceedings of the Snellius-II Symposium. Netherland Journal of Sea Research*, 25: 431–447.

Australia's Southern Shelf

Collette, B., S.-K. Chang, A. Di Natale, W. Fox, M. Juan Jorda, N. Miyabe, R. Nelson, Y. Uozumi and S. Wang. 2011. *Thunnus maccoyii*. The IUCN Red List of Threatened Species. Version 2015.1. www.iucnredlist.org. Accessed 03 June 2015.

Gill, P.C., M.G. Morrice, B. Page, R. Pirzl, A.H. Levings, and M. Coyne. 2011. Blue whale habitat selection and within-season distribution in a regional upwelling system off southern Australia. *Marine Ecology Progress Series* 421: 243–263.

Herzfeld, M. 1997. The annual cycle of sea surface temperature in the Great Australian Bight. *Progress in Oceanography* 39(1): 1–27.

Kämpf, J., M. Doubell, D. Griffin, R.L. Matthews, and T.M. Ward. 2004. Evidence of a large seasonal coastal upwelling system along the southern shelf of Australia. *Geophysical Research Letters* 31: L09310. doi:10.1029/2003GLO19221.

Kämpf, J. 2010. On the preconditioning of coastal upwelling in the eastern Great Australian Bight. *Journal of Geophysical Research*, 115, C12071: 11. doi:10.1029/2010JC006294.

Kämpf, J. 2015. Phytoplankton blooms on the western shelf of Tasmania: evidence of a highly productive ecosystem. *Ocean Science* 11: 1–11. doi:10.5194/os-11-1-2015.

Kennedy, J. 1999. A dynamic model of cooperative and non-cooperative harvesting of Southern Bluefin Tuna with an open access fringe, paper presented at the *World Conference on Natural Resource Modeling*, Saint Mary's University, Halifax, Canada, 23–25 June.

Lewis, R.K. 1981. Seasonal upwelling along the southeastern coastline of South Australia. *Australian Journal of Marine and Freshwater Research* 32: 843–854.

McClatchie, S., J.F. Middleton, and T.M. Ward. 2006. Water mass analysis and alongshore variation in upwelling intensity in the eastern Great Australian Bight. *Journal of Geophysical Research* 111(C8): C08007. doi:10.1029/2004JC002699.

Middleton, J.F., and J.A.T. Bye. 2007. A review of the shelf-slope circulation along Australia's southern shelves: Cape Leeuwin to Portland. *Progress in Oceanography* 75(1): 1–41.

Ridgway, K.R., and S.A. Condie. 2004. The 5500-km-long boundary flow off western and southern Australia. *Journal of Geophysical Research* 109(4): c04017. doi:10.1029/2003JC001921.

Rochford, D.J. 1977. *A review of possible upwelling situation off Port MacDonald, South Australia*. CSIRO Australian Division Oceanography Report No. 81, 4 pp.

Van Ruth, P.D., G.G. Ganf, and T.M. Ward. 2010a. Hot spots of primary productivity: an alternative interpretation to conventional upwelling models. *Estuarine, Coastal and Shelf Science* 90: 142–158.

Van Ruth, P.D., G.G. Ganf, and T.M. Ward. 2010b. The influence of mixing on primary productivity: a unique application of classical depth theory. *Progress in Oceanography* 85: 224–235.

Around New Zealand

Bowman, M.J., S.M. Chiswell, P.P. Lapcnnas, R.A. Murtagh, B.A. Foster, V. Wilkinson and W Battaerd. 1983. Coastal upwelling, cyclogenesis and squid fishing near Cape Farewell, New Zealand. In *Coastal Oceanography*, ed. H. Gade. Plenum Press.

Bradford, J.M. 1983. Physical and chemical oceanographic observations off Westland, New Zealand, June 1979. *New Zealand Journal of Marine and Freshwater Research* 17: 71–81.

Bradford, J.M., and P.E. Roberts. 1978. Distribution of reactive phosphorous and plankton in relation to upwelling and surface circulation around New Zealand. *New Zealand Journal of Marine and Freshwater Research* 12: 1–15.

Brodie, J.W. 1960. Coastal surface currents around New Zealand. *New Zealand Journal of Geology and Geophysics* 3: 235–252.

Foster, B.A., and W.R. Battaerd. 1985. Distribution of zooplankton in a coastal upwelling in New Zealand. *New Zealand Journal of Marine and Freshwater Research* 19: 213–226.

Garner, D.M. 1954. Sea surface temperature in the Southwest Pacific Ocean from 1949 to 1952. *New Zealand Journal of Science and Technology, Series B* 36: 285–303.

Longdill, P.C., R.T. Healy, and K.P. Black. 2008. Transient wind-driven coastal upwelling on a shelf with varying width and orientation. *New Zealand Journal of Marine and Freshwater Research* 42(2): 181–196. doi:10.1080/00288330809509947.

Shlrtcliffe, T.G.L., M.I. Moore, A.G. Cole, A.B. Viner, R. Baldwin, and B. Chapman. 1990. Dynamics of the Cape Farewell upwelling plume, New Zealand. *Journal of Marine and Freshwater Research* 24(4): 555–568. doi:10.1080/00288330.1990.9516446.

Stanton, B.R. 1976. Circulation and hydrology off the west coast of the South Island, New Zealand. *New Zealand Journal of Marine and Freshwater Research* 19: 445–467.

Zeldis, J.R., R.A. Walters, M.J.N. Greig, and K. Image. 2004. Circulation over the northeastern New Zealand continental slope, shelf and adjacent Hauraki Gulf, during spring and summer. *Continental Shelf Research* 24: 543–561.

Northern Indian Ocean

Baars, M.A., P.H. Schalk, and M.J.W. Veldhuis. 1998. Seasonal fluctuations in plankton biomass and productivity in the ecosystems of the Somali Current, Gulf of Aden, and Southern Red Sea. In *Large Marine Ecosystems of the Indian Ocean: Assessment, Sustainability, and Management*, ed. K. Sherman, E.N. Okemwa, and M.J. Ntiba, 143–174. Malden, MA: Blackwell Scientific.

Bakun, A., C. Roy, and S. Lluch-Cota. 1998. Coastal upwelling and other processes regulating ecosystem productivity and fish production in the western Indian Ocean. In *Large Marine Ecosystems of the Indian Ocean: Assessment, Sustainability, and Management*, ed. K. Sherman, E.N. Okemwa, and M.J. Ntiba, 103–141. MA: Blackwell Scientific. Malden.

Beal, L.M., and K.A. Donohue. 2013. The Great Whirl: Observations of its seasonal development and interannual variability. *Journal of Geophysical Research* 118: 1–13. doi:10.1029/2012JC008198.

Bobzin, E. 1922. Vergleichende Betrachung des Klimas und der kalten Auftriebströmungen an der südwest-afrikanischen und südarabischen Küste. *Deutsche Übersee Meteorologische Beobachtungen* 23: 1–18.

Bottero, J.S. 1969. *An analysis of upwelling off the South-East Arabian Coast during the Summer Monsoon*. M.Sc.Thesis: Oregon State University.

Bruce, J.G. 1974. Some details of upwelling off the Somali and Arabian coasts. *Journal of Marine Research* 32: 419–423.

Currie, R.J. 1964. A fertile sea. *The Geographic Magazine* 37: 198–211.

Currie, R.J. 1992. Circulation and upwelling off the coast of South-East Arabia. *Oceanologica Acta* 15: 43–60.

Currie, R.J., A.E. Fisher, and P.M. Hargreaves. 1973. Arabian Sea upwelling. In *The biology of the Indian Ocean*, ed. B. Zeitschel, and S.A. Gerlach, 37–52. New York: Springer.

de Vos, A., C.B. Pattiaratchi, and R.G. Harcourt. 2014a. Inter-annual variability in blue whale distribution off southern Sri Lanka between 2011 and 2012. *Journal of Marine Science and Engineering* 2: 534–550.

de Vos, A., C.B. Pattiaratchi, and E.M.S. Wijeratne. 2014b. Surface circulation and upwelling patterns around Sri Lanka. *Biogeosciences* 11: 5909–5930. doi:10.5194/bg-11-5909-2014.

du Vall, K., S. Ingle, J. Snider, and S.F. DiMarco. 2011. Cabled ocean observatories in the Sea of Oman and Arabian Sea. *Oceans* 2011: 1–6.

FAO. 1997. *Review of the State of the World Fishery Resources: Marine Fisheries*. In 2 Lanternfishes: a potential fishery in the Northern Arabian Sea? www.fao.org/docrep/003/w4248e/w4248e34.htm. Accessed 4 May 2016.

FAO. 2003. Trends in oceanic captures and clustering of large marine ecosystems—Two studies based on the FAO capture database. In *FAO Fisheries Technical Paper 435*, 71 pp.

Fischer, J., F. Schott, and L. Stramma. 1996. Currents and transport of the Great Whirl-Socotra Gyre system during the summer monsoon, August 1993. *Journal of Geophysical Research* 101: 3573–3587.

Hitchcock, G.L., and D.B. Olson. 1992. NE and SW monsoon conditions along the Somali coast during 1987. In *Oceanography of the Indian Ocean*, ed. B.N. Desai, 583–593. Goa, India: National Institute of Oceanography.

Honjo, S., J. Dymond, W. Prell, and V. Ittekkot. 1999. Monsoon-controlled export fluxes to the interior of the Arabian Sea. *Deep-Sea Research II* 46: 1859–1902.

Jayaram, C., N. Chacko, K.A. Joseph, and A.N. Balchand. 2010. Interannual variability of upwelling indices in the Southeastern Arabian Sea: A satellite based study. *Ocean Science Journal* 45(1): 27–40.

Kortzinger, A., and J.C. Duinker. 1997. Strong CO_2 emissions from the Arabian Sea during South-West Monsoon. *Geophysical Research Letters* 24: 1763–1766.

Kumar, S.P., M. Madhupratap, M.D. Kumar, M. Gauns, P.M. Muraleedharan, V.V.S.S. Sarma and S.N. D'Souza. 2000. Physical control of primary productivity on a seasonal scale in central and eastern Arabian Sea. In *Proceedings of the Indian Academy of Science (Earth Planetary Sciences)*, 109, No. 4, 433–441.

Luis, A.J., and H. Kawamura. 2004. Air-sea interaction, coastal circulation and biological production in the eastern Arabian Sea: A review. *Journal of Oceanography* 60: 205–218.

Mantoura, R.F., C.S. Law, N.J.P. Owens, P.H. Burkill, E.M.S. Woodward, R.J.M. Howland, and C.A. Llewellyn. 1993. Nitrogen biogeochemical cycling in the northwesterrn Indian Ocean. *Deep-Sea Research II* 40: 651–671.

Marra, J., and R.T. Barber. 2005. Primary productivity in the Arabian Sea: a synthesis of JGOFS data. *Progress in Oceanography* 65: 159–175.

Morrison, J.M., L.A. Codispoti, S. Gaurin, B. Jones, V. Manghnani, and Z. Zheng. 1998. Seasonal variation of hydrographic and nutrient fields during the US JGOFS Arabian Sea process study. *Deep-Sea Research II* 45: 2053–2101.

Morrison, J.M., L.A. Codispoti, S.L. Smith, K. Wishner., C. Flagg, W.D. Gardner, S. Gaurin, S.W. A. Naqvi, V. Manghnani, L. Prosperie and J.S. Gundersen. 1999. The oxygen minimum zone in the Arabian Sea during 1995. *Deep-Sea Research II*, 46: 1903–1931.

Naqvi, S.W.A. 1991. Geographical extent of denitrification in the Arabian Sea in relation to some physical processes. *Oceanologica Acta* 14: 281–290.

Naqvi (2006) Oxygen deficiency in the north Indian Ocean. *Gayana (Concepción), 70* (Suplemento 1), 70: 53–58. doi:10.4067/S0717-65382006000300011, http://www.scielo.cl/scielo.php?script=sci_arttext&pid=S0717-65382006000300011. Accessed 15 April 2016.

Naqvi, S.W.A., R.J. Noronha, M.S. Shailaja, K. Somasundar, and R. Sen Gupta. 1992. Some aspects of the nitrogen cycling in the Arabian Sea. In *Oceanography of the Indian Ocean*, ed. B.N. Desai, 285–311. New Delhi: Oxford and IBH Publishing.

Naqvi, S.W.A., D.A. Jayakumar, P.V. Narvekar, H. Naik, V.V.S.S. Sarma, W. D'Souza, S. Joseph, and M.D. George. 2000. Increased marine production of N_2O due to intensifying anoxia on the Indian continental shelf. *Nature* 408: 346–349.

Okemwa, E. 1998. Application of the Large Marine Ecosystem concept to the Somali Current. In *Large Marine Ecosystems of the Indian Ocean: Assessment, sustainability, and management*, ed. K. Sherman, E.N. Okemwa, and M.J. Ntiba, 73–99. Malden, MA: Blackwell Scientific.

Paulmier, A., and D. Ruiz-Pino. 2009. Oxygen minimum zones (OMZs) in the modern ocean. *Progress in Oceanography* 80: 113–128. doi:10.1016/j.pocean.2008.08.001.

Puff, A. 1890. *Das kalte Auftriebwasser an der Ostseite des Nordatlantischen und der Westseite der Nordindischen Ozeans*. Dissertation, University of Marburg, 99 pp.

Resplandy, L., M. Lévy, G. Madec, S. Pous, O. Aumont, and D. Kumar. 2011. Contribution of mesoscale processes to nutrient budgets in the Arabian Sea. *Journal of Geophysical Research* 116: C11007. doi:10.1029/2011JC007006.

Royal Society. 1963. *International Indian Ocean expedition RRS Discovery cruise 1. South East Arabian Upwelling Region*. London: Cruise Report, Royal Society. 24 pp.

Sabine, C.L., R. Wanninkhof, R.M. Key, C. Goyet, and F.J. Millero. 2000. Seasonal CO_2 fluxes in the tropical and subtropical Indian Ocean. *Marine Chemistry* 72: 33–53.

Sanjeevan, V.N., P. Jasmine, B.R. Smitha, T. Ganesh, P. Sabu and T. Shanmugaraj. 2009. Eastern Arabian Sea marine ecosystems. In *MECOS'09—International symposium on marine ecosystems challenges and opportunities*, 9–12 February 2009 Cochin, India.

Sarma, V.V.S.S., A. Lenton, R.M. Law, N. Metzl, P.K. Patra, S. Doney, I.D. Lima, E. Dlugokencky, M. Ramonet, and V. Valsala. 2013. Sea–air CO_2 fluxes in the Indian Ocean between 1990 and 2009. *Biogeosciences* 10: 7035–7052. doi:10.5194/bg-10-7035-2013.

Schott, F. 1983. Monsoon response of the Somali Current and associated upwelling. *Progress in Oceanography* 12: 357–381.

Schott, F., J. Fischer, U. Garternicht, and D. Quadfasel. 1997. Summer monsoon response of the Northern Somali Current, 1995. *Geophysical Research Letters* 24: 2565–2568.

Schott, F., M. Dengler, and R. Schoenefeldt. 2002. The shallow overturning circulation of the Indian Ocean. *Progress in Oceanography* 53: 57–103.

Schott, F.A., and J.P. McCreary. 2001. The monsoon circulation of the Indian Ocean. *Progress in Oceanography* 51: 1–123.

Shimmield, G.B., N.B. Price and T.F. Pedersen. 1990. The influence of hydrography, bathymetry and productivity on sediment type and composition of the Oman margin and in the northwest Arabian Sea. In *The Geology and Tectonics of the Oman Region*, ed. A. Robertson, M. Searle and S. Riese. Geological Society Special Publication 49, 759–769.

Smith, R.L., and J.S. Bottero. 1977. On upwelling in the Arabian Sea. In *A voyage of discovery*, ed. M.V. Angel, 291–304. Supplement: Deep-Sea Research.

Smith, S.L. 2001. Understanding the Arabian Sea: Reflections on the 1994–1996 Arabian Sea expedition. *Deep-Sea Research II* 48: 1385–1402.

Smith, S., M. Roman, K. Wishner, M. Gowing, L. Codispoti, R. Barber, J. Marra, I. Prusova, and C. Flagg. 1998. Seasonal response of zooplankton to monsoonal reversals in the Arabian Sea. *Deep-Sea Research II* 45: 2369–2404.

Swallow, J.C., and J.G. Bruce. 1966. Current measurements off the Somali coast during the south west monsoon of 1964. *Deep-Sea Research* 13: 861–888.

Swallow, J.C., and M. Fieux. 1982. Historical evidence for two gyres in the Somali Current. *Journal of Marine Research* 40(Supplement): 747–755.

Talley, L.D. 2013. *Hydrographic Atlas of the World Ocean Circulation Experiment (WOCE). Volume 4: Indian Ocean*, ed. M. Sparrow, P. Chapman and J. Gould. International WOCE Project Office, Southampton, U.K.
Van der Elst, R., B. Everett, N. Jiddawi, G. Mwatha, P.S. Afonso, and D. Boulle. 2005. Fish, fishers and fisheries of the Western Indian Ocean: Their diversity and status. A preliminary assessment. *Philosophical Transactions of the Royal Society* 363: 263–284.
Vinayachandran, P.N., P. Chauhan, M. Mohan, and S. Nayak. 2004. Biological response of the sea around Sri Lanka to summer monsoon. *Geophysical Research Letters* 31: L01302. doi:10.1029/2003GL018533.
Wang, Z., S.F. DiMarco, A.E. Jochens, and S. Ingle. 2013. High salinity events in the northern Arabian Sea and Sea of Oman. *Deep-Sea Research I* 74: 14–24.
Wishner, K.F., M.M. Gowing, and C. Gelfman. 1998. Mesozooplankton biomass in the upper 1000 m in the Arabian Sea: overall, seasonal and geographic patterns, and relationship to oxygen gradients. *Deep-Sea Research II* 45: 2405–2432.
Wyrtki, K. 1973. *Physical oceanography of the Indian Ocean: The biology of the Indian Ocean*, ed. B. Zeitzschel, 18–36. Springer, Berlin.

Gulf of Mexico

Feng, Y., S.F. DiMarco, and G.A. Jackson. 2012. Relative role of wind forcing and riverine nutrient input on the extent of hypoxia in the northern Gulf of Mexico. *Geophysical Research Letters* 39: L09601. doi:10.1029/2012GL051192.
Feng, Y., K. Fennel, G.A. Jackson, S.F. DiMarco, and R.D. Hetland. 2014. A model study of the response of hypoxia to upwelling-favorable wind on the northern Gulf of Mexico shelf. *Journal of Marine Systems* 131: 63–73.
Rabalais, N.N., R.E. Turner, B.K. Sen Gupta, D.F. Boesch, P. Chapman, and M.C. Murrell. 2007. Hypoxia in the northern Gulf of Mexico: Does the science support the plan to reduce, mitigate, and control hypoxia? *Estuaries and Coasts* 30: 753–772.
Liu, K.-K., and N. Dittert. 2010. Web-based electronic supplements, Appendix C. In *Carbon and Nutrient Fluxes in Continental Margins,* ed. K.-K. Liu, L. Atkinson, R.A. Quiñones and L. Talaue-McManus. Springer, Berlin. http://cmtt.pangaea.de/. Accessed 11 June 2015.
Obenour, D.R., D. Scavia, N.N. Rabalais, R.E. Turner, and A.M. Michalak. 2013. Retrospective analysis of midsummer hypoxic area and volume in the northern Gulf of Mexico, 1985–2011. *Environmental Science and Technology* 47: 9808–9815.
Zavala-Hidalgo, J., A. Gallegos-Garcia, B. Martinez-Lopez, S.L. Morey, and J.J. O'Brien. 2006. Seasonal upwelling on the western and southern shelves of the Gulf of Mexico. *Ocean Dynamics* 56: 333–338.

Caribbean Sea

Corredor, J.E. 1979. Phytoplankton response to low level nutrient enrichment through upwelling in the Colombian Caribbean Basin. *Deep-Sea Research* 26A: 731–741.
Gordon, A.L. 1967. Circulation of the Caribbean Sea. *Journal of Geophysical Research* 72: 6207–6223.
Herrera, L., and G. Febres-Ortega. 1975. Kinematics of the wind-generated velocity field in the surface waters off eastern *Venezuela, Caribbean Sea. Bulletin of the Oceanographic Institute, University of Oriente*, 14(2): 165–186.
Liu, K.-K., and N. Dittert. 2010. Web-based electronic supplements, Appendix C. In *Carbon and Nutrient Fluxes in Continental Margins*, ed. K.-K. Liu, L. Atkinson, R.A. Quiñones and L. Talaue-McManus. Springer, Berlin. http://cmtt.pangaea.de/. Accessed 11 June 2015.

Muller-Karger, F.E., and R. Aparicio. 1994. Mesoscale processes affecting phytoplankton abundance in the southern Caribbean Sea. *Continental Shelf Research* 14(2/3): 199–221. doi:10.1016/0278-4343(94)90013-2.

Muller-Karger, F., R. Varela, R. Thunell, Y. Astor, H.Y. Zhang, R. Luerssen, and C.M. Hu. 2004. Processes of coastal upwelling and carbon flux in the Cariaco Basin. *Deep-Sea Research* 51: 927–943. doi:10.1016/j.dsr2.2003.10.010.

Richards, F. 1960. Some chemical and hydrographic observations along the north coast of South America. *Deep-Sea Research* 7: 163–182. doi:10.1016/0146-6313(60)90023-X.

Rueda-Roa, D.T., and F.E. Muller-Karger. 2013. The Southern Caribbean Upwelling System: Sea surface temperature, wind forcing and chlorophyll concentration patterns. *Deep-Sea Research* 78: 102–114.

Ruiz-Ochoa, M., E. Beier, G. Bernal, and E.D. Barton. 2012. Sea surface temperature variability in the Colombian Basin, Caribbean Sea. *Deep-Sea Research I* 64: 43–53. doi:10.1016/j.dsr.2012.01.013.

Stromme, T., and G. Saetersdal. eds. 1989. *Final report. Surveys of the fish resources in the shelf areas between Suriname and Colombia*. NORAD,FAO/UNDPGLO82/001, Institute of Marine Research. Bergen, Norway, Cruise Reports Dr. Fridtjof Nansen. 139 pp. http://www.fao.org/WAIRDOCS/FNS/X6078E/x6078e00.htm. Accessed 1 Apr 2016.

Brazil

Bakun, A., and R.H. Parrish. 1991. Comparative studies of coastal pelagic fish reproductive habitats: the anchovy (*Engraulis anchoita*) of the southwestern Atlantic. *ICES Journal of Marine Science* 48: 343–361.

Campos, P.C., O.O. Moller Jr., A.R. Piola, and E.D. Palma. 2013. Seasonal variability and coastal upwelling near Cape Santa Marta (Brazil). *Journal of Geophysical Research* 118: 1420–1433. doi:10.1002/jgrc.20131.

Castelão, R.M., and J.A. Barth. 2006. The relative importance of wind strength and along-shelf bathymetric variations on the separation of a coastal upwelling jet. *Journal of Physical Oceanography* 36: 412–425.

FAO. 2003. Trends in oceanic captures and clustering of Large Marine Ecosystems—two case studies based on the FAO Capture Database. *FAO Fisheries Technical Paper*, 435.

Gaeta, A.S., J.Á. Lorenzetti, L.B. Miranda, S.M.M. Susini-Ribeiro, M. Pompeu, and C.E.S. De Araujo. 1999. The Victoria eddy and its relation to the phytoplankton biomass and primary productivity during the austral Fall of 1995. *Archive of Fishery and Marine Research* 47(2/3): 253–270.

Matsuura, Y. 1998. Brazilian sardine (*Sardinella brasiliensis*) spawning in the southeast Brazilian Bight over the period 1976-1993. *Revista de Biología Marina y Oceanografía* 46(1): 33–43.

Mazzini, P.L.F., and J.A. Barth. 2013. A comparison of mechanisms generating vertical transport in the Brazilian coastal upwelling regions. *Journal of Geophysical Research* 118: 5977–5993. doi:10.1002/2013JC008924.

Paiva, M.P. ed. 1997. Recursos pesqueiros estuarinos e marinhos do Brasil. *Avaliação do potencial sustentável de recursos vivos na zona econômica exclusiva*. Universidade Federal do Ceará, Fortaleza: 286 pp.

Stech, J.L., and J.A. Lorenzzetti. 1992. The response of the South Brazil Bight to the passage of wintertime cold fronts. *Journal of Geophysical Research* 97(6): 9507–9520.

UNEP (2004) Marques, M., B. Knoppers, A.E. Lanna, P.R. Abdallah and M. Polette. 2004. *Brazil Current, GIWA Regional Assessment 39*. Kalmar, Sweden: University of Kalmar.

Eurafrican Mediterranean Sea

Agostini, V., and A. Bakun. 2002. Ocean triads in the Mediterranean Sea: physical mechanisms potentially structuring reproductive habitat suitability (with example application to European anchovy, *Engraulis encrasicolus*). *Fisheries Oceanography* 11: 129–142.

Bakun, A. 1996. *Patterns in the Ocean: Ocean processes and marine population dynamics*. California, USA: University of California Sea Grant, San Diego, in cooperation with Centro de Investigaciones Biolo´gicas de Noroeste, La Paz, Baja California Sur, Mexico, 323 pp.

Bakun, A., and V. Agostini. 2001. Seasonal patterns of wind-driven upwelling/downwelling in the Mediterranean Sea. *Scientia Marina* 65(3): 243–257.

d'Ortenzio, F., D. Antoine and S. Marullo. 2008. Satellite-driven modeling of the upper ocean mixed layer and air–sea CO2 flux in the Mediterranean Sea. *Deep-Sea Research*, 55: 405–434.

Goffredo, S., and Z. Dubinsky (eds.). 2014. *The Mediterranean Sea: Its history and present challenges*. New York: Springer. 678 pp.

Siokou-Frangou, I., U. Christaki, M.G. Mazzocchi, M. Montresor, M. Ribera d'Alcalá, D. Vaqué and A. Zingone. 2010. Plankton in the open Mediterranean Sea: a review. *Biogeosciences*, 7: 1543–1586. doi:10.5194/bg-7-1543-2010.

Stergiou, K.I., and E.D. Christou. 1996. Modelling and forecasting annual fisheries catches: comparison of regression, univariate and multivariate time series methods. *Fishery Research* 25: 105–138.

Chapter 9
Other Important Upwelling Systems

Abstract This chapter describes other significant upwelling systems of the world oceans that were not covered in previous chapters. This includes wind-driven upwelling in the open ocean (e.g., equatorial upwelling) and upwelling domes, shelf-break upwelling induced by boundary currents and their mesoscale eddies, nutrient upwelling in tidal-mixing fronts, and ice-edge upwelling.

Keywords Equatorial upwelling · Upwelling domes · Miscellaneous upwelling mechanisms · Shelf-break upwelling · Tidal-mixing fronts · Ice-edge upwelling

The fishermen know that the sea is dangerous and the storm terrible,
but they have never found these dangers sufficient reason for remaining ashore.
Vincent Van Gogh (1853–1990)

9.1 Introduction

This book has concentrated on regions where upwelling along coastal boundaries is important. Apart from the classical wind-driven upwelling regions, however, other upwelling regions of relatively high productivity occur in the world ocean, associated with different physical generation mechanisms. One example is current-driven upwelling on the Patagonian Shelf on the southern Atlantic coast of South America; this creates a Class II, moderately productive Large Marine Ecosystem (150–300 g C/m^2/year) in the absence of upwelling-favourable coastal winds. This chapter provides a summary of the most important of these other marine ecosystems, but we have deliberately not gone into as great an amount of detail as with the coastal regimes and have concentrated on the physical mechanisms involved. Readers are encouraged to consult the references for more information on these systems (Fig. 9.1).

© Springer International Publishing Switzerland 2016

J. Kämpf and P. Chapman, *Upwelling Systems of the World*,
DOI 10.1007/978-3-319-42524-5_9

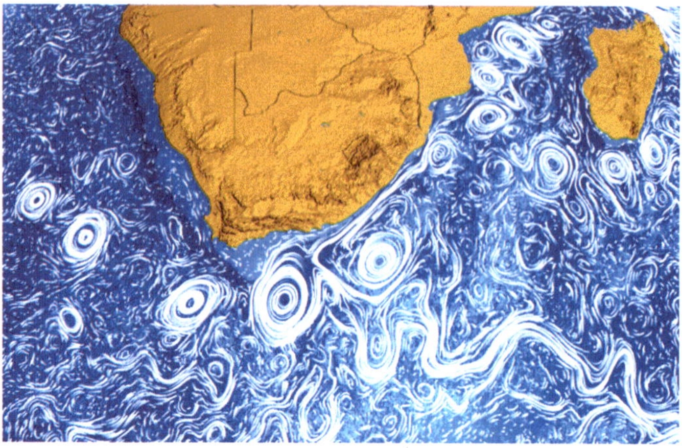

Fig. 9.1 NASA's visualization of oceanic currents around South Africa (slightly modified). Image source: http://svs.gsfc.nasa.gov/cgi-bin/details.cgi?aid=3827 [accessed on 6 April 2016]

9.2 Southern Ocean Upwelling

The upwelling of deep water along the Antarctic Divergence of the Southern Ocean is by far the biggest upwelling system on Earth (Fig. 9.2). It is created from a divergence of Ekman-layer transport in the surface ocean driven by the meridional variation of the strength of the Westerlies between 55°–65° S. Additional energy is provided where the frontal systems (the Antarctic Circumpolar Current and the Polar Front) cross large topographic features such as the mid-ocean ridges. This generates an increase in potential vorticity at these sites, which is dissipated through the formation of meanders and eddies that increase local upwelling (Moore and Abbott 2002). The total upwelling flux of deep water in the Southern Ocean is about 29 Sv (Talley 2013). Despite this widespread upwelling in the Southern Ocean (Fig. 9.3) of nutrient-rich water, containing about 32–34 μM/kg nitrate, 2.2–2.4 μM/kg phosphate, and 80–120 μM/kg silicate (data from WOCE Atlas series), photosynthesis by marine phytoplankton and the associated production of carbon are thought to be currently limited by the availability of iron (Martin 1990; Kumar et al. 1995). Hence, most of the deep water upwelling region constitutes a high-nutrient, low-chlorophyll (HNLC) region of very low primary productivity.

Using incubation experiments, Church et al. (2000) found that bacterial growth in this region was constrained primarily by the availability of dissolved organic matter. Bacterial growth in the subtropical front, subantarctic zone, and subantarctic front all responded most favourably to additions of dissolved free amino acids or glucose plus ammonium. Unlike similar experimental results in other HNLC regions (subarctic and equatorial Pacific), where additions of iron alone caused

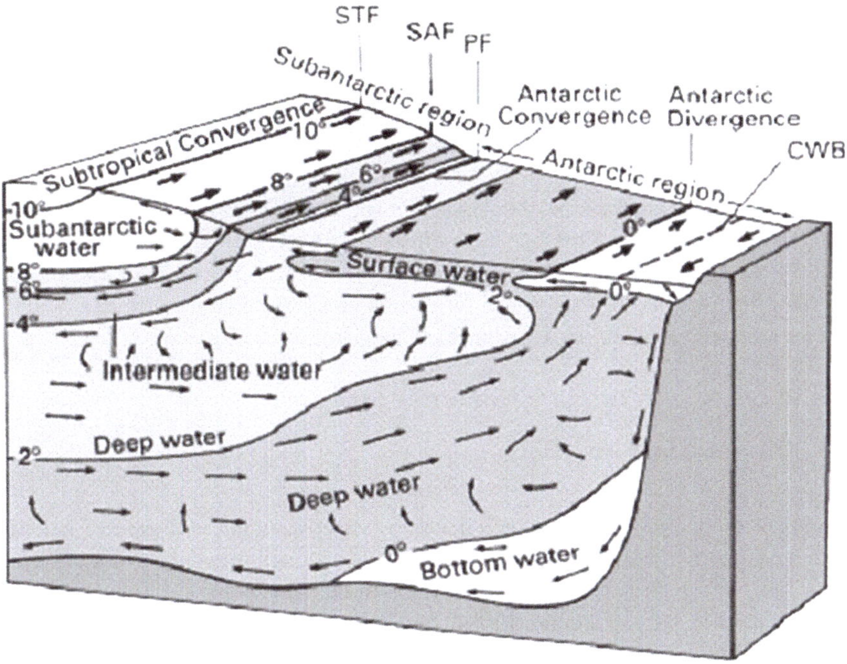

Fig. 9.2 Schematic of the water circulation and associated water masses in the Southern Ocean (from Tomczak and Godfrey 2003). *STF* shows the Subtropical Front, *SAF* the Sub-Antarctic Front, and *PF* the Polar Front. *CWB* is the Continental Water Boundary

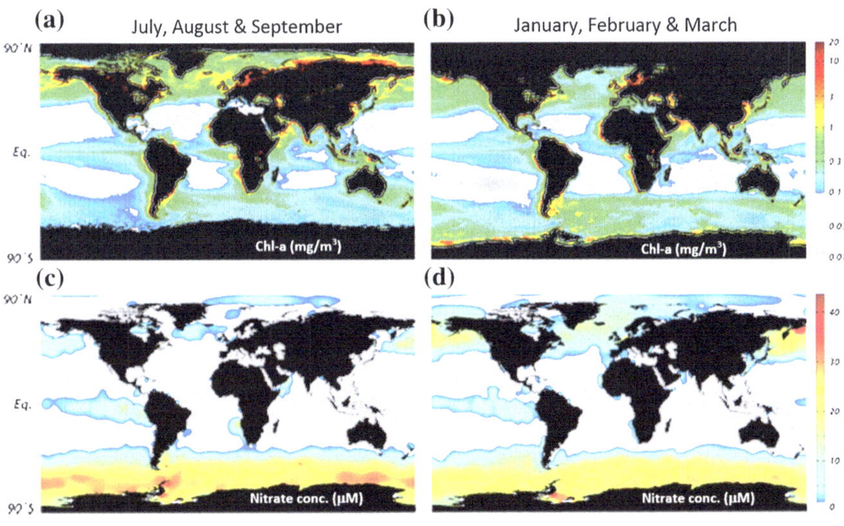

Fig. 9.3 Composite global ocean maps of concentrations of satellite-derived chlorophyll-*a* (**a**, **b**) and ship sampled nitrate (**c**, **d**). The *left* (*right*) panels show northern (southern) hemisphere summer. *Gray* regions denote low values. Data shown are available from NASA (http://oceancolor.gsfc.nasa.gov/) and NOAA's National Oceanographic Data Center (www.nodc.noaa.gov/)

increases in phytoplankton productivity, additions of glucose plus iron were required to produce substantial increases in rates of bacterial growth and biomass accumulation and the addition of iron alone resulted in relatively little change in bacterial growth rates.

The relatively low primary production of large areas of the Southern Ocean (on a per square metre basis) results in a short food chain, with Antarctic krill (*Euphausia superba*) having evolved as the keystone species within the Southern Ocean ecosystem (Reid 2001). Krill provide a direct link in the marine food web between primary producers and upper trophic levels (Marr 1962; Croxall et al. 1999), and are the principle diet of many large carnivores including invertebrates (e.g. squid) and vertebrates (e.g. fish, birds, whales) (Cuzin-Roudy 2000).

9.3 Equatorial Upwelling

Thanks to the Coriolis effect, which becomes effective within a degree of latitude from the equator, the trade winds create divergence along the equator that transports away from it in both hemispheres. This leads to the upwelling of nutrient-enriched water from a depth of around 100 m of the equatorial undercurrent to the sea surface (Wyrtki 1981), where it enhances primary production. This is particularly obvious in the Pacific Ocean, where the plume of enhanced surface nitrate stretches from South America to beyond the dateline (Fig. 9.3). The Ekman drift implies a distinct spatial zonation of trophic levels in both hemispheres. Phytoplankton concentrations peak closest to the equator, the zone of highest fish production forms some 400 km away from the equator and the zone of enhanced zooplankton abundance is located between the two (Fig. 9.4). Equatorial fisheries mainly target large pelagic fish (tuna, marlin and sharks).

Equatorial upwelling occurs in the Pacific, subject to ENSO events, and the Atlantic Oceans. The wind field that drives the equatorial Indian Ocean is unique in that its annual mean is weak and westerly, and its annual cycle has a strong semi-annual component (see Sect. 8.3). Consequently, the Indian Ocean equatorial currents are much more variable and quite different in character from those in the other oceans, and as a result, persistent equatorial upwelling does not develop here (Schott and McCreary 2001).

A number of important field experiments were conducted in the equatorial Pacific between 1990 and 1997 by researchers from Australia, France, Japan and the U.S. under the auspices of the Joint Global Ocean Flux Study (JGOFS). While it is impossible to discuss all the results from these experiments, which fill several volumes of scientific papers, we summarize some key findings, and readers are referred to the primary literature for more details of the work. JGOFS research concentrated on carbon cycling in the equatorial Pacific, including physical and chemical controls, the biological response and interactions with the atmosphere and export to the deep ocean, andcaptured both El Niño and La Niña conditions (Murray et al. 1994, 1995, 1997).

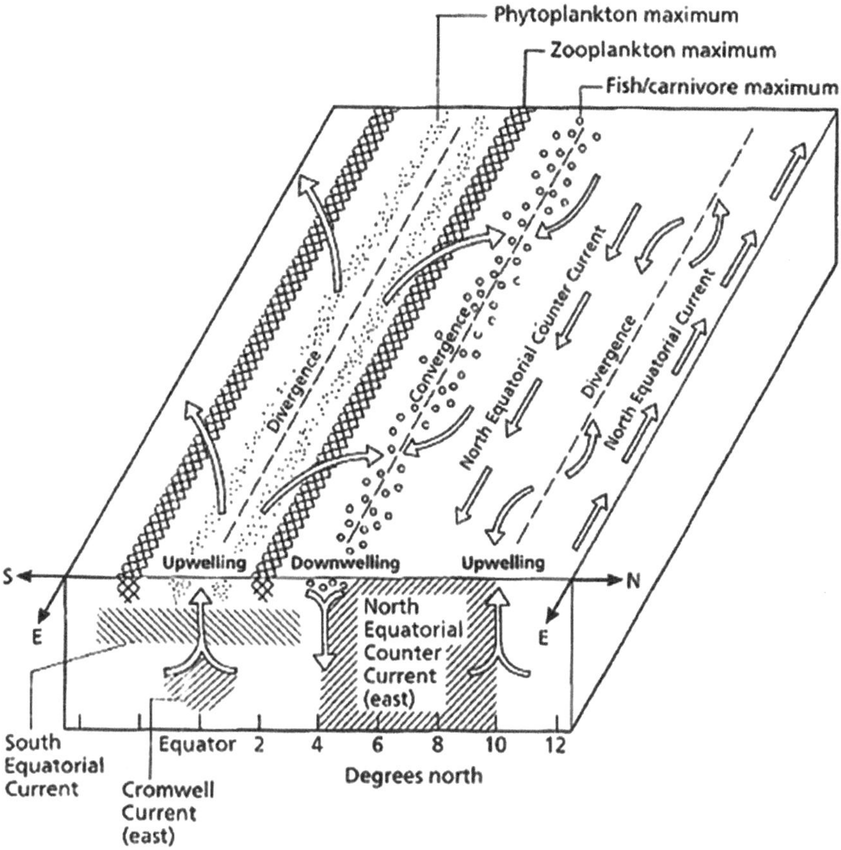

Fig. 9.4 Zonation of marine trophic levels in relation to currents and upwelling/downwelling features, shown for the Pacific Ocean (from Mann and Lazier 1996)

The upper-ocean circulation in the equatorial Atlantic and Pacific has a clear zonal structure and generally consists of westward surface flows and an eastward undercurrent found at depths below ~100 m. The equatorial undercurrent provides the source of upwelling water, but although enriched in nutrients compared with coastal upwelling regimes (e.g., ~5 μM/L nitrate; Fig. 9.5), it is still relatively low in nutrients. The strength of equatorial upwelling depends largely on the magnitude of the trade winds, although wind variability in the western Pacific and tropical instability waves with periods of 20–30 days can modulate this as they determine the depths from which water upwells farther east (Murray et al. 1994; Foley et al. 1997), Wyrtki (1981) estimated a mean upwelling rate of 50×10^6 m^3/s for the Pacific. Seasonal variations in equatorial upwelling are associated with seasonal shifts of the location of the Inter-Tropical Convergence Zone (ITCZ) (see Fig. 2.13).

The western wedge of upwelling in the equatorial Pacific is clearly marked by a salinity front (e.g., Kuroda and McPhaden 1993), which separates the warm, fresh

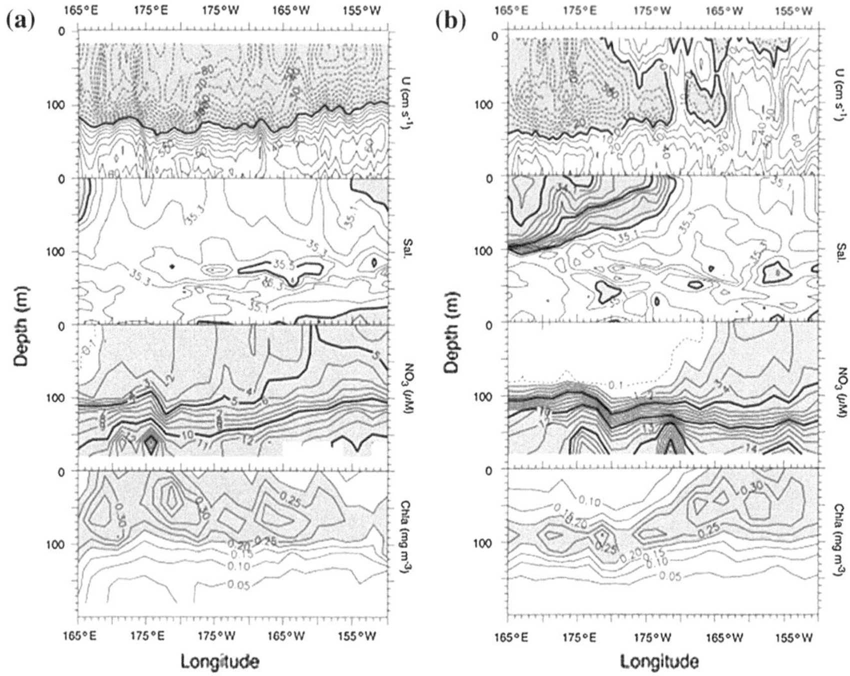

Fig. 9.5 Results from the equatorial leg of **a** the ALIZÉ 2 cruise in January–February 1991 and **b** the FLUPAC cruise in October 1994 in the equatorial Pacific Ocean. Shown are: zonal current (U, the shaded areas are westward flow), salinity (Sal., shaded area <35), nitrate (NO_3, shaded >1 μM/L) and chlorophyll-*a* (Chl*a*, shaded >0.2 mg/m^3) zonal sections (from Rodier et al. 2000)

pool of water to the west from colder and saltier upwelled waters to the east (Fig. 9.5). On interannual timescales, zonal displacements of the salinity front are closely correlated with the Southern Oscillation Index (SOI) (Delcroix and Picaut 1998). Upwelling in the east of the basin is enhanced through the general eastward shoaling of the equatorial pycnocline and nutricline (see Fig. 3.2). Like much of the Southern Ocean, the upwelling zone in the equatorial Pacific is known as a High Nutrient–Low Chlorophyll (HNLC) area (Minas et al. 1986), which implies that chlorophyll is relatively low compared to available nutrient levels and suggests some form of nutrient limitation.

Most of the production appears to be dependent on nano- or picoplankton, that take up ammonia rather than nitrate and are fed on by protozoans, with the production of considerable amounts of dissolved organic carbon that are recycled within the upper 200 m of the water column (Murray et al. 1994; Landry et al. 1995). While grazing pressure is one control of phytoplankton biomass (Loukos et al. 1997), export production is low in this region, with very little being transported into the deep ocean and the majority of the DOC pool that survives local recycling is advected out of the upwelling zone.

Rodier et al. (2000) discuss field data from two cruises covering the equatorial Pacific from 165° E to 150° W. The first cruise (ALIZÉ 2, Fig. 9.5a) took place during a non-El Niño period (3 January–5 March 1991). In contrast, the second cruise (FLUPAC, Fig. 9.5b) took place during a moderate El Niño event (23 September–29 October 1994). The ALIZÉ 2 data show persistent upwelling along most of the study region, with nitrate levels in the surface mixed layer that have increased from <1–5 μM over the length of the transect (Fig. 9.5a). Primary production was supported throughout the surface mixed layer, although chlorophyll concentrations were only ~0.3 mg/m^3, which is an order of magnitude lower compared with major coastal upwelling regions.

The FLUPAC transect (Fig. 9.5b) shows a dramatically different situation. Here we can see the existence of a sharp salinity front at ~173° W separating the warm pool to the west from the upwelling regime to the east. Nitrate levels within the warm pool are extremely low (~0.1 μM), but nutrient fluxes are still sufficient to create sub-surface phytoplankton blooms at a depth of ~50–100 m where chlorophyll concentrations peak at ~0.25 mg/m^3. In contrast, the upwelling regime has nitrate levels of 2–4 μM and chlorophyll concentrations of ~0.3 mg/m^3 throughout the surface layer. Interestingly, there was a substantial spatial mismatch of ~500 km between the location of the salinity front and that of the nutrient/chlorophyll front (Fig. 9.6). Rodier et al. (2000) argue that this zonal separation arises from biological processes of nitrate uptake and grazing, but their model could not explain why no such offset was observed for zooplankton biomass.

While tropical instability waves can control the strength of upwelling, as suggested above, they can also affect production through their delivery of nutrients (Foley et al. 1997). This includes not only macronutrients, but also iron, which is supplied by the Equatorial Undercurrent (Gordon et al. 1997). As the Equatorial Undercurrent tends to be deeper and slower during El Niño events than at other times, this will reduce the supply of iron and be another constraint on primary production and export during these periods (Murray et al. 1995).

Equatorial upwelling in the Pacific makes the domain both a source of CO_2 to the atmosphere and an internal carbon sink due to its increased primary production (Murray et al. 1994). Given the vast size of the upwelling region, the equatorial Pacific is a major source of CO_2 to the atmosphere, estimated at up to about 1 G tonnes/yr (Tans et al. 1990), and it contributes to >18 % of global new biological production (Chavez and Toggweiler 1995). Due to differential upwelling, the surface salinity front corresponds to a pCO_2 front (see Fig. 9.6; Inoue et al. 1996; Boutin et al. 1999). Accordingly, ENSO events strongly control CO_2 exchanges with the atmosphere over the entire equatorial Pacific region; that is, El Niño events transiently reduce the CO_2 flux to the atmosphere substantially because the upwelling water contains less CO_2 and not because of changes in the biology (Fig. 9.7; Feely et al. 2002). As an example, the 1992 El Niño reduced the flux to only about 0.3 G ton/year (Feely et al. 1995), but following a return to more normal conditions, the flux increased again to 0.6 G tonnes in 1993 and 0.7 G tonnes in 1994 (Feely et al. 1997). The spreading of low salinity water from the western

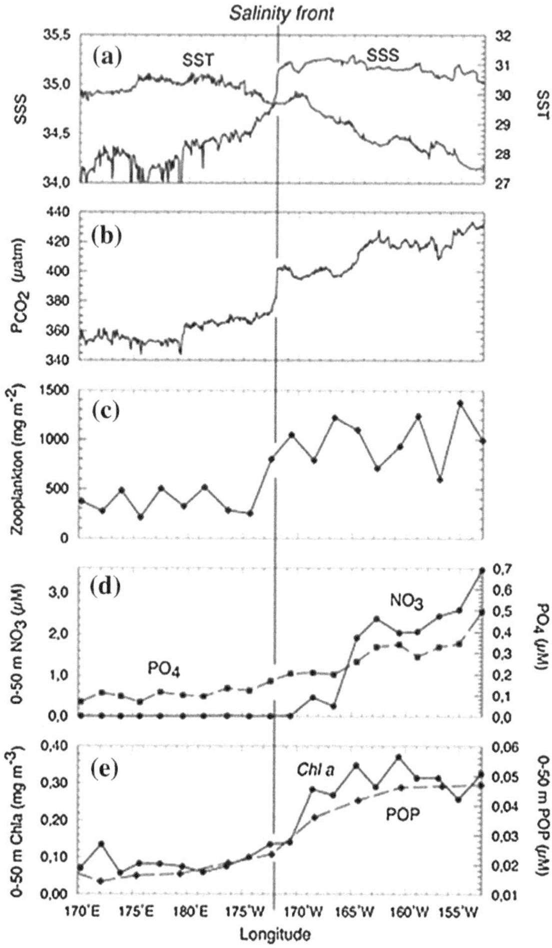

Fig. 9.6 Details of the hydrographic and planktonic features in the upper layer during FLUPAC cruise: **a** SST, SSS, **b** pCO_2, **c** 0–100 m zooplankton biomass in ash-free dry weight, **d** mean NO_3 and PO_4 and **e** mean chlorophyll-*a* (Chl *a*) and POP (Particulate organic phosphorous) in the 0–50 m upper layer. SSS, SST and pCO2 data correspond to underway measurements, whereas nutrient, phytoplankton and zooplankton data were obtained every two degrees of longitude (from Rodier et al. 2000)

Pacific at the ocean surface can also reduce sea-air transfer of CO_2 by capping the higher concentrations deeper in the water column; this happened from 2° N–2° S in 1992 (Murray et al. 1994).

It should be noted that Ryan et al. (2002) describe an unusual large-scale accumulation of phytoplankton during the 1998 transition from El Niño to La Niña. Interestingly, this bloom started as a sub-surface bloom in the Equatorial Undercurrent, spread eastward over a distance of 4500 km and then dispersed north and south of the equator over several hundred kilometres due to advection by tropical instability waves. Ryan et al. (2002) describe the different factors that contributed to the formation of this massive phytoplankton bloom including nutricline shoaling, along-isopyncal upwelling, turbulent vertical mixing and wind-driven upwelling.

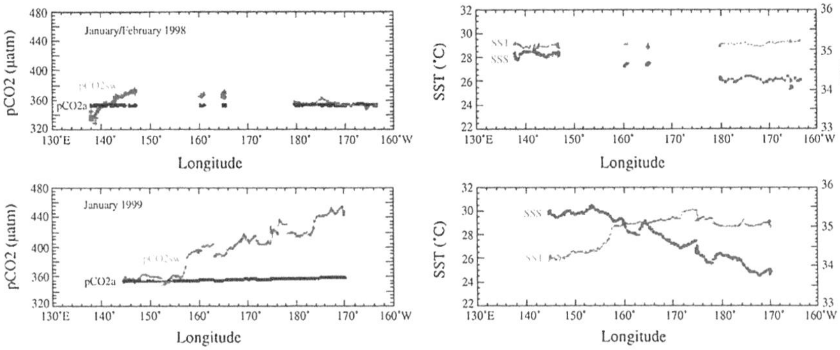

Fig. 9.7 Longitudinal distribution of pCO_2 (*left*) and SST and SSS (*right*) along the equator in January/February 1998, and January 1999 (from Feely et al. 2002)

Nitrogen fixation is the use of nitrogen gas (N_2) as the source of nitrogen atoms for growth, rather than nitrate (NO_3^-) or ammonium (NH_4^+), which are more convenient for phytoplankton to use because less energy expenditure is required. While nitrogen fixation is commonly found in regions where dissolved inorganic nitrogen concentrations are very low, such as the oligotrophic gyres in the centres of the main ocean basins (Fig. 9.3), recent observational findings challenge the long-held paradigm that oligotrophic gyres have the highest nitrogen fixation rates. In the equatorial Atlantic, Subramaniam et al. (2013) derived nitrogen fixation rates (up to 400 $\mu M/m^3/day$) that far exceed those typically measured in subtropical gyres (up to 200 $\mu M/m^3/day$) or in other upwelling regions (up to 85 $\mu M/m^3/day$). Hence, equatorial waters of the Atlantic support surprisingly extensive diazotrophic (nitrogen fixating) communities, and it seems that the understanding of nitrogen and carbon cycles in equatorial regions is far from complete.

9.4 Upwelling Domes

Upwelling domes are quasi-stationary circular ocean-ocean features having a diameter of 300–600 km in which nutrient-rich subsurface water is brought closer to the euphotic zone. They can clearly by identified in satellite images of sea-surface temperature and ocean colour (chlorophyll-*a*). The classic example of an upwelling dome is the Costa Rica Dome at 7.5°–9.0° N, 87°–90° W in the eastern tropical Pacific (Wyrtki 1964; Fiedler 2002) (Fig. 9.8). This dome forms in a region where the Equatorial Countercurrent (ECC) turns north into the Costa Rican Coastal Current and then westward into the North Equatorial Current. It is most prominent when the ECC is strongest in late summer and in autumn, but it exists at all seasons, even when the ECC is almost absent.

Mazeika (1967) described upwelling domes in the eastern tropical Atlantic. From January to April (austral summer) the Angola Dome forms off Angola.

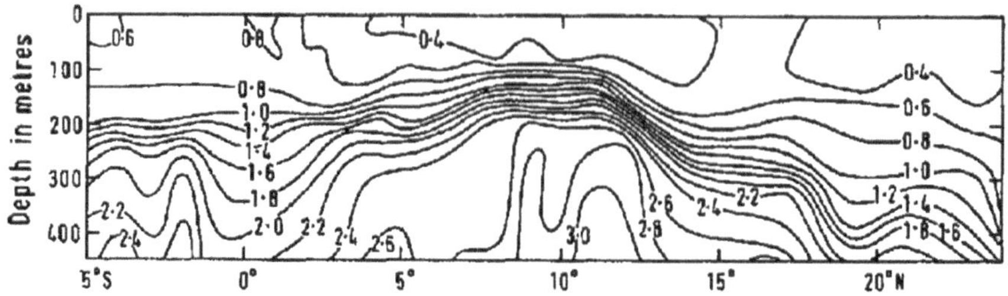

Fig. 9.8 Section of the Costa Rica Dome, in phosphorous (mg-atoms/L). Taken from Cromwell (1958)

Similarly from July to September (boreal summer), the Guinea Dome forms southwest of Dakar. It is formed as the Equatorial Countercurrent turns into the North Equatorial Current; in other words, it is a seasonal analogue of the Costa Rica Dome. Mazeika (1967) also showed that another seasonal upwelling dome forms during austral summer months just south of the equator between 8° W and 3° E. Wyrtki (1962) reported the existence of an upwelling dome in the tropical eastern Indian Ocean—the Sumbawa Dome—that forms off Java and Sumbawa from May to September during the southeasterly monsoon, while more recently the existence of a similar dome near the Seychelles has been discussed by Xie et al. (2002) and Yokoi et al. (2008).

These large upwelling domes are all associated with the eastern edge of equatorial currents where they turn polewards after hitting the continental shelf. On a smaller scale, any cold-core, cyclonic eddy will also induce upwelling. While open ocean eddies are ephemeral and erratic, there are quasi-stationary eddies near the coast that may form or be trapped by topography. One such example comes from the Gulf of Mexico, where Belabbassi et al. (2005) have shown that such a quasi-stationary eddyin vicinity of the De Soto Canyon, east of the Mississippi delta, where it supplies nutrients to the generally oligotrophic shelf of the northeast Gulf of Mexico.

The Costa Rica Dome, and presumably also the other upwelling domes, are important habitats for dolphins and migratory megafauna such as blue whales. However, the Costa Rica Dome is located in an overall productive region and does not necessarily coincide with the region of the highest phytoplankton productivity. For example, an oceanographic survey of the eastern tropical Pacific Ocean in August–November 1990 observed higher productivity of >600 mg C/m^2/day along the divergence of the Equatorial Countercurrent to the west of the Costa Rica Dome than in the dome itself (Fiedler et al. 1991). Also, in recent decades, the jumbo flying squid (*Dosidicus gigas*) has evolved as one of the most important species of cephalopod fisheries in the eastern Pacific (Chen et al. 2014). The highest abundance of this species usually occurs in the Californian, Peruvian, and Chilean upwelling systems including the Costa Rica Dome. Chen et al. (2013) proposed that the Costa Rica Dome and adjacent waters serve as a potential spawning ground for this species. There is an international effort to enhance the protection level of the Costa Rica Dome, but this effort is complicated by the fact that, owing to seasonal variations, its location is sometimes completely in international waters of the "high seas" while at other times it includes the Exclusive Economic Zones of five Central American countries (Hoyt 2011).

9.5 Current-Driven Upwelling in Western Boundary Currents

9.5.1 Overview

Western boundary currents can create upwelling by various physical mechanisms, such as:

- Dynamic uplift controlled by the overall strength of the boundary current (e.g., Middleton et al. 1994),
- Creation of geostrophic turbulence (mesoscale eddies),
- Convergence of transports in the bottom Ekman layer. (e.g., McClain et al. 1984), and
- Interaction with irregular topography and the creation of stationary topographic waves (e.g., Kämpf 2012).

Western boundary currents of subtropical gyres (the Gulf Stream, Brazil Current, Agulhas Current, Somali Current, Kuroshio Current, and East Australian Current) are dynamically distinct from those of subpolar gyres (East Greenland Current), Malvinas Current, Oyashio Current, Kamchatka Current, and Labrador Current Subtropical western boundary currents have two special characteristics that make them favourable for inducing upwelling. The first characteristic is that in regions in which they interact with the seafloor they create shoreward veering in their bottom Ekman layer and a convergence of bottom Ekman-layer transports on their shoreward side. The second characteristic is that their currents are directed opposite to the propagation direction of free topographic Rossby waves, which can lead to the creation of localized upwelling flows across the shelf break in interaction with bathymetric undulations such as submarine canyons (Kämpf 2012).

In contrast, western boundary currents of subpolar gyres do not possess these characteristics. Theoretically, they should create downwelling as a result of Ekman-layer divergence on their shoreward side and they cannot create stationary topographic waves given that wave disturbances travel in the same direction as the mean flow. The Leeuwin Current, which flows south along the west coast of Australia because of the sea level pressure gradient created by the Indonesian throughflow (Godfrey and Ridgway 1985), also belongs to this type of boundary current. Despite this theoretical argument, however, it should be noted that the Malvinas Current on the Patagonian Shelf, which is a subpolar western boundary current, creates highly productive shelf-break upwelling and a lucrative fishing ground. Possible mechanisms for this phenomenon are given below.

9.5.2 Western Boundary Currents of Subtropical Gyres

Western boundary currents of subtropical gyres, such as the Gulf Stream, are characterized by shoreward flows in their bottom friction layer. Given that this shoreward flow is attached to the boundary current and does not extend to the coast, it is rather the convergence of near-bottom Ekman flows that creates a localized vertical uplift of near-bottom water. This type of upwelling can be continuous in regions where the current is located permanently on the continental shelf or it can be transient in regions where the current moves occasionally onto the shelf. This episodic current-driven upwelling associated with meanders of a western boundary current has been observed on the South Carolina Shelf (Lee et al. 1989) and in the South Brazil Bight (Campos et al. 2000).

Upwelling can also result from bathymetric interactions. The Gulf Stream, for instance, creates local upwelling in a region where it is deflected offshore by the Charleston Bump, which includes an underwater ridge and trough feature on which the seafloor rises from 700 to 400 m within a relatively short distance (Brooks and Bane 1978). Numerous submarine canyons cut into the shelf break of the New England and Mid-Atlantic continental shelves of the U.S. east coast. These shelf-break submarine canyons, such as the Hudson, Wilmington, Lydonia, Oceanographer and Norfolk Canyons, are known to be hotspots of enhanced physical-biological interaction and biological productivity (Greene et al. 1988; Hooker et al. 1999; Waring et al. 2001; Ryan et al. 2005; Levin and Sibuet 2012).

Consistent with the theory of stationary canyon upwelling (Kämpf 2012), warm core rings of the Gulf Stream, which are associated with northward currents on the continental shelf, can strongly influence the magnitude and direction of the flow inside the canyons (Butman 1986). In particular, the formation of a "cold pool" of nutrient-rich water on the shelf (Houghton et al. 1982) can promote nutrient exchange and biological production (Church et al. 1984). This is similar to what happens in the eastern the Great Australian Bight Australia, where a cold pool forms southwest of Kangaroo Island and upwelling is enhanced by the presence of a canyon (Kämpf 2010; see Sect. 8.2.4). Hence, there is some evidence for the theory that subtropical western boundary currents initiate localized canyon upwelling, but more research is required to confirm this.

It is widely known that the Agulhas Current, which is the western boundary current of the southern subtropical Indian Ocean, creates widespread upwelling through its interaction with the Agulhas Bank (Fig. 9.9). The inshore regions of the central and eastern Agulhas Bank are characterized by intermittent coastal upwelling events, particularly downwind of prominent capes (Schumann et al. 1988). Wind-driven upwelling occurs predominantly in austral summer and autumn. The Agulhas Current and its meanders can also trigger episodic shelf-break upwelling in the eastern sector of the Agulhas Bank throughout the year that may persist for a number of weeks. Ecologically more important, however, is the upwelling onto the shelf that occurs near 26° E; this forms a quasi-permanent pool of cold water along

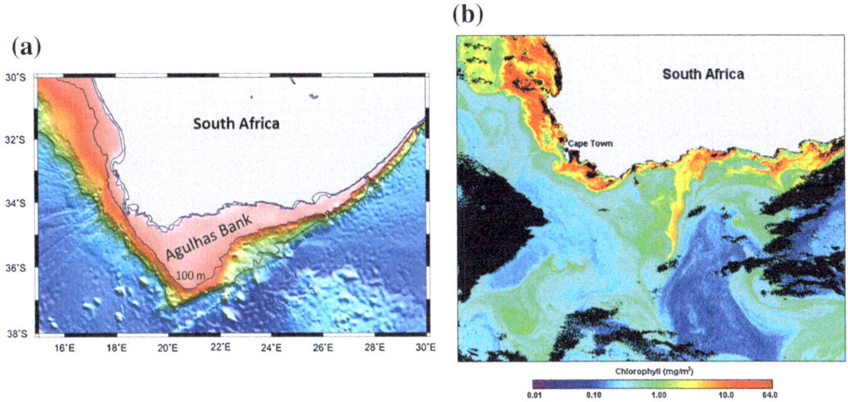

Fig. 9.9 **a** Bathymetry of the Agulhas Bank, Taken from Liu and Dittert (2010). **b** Satellite image of SeaWiFS chlorophyll-*a* concentration showing upwelling filaments on the Agulhas Bank and within the Benguela upwelling system on the west coast of South Africa. Data source: NASA

the centre of the Agulhas Bank that moves slowly westwards until it leaves the bank between 20° and 22° E (Largier et al. 1992; Lutjeharms 2006). This is visible in summer and contributes to the intense thermocline (<10 °C in 5–10 m) that develops on the bank at this time (see also Fig. 7.9). In winter, it dissipates because of deep mixing. The distribution of copepod diversity (Fig. 9.10) clearly reflects the complex flow dynamics in the region.

DiMarco et al. (2000) also identified upwelling on the continental shelf of Madagascar, confirmed later by hydrographic data (Machu et al. 2002), which may

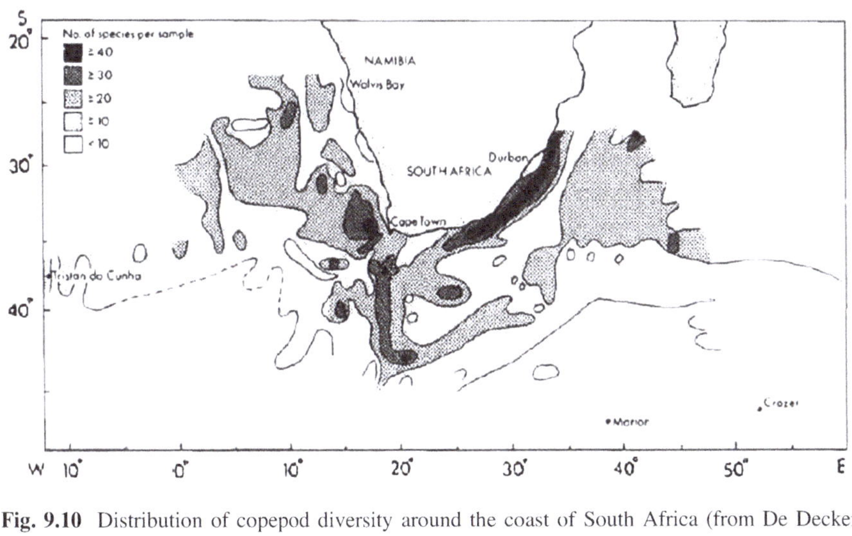

Fig. 9.10 Distribution of copepod diversity around the coast of South Africa (from De Decker 1984)

occur in the same manner, although whether this upwelling is predominantly current-driven by the East Madagascar Current wind-driven, or some combination of both, remains unknown.

Other regions where current-driven shelf-break upwelling is known to occur include the Taiwan Warm Current, a branch of the Kuroshio Current that creates shelf-break upwelling in the East China Sea (Hu et al. 1980; Zhu 2003), the East Australian Current which creates shelf-break upwelling near Frazer Island known as the Southeast Frazer Island Upwelling System (Brieva et al. 2015), and the flow through Yucatan Strait, which causes upwelling along the eastern edge of Campeche Bank (Cochrane 1969).

9.5.3 Western Boundary Currents of Subpolar Gyres

The Malvinas Current (Fig. 9.11), previously called the Falkland Current, has a high productivity associated with persistent shelf-break upwelling (Fig. 9.12a). The food supply in the Malvinas region is so plentiful and reliable that elephant seals, which breed and moult on the Argentinean shoreline, cross the ~400 km wide Patagonian shelf to feed there (Campagna et al. 1998). Unsurprisingly, this region hosts one of the largest fisheries in the Southern Hemisphere. Night-time satellite pictures, for example, routinely show a dense congregation of squid fishing vessels whose illumination rivals the urban centres of Buenos Aires, Argentina, and Montevideo, Uruguay (Fig. 9.12b). Local productivity means that the Patagonian Shelf is classified as a moderately productive (Class II) Large Marine Ecosystem (150 g C/m^2/yr).

The persistent shelf-break upwelling along the edge of the Patagonian Shelf, has puzzled oceanographers for decades, given that the Malvinas Current is a "downwelling" current; that is, its bottom Ekman layer induces a seaward, downslope flow. Due to its flow direction, it can also not support the creation of stationary topographic waves. Moreover, the winds in the Patagonia region are not upwelling favourable, tidal mixing is relatively small in the shelf-break region, and the Malvinas Current does not show the eddy shedding and meandering that drive the upwelling of other western boundary systems (e.g., Campos et al. 2000).

Instead, process-oriented hydrodynamic modelling studies suggest that the shelf-break upwelling on the Patagonian Shelf is driven by flow convergence either in the bottom Ekman layer (Benthuysen et al. 2015) or just above this layer (Matano and Palma 2008). Let us consider this process in more detail to avoid confusion. Although the bottom Ekman flow is directed in an offshore direction, the offshore Ekman transport decreases in the same direction. This decrease constitutes a flow convergence triggering upward flow and, hence, upwelling. While this process takes place in vicinity of the shelf break here, it is akin to the traditional wind-driven upwelling process where the bottom Ekman layer also experiences a flow convergence and creates upwelling as it approaches the coast.

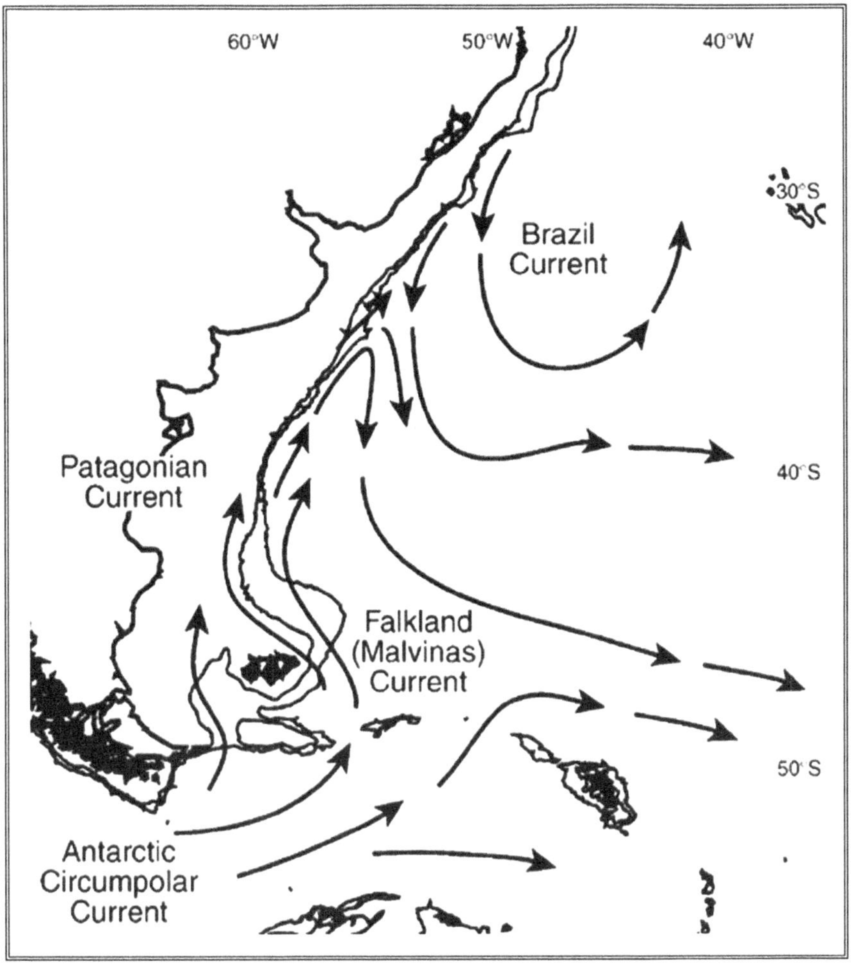

Fig. 9.11 The major surface currents in the southwest Atlantic. The 200 and 1000 m depth contours are also shown (from Anderson and Rodhouse 2001)

While the Oyashio Current, which is the western boundary current of the sub-polar gyre in the North Pacific, does not experience noticeable upwelling along its path, it also supports a highly productive ecosystem. Most of its fauna are sub-polar fish species, but it is also an important summer feeding ground for subtropical migrants such as the Japanese sardine, Japanese anchovy, Pacific saury, mackerels, Japanese common squid, whales and seabirds (Sakurai 2007). The reason for this productivity is that the current receives high concentrations of nutrients from the Western Subarctic Gyre (WSG) (Sakurai 2007). The WSG is itself the consequence of a wind-stress curl that creates extensive open-ocean upwelling in the gyre's centre and a cyclonic circulation around it.

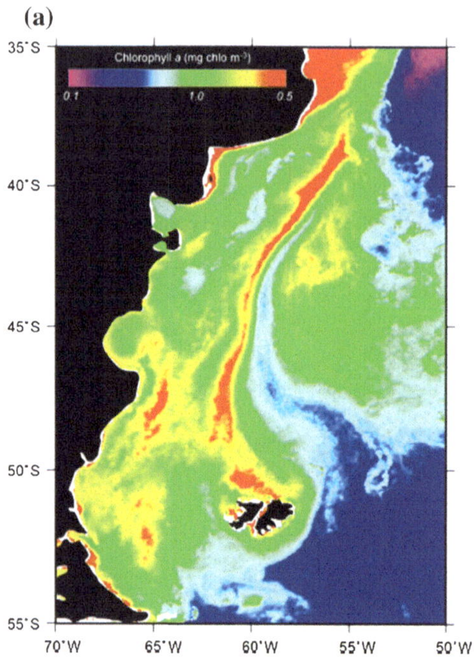

Fig. 9.12 **a** Summer values of surface chlorophyll-*a* derived from SeaWiFS (mg/m^3) in the Patagonia region; **b** nighttime lights in the southwestern Atlantic. Lights over the Patagonian shelf (shown in *white*) are generated by squid fishing vessels (*inset*) (after Rodhouse et al. 2001). Taken from Matano and Palma (2008)

It should be noted that the Oyashio Current creates meanders and eddies and hence open-ocean upwelling and enhanced primary production in the confluence region where it interacts with the Kuroshio Current to form the polar front (Tomczak and Godfrey 2003; Kimura et al. 2000). Mesoscale eddies of the Leeuwin Current which is also a "downwelling" current, similarly create upwelling and regionally important ecosystems through their interaction with the Perth Canyon off southwestern Australia (Rennie et al. 2009).

9.6 Other Current-Driven Upwelling Systems

9.6.1 The Green Belt of the Bering Sea

In 1899, Frederic Lucas provided one of the first descriptions of the region along the edge of the continental shelf of the Bering Sea that is now commonly known as the "green belt," a synonym for a highly productive habitat. Lucas (1899) stated that:

> The most frequent feeding grounds, as indicated by the logs of the pelagic (fur) sealers, lie from 75 to 150 miles to the southward and eastward and to the northward and westward of the Pribilofs (Islands), some little distance outside the 100-fathom (180-m) line, or where the bottom of the sea dips abruptly downward from 500 to 5,000 ft.

Physical processes at the shelf edge, such as intense tidal mixing, mesoscale frontal eddies and canyon-flow interaction in the Bering Slope Current (Fig. 9.13) are believed to bring nutrients into the euphotic zone to initiate enhanced primary and secondary production (Springer et al. 1996). This creates a narrow corridor in which fish and squid concentrate because of favourable feeding conditions. This large abundance of zooplankton, fish and squid, in turn, attracts large numbers of marine birds and mammals (Springer et al. 1996). Overall, primary productivity in the Bering Sea ranges between 175 and 275 g C/m^2/yr, which makes the region a moderately productive (Class II) Large Marine Ecosystem.

It should be noted that, technically, the Bering Slope Current is a "downwelling" current similar the Malvinas Current. It is therefore possible that the same physical mechanisms create shelf-break upwelling in the Bering Sea as those operating on the edge of the Patagonian Shelf. In addition, small mesoscale eddies of 5–30 km diameter also are common at the shelf edge of the Bering Sea and likely are capable of pumping nutrients from depth into the euphotic zone to maintain their own production base (Falkowski et al. 1991).

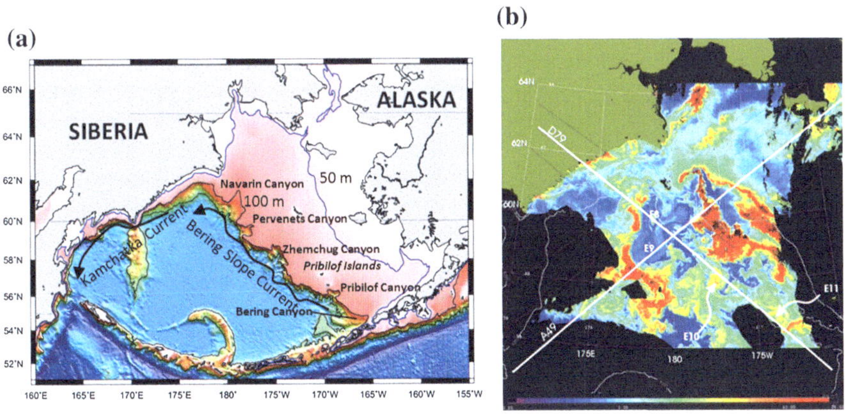

Fig. 9.13 **a** Bathymetry of the Bering Sea, also showing major shelf-break canyons and boundary currents. Modified from Liu and Dittert (2010). **b** Satellite image of SeaWiFS chlorophyll-*a* concentration for 18 June 2001. The chlorophyll concentration scale ranges from 0.3 to 30 mg/m^3 (from Okkonen et al. 2004)

9.6.2 The Grand Banks of Newfoundland

The Grand Banks off Newfoundland once supported one of the largest fisheries in the world and commercial fishing here began in the 16th century (Prowse 1895). The fishery provided a relatively stable livelihood to island residents for five centuries, making Canada one of the world's leading fishing economies. During the peak, from 1950 to 1970, cod catches in the northwest Atlantic, centered over the Grand Banks, amounted to more than one million tonnes per annum, with the largest catches in the 1.5–2.0 million tonnes range taken during the 1960s (Fig. 9.14). However, the combination of the following factors has led inexorably to a rapid decline during the 1970s, so that by the mid-1990s the former great shoals of cod had vanished from the Grand Banks:

(a) improved fishing technology,
(b) the removal of large quantities of bycatch, notably the capelin that provided the food supply for the cod as well as smaller cod that were below marketable size (Myers et al. 1997),
(c) fishing pressure from foreign fleets, and
(d) mismanagement of the resource by allowing too high a catch,

As stocks dwindled across the North Atlantic there were fears that overfishing might drive cod to complete extinction. In 1992, the Canadian federal government took action by imposing a moratorium on cod fishing, although this was too little and too late for the Grand Banks fishery. Although a recreational and artisanal fishery now exists on the Grand Banks, with annual catches of about 5,500 tonnes, the moratorium on commercial cod fishing remains, and even after two decades stocks have not recovered (Shelton et al. 2006; Hilborn and Litzinger 2009). As the cod stock declined, shrimp and crab flourished, and cod fishers have turned to exploit these species. Note that, in contrast to the northwestern stock, cod catches in

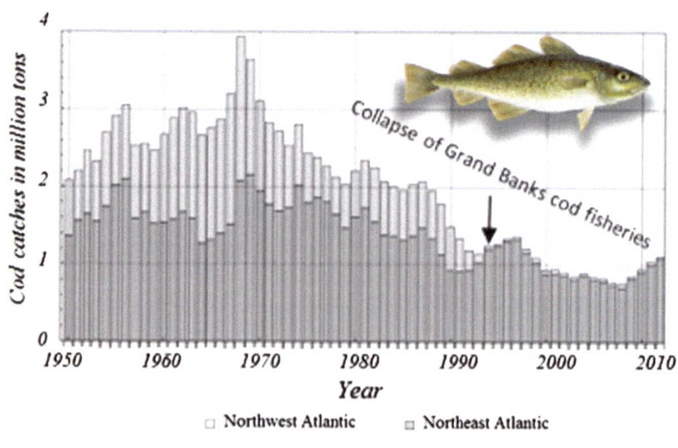

Fig. 9.14 Cod catches for the northwest and northeast Atlantic from 1950–2012. Data from FAO

the northeastern Atlantic have remained relatively stable at around 1 million tonnes for the last 25 years.

The Grand Banks is a shallow plateau of 400 km by 400 km in horizontal extent and has a depth of between 100 and 200 m (Fig. 9.15a). The region is mainly influenced by the western boundary current of the Labrador Sea—the *Labrador Current*—that continuously transports cold, low-salinity and nutrient-rich sub-polar surface water across the Grand Banks. The Gulf Stream extension, the *North Atlantic Current* is located farther to the east, but it creates recirculation patterns and water mass exchanges in interaction with the Labrador Current on the eastern and southern shelf of the Grand Banks (Krauss et al. 1990).

As pointed out in Chap. 1, light and nutrients are two of the most important factors that support or limit phytoplankton growth and biomass. On one hand, the density stratification of the water column within the euphotic layer favours

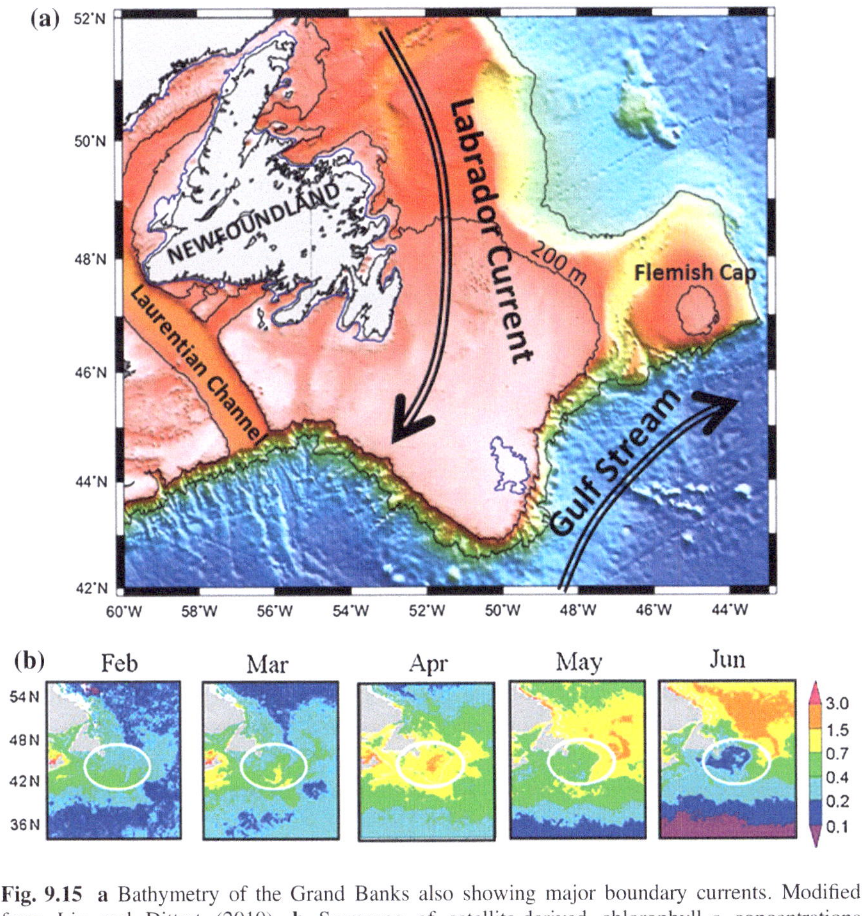

Fig. 9.15 **a** Bathymetry of the Grand Banks also showing major boundary currents. Modified from Liu and Dittert (2010). **b** Sequence of satellite-derived chlorophyll-*a* concentrations (SeaWiFS), taken from Zhao et al. (2013)

phytoplankton photosynthesis and inhibits transport of phytoplankton into the deep dark layers. On the other hand, pronounced density stratification hinders upward entrainment of nutrients into the euphotic layer. The light limitation is typical of high and temperate latitudes in winter. Off Canada's east coast, spring phytoplankton blooms occur generally during March–May every year (Fig. 9.15b). On the eastern Grand Banks the blooms usually start during March or April encompassed by the shelf edge, and then spread inwards (westward) and northward with the seasonal progression, until eventually covering the Grand Banks and the entire northeast Newfoundland Shelf in May (e.g., Wu et al. 2007).

Sverdrup (1953) suggested that for a spring bloom to occur, the thickness of the surface mixed layer must be shallower than the critical depth over which integrated production is equal to integrated respiration. The analysis by Zhao et al. (2013) confirms Sverdrup's "critical depth" hypothesis for the Grand Banks, where the timing of blooms is determined by the timing of shoaling of the surface mixed-layer due to ice melting, which is influenced by insolation, sea-surface temperature and wind-stress magnitude.

There are, of course, several other factors that affect production, particularly at frontal regions or over the shelf, as shown by studies elsewhere in the northwest Atlantic. These include shelf-break upwelling and interaction with meanders and mesoscale eddies. For more details on earlier work, see the review by Smith and Sandström (1988), or the recent book by Simpson and Sharples (2012).

9.6.3 The Guinea Current Upwelling System

Seasonal coastal upwelling occurs along the equatorward-facing coast of western central Africa off the Ivory Coast and Ghana from July to October (Bakun 1978) (Figs. 6.1 and 9.16). The region is similar to eastern ocean boundary current upwelling systems in terms of the appearance of cold surface waters near the coast, productive coastal fisheries, and a zone of low precipitation. However, it differs from other coastal upwelling regions in several important aspects. These include the zonal rather than meridional orientation of the coast, the influence of a swift coastal current, the *Guinea Current* relatively close vicinity to the equator (500–600 km), and exposure to very weak winds.

While Bakun (1978) could show that coastal winds in the Gulf of Guinea could support offshore Ekman transport year-round, the winds were generally weak and inconsistent with the pronounced seasonal variation of sea-surface temperatures. Moreover, a time-series of sea surface temperature at various coastal stations revealed wave-type upwelling events with periods of ~ 14 days that were unrelated to coastal winds and propagated from east to west, as shown in Fig. 9.17. These temperature variations are caused by a reversal of the currents on the shelf, periodic lifting of the thermocline, and advection of nutrient-rich water towards the coast (Tomczak and Godfrey 2003). Hence, the observed upwelling events are the clear signature of intense internal wave motions that are only created during boreal

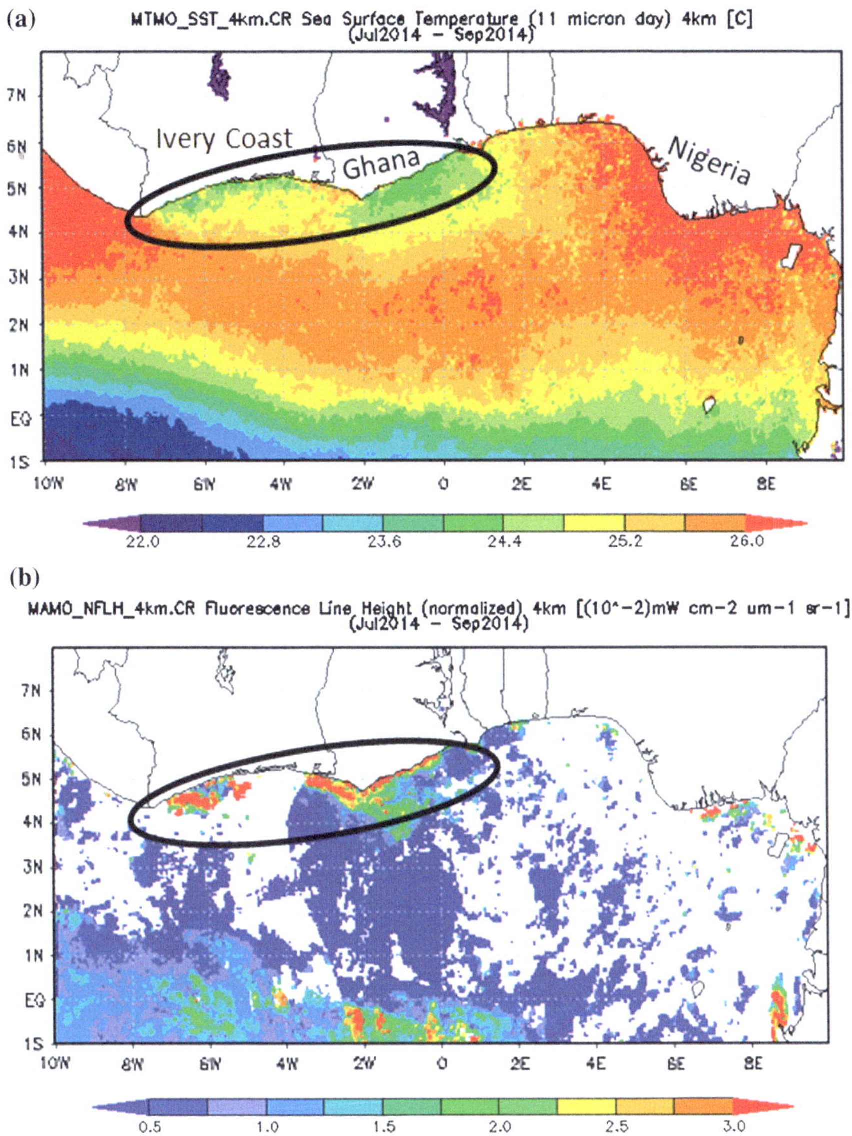

Fig. 9.16 Evidence of coastal upwelling off the coasts of Ivory Coast and Ghana in terms of **a** sea surface temperatures and **b** fluorescence line height (a measure of chlorophyll-*a* concentration), shown for the period from July to September 2014. *Source* Giovanni (NASA)

summer months. Using a simplified hydrodynamic model, O'Brien et al. (1978) suggested that the Guinea Current upwelling is the signature of topographic Rossby waves following on from equatorial Rossby waves that propagate eastward along the equator and then turn to follow the shelf once they hit the eastern boundary.

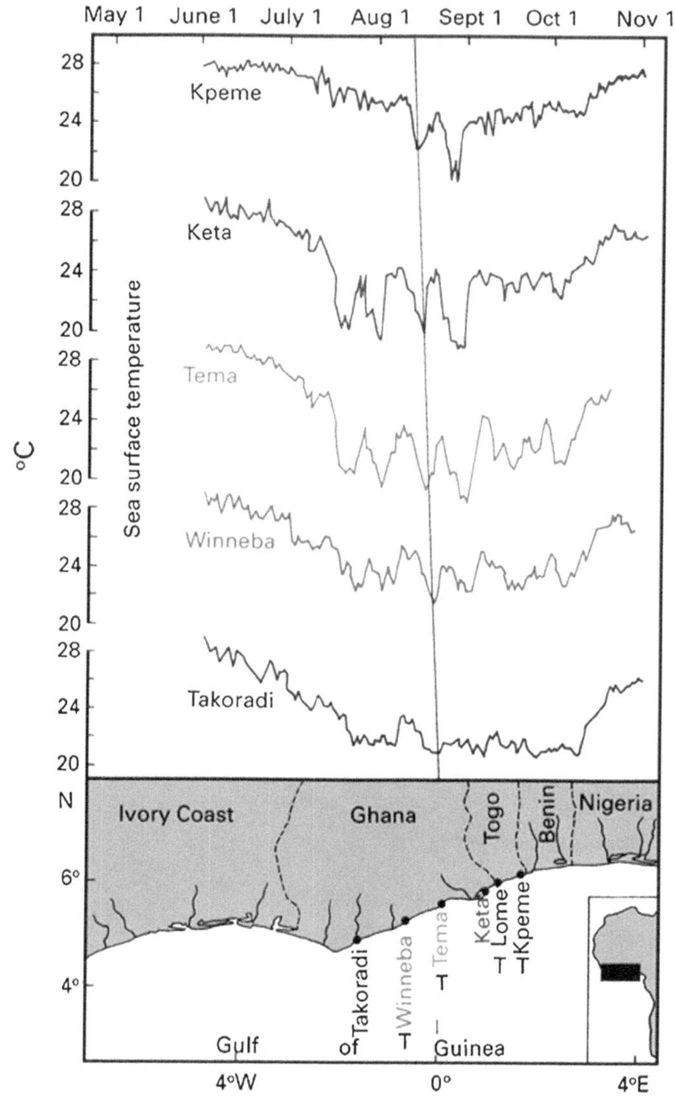

Fig. 9.17 Sea surface temperature (°C) as observed in 1994 at various locations in the Gulf of Guinea. Taken from Tomczak and Godfrey (2003), adapted from Moore et al. (1978)

Another possibility is the interaction of the Guinea Current with variable bathymetry, not considered by O'Brien et al. (1978), that could initiate stationary topographic Rossby waves (Kämpf 2012). As a result of the intense upwelling during summer, the Guinea Current upwelling system is one of the world's most productive ecosystems (>300 g C/m^2/yr).

9.6.4 *Island-Induced Upwelling*

In Fig. 2.10, a satellite image of the eastern equatorial Pacific shows high chlorophyll concentrations in the upwelling system off the coast of Peru. The picture also shows high concentrations to the west of the Galapagos Islands, which are almost 2,000 km off the South American mainland. What causes this high chlorophyll concentration so far offshore? The answer is topographical upwelling (Houvenaghel 1978) caused by the eastward flowing Equatorial Undercurrent hitting the steep topography on the western side of the islands. This brings the nutrient-rich water from the undercurrent to the surface, where it mixes with the westward flowing South Equatorial Current (Feldman 1986), and allows a much larger phytoplankton biomass to develop than is found further west along the equator (see Sect. 9.3). Similar effects are found at other island groups near the equator [e.g., the Seychelles in the Indian Ocean, or Jarvis Island at 160° W in the Pacific (Gove et al. 2006)]. The effect of the Canary Islands on productivity off the coast of North Africa has already been mentioned (see Chap. 6), but here it is the eddies and filaments downstream of the islands that are important, rather than the upstream effects, as is also the case for islands in the Coral Sea in the western Pacific (e.g., Rissik and Suthers 2000). Other mechanisms that can induce greater productivity in the vicinity of islands include nutrient runoff from the land, and tidal mixing (Feldman 1986).

9.7 Tidal-Mixing Ecosystems

Tidal mixing is the process of tidally created turbulence that induces vertical mixing in the water column. The efficiency of the tidal-mixing process depends on the existing density stratification and the strength of tidal currents which varies with total water depth. Given the bathymetric variations in a region, the tidal-mixing process can create a *tidal-mixing front* (also called *shallow-sea front*) which separates well-mixed water on one side from stratified water on the other side. Tidal-mixing fronts are generally highly productive zones for several reasons. The first is that tidal currents can induce short-term upwelling episodes ($\sim$2–3 h) characterized by enhanced upward entrainment of nutrients during the tidal cycle (Franks and Chen 1996). Secondly, tidal-mixing fronts also induce a surface flow convergence (and downwelling) along their front axis, which can trap buoyant organic matter and retain larvae (Lough and Manning 2001). Thirdly, upwelling has been observed on the mixed side of a tidal front (Lough and Manning 2001), which can lead to nutrient recycling from the seafloor. And finally, tidal-mixing fronts also create swift geostrophic flows along the front that can shed mesoscale eddies with cyclonic eddies inducing localized upwelling along their path.

Among the most productive tidal-mixing ecosystems are the Irish Sea (e.g., Simpson and Pingree 1978; Simpson 1981), Georges Bank in the northwestern Atlantic (e.g., Backus 1987), the East China and Yellow Seas (e.g., Chen et al. 1994), and the continental shelf around the Faroe Islands (Gaard 2000). Tidal fronts also exist in Sea of Okhotsk (Zhabin and Dubina 2013), the eastern North Sea (Munk 1993), Bass Strait (Australia) (Baines and Fandry 1983), and other shelf seas and estuaries.

9.8 Ice-Edge Upwelling

In polar and sub-polar regions sea ice plays an important role in both biotic and abiotic cycles. It can accumulate nutrients and metals from atmospheric deposition, river discharges and oceanic exchanges, and release them into the surface ocean when the ice melts (Tovar-Sánchez et al. 2010; and references therein). As already discussed in the section on Newfoundland's Grand Banks, meltwater also provides a stable density stratification within the euphotic zone, which supports phytoplankton growth. Hence, Marginal Ice Zones (MIZ) are generally highly productive ecosystems.

Two separate methods for producing upwelling are found in these regions. Coastal winds can create the classical upwelling either in ice-free shelf waters inshore of multi-year pack ice, or along the edges of ice shelves and large ice sheets in the same way that winds create upwelling along coasts (e.g., Buckley et al. 1979). In contrast to coasts, however, the geometry of ice shelves can undergo substantial temporal variations as a result of seasonal changes in insolation and thermodynamic interactions with currents, and there can also be abrupt changes due to sudden breakups (e.g., Vincent et al. 2001). In addition, ice shelves and their ecologic systems can also be modified by oceanic processes such as tidal mixing.

Shelf-break upwelling is observed in all seasons in both the Alaskan and Canadian Beaufort Seas. It is most common in the autumn and winter months when Aleutian low pressure systems, passing to the south, result in easterly winds along the northern slope of Alaska and Canada. Under such conditions the normally eastward-flowing shelf-break jet of water from the Pacific reverses to flow towards the west, and water from the interior halocline is brought onto the shelf (e.g., Williams et al. 2006, Pickart et al. 2009; Schulze and Pickart 2012).

Carmack and Chapman (2003) explored the situation in which the shelf waters are ice free and multi-year ice is found at some distance offshore. They refer to a time-series of ice conditions on the Canadian Shelf from 1968 to 2000 (compiled by the Canadian Ice Service) that shows considerable interannual variability both in ice extent and in the duration of open water. In particular, mean conditions during mid-summer (August–September) show an ice edge that retreats past the shelf break for a significant portion of the summer in about half of the years. Furthermore, Carmack and Chapman (2003) refer to wind observations based on 6-hourly NCAR/NCEP reanalysis data showing that, for the Canadian Shelf, the wind is upwelling-favourable about one third of the time. In this situation, upwelling-favourable longshore winds

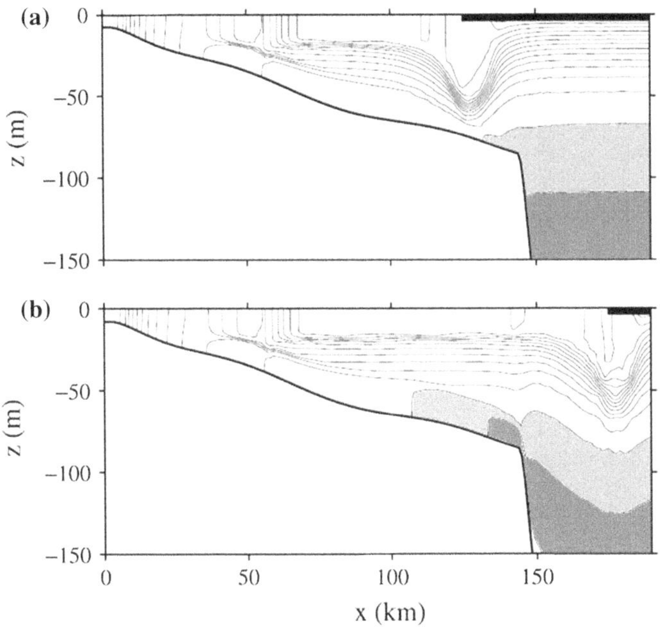

Fig. 9.18 Cross-shelf salinity field from the numerical model with an upwelling-favourable wind stress of 0.2 Pa blowing parallel to the coast for 7 days. A stationary ice edge, denoted by the *black bar* at the ocean surface, is located **a** 125 km and **b** 175 km from the coast. *Lines* are salinity contours. Taken from Carmack and Chapman (2003)

lead to the formation of an upwelling cell on the continental shelf, but trigger downwelling along the ice edge (Fig. 9.18). More importantly, it appears that the water-mass exchange across the shelf break decreases considerably if the ice edge is located inshore of the shelf break (Fig. 9.18a), whereas intense shelf-break upwelling develops once the ice edge retreats past the shelf break (Fig. 9.18b). Carmack and Chapman (2003) suggest that overall ice-edge retreat under the global warming scenario can have dramatic consequences for cross-shelf exchanges and biological productivity in the Arctic Ocean.

On the other hand, wind can also create upwelling directly along the ice edge if it creates an Ekman transport directed away from it. For instance, Mundy et al. (2009) report such an ice-edge upwelling event in a shallow Arctic bay of the Canadian Beaufort Sea that led to a 3-week long phytoplankton bloom, whereby 22 % of the bloom was derived from under-ice production. Shelf-break upwelling has also been observed in the Bering Sea (Alexander and Niebauer 1981), in conjunction with extensive under-ice blooms in the Chukchi Sea (Spall et al. 2014), and in other polar regions. McPhee et al. (2005) observed upwelling of Arctic pycnocline water associated with shear motion of sea ice.

9.9 Summary

This chapter gave an overview of other important upwelling systems that are not associated with the classical wind-driven coastal upwelling process. In the end, we have described most upwelling systems that are associated with Large Marine Ecosystems of the world (see Fig. 1.1). The other branch of highly productive marine ecosystems, not described in this book, are those associated with estuaries and nutrient inputs from rivers.

References

Alexander, V., and H.J. Niebauer. 1981. Oceanography at the eastern Bering Sea ice edge zone in spring. *Limnology and Oceanography* 26: 1111–1125.

Anderson, C.I.H., and P.G. Rodhouse. 2001. Life cycles, oceanography, and variability: Ommastrephid squid in variable oceanographic environments. *Fisheries Research* 54: 133–143.

Backus, R.H. (ed.). 1987. *Georges Bank*. Cambridge, MA: MTT Press. 593 pp.

Baines, P.G., and C.B. Fandry. 1983. Annual cycle of the density field in Bass Strait. *Australian Journal of Marine and Freshwater Research* 34: 143–153.

Bakun, A. 1978. Guinea Current upwelling. *Nature* 271: 147–150.

Belabbassi, L., P. Chapman, W.D. Nowlin Jr., A.E. Jochens, and D.C. Biggs. 2005. Summertime nutrient supply to near surface waters of the northeastern Gulf of Mexico: 1998, 1999 and 2000. *Gulf of Mexico Science* 23(2): 137–160.

Benthuysen, J., L.N. Thomas, and S.J. Lentz. 2015. Rapid generation of upwelling at a shelf break caused by buoyancy shutdown. *Journal of Physical Oceanography* 45: 294–312.

Boutin, J., J. Etcheto, Y. Dandonneau, and D.C.E. Bakker. 1999. Satellite sea surface temperature: A powerful tool for interpreting in situ pCO_2 measurements in the equatorial Pacific Ocean. *Tellus* 51B: 490–508.

Brieva, D., J. Ribbe, and C. Lemckert. 2015. Is the East Australian Current causing a marine ecological hot spot and an important fishery near Fraser Island, Australia? *Estuarine Coastal and Shelf Science* 153: 121–134. doi:10.1016/j.ecss.2014.12.012.

Brooks, D.A., and J.M. Bane Jr. 1978. Gulf Stream deflection by a bottom feature off Charleston, South Carolina. *Science* 201: 1225–1226.

Buckley, J.R., T. Gammelsrød, J.A. Johannessen, O.M. Johannessen, and L.P. Røed. 1979. Upwelling: Oceanic structure at the edge of the Arctic ice pack in winter. *Science* 203: 165–167.

Butman, B. 1986. North Atlantic slope and canyon study. Technical Report 88–27A, U.S. Department of Interior Geological Survey.

Campagna, C., F. Quintana, B.J. Le Boeuf, S.B. Blackwell, and D.E. Crocker. 1998. Diving behaviour and foraging ecology of female southern elephant seals from Patagonia. *Aquatic Mammals* 24: 1–11.

Campos, E.J.D., D. Velhote, and I.C.A. Da Silveira. 2000. Shelf break upwelling driven by Brazil Current cyclonic meanders. *Geophysical Research Letters* 27: 751–754. doi:10.1029/1999GL010502.

Carmack, E., and D.C. Chapman. 2003. Wind-driven shelf/basin exchange on an Arctic shelf: the joint roles of ice cover extent and shelf-break bathymetry. *Geophysical Research Letters* 30 (1778). doi:10.1029/2003GL017526.

Chavez, F.P., and J.R. Toggweiler. 1995. Physical estimates of global new production: The upwelling contribution. In: *Upwelling in the Ocean: Modern Processes and Ancient Records*, ed. C.P. Summerhayes, K.-C. Emeis, M.V. Angel, R.L. Smith and B. Zeitschel. Chichester: Wiley.

Chen, C.S., R.C. Beardsley, R. Limeburner, and K. Kim. 1994. Comparison of winter and summer hydrographic observations in the Yellow and East China Seas and adjacent Kuroshio during 1986. *Continental Shelf Research* 14(7–8): 909–929. doi:10.1016/0278-4343(94)90079-5.

Chen, X.J., J.H. Li, B.L. Liu, Y. Chen, G. Li, Z. Fang, and S.Q. Tian. 2013. Age, growth and population structure of Jumbo flying squid, *Dosidicus gigas*, off the Costa Rica Dome. *Journal of the Marine Biological Association of the UK* 93(2): 567–573.

Chen, X.J., J.H. Li, B.L. Liu, G. Li, and H. Lu. 2014. Fishery biology of jumbo flying squid *Dosidicus gigas* off Costa Rica Dome. *Journal of Ocean University of China* 13(3): 485–490. doi:10.1007/s11802-014-2165-9.

Church, M.J., D.A. Hutchins, and H.W. Ducklow. 2000. Limitation of bacterial growth by dissolved organic matter and iron in the Southern Ocean. *Applied Environmental Microbiology* 66(2): 455–466.

Church, T.M., C.N.K. Mooers, and A.D. Voorhis. 1984. Exchange processes over a middle Atlantic bight shelfbreak canyon. *Estuarine, Coastal and Shelf Science* 19: 393–411.

Cochrane, J.D. 1969. Water and circulation on Campeche Bank in May. Papers in Dedication to Professor Michitaka Uda, 1969. *Special Bulletin of the Japanese Society for Fisheries Oceanography* 123–129.

Cromwell, T. 1958. Thermocline topography, horizontal currents and "ridging" in the eastern tropical Pacific. *Bulletin "of the" Inter-American Tropical Tuna Commission* 111: 135–164.

Croxall, J.P., K. Reid, and P.A. Prince. 1999. Diet, provisioning and productivity responses of marine predators to differences in availability of Antarctic krill. *Marine Ecology Progress Series* 177: 115–131.

Cuzin-Roudy, J. 2000. Seasonal reproduction, multiple spawning, and fecundity in northern krill, *Meganyctiphanes norvegica*, and Antarctic krill, *Euphausia superba*. *Canadian Journal of Fisheries and Aquatic Sciences* 57: 6–15.

De Decker, A.H.B. 1984. Near-surface copepod distribution in the southwestern Indian and south-eastern Atlantic Ocean. *Annals of the South African Museum* 93: 303–370.

Delcroix, T., and J. Picaut. 1998. Zonal displacement of the western equatorial Pacific Fresh pool. *Journal of Geophysical Research* 103: 1087–1098.

DiMarco, S.F., P. Chapman, and W.D. Nowlin. 2000. Satellite observations on the continental shelf south of Madagascar. *Geophysical Research Letters* 27(24): 3965–3968.

Falkowski, P.G., D. Zieman, Z. Kolber, and P.K. Bienfang. 1991. Role of eddy pumping in enhancing primary production in the ocean. *Nature* 352: 55–58.

Feely, R.A., R. Wanninkhof, C.E. Cosca, P.P. Murphy, M.F. Lamb, and M.D. Steckley. 1995. CO2 distributions in the equatorial Pacific during the 1991–1992 ENSO event. *Deep-Sea Research II* 42: 365–386.

Feely, R.A., R. Wanninkhof, C. Goyet, D.E. Archer, and T. Takahashi. 1997. Variability of CO_2 distributions and sea-air fluxes in the central and eastern equatorial Pacific during the 1991–94 El Niño. *Deep-Sea Research II* 44: 1851–1867.

Feely, R.A., J. Boutin, C.E. Cosca, Y. Dandonneau, J. Etcheto, H.Y. Inoue, M. Ishii, C. Le Quéré, D.J. Mackey, M. McPhaden, N. Metzl, A. Poisson, and R. Wanninkhof. 2002. Seasonal and interannual variability of CO_2 in the equatorial Pacific. *Deep Sea Research II* 49(13–14): 2443–2469. doi:10.1016/S0967-0645(02)00044-9.

Feldman, G.C. 1986. Patterns of phytoplankton production around the Galapagos Islands. In *Tidal Mixing and Plankton Dynamics*, ed. M.J. Bowman, C.M. Yentsch, and W.T. Peterson, 77–106. New York: Springer.

Fiedler, P.C. 2002. The annual cycle and biological effects of the Costa Rica Dome. *Deep-Sea Research I* 49: 231–338.

Fiedler, P.C., V. Philbrick, and F.P. Chavez. 1991. Oceanic upwelling and productivity in the eastern Tropical Pacific. *Limnology and Oceanography* 36: 1834–1850.

Foley, D.G., T.D. Dickey, M.J. McPhaden, R.R. Bidigare, M.R. Lewis, R.T. Barber, S.T. Lindley, C. Garside, D.V. Manovt, and J.D. McNeill. 1997. Longwaves and primary productivity variations in the equatorial Pacific at 0°, 140° W. *Deep-Sea Research II* 44: 1801–1826.
Franks, P.J.S., and C. Chen. 1996. Plankton production in tidal fronts: A model of Georges Bank in summer. *Journal of Marine Research* 54: 631–651.
Gaard, E. 2000. Seasonal abundance and development of the copepod *Calanus finmarchicus* in relation to phytoplankton and hydrography on the Faroe shelf. *ICES Journal of Marine Science* 57: 1605–1611.
Godfrey, J.S., and K.R. Ridgway. 1985. The large-scale environment of the poleward flowing Leeuwin Current, Western Australia: Longshore steric height gradients, wind stresses, and geostrophic flow. *Journal of Physical Oceanography* 15: 481–495.
Gordon, R.M., K.H. Coale, and K.S. Johnson. 1997. What controls dissolved iron concentrations in the world ocean? *Marine Chemistry* 57: 137–161.
Gove, J.M., M.A. Merrifield, and R.E. Brainard. 2006. Temporal variability of current-driven upwelling at Jarvis Island. *Journal of Geophysical Research* 111. doi:10.1029/2005JC003161.
Greene, C.H., P.H. Wiebe, J. Burczynski and M.J. Youngbluth. 1988. Acoustical detection of high-density krill demersal layers in the submarine canyons off Georges Bank. *Science* 241, 4863 (071988): 359–361.
Hilborn, R., and E. Litzinger. 2009. Causes of decline and potential for recovery of Atlantic cod populations. *The Open Fish Science Journal* 2: 32–38.
Hooker, S.K., H. Whitehead, and S. Gowans. 1999. Marine protected area design and the spatial and temporal distribution of cetaceans in a submarine canyon. *Conservation Biology* 13(3): 592–602.
Houghton, R.W., R. Schiltz, R.C. Beardsley, B. Butman, and J.L. Chamberin. 1982. Middle Atlantic cold pool: Evolution of the temperature structure during summer 1979. *Journal of Physical Oceanography* 12: 1019–1029.
Houvenaghel, O.T. 1978. Oceanographic conditions in the Galapagos Archipelago and their relationships with life on the islands. In *Upwelling Ecosystems*, ed. R. Boje, and M. Tomczak, 181–202. New York: Springer.
Hoyt, E. 2011. *Marine protected areas for whales, dolphins and porpoises: A world handbook for cetacean habitat conservation and planning*. London & Washington DC: Earthscan. 448 pp.
Hu, D.X., L.H. Lu, Q.C. Xiong, Z.X. Ding, and S.X. Sun. 1980. Study on the upwelling of Zhejiang coastal waters. *Chinese Science Bulletin* 3: 131–133.
Inoue, H.Y., M. Ishii, H. Matsueda, M. Aoyama, and I. Asanuma. 1996. Changes in longitudinal distribution of the partial pressure of CO_2 (pCO_2) in the central and western equatorial Pacific, west of 160 W. *Geophysical Research Letters* 23: 1781–1784.
Kämpf, J. 2010. On the preconditioning of coastal upwelling in the eastern Great Australian Bight. *Journal of Geophysical Research* 115(C12071): 11. doi:10.1029/2010JC006294.
Kämpf, J. 2012. Lee effects of localized upwelling in a shelf-break canyon. *Continental Shelf Research* 42: 78–88. doi:10.1016/j.csr.2012.05.005.
Kimura, S., H. Nakata, and Y. Okazaki. 2000. Biological production in meso-scale eddies caused by frontal disturbances of the Kuroshio Extension. *ICES Journal of Marine Science* 57: 133–142. doi:10.1006/jmsc.1999.0564.
Krauss, W., R.H. Kaese, and H.H. Hinrichsen. 1990. The branching of the Gulf Stream southeast of the Grand Banks. *Journal of Geophysical Research* 95(C8): 13089–13103.
Kumar, N., R.F. Anderson, R.A. Mortlock, P.N. Froelich, P. Kubik, B. Dittrich-Hannen, and M. Suter. 1995. Increased biological productivity and export production in the glacial Southern Ocean. *Nature* 378: 675–680.
Kuroda, Y., and M. McPhaden. 1993. Variability in the western equatorial Pacific Ocean during Japanese Pacific Climate Study cruises in 1989 and 1990. *Journal of Geophysical Research* 98: 4747–4759.
Landry, M.R., J. Constantinou, and J. Kirshtein. 1995. Microzooplankton grazing in the central equatorial Pacific during February and August, 1992. *Deep-Sea Research II* 42: 657–671.

Largier, J.L., P. Chapman, W.T. Peterson, and V.P. Swart. 1992. The western Agulhas Bank: Circulation, stratification and ecology. *South African Journal of Marine Science* 12: 319–339.

Lee, T., N.E. Williams, J. Wang, and R. Evans. 1989. Response of South Carolina continental shelf waters to wind and Gulf Stream forcing during winter of 1986. *Journal of Geophysical Research* 94: 10715–10754.

Levin, L.A., and M. Sibuet. 2012. Understanding continental margin biodiversity: A new imperative. *Annual Review of Marine Science* 4: 79–112.

Liu, K.-K., and N. Dittert. 2010. Web-based electronic supplements, Appendix C. In: *Carbon and nutrient fluxes in continental margins*, ed. K.-K. Liu, L. Atkinson, R.A. Quiñones and L. Talaue-McManus. Berlin: Springer. http://cmtt.pangaea.de/. Assessed 11 June 2015.

Lough, R.G., and J.P. Manning. 2001. Tidal-front entrainment and retention of fish larvae on the southern flank of Georges Bank. *Deep Sea Research II* 48: 631–644.

Loukos, H., B. Frost, D.E. Harrison, and J.W. Murray. 1997. An ecosystem model with iron limitation of primary production in the equatorial Pacific at 140° W. *Deep-Sea Research II* 44: 2221–2249.

Lucas, F.A. 1899. In *The Fur Seals and Fur-Seal Islands of the North Pacific Ocean*, ed. D.S. Jordan, 59–68. Washington, DC: Government Printing Office.

Lutjeharms, J.R.E. 2006. *The Agulhas Current*. Heidelberg, New York: Springer. 330 pp.

Machu, E., J.R.E. Lutjeharms, A.M. Webb, and H.M. Van Aken. 2002. First hydrographic evidence of the southeast Madagascar upwelling cell. *Geophysical Research Letters* 29(21). doi:10.1029/2002GL015381.

Mann, K.H., and J.R.N. Lazier. 1996. *Dynamics of marine ecosystems: Biological-physical interactions in the oceans*, 2nd ed, 662. Oxford: Blackwell Science Limited.

Marr, J.W.S. 1962. The natural history and geography of the Antarctic krill (Euphausia superba). *Discovery Reports* 32: 33–464.

Martin, J.H. 1990. Glacial-interglacial CO2 change; the iron hypothesis. *Paleoceanography* 5: 1–13.

Matano, R.P., and E.D. Palma. 2008. On the upwelling of downwelling currents. *Journal of Physical Oceanography* 38: 2482–2500. doi:10.1175/2008JPO3783.1.

Mazeika, P.A. 1967. Thermal domes in the eastern tropical Atlantic Ocean. *Limnology and Oceanography* 12: 537–539.

McClain, C.R., L.J. Pietrafesa, and J.A. Yoder. 1984. Observations of Gulf Stream-induced and wind-driven upwelling in the Georgia Bight using ocean color and infrared imagery. *Journal of Geophysical Research* 89(C3): 3705–3723.

McPhee, M.G., R. Kwok, R. Robins, and M. Coon. 2005. Upwelling of Arctic pycnocline associated with shear motion of sea ice. *Geophysical Research Letters* 32: L10616. doi:10.1029/2004GL021819.

Middleton, J.H., P. Coutis, D.A. Griffin, A. Macks, A. McTaggart, M. Merrifield, and G.D. Nippard. 1994. Circulation and water mass characteristics of the southern Great Barrier Reef. *Australian Journal of Marine and Freshwater Research* 45: 1–18.

Minas, H.J., M. Minas, and T.T. Packard. 1986. Productivity in upwelling areas deduced from hydrographic and chemical fields. *Limnology and Oceanography* 31: 1182–1206.

Moore, J.K., and M.R. Abbott. 2002. Surface chlorophyll concentrations in relation to the Antarctic Polar Front: Seasonal and spatial patterns from satellite observations. *Journal of Marine Systems* 37: 69–86.

Moore, D.W., P. Hisard, J. McCreary, J. Merle, J.J. O'Brien, J. Picaut, J.M. Verstraete, and C. Wunsch. 1978. Equatorial adjustment in the eastern Atlantic. *Geophysical Research Letters* 5: 637–640.

Mundy, C.J., M. Gosselin, J. Ehn, Y. Gratton, A. Rossnagel, D.G. Barber, J. Martin, J.E. Tremblay, M. Palmer, K.R. Arrigo, G. Darnis, L. Fortier, B. Else, and T. Papakyriakou. 2009. Contribution of under-ice primary production to an ice-edge upwelling phytoplankton bloom in the Canadian Beaufort Sea. *Geophysical Research Letters* 36: L17601. doi:10.1029/2009gl038837.

Munk, P. 1993. Differential growth sprat *Sprattus sprattus* across a tidal front in the eastern North Sea. *Marine Ecology Progress Series* 99(1–2): 17–29.

Murray, J.W., R.T. Barber, M.R. Roman, M.P. Bacon, and R.A. Feely. 1994. Physical and biological controls on carbon cycling in the equatorial Pacific. *Science* 266: 58–65.

Murray, J.W., E. Johnson, and C. Garside. 1995. A U.S. JGOFS process study in the equatorial Pacific (EqPac): Introduction. *Deep-Sea Research II* 42: 275–293.

Murray, J.W., R. LeBorgne, and Y. Dandonneau. 1997. JGOFS studies in the equatorial Pacific. *Deep-Sea Research II* 44: 1759–1763.

Myers, R.A., J.A. Hutchings, and N.J. Barrowman. 1997. Why do fish stocks collapse? The example of cod in Atlantic Canada. *Ecological Applications* 7: 91–106.

O'Brien, J.J., D. Adamec, and D.W. Moore. 1978. A simple model of upwelling in the Gulf of Guinea. *Geophysical Research Letters* 5: 641–644.

Okkonen, S.R., G.M. Schmidt, E.D. Cokelet, and P.J. Stabeno. 2004. Satellite and hydrographic observations of the Bering Sea 'Green Belt'. *Deep Sea Research II* 51: 1033–1051.

Pickart, R.S., G.W.K. Moore, D.J. Torres, P.S. Fratantoni, R.A. Goldsmith and J. Yang. 2009. Upwelling on the continental slope of the Alaskan Beaufort Sea: storms, ice, and oceanographic response. *Journal of Geophysical Research* 114(C00A13). doi:10.1029/2208JC005009.

Prowse, D.W. 1895. *A History of Newfoundland*. London: MacMillan and Co.

Reid, K. 2001. Growth of Antarctic krill *Euphausia superba* at South Georgia. *Marine Biology* 138: 57–62.

Rennie, S., R.D. McCauley, C. Pattiaratchi, C. Hanson, C. Burton, J. Bannister, C. Jenner, and M.-N. Jenner. 2009. Physical properties and processes in the Perth Canyon, Western Australia: Links to water column production and seasonal pygmy blue whale abundance. *Journal of Marine Systems* 77: 21–44.

Rissik, D., and I.M. Suthers. 2000. Enhanced feeding by pelagic juvenile myctophid fishes within a region of island-induced flow disturbance in the Coral Sea. *Marine Ecology Progress Series* 203: 263–273.

Rodhouse, P.G., C.D. Elvidge, and P.N. Trathan. 2001. Remote sensing of the global light-fishing fleet: An analysis of interactions with oceanography, other fisheries and predators. *Advances in Marine Biology* 39: 261–303.

Rodier, M., G. Eldin, and R. Le Borgne. 2000. The western boundary of the equatorial Pacific upwelling: Some consequences of climate variability on hydrological and planktonic properties. *Journal of Oceanography* 56: 463–471.

Ryan, J.P., P.S. Polito, P.G. Strutton, and F.P. Chavez. 2002. Unusual large-scale phytoplankton blooms in the equatorial Pacific. *Progress in Oceanography* 55(3–4): 263–285. doi:10.1016/S0079-6611(02)00137-4.

Ryan, J.P., F.P. Chavez, and J.G. Bellingham. 2005. Physical-biological coupling in Monterey Bay, California: Topographic influences on phytoplankton ecology. *Marine Ecology Progress Series* 287: 23–32.

Sakurai, Y. 2007. An overview of the Oyashio ecosystem. *Deep Sea Research II* 54: 2526–2542.

Schott, F., and J.P. McCreary. 2001. The monsoon circulation of the Indian Ocean. *Progress in Oceanography* 51: 1–123.

Schulze, L.M., and R.S. Pickart. 2012. Seasonal variation of upwelling in the Alaskan Beaufort Sea: impact of sea ice cover. *Journal of Geophysical Research* 107. doi:10.1029/2012JC007985.

Schumann, E.H., G.J.B. Ross, and W.S. Goschen. 1988. Cold water events in Algoa Bay and along the Cape south coast, South Africa, in March/April 1987. *South African Journal of Science* 84: 579–583.

Shelton, P.A., A.F. Sinclair, G.A. Chouinard, R. Mohn, and D.E. Duplisea. 2006. Fishing under low productivity conditions is further delaying recovery of Northwest Atlantic cod (*Gadus morhua*). *Canadian Journal of Fisheries and Aquatic Sciences* 63: 235–238. doi:10.1139/f05-253.

Simpson, J.H. 1981. The shelf-sea fronts: Implications of their existence and behaviour. *Philosophical Transactions of the Royal Society of London* A302: 531–546.

Simpson, J.H., and R.D. Pingree. 1978. Shallow sea fronts produced by tidal stirring. In *Oceanic Fronts in Coastal Processes*, ed. M.J. Bowman, and W.E. Esaias, 29–42. Berlin: Springer.

Simpson, J.H., and J. Sharples. 2012. *Introduction to the Physical and Biological Oceanography of Shelf Seas*. Cambridge, U.K.: Cambridge University Press.

Smith, P.C., and H. Sandström. 1988. Physical processes at the shelf edge in the northwest Atlantic. *Journal of Northwest Atlantic Fishery Science* 8: 5–13.

Spall, M.A., R.S. Pickart, E.T. Brugler, G.W.K. Moore, L. Thomas, and K.R. Arrigo. 2014. Role of shelfbreak upwelling in the formation of a massive under-ice bloom in the Chukchi Sea. *Deep-Sea Research II* 105: 17–29. doi:10.1016/j.dsr2.2014.03.017.

Springer, A.M., C.P. McRoy, and M.V. Flint. 1996. The Bering Sea green belt: Shelf-edge processes and ecosystem production. *Fisheries Oceanography* 5: 205–223.

Subramaniam, A., C. Mahaffey, W. Johns, and N. Mahowald. 2013. Equatorial upwelling enhances nitrogen fixation in the Atlantic Ocean. *Geophysical Research Letters* 40: 1766–1771. doi:10.1002/grl.50250.

Sverdrup, H.U. 1953. On conditions for the vernal blooming of phytoplankton. *Journal du Conseil / Conseil Permanent International pour l'Exploration de la Mer* 18(3): 287–295. doi:10.1093/icesjms/18.3.287.

Talley, L.D. 2013. Closure of the global overturning circulation through the Indian, Pacific and Southern Oceans: Schematics and transports. *Oceanography* 26: 80–97.

Tans, P.P., I.Y. Fung, and T. Takahashi. 1990. Observational constraints on the global atmospheric CO2 budget. *Science* 247: 1431–1438.

Tomczak, M., and J.S. Godfrey. 2003. *Regional oceanography: An introduction*, 2nd edn. Daya Publishing House, 390 pp.

Tovar-Sánchez, A., C.M. Duarte, J.C. Alonso, S. Lacorte, R. Tauler, and Galbán-Malagón. 2010. Impacts of metals and nutrients released from melting multiyear Arctic sea ice. *Journal of Geophysical Research* 115: C07003. doi:10.1029/2009jc005685.

Vincent, W.F., J.A.E. Gibson, and M.O. Jeffries. 2001. Ice shelf collapse, climate change, and habitat loss in the Canadian high Arctic. *Polar Record* 37(201): 133–142.

Waring, G.T., T. Hamazaki, D. Sheehan, G. Wood and S. Baker. 2001. Characterization of beaked whale (Ziphiidae) and sperm whale (*physeter macrocephalus*) summer habitat in shelf-edge and deeper waters off the northeast U. S. *Marine Mammal Science* 17(4): 703–717.

Williams, W.J., E.C. Carmack, K. Shimada, H. Melling, K. Aagaard, R.W. Macdonald, and R.G. Ingram. 2006. Joint effects of wind and ice motion in forcing upwelling in Mackenzie Trough, Beaufort Sea. *Continental Shelf Research* 26: 2351–2366.

Wu, Y.S., I.K. Peterson, C.L. Tang, T. Platt, S. Sathyendranath, and C. Fuentes-Yaco. 2007. The impact of sea-ice on the spring bloom on the Labrador and Newfoundland Shelves. *Journal of Plankton Research* 29: 509–514.

Wyrtki, K. 1962. The upwelling in the region between Java and Australia during the southeast monsoon. *Australian Journal of Marine and Freshwater Research* 13(3): 217–225.

Wyrtki, K. 1964. Upwelling in the Costa Rica Dome. *Fishery Bulletin* 63(2): 355–372.

Wyrtki, K. 1981. An estimate of equatorial upwelling in the Pacific. *Journal of Physical Oceanography* 11(9): 1205–1214.

Xie, S.-P., H. Annamalai, F.A. Schott, and J.P. McCreary. 2002. Structure and mechanisms of south Indian Ocean climate variability. *Journal of Climate* 15: 864–878.

Yokoi, T., T. Tozuka, and T. Yamagata. 2008. Seasonal variation of the Seychelles Dome. *Journal of Climate* 21: 3740–3754.

Zhabin, I.A., and V.A. Dubina. 2013. The structure of the tidal mixing front in the region of Shantar Islands (Sea of Okhotsk) according to data of satellite observations. *Izvestiya, Atmospheric and Oceanic Physics* 48(9): 999–1005.

Zhao, H., G. Han, and D. Wang. 2013. Timing and magnitude of spring bloom and effects of physical environments over the Grand Banks of Newfoundland. *Journal of Geophysical Research* 118: 1385–1396. doi:10.1002/jgrg.20102.

Zhu, J. 2003. Dynamic mechanism of the upwelling on the west side of the submerged river valley off the Changjiang mouth in summertime. *Chinese Science Bulletin* 48(24): 2754–2758.

Chapter 10
Comparison, Enigmas and Future Research

Abstract This chapter summarizes the findings of scientific research that compare major eastern boundary upwelling ecosystems, highlights the significance of upwelling ecosystems in Southeast Asian waters, and lists enigmas and open questions as motivation for future research.

Keywords Upwelling • Eastern boundary currents • Comparison • Carbon fluxes • Enigmas • Open questions

> Out in the dark it throbs and glows - The wide, wild sea that no man knows!
> The wind is chill, the surge is white - And I must sail that sea tonight.
> William Winter (1836–1917) (from On the Verge, 1900)

10.1 Overview

Upwelling systems have been the focus of scientific research for many decades. The strong coupling between atmospheric forcing, oceanic currents, biogeochemical cycles and food web dynamics has been the subject of various multidisciplinary scientific studies which have significantly improved our understanding of such systems. Moreover, the need for sustainable management of global marine resources has led to the establishment of freely available and shared global data bases (e.g., FAO fishery data, NOAA satellite data, among many others) and science-based management frameworks. In this era of changing climate, the need for monitoring and further advances in the scientific understanding of processes inherent to upwelling systems is becoming progressively pressing (Fig. 10.1).

Some readers may think that the era of scientific research and in situ field programs has now been exhausted and that we already know enough to be able to manage marine ecosystems in a sustainable way and to mitigate the effects of global warming. This is far from the truth. Despite satellite remote sensing and the establishment of global data bases of physical and biological data such as the World

© Springer International Publishing Switzerland 2016

J. Kämpf and P. Chapman, *Upwelling Systems of the World*,
DOI 10.1007/978-3-319-42524-5_10

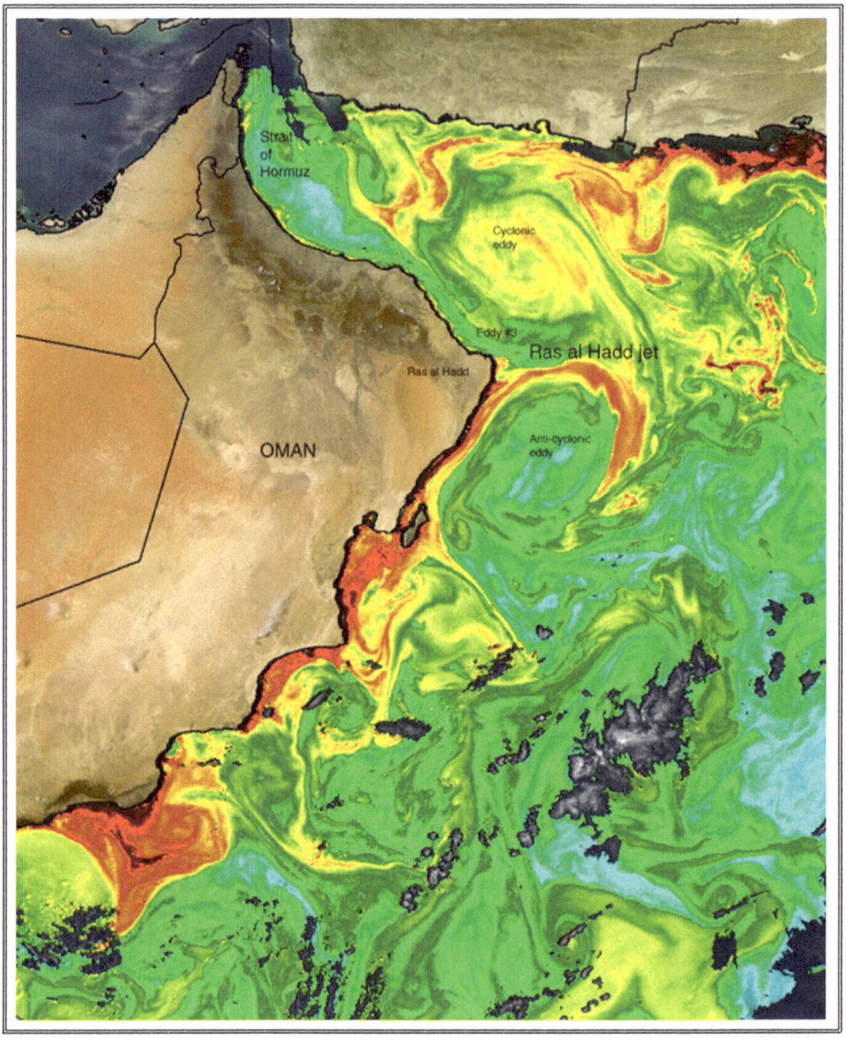

Fig. 10.1 Upwelling filaments off the coast of Oman seen in satellite-derived chlorophyll data. Image source: NASA

Ocean Atlas (Conkright et al. 2002) or the Census of Marine Life (e.g., Ausubel 1999; Yarincik and O'Dor 2005; see http://www.coml.org/), sophisticated observing systems such as the global ARGO float system (Roemmich et al. 2004; see http://www.argo.ucsd.edu/) and synthetic models, information on oceanic properties and their variability below the sea surface is highly fragmented both in time and space, particularly over the continental shelves, and the interpretation of such data is difficult and often inconclusive. To this end, the understanding of physical and biogeochemical processes and the food web dynamics in many key upwelling regions, such as the warm waters of the Indonesian Seas that are central to the formation of El Niño events (Kida and Richards 2009; and references therein) is still at a rudimentary level.

Globalization has also led to the establishment of potential global management frameworks of marine resources. Most noticeable are those based on Large Marine Ecosystems (LMEs) (Sherman 1993; Sherman and Hempel 2008; see Fig. 1.1), which compare the total biological productivity of different marine ecosystems based on their annual depth-integrated phytoplankton carbon fixation. The LME approach to the assessment and management of marine resources for monitoring and assessing changing conditions in ecosystems is based on five modules: productivity, fish and fisheries, pollution and ecosystem health, socioeconomics, and governance (Duda and Sherman 2002).

While the LME approach has clear merits as it provides a global framework for the management of important marine ecosystems, its underlying scientific methodology can certainly be improved. One shortcoming, for instance, is that the reported productivity of a LME depends on the defined areal extent of the system and that the core classification is based on annual values. This leads to a perceived bias towards the big four coastal upwelling systems and there is a risk that seasonal upwelling systems are overlooked despite their ecological significance. One example already mentioned is the South Australian upwelling system in the Great Australian Bight (see Chap. 8), which despite relatively low productivity plays a vital role in the life cycle of southern bluefin tuna (*Thunnus maccoyii*). Perhaps there should be another classification addressing small but ecologically highly significant marine ecosystems. Another problem is that the productivity algorithm does not capture subsurface phytoplankton blooms (see Fig. 6.17), which are an important biochemical contributors to total productivity (e.g., Richardson et al. 2000). Only a few regions supporting subsurface algal blooms, such as those off Bermuda and Hawaii and the extensive upwelling region of the equatorial Pacific Ocean have been studied systematically. Hence, the next important step in this context is to routinely equip more ARGO floats with fluorometers that will allow us to see how widespread and productive they really are (see Xing et al. 2011).

Scientists have argued for many years about how to define the precise position and areal extent of bio-regions. For example, Briggs (1974, 1995) focused on a system of coastal and shelf provinces defined by their degree of endemism. This strong taxonomic focus and clear definition have led to relatively widespread adoption of Briggs' system; it was used, for example, by Hayden et al. (1984) with minor amendments. Adey and Steneck (2001) provided independent verification of many of Briggs' subdivisions. In another systematic approach, Longhurst (1998) developed a two-tier classification of pelagic systems of biomes and biogeochemical provinces. These subdivisions were based on a detailed array of oceanographic factors including a global database of chlorophyll profiles.

Unlike the classifications of Briggs and Longhurst, LMEs represent an expert-derived system, but one without a rigorous, replicable core definition (Spalding et al. 2007). Instead, LMEs are loosely defined as relatively large regions (>200,000 km^2), characterized by distinct bathymetry, hydrography, productivity, and trophically dependent populations (see www.lme.noaa.gov/Portal/). Spalding et al. (2007) recently proposed a revised bioregionalization of coastal and shelf areas for the entire world ocean, which includes 12 realms, 62 provinces, and 232 ecoregions, including a substantial area (an additional 10 %) not covered by LMEs. Whether such a complex but allegedly improved bioregionalization will be adopted at an international level remains to be seen; we note that many countries have also introduced smaller management units of marine parks, reserves and sanctuaries, some of which are not always based on scientific evidence but which may at least provide a potential refuge within larger areas of exploitation. It must be said, however, that this is not always the case. For example, the marine reserve in the central equatorial Pacific, the Pacific Remote Islands Marine National Monument, proclaimed in 2009 by U.S. President G.W. Bush and expanded in 2014 by President Obama protects a vast area of reefs and atolls that cover over 1.2 million km^2, while others protect spawning grounds for particular species.

This chapter is organized as follows. First, we summarize scientific work that compared the four major eastern boundary current upwelling systems; i.e., upwelling in the Benguela, California, Iberia/Canary and Humboldt ecosystems. The total fish production of the big four upwelling systems is then compared to that of Southeast Asian waters. Finally, we present a list of open questions and enigmas as a motivation of future research.

10.2 The Big Four Coastal Upwelling Systems Compared

10.2.1 Introduction

There are a number of processes and features of importance that are common to the upwelling dynamics in all, or most of these upwelling regions. These include: the rates of nutrient supply and possible nutrient limitation; the influence of the large-scale oceanic circulation; habitat zonation (by which we mean the positioning of upwelling centres and upwelling shadows, spawning and nursery grounds); how system parameters (i.e., thermocline depth, coastal wind forcing, ecologic adjustment) vary with time; the importance of relaxation periods to production; the role of topographic irregularities (e.g., capes, islands, submarine canyons); the offshore export and leakage of nutrients and biomass (i.e., carbon fluxes); how well the biota can cope with rapidly changing conditions; and effects associated with low-oxygen environments.

There a many different ways of comparing upwelling systems: on the basis of their atmospheric drivers (i.e. upwelling index), their physical characteristics (e.g. SST anomalies, upwelling rates, nutrient fluxes), their estimated rates of carbon fixation linked to the concept of Large Marine Ecosystems, their biological responses (e.g. phytoplankton and zooplankton variability), or their marine food web composition and interaction. All these measures are valuable as they help us to understand how marine ecosystems function, while also increasing our knowledge of how they may respond to environmental changes (e.g. global warming) or industrial interference (e.g. commercial fishing).

Sverdrup et al. (1942) provided an early temperature-based comparison of the four main upwelling systems. The recognition of their significance for fish production (Ryther 1969) led to the Coastal Upwelling Ecosystem Analysis (CUEA) program during the 1970s, which undertook measurements in the upwelling systems of California, northwest Africa and Peru. Barber and Smith (1981) published a comparative synthesis of the research findings. Several newer comparisons of these upwelling regions can be found in the literature; the most recent ones are those by Carr and Kearns (2003), Mackas et al. (2006) and Chavez and Messié (2009). Before we discuss differences, let us briefly summarize the key features of the big four coastal upwelling systems.

The Humboldt Current (Peru/Chile) coastal upwelling system (Fig. 10.2; see Chap. 5) extends from the cold waters of southern Chile (~45° S) to the warm tropical waters at the Equatorial Front off northern Peru (~4° S). Upwelling off Peru (4°–16° S) is highly productive year-round, while a similarly highly productive seasonal summer upwelling occurs off southern Chile (30°–40° S) based on

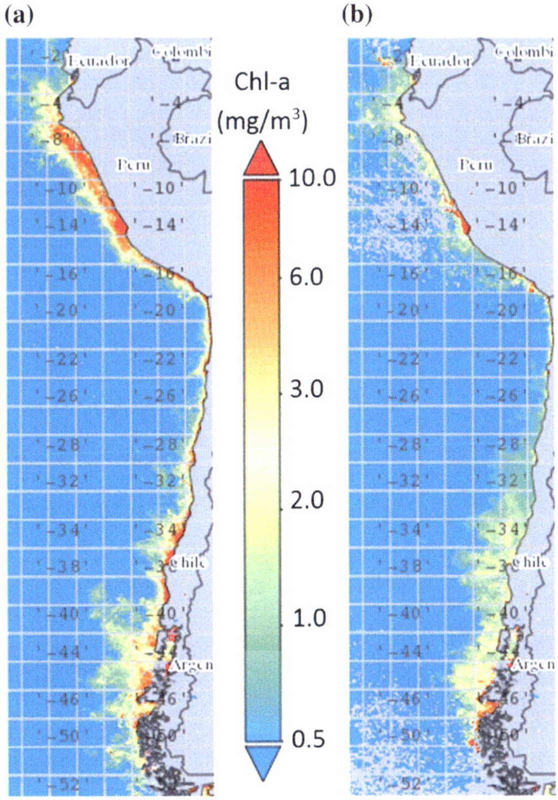

Fig. 10.2 Example of **a** summer and **b** winter season distributions of MODIS Aqua (4 km resolution) surface chlorophyll pigment concentrations in the Humboldt Current upwelling system. Summer (winter) refers to the period Dec–Feb 2010/11 (July–August 2010). *Source* Giovanni (NASA), Version 4.19, http://giovanni.gsfc.nasa.gov/giovanni/

specific upwelling cells. An upwelling shadow of moderate to low productivity exists off northern Chile and southern Peru (18°–26° S). The greatest upwelling, which is responsible for the enormous production of anchoveta in the region (e.g., Chavez et al. 2008), occurs off central and northern Peru. Upwelling within the Humboldt Current system is strongly modulated by the El Niño Southern Oscillation, which affects the amount of nutrients that are upwelled and thus controls primary productivity.

The Benguela Current coastal upwelling system extends from the southern tip of Africa (~35° S) to about 14° S off Angola (Fig. 10.3; see Chap. 7). Year-round upwelling occurs from 15°–27° S, while farther south there is a seasonal summer maximum. There are a number of individual upwelling cells and the upwelling system in the south is highly fragmented. The major difference between this and the other three major upwelling systems is the influence of the warm Agulhas Current and its retroflection at the southern end; the other three systems have cooler water at their poleward extremities.

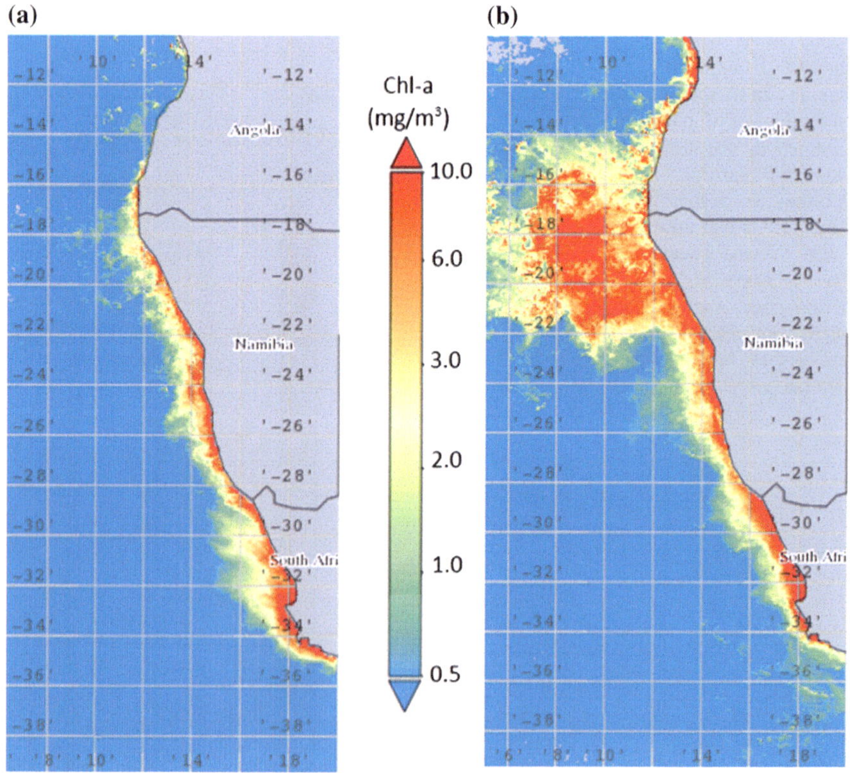

Fig. 10.3 Same as Fig. 10.2, but for the Benguela Current upwelling system. Panel **a** refers to austral summer; panel **b** to winter

The Canary Current coastal upwelling system includes two distinct upwelling regions, the Iberian upwelling region and the system off northwest Africa (Fig. 10.4; see Chap. 6). These are separated by the gap in the Gulf of Cadiz where water flows into and out of the Mediterranean Sea and separates the connectivity between the Iberian and Moroccan systems. Upwelling is relatively weak and seasonal along the Iberian Peninsula (~36°–44° N) and off Morocco (~25°–36° N), with maximum upwelling occurring in late summer and summer respectively. Weak seasonality is also found further south between 20°–25° N, but upwelling becomes quasi-continuous with a spring maximum within the 10°–20° N latitudinal band (Lathuilière et al. 2008). Dust deposition appears to play an important role in the productivity of the Canary Current coastal upwelling system and, to a lesser extent, in the Benguela region.

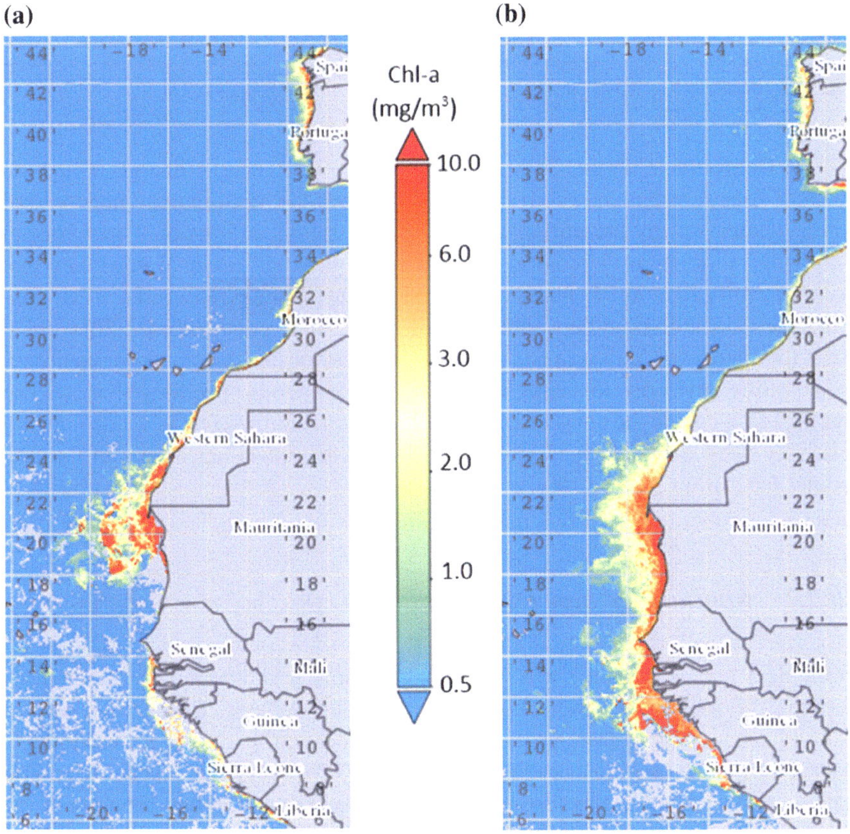

Fig. 10.4 Example of **a** summer and **b** winter season distributions of MODIS Aqua (4 km resolution) surface chlorophyll pigment concentrations in the Canary/Iberia Current upwelling system. Summer (winter) refers to the period July–August 2010 (Dec–Feb 2010/11). *Source* Giovanni (NASA), Version 4.19, http://giovanni.gsfc.nasa.gov/giovanni/

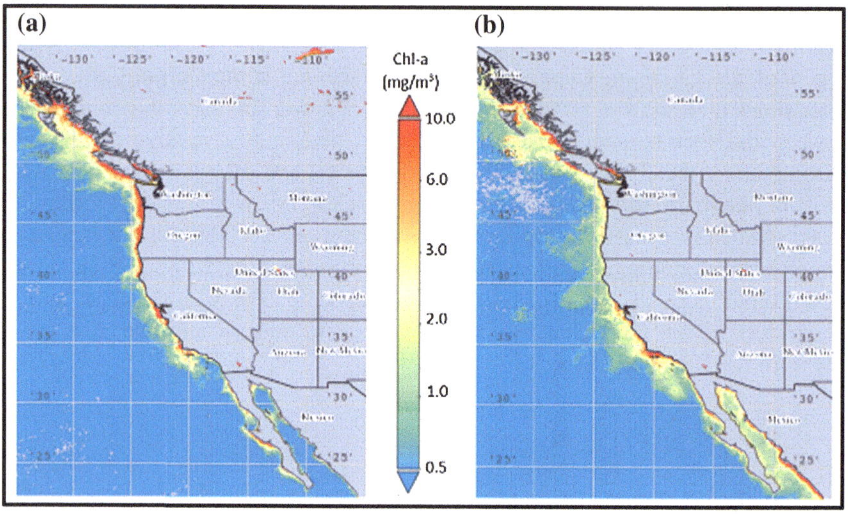

Fig. 10.5 Same as Fig. 10.4, but for the California Current upwelling system

The California Current Coastal Upwelling System (Fig. 10.5, see Chap. 4) contains four sub-regions. From the north these are the Washington/Oregon coast where winter storms are strong and freshwater runoff is important in terms of supplying nutrients to the coastal region. This region has seasonal upwelling that occurs mostly in summer. Further south, along the central Californian shelf, conditions are much drier and upwelling occurs during spring and summer. This region abuts the Southern California Bight, which, because of the change in orientation of the coastline, gives rise to a large and relatively continuous upwelling shadow. The fourth region covers the coast of Baja California, where upwelling is year-round but relatively weak (Mackas et al. 2006). The strongest upwelling occurs in the latitudinal band from 34°–44° N.

10.2.2 Similarities and Differences

In the next few sections we will discuss some specific similarities and differences between the four main eastern boundary upwelling systems. Chavez and Messié (2009) list a number of features that describe the mean properties of a 150 km wide coastal strip for a band of coastline that covers 10° of longitude (Table 10.1). The specific range of latitude is listed for each system. The intriguing thing about the details in this table is that physical features, such as mean wind speed, mixed layer depth, light intensity, and even average nitrate concentration at 60 m depth, are all very similar in the four systems, varying only by a factor of about 1.5, but the biology varies by a factor of four.

Table 10.1 Comparison of upwelling parameters for major coastal upwelling systems

Parameter	California 34°–44° N	Peru 16°–6° S	NW Africa 12°–22° N	Benguela 28°–18° S
Coriolis parameter (s^{-1})	9.2×10^{-5}	-2.8×10^{-5}	4.2×10^{-5}	-5.7×10^{-5}
Average wind speed (m/s)	7.8	5.7	6.8	7.2
Upwelled volume (10^6 m^3/s)	1.0	1.6	1.4	1.5
Average [NO_3] at 60 m (μM/L)	14.9	16.8	19	16.9
Potential new production (g C/m^2/ yr)	323	566	539	517
SeaWiFS primary production (g C/m^2/ yr)	479	855	1213	978
f ratio[1] (unitless)	67 %	66 %	44 %	53 %
1 mg m^{-3} Chl extension (km)	49	120	164	160
Turbulence (m^3/ s^3)	610	225	371	444
Mixed layer depth (m)	43	31	48	45
PAR (E m^2/ day)	34	43	48	43
Dust deposition (g m^2/ yr)	0.3	0.3	33.1	11.7

See Chavez and Messié (2009) for more details. Note that satellite-measured primary production generally exceeds potential new production (estimated from nitrate fluxes) given the additional nutrient regeneration (see Chap. 1)

[1]The *f* ratio is calculated from the ratio between potential new production and SeaWiFS primary production

Mackas et al. (2006) showed that broad-scale seasonal variations in phytoplankton productivity and biomass in the four upwelling regions are closely related to seasonal and alongshore variation of offshore Ekman transport. In summer, well-defined areas of elevated phytoplankton concentrations develop next to the coast (see Figs. 10.2, 10.3, 10.4 and 10.5), while open-ocean regions devoid of upwelling/mixing mechanisms experience a decrease due to stratification and nutrient depletion. During winter, the phytoplankton concentration increases slightly in offshore waters due to enhanced wind-driven mixing.

One thing that all four systems have in common is their inherent short-term variability. Upwelling episodes are often short-lived and vary in both time and space, while eddies and fronts move within each system, creating changes in nutrient concentrations on short time scales. These changes can lead to biological effects such as a mismatch between predators and prey, or the advection of organic material out of the productive inshore system–one example of this is the reported loss of a large proportion of fish larvae from the southern Benguela following a "close encounter" with an Agulhas ring (Duncombe Rae et al. 1992). It has been suggested that coastal upwelling systems that show the most variability will tend to have more such mismatches than those that are more stable. For example, Bakun and Weeks (2008) compared the southern Benguela with the Humboldt current system and concluded that the increased stability in the latter enhanced trophic transfer efficiency.

The main differences between the eastern margins of the Atlantic and Pacific are associated with:

(a) the relative size of oceanic basins, which controls the depth of the thermocline in the eastern part of the basin. It is relatively shallow off Peru, which allows high nutrient concentrations to upwell except during El Niño episodes when the thermocline deepens;
(b) the sub-surface circulation of the ocean, which results in the sub-pycnocline water of the Pacific having a lower oxygen content and a higher concentration of macronutrients than the Atlantic; and
(c) different levels of exposure to influences from polar oceans and continents.

Despite the overall higher concentration of macronutrients found in old sub-pycnocline waters of the Pacific, all four upwelling systems have very similar nutrient concentrations at 60-m depth (see Table 10.1). This surprising similarity is explained by the enhanced nitrate loss due to denitrification in the extremely oxygen-poor waters in the eastern equatorial Pacific (see Sect. 1.5.3), which is preconditioned by the large-scale circulation and intensified during the decomposition of organic matter as it is transported along poleward undercurrents (Chavez and Messié 2009). The enhanced denitrification off Peru, which is an important component of the ocean's nitrate budget (Codispoti et al. 2001), keeps nitrate levels in this upwelling system at a lower level than expected. While there is an extensive region of low oxygen concentration in the eastern tropical South Atlantic (Wattenberg 1938; Chapman and Shannon 1985), denitrification is not found in the open ocean but only associated with coastal upwelling close to and over the shelf.

10.2.3 Overall Productivity

Potential new production can be derived from nitrate fluxes based on nitrate levels at 60-m depth in conjunction with vertical upwelling rates inferred from (i) offshore Ekman transport and (ii) Ekman pumping (Messié et al. 2009). Two important points need to be noted here. First, it turns out the Ekman pumping due to wind-stress curl cannot be neglected and generally contributes >25 % to the total nitrate flux. Second, winds are weaker off Peru as compared to the other regions (see Table 10.1), which amplifies when converted to wind stress (which is proportional to the square of wind speed) or turbulence (which is proportional to the cube of wind speed). Despite this reduced wind stress, the Peru system still experiences relatively strong upwelling due to its close proximity to the equator and a small Coriolis parameter, which amplifies Ekman transports (see Chap. 2), although this has much less effect at the southern end of the system in southern Chile as there is an inverse relationship between the Coriolis parameter and total upwelling.

To this end, both the total upwelled volume and potential new production are very similar for the Peru, northwest Africa and Benguela upwelling systems, but California lags behind with 30–40 % smaller values. This difference is also evident from primary production estimated with SeaWiFS satellite data (Chavez and Messié 2009). Note that northwest Africa shows the highest SeaWiFS production estimate, which can be explained by enhanced iron supply resulting from dust deposition from the Sahara (see Chap. 6), but which may also result from interference from airborne dust in the atmospheric corrections applied to the satellite signal (Chavez and Messié 2009).

Table 10.2 shows previous estimates of phytoplankton production rates for the big four upwelling regions, separated into latitudinal bands and indicating the months of highest production. These numbers are based on 3 years of measurements of chlorophyll calculated from SeaWiFs satellite data (Carr and Kearns 2003). The observations show interannual variability caused by large-scale ocean effects, such as El Niño-La Niña, as well as considerable seasonal and regional variability, which depends to some extent on changes in offshore Ekman transport rates (Thomas et al. 2001). Overall, however, despite higher seasonal rates in limited latitudinal ranges, all four regions show theoretical mean annual primary production rates of 2–3 g Cm^2/day.

Carr and Kearns (2003) also discuss the chlorophyll response in the big four upwelling systems to different phosphate levels in upwelled water (Fig. 10.6). For

Table 10.2 Production in the eastern boundary coastal upwelling regions

Region	Area (10^3 km^2)	Production rate (g C/m^2/day)
Benguela (36°–35° S)	6.7	1.0 (2; OND)
(35°–33° S)	22.2	3.4 (4; OND)
(33°–29° S)	63.3	4.8 (5; JFM)
(29°–24° S)	112.1	3.9 (5; JFM)
(24°–19° S)	72.2	4.9 (5; JFM)
(19°–15° S)	24.4	4.0 (4.5; JFM)
California (40°–48° N)	44.4	2.2 (4; JAS)
(34°–40° N)	23.3	2.2 (3; AMJ)
(32°–34° N)	15.5	1.3 (4; AMJ)
(26°–32° N)	90.5	2.5 (4; AMJ)
(24°–26° N)	28.9	2.7 (4; AMJ)
Canary (35°–42° N)	50.5	2.2 (3; JAS)
(33°–35° N)	17.8	2.0 (2; JAS)
(27°–33° N)	54.9	4.0 (5; JAS)
(20°–27° N)	75.5	4.5 (5; JAS)
(11°–20° N)	82.1	5.0 (6; AMJ)
(6°–11° N)	122.1	4.0 (4; OND)
Humboldt (40°–33° S)	9.7	3.3 (4; OND)
(33°–27° S)	10.0	2.1 (3; JFM)
(27°–18° S)	15.0	3.2 (3; OND)
(18°–15° S)	14.4	3.2 (4; JFM)
(15°–5° S)	86.3	3.9 (5; JFM)

Area denotes the area of both the inner (<100 m) and outer shelf (<500 m) for each region. In the column of production rates, figures in parentheses denote maximum rates in g C/m^2/day and the quarter of the year when this occurs (JFM, AMJ, JAS, OND). Based on data in Carr and Kearns (2003)

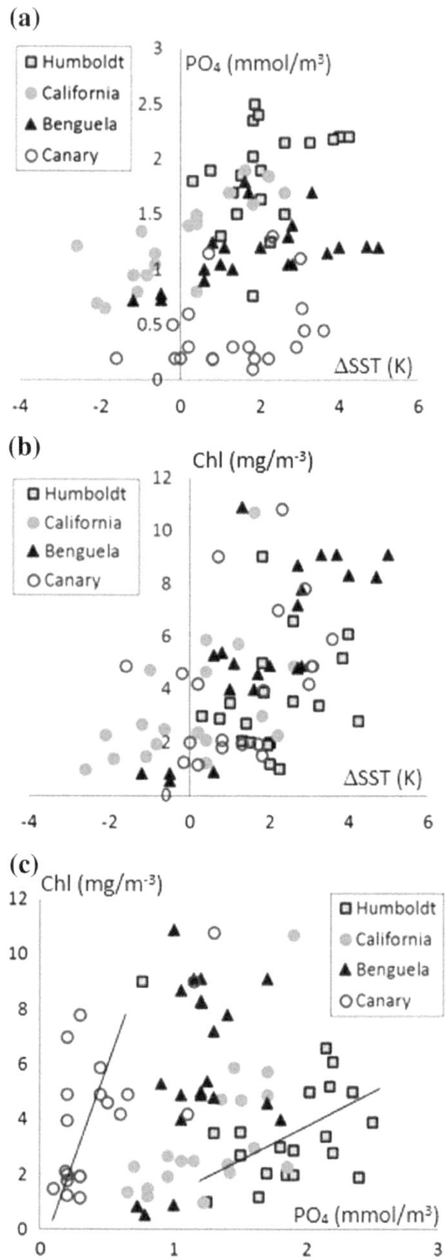

Fig. 10.6 Relations between SST anomalies (positive values refer to upwelling of colder water), phosphate (PO_4) concentrations and chlorophyll levels for the four major upwelling regions (reproduced from data shown by Carr and Kearns 2003)

similar temperature anomalies there are substantial differences in phosphate levels among the upwelling regions (Fig. 10.6a). The Canary Current upwelling system has the lowest phosphate levels, followed by the Benguela, California and the

Humboldt upwelling systems, and the order essentially agrees with the mean oxygen concentration, with the lowest oxygen levels (in the Humboldt and Benguela Current systems) coinciding with high phosphate concentrations and the Canary Current system, which has the highest oxygen levels, having the lowest phosphate concentration.

The California Current system is somewhat odd in this respect, as it has higher phosphate concentrations than expected on the basis of its oxygen concentration. This is for two reasons, higher phosphate concentrations in the source water (the Equatorial Undercurrent) and additional sources from the Columbia River and the Strait of Juan de Fuca. Despite these differences, the chlorophyll responses among the upwelling regions are quite similar, with all four regions showing maximum chlorophyll-*a* concentrations when the SST anomalies are highest (Fig. 10.6b). Based on the data (Fig. 10.6c), Carr and Kearns (2003) concluded that the phosphate utilization in the Canary Current is greater than in the Humboldt Current. This is consistent with the results of Minas et al. (1986), who suggested that nearshore grazing pressure prevented optimal nutrient utilization in the Humboldt Current upwelling system. Other scientists, however, have argued that the difference in nutrient utilization could also be due to the availability of iron (e.g., Fung et al. 2000) associated with different contributions by aeolian dust deposition and benthic or riverine supplies.

Despite similar upwelling-induced nitrate fluxes and primary production rates, the Peruvian system stands out for its extremely high production, which is known as the Peruvian puzzle (see Sect. 5.9). Under the assumption of 10 % energy loss per trophic level, Chavez and Messié (2009) conclude that there are approximately 2.5 trophic levels between primary production and fish off Peru, 3 off California and 3. 5–4 within the Benguela and northwest Africa systems, implying higher transfer efficiencies off Peru and reinforcing the central role of anchoveta in trophic interactions in this region. The reason for this difference in efficiencies is still unknown, given that unlike sardines, which obtain most of their energy from phytoplankton, anchoveta are thought to gain most of their energy from zooplankton, primarily euphausiids and large copepods (Espinoza and Bertrand 2008). To this end, Chavez and Messié (2009) suggest other possible explanations of this discrepancy including the presence of a shallow and intense oxygen minimum zone off Peru (see Table 10.1) that leads to distinct zooplankton-prey interactions, the generally lower mean wind speeds that provide optimum conditions (Table 10.1; see also Sect. 2.4), and a possible periodic reset of the food web after strong El Niño events which keeps predators from becoming dominant and provides effective top-down control.

Another important factor that governs overall productivity is the relaxation time between upwelling events. This is particularly important for regions where upwelling-favourable wind stress is not continuous, but depends on the passage of short-period weather events. An example of this is the southern Benguela system, which is influenced by the passage of individual weather systems on a roughly fortnightly basis during the summer (see Chap. 7). Upwelling-favourable winds typically blow for 3–4 days, followed by relaxation periods that allow time for blooms to develop close to the coast, so that the new organic matter produced is not dispersed out to sea before higher trophic levels can develop and consume it. Wider shelves and upwelling shadow zones, such as in the southern California Bight, also contribute to retaining more of the primary production, as they provide a larger area within which the bloom can remain coherent.

10.2.4 Seasonal Variations

Chlorophyll-*a* concentrations in the highly productive latitude band off California closely follow the seasonal cycle of upwelling due to Ekman transport and Ekman pumping with a maximum in boreal summer months (May–July; Fig. 10.7a). Chlorophyll concentrations vary seasonally between 2 and 5 mg/m^3. Off northwest Africa, both upwelling fluxes and chlorophyll-*a* levels have pronounced maxima between January and May, and chlorophyll-*a* levels are extremely high at 7 mg/m^3 (likely due to dust deposition; Fig. 10.7b). From May to July, chlorophyll levels

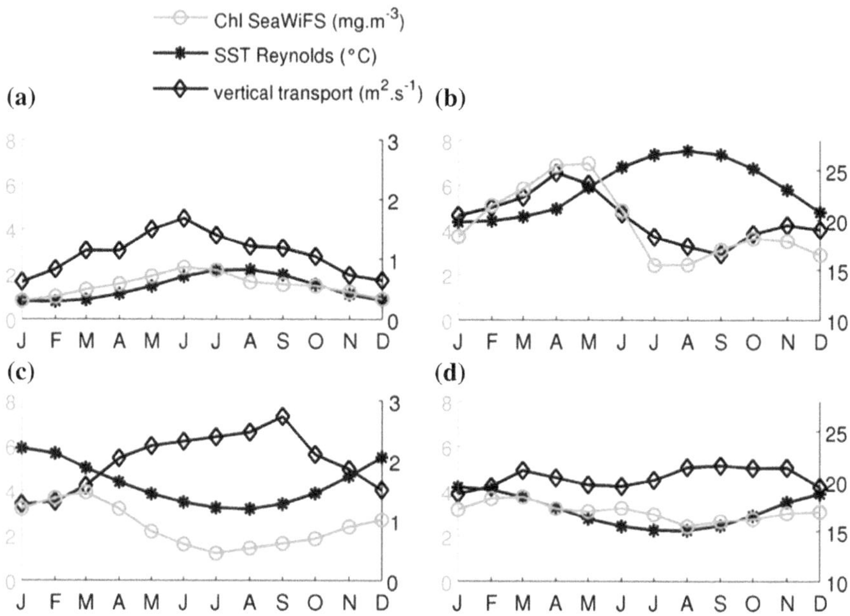

Fig. 10.7 Seasonal cycles of chlorophyll concentration, sea surface temperature and total vertical transport (Ekman transport plus Ekman pumping) for the four eastern boundary upwelling systems for the specific 10° latitude bands, given in Table 10.1, up to 150 km offshore. California and northwest Africa show the expected positive relationship between transport and chlorophyll. The Benguela is not highly seasonal, while Peru has an inverse relationship with SST. SST values are monthly from October 1981 to February 2007 (from Reynolds et al. 2007); chlorophyll is SeaWiFS climatology; winds are QuikSCAT weekly from July 1999 to July 2008. Taken from Chavez and Messié (2009)

drop significantly to ~2 mg/m^3, which coincides with the onset of thermal stratification and nutrient depletion, commonly seen in the North Atlantic after the spring phytoplankton bloom (Chavez and Messié 2009; see Chap. 3).

The most surprising result is seen off Peru (Fig. 10.7c), where the upwelling forcing is strongest in austral winter but the highest phytoplankton concentrations develop during austral summer (see Sect. 5.9; Mackas et al. 2006; Chavez and Messié 2009). Winter mixing and iron limitation have been suggested as explanations for this unexpected feature (Calienes et al. 1985; Chavez et al. 2008; Echevin et al. 2008; Friederich et al. 2008). The situation in the chosen region of the northern Benguela is similar, with a slight summer maximum in chlorophyll levels that does not coincide with the greatest vertical transport, although summer upwelling is higher than in autumn (Fig. 10.7d).

10.2.5 Large-Scale Setting

As pointed out by Chavez and Messié (2009) the upwelling ecosystems of eastern boundary currents, which are associated with the subtropical gyres of the Atlantic and Pacific, owe their existence to the large-scale structure of the thermocline which separates nutrient-poor surface waters from nutrient-enriched deeper water. Given the wind-driven dynamics that drive subtropical gyres, the thermocline shoals towards the east, which lifts the nutricline across the shelf break and onto the continental shelf (see Fig. 3.2). Without this preconditioning process, which is driven by the large-scale wind stress curl, the coastal upwelling process would be far less efficient at delivering nutrients to the euphotic zone. This large-scale view also highlights the significance of intra-decadal variability associated with variations of the large-scale wind-driven circulation.

From this, it is also clear that north–south variations of thermocline doming and weather-driven mixing at higher latitudes make latitude an important characteristic of each major upwelling system (Fig. 10.8). At a reference distance of 1000 km offshore, the nutricline shoals near the equator, deepens in the subtropical gyres and then shoals again near the poleward margin of the subtropical gyres in the regions subject to the influence of westerly winds. The shoaling near the equator contributes to enhanced nutrient supply in the regions where the Humboldt Current and Canary Current upwelling systems impinge on the equator (see Figs. 10.2 and 10.4). At around 20–30° N/S, the equatorward-flowing subtropical gyres, despite their deeper nutriclines, still provide sufficiently high nitrate concentrations to bottom waters on the shelf, close to the Southern California Bight, in southern Peru and northern Chile, around Morocco and Namibia. At higher latitudes at the northern or southern limits of the upwelling systems, enhanced vertical mixing in subtropical frontal zones and by weather patterns of the westerlies further enhance biological productivity.

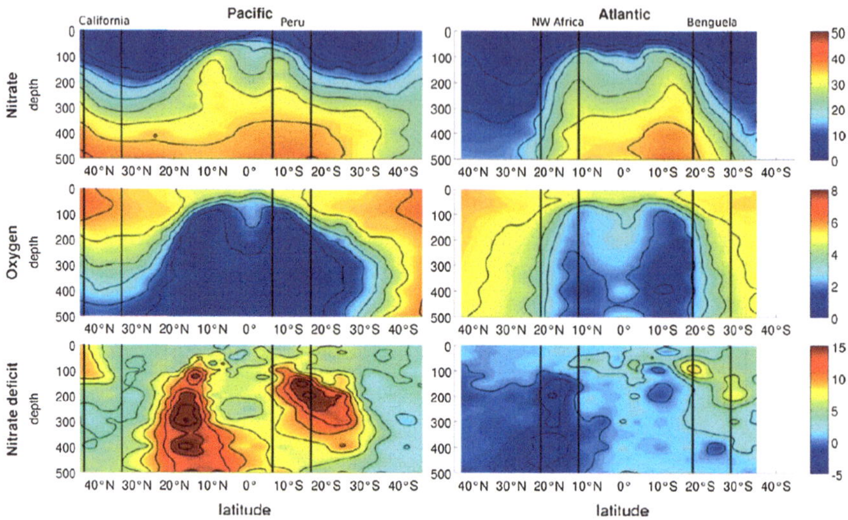

Fig. 10.8 North–south section of nitrate (μM/L), oxygen (mL/L), and nitrate deficit (μM/L) calculated as 16 $[PO_4^{3-}]-[NO_3^-]$ in the Pacific and Atlantic Oceans following the coast 1000 km offshore of the eastern boundary. Black lines correspond to the latitudes of vertical sections (not shown). Data are from the World Ocean Atlas 2005 (Garcia et al. 2006a, b). Note the strong correlation between hypoxia and nitrate deficit in the Pacific section. Taken from Chavez and Messié (2009)

10.2.6 Air-Sea Carbon Fluxes

Air-sea carbon fluxes are driven by the difference in partial gas pressures across the sea surface (see Sect. 1.6). The partial gas pressure of CO_2 in upwelling systems is controlled and influenced by the upwelling of cold and carbon-rich sub-pycnocline water, which tends to increase surface pCO_2 values, and photosynthetic primary production, which tends to reduce it. Productivity in turn is influenced by many factors including temperature, nutrient limitation and complex food-web dynamics and biogeochemical cycles (see Chap. 1). Theoretically there should be a difference in air-sea carbon fluxes between upwelling centres near the coast, which tend to be sources at least some of the time, and waters farther offshore, which are influenced by the export of organic matter via mesoscale eddies and filaments (see Fig. 10.1) and which tend to be sinks. While in situ observations suggest that net CO_2 fluxes vary substantially among the big four coastal upwelling systems (Table 10.3), the overall net flux may not be very large, except in the Humboldt Current system, and there is considerable scatter in the estimates, which depend on a number of assumptions, not the least of which is the area affected (see Table 10.3).

Offshore of Oregon and California, the region is either a slight CO_2 sink (Hales et al. 2005) or neutral (Chavez et al. 2007), although Pennington et al. (2010) suggest that inshore waters (<60 km from the coast) act as a CO_2 source to the atmosphere, but as a sink farther offshore. In contrast, off Peru (Friederich et al. 2008) and central

Table 10.3 Estimated source and sink regions for carbon dioxide in eastern boundary systems

Region	Source	Sink	Reference
California Current			
Inshore (<60 km)	<1		Pennington et al. (2010)
60–300 km		−0.3	Pennington et al. (2010)
Humboldt Current			
18–27° S	0.3–5.2		Quiñones et al. (2010)
27–33° S	<1.94	<−0.3	Quiñones et al. (2010)
Northwest Africa			
	0–1.0		Takahashi et al. (2002)
		"Small"	Fischer et al. (2010)
18–23° N	<3.0 (spring)		Gonzales-Davila and Santana-Casiano (2013)
18–23° N	0.65 (annual)		Gonzales-Davila and Santana-Casiano (2013)
Iberian system			
Shelf	0.35–2.4		Alvarez-Salgado et al. (2010)
Offshore	0–1.8		Alvarez-Salgado et al. (2010)
Benguela Current			
North of 20° S	0.15–1.5		Santana-Casiano et al. (2009)
North of 20° S	0.82		Monteiro (2010)
South of 20° S		−0.71 to −2.4	Santana-Casiano et al. (2009)
South of 20° S		0	Monteiro (2010)

Sources and sinks given in mole C/m^2/yr

and northern Chile (Torres et al. 2002; Paulmier et al. 2008), measured CO_2 fluxes suggest a strong flux to the atmosphere, as found also by Quiñones et al. (2010). Santana-Casiano et al. (2009) derived a similar pattern in the Benguela Current system, with the ocean being a sink at higher latitudes but outgassing to the atmosphere at lower latitudes. Monteiro (2010), however, while agreeing that the northern Benguela is a significant source, calculated that the southern Benguela was effectively in balance with no net flux in either direction. Off northwest Africa, CO_2 measurements indicate that this region too is a weak source of CO_2 (e.g., Gonzales-Davila and Santana-Casiano 2013), while the higher-latitude Iberian Peninsula acts as a CO_2 sink (Arístegui et al. 2009). Chavez and Messié (2009) attribute these differences in net CO_2 fluxes to varying rates of denitrification and ambient levels of preformed nitrate (nitrate levels of the source water mass).

Given that some regions experience large seasonal and inter-annual variations, the complexity of physical and biogeochemical processes that determine the resultant air-sea fluxes of CO_2 and the vast range of temporal and spatial scales involved, it is impossible at present to define the fluxes any more precisely or to predict likely future changes in CO_2 fluxes in these upwelling regions with any confidence.

10.2.7 Multi-decadal Variability and Global Trends

Multi-decadal fluctuations of the large-scale thermocline/nutricline have been linked to regime shifts in the ecosystems of the four main coastal upwelling systems (see Sect. 3.5.2). Historically, when the upwelling systems experienced warming, sardines increased in abundance. Conversely anchovies have been favoured during cooler periods (Chavez et al. 2003; and references therein). Longer sedimentary records document even more dramatic biogeochemical and ecological changes coinciding with glacial maxima and ice ages; during glacial periods both denitrification and the area affected by oxygen minimum zones decrease (Galbraith et al. 2004). Even the Little Ice Age shows up in the sediment record as a period of reduced productivity in the northern Humboldt Current system off Peru, with low abundances of both sardine and anchovy that only recovered after the end of this cool period in about 1820 (Gutiérrez et al. 2009).

Chavez and Messié (2009) point to the conundrum that arises from the fact that warm interglacial periods had more productive major upwelling regions, which is in contrast to evidence that warm phases of the Pacific Decadal Oscillation are associated with lowered ecosystem productivity in the eastern Pacific. The question then arises as to whether global warming will lead to decreased or increased productivity. As we have seen, Bakun (1990) proposed that increased land-sea temperature differences under the global-warming scenario will lead to stronger upwelling-favourable winds, which appears to be the case for the California, Humboldt and Benguela upwelling systems (Sydeman et al. 2014), but this is still a subject of intense discussion.

A time-series of SeaWiFS chlorophyll-*a* data from 1998 to 2007 indicates that in three of the four major upwelling systems (northwest Africa is an exception), chlorophyll levels experienced a slight increase over the decadal record (Fig. 10.9; Chavez and Messié 2009). Simultaneously, SST trends in the Peru and California systems have experienced slight cooling. Nevertheless, this interpretation has to be treated with suspicion, given that the positive chlorophyll trends for Peru and California are strongly biased by the absence of strong summer peaks at the start of the time-series from 1998–2000, following the strong 1997/98 El Niño. This bias is also consistent with the alleged cooling trend for Peru and California, which experienced a transient warm water phase initiated by the El Niño event. The length of this ocean colour time series is, to date, simply not long enough for us to extract statistically reliable trends that are unbiased by natural climate variability. However, the longer time series for the period 1965–2004 off Peru (see Fig. 5.9; Gutiérrez et al. 2011) shows a statistically more reliable gradual increase, but, again, this feature could by the result of inter-decadal variability (i.e., the Pacific Decadal Oscillation), longer-term variations of the nutricline structure influencing cross-shelf nutrient fluxes, and/or shifts in food-web dynamics. This example underpins the great significance and importance of multi-decadal time series for evaluating long-term trends.

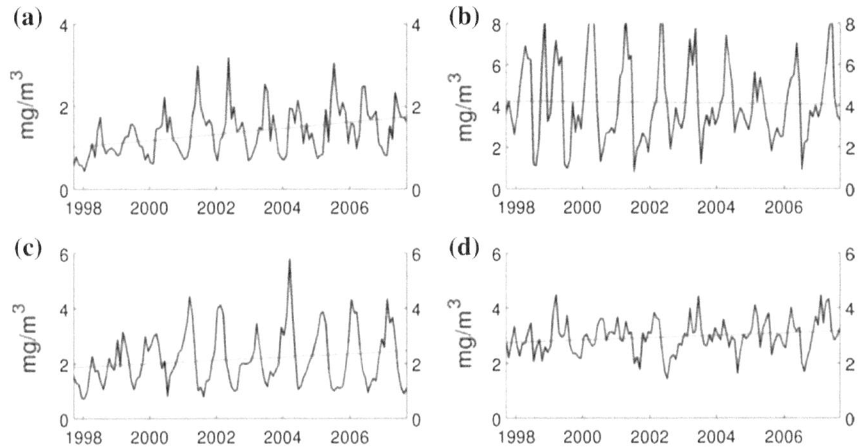

Fig. 10.9 Time series of SeaWiFS chlorophyll for each eastern boundary current upwelling system. The trend line shows increasing concentrations of chlorophyll in the Benguela, California, and Peru systems, and small decreases in northwest Africa. The data are SeaWiFS full resolution monthly (9 km) averaged on a 0.25° by 0.25° grid (geometric mean) and then within each 10° latitude band up to 150 km offshore. Taken from Chavez and Messié (2009). **a** California: +0.073 mg/m^3/yr. **b** NW Africa: −0.018 mg/m^3/yr. **c** Peru: +0.058 mg/m^3/yr. **d** Benguela: +0.041 mg/m^3/yr

10.2.8 Fisheries

From the data in Table 10.4, it can be seen that the Humboldt Large Marine Ecosystem produces far more fish than the other three major upwelling systems combined. Its fish production is twice that of the Canary LME and the difference is explained by the large anchoveta catches. The Humboldt and Canary systems combined account for almost 90 % of the fish caught in the four major upwelling systems.

Despite the large fisheries production of the eastern boundary current upwelling systems, the world's most productive fisheries are found in tropical and subtropical Asian shelf seas, namely the South China Sea, the East China and Yellow Seas, and the Bay of Bengal, some of which receive large nutrient inputs from river

Table 10.4 Average catches (in 1000 tonnes per annum) for the big four upwelling systems during the period 1990–2010

Region	Total	Anchovies	Herring-like	Perch-like	Cod-like	Other[a]
Humboldt	14,183	8,774	1,659	2,383	412	955
Canary	6,662	74	1,302	1,125	21	4,140
Benguela	1,653	215	309	562	263	304
California	884	12	169	158	184	361
Sum	**23,382**	**9,075**	**3,439**	**4,228**	**880**	**5,760**

Data from http://www.seaaroundus.org/

[a]Includes estimates of bycatch/unreported data

Table 10.5 Catches (in 1000 tonnes per annum) for selected Asian shelf seas for 2010

Region	Total	Anchovies	Herring-like	Perch-like	Cod-like	Other[a]
Yellow Sea	3,800	220	117	1,337	87	2,039
Sea of Japan	2,100	124	78	408	401	1,089
Kuroshio	1,750	137	45	561	60	947
East China Sea	6,750	519	166	2,845	3	3,217
South China Sea	12,900	626	572	5,800	0	5,902
Indonesian Seas	2,850	134	222	1251	0	1,243
Gulf of Thailand	3,450	239	209	1,732	0	1,270
Bay of Bengal	7,800	306	772	3045	3	3,674
Sulu-Celebes Sea	1,760	39	271	631	0	819
Sum	**43,160**	**2,344**	**2,451**	**17,610**	**554**	**20,200**

Data from http://www.seaaroundus.org/
[a]Includes estimates of bycatch/unreported data

discharges. Altogether, these regions produce almost twice as much fish as the big four upwelling regions combined, although one should note that the fishery data are very uncertain due to unreported catches and possibly over-reporting by Chinese sources. In a manner similar to what has happened in the major coastal upwelling systems, total fish catches have stagnated in recent decades, which is a strong indication that a critical catch limit has been reached (Table 10.5).

For completeness, we show seasonal distributions of normalized fluorescence line height as a measure of phytoplankton concentrations (Fig. 10.10) for Southeast

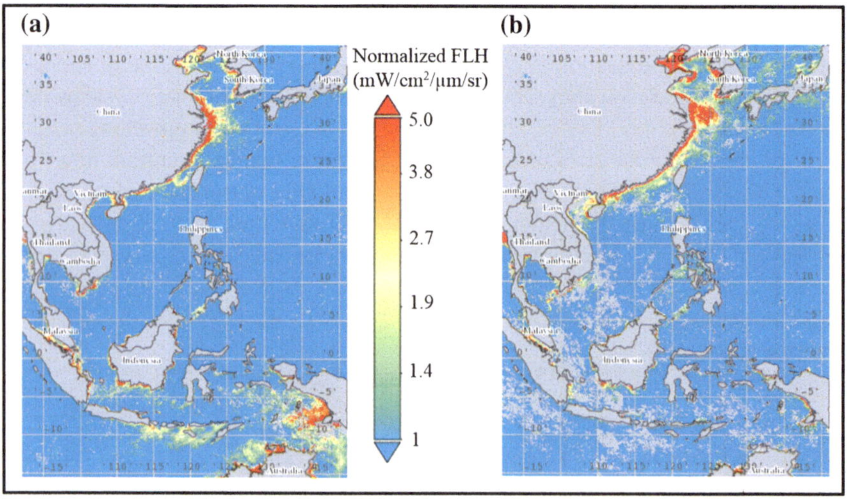

Fig. 10.10 Example of **a** boreal winter and **b** boreal summer season distributions of MODIS Aqua (4 km resolution) surface chlorophyll pigment concentrations for southeast Asian waters. Winter (summer) refers to the period Dec–Feb 2010/11 (July–August 2010). *Source* Giovanni (NASA), Version 4.19, http://giovanni.gsfc.nasa.gov/giovanni/

Asian waters, noting that this method eliminates potential biases from suspended sediment and coloured dissolved organic matter (see Sect. 2.1.7). Details on individual upwelling systems in this region are given in Chap. 8.

10.3 Research Gaps and Enigmas

10.3.1 Overview

The aim of this section is to provide readers with a summary of key gaps in the understanding of upwelling systems as a guide for future research directions. As far as humans are concerned, the most important role of upwelling systems is their production of enormous amounts of biological resources at higher trophic levels, such as fisheries. This production of carbon is fueled by nutrient fluxes and strongly interlinked with the atmospheric carbon budget and the ocean's biogeochemistry. Despite intensive past research, going back at least to the 1920s, there are still many unresolved scientific questions regarding upwelling systems. One of the largest, and perhaps the most important unknown, is whether human behaviour will change and encourage a transition to a world of zero anthropogenic carbon emissions, population growth, management of pollution, and changes in the exploitation of marine resources. This, however, is a social and political problem that is well beyond the scope of this publication.

10.3.2 Ocean Acidification and Expanding OMZs

One of the oceanography's leading questions is how the ecology of upwelling systems will change and adapt under the scenario of global climate change. While many aspects of relevance to upwelling systems, such as future wind changes, cannot be predicted with certainty, the clearest and best documented global-warming signal is the shoaling of carbonate mineral saturation horizons and the expansion of oxygen minimum zones (OMZs).

Changes in carbonate chemistry mostly impact coral reefs, which are integral parts of some upwelling systems such as those in the Caribbean Sea and southeast Asian waters, as well as any other organisms, such as oysters and other shellfish, that use carbonate to build their shells. Carbon dioxide levels are certainly increasing, which is reducing the pH of the surface ocean, and the rate at which the ocean can accept additional atmospheric CO_2 is slowing down. How will this affect upwelling systems? At present, given the large uncertainty in CO_2 flux estimates associated with upwelling zones (Sect. 10.2.6), we cannot answer this question.

Hypoxia is of relevance to many important upwelling systems in the Pacific (Peru-Chile and California), the Indian Ocean (Somalia) and the Atlantic (northern Benguela). The marine biology and ecology of open-ocean hypoxic environments is a relatively new research field and there are still many open questions. Diaz (2001), Levin (2003), and Diaz and Rosenberg (2008) give reviews of earlier research findings, although these are concentrated on coastal hypoxia, and point out that many scientific questions remain unanswered concerning controls on faunal standing stocks in OMZs, and the physiological, enzymatic, metabolic, reproductive and molecular adaptations that permit benthic animals to live in them. We should note here that there is a considerable difference between hypoxia in upwelling systems, where the fauna have evolved mechanisms to cope with low and variable oxygen concentrations, and other coastal regions subject to sudden increases in oxygen demand, where they have not. Sulphate reducing bacteria, such as *Beggiatoa*, are one result of overwhelming the system with organic carbon that uses up all the dissolved oxygen during respiration, but these are not dominant in most upwelling systems, and it is unclear how oxygen levels in coastal upwelling systems will change with continued global warming. Ekau et al. (2010) and Stramma et al. (2010) discuss potential impacts of hypoxia on the structure and processes in pelagic communities including zooplankton, macro-invertebrates and fish. These authors conclude that there will be habitat compression in coastal regions as the depth of the upper surface of the OMZs become shallower, leading to possible shifts in faunal distribution and a decrease in biodiversity in hypoxic areas that will reduce ecosystem resilience and resistance and may decrease its function in nutrient cycling processes.

In this vein, Chu and Tunnicliffe (2015) discuss oxygen limitations on marine animal distributions in the northeast Pacific and report on the collapse of epibenthic community structure, due mainly to migration of motile species out of the hypoxic zone, during episodes of shoaling hypoxia. While their site (Saanich Inlet) is not part of the California Current upwelling system, it is situated on the southern end of Vancouver Island and so affected by the same water masses that affect the northern portion of the system. Interestingly, the authors noted that in situ oxygen changes were not congruent with lethal and sublethal hypoxia thresholds calculated across the literature for major taxonomic groups, which indicates that research biases towards laboratory studies on Atlantic species are not globally applicable, while they agreed with the general conclusions of Ekau et al. (2010) and Stramma et al. (2010). Chu and Tunnicliffe (2015) suggested that because organisms react differently to low oxygen depending on the temperature and salinity in a given area, region-specific hypoxia thresholds are necessary to predict future impacts of deoxygenation on marine biodiversity. This example illustrates the significance of future hypoxia research.

10.3.3 Lack of Systematic Monitoring

While the Californian Current upwelling system has been systematically studied over many decades thanks to the CALCOFI program, most other upwelling systems remained underexplored. For examples, the Canary Current Upwelling System is the least explored of the four major coastal upwelling systems, and relatively little is known about ecological features of the seasonal upwelling system of the Somali Current in the Indian Ocean. Most striking is the lack of systematic observations to date in east and southeast Asian waters which produce almost twice as much fish as the big four upwelling systems combined, although Chinese researchers are increasing their research effort in the South China Sea and East China Sea. Hence, there are substantial research gaps for many upwelling systems worldwide.

10.3.4 Uncertainty of Future Continental Runoff

Some upwelling regions are strongly influenced by nutrient inputs from rivers. This includes the major coastal upwelling systems off Washington/Oregon and Portugal and some important seasonal upwelling systems such as those in the South China Sea. Although various climate models predict future changes in average river flows (and temperatures) under different climate scenarios (van Vliet et al. 2013), these studies do not include changes in human intervention factors (e.g., water extraction for agriculture and human consumption). To this end, changes in both flow rates and water quality of river discharges reaching individual upwelling regions remain highly uncertain.

10.3.5 Global Warming Versus Geological Records

As stated earlier (Sect. 10.2.7), there is evidence on geological timescales that, during glacial maxima and the Little Ice Age, oxygen minimum zones retreated and denitrification including microbial processes removing fixed nitrogen was reduced. During the Little Ice Age, core samples show that the productivity of the Humboldt Current upwelling system decreased at all trophic levels (Gutiérrez et al. 2009) leading to very low abundances of both anchovies and sardines in the northern region off Peru. On the other hand, warm interglacial periods were associated with more productive eastern boundary current upwelling systems (Chavez and Messié 2009), and the question arises as to whether the anthropogenic global warming signal together with expanding oxygen minimum zones will increase overall productivity, as occurred during warm interglacial periods, or decrease productivity, as seen during strong El Niño events or during warm phases of the PDO).

10.3.6 Zooplankton

While most of the current knowledge of the functioning of upwelling systems has been derived from physical data, nutrient fluxes, phytoplankton distributions and fishery data, little is known about the intermediate trophic level of zooplankton dynamics and differences in zooplankton composition and behaviour between upwelling systems.

A few species of euphausiids (krill) grow to 6–15 cm in length and can live up to 10 years. Krill play a critical role in many marine food webs, feeding on phytoplankton and smaller zooplankton, and consumed by whales, seals, penguins, squid and fish. Most krill are found in large schools or swarms that can stretch over kilometres. While krill are known to play an important role in upwelling food webs and biogeochemical cycles, particularly in the Antarctic, information on euphausiids is severely lacking (Chavez and Messié 2009). This lack of knowledge is due to the long lifespan of this mobile species, which makes the prediction of longer-term migration paths largely impossible. Nevertheless, the importance of krill is seen in many upwelling systems such as that off southern Australia (Kämpf et al. 2004) which attract large numbers of blue whales and other whale species during the upwelling season (Gill et al. 2011).

One question of relevance is whether zooplankton structure and abundance is controlled by "top down" or "bottom up" processes. Rykaczewski and Checkley (2008) have suggested that the local upwelling rate determines both zooplankton size structure and population density in coastal upwelling systems, with faster upwelling rates favouring larger plankton and anchovies and slower upwelling rates favouring smaller plankton and sardines. As the wind stress is highest off California, this suggests that the largest zooplankton should occur there. While such "bottom up" control appears to be generally true, a "top down" control from anchoveta can predominate on local scales (Ayon et al. 2008).

10.3.7 Interconnections of Biomes

Little is known about the connectivity between upwelling biomes. While there can be a certain degree of synchronicity in ecosystem variations due to large-scale climate variability and climate change, little is known about the migration of longer-lived marine species such as euphausiids or forage fish (salmon, sardine, tuna) between upwelling systems, or if it even occurs. Satellite tagging studies, such as those on Pacific predators (Fig. 10.11), are the key in the understanding of this connectivity.

10.3.8 Role of Fish in Carbon Fluxes

Walsh (1981) and Bakun and Weeks (2004) suggested that large populations of small pelagic fish in upwelling systems can play a role in carbon fluxes. Walsh (1981) argued that the collapse of the anchovy fish stock in the early 1970s would intensify anoxia off Peru via enhanced vertical carbon fluxes on the shelf. Bakun

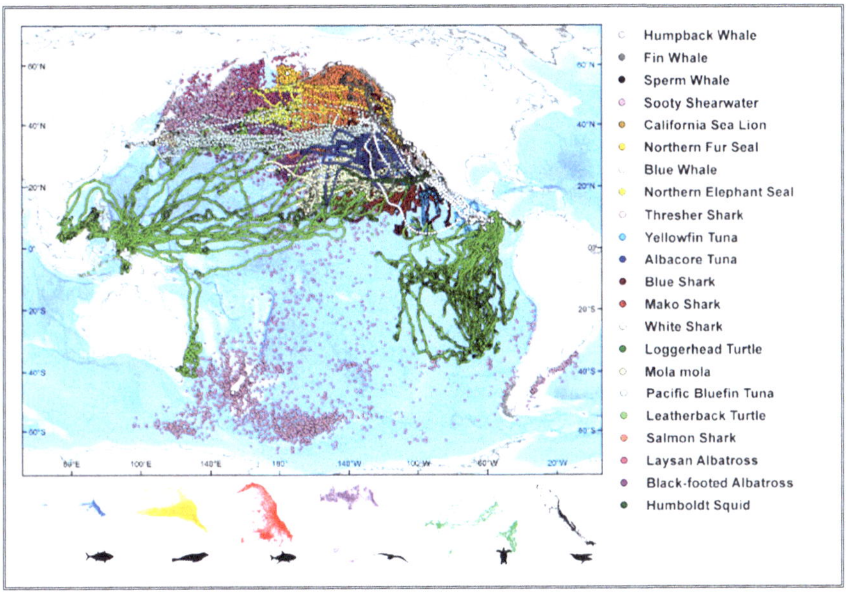

Fig. 10.11 Census of Marine Life Tagging of Pacific Predators (TOPP). Positions of all TOPP animals, color-coded based on species group: *blue* tunas (yellowfin, bluefin and albacore), *orange* pinnipeds (northern elephant seals, California sea lions and northern fur seals), *red* sharks (salmon, white, blue, common thresher and mako), *purple* seabirds (Laysan and black-footed albatrosses and sooty shearwaters), *green* sea turtles (leatherback and loggerhead) and *black* cetaceans (blue, fin, sperm and humpback whales). Taken from Block et al. (2011)

and Weeks (2004) similarly suggested that variations in the abundance of sardines in the Benguela might be responsible for episodic anoxia and the production of hydrogen sulphide through increased deposition of phytoplankton carbon at times of low sardine abundance. Neither of these hypotheses has been confirmed, and they are difficult to study given that other factors (e.g., changes in coastal winds or climate variability) often dominate the changes in oxygen levels in coastal waters.

10.4 Future Research

Given that many important parameters and processes, such as changes in the microbial loop, species composition, sub-surface phytoplankton production, zooplankton composition, etc. cannot be derived from satellite observations, in situ observations within upwelling systems will continue to be needed for the foreseeable future. This requires the coordination of existing in situ monitoring programs and the development of new ones. Hence, this epoch of the anthropocene should be accompanied by an intensification of observational efforts in order to understand the dramatic changes in marine ecosystems that are expected to dominate the upcoming decades.

While most of the basic physical mechanisms governing upwelling regions are reasonably well understood by now, there are many aspects inherent to these systems that are barely known, in particular how they will respond to anthropogenic global warming and associated changes of the carbonate system. Even more basic questions, such as what percentage of organic carbon production is transferred to the sediments in each upwelling system, or the size of the zooplankton populations in the different systems, have very large error bars on their estimates. The enigmas listed above also provide many potential research opportunities, although these are biased towards the four large eastern boundary upwelling systems. Our knowledge of even the basic controls and parameters within the other upwelling systems covered in this book is even more uncertain.

Finally, we return to the importance of upwelling systems to humans and perhaps the most important reason for continued research on them. The harvesting of marine food resources from upwelling systems is substantial, and the economic and social benefits that result are extremely significant for many developed and developing countries. It is currently unknown how these benefits will change in the global warming scenario, and that is why continued and intensified scientific effort has never been as important as it is for this generation.

References

Adey, W.H., and R.S. Steneck. 2001. Thermogeography over time creates biogeographic regions: A temperature/space/time-integrated model and an abundance-weighted test for benthic marine algae. *Journal of Phycology* 37: 677–698.

Alvarez-Salgado, X.A., A.V. Borges, F.G. Figueras, and L. Chou. 2010. Iberian margin: The rias. In *Carbon and nutrient fluxes in continental margins*, ed. K.-K. Liu, L. Atkinson, R. Quiñones, and L. Talaue-McManus, 103–120. Berlin: Springer.

Arístegui, J., E.D. Barton, X.A. Álvarez-Salgado, A.M.P. Santos, F.G. Figueiras, S. Kifani, S. Hernández-León, E. Mason, E. Machu, and H. Demarcq. 2009. Subregional ecosystem variability in the Canary Current upwelling. *Progress in Oceanography* 83(1): 33–48. doi:10.1016/j.pocean.2009.07.031.

Ausubel, J. 1999. Toward a census of marine life. *Oceanography* 12: 4–5.

Ayon, P., G. Swartzman, A. Bertrand, M. Gutierrez, and S. Bertrand. 2008. Zooplankton and forage fish species off Peru: large-scale bottom-up forcing and local-scale depletion. *Progress in Oceanography* 79: 208–214.

Bakun, A. 1990. Global climate change and intensification of coastal ocean upwelling. *Science* 247(4939): 198–201.

Bakun, A., and S.J. Weeks. 2004. Greenhouse gas buildup, sardines, submarine eruptions, and the possibility of abrupt degradation of intense marine upwelling ecosystems. *Ecology Letters* 7: 1015–1023.

Bakun, A., and S.J. Weeks. 2008. The marine ecosystem off Peru: what are the secrets of its fishery productivity and what might its future hold? *Progress in Oceanography* 79: 290–299.

Barber, R.T., and R.L. Smith. 1981. Coastal upwelling ecosystems. In *Analysis of marine ecosystems*, ed. A. Longhurst, 31–68. Academic Press.

Block, B.A., I.D. Jonsen, S.J. Jorgensen, et al. 2011. Tracking apex marine predator movements in a dynamic ocean. *Nature* 475: 86–90. doi:10.1038/nature10082.

Briggs, J.C. 1974. *Marine zoogeography*. New York: McGraw-Hill.

Briggs, J.C. 1995. *Global biogeography*. Amsterdam: Elsevier.

Calienes, R., O. Guillén, and N. Lostaunau. 1985. Variabilidad espacio-temporal de clorofila, produccion primaria y nutrientes frente a la costa peruana. *Boletín del Instituto del Mar del Perú* 10: 1–44.

Carr, M.-E., and E.J. Kearns. 2003. Production regimes in four eastern boundary current systems. *Deep-Sea Research II* 50: 3199–3221.

Chapman, P., and L.V. Shannon. 1985. The Benguela ecosystem. Part II. Chemistry and related processes. *Oceanography and Marine Biology Annual Reviews* 23: 183–251.

Chavez, F.P., J.P. Ryan, S.E. Lluch-Cota, and C.M. Ñiquen. 2003. From anchovies to sardines and back: Multidecadal change in the Pacific Ocean. *Science* 299(5604): 217–221.

Chavez, F.P., T. Takahashi, W.-J. Cai, G. Friederich, B. Hales, R. Wanninkhof and R. Feely. 2007. Coastal oceans. In *The First State of the Carbon Cycle Report (SOCCR): The North American Carbon Budget and Implications for the Global Carbon Cycle*, ed. A. King, I. Dilling, G. Zimmerman, D. Fairman, R. Houghton, G. Marland, A. Rose and T. Wilbanks, A, Report by the US Climate Change Science Program and the Subcommittee on Global Change Research, 157–166. National Oceanic and Atmospheric Administration, National Climatic Data Center, Asheville, NC, USA.

Chavez, F.P., A. Bertrand, R. Guevara, P. Soler, and J. Csirke. 2008. The northern Humboldt Current System: Brief history, present status and a view towards the future. *Progress in Oceanography* 79(2–4): 95–105.

Chavez, F.P., and M. Messié. 2009. A comparison of eastern boundary upwelling ecosystems. *Progress in Oceanography* 83(1–4): 80–96. doi:10.1016/j.pocean.2009.07.032.

Chu, J.W.F., and V. Tunnicliffe. 2015. Oxygen limitations on marine animal distributions and the collapse of epibenthic community structure during shoaling hypoxia. *Global Change Biology* 21: 2989–3004. doi:10.5061/dryad.1p55v.

Codispoti, L.A., J.A. Brandes, J.P. Christensen, A.H. Devol, S.W.A. Naqvi, H.W. Paerl and T. Yoshinari. 2001. The oceanic fixed nitrogen and nitrous oxide budgets: Moving targets as we enter the anthropocene? *Scientia Marina*, 65 (S2): 85–105.

Conkright, M.E., et al. 2002. *World Ocean Atlas 2001*, Volume 1: Introduction. Technical Report, NOAA, Atlas NESDIS 42, 160 pp.

Diaz, R. 2001. Overview of hypoxia around the world. *Journal of Environmental Quality* 30: 275–281.

Diaz, R.J., and R. Rosenberg. 2008. Spreading dead zones and consequences for marine ecosystems. *Science* 321: 926–929.

Duda, A., and K. Sherman. 2002. A new imperative for improving management of large marine ecosystems. *Ocean and Coastal Management* 45: 797–833.

Duncombe Rae, C.M., A.J. Boyd, and R.J.M. Crawford. 1992. "Predation" of anchovy by an Agulhas ring: a possible contributory cause for the very poor year class of 1989. *South African Journal of Marine Science* 12: 167–173.

Echevin, V., O. Aumont, J. Ledesma, and G. Flores. 2008. The seasonal cycle of surface chlorophyll in the Peruvian upwelling system: A modelling study. *Progress in Oceanography* 79(2–4): 167–176.

Ekau, W., H. Auel, H.-O. Portner, and D. Gilbert. 2010. Impacts of hypoxia on the structure and processes in pelagic communities (zooplankton, macro-invertebrates and fish). *Biogeosciences* 7: 1669–1699.

Espinoza, P., and A. Bertrand. 2008. Revisiting Peruvian anchovy (*Engraulis ringens*) trophodynamics provides a new vision of the Humboldt Current system. *Progress in Oceanography* 79(2–4): 215–227.

Fischer, G., S. Neuer, R. Davenport, P. Helmke, R. Schlitzer, O. Romero, V. Ratmeyer, B. Donner, T. Freudenthal, H. Meggers, and G. Wefer. 2010. The Northwest African margin. In

Carbon and nutrient fluxes in continental margins, ed. K.-K. Liu, L. Atkinson, R. Quiñones, and L. Talaue-McManus, 78–103. Berlin: Springer.
Friederich, G.E., J. Ledesma, O. Ulloa, and F.P. Chavez. 2008. Air–sea carbon dioxide fluxes in the coastal southeastern tropical Pacific. *Progress in Oceanography* 79(2–4): 156–166.
Fung, I., S.K. Meyn, I. Tegen, S. Doney, J. John, and J. Bishop. 2000. Iron supply and demand in the upper ocean. *Global Biogeochemical Cycles* 14(1): 281–295. doi:10.1016/j.dsr2.2003.07.015.
Galbraith, E.D., M. Kienast, T.F. Pedersen and S.E. Calvert. 2004. Glacial-interglacial modulation of the marine nitrogen cycle by high latitude O_2 supply to the global thermocline. *Paleoceanography*, 19, PA4007. doi:10.1029/2003PA001000(4).
Garcia, H.E., R.A. Locarnini, T.P. Boyer and J.I. Antonov. 2006a. *World Ocean Atlas 2005*. In Dissolved oxygen, apparent oxygen utilization, and oxygen saturation, ed. S. Levitus, NOAA Atlas NESDIS 63, vol. 3, 342 pp. US Government Printing Office, Washington, DC.
Garcia, H.E., R.A. Locarnini, T.P. Boyer and J.I. Antonov. 2006b. *World Ocean Atlas 2005*. In Nutrients (phosphate, nitrate, silicate), ed. S. Levitus, NOAA Atlas NESDIS 63, vol. 4, 396 pp. US Government Printing Office, Washington, DC.
Gill, P.C., M.G. Morrice, B. Page, R. Pirzl, A.H. Levings, and M. Coyne. 2011. Blue whale habitat selection and within-season distribution in a regional upwelling system off southern Australia. *Marine Ecology Progress Series* 421: 243–263.
Gonzales-Davila, M., and J. Santana-Casiano. 2013. Seasonal air-sea CO_2 fluxes off northwest Africa at the Mauritanian region. AGU Fall Meeting, abstract #OS51A-1637.
Gutiérrez, D., A. Sifeddine, D.B. Field, L. Ortlieb, G. Vargas, F. Chavez, F.P. Velazco, V. Ferriera, P. Tapia, R. Salvatteci, H. Boucher, M.C. Morales, J. Valdes, J.-L. Reyss, A. Campusano, M. Boussafir, M. Mandang-Yogo, M. Garcia, and T. Baumgartner. 2009. Rapid reorganization in ocean biogeochemistry off Peru towards the end of the Little Ice Age. *Biogeosciences* 6: 835–848.
Gutiérrez, D., I. Bouloubassi, A. Sifeddine, S. Purca, K. Goubanova, M. Graco, D. Field, L. Méjanelle, F. Velazco, A. Lorre, R. Salvatteci, D. Quispe, G. Vargas, B. Dewitte, and L. Ortlieb. 2011. Coastal cooling and increased productivity in the main upwelling zone off Peru since the mid twentieth century. *Journal of Geophysical Research* 38: L07603. doi:10.1029/2010GL046324.
Hales, B., T. Takahashi, and L. Bandstra. 2005. Atmospheric CO_2 uptake by a coastal upwelling system. *Global Biogeochemical Cycles* 19(GB1009): 2004G. doi:10.1029/B002295.
Hayden, B.P., G.C. Ray, and R. Dolan. 1984. Classification of coastal and marine environments. *Environmental Conservation* 11: 199–207.
Kämpf, J., M. Doubell, D. Griffin, R.L. Matthews, and T.M. Ward. 2004. Evidence of a large seasonal coastal upwelling system along the southern shelf of Australia. *Geophysical Research Letters* 31: L09310. doi:10.1029/2003GLO19221.
Kida, S., and K.J. Richards. 2009. Seasonal sea surface temperature variability in the Indonesian Seas. *Journal of Geophysical Research*, 114. doi:10.1029/2008JC005150.
Lathuilière, C., V. Echevin, and M. Lévy. 2008. Seasonal and intraseasonal surface chlorophyll-a variability along the northwest African coast. *Journal of Geophysical Research* 113: C05007. doi:10.1029/2007JC004433.
Levin, L.A. 2003. Oxygen minimum zone benthos: Adaptation and community response to hypoxia. *Annual Reviews of Oceanography and Marine Biolology* 41: 1–45.
Longhurst, A. 1998. *Ecological geography of the sea*. San Diego: Academic Press.
Mackas, D. L., P.T. Strub, A. Thomas and V. Montecino. 2006. Eastern ocean boundaries. In *The Sea*, vol. 14a, ed. A.R. Robinson and K.H. Brink, pp. 21–59. Harvard University Press.
Messié, M., J. Ledesma, D.D. Kolber, R.P. Michisaki, D.G. Foley, and F.P. Chavez. 2009. Potential new production estimates in four eastern boundary upwelling ecosystems. *Progress in Oceanography* 83(1–4): 151–158. doi:10.1016/j.pocean.2009.07.018.
Minas, H.J., M. Minas, and T.T. Packard. 1986. Productivity in upwelling systems deduced from hydrographic and chemical fields. *Limnology and Oceanography* 31: 1182–1206.

Monteiro, P.M.S. 2010. The Benguela current system. In *Carbon and nutrient fluxes in continental margins*, ed. K.-K. Liu, L. Atkinson, R.A. Quiñones, and L. Talaue-McManus, 65–78. Berlin: Springer.
Paulmier, A., D. Ruiz-Pino, and V. Garcon. 2008. The oxygen minimum zone (OMZ) off Chile as intense source of CO_2 and N_2O. *Continental Shelf Research* 28(20): 2746–2756.
Pennington, J.T., G.E. Friederich, C.G. Castro, C.A. Collins, W.W. Evans, and F.P. Chavez. 2010. The northern and central California coastal upwelling system. In *Carbon and nutrient fluxes in continental margins*, ed. K.-K. Liu, L. Atkinson, R. Quiñones, and L. Talaue-McManus, 65–78. Berlin: Springer-Verlag.
Quiñones, R.A., M.H. Gutierrez, G. Daneri, D.G. Aguilar, H.E. Gonzales, and F.P. Chavez. 2010. The Humboldt current system. In *Carbon and nutrient fluxes in continental margins*, ed. K.-K. Liu, L. Atkinson, R. Quiñones, and L. Talaue-McManus, 44–64. Berlin: Springer.
Reynolds, R.W., T.M. Smith, C. Liu, D.B. Chelton, K.S. Casey, and M.G. Schlax. 2007. Daily high-resolution-blended analyses for sea surface temperature. *Journal of Climate* 20(22): 5473–5496.
Richardson, K., A.W. Visser, and F.B. Pedersen. 2000. Subsurface phytoplankton blooms fuel pelagic production in the North Sea. *Journal of Plankton Research* 22: 1663–1671.
Roemmich, D., S. Riser, R. Davis, and Y. Desaubies. 2004. Autonomous profiling floats: workhorse for broad-scale ocean observations. *Marine Technology Society Journal* 38: 21–29.
Rykaczewski, R.R., and D.M. Checkley. 2008. Influence of ocean winds on the pelagic ecosystem in upwelling regions. *Proceedings of the National Academy of Sciences* 105: 1965–1970.
Ryther, J.H. 1969. Photosynthesis and fish production in the sea. *Science* 166(3901): 72–76.
Santana-Casiano, J.M., M. Gonzales-Davil, and I.R. Ucha. 2009. Carbon dioxide fluxes in the Benguela upwelling system during winter and spring. A comparison between 2005 and 2006. *Deep-Sea Research II* 46: 533–541.
Sherman, K, ed. 1993. *Emerging theoretical basis for monitoring the changing states (health) of large marine ecosystems*. Summary report of two workshops: 23 April 1992, National Marine Fisheries Service, Narragansett, Rhode Island, and 11–12 July 1992, Cornell University, Ithaca, New York. NOAA Technical Memorandum NMFS-F/NEC-100.
Sherman, K., and G. Hempel, eds. 2008. *The UNEP large marine ecosystem report: A perspective on changing conditions in LMEs of the world's Regional Seas*. UNEP Regional Seas Report and Studies No. 182. United Nations Environment Programme. Nairobi, Kenya.
Spalding, M.D., H.E. Fox, G.R. Allen, N. Davidson, Z.A. Ferdana, M. Finlayson, B.S. Halpern, M.A. Jorge, A. Lombana, S.A. Lourie, K.D. Martin, E. McManus, J. Molnar, C.A. Recchia, and A. Robertson. 2007. Marine ecoregions of the World: a bioregionalization of coastal and shelf areas. *BioScience* 57(7): 573–583.
Stramma, L., S. Schmidko, L.A. Levin, and G.C. Johnson. 2010. Oxygen ocean minima expansions and their biological impacts. *Deep-Sea Research I* 57: 587–595.
Sverdrup, H.U., M.W. Johnson and R.H. Fleming. 1942. *The oceans, their physics, chemistry, and general biology*. Prentice-Hall Inc.
Sydeman, W. J., M. García-Reyes, D.S. Schoeman, R.R. Rykaczewski, S.A. Thompson, B.A. Black and S.J. Bograd. 2014. Climate change and wind intensification in coastal upwelling ecosystems. *Science* 345, 77–80. doi:10.1126/science.1251635.
Takahashi, T., C. Sutherland, C. Sweeney, A. Poisson, N. Metzl, B. Tilbrook, N. Bates, R. Wanninkhof, R.A. Feely, C. Sabine, J. Olafsson, and Y. Nojiri. 2002. Global sea-air CO_2 flux based on climatological surface ocean pCO_2 and seasonal biological and temperature effects. *Deep-Sea Research II* 49: 1601–1622.
Thomas, A.C., M.E. Carr, and P.T. Strub. 2001. Chlorophyll variability in eastern boundary currents. *Geophysical Research Letters* 28(18): 3421–3424.
Torres, R., D. Turner, J. Rutllant, M. Sobarzo, T. Antezana, and H.E. Gonzalez. 2002. CO_2 outgassing off central Chile (31–30 S) and northern Chile (24–23 S) during austral summer 1997: the effect of wind intensity on the upwelling and ventilation of CO_2-rich waters. *Deep-Sea Research I* 49: 1413–1429.

van Vliet, M.T.H., W.H.P. Franssen, J.R. Yearsley, F. Ludwig, I. Haddeland, D.P. Lettenmaier, and P. Kabat. 2013. Global river discharge and water temperature under climate change. *Global Environmental Change* 23: 450–464. doi:10.1016/j.gloenvcha.2012.11.002.

Walsh, J.J. 1981. A carbon budget for overfishing off Peru. *Nature* 290: 300–304.

Wattenberg, H. 1938. Die Verteilung des Sauerstoffs im Atlantischen Ozean. *Wissenschaftliche Ergebnisse der Deutschen Atlantischen Expedition der Meteor, 1925–1927* 9: 1–132.

Xing, X., A. Morel, H. Claustre, D. Antoine, F. D'Ortenzio, A. Poteau, and A. Mignot. 2011. Combined processing and mutual interpretation of radiometry and fluorimetry from autonomous profiling Bio-Argo floats: Chlorophyll a retrieval. *Journal of Geophysical Research* 116: C06020. doi:10.1029/2010JC006899.

Yarincik, K., and R. O'Dor. 2005. The census of marine life: goals, scope and strategy. *Scientia Marina*, 69. doi:10.3989/scimar.2005.69s1201.

Index

© Springer International Publishing Switzerland 2016

J. Kämpf and P. Chapman, *Upwelling Systems of the World*,
DOI 10.1007/978-3-319-42524-5

The manufacturer's authorised representative in the EU is Springer Nature Customer Service Centre GmbH, Europaplatz 3, 69115 Heidelberg, Germany. If you have any concerns regarding our products, please contact ProductSafety@springernature.com

Printed and bound by CPI Group (UK) Ltd, Croydon, CR0 4YY
05/10/2026
02228127-0014

Springer Tracts in Advanced Robotics

Volume 13

Editors: Bruno Siciliano · Oussama Khatib · Frans Groen

W. Chung

Nonholonomic Manipulators

With 90 Figures

Springer

M
X RW 1461

Professor Bruno Siciliano, Dipartimento di Informatica e Sistemistica, Università degli Studi di Napoli Federico II, Via Claudio 21, 80125 Napoli, Italy, email: siciliano@unina.it

Professor Oussama Khatib, Robotics Laboratory, Department of Computer Science, Stanford University, Stanford, CA 94305-9010, USA, email: khatib@cs.stanford.edu

Professor Frans Groen, Department of Computer Science, Universiteit van Amsterdam, Kruislaan 403, 1098 SJ Amsterdam, The Netherlands, email: groen@science.uva.nl

STAR (Springer Tracts in Advanced Robotics) has been promoted under the auspices of EURON (European Robotics Research Network)

Author

Dr. Woojin Chung

Korea Institute of Science and Technology (KIST)
Intelligent Robotics Research Center
Sungbuk-ku, Hawolgok-dong 39-1
Seoul 136-791, Korea

Ausgesondert

Fachhochschule
Bonn-Rhein-Sieg

04/11133

Hochschul- und
Kreisbibliothek
Bonn-Rhein-Sieg

ISSN 1610-7438

ISBN 3-540-22108-5 Springer Berlin Heidelberg New York

Library of Congress Control Number: 2004108313

This work is subject to copyright. All rights are reserved, whether the whole or part of the material is concerned, specifically the rights of translation, reprinting, reuse of illustrations, recitation, broadcasting, reproduction on microfilm or in other ways, and storage in data banks. Duplication of this publication or parts thereof is permitted only under the provisions of the German Copyright Law of September 9, 1965, in its current version, and permission for use must always be obtained from Springer-Verlag. Violations are liable to prosecution under German Copyright Law.

Springer is a part of Springer Science+Business Media

springeronline.com

© Springer-Verlag Berlin Heidelberg 2004
Printed in Germany

The use of general descriptive names, registered names, trademarks, etc. in this publication does not imply, even in the absence of a specific statement, that such names are exempt from the relevant protective laws and regulations and therefore free for general use.

Typesetting: Digital data supplied by author.
Data-conversion and production: PTP-Berlin Protago-TeX-Production GmbH, Germany
Cover-Design: design & production GmbH, Heidelberg
Printed on acid-free paper 62/3020Yu - 5 4 3 2 1 0

Editorial Advisory Board

EUROPE

Herman Bruyninckx, KU Leuven, Belgium
Raja Chatila, LAAS, France
Henrik Christensen, KTH, Sweden
Paolo Dario, Scuola Superiore Sant'Anna Pisa, Italy
Rüdiger Dillmann, Universität Karlsruhe, Germany

AMERICA

Ken Goldberg, UC Berkeley, USA
John Hollerbach, University of Utah, USA
Lydia Kavraki, Rice University, USA
Tim Salcudean, University of British Columbia, Canada
Sebastian Thrun, Carnegie Mellon University, USA

ASIA/OCEANIA

Peter Corke, CSIRO, Australia
Makoto Kaneko, Hiroshima University, Japan
Sukhan Lee, Sungkyunkwan University, Korea
Yangsheng Xu, Chinese University of Hong Kong, PRC
Shin'ichi Yuta, Tsukuba University, Japan

To Yoola and my parents, I owe everything.

Foreword

At the dawn of the new millennium, robotics is undergoing a major transformation in scope and dimension. From a largely dominant industrial focus, robotics is rapidly expanding into the challenges of unstructured environments. Interacting with, assisting, serving, and exploring with humans, the emerging robots will increasingly touch people and their lives.

The goal of the new series of Springer Tracts in Advanced Robotics (STAR) is to bring, in a timely fashion, the latest advances and developments in robotics on the basis of their significance and quality. It is our hope that the wider dissemination of research developments will stimulate more exchanges and collaborations among the research community and contribute to further advancement of this rapidly growing field.

The monograph written by Woojin Chung is an evolution of the Author's Ph.D. dissertation. The work builds upon an increasing interest in nonholomic mechanical systems which have attracted several researchers in control and robotics. A key feature of the work is the possibility to exploit nonholonomic theory to design innovative mechanical systems with a reduced number of actuators without reducing the size of their controllable space. The volume offers a comprehensive treatment of the problem from the theoretical development of the various control schemes to prototyping new types of manipulators, while testing their performance by simulation and experiments in a number of significant cases.

One of the first focused books on nonholomic manipulators, this title constitutes a fine addition to the series!

Naples, Italy
April 2004

Bruno Siciliano
STAR Editor

Preface

Recently the nonholonomic mechanical systems have received much attention in the field of robotics and control engineering. The scope of this book is about the definitions and developments of new nonholonomic machines which are designed on the basis of nonlinear control theory for nonholonomic mechanical systems. So far, many useful research achievements had been accumulated for the kinematic analysis and development of control schemes for driftless nonholonomic systems. Control theoretic strategies based on the differential geometric framework have achieved remarkable progresses. However, previous works on the nonholonomic mechanical systems have been carried out mainly focusing on the development of control strategies for known existing nonholonomic machines.

The starting point of this book is to explore the possibility of innovative and useful mechanisms from the nonlinear control theoretic background. While previous publications have assumed the nonholonomic systems to have given and developed theory for these systems, this book points out a new direction where the nonholonomic theory is used to design controllable systems.

The specific goal of this study is to design and to control a manipulator which consists of n revolute joints using only two actuators, by exploiting the unique feature of nonholonomic systems. This fact will cause a revolutionary change in mechanical design, especially, when it is essential to reduce the number of actuators without reducing the dimension of the reachable configuration space.

This book is based on my Ph.D. dissertation written under the supervision of Professor Yoshihiko Nakamura at Department of Mechano-Informatics, the University of Tokyo. Professor Nakamura has inspired me to do my best. My personal development have benefited immensely from not only his scholarly attitude to research, but also his own enthusiastic manner to life.

I would like to thank Professor Ken-ichi Yoshimoto, who was in charge of Mechanisms and Control Laboratory, for guiding me through research efforts and fruitful discussions.

I would like to thank Doctor Ole Jacob Sørdalen for the part of this book, for leading me to interesting nonholonomic world and valuable comments and advices. I acknowledge Doctor Tetsuro Yabuta, Doctor Ken Tsujimura in NTT Co., Ltd., and Mr. Kazuhusa Noda, Mr. Osamu Muraki in Oshima Prototype Engineering Co., Ltd. for their supports and collaboration in building the prototypes. I gratefully acknowledge POSCO Scholarship Society for their financial support and good service during my stay in Japan.

I am very grateful to the research staffs and students of the Mechanisms and Control Laboratory in University of Tokyo. I am specially grateful to Hideaki Ezaki for his great contribution in building an experimental setup. I would like to thank colleagues and students at the Intelligent Robotics Research Center, Korea Institute of Science and Technology.

This research was supported in part by NTT Co., Ltd., Japan Society of the Promotion of Science, the Center of Maritime Control Systems at NTH/SINTEF and the Grant in Aid of Scientific Research from the Ministry of Culture, Sports and Education (General Research (b)04452153 and (b)07455110).

I wish to thank Professor Bruno Siciliano and Dr. Thomas Ditzinger for their patience and support during the preparation of the manuscript. I would like to thank my student Myoungkuk Park and my friends Hansung Kim and Jaeheum Han for their support.

Finally, I am very grateful to my parents for their encouragement and support throughout my life, and especially to my wife Yoola Shin, who is always a friendly partner.

Seoul, Korea
April 2004

Woojin Chung
Korea Institute of Science and Technology

Contents

1

Introduction

1.1 Nonholonomy

Physical systems are classified into linear and nonlinear ones. Nonlinear systems can be further divided into holonomic and nonholonomic systems. *Holonomy* and *nonholonomy* are key concepts for such classification and describes the essential difference of behaviors of physical systems. In this section, definitions and several examples of the holonomic and the nonholonomic systems are presented.

If a constraint is written by the following form, it is called *holonomic.*

$$f(q,t) = 0 \tag{1.1}$$

where q is the generalized coordinate of the system and t represents time. *Generalized coordinates* are used to locate a system with respect to a reference frame. A set of generalized coordinates may include cartesian or spherical coordinate, but also include lengths or angles which can be chosen conveniently to describe the system. A set of generalized coordinates is said to be *complete* if the values of the coordinates corresponding to an arbitrary geometrically admissible configuration of the system are sufficient to fix the location of all parts of the system [13]. In this book, the term *configuration variable* will be alternatively used as the same concept as the generalized coordinate. For the robot manipulator consisting of n revolute joints, a set of joint angles θ_i, where $i \in \{1, \cdots, n\}$, are the generalized coordinates.

If all the constraints of the system are holonomic, the system is said to be holonomic. Equation (1.1) always reduces the dimension of the system, *i.e.*, if there are m holonomic constraints of q_i, where $i \in \{1, \cdots, n\}$, the number of independent coordinates are $n - m$.

A well-known example of holonomy is a system consisting of mass particles. For a single particle, there are three generalized coordinates to specify current configuration. If another particle is connected to it with a rigid bar, then there are six coordinates, one holonomic constraint. (*i.e.* The length of a bar doesn't change.) Therefore the number of configuration variables is five. In the same way, there are nine coordinates, three holonomic constraints and six configuration variables for a rigidly connected three-particle system. Consequently, there are six generalized coordinates to describe the configuration of a rigid body consisting of n particles, where $n \geq 3$. This result illustrates the reason why six components are needed to describe rigid body motions.

When a conventional robot manipulator is said to be holonomic, it explains the geometric constraints between joint angles and the configuration of the end-effector. If the end-effector location and orientation are denoted by Y, the constraint is described as $Y = F(X)$ which is the form in eq. (1.1), where X is a set of generalized coordinates representing joint configurations. It is obvious that steering the holonomic system requires as many inputs as the number of independent configuration variables.

If a constraint cannot be expressed in the form of eq. (1.1), it is *nonholonomic*. For instance, an inequality condition is nonholonomic. A typical example is the position of a molecule moving inside a soccer ball. The constraint is represented as $x^2 + y^2 + z^2 \leq r^2$, where $\{x, y, z\}$ is a position and r is a radius of the ball.

The nonholonomic constraints of mechanical systems are expressed by the following equation in most cases.

$$f(q, \dot{q}, \ddot{q}, t) = 0 \tag{1.2}$$

If the constraint in eq. (1.2) cannot be reduced to the form of eq. (1.1) (*i.e.* if it is nonintegrable), the system is nonholonomic.

The constraint of the following equation will be considered in this book:

$$f(q, \dot{q}) = 0 \tag{1.3}$$

Equation (1.3) is a kinematic constraint while eq. (1.2) represents a dynamic constraint. Control issues of nonholonomic systems under dynamic constraints can be found in [6].

Equation (1.3) can be understood as a velocity constraint of the system at the given configuration. The nonintegrable constraint does not necessarily reduce the number of generalized coordinates. There arises a possibility to steer the generalized coordinates using less number of inputs, which is unlikely in the holonomic systems. A system whose control vector has a lower dimension than the configuration vector is called a *underactuated* system.

One of the examples is a two-wheeled mobile robot as shown in fig. 1.1. Its kinematic model is given as follows:

$$\begin{bmatrix} \dot{x} \\ \dot{y} \\ \dot{\theta} \end{bmatrix} = \begin{bmatrix} cos\theta & 0 \\ sin\theta & 0 \\ 0 & 1 \end{bmatrix} \begin{bmatrix} v \\ \omega \end{bmatrix} \tag{1.4}$$

Eliminating an input v in eq. (1.4) leads to the following nonholonomic constraint.

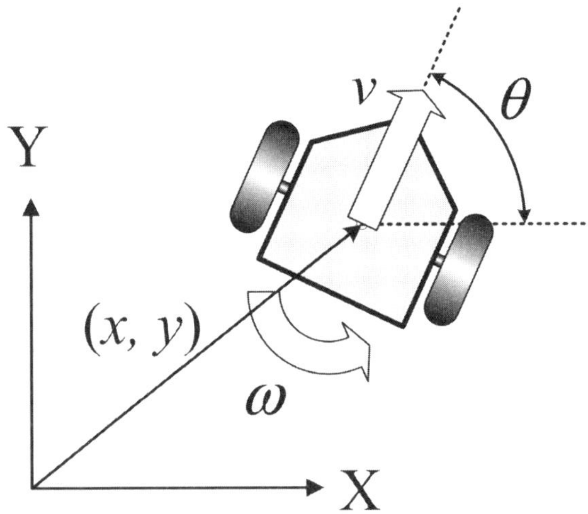

Fig. 1.1. A two-wheeled mobile robot.

$$\dot{x}sin\theta - \dot{y}cos\theta = 0 \tag{1.5}$$

Equation (1.5) is nonintegrable and implies no-side slip condition under the assumption of rolling contact. There are only two inputs in total, but we know from our daily experiences that a simple car like eq. (1.5) is controllable in the three dimensional configuration space.

In holonomic systems, the number of degrees of freedom is same to the number of generalized coordinates. However, it is no longer valid for nonholonomic systems. *The number of degrees of freedom* is defined as the number of coordinates which are used to specify the configuration of the system subtracted by the number of independent equations of constraint [19]. Therefore, the number of degrees of freedom of the two-wheeled mobile robot is two.

It can be easily checked that the first order linear approximation of the nonlinear system in eq. (1.4) is uncontrollable. Therefore, nonlinear approaches should be taken to control the system.

Constraints are further classified according as the equation of constraint contain time as an explicit variable, which is called rheonomous, or not explicitly dependent on time, which is scleronomous [18, 19]. Constraints in eqs. (1.1) and (1.2) are written as rheonomous for generality.

By exploiting nonlinearity caused by rolling contacts, various mechanisms have been designed. One of such applications is the kinematic integrating machine. Fig. 1.2 is the Henrici-Coradi harmonic analyzer [54]. Another model of harmonic analyzer was studied by [58]. If we move the tip along the function $y = f(x)$, then the motion of the tip causes rotations of the wheels which are in contact with the balls due to rolling contacts. The resultant rotation angle of a wheel corresponds to the coefficient of Fourier series expansion of $f(x)$. That is synthesized by the nonlinear mechanical structures designed in it. The

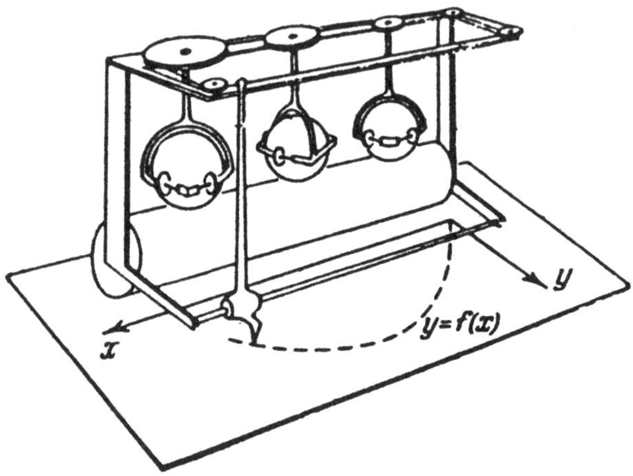

Fig. 1.2. The Henrici-Coradi harmonic analyzer [54].

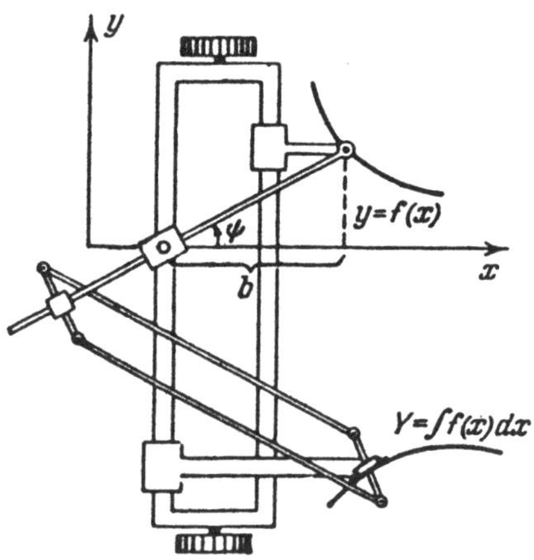

Fig. 1.3. The Abdank-Abakanowicz integraph [54].

harmonic analyzer in fig. 1.2 seems somewhat mysterious but the mechanism can be easily understood using the kinematical analysis in section 2.2. Fig. 1.3 shows the Abdank-Abakanowicz integraph, which performs integration of given function $f(x)$ using rolling contact and special linkage design.

In this book, we develop theories to design and to control nonholonomic underactuated manipulators and to build their prototypes. The nonholonomic gear is a key component of designing underactuated manipulators. It is a special type of the velocity transmission and utilizes rolling contact between a ball and wheels. Its principle of the velocity transmission will be presented in section 2.2.

1.2 Previous and concurrent works

1.2.1 Various examples of nonholonomic systems

One of the typical examples of nonholonomic constraints is the motion of rigid bodies with rolling constraint. Li and Canny [34] discussed controllability of the two spheres in contact. Under the assumption of no relative motions with respect to the surface normal axis at the contact point, it was shown that the two spheres can be reoriented with rolling, if two spheres have different radii. A motion planning scheme for a sphere rolling on the plane was proposed.

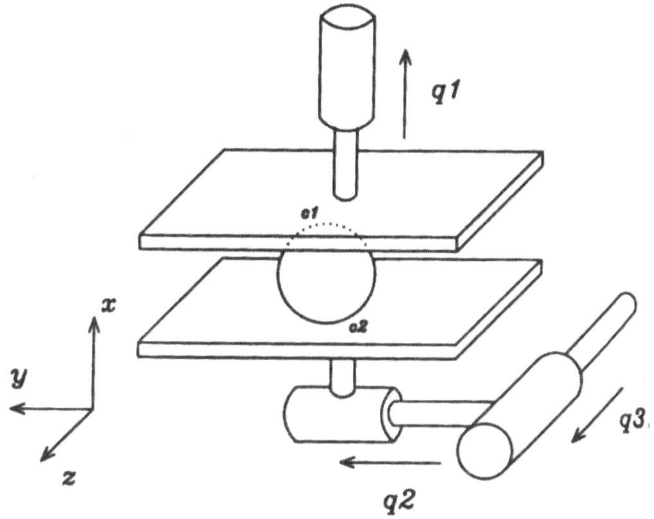

Fig. 1.4. Dexterous manipulation by by rolling in [5].

Bicchi and Sorrentino [5] studied controllability of the simple mechanism consists of two planar finger and an object. Fig. 1.4 shows the experimental setup. Under the assumption of regularity and convexity, a tracking control scheme in two dimensional subspace out of 5 dimensional configuration space was successfully tested on a spherical object. This work was extended to the irregular object by Marigo, Chitour and Bicchi [40]. Reachable configurations of the polyhedral parts by rolling are studied and a path finding algorithm was presented.

Manipulating an object by pushing on a plane is an interesting nonholonomic problem and was studied by Lynch and Mason [37, 36]. Motions are limited by the applicable force and friction, therefore controllability can be referred to as a nonholonomic problem of positioning and reorienting an object. Fig. 1.5 shows a result of successful motion planning. A path planning

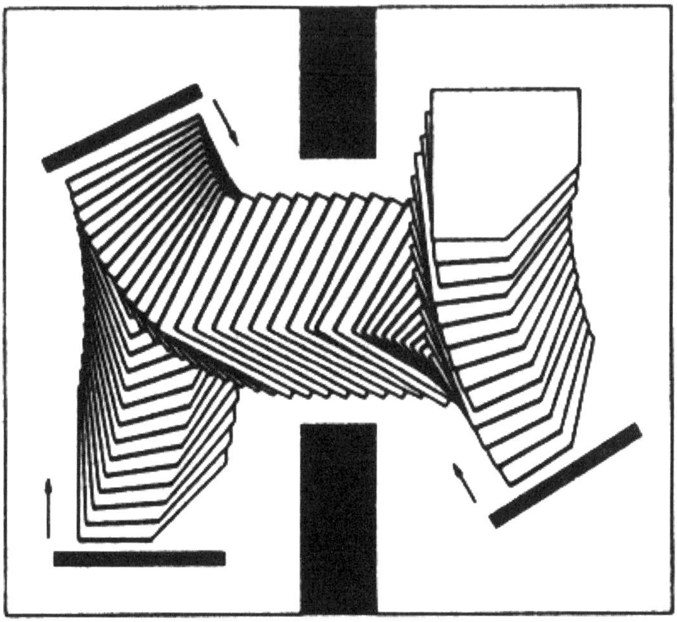

Fig. 1.5. Manipulating an object by pushing in [37].

problem of a disk among the obstacles was studied by Agarwal, Latombe, Motwani and Raghavan [1].

Angular momentum conservation law of a free-flying space robot is a nonholonomic constraint. For a spacecraft equipped with a robot manipulator, nonintegrability enables reorienting a robot just using an internal motion, which is different from an orientation control with control input such as thrusters, for example, see [25]. Under the same framework, motion control of a falling cat can be discussed.

Nakamura and Mukherjee [49] proposed a control scheme of a space robot using a bidirectional approach. A space robot can be stabilized to an equilibrium manifold instead of a stationary goal point. Two space robots, one is at the start configuration and the other is at the goal configuration, were controlled to the manifold based on Lyapunov's direct method. Complete construction of path planning was made by combining feedback and open loop control. The concept of nonholonomic redundancy was presented also in [49] and it was further studied in [50]. Nonholonomic redundancy implies that there are many possible alternative configurations satisfying the same boundary conditions of the end-effector. Nonholonomic redundancy was shown to be useful to avoid joint limits and obstacles. Nakamura and Suzuki [82, 52] successfully planned the end-effector motion without changing the orientation of the base body using spiral motions.

Underwater vehicles may have less control inputs than the dimension of the reachable configuration space. Nakamura and Savant [51] proposed a globally stable nonlinear tracking control law of the autonomous underwater vehicle. Under the non-zero reference velocity assumption, 6 generalized coordinates can be controlled using only 4 inputs.

Designing new mechanisms using the nonholonomic constraints is carried out in various way.

Peshkin, Colgate, Moore and Wannasuphoprasit [59, 11] proposed a passive robot which can display programmable constraints. A continuously variable transmission, which is similar to the nonholonomic gear in section 2.2, couples two velocities by an adjustable ratio. This study is a good example that nonholonomic constraints can be used to create new helpful mechanical systems, which is a major goal of this book. Wannasuphoprasit, Gillespie, Colgate and Peshkin [95] recently proposed a passive mobile robot whose path is programmable using the rolling constraints between tires and the ground.

Another work of using nonholonomic constraints to the mechanical design was done by Luo and Funaki in [35]. A cartesian nonholonomic table can be driven by two actuators and spherical elements, which is also similar to the nonholonomic gear. The implementation of the table was made and experimentally proved to be useful in [39].

A number of studies have been carried out for nonholonomic systems including those introduced in this section. Works related to wheeled mobile robots and its applications are focused on mainly because they are most directly related to the study of this book.

1.2.2 Open loop strategies

Laumond [27] proved that a car-like robot is controllable even when the steering angle is limited, which can be described as an inequality constraint. This work was extended by Laumond and Siméon [32] for the mobile robot towing one trailer, by showing the existence of path connecting initial and final configurations. Controllability of a mobile robot with n trailer was shown in [28, 29]. There have been a number of motion planning schemes on the specific low dimensional model based on a heuristic approaches, for example, [14, 62, 33].

On the other hand, control theoretic approaches were made based on differential geometric framework for general nonholonomic systems. Murray and Sastry [47] proposed a chained form, which is a canonical form for a class of driftless nonholonomic systems including a mobile robot with n trailer system. In [46, 47], sinusoidal inputs at integrally related frequencies were used to steer a chained form system, based on Fourier series techniques. By applying the proposed scheme, chained form variables are steered to their desired values step by step. Sufficient conditions for chained form conversion were presented,

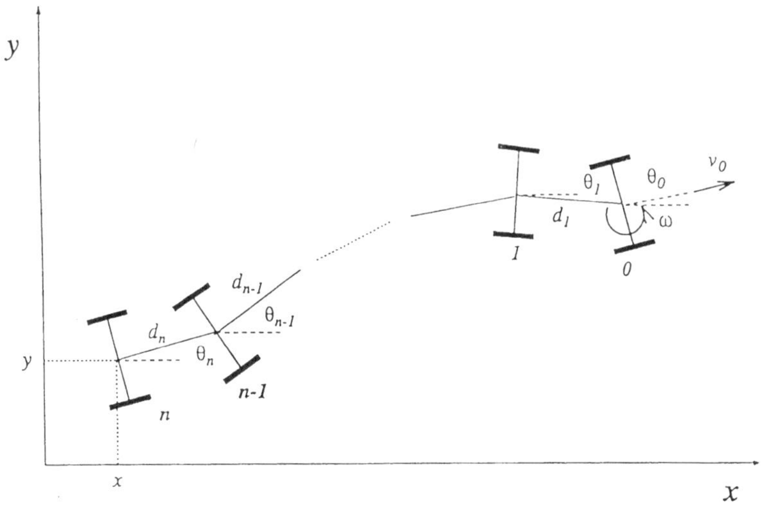

Fig. 1.6. Configuration of a car with n trailers in [73].

which were applicable for a car with one trailer system. The algorithm failed when additional trailers were added.

Sørdalen [72, 73] solved this problem successfully by putting the coordinate frame on the last trailer and modifying kinematic equations. Appropriate transformation of coordinates and inputs enabled chained form conversion locally in orientation and globally in position. Fig. 1.6 illustrates a car with n trailer system. Owing to the simple structure of chained form, many useful control schemes were developed and they can be applied to control systems which satisfy chained form convertibility.

Tilbury, Murray and Sastry [88, 89] proposed useful path planners for the chained form. Instead of time-consuming sinusoids in [47], sinusoidal inputs which can steer a system all at once were presented. Time polynomial can be an alternative choice. Both methods are common in which inputs are given by solving algebraic equations under the input parameterization.

Monaco and Normand-Cyrot [44] proposed an algorithm of steering nonlinear systems using multirate digital control under the admissible boundary conditions. Piecewise constant inputs can steer chained system exactly because chained system is nilpotent. It was further studied by Giamberardino,Monaco and Normand-Cyrot [16] for the feedback nilpotenzable system.

Lafferriere and Sussmann [26] proposed a steering algorithm for driftless systems. By extending a system with higher order Lie brackets of inputs, an arbitrary path in the configuration space can be generated. It gives exact solutions for nilpotent systems and approximate for non-nilpotent systems. However, it is difficult to be implemented for high dimensional systems because concatenations of piecewise constant inputs were used to generate higher order

Lie bracketing motions. Sussmann and Liu [81] used high frequency sinusoids to approximate any path by admissible trajectories. If the input frequency goes to infinity, planned trajectory converges to the desired trajectory. Tilbury, Laumond, Murray, Sastry and Walsh [87] made a synthesis of [81] and [47] for an example of a mobile robot with two trailers.

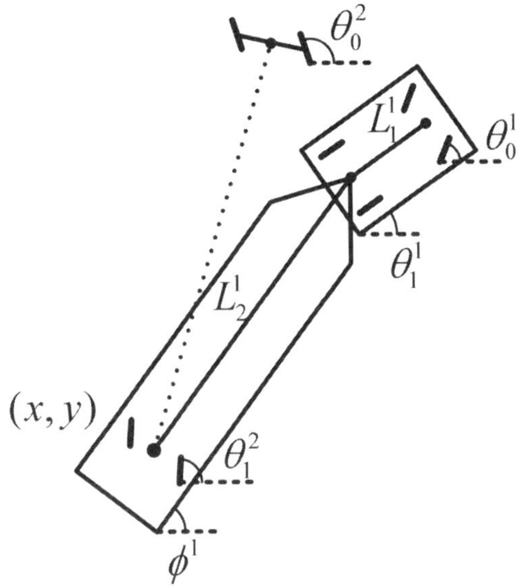

Fig. 1.7. A sketch of the fire truck system in [90].

In addition to a car with n trailer problem, a vehicle train consists of several steerable and passive trailers was studied by Tilbury, Sørdalen, Bushnell and Sastry [90, 91]. In order to convert the system into multi-input chained form, dynamic extension by adding the virtual axles was carried out, which results in a dynamic feedback. A simplest example is the fire truck which has a steerable wheel on the trailer as shown in fig. 1.7.

Goursat normal form [8, 45, 89] can be thought as another canonical form. Main advantage of using Goursat normal form is that conditions for conversion is straightforward and there exists an algorithm for finding necessary coordinate transformation. Since Goursat normal form can be converted to the chained form, controllers to the chained form are applied in practice.

Power form [41, 17] , which is dual of the chained form, is another canonical form. Since power form can be transformed to the chained form, there are no fundamental difference. As stated by Samson [70], two forms present complementary advantages and drawbacks. A kinematic model of a car with n trailer system represented by cartesian coordinates is closer to the chained form and it is the reason why chained form systems are mainly considered in this book.

1.2.3 Closed loop strategies

Feedback control of a driftless control system is known to be difficult because of the limitations given by Brockett's theorem [7]. In [7], necessary conditions for the existence of continuous static state feedback, which makes the origin asymptotically stable, are presented. Since driftless nonholonomic systems considered in this book does not satisfy the condition, a controller stabilizing to a stationary point should be time-variant, discontinuous or combination of the two. Other approaches are to design a controller which stabilize a system to a manifold or moving target point(*i.e.* path tracking), instead of stabilizing to the stationary point.

Walsh, Tilbury, Sastry, Murray and Laumond [93, 94] proposed a control law which stabilizes a system to a desired feasible trajectory, instead of a point. By computing a linearization of the system about the nominal trajectory, linear time-varying feedback law [10] can be applied. A simulation result shows that a four-wheeled car can be easily steered to a desired circular path.

Sampei [64, 67] proposed time-state control form and feedback control strategies. A class of driftless nonholonomic systems (including chained form) can be converted to the time-state control form by appropriate coordinate and input transformations. By separating one variable from the original system and driving it monotonically, subsystem can be stabilized using smooth, static feedback, because linearized subsystem is controllable. Roughly speaking, this controller stabilizes the system to an one dimensional manifold in the configuration space. As examples, control of a simple space robot [66], a four-wheeled mobile robot [69] and position control of a ball between two parallel plates [68] were presented. Since one variable should be controlled separately, it is difficult to specify the convergence rate.

Imura, Kobayashi and Yoshikawa [21, 22] proposed time-varying stabilizer for a chained form to a reference path which converges to a desired point exponentially. Using this approach, the number of switching points can be given explicitly.

Many studies have been carried out on designing time-varying controller.

Coron [12] showed driftless systems can be stabilized using smooth, periodic, time-varying feedback.

Pomet [61] proposed a systematic procedure to construct a time-varying controller which makes the origin asymptotically stable. Explicit controller design was presented using the classical Lyapunov approach.

One of the drawbacks of [61] was the slow convergence rate because smooth controller shows just polynomial convergence. In order to solve it, Pomet, Thuilot, Bastin and Campion [60] proposed a hybrid strategy to control a cart. To obtain a fast convergence rate, time-invariant feedback was applied outside the neighborhood of the origin, then time-varying control was applied inside a neighborhood.

A time-varying controller to the chained form was proposed by Samson [70]. Samson's controller is designed based on explicit Lyapunov technique using the Lyapunov-like function which excludes one chained form variable z_1. A linear coordinate transformation to convert chained form to the skew-symmetric chained form was presented to guarantee asymtotic stability, even input $v_1(t)$ passes through zero. Proposed controller was shown to be useful for both path-following and point stabilization for the example of the car with trailers.

Teel, Murray and Walsh [86] proposed a time-varying controller which globally asymptotically stabilizes a system in power form, which is equivalent to the chained form. It utilizes sinusoids at integrally related frequencies, which are related in part to [47, 81].

Another approach to design a point stabilizing controller is to use discontinuous feedback. A major disadvantage of smooth controller is that it cannot have an exponential convergence [20].

Canudas de Wit and Sørdalen [9] proposed a piecewise continuous, exponentially convergent feedback for a two-wheeled (i.e. no lower bound on its turning radius) mobile robot. Coordinate transformations using a circle was a powerful tool in the analysis of convergence properties and controller design. This method was extended to the path following problem [75] with the same convergence rate. This work was extended to the chained form by Sørdalen and Egeland as a time-varying, piecewise smooth, κ-exponentially convergent feedback [76, 77]. Application to control of a car with n trailer problem was presented in [79]. A coordinate transformation to shift an equilibrium point to the arbitrary desired point was proposed also in [79]. The controller switches feedback gain at discrete instants of time and smooth between switching instants.

Astolfi [2, 3, 4] proposed a time-invariant, discontinuous feedback with exponential convergence for a mobile robot and chained form systems. Under the uniform convergence of z_1, all the other variables can be stabilized using the technique similar to gain scheduling. Singularity occurs at $z_1 = 0$ but it can be avoided by appropriate selection of initial conditions or intermediate points.

Narikiyo [53] also proposed static, discontinuous feedback excluding origin. For the unicycle model, it is same to Astolfi's. However, since the way of controller design is different, it may have different forms for the high dimensional model.

M'Closkey and Murray [42, 43] proposed a systematic procedure to extend smooth, time-varying asymptotic stabilizers to homogeneous, exponential stabilizers. Feedback is continuous, smooth away from the origin and non-Lipschitz to achieve ρ-exponential convergence rate. It can be extended to dynamic, torque input system instead of velocity inputs.

1.3 Contributions and outline of this book

Recently the nonholonomic mechanical systems have received much attention in the field of robotics and control engineering as described in the previous section. As illustrated in section 1.1, nonholonomic systems in robotics can be mainly divided into two categories according to the constraint equations. One is a kinematic constraint and the other is a dynamic constraint. A system under the kinematic constraint is called first order nonholonomic system, which is mainly considered in this book. The constraint equations are represented by a function of generalized coordinates and first time derivative of the coordinates.

The scope of this book is the definitions and developments of new nonholonomic machines which are designed on the basis of nonlinear control theory for nonholonomic mechanical systems. So far, many useful research achievements were accumulated for the kinematic analysis and development of control schemes for driftless nonholonomic systems. Control theoretic strategies based on the differential geometric framework achieved remarkable progresses. However, previous works on the nonholonomic mechanical systems have been carried out mainly focusing on the development of control strategies for known existing nonholonomic machines.

The starting point of this study is to explore the possibility of innovative and useful mechanisms from the nonlinear control theoretic background. While previous publications have assumed the nonholonomic systems are given and developed theory for these systems, this book points out a new direction where the nonholonomic theory is used to design controllable systems. The new mechanisms have several advantages which were unobtainable with conventional mechanism design.

Nonholonomic systems are typically controllable in a configuration space of higher dimension than the input space. This fact implies that a large number of generalized coordinates can be controlled by a few control inputs. The specific goal of this study is to design and to control a manipulator which consists of n revolute joints using only two actuators, by exploiting the unique feature of nonholonomic systems. This fact will cause a revolutionary change in mechanical design, especially, when it is essential to reduce the number of actuators without reducing the dimension of the reachable configuration space.

The outline and major contributions of this book are as follows:

1. **Design of the nonholonomic manipulator**
   It is shown how a manipulator with n revolute joints can be controlled by using only two actuators and a special type of mechanical transmissions. All the joints are passive and there are no devices which require additional control inputs. The nonhonolomic manipulator does not just imply reducing the number of actuators. A manipulator driven by a hydraulic actuator with several valves can be imagined as an example of a

system with one actuator. However, it is essentially different from the nonholonomic manipulator, because the hydraulic actuator system requires as many control inputs to the manipulator joints as the conventional robot manipulator. The number of control inputs of the nonholonomic manipulator is always two. Therefore, it is a completely new underactuated mechanism. The nonholonomic gear is designed to transmit input angular velocities to manipulator joints from the actuator which is located at the base. The nonholonomic gear is a special type of velocity transmission whose gear ratio is represented by trigonometric functions. Nonholonomic kinematic constraints are created by rolling contacts between a ball and wheels, which transmit angular velocities. For the nonholonomic manipulator, the nonholonomic gear is used as a single input, multiple output CVT (Continuously Variable Transmission). A unique mechanical design is presented, and the nonlinear kinematic model is derived. Then, the designed nonholonomic manipulator is shown to be completely controllable. Conversion of the kinematic model to the chained form is derived under certain conditions which can be easily checked. Once the kinematic model can be represented by the chained form, then existing various control laws for the chained form can be used to stabilize or to steer the manipulator to any given configuration in the joint space.
This part of work is described in chapter 2.

2. **Prototyping and control of the nonholonomic manipulator**
The nonholonomic manipulator was theoretically designed from the viewpoint of kinematic constraints and nonlinear control. There are significant issues on mechanical implementation and prototyping. Prototyping was carried out on the basis of mechanical and control point of view. Various important issues in mechanical design are discussed and the analysis and computation of driving force of the nonholonomic gears are presented. Since the nonholonomic manipulator is designed to satisfy chained form convertibility, any of the proposed controllers to the chained form can be applied. Nonlinear controllers including both open-loop ones and feedback ones are successfully applied to the prototype, which illustrate the usefulness of design of the nonholonomic manipulator. Some practical control strategies for real implementations are proposed. A proposed time scaling is a powerful tool to deal with actuator saturations or high joint accelerations. By using the proposed time scaling scheme, any of the existing open-loop controller can be applied in order to generate physically feasible motions. A practical feedback control strategy for a chained system by a pseudo-linearization is proposed. It does not require large number of switching stages and the equations of control law are extremely simple. One generalized coordinate of the kinematic model is specially included as

an internal parameter for control. It is shown that the parameter greatly contributes to the application of pseudo-linearization.
This part of work is described in chapter 3.

3. **Design of the chained form manipulator**
It is desirable to design a mechanism taking account of the available control strategies. Therefore, designing a system which satisfies chained form convertibility is significant. Although there are other canonical forms such as the Goursat normal form or the power form, the chained form is close to real physical systems and control laws are most easily developed. It is the reason why the scope of this study is limited to the chained form. There are many possible alternatives of designing controllable underactuated manipulators with two actuators using the nonholonomic gear. The nonholonomic manipulator was designed focusing on the mechanical simplicity. As a result, several problems were faced upon, when we tried to control a manipulator with a large number of joints. Control performances are greatly affected by mathematical properties of chained form conversion equations, which are highly dependent on the mechanical structure of the system. Such problems are clarified by numerical computation results. Solving mechanical and mathematical problems is the goal of a new design. The chained form manipulator is successfully designed to satisfy control simplicity. Although we have to add a little complication to mechanical structures, the chained form manipulator has a well-conditioned kinematic model even for the high dimensional model. Mechanical design of the chained form manipulator is also established. Simulation results show that the new design satisfies control simplicity for the high dimensional model.
This part of work is described in chapter 4.

4. **Prototyping and control of the chained form manipulator**
For the experimental verification of the chained form manipulator, a prototype of the chained form manipulator, which consists of five joints, is fabricated. If mechanical capacities of actuators and nonholonomic gears are sufficiently high enough to bear inertial torques, there is no limit on the number of joints. Although the presented prototypes of underactuated manipulators are planar structure, it can be easily extended to a spatial structure if the frictional force at the nonholonomic gear is sufficiently large enough to overcome gravity. The efficient motion planning scheme is developed based on the sinusoidal inputs to the chained form. Holonomic reference path of robot manipulator can be generated using a conventional robot motion planner. Then, the motion is approximated by feasible nonholonomic path for the chained form manipulator by applying the proposed scheme. Smooth and efficient motions are obtained

in simulations. It is clarified that the significant factor to improve actual performance is to consider sensitivity of the planned motion with respect to positional errors. Although the open loop controller cannot compensate various errors or disturbances, the effect of errors can be reduced through the presented analysis. The investigation of the effect of positional error is specified through the computed trajectories using sinusoidal inputs under the existence of the initial error of joint angles. The result of analysis clearly provides a solution for the problem how the sensitivity can be reduced. Presented experimental result shows that the decreased sensitivity greatly contributes to actual performances.
This part of work is described in chapter 5.

Chapter 6 concludes the whole work and provides some remarks.

This book presents two types of underactuated manipulators. The two underactuated manipulators have common features as follows:

1. Both manipulators are under kinematic nonholonomic constraints. Nonholonomic gears are exploited in order to create nonholonomic constraints.
2. Nonlinear kinematic models can be converted to chained form.
3. Only two actuators are required in order to drive n revolute joints.

However, kinematic models are totally different. The kinematic model of the chained form manipulator is well-conditioned, which provides significant advantages in order to improve control performance. Chained form conversion equations and nonlinear kinematic model remains simple regardless of the number of joints. This fact guarantees good control performance of the chained form manipulator. The kinematic well-condition implies that the mapping between the chained form variables and joint angles are simple and straight forward. Therefore, manipulator motion can be easily estimated and controlled by designing appropriate controller for the chained form variables. For the case of a nonholonomic manipulator, joint trajectories were highly oscillatory even if the computed trajectory in the chained form variable space was smooth. A kinematic ill-condition of a nonholonomic manipulator does not imply that it is an extreme example of poor design. A car with n trailer system also possesses ill-conditioned conversion equations. Therefore, it is difficult to control with a high dimensional model. In order to achieve well-conditioned kinematic model, a mechanical structure is totally redesigned in the chained form manipulator. In addition, joint actuating torques can be efficiently delivered from an actuator to joint, by using the improved mechanical structure of the chained form manipulator.

2

Design of the nonholonomic manipulator

2.1 Introduction

Recently the nonholonomic mechanical systems have received much attention in the field of robotics and control engineering. Nonholonomic systems are typically controllable in a configuration space of higher dimension than the input space. Therefore, behaviors of the nonholonomic systems are often found to be rich, even if there are a few control inputs.

Previous works on the nonholonomic mechanical systems have been carried out mainly focusing on the analysis and development of control strategies. In this chapter, design of an innovative and useful mechanism is explored from the nonlinear control theoretic viewpoint. Nonholonomic constraints are exploited to design a n-joint manipulator which can be controlled by using only two actuators. This is achieved by introducing a nonholonomic gear consisting of a ball and wheels at the joints instead of actuators or any other controlled devices like brakes or clutches. Nonholonomic gears transmit velocities from the actuator inputs to the passive joints.

It is shown that the resulting kinematic model is globally, completely controllable. The system is designed with a triangular structure, then a conversion into chained form is determined straightforward. Once the system is on a chained form, previously proposed control laws can be used to control the manipulator. For example, to obtain asymptotic stability about a given configuration in the joint space, stabilizers for chained form like [61, 86] can be used, or [76] to obtain exponential convergence. Path planners for chained form can be found in [88].

For some applications, where only limited resources are available, it would be essential to reduce the number of actuators and to reduce the cost and weight without reducing the reachable configuration space.

2.2 The nonholonomic gear

Nonholonomic constraints of the wheeled mobile robots are due to the rolling contacts between tires and the ground. To drive the passive joints exploiting nonholonomic constraints, it is needed to contrive some nonlinear mechanical elements inside the robot manipulator. In order to transmit velocities we design a gear as illustrated in fig. 2.1.

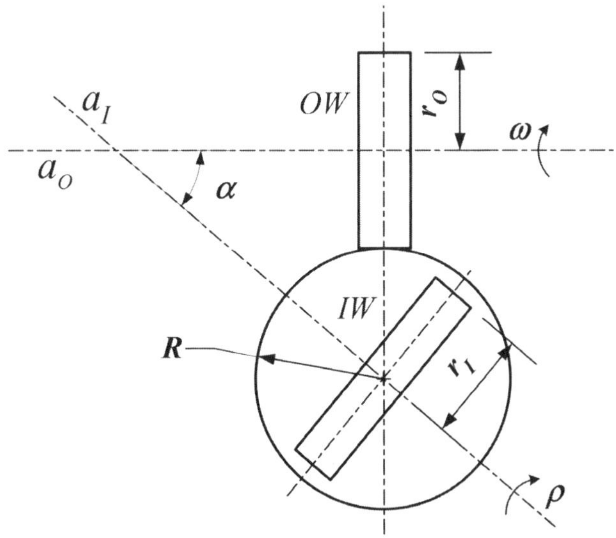

Fig. 2.1. Illustration of the nonholonomic gear seen from above.

The basic components of this gear are a ball with radius R and two wheels IW and OW, indicating Input Wheel and Output Wheel, with radii r_I and r_O. The velocity constraints of the ball are only due to point contacts with the wheels. Wheel IW rotates about a fixed axis a_I with a given angular velocity ρ which makes the ball rotate. Point contact without slipping is assumed between wheel IW and the ball. Wheel OW is driven by the ball and rotates about an axis a_O. Point contact without slipping is assumed. Wheel IW is in contact with the ball at the north pole, while wheel OW is on the equator such that its axis a_O lies in the plane of the equator. The angle between a_I and a_O is denoted by α.

When a ball and a wheel make a contact and rotate without slipping, the rolling-without-slipping constraint is not sufficient to uniquely determine the rotation of the ball. The rotation of the ball has two degrees of freedom, and the axis of rotation of the ball can lie in a plane involving the wheel axis and center of the ball. We call such a plane a constraint plane. If the ball makes contact with two wheels with different constraint planes, the rotation of the ball has only one degree of freedom. When the axes of the wheels are fixed, the ball rotates about the intersection of the two constraint planes. At this configuration, the ball rotates about axis A which is parallel to a_I and passes through the center. When wheel IW rotates with ρ, the angular velocity Ω of the ball about axis A is given by

$$\Omega = -\frac{r_I}{R}\rho \tag{2.1}$$

The angular velocity ω of wheel OW is then given by

$$\omega = -\Omega\frac{R}{r_O}\cos\alpha \tag{2.2}$$

Combining eqs. (2.1) and (2.2) yields

$$\omega = \rho\frac{r_I}{r_O}\cos\alpha \tag{2.3}$$

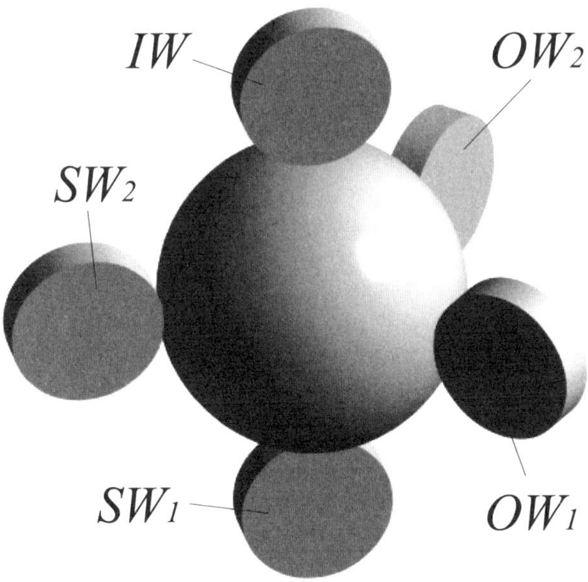

Fig. 2.2. The nonholonomic gear at joint i with supporting wheels.

In addition to these three main components, additional supporting mechanisms are needed to maintain the contact between the ball and the wheels, and to assure the rotation of the ball. The analysis of constraint planes provides a suggestion to design supporting mechanism of the ball. The mechanism should support the ball without adding further constraints. A wheel that shares the constraint plane with OW, for example, adds no additional constraints as long as its axis does not move relative to the axis of OW. Therefore, we can add any number of wheels on the equator by fixing their axes to the constraint plane of OW and to the axis of OW. This fact implies a nonholonomic gear can have multiple output wheels. Similarly, a wheel that shares the constraint plane with IW, located on the south pole can be added.

Fig. 2.2 shows a 3-dimensional view of a nonholonomic gear at joint i with supporting wheels. SW indicates Supporting Wheel. The rotation axes of IW and SW_1 are fixed to link $i-1$. The rotation axes of OW_1,OW_2 and SW_2 are fixed to link i. Therefore, the nonholonomic gear in fig. 2.2 consists of an input wheel IW, two output wheels OW_1 and OW_2, and two supporting wheels SW_1 and SW_2. Suppose that OW_2 in fig. 2.2 corresponds to OW in fig. 2.1. By the similar kinematic analysis, it is clear that the angular velocity of OW_1 in fig. 2.2 is given by the following equation.

$$\omega = \rho \frac{r_I}{r_O} \sin \alpha \tag{2.4}$$

where α corresponds to the angle of joint i. From eqs. (2.3) and (2.4), it is obvious that the nonholonomic gear is a continuously variable transmission (CVT) which is composed of single input and multiple output wheels. By changing the roles of input and output wheels, gear ratio are functions of $\frac{1}{cos\alpha}$ or $\frac{1}{sin\alpha}$. Therefore, gear ratios of the nonholonomic gear are represented by trigonometric functions.

When α in eq. (2.3) changes, wheels OW_1,OW_2,SW_2 and the ball rotate about the axis connecting the two poles. But even if $\dot{\alpha}$ is nonzero, the axis of rotation of the ball is axis A with respect to the coordinate frame fixed to link i, therefore eqs. (2.3) and (2.4) are still valid.

2.3 Theoretical design of the nonholonomic manipulator

2.3.1 Joint driving mechanism

Assume that the gear presented in the previous section is located at joint 1. We use one input wheel IW_1 and two output wheels, $OW_{1,1}$ and $OW_{2,1}$ where the last index indicates the joint. Notice that the nonholonomic gear is used as a velocity transmission which is composed of single input and multiple output wheels. Fix axis $a_{I,1}$ of IW_1 to the base and the two axes of $OW_{1,1}$ and $OW_{2,1}$ to link 1 with orientations $\alpha_{1,1} \equiv \theta_1$ and $\alpha_{2,1} \equiv \theta_1 - \frac{\pi}{2}$ with respect to axis $a_{I,1}$. We denote the angular velocities of $OW_{1,1}$ and $OW_{2,1}$ by $\omega_{1,1}$ and $\omega_{2,1}$, respectively. Let the inputs u_1 and u_2 be the angular velocities of joint 1 and the input wheel IW_1 as follows:

$$\dot{\theta}_1 = u_1, \qquad \rho_1 = u_2 \tag{2.5}$$

By using mechanical transmissions like e.g. shafts and gears, or belts and pulleys, $OW_{2,1}$ drives the second joint as

$$\dot{\theta}_2 = \eta_{2,2}\,\omega_{2,1} = \eta_{2,2}\frac{r_{I,1}}{r_{O,1}}\sin\theta_1\;\rho_1$$

since $\alpha_{2,1} = \theta_1 - \frac{\pi}{2}$ where $\eta_{2,2}$ is a gear ratio. We assume that $OW_{1,1}$ and $OW_{2,1}$ have the same radius $r_{O,1}$ for simplicity. Locate another similar gear at joint 2. By using mechanical transmissions from wheel $OW_{1,1}$ let the angular velocity ρ_2 of IW_2 be given by

$$\rho_2 = \eta_{1,2}\,\omega_{1,1} = \eta_{1,2}\frac{r_{I,1}}{r_{O,1}}\cos\theta_1\;\rho_1$$

since $\alpha_{1,1} = \theta_1$ where $\eta_{1,2}$ is a gear ratio. By locating such a gear at each joint i for $i \in \{1, \ldots, n-2\}$ we get

$$\dot{\theta}_{i+1} = \eta_{2,i+1}\frac{r_{I,i}}{r_{O,i}}\sin\theta_i\;\rho_i \tag{2.6}$$

$$\rho_{i+1} = \eta_{1,i+1}\frac{r_{I,i}}{r_{O,i}}\cos\theta_i\;\rho_i \tag{2.7}$$

The coupling between the consecutive joints is illustrated in fig. 2.3.

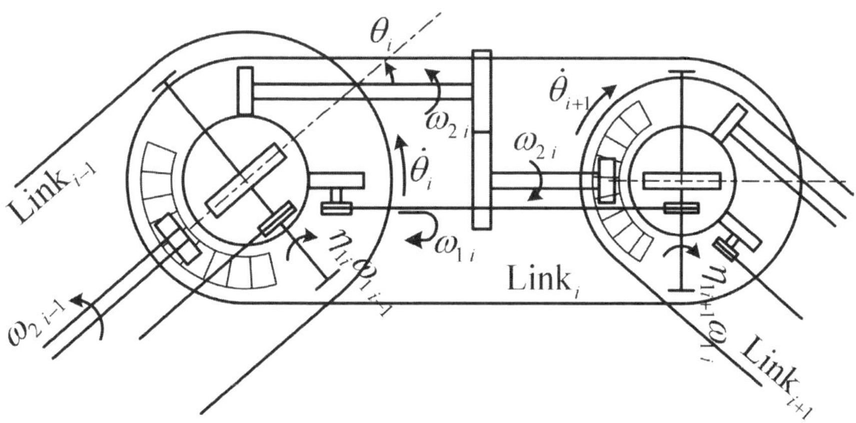

Fig. 2.3. The placement of the nonholonomic gear at joint $i-1$.

An example of four-joint planar nonholonomic manipulator is illustrated in fig. 2.4. The direction of the axis of a revolute joint $i+1$ can be arbitrary with respect to the axis of a joint i by designing the mechanical transmissions between the joints appropriately. Therefore, the manipulator is not restricted to be planar.

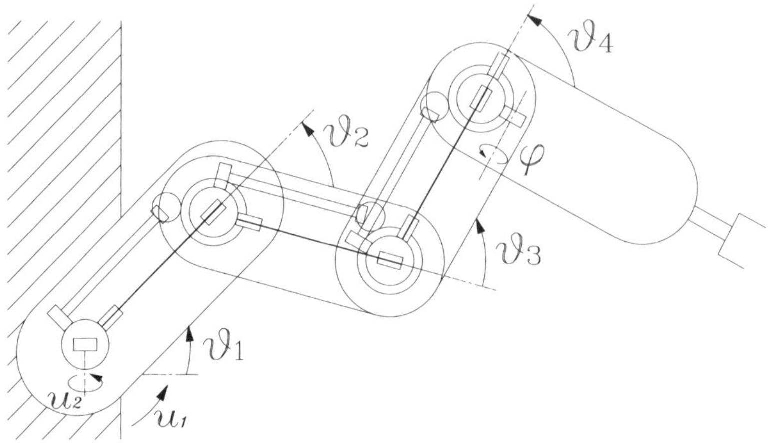

Fig. 2.4. Illustration of a four-joint manipulator and the velocity transmissions.

2.3.2 Kinematic model of the nonholonomic manipulator

In addition to the joint angles, the orientation of one of the output wheels at the joint $n-1$, which is denoted by φ, is added to the kinematic model. By taking φ as a control parameter, kinematic model of the nonholonomic manipulator can be converted to the chained form. Notice that φ is not a control input and the details of φ will be presented in section 3.5. Accordingly, there are $n-1$ joints for n dimensional kinematic model.

From (2.5)– (2.7) and adding φ, the following kinematic model in the configuration space S^n is obtained:

$$\dot{\theta}_1 = u_1 \tag{2.8}$$

$$\dot{\theta}_i = k_i \sin\theta_{i-1} \prod_{j=1}^{i-2} \cos\theta_j u_2, \quad i \in \{2, \ldots, n-1\} \tag{2.9}$$

$$\dot{\varphi} = k_n \prod_{j=1}^{n-1} \cos\theta_j u_2 \tag{2.10}$$

where

$$k_i \triangleq \eta_{2,i} \prod_{j=1}^{i-1} \eta_{1,j} \frac{r_{I,j}}{r_{O,j}}, \qquad k_n \triangleq \prod_{j=1}^{n-1} \eta_{1,j} \frac{r_{I,j}}{r_{O,j}} \tag{2.11}$$

where $\eta_{1,1} = 1$.

It is noteworthy that this kinematic model has a similar structure as the kinematic model of a car with $n-3$ trailers, but not identical. The angle θ_1 can be thought of as the steering angle and θ_{n-1} and φ can be thought of as the y- and x-position, respectively, of the last trailer.

The system's nonholonomic constraints are due to the rolling contact between the balls and the wheels. The $n-2$ constraints in the joint space are given from eqs. (2.8)–(2.10) by

$$k_i \sin\theta_{i-1}\dot{\varphi} - k_n \prod_{j=i-1}^{n-1} \cos\theta_j \, \dot{\theta}_i = 0, \quad i \in \{2,\dots,n-1\} \tag{2.12}$$

An alternative formulation of the kinematics eqs. (2.8)–(2.10) is

$$\dot{\theta}_1 = u_1 \tag{2.13}$$
$$\dot{\theta}_i = k_i \sin\theta_{i-1} v_{i-1} \qquad i \in \{2,\dots,n-1\}$$
$$\dot{\varphi} = k_n \cos\theta_{n-1} v_{n-1} \tag{2.14}$$

where

$$v_{i-1} = \prod_{j=1}^{i-2} \cos\theta_j u_2 = C_1^{i-2}(\Theta_{i-2}) u_2$$

where $\Theta_{i-2} \triangleq [\theta_1,\dots,\theta_{i-2}]^T$ and

$$C_1^{i-2}(\Theta_{i-2}) \triangleq \prod_{j=1}^{i-2} \cos\theta_j$$

By setting configuration variables $\xi \triangleq [\theta_1,\dots,\theta_{n-1},\varphi]^T$, the kinematic model can be represented using input vectors as following equations.

$$\dot{\xi} = \boldsymbol{V_1} u_2 + \boldsymbol{\Omega_1} u_1 \tag{2.15}$$

$$\boldsymbol{V_1} = [0, k_2 s_1, k_3 s_2 C_1^1, \dots, k_{n-1} s_{n-2} C_1^{n-3}, k_n c_{n-1} C_1^{n-2}]^T \tag{2.16}$$

$$\boldsymbol{\Omega_1} = [1, 0, \dots, 0]^T \tag{2.17}$$

where $s_i \triangleq \sin\theta_i$ and $c_i \triangleq \cos\theta_i$. This formulation of the kinematic model will be useful in the following section to show that the system is controllable in the whole joint space.

2.4 Controllability

Controllability of eqs. (2.8)–(2.10) will be shown along the same lines as in [28] and [74]. To show controllability we introduce the accessibility algebra $\mathcal{C}$ for eqs. (2.8)–(2.10) which is the smallest subalgebra of $V^\infty(M)$ (the Lie algebra of smooth vector fields on the configuration manifold S^n, [56]) that contains the input vector fields $\boldsymbol{V_1}$ and $\boldsymbol{\Omega_1}$, in eqs. (2.16)–(2.17). The accessibility distribution C of eqs. (2.8)–(2.10) is given by

$$C(q) = span\,\{v(q) \mid v \in \mathcal{C}\}, \quad q \in S^n$$

From [56] p. 83 we have the following theorem:

Theorem 2.1. *Assume that*

$$\forall q \in S^n \qquad \dim C(q) = n$$

Then, the system in eqs. (2.8)–(2.10) is completely controllable.

Let $[\boldsymbol{\Omega_j}, \boldsymbol{V_j}]$ denote the Lie bracket of the vector fields $\boldsymbol{\Omega_j}$ and $\boldsymbol{V_j}$. We introduce the following vector fields

$$\boldsymbol{Y_1} = [\boldsymbol{\Omega_1}, \boldsymbol{V_1}] \tag{2.18}$$

$$\boldsymbol{V_{i+1}} = \cos\theta_i \boldsymbol{V_i} - \sin\theta_i \boldsymbol{Y_i} \tag{2.19}$$

$$\boldsymbol{\Omega_{i+1}} = (\sin\theta_i \boldsymbol{V_i} + \cos\theta_i \boldsymbol{Y_i})\frac{1}{k_{i+1}} \tag{2.20}$$

$$\boldsymbol{Y_{i+1}} = [\boldsymbol{\Omega_{i+1}}, \boldsymbol{V_{i+1}}] \tag{2.21}$$

where $i \in \{1, \ldots, n-1\}$.

Lemma 2.2. *Let the vector fields $\boldsymbol{V_i}$, $\boldsymbol{\Omega_i}$, and $\boldsymbol{Y_i}$ be iteratively defined for $i \in \{1, \ldots, n\}$ by eqs. (2.16)–(2.17) and (2.18)–(2.21). These vector fields have then the following structure*

$$\boldsymbol{V_i} = [\boldsymbol{0_i}, \boldsymbol{v_i^n}]^T \tag{2.22}$$

$$\boldsymbol{\Omega_i} = [\boldsymbol{0_{i-1}}, 1, \boldsymbol{0_{n-i}}]^T \tag{2.23}$$

$$\boldsymbol{Y_i} = [\boldsymbol{0_i}, \boldsymbol{y_i^n}]^T \tag{2.24}$$

where

$$\boldsymbol{0_i} \triangleq [0, 0, \ldots, 0], \quad dim\, \boldsymbol{0_i} = i \tag{2.25}$$

$$\boldsymbol{v_i^n} \triangleq [k_{i+1}\sin\theta_i, \cos\theta_i \boldsymbol{v_{i+1}^n}], \quad \boldsymbol{v_{n-1}^n} \triangleq k_n \cos\theta_{n-1} \tag{2.26}$$

$$\boldsymbol{y_i^n} \triangleq [k_{i+1}\cos\theta_i, -\sin\theta_i \boldsymbol{v_{i+1}^n}] \tag{2.27}$$

for $i \in \{1, \ldots, n-2\}$ and

$$\boldsymbol{V_{n-1}} = [\boldsymbol{0_{n-1}}, k_n \cos\theta_{n-1}]^T \tag{2.28}$$

$$\boldsymbol{\Omega_{n-1}} = [\boldsymbol{0_{n-2}}, 1, 0]^T \tag{2.29}$$

$$\boldsymbol{Y_{n-1}} = [\boldsymbol{0_{n-1}}, -k_n \sin\theta_{n-1}]^T \tag{2.30}$$

$$\boldsymbol{V_{n-1}} = [\boldsymbol{0_{n-1}}, k_n]^T \tag{2.31}$$

Proof: The proof will be given by induction. Assume that the vector fields $\boldsymbol{V_i}$, $\boldsymbol{\Omega_i}$, and $\boldsymbol{Y_i}$ are given by eqs. (2.22)–(2.24) for an $i \in \{1, \ldots, n-2\}$. We find from eqs. (2.19) and (2.22)–(2.24) that

$$\begin{aligned} \boldsymbol{V_{i+1}} &= \cos\theta_i \boldsymbol{V_i} - \sin\theta_i \boldsymbol{Y_i} \\ &= [\boldsymbol{0_i}, \cos\theta_i \boldsymbol{v_i^n} - \sin\theta_i \boldsymbol{y_i^n}]^T \\ &= [\boldsymbol{0_i}, k_{i+1}(c_i s_i - s_i c_i), c_i^2 \boldsymbol{v_{i+1}^n} + s_i^2 \boldsymbol{v_{i+1}^n}]^T \\ &= [\boldsymbol{0_{i+1}}, \boldsymbol{v_{i+1}^n}]^T \end{aligned} \tag{2.32}$$

$$\begin{aligned} \boldsymbol{\Omega_{i+1}} &= (\sin\theta_i \boldsymbol{V_i} + \cos\theta_i \boldsymbol{Y_i}) \frac{1}{k_{i+1}} \\ &= (\sin\theta_i [\boldsymbol{0_i}, \boldsymbol{v_i^n}]^T + \cos\theta_i [\boldsymbol{0_i}, \boldsymbol{y_i^n}]^T) \frac{1}{k_{i+1}} \\ &= [\boldsymbol{0_i}, 1, \boldsymbol{0_{n-i-1}}]^T \end{aligned} \tag{2.33}$$

$$\begin{aligned} \boldsymbol{Y_{i+1}} &= [\boldsymbol{\Omega_{i+1}}, \boldsymbol{V_{i+1}}] = \frac{\partial \boldsymbol{V_{i+1}}}{\partial q} \boldsymbol{\Omega_{i+1}} - \frac{\partial \boldsymbol{\Omega_{i+1}}}{\partial q} \boldsymbol{V_{i+1}} \\ &= \frac{\partial \boldsymbol{V_{i+1}}}{\partial \theta_{i+1}} \end{aligned} \tag{2.34}$$

where $q \triangleq [\theta_1, \theta_2, \ldots, \theta_{n-1}, \varphi]^T$. We find from eqs. (2.32) and (2.26)

$$\begin{aligned} \boldsymbol{Y_{i+1}} &= \frac{\partial \boldsymbol{V_{i+1}}}{\partial \theta_{i+1}} \\ &= [\boldsymbol{0_{i+1}}, \boldsymbol{y_{i+1}^n}]^T \end{aligned}$$

We find from eqs. (2.16) and (2.17) that

$$\boldsymbol{Y_1} = [\boldsymbol{\Omega_1}, \boldsymbol{V_1}] = \frac{\partial \boldsymbol{V_1}}{\partial \theta_1} = [0, \boldsymbol{y_1^n}]^T$$

Hence, (2.24) is satisfied for $i = 1$. The proof of (2.22)–(2.24) is then completed by noting from (2.16) and (2.17) that (2.22) and (2.23) are satisfied for $i = 1$. The proof of (2.28)–(2.31) follows from a simple calculation using (2.19)–(2.21).

We see from this lemma that the vector fields $\boldsymbol{V_i}$ and $\boldsymbol{\Omega_i}$ as defined by (2.18)–(2.21) have the same structure with respect to the sub-system consisting of links i through n as the input vector fields $\boldsymbol{V_1}$ and $\boldsymbol{\Omega_1}$ have with respect to the complete system. Therefore, using the inputs to e.g. generate a motion in $\boldsymbol{\Omega_i}$ direction makes joint i rotate.

The following theorem states that the nonholonomic manipulator is controllable.

Theorem 2.3. *The kinematic model of a nonholonomic manipulator as given by eqs. (2.8)–(2.10) is controllable.*

Proof: From (2.22)–(2.24), (2.26)–(2.27), (2.28)–(2.31) we have

$$\det\left[\boldsymbol{\Omega}_1, \ldots, \boldsymbol{\Omega}_{n-1}, \boldsymbol{V_{n-1}}\right] = k_n \neq 0$$

Therefore,

$$\dim\left\{\boldsymbol{\Omega}_1, \ldots, \boldsymbol{\Omega}_{n-1}, \boldsymbol{V_{n-1}}\right\} = n, \quad \forall q \in \mathbb{R}^n$$

From the construction of $\boldsymbol{\Omega}_i$ and $\boldsymbol{V_{n-1}}$, Eqs. (2.18)–(2.21), it follows that the system is controllable according to Theorem 2.1.

2.5 Conversion to the chained form

2.5.1 Triangular structure and chained form conversion

In order to control the nonholonomic manipulator, the kinematic model eqs. (2.8)–(2.10) will be converted into a chained form implying that existing control laws for chained form can be applied. Conversion of the kinematic model of a car with n trailers into a chained form was presented in [72] and an exponentially convergent stabilizer was proposed in [76]. Chained form was introduced in [47]. The chained form considered here is given by

$$\dot{z}_1 = v_1 \tag{2.35}$$

$$\dot{z}_2 = v_2 \tag{2.36}$$

$$\dot{z}_i = z_{i-1} v_1, \quad i \in \{3, \ldots, n\} \tag{2.37}$$

To convert the kinematic model into a chained form, we present the following theorem on the conversion of triangular systems into a chained form. First, denote

$$\underline{x}_i \triangleq [x_i, \ldots, x_n]^T$$

$$\underline{f}_i(\underline{x}_{i-1}) \triangleq [f_i(\underline{x}_{i-1}), \ldots, f_n(\underline{x}_{n-1})]^T$$

Theorem 2.4. *Let a driftless, two-input system be given by*

$$\dot{x}_1 = u_1 \tag{2.38}$$

$$\dot{x}_2 = u_2 \tag{2.39}$$

$$\dot{x}_i = f_i(\underline{x}_{i-1}) u_1, \quad i \in \{3, \ldots, n\} \tag{2.40}$$

where $f_i(\cdot)$ is a smooth function. Assume that at a configuration $x = p$ on the configuration manifold

$$\frac{\partial f_i(\underline{x}_{i-1})}{\partial x_{i-1}}|_{x=p} \neq 0, \quad \forall i \in \{3, \ldots, n\} \tag{2.41}$$

Then, a coordinate transformation $z = h(x)$ and an input feedback transformation $v = g(x)u$ converting eqs. (2.38)–(2.40) into the chained form in eqs. (2.35)–(2.37) in a neighborhood of $x = p$ is given by

$$z_n = h_n(x_n) \tag{2.42}$$

$$z_i = h_i(\underline{x}_i) \triangleq \frac{\partial h_{i+1}(\underline{x}_{i+1})}{\partial \underline{x}_{i+1}} \underline{f}_{i+1}(\underline{x}_i) \tag{2.43}$$

$$z_1 = h_1(x_1) \triangleq x_1 \tag{2.44}$$

where $i \in \{2, \ldots, n-1\}$ and $h_n(x_n)$ is any smooth function such that

$$\frac{\partial h_n(x_n)}{\partial x_n}|_{x=p} \neq 0$$

and

$$v_1 = u_1 \tag{2.45}$$

$$v_2 = \frac{\partial h_2(\underline{x}_2)}{\partial \underline{x}_3} \underline{f}_3(\underline{x}_2) u_1 + \frac{\partial h_2(\underline{x}_2)}{\partial x_2} u_2 \tag{2.46}$$

Proof [79]: The proof consists of two parts. First, it will be shown that the transformations in eqs. (2.42)–(2.44) and (2.45)–(2.46) result in the chained form in eqs. (2.35)–(2.37). Then, it will be shown that these transformations are diffeomorphic.

Assume that z_i is given by eqs. (2.42)–(2.43) for all $i \in \{1, \ldots, n\}$. Then, from eqs. (2.38)–(2.40) and (2.45) we get

$$\dot{z}_i = \frac{\partial h_i(\underline{x}_i)}{\partial \underline{x}_i} \dot{\underline{x}}_i = \frac{\partial h_i(\underline{x}_i)}{\partial \underline{x}_i} \underline{f}_i(\underline{x}_{i-1}) u_1 = z_{i-1} v_1$$

From eqs. (2.43), (2.38)–(2.40), and (2.46) we get

$$\dot{z}_2 = \frac{\partial h_2(\underline{x}_2)}{\partial \underline{x}_2} \dot{\underline{x}}_2 = \frac{\partial h_2(\underline{x}_2)}{\partial \underline{x}_3} \underline{f}_3(\underline{x}_2) u_1 + \frac{\partial h_2(\underline{x}_2)}{\partial x_2} u_2 = v_2$$

From eqs. (2.38), (2.44), and (2.45) it follows readily that $\dot{z}_1 = v_1$.

To show that the transformation $z = h(x)$ from eqs. (2.42)–(2.44) is diffeomorphic, we study the jacobian $J(x)$,

$$J(x) \triangleq \frac{\partial h(x)}{\partial x}$$

where $x = [x_1, \ldots, x_n]^T$. Because of the triangular structure of $h(x)$, $J(x)$ is non-singular if and only $J_{ii}(x) \neq 0$ for all $i \in \{1, \ldots, n\}$. From eq. (2.43) we have that for $i \in \{2, \ldots, n-1\}$

$$\begin{aligned} h_i(\underline{x}_i) &= \frac{\partial h_{i+1}(\underline{x}_{i+1})}{\partial \underline{x}_{i+1}} \underline{f}_{i+1}(\underline{x}_i) \\ &= \frac{\partial h_{i+1}(\underline{x}_{i+1})}{\partial x_{i+1}} f_{i+1}(\underline{x}_i) + \frac{\partial h_{i+1}(\underline{x}_{i+1})}{\partial \underline{x}_{i+2}} \underline{f}_{i+2}(\underline{x}_{i+1}) \end{aligned}$$

Hence,

$$J_{ii}(\underline{x}_i) = \frac{\partial h_i(\underline{x}_i)}{\partial x_i} = \frac{\partial h_{i+1}(\underline{x}_{i+1})}{\partial x_{i+1}} \frac{\partial f_{i+1}(\underline{x}_i)}{\partial x_i}$$
$$= J_{i+1,i+1} \frac{\partial f_{i+1}(\underline{x}_i)}{\partial x_i}$$

$J_{nn} = \frac{\partial h_n(x_n)}{\partial x_n}$ is assumed to be non-zero in a neighborhood of $x = p$. It then follows by induction that $J_{ii}(\underline{x}_i) \neq 0$ in a neighborhood of $x = p$ if and only if

$$\frac{\partial f_{i+1}(\underline{x}_i)}{\partial x_i}|_{x=p} \neq 0$$

for $i \in \{2, \ldots, n-1\}$. From eq. (2.44) we see that $J_{11}(x) = 1$. From The Inverse Function Theorem it then follows that $z = h(x)$ is diffeomorphic if eq. (2.41) is satisfied. From eqs. (2.45)–(2.46) we then see that the input transformation is diffeomorphic if

$$\frac{\partial h_2(\underline{x}_2)}{\partial x_2}|_{x=p} \neq 0$$

2.5.2 Chained form conversion of the nonholonomic manipulator

The kinematic model of the nonholonomic manipulator eqs. (2.8)–(2.10) can be put locally on the following form:

$$\dot{\theta}_1 = u_1 \tag{2.47}$$
$$\dot{\varphi} = \mu_2 \tag{2.48}$$
$$\dot{\theta}_i = k_i \tan\theta_{i-1} \frac{1}{k_n \prod_{j=i}^{n-1} \cos\theta_j} \mu_2, \quad i \in \{2, \ldots, n-1\} \tag{2.49}$$

$$\mu_2 \triangleq k_n \prod_{j=1}^{n-1} \cos\theta_j u_2$$

This system has the structure as given by eq. (2.40). From Theorem 2.4 it follows that eqs. (2.8)–(2.10) is convertible to a chained form:

Corollary 2.5. *The kinematic model in eqs. (2.8)–(2.10) is convertible into chained form in the subspace* $\theta_i \in (-\frac{\pi}{2}, \frac{\pi}{2})$, $i \in \{1, \ldots, n-1\}$. *A coordinate transformation and an input feedback transformation is given by eqs. (2.42)–(2.46) with* $x = [\varphi, \theta_1, \theta_2, \ldots, \theta_{n-1}]^T$ *and*

$$f_{i+1}(\underline{x}_i) = k_i \tan\theta_{i-1} \frac{1}{k_n \prod_{j=i}^{n-1} \cos\theta_j}$$

Proof: The modified kinematic model in eqs. (2.47)–(2.49) is valid for $\theta_i \in (-\frac{\pi}{2}, \frac{\pi}{2})$ where $i \in \{1, \ldots, n-1\}$. The result then follows readily from Theorem 2.4 and eqs. (2.47)–(2.49) since for $\theta_{i-1} \in (-\frac{\pi}{2}, \frac{\pi}{2})$ and for all $i \in \{2, \ldots, n-1\}$

$$\frac{\partial f_{i+1}(\underline{x}_i)}{\partial x_i} = k_i \frac{1}{\cos^2 \theta_{i-1}} \neq 0$$

It can be easily checked that the kinematic model in eqs. (2.47)–(2.49) cannot be converted to the chained form without a control parameter φ. Although it has a triangular structure as in eqs. (2.38)–(2.40), the condition in eq. (2.41) cannot be satisfied. Furthermore, φ plays a significant role in controller design, which will be discussed in section 3.5.3.

2.6 Conclusion

Using the nonholonomic gears introduced here, the nonholonomic manipulator with n revolute joints is made completely controllable in $T^n = S^1XS^1X\ldots XS^1$ with only two actuators. For some applications, where only limited resources are available, it would be essential to reduce the number of actuators and to reduce the cost and weight without reducing the reachable configuration space.

It is shown how a manipulator can be designed if it is preferred to have a few actuators. The basic idea is to introduce nonholonomic constraints in the design and to exploit the property that such constraints are nonintegrable in order to increase the dimension of the reachable space. The nonholonomic gear can be employed not only to build underactuated manipulators but also to be used in many other applications as a continuously variable transmission having a nonlinear gear ratio of trigonometric functions.

While previous publications have assumed the nonholonomic systems are given and developed theory for these systems, this book points out a new direction where the nonholonomic theory is used to design controllable and stabilizable systems.

To stabilize the system, the system is designed with a triangular structure so that the kinematic model can be converted locally into chained form due to a theorem presented here. Existing open- and closed-loop controllers for chained form can then be used to control the nonholonomic manipulator.

Since there are only two velocity inputs, there are only two degrees of freedom and an n-dimensional trajectory for the joints can not be followed in general. Therefore, if exact tracking is required in a space of higher dimension than two, more than two actuators are needed. However, since the nonholonomic manipulator is controllable, any path in the n-dimensional joint space can be followed with any specified accuracy. Obstacles in the workspace can therefore easily be avoided.

3

Prototyping and control of the nonholonomic manipulator

3.1 Introduction

Since a nonholonomic manipulator was theoretically designed from the viewpoint of kinematic constraints and nonlinear control, mechanical implementation and prototyping are significant in practice. Since joint actuating torques are transmitted through the nonholonomic gears using frictional forces, a prototype should be carefully designed to assure the principle of velocity transmission, which was presented in section 2.3.

The nonholonomic gear, introduced in section 2.2, can be used as a single input, multiple output CVT (continuously variable transmission). Exploiting mechanical elements similar to the nonholonomic gear, many applications can be found. For example, kinematic integrating machines [54, 58] can be designed using the continuously variable gear ratio which is represented by trigonometric functions. Another interesting mechanism is a passive haptic display [59]. A key component is a programmable CVT which adjusts the ratio of angular velocities of two wheels. The nonholonomic gear-like mechanical components could find its practical applications in various ways.

Although controllability can be proven as shown in section 2.4, control problem is still not easy because of its nonlinearlity. Furthermore, since there is no smooth, static state feedback [7], stabilizer design is another important problem. Since the nonholonomic manipulator satisfies chained form convertibility as in section 2.5, many proposed controllers can be applied.

In this chapter, various important issues in mechanical design are discussed in section 3.2. Section 3.3 illustrates the analysis and computation of driving force of the nonholonomic gears towards practical applications. A prototype of the nonholonomic manipulator is introduced in section 3.4. Some practical control schemes including both open loop ones and feedback ones are presented in section 3.5. Experimental results are given with the fabricated prototype.

3.2 Issues in mechanical design

3.2.1 Traction drive and friction drive

There are two strategies of power transmission using frictional forces between solid bodies, depending on the use of traction oil. One is friction drive with traction oil, and the other is friction drive without traction oil. In this paper, in order to distinguish the two, we call the former "traction drive" and the latter "friction drive".

In traction drive, the elasto-hydraulic shear stress of the traction fluid between two solid bodies is utilized to transmit power. Viscosity of the traction oil increases as the pressure goes up. Under the high pressure, the fluid film of the traction oil can bear large frictional force. A coefficient of friction is typically 0.1, and macroscopic slip occurs at any time. Since there are no direct contact between the solid bodies, the frictional wear is not a problem. Therefore, traction drive is frequently used for industrial applications such as toroidal or spherical CVTs, for example, see [38, 85, 48]. The drawbacks of traction drive are: (1) temperature must be carefully controlled, (2) a transmission must be sealed because of the traction oil, which results in complicated and heavy mechanical structures.

On the other hand, the friction drive utilizes frictional force between two surfaces of solid bodies in direct contact. Friction drive is one of the hot issues in the field of tribology, because precise positioning can be accomplished avoiding backlashes [57, 24, 83, 84]. However, the mechanism of slip has not been made clear and it is a major obstacle for practical applications. One of the advantages of this system is that a coefficient of friction is higher than that of the traction drive. Furthermore, sealing and temperature control is unnecessary. Frictional wear of surfaces due to direct contact is a major problem.

For prototyping, friction drive was chosen because a priority of design was placed on the mechanical simplicity.

3.2.2 Supporting mechanism and adjusting spring force

In order to assure frictional force at a contact zone, Appropriate surface normal forces should be applied at the contact zone between a ball and wheels. Two cantilever springs, which make the mechanical structure simple, are employed at each nonholonomic gear. One is for the supporting (not the input) wheel at the south pole, and the other is for the supporting (not the output) wheel at the equator. Locations of wheels of the prototype are same to fig. 2.2. In fig. 3.1, a supporting wheel with a cantilever spring is presented.

The cantilever spring drawn with dotted line represents its original shape. By inserting a spacer (a piece of metal slice) whose thickness is equal to δ, surface normal force is generated. Cantilever spring drawn with solid line represents its deformed shape.

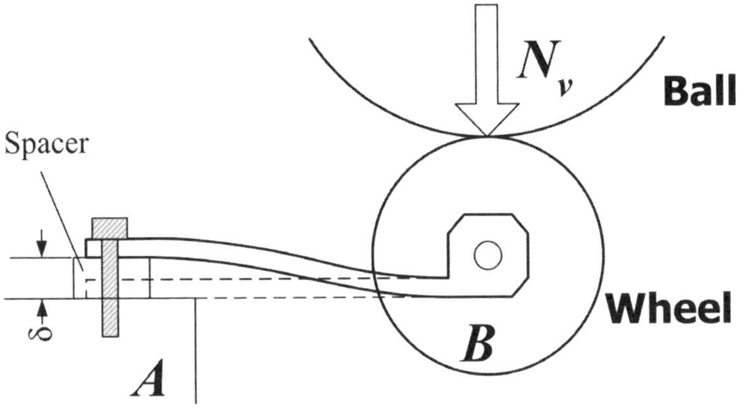

Fig. 3.1. A supporting wheel with a cantilever spring

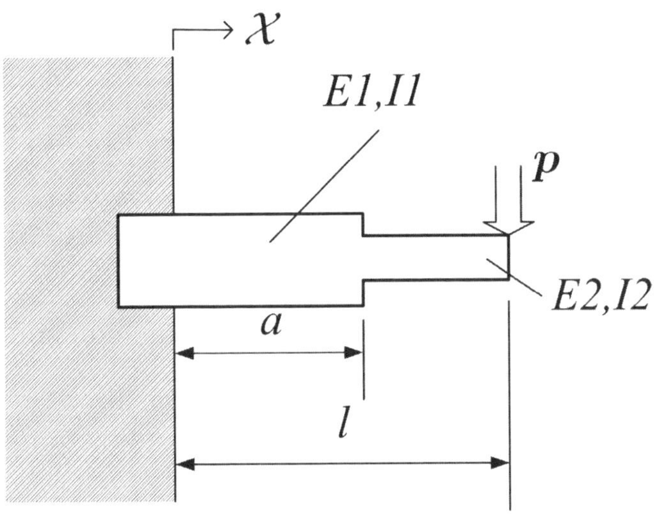

Fig. 3.2. Model of a cantilever spring

Since spring force is closely related to the limit of the power transmission, we need to measure spring force and adjust it. In fig. 3.2, the model of a cantilever spring is presented.

The relation of deflection δ and spring force P is given by following equation.

$$\delta = \frac{P}{3}[\frac{1}{E_1 I_1}[l^3 - (l-a)^3] + \frac{1}{E_2 I_2}(l-a)^3] \tag{3.1}$$

From eq. (3.1), it is clear that spring force is proportional to the deflection δ. Accordingly, once we have prepared several thicknesses of spacers, then, by choosing many different combinations, the spring force can be widely changed.

Machining errors of mechanical components might cause a difference between the real deflection δ and the predicted one. To set an accurate spring force, strain gauges were attached to each cantilever and calibrations were carried out. Resultant spring forces can be adjusted from the strain gauge signal after constructing all the mechanical components.

Determining stiffness of a cantilever is another problem. If stiffness of a cantilever is too high, then the thicknesses of spacers should be changed minutely to adjust spring force. This would require very precise machining and great expense. Conversely, if the stiffness is too low, sufficient forces would be unobtainable. Selecting appropriate stiffness, dimensions were carefully designed using the data set of surface normal force, which will be discussed in section 3.3.

3.2.3 Other issues related to mechanical design

Forces acting around the ball cannot be specified because the nonholonomic gear is a statically indeterminate system. To estimate the limit of power transmission of a nonholonomic gear, experimental analysis is needed. Transmitted power can be computed in experiments. Once angular velocities and accelerations of all the links are measured, then inverse dynamics can be solved to obtain transmitted force/moment. On the other hand, slipping can be measured by observing the velocities of two motors and those of joints. Comparing the transmitted force/moment and the slipping velocity, the limit of power transmission can be estimated. The resonance can be investigated by the experiment of frequency response.

3.3 Analysis and computation of driving force

In order to transmit angular velocity by friction drive, large frictional forces are desirable between two rotating bodies. Frictional force depends on the surface normal force and a coefficient of friction. The nominal value of a coefficient of friction is determined by material property. Although surfaces can be made rough by sand blasting, its effect on a coefficient of friction is not much under the high pressure. However, applying too large a surface normal force causes yielding and plastic deformation of materials.

General solution to the contact problem of two elastic solids is well known as the Hertz theory of elastic contact. Let the coordinate frames of the contact zone be as in fig. 3.3.

According to the analysis, the following general results are found in the literature [92].

1. The contact zone is bounded by an ellipse, of semiaxes a and b.
2. The normal pressure distribution over this area is given by following equation.

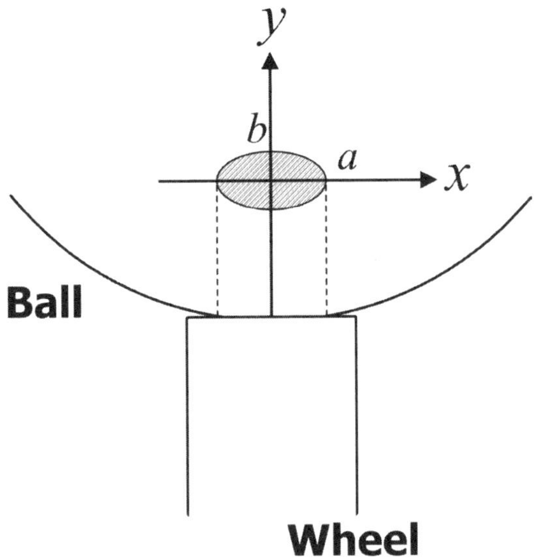

Fig. 3.3. Configuration of a coordinate frame

$$p = p_0(1 - (\frac{x}{a})^2 - (\frac{y}{b})^2)^{1/2} \tag{3.2}$$

The maximum pressure p_0 occurs above the origin of a coordinate frame.

3. The dimensions a and b of the ellipse of contact increase directly as $N^{1/3}$, where N is the surface normal force at the contact zone.

On the stress distribution over the entire stress field, the following results are known [80, 15].

1. Cylinder and cylinder contact
   - The shape of a contact zone is rectangular.
   - $\tau_{max} = 0.31 p_0$ at $z = 0.47a$
2. Sphere and sphere contact
   - The shape of a contact zone is circular.
   - $\tau_{max} = 0.30 p_0$ at $z = 0.79a$

   where τ_{max} is a maximum shear stress over the entire stress field.
3. If the frictional coefficient is 1/9, for example, and frictional force present, then maximum shear stress increases approximately 43% from the case without frictional force.

Since a nonholonomic gear is composed of a ball and wheels, the shape of the contact zone is elliptical. The maximum shear stress τ_{max} is assumed to be in the range from $0.3p_0$ to $0.31p_0$. Furthermore, the frictional force is acting on the contact zone with the frictional coefficient 0.1, τ_{max} increases nearly 43%. For the prototype nonholonomic manipulator designed, radii of

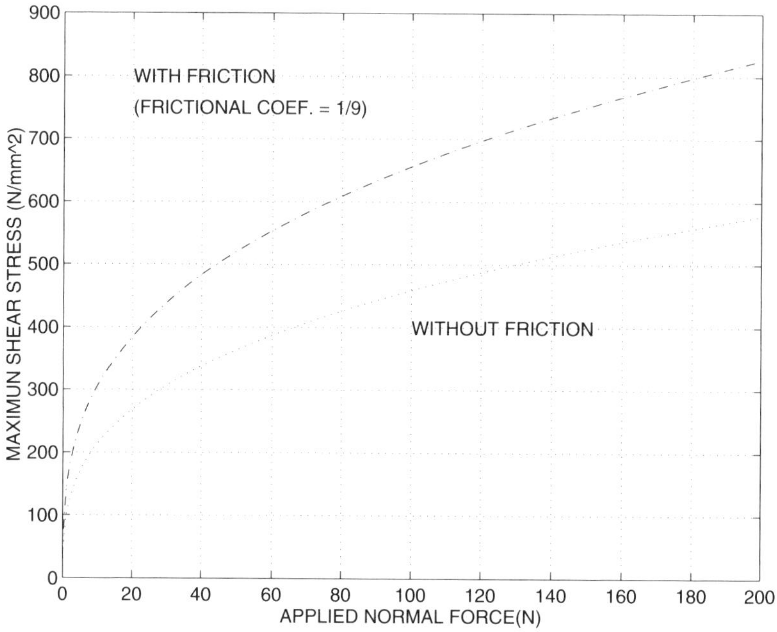

Fig. 3.4. Applied normal force versus Maximum shear stress

a ball and a wheel are $19mm$ and $10mm$. Considering these conditions, the maximum shear stress is presented as a function of the applied normal force in fig. 3.4.

The shear strength of material used (steel SUJ2) is approximately $700N/mm^2$. The frictional coefficient was assumed to be 0.1. From fig. 3.4, we conservatively see that we can apply up to $120N$ of surface normal force using a cantilever spring.

3.4 Prototype of the nonholonomic manipulator

According to the design concept presented in previous sections, a prototype of the nonholonomic manipulator was fabricated as shown in fig. 3.5. The prototype has four revolute joints, whose kinematic model corresponds to five dimensional chained form, as illustrated in section 2.3. It is a planar structure to avoid gravity. If sufficient joint actuating torques can be obtained, it can be designed with a spatial structure having arbitrary directions of joint axes.

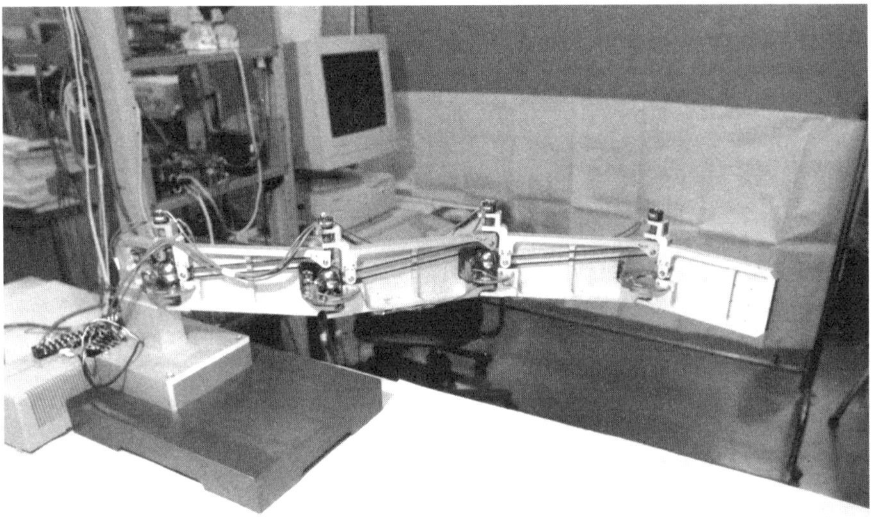

Fig. 3.5. Prototype of the nonholonomic manipulator

There are two actuators in total and both are located at the base. There are many possible alternatives in determining the location of actuators. Putting actuators on the base is desirable because there are only mechanical components on the robot arm, which results in light weight and simple structure of the arm.

In order to obtain high joint actuating torques, gear ratio k_i in eq. (2.9) was chosen to be 1/10, where $i \in \{2, 3, 4\}$. A DC motor($8.6W$) drives first joint, which corresponds to an input u_1 in eq. 2.8. A gear hat with gear ratio 1/50 was attached to the motor axis. For driving an input wheel of the first joint which is an input u_2 in eq. (2.9), a brushless DC motor($100W$) equipped with a gear hat whose gear ratio is 1/36, was employed. Both actuators have its own controllers in which velocity feedback loops. Therefore, velocity commands are sent to the controllers by open loop.

All the joints are equipped with potentiometers to read joint angles. These are not used only for the feedback control, but also for monitoring slip at non-holonomic gears. By comparing the joint angles computed by the numerical integration using input velocities with actual joint angles, slip can be detected.

Fig. 3.6 and Fig. 3.7 show the nonholonomic gear at the second joint. Belt transmission to drive the input wheel of the second joint is presented in fig. 3.8. A joint driving mechanism using a shaft and a gear train is shown in fig. 3.9.

Manipulator links are made with aluminum alloy (52S), 300mm in lengths. Approximate weight of each link after constructing all the other mechanical components is 1.1 kg, which can be hardly achieved if actuators are placed to each link as in conventional robot manipulators.

Fig. 3.6. Prototype of the nonholonomic gear

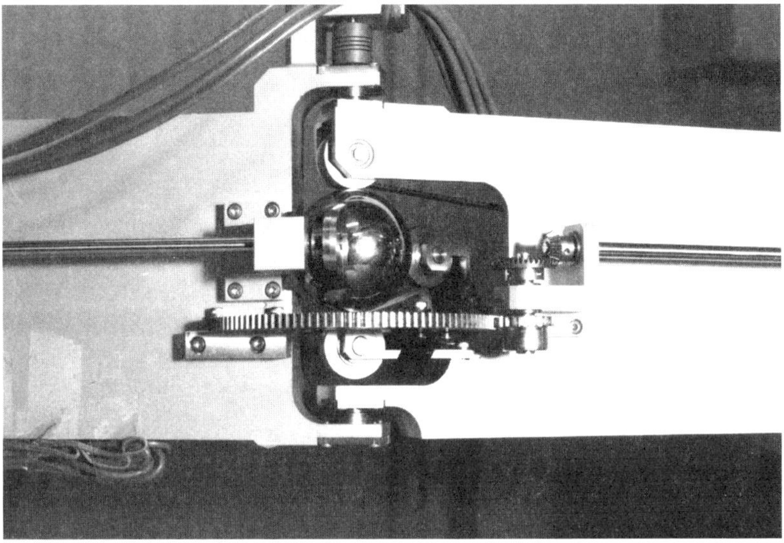

Fig. 3.7. Prototype of the nonholonomic gear seen from the other side

Fig. 3.8. Mechanism of driving the input wheel of the nonholonomic gear

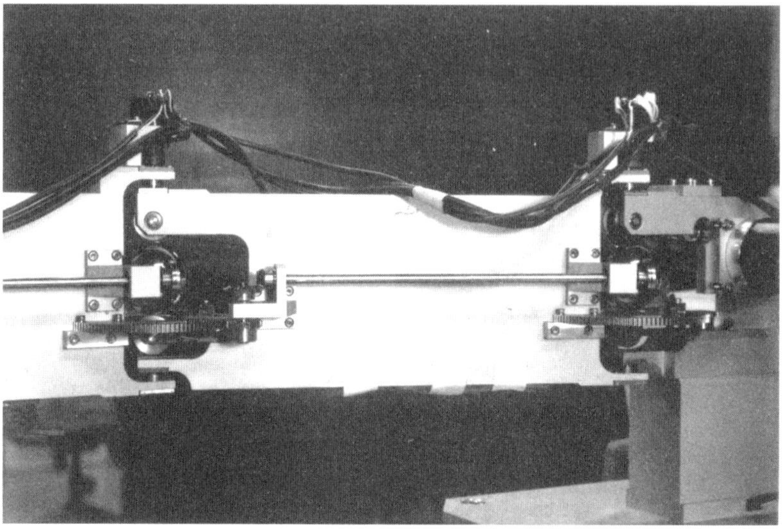

Fig. 3.9. Joint driving mechanism

3.5 Control of the nonholonomic manipulator

There are two major approaches in control of nonholonomic systems. One is open loop motion planning and the other is feedback control. Control inputs can be easily computed by applying the open loop strategies, which provide simple and practical solutions in many cases. However, open loop controller cannot deal with uncertainties and disturbances. Feedback controllers show complementary advantages and disadvantages of the open loop controllers. More detailed explanations will be given in chapter 5.

In this section, one open loop and two feedback controllers are experimentally tested. The first approach is to employ time polynomial inputs in [89]. The second is the feedback control with exponential convergence in [76]. The first and the second experiment show that any of the previously proposed controllers to the chained form can be applied to control the nonholonomic manipulator. Finally, the feedback control by pseudo-linealization is proposed in order to provide a practical control solution of the nonholonomic manipulator.

In simulations and experiments, control schemes are applied to the three joint manipulator model as shown in fig. 3.10. Correspondence of simulations and experiments in this section is presented in table 3.1.

For the four joint model, some mathematical and mechanical problems were faced with. Although there is no limitation on the number of the controllable joint, a kinematic model becomes ill-conditioned as the number of

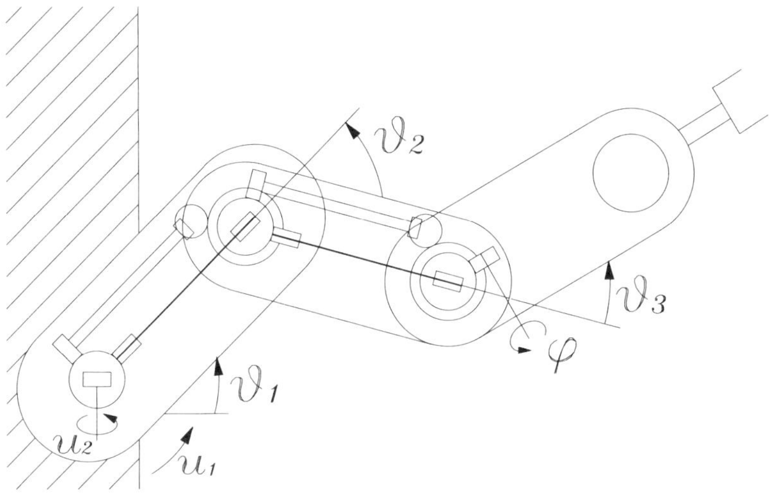

Fig. 3.10. The four dimensional nonholonmic manipulator with three joints.

the joint increases. This problem will be further discussed in 4.2. Therefore, the scope of experiments will be limited to the three joint model, which corresponds to the four dimensional chained form system. φ, which is a rotation angle of the output wheel of the third joint, is a control parameter and free boundary conditions can be given.

Table 3.1. Correspondence of simulations and experiments for the nonholonomic manipulator.

Simulation 1	no experiment
Simulation 1 modified by time scaling	Experiment A
Simulation 3	Experiment B
Simulation 4	Experiment C

3.5.1 Open loop control

There are many existing open loop controllers to the chained form. Sinusoidal inputs [47, 89], time polynomial inputs [89] and piecewise constant inputs [44] are typical examples. Such control schemes are common in that inputs are given by solving algebraic equations using the input parameterization. Any of such control laws can be applied to control the nonholonomic manipulator. In this section, time polynomial inputs are applied because inputs are easily obtained by solving simple algebraic equations and resultant trajectories are smooth.

The inputs to the chained form (2.35)–(2.37) are given by time polynomials as follows:

$$v_1 = b_0 \tag{3.3}$$
$$v_2 = c_0 + c_1 t + \cdots + c_{n-2} t^{n-2} \tag{3.4}$$

To steer the state z from $z(0)$ to $z(T)$ in the finite time T, assuming that $z_1(0) \neq z_1(T)$, the coefficients are given as follows:

$$b_0 = \frac{z_1(T) - z_1(0)}{T}$$

$$M \begin{bmatrix} c_0 \\ c_1 \\ \vdots \\ c_{n-2} \end{bmatrix} + N \begin{bmatrix} z_2(0) \\ z_3(0) \\ \vdots \\ z_n(0) \end{bmatrix} = \begin{bmatrix} z_2(T) \\ z_3(T) \\ \vdots \\ z_n(T) \end{bmatrix}$$

where

$$M_{i,j} \triangleq \frac{{b_0}^{i-1}(j-1)!}{(i+j-1)!} T^{i+j-1}$$
$$N_{i,j} \triangleq \begin{Bmatrix} 0 & , & i < j \\ \frac{{b_0}^{i-j}}{(i-j)!} T^{i-j} & , & i \geq j \end{Bmatrix}$$

If $z_1(0) \neq z_1(T)$, M is nonsingular. Since free boundary conditions can be given to z_1, M can be always made nonsingular.

The joint and motor trajectories are computed. In fig. 3.11, joint angles versus time angles are plotted (simulation 1). Fig. 3.12 shows input angular velocities of two actuators. Manipulator motions in $\mathbb{R}^2$ are shown in fig. 3.13. Fig. 3.14 shows the motion with respect to time which corresponds to z axis. The total time T was set to 10 seconds. The gear ratios were selected as $[k_2, k_3] = [1/10, 1/10]$ which are same to the prototype. Joint angles of initial configuration $\theta(0)$ and desired configuration θ_d are $\theta(0) = [-20, -20, -20](deg)$, $\theta_d = [20, 20, 20](deg)$. From fig. 3.11, it is clear that the joint angles reach their desired values.

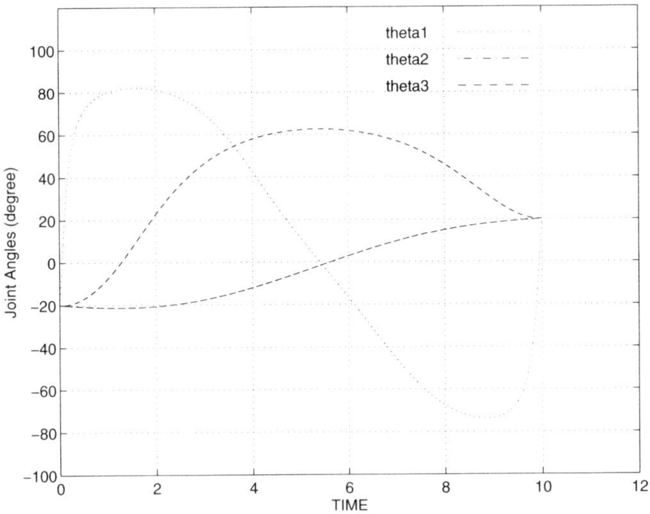

Fig. 3.11. Joint angles versus time with polynomial inputs: simulation 1.

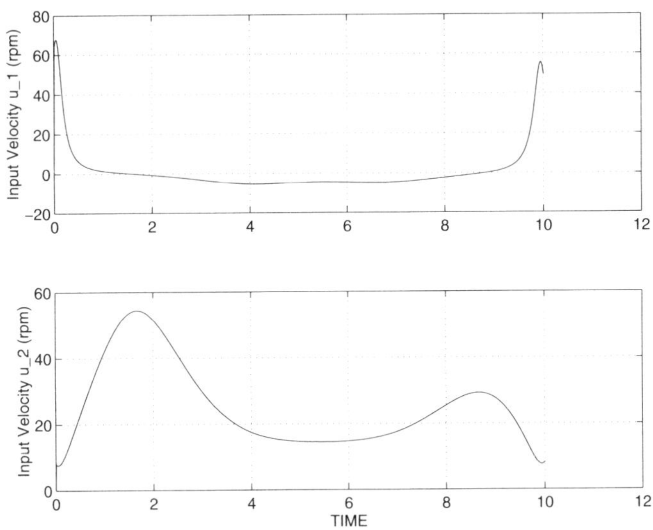

Fig. 3.12. Input angular velocities versus time with polynomial inputs: simulation 1.

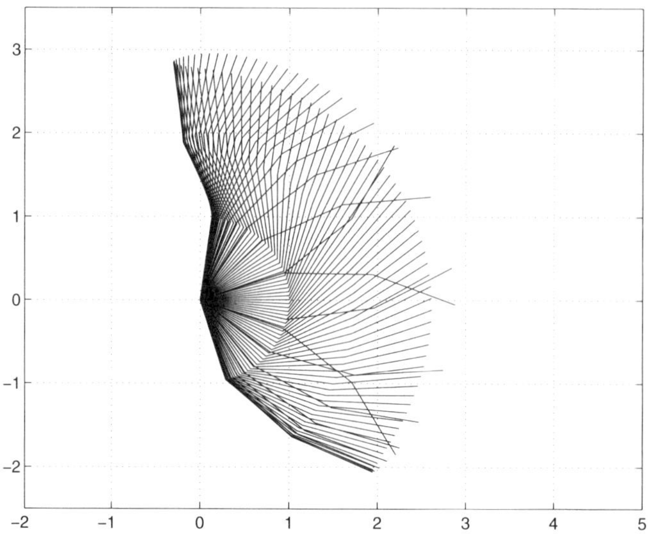

Fig. 3.13. Manipulator motion in $\mathbb{R}^2$ with polynomial inputs: simulation 1.

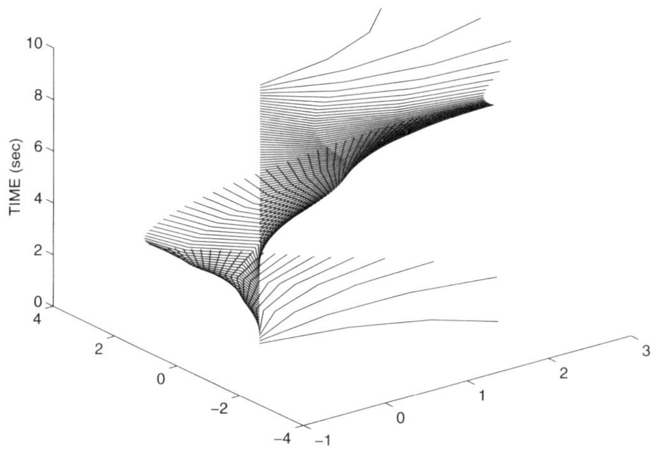

Fig. 3.14. Manipulator motion in $\mathbb{R}^2 \times t$ with polynomial inputs: simulation1.

Although motions are successfully planned in simulations, the nonholonomic manipulator has mechanical limitations in practice. Trajectories are physically infeasible because of actuator saturations or high joint accelerations exceeding the limitations of nonholonomic gears.

A time scaling is a powerful tool to deal with this problem. Suppose that a new time scale $\hat{t}$ is given as a following equation:

$$\hat{t} = S_f \;\; t \tag{3.5}$$

where S_f is a scaling factor which is assumed to be a constant during a time interval $t \in [0, t]$. From eq. (2.12), it is obvious that the scaling factor S_f does not appear in the constraint equation eq. (3.5) after time scaling. Equation (2.15) can be simply rewritten as $\dot{\xi} = g(\xi)u$. Since $d\hat{t} = S_f \;\; dt$, following equations are obtained.

$$\frac{d\hat{\xi}}{d\hat{t}} = g(\hat{\xi})\frac{1}{S_f}u = g(\hat{\xi})\hat{u} \tag{3.6}$$

where $\hat{\xi}$ and $\hat{u}$ are scaled coordinates and inputs, respectively. Since $\hat{u} = \frac{1}{S_f}u$ from eq. (3.6), it is clear that the time scaling of the nonholonomic manipulator implies the scaling of inputs. Since $\xi(t) = \hat{\xi}(S_f t) = \hat{\xi}(\hat{t})$, trajectories in the joint space does not change. Therefore, time scaling can be applied to plan a physically feasible path for any open loop controller. Another expression of time scaling can be found in [78] for the polynomial inputs.

Here, time scaling is applied to saturate an input angular velocity u_1, *i.e.*:

$$u_1 = sat(u_1, u_{1,sat}) \tag{3.7}$$

where

$$\text{sat}(p, K) \triangleq \left\{ \begin{array}{ccc} p & , & |p| < K \\ K\text{sgn}(p) & , & |p| \geq K \end{array} \right\}$$

The velocity limit of u_1 is set first, and trajectories computed by time polynomial inputs are modified by time scaling. Fig. 3.15 shows the joint angle trajectories after the time scaling (simulation 2). Scaled inputs are presented in fig. 3.16. The absolute value of the desired velocity limit of $u_{1,sat}$ was selected as $0.3141(rad/sec) = 3(r/min)$. The total time need to reach the goal was 22 seconds, it took 12 seconds longer than the case of the standard time polynomial inputs. The same motion was experimentally tested by applying velocity control to two motors for their predetermined trajectories (experiment A). Fig. 3.17 illustrates the results of experiment.

The index of error of the joint angles is defined as follows:

$$E = \sum_{i=1}^{3} |\theta_{i,f} - \theta_{i,d}| \tag{3.8}$$

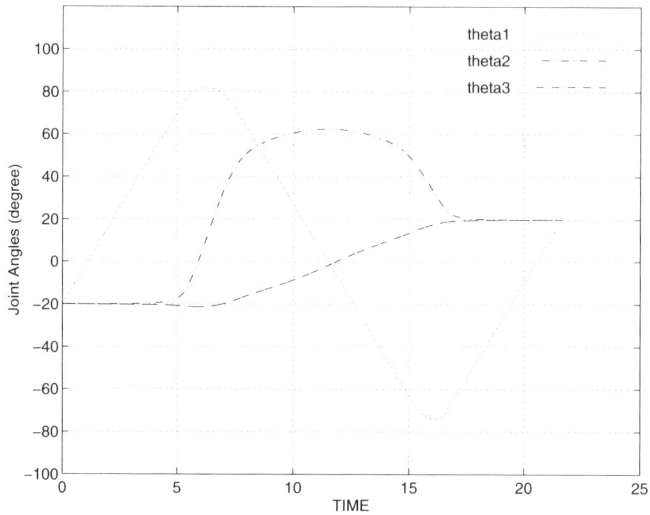

Fig. 3.15. Joint angles versus time with time scaled polynomial inputs: simulation 2.

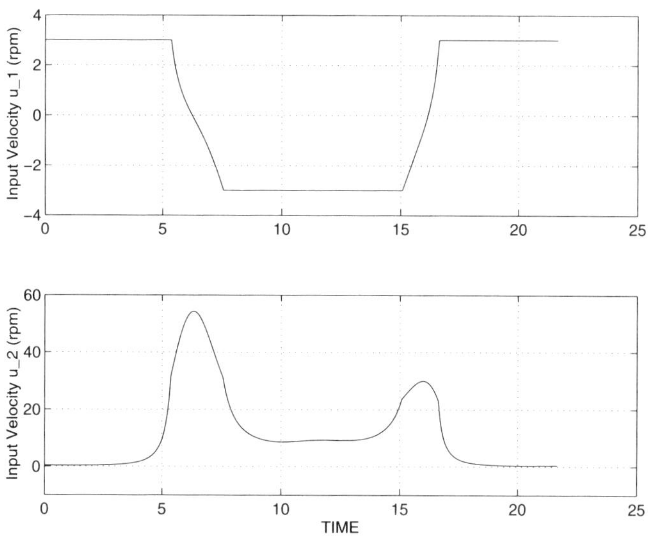

Fig. 3.16. Input angular velocities versus time with time scaled polynomial inputs: simulation 2.

where $\theta_{i,f}$ is a joint angle after completing control, $\theta_{i,d}$ is a desired joint angle. At the end of experiment A, E was $5deg$. The backlash at the spur gear and the low stiffness of the long shaft would have caused the error.

Measured input velocities are plotted in fig. 3.18 and it well matches with fig. 3.16. Therefore, it is obvious that the velocity of actuators are precisely controlled. A resultant motion of the prototype nonholonomic manipulator is shown in fig. 3.19.

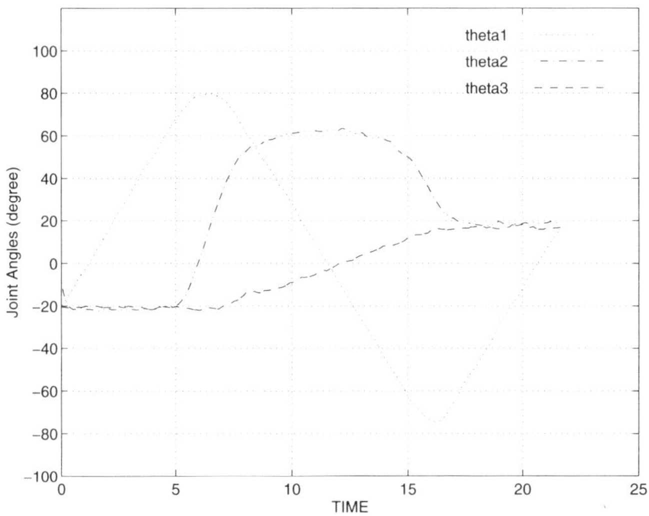

Fig. 3.17. Experimented joint angles versus time with time scaled polynomial inputs: experiment A.

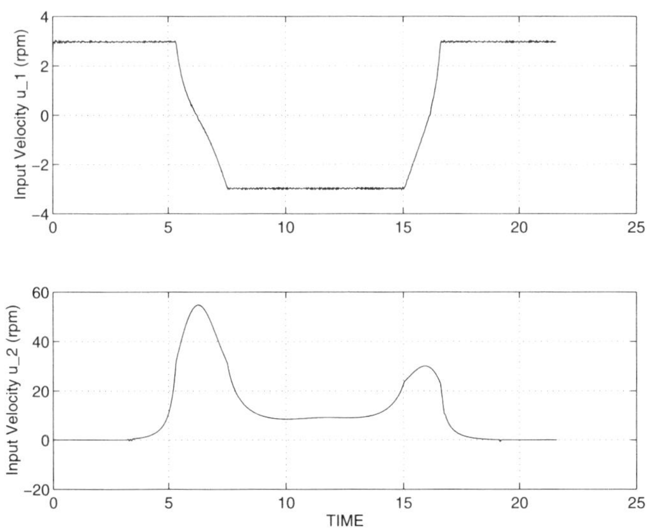

Fig. 3.18. Experimented input angular velocities of two motors versus time with time scaled polynomial inputs: experiment A.

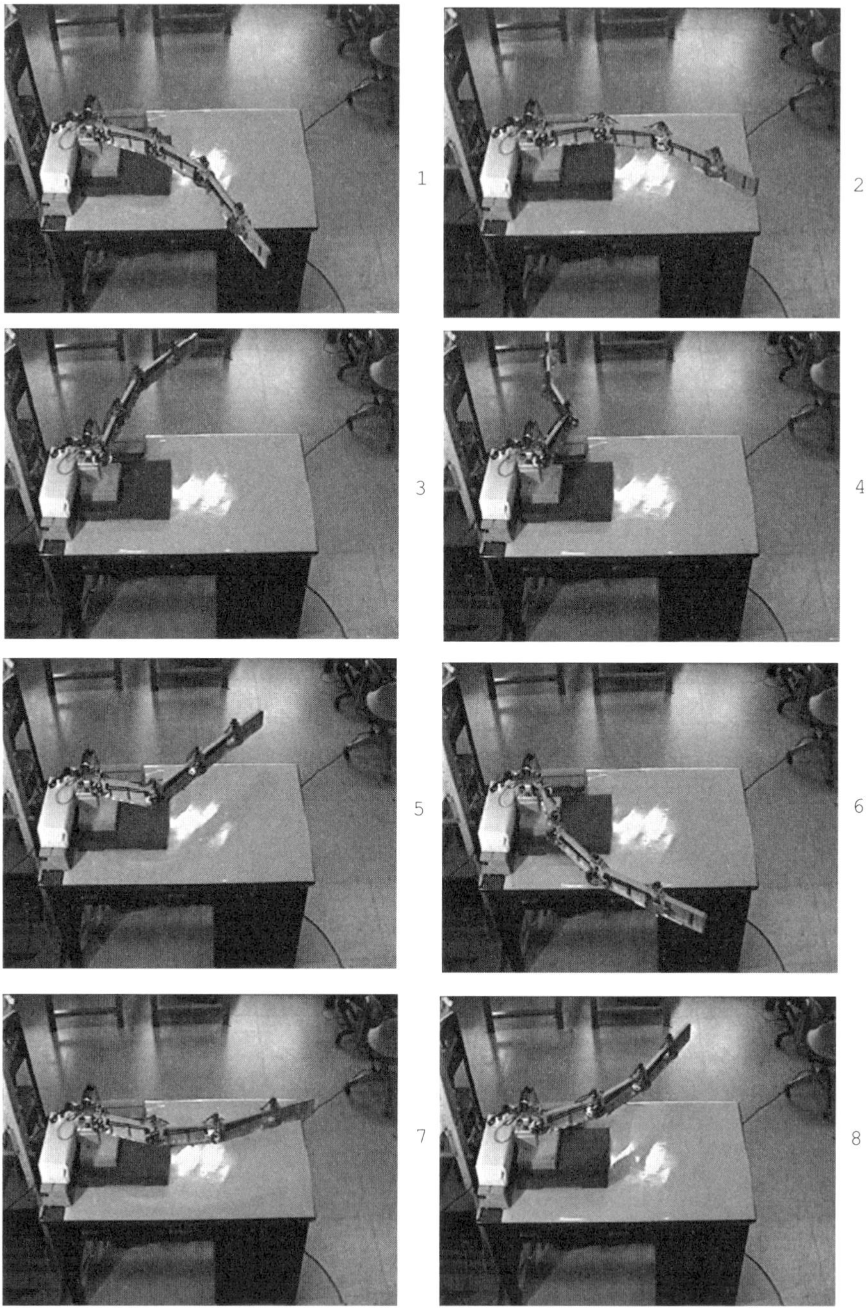

Fig. 3.19. A resultant motion of the nonholonomic manipulator: experiment A.

3.5.2 Feedback control with exponential convergence

Feedback control would be significant to live with various uncertainties. Feedback control of a class of system, including driftless nonholonomic systems is known to be difficult because of the limitations given by Brockett's theorem [7], *i.e.* there is no smooth, static state feedback law. Therefore, many approaches were made without violating Brockett's necessary conditions. Typical examples are time dependent feedback law by Samson [70] or Pomet [61]. Since these controllers are smooth, convergence rate is slow [20]. Introduction of such feedback control laws and comparison by simulations can be found in [65].

In this section, the strategy from [73, 76] resulting in a feedback control with exponential convergence is applied. The control scheme is piecewise continuous and time dependent. The feedback control is given by following equations:

$$v_1 = \gamma(z(t_i))\, l(t) \tag{3.9}$$

$$v_2 = -\lambda_2(z_2 - z_2^d) + \dot{z}_2^d \tag{3.10}$$

$$z_2^d \triangleq (-\lambda_3(z_3 - z_3^d) + \dot{z}_3^d)/\gamma$$

$$z_3^d \triangleq (-\lambda_4 z_4)/\gamma$$

where $l(t) = (1 - \cos\omega t)/2$. $\omega = \frac{2\pi}{T}$ and $T = t_{i+1} - t_i$ are supposed to be constants. The parameter $\gamma(\cdot)$ is given as follows

$$\gamma = \text{sat}\left(-\left[z_1(t_i) + \text{sgn}(z_1(t_i))G(\|q(t_i)\|)\right]\beta, K\right) \tag{3.11}$$

where $q = [z_2, \ldots, z_n]^T$ and

$$G(\|q(t_i)\|) \triangleq \kappa\|q(t_i)\|^{\frac{1}{2(n-2)}} \tag{3.12}$$

$$\beta \triangleq 1/\int_{t_i}^{t_{i+1}} l(\tau)d\tau \tag{3.13}$$

$$\text{sgn}(p) = \begin{cases} 1, & p \geq 0 \\ -1, & p < 0 \end{cases} \tag{3.14}$$

where κ is a positive constant.

In the simulation (simulation 3), the instants of time t_i where $l(\xi(t_i))$ may switch is given by the set $\{0, 2\pi, 4\pi, 6\pi, \ldots\}$, *i.e.* $\omega = 1$, where t_0 is taken to zero. The controller parameters were chosen as $\lambda_2 = \lambda_3 = \lambda_4 = 0.5$, $\kappa = 1.5$, and $K = 10$.

In fig. 3.20, simulated joint angles are plotted (simulation 3). Joint angles of initial configuration $\theta(0)$ and desired configuration θ_d are $\theta(0) = [10, 10, 10](deg)$ $\theta_d = [0, 0, 0](deg)$. From fig. 3.20, it is clear that joint angles converge to the origin with exponential convergence.

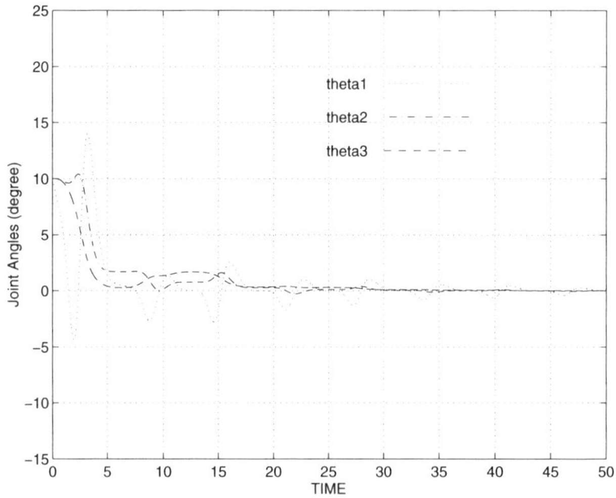

Fig. 3.20. Joint angles versus time with the exponentially convergent feedback law: simulation 3.

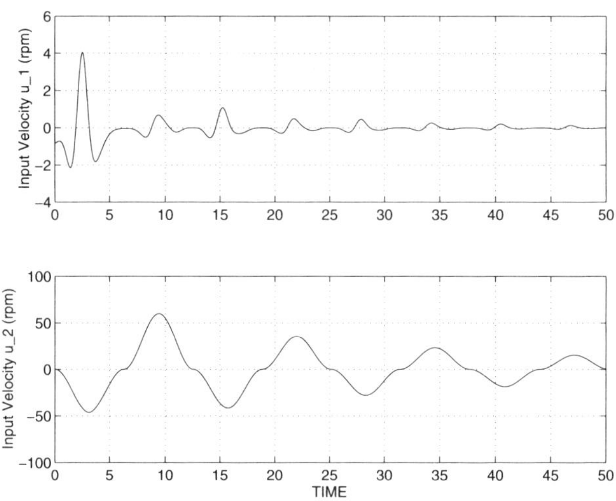

Fig. 3.21. Input angular velocities of two motors versus time with the exponentially convergent feedback law: simulation 3.

Experiment was carried out under the same condition (experiment B). Fig. 3.22 shows the result. We set the condition of terminating the control as $E < 3°$ in (3.8). Profiles of the joint angle trajectories are similar to that of the first 23 seconds of fig. 3.20. A small difference of the magnitude of oscillation might be caused by experimental errors such as the effect of joint backlashes, slippage at the nonholonomic gear and so forth.

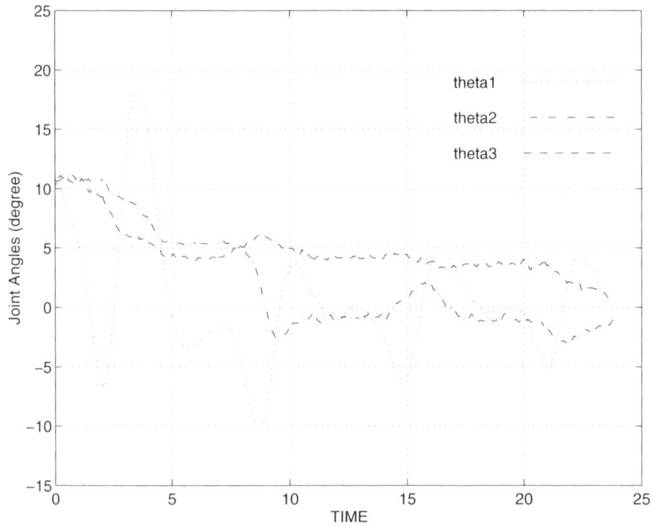

Fig. 3.22. Joint angles versus time with the exponentially convergent feedback law: experiment B.

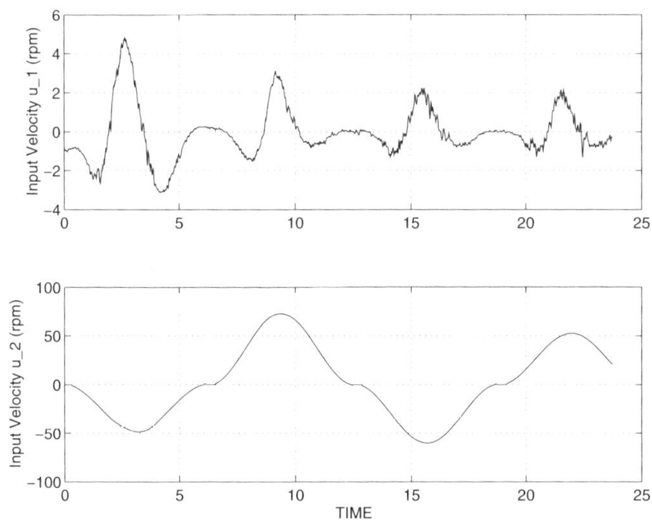

Fig. 3.23. Input angular velocities of two motors versus time with the exponentially convergent feedback law: experiment B.

3.5.3 Feedback control by pseudo-linearization

The main drawback of the feedback controller presented in the previous section is that the number of switching is quite a lot, which results in energy consuming and oscillatory motions. Furthermore, feedback equations become highly complicated as the dimension of a system increases. In this section, a simple feedback strategy is developed towards practical implementations.

If a constant a_0 is set to the input v_1 of the chained form of eqs.(2.35) –(2.37), then the system is described as follows:

$$\dot{z_1}(t) = a_0 \tag{3.15}$$

$$\begin{bmatrix} \dot{z_2}(t) \\ \dot{z_3}(t) \\ \vdots \\ \dot{z_n}(t) \end{bmatrix} = A \begin{bmatrix} z_2(t) \\ z_3(t) \\ \vdots \\ z_n(t) \end{bmatrix} + \begin{bmatrix} 1 \\ 0 \\ \vdots \\ 0 \end{bmatrix} v_2 \tag{3.16}$$

$$A_{i,j} \triangleq \begin{Bmatrix} 0 \ , & i-j \neq 1 \\ a_0 \ , & i-j = 1 \end{Bmatrix}$$

Equation (3.16) is a state equation of a linear system. Therefore, linear system control theory can be applied. z_1 in eq. (3.15) is controlled with open loop. Since z_1 corresponds to a parameter φ in fig. 3.10, arbitrary boundary conditions of z_1 can be given.

So far, the goal point of stabilization was the origin, which is an equilibrium point. To stabilize the system to the arbitrary configuration, an equilibrium point should be shifted to the desired point. However, the arbitrary change of its location of eq. (3.16) is not easy with the standard linear control theory.

It is not easy to stabilize the system in (3.16) to the non-zero desired configuration with the standard linear control theory. Fortunately, nonlinear coordinate transformation to shift an equilibrium point of the chained system is developed in [79], as following equations:

$$z_i^r \triangleq z_i^d + \sum_{j=2}^{i-1} z_j^d \frac{1}{(i-j)!}(z_1 - z_1^d)^{i-j} \tag{3.17}$$

$$\bar{z}_i = z_i - z_i^r, \quad i \in \{1, \cdots, n\} \tag{3.18}$$

where z_i^d is a desired point. This coordinate transformation is valid for any control strategy to control the chained system to the origin. The equilibrium point of the closed loop system can be shifted to the arbitrary desired configuration z_i^d, using the transformation in (3.17)–(3.18). In order to use coordinate transformations in (3.17)–(3.18), $z_i (i \in \{1, \cdots, n\})$ should be controlled to the origin. The pseudo-linearized system of (3.16) can be stabilized to the origin easily by a state feedback. By setting the proper a_0, z_1 goes to 0 with linear convergence. This determines a finite time to control.

In fig. 3.24, joint angles versus time resulted from the feedback control by a pseudo-linearization are plotted (simulation 4). Joint angles of initial configuration $\theta(0)$ and desired configuration θ_d are $\theta(0) = [10, 10, 10](deg)$, $\theta_d = [-10, -10, -10](deg)$. $v_1 = a_0$ was $-0.3272(rad/sec)$ and finite time of convergence was 20 seconds. Feedback gains were $[g_1, g_2, g_3]=$ $[3.7131, -5.7865, 3.1623]$. It can be seen that the joint angles reach the desired values after the finite time. Fig. 3.25 shows input velocities of two actuators. Since a prototype has a velocity saturation at $u_1 = 5r/min$, u_2 should be scaled with the same rate. As described in section 3.5.1, resultant trajectories in the joint space under the existence of velocity saturation is same to the trajectories without a velocity saturation.

An experiment was carried out under the same condition (experiment C). Fig. 3.26 shows experimental joint angles. Measured inputs are presented in fig. 3.27. Since there were some noise on the signal from the potentiometers, high frequency noise can be seen in inputs. The index of errors at the 20 seconds was $E= 1deg$. Profiles of the joint angle trajectories are almost same as those of fig. 3.24. Since the control scheme introduced here requires only simple computation, it can be used for a nonholonomic manipulator with large number of joints. A resultant motion of the prototype nonholonomic manipulator is shown in fig. 3.28.

Similar approaches to the pseudo-linearization were made previously. For example, Samson pointed out this idea in [70]. Sampei [64, 67] proposed a general formulation of pseudo-linearization which can be extended to a class of driftless systems including chained form. Astolfi's controller [2, 3, 4] is designed based on the idea of uniform convergence of z_1.

However, the strategy introduced in this section is especially useful for the nonholonomic manipulator because z_1 is designed as a parameter, which cannot be achieved for other systems that can be converted into the chained form. For the car with n trailer system, z_1 corresponds to the x coordinate of the last trailer. Therefore, z_1 should be carefully dealt with. Convergence analysis of Sampei's controller has not been made clear because z_1 is controlled independently. For the nonholonomic manipulator, only a subsystem in (3.16) should be considered and its convergence rate is exponential.

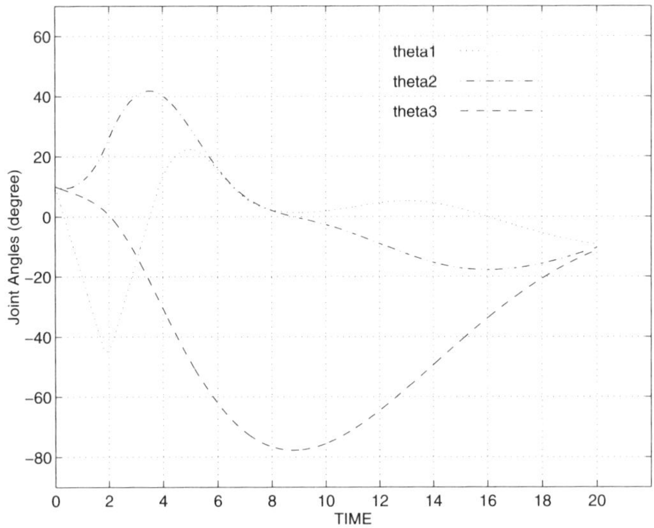

Fig. 3.24. Joint angles versus time with the feedback control by pseudo-linearization: simulation 4.

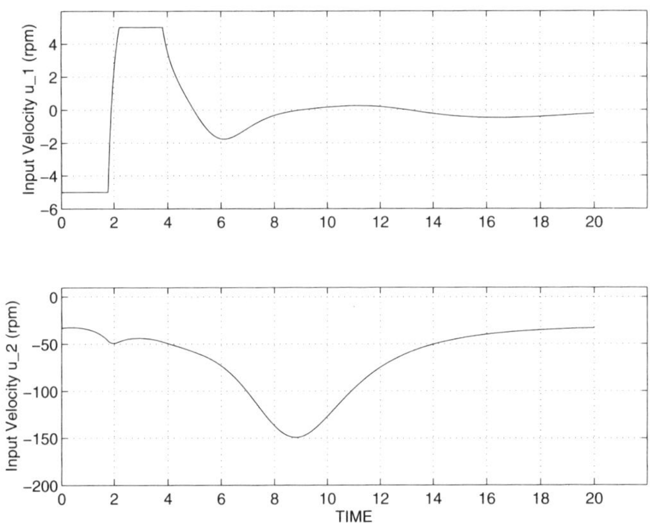

Fig. 3.25. Input angular velocities of two motors versus time with the feedback control by pseudo-linearization: simulation 4.

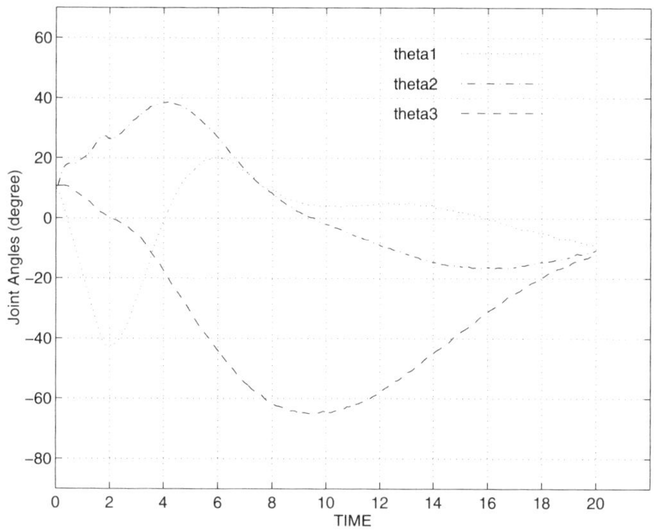

Fig. 3.26. Joint angles versus time with the feedback control by pseudo-linearization: experiment C.

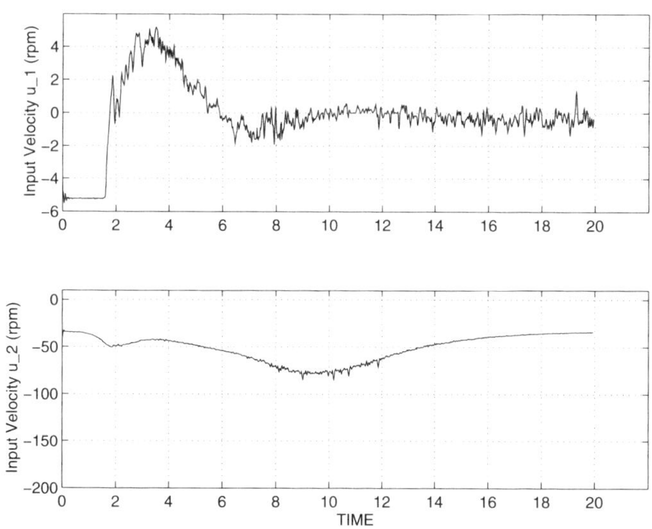

Fig. 3.27. Input angular velocities of two motors versus time with the feedback control by pseudo-linearization: experiment C.

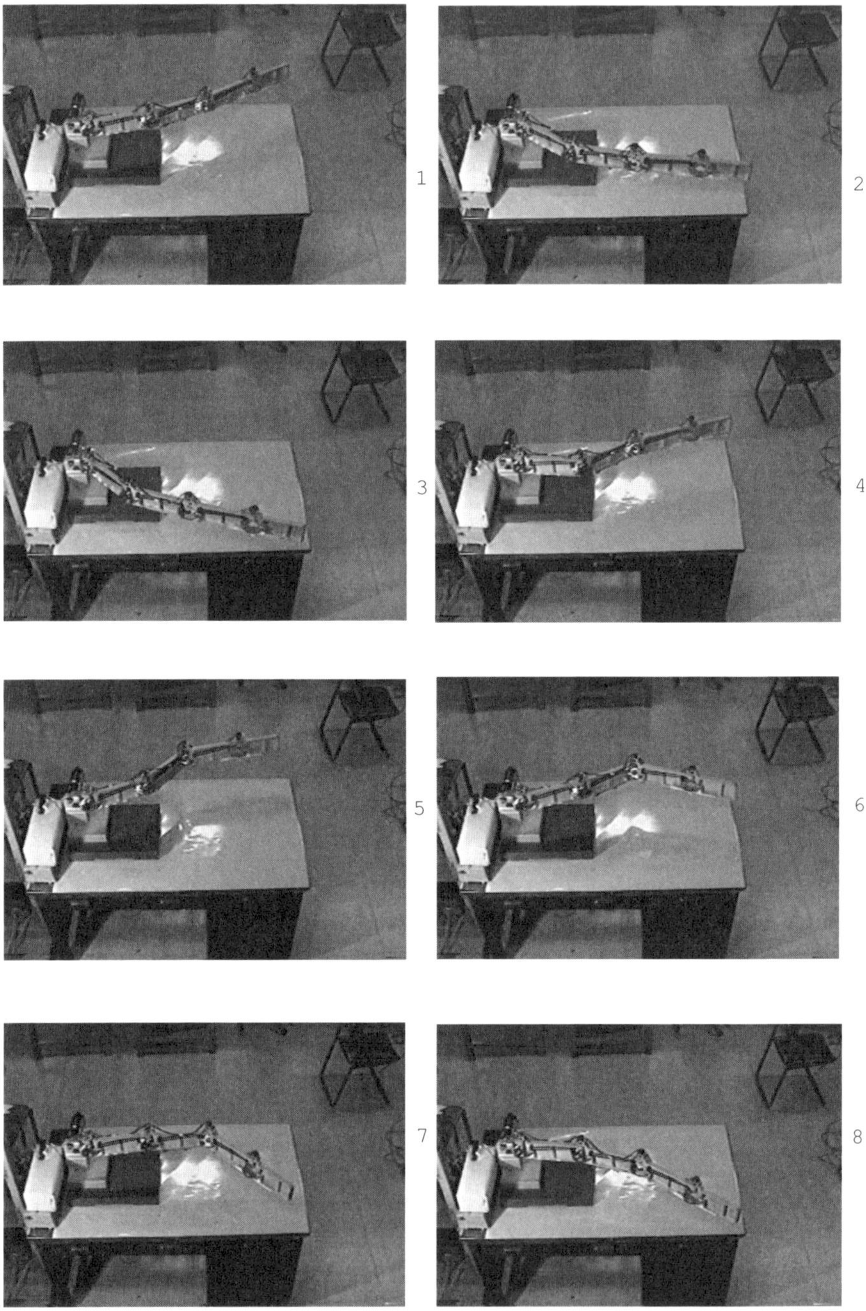

Fig. 3.28. A resultant motion of the nonholonomic manipulator: experiment C.

3.6 Conclusion

The nonholonomic manipulator is designed from the viewpoint of kinematics and velocity constraints. However, practical control performances are dependent upon the system dynamics. If there exists discontinuity of the input angular velocity, the joint velocity becomes infeasible due to the violation of the kinematic constraints. The discontinuity of the joint velocity requires high joint torque which exceeds the mechanical limitation of the nonholonomic gear. Therefore, the input of the nonholonomic manipulator should be continuous in order to guarantee kinematic constraints. Once the smooth manipulator motion is obtained, joint torques of the nonholonomic manipulator can be computed by applying the conventional inverse dynamics. The maximum joint torque is limited by the frictional forces of the nonholonomic gear. Therefore, feasible manipulator motions should be carefully planned by taking the joint torque limitations into account.

Various issues in mechanical design of a nonholonomic manipulator were discussed. Analysis and computations on the mechanical design were carried out and applied to the fabrication of the prototype.

By simulations and experiments, it was shown that any of the existing controllers to the chained form can be applied to control the nonholonomic manipulator. Experimental results proved the usefulness of design of the nonholonomic manipulator and applied control schemes. A time scaling technique is shown to plan physically feasible paths. A practical feedback strategy by pseudo-linearization was proposed and shown to satisfy control simplicity.

4

Design of the chained form manipulator

4.1 Introduction

Nonholonomic underactuated systems are typically modelled as highly nonlinear ones, which becomes obvious as the dimension of the system increases. To the author's knowledge, so far there is no canonical form for the general nonholonomic systems, for example, free flying space robots. Therefore, even a proof of the controllability is made based on numerical computations, for example, see [49]. Furthermore, a motion planning requires high computational costs in dealing with complicated equations, for example see [82, 52].

Fortunately, chained form conversion provided powerful tools to control a class of systems including wheeled mobile robots. Systems which can be converted to the chained form can be controlled easily with low computational costs. Sørdalen showed conversion of a mobile robot with n trailer system to the chained form [72].

Goursat normal form [8, 45, 89] can be thought as another canonical form. Main advantage of using Goursat normal form is that conditions for conversion is straight forward and there exists an algorithm for finding necessary coordinate transformation. Since Goursat normal form can be converted to the chained form, controllers to the chained form are applied in practice.

Power form [41, 17], which is dual of the chained form, is another canonical form. Since power form can be transformed to the chained form, there are no fundamental difference. As Stated by Samson [70], two forms present complementary advantages and drawbacks. A kinematic model of a car with n trailer system represented by cartesian coordinates is closer to the chained form and it is the reason why chained form system is mainly considered in this dissertation.

There are many possible alternatives of designing active underactuated manipulators with only two actuators using the nonholonomic gear. The nonholonomic manipulator was designed focusing on the mechanical simplicity. As a result, several problems were faced with in a high dimensional model. Only three joint model were considered for the case of the prototype nonholonomic manipulator.

In this chapter, it is clarified that such problems are caused by mathematical properties of a kinematic model and the mechanical structure of power transmission. Mathematical problems related to highly complicated chained form conversion equations are investigated by numerical computations. Such problems are design requirements of a new underactuated manipulator consisting of large number of joints.

The chained form manipulator is designed to achieve control simplicity, although a little complication should be added to mechanical structures. The improvement of mathematical properties of chained form transformation equation is successfully carried out with the chained form manipulator. This fact does not imply that the nonholonomic manipulator is an extreme example of poor design. The car with n trailer system [29], for example, has ill conditioned conversion equations. Although chained form conversion is carried out diffeomorphically, ill conditions of conversion equations make control extremely difficult. Mechanical design and a kinematic model are presented according to the design requirements. Results of numerical computation of coordinate transformation and a simulation show that the chained form manipulator has a well conditioned kinematic model even for the high dimensional model.

4.2 Problems of the nonholonomic manipulator

4.2.1 Numerical computation of the coordinate transformation

A validity of chained form conversion is guaranteed by mathematical proof of diffeomorphism as presented in section2.5. However, control performances of the system are greatly affected by the properties of conversion equations.

For the nonholonomic manipulator, $f_i(\cdot)$ in eq.(2.40) is given as a following equation.

$$f_i(\cdot) = k_i \tan\theta_{i-1} \frac{1}{k_n \prod_{j=i}^{n-1} \cos\theta_j} \tag{4.1}$$

By substituting $f_i(\cdot)$ of eq. (4.1) into the eq.(2.43), the coordinate transformations are determined. It is obvious that the coordinate transformations become highly complicated as the number of joints increases. This fact results in a computational inefficiency and local ill-condition. A speed of computation is not a dominant problem. A more significant factor is properties of the conversion equations.

To investigate the feature of nonlinear mapping, 1/4 of a unit sphere in the joint space is mapped into z space. Fig. 4.1 shows the computed results. A quadrant of a sphere ($r = 30deg$) in the joint space is in the left top. A mapped image in z space ($r = 30deg$) is in the right top. The left and right bottoms are the images of $r = 60deg$, $r = 80deg$ respectively. Three axes represent θ_1, θ_2 and θ_3 in the joint space and z_2, z_3 and z_4 in z space, respectively.

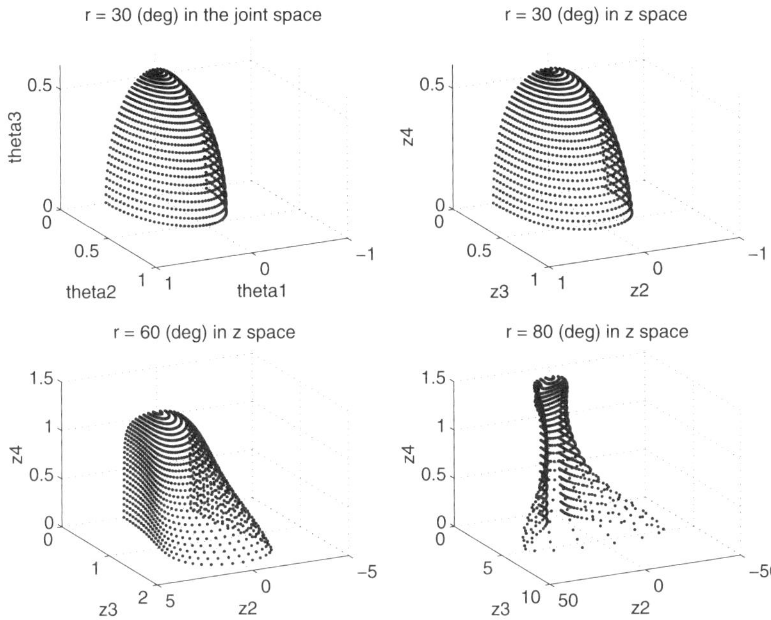

Fig. 4.1. Mapping of a quadrant of sphere from θ to z space.

It is clear that regions nearby origin show strong similarity between two spaces. As the radius of a sphere increases, the unit sphere is partially scaled up and down. This fact implies that the path in the joint space can be mapped into highly distorted path in z space. This kind of distortion becomes larger in the region which is far from the origin. Another problem is that a small computational error in the joint space possibly enlarged in the region which is far from the origin. Numerical problem such as a divergence occurred in computing a trajectory of the nonholonomic manipulator by a numerical integration when the joint trajectory goes close to $\pm 90deg$ which is a singularity of coordinate transformation. These problems are caused by a strong nonlinearity and a complexity of transformation equations and become serious for high dimensional models.

Fig. 4.2 illustrates that the controller is implemented using the chained form conversion, which are nonlinear input and coordinate transformations. Detailed conversion equations were described in section 2.5.

4.2.2 Simulated motion of the 5 joint nonholonomic manipulator

In this section, a joint trajectory of the nonholonomic manipulator is computed to clarify the drawback of the ill-conditioned conversion equations. Among the many open loop controller, sinusoidal inputs [88] are applied to

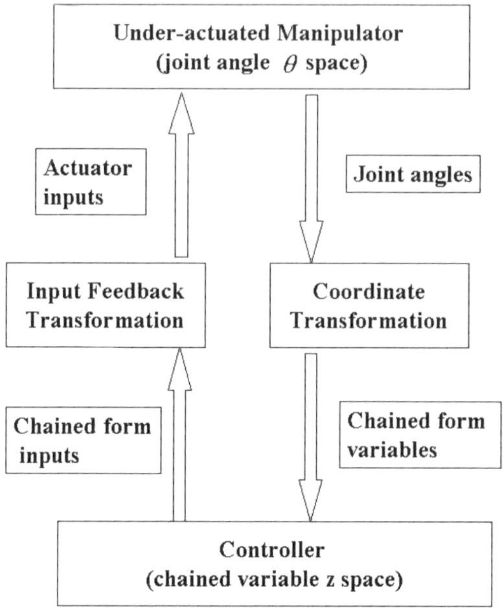

Fig. 4.2. Control of the chained form manipulator using the chained form transformation.

compute a joint trajectory. Details on the motion planning scheme will be presented in section 5.3.

In a simulation, boundary conditions are given as: $\theta_i(0) = -30deg$, $\theta_i^d = 30deg$ ($i \in \{1, \cdots, 5\}$) in the joint space, which corresponds to $z(0) = [0 \ -49.68 \ -6.431 \ -1.442 \ -0.6667 \ -0.5236]$, $z^d = -z(0)$. Fig. 4.3 shows normalized trajectories of z_i. It can be seen that z_i's move to their desired values smoothly. Joint trajectories corresponding to fig. 4.3 are computed using inverse coordinate transformations and presented in fig. 4.4. On the contrary to the smooth trajectories in fig. 4.3, it is clear that the joint trajectories are highly oscillatory, which is hard to implement. It is natural to conclude that the ill-conditioned mapping as shown in fig. 4.1 resulted in highly distorted joint trajectories. Ill-conditions become serious as the number of joint increases, which results in poor control performance of the nonholonomic manipulator with many joints. A presented result in fig. 4.4 is one of the "successful" examples. For most cases, physically infeasible paths were obtained depending on boundary conditions and the choice of a controller.

The coordinate transformation z_i in eq.(2.43) is iteratively determined from the choice of h_n in eq.(2.42). Input transformation in eq.(2.45) and (2.46)

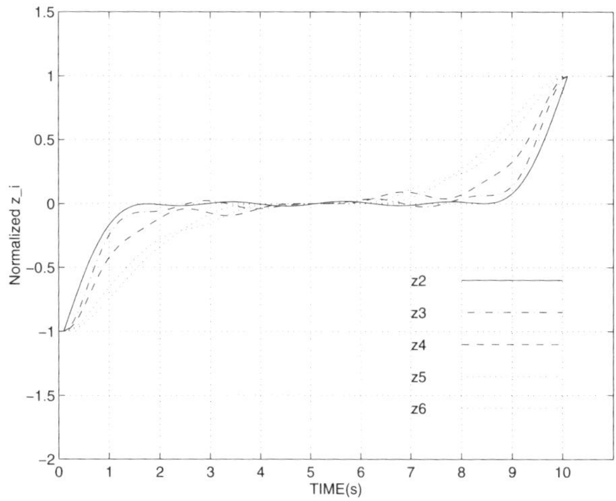

Fig. 4.3. Normalized chained form variable z versus time with sinusoidal inputs. (Simulated result)

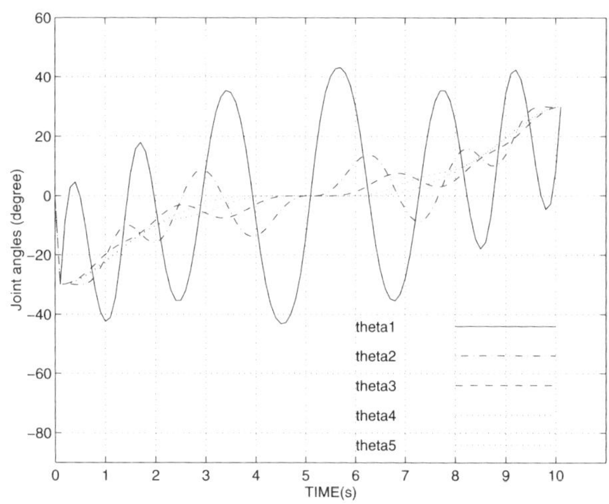

Fig. 4.4. Simulated joint angles of the nonholonomic manipulator.

are dependent on h_n also. Conversion equations are not unique depending on the choice of h_n in eq.(2.42) and h_1 in eq.(2.44). However, even if h_n is chosen as a simplest form, conversion equations remain complicated, which can be checked easily.

$f_i(\cdot)$ in eq.(2.40) is determined by the kinematic structure of the system. From eq.(2.43), it is clear that $h_i(\underline{x}_i)$ is affected by the selection of $f_i(\cdot)$. Since $f_i(\cdot)$ implies joint driving mechanism using the nonholonomic gear, an appropriate choice of $f_i(\cdot)$ makes kinematic structure simple. It can be concluded that simplicity of the transformation equation is important and should be well considered in mechanical design. It is much more significant for the nonholonomic manipulator with a large number of joints.

4.2.3 Mechanical problem related to the power transmission

The nonholonomic gear utilizes frictional force to transmit joint driving force. Therefore, limits of transmitted power is relatively low compared with typical mechanical components such as gears or shafts. For the nonholonomic manipulator, driving torque of joint i is transmitted from the actuator through $i-1$ nonholonomic gears. Fig. 4.5 illustrates power transmission of the nonholonomic manipulator. All the joint driving torques of joint i ($i \in \{2, \cdots, n-1\}$) are applied to the nonholonomic gear at the first joint, which confines the number of joints in practice.

The modular structure is advantageous in that it can be easily reconstructed. However, the structure of cascade concatenation of the nonholonomic gears is unfavorable from the viewpoint of power transmission. Although the number of joint can be arbitrary in theory, it is limited by this kind of mechanical drawback. Together with mathematical problems, mechanical problems should be considered in manipulator design.

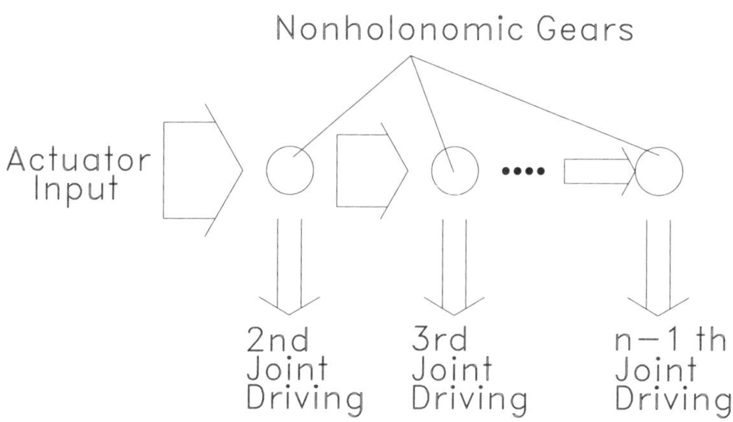

Fig. 4.5. Power flow of the nonholonomic manipulator

4.3 Design of the chained form manipulator

4.3.1 Design of the main power train

At first, the power transmitting mechanism as shown in fig. 4.6 is employed to solve the mechanical problem described in section 4.2.3. A distinguishing

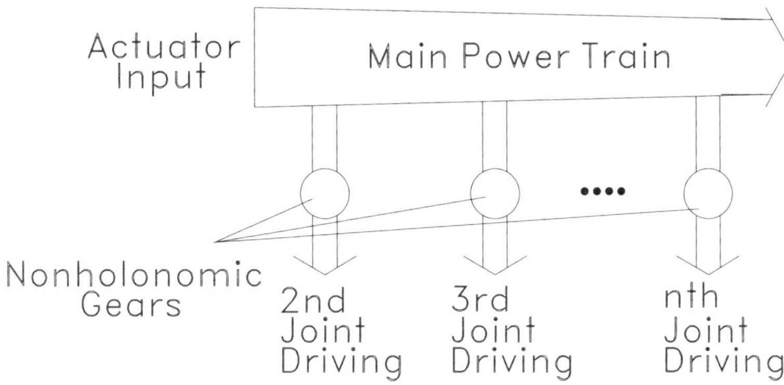

Fig. 4.6. Power flow of the chained form manipulator

characteristic is to use a main power train which is placed along the manipulator. By using a shaft as a main power train, joint driving torques are not concentrated on a nonholonomic gear, which is different from the case of the nonholonomic manipulator. Therefore, high dimensional model can be easily built from the viewpoint of power transmission. Angular velocities from the main power train are transmitted to each joints through nonholonomic gears, which provides nonlinear kinematic structure.

It is obvious that a rigid shaft cannot be used as a power train because of a manipulator motion. In addition, it is desirable to transmit inputs uniformly to all the joints independent of joint motions, which makes a kinematic model simple. By using a mechanism presented in this section, a manipulator can have a spatial structure without any restriction. However, the scope of this section will be limited to the planar structure for simplicity.

In fig. 4.7, a schematic diagram of velocity transmission using universal joints (Hooke's joints) is presented [23]. As well known, ω_B, the angular velocity of the shaft B is given as following equation.

$$\omega_B = \frac{\cos\varphi}{1 - \sin^2\theta_A \cdot \sin^2\varphi}\omega_A \tag{4.2}$$

where θ_A denotes the rotation angle of a shaft A. From eq. (4.2), it is clear that the transmitted velocity with one universal joint varies periodically. Therefore, another universal joint is connected serially to eliminate a periodic variation. To summarize, ω_C is equal to ω_A in fig. 4.7. In practice, point D in fig. 4.7 is placed on the joint axis. Accordingly, the length of a shaft B is changed depending on the joint angle. This problem can be solved by using a sliding link which transmits rotations. The shafts A and C are placed to link $i-1$ and i, respectively.

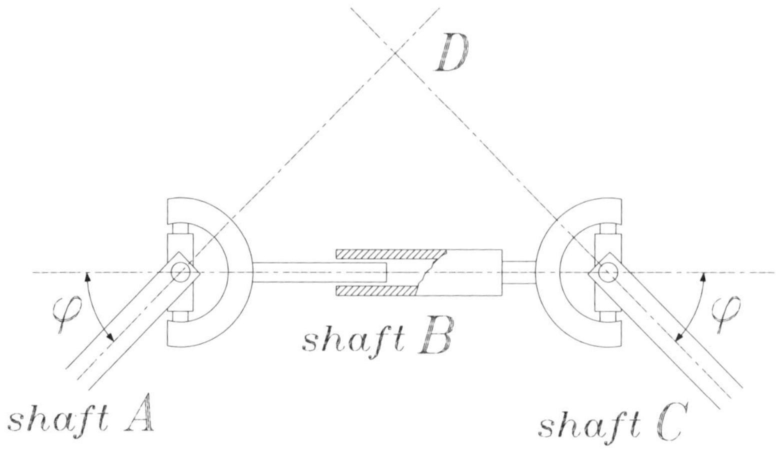

Fig. 4.7. Velocity transmission using universal joints.

4.3.2 Joint driving of the chained form manipulator

In order to create nonlinear structures of the chained form manipulator, the nonholonomic gears are used as the nonholonomic manipulator. Remind that if the nonholonomic gear is placed to the joint i, its gear ratio can be $\sin\theta_i$, $\cos\theta_i$, $\frac{1}{\cos\theta_i}$ or $\frac{1}{\sin\theta_i}$ depending on the placement of wheels when $\theta_i = 0$.

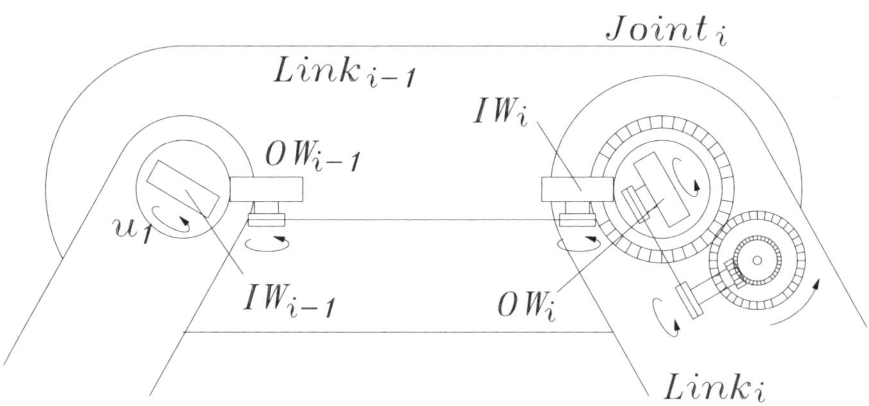

Fig. 4.8. Joint driving mechanism of the chained form manipulator

Fig. 4.8 shows a joint driving mechanism of the chained form manipulator. At first, the joint $i-1$ is considered, where $i \in \{2, \ldots, n-2\}$ if the number of joints are $n-2$. An input wheel IW_{i-1} and an output wheel OW_{i-1} are in contact with a ball at the north pole and on the equator of a ball respectively. The axes of rotation of IW_{i-1} and OW_{i-1} are fixed to the links $i-2$ and $i-1$

respectively. If IW_{i-1} is driven with a input angular velocity $\omega_{IW_{i-1}} = u_1$ by the main power train, then the output becomes as a following equation:

$$\omega_{OW_{i-1}} = sin\theta_{i-1}u_1 \tag{4.3}$$

Then $\omega_{OW_{i-1}}$ is transmitted to the IW_i by mechanical components such as shafts and a set of bevel gears. At joint i, IW_i is in contact with a ball on the equator, which is contrary to joint $i-1$. The axes of rotation of IW_i and OW_i are fixed to the links $i-1$ and i respectively. Output is obtained from OW_i on the north pole and its angular velocity is represented as following equations.

$$\omega_{OW_i} = \frac{1}{cos\theta_i}\omega_{IW_i} = \frac{1}{cos\theta_i}\omega_{OW_{i-1}} \tag{4.4}$$

Finally, ω_{OW_i} is transmitted to drive joint i through belts, a shaft, a set of bevel gear and a set of spur gear in sequence. In order to obtain high joint actuating torques, angular velocities are reduced with a gear ratio k_g.

To summarize this principle of transmission, a following equation of the joint angular velocity is obtained:

$$\dot{\theta}_i = k_g\frac{sin\theta_{i-1}}{cos\theta_i}u_1 \tag{4.5}$$

For the nonholonomic manipulator, the nonholonomic gear was used as a single input, double output transmission. On the other hand, the nonholonomic gear in the chained form manipulator possesses single input and single output. First joint is driven by the other actuator input u_2 directly, which is represented as $\dot{\theta}_1 = u_2$. There are two nonholonomic gears to drive joint i.

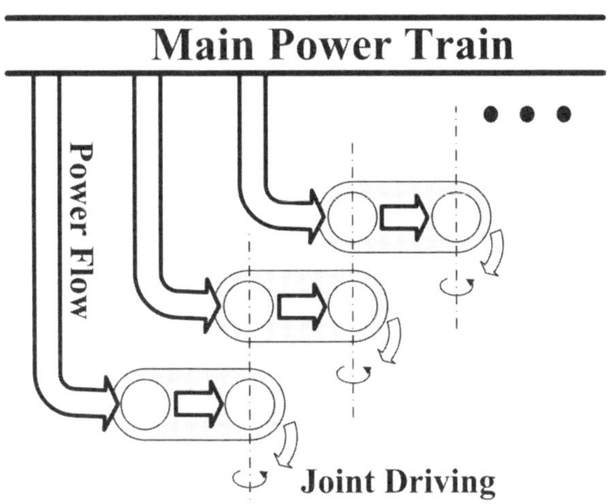

Fig. 4.9. Serial concatenation of manipulator links.

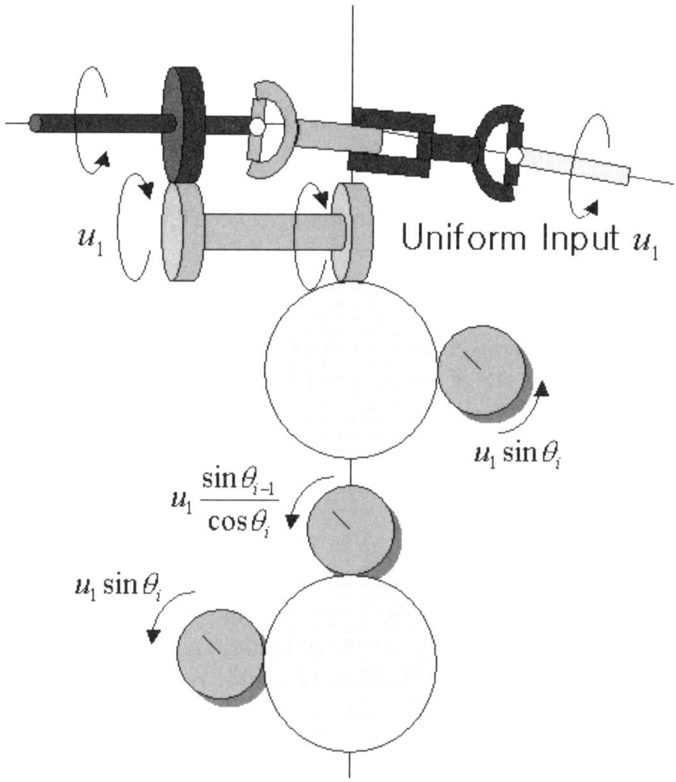

Fig. 4.10. Structure of the joint i seen from side.

One's gear ratio is $sin\theta_{i-1}$ and the other's is $\frac{1}{cos\theta_i}$. This structure is shown in fig. 4.9.

Fig. 4.10 shows the mechanical structure of joint i. As explained in section 4.3.1, sliding link mechanism with universal joints is exploited to transmit u_1. The nonholonomic gear placed above is used as a transmission having a gear ratio $sin\theta_i$. The nonholonomic gear placed below is used to drive joint i with a gear ratio $\frac{1}{cos\theta_i}$. Accordingly, two nonholonomic gears should be placed along the joint axis at joint i. Therefore mechanical structure of the chained form manipulator becomes a little complicated than that of the nonholonomic manipulator.

4.4 Kinematic model

From design of the chained form manipulator in previous section, a following kinematic model is obtained:

$$\dot{\varphi}_1 = k_g u_1 \tag{4.6}$$
$$\dot{\theta}_1 = u_2 \tag{4.7}$$
$$\dot{\theta}_i = k_g \frac{sin\theta_{i-1}}{cos\theta_i}\ u_1 \qquad i \in \{2, \ldots, n-2\} \tag{4.8}$$
$$\dot{\varphi}_2 = k_g sin\theta_{n-2}\ u_1 \tag{4.9}$$

A kinematic model consists of $n-2$ joints which corresponds to n dimensional chained form system. From eqs. (4.6)–(4.9), it is obvious that the kinematic model is described as simple equations regardless of the number of joints, which was unobtainable in the nonholonomic manipulator. Transmitting u_1 through a power train contributes not only to solve the mechanical problem but also to make the kinematic model simple.

Proof of controllability is straightforward by investigating LARC (Lie Algebra Lank Condition), which can be made with the same procedure presented in section 2.4. Notice that the chained form manipulator is controllable in the joint subspace $\theta_i \in (-\frac{\pi}{2}, \frac{\pi}{2})$. Although chained form conversions are carried out locally for both cases, the nonholonomic manipulator is controllable in the whole joint space, which is different from the chained form manipulator.

By rewriting the input u_1 as $\hat{u}_1 = k_g u_1$ in eqs. (4.6)–(4.9), a kinematic model has a triangular structure in (2.38)–(2.40). To obtain coordinate transformation equations, $h_n = \varphi_2$ and $h_1 = \varphi_1$ are assumed in (2.42) and (2.44) respectively. By substituting f_i from eqs. (4.8) and (4.9) to (2.42)–(2.46), coordinate and input transformation equations are given as follows:

$$z_1 = \varphi_1 \tag{4.10}$$
$$z_i = sin\theta_{i-1} \qquad i \in \{2, \ldots, n-1\} \tag{4.11}$$
$$z_n = \varphi_2 \tag{4.12}$$
$$v_1 = k_g u_1 \tag{4.13}$$
$$v_2 = cos\theta_1 u_2 \tag{4.14}$$

where z_1 and z_n are not joint angles, but control parameters.

It is known that systems which can be converted to the chained form are flat. A system is said to be flat if the following conditions are satisfied [63].

1. there exists a finite set $y = y(y_1, \cdots, y_n)$ of variables which are differentially independent, i.e., not related by any differential equations. $y = y(y_1, \cdots, y_n)$ are called flat or linearizing outputs.
2. the y_i's are differential functions of the system variables, i.e., are functions of the system variables (state, input, $\cdots$) and a finite number of their derivatives.

3. Any system variable is a differential function of the y_i's, i.e., is a function of the y_i's and of a finite number of their derivatives.

It can be easily checked the chained form manipulator satisfies above sufficient conditions. Flat outputs are z_1 and z_n in eqs. (4.6)–(4.9). For details on the flatness and motion planning, see, for example, [63, 55]. By putting z_1 and z_n as parameters, free boundary conditions can be given, which provides various advantages to control z_i $(i \in \{2, \ldots, n-1\}$). For example, feedback control using a pseudo-linearization in section 3.5 can be applied. A distinguishing characteristic is that both of flat outputs are parameters in the chained form manipulator. For the case of the nonholonomic manipulator, only z_1 was a parameter. As a result, the chained form manipulator consists of $n-2$ revolute joints in eqs. (4.6)–(4.9) are converted to the n dimensional chained form. The gear ratio k_g in eq. (4.13) implies scaling of inputs.

4.5 Advantages of the chained form manipulator

4.5.1 Numerical mapping of the conversion equations

It is clear that the conversion equations of the chained form manipulator are simple regardless of the number of joints. To investigate the advantage of mapping in eqs. (4.6)–(4.9), a computation was carried out under the same conditions as in fig. 4.1. The result of computed mapping is shown in fig. 4.11. It is also clear that the shape of sphere shows strong similarity even in the region far from the origin. This fact guarantees a well-conditioned mapping between two spaces, which provides a complete and fundamental solution to the mathematical problems of the nonholonomic manipulator presented in section 4.2. Moreover, inverse transformation is expressed explicitly as $\theta_i = arcsinz_i$ where $i \in \{1, \dots, n-1\}$. Therefore, if trajectories are given in z space, trajectories in joint space can be obtained directly. Therefore, manipulator motions can be easily estimated from the trajectory in z space.

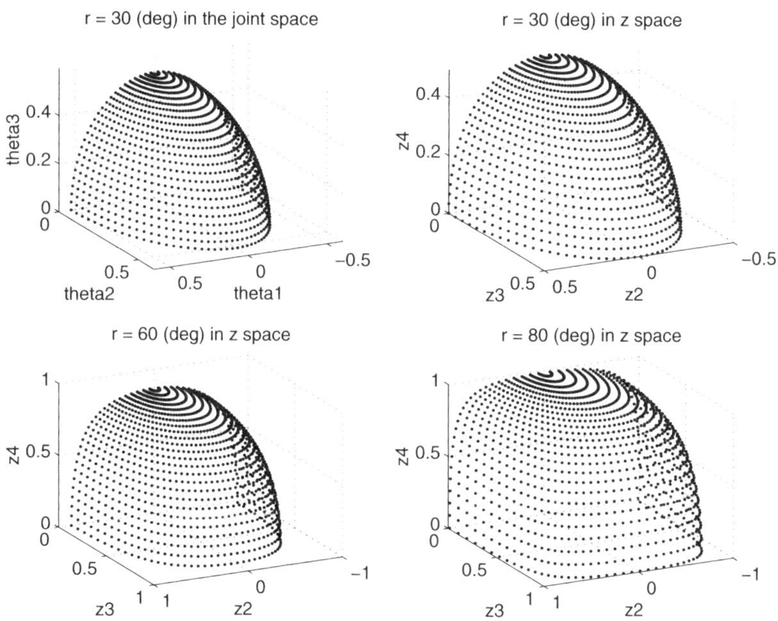

Fig. 4.11. Mapping of a quadrant of sphere from θ to z space for the chained form manipulator.

4.5.2 Simulated motion of the 6 joint chained form manipulator

In order to show the advantage of well-conditioned conversion equations, a simulation is carried out for the 6 joint chained form manipulator. Similar to the simulation in section 4.2, sinusoidal inputs are exploited to plan a trajectory. Boundary conditions are given as $\theta_i(0) = \varphi_2(0) = -30deg$, $\theta_i^d = \varphi_2^d = 30deg$ $(i \in \{1, \cdots, 6\})$, which corresponds to $z_i(0) = -0.5$, $z_i^d = -z_i(0)$ $(i \in \{2, \cdots, 7\})$ in the z space. Free boundary conditions are φ_1 and φ_2, and $\varphi_1(0) = \varphi_1^d = 0$ is given for simplicity. The details on the motion planning strategies will be presented in section 5.3.

Fig. 4.12 presents simulated trajectories of z. It can be seen that z_i's started from initial configurations are smoothly steered to the desired configurations. Inverse coordinate transformations are carried out to map the trajectory to the joint space and presented in fig. 4.13. Joint trajectories are quite smooth and show strong similarity with fig. 4.12, which is obvious from the simple transformation equations in eqs. (4.10)–(4.12).

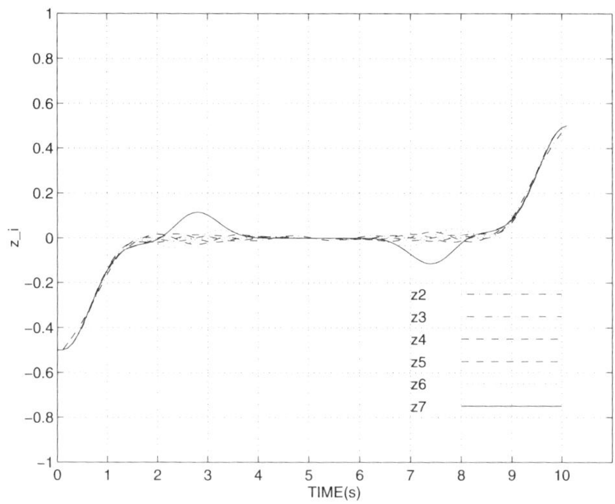

Fig. 4.12. Chained form variable z versus time with sinusoidal inputs.

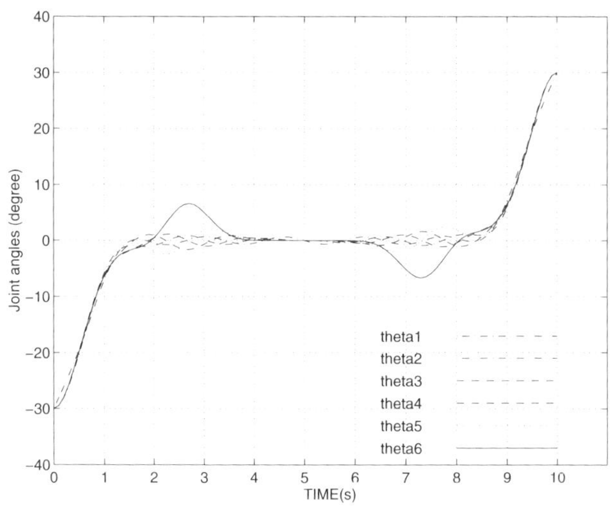

Fig. 4.13. Simulated joint angles of the chained form manipulator.

4.6 Conclusion

Problems of the nonholonomic manipulator was clarified, then design requirements of the chained form manipulator was presented in the light of mathematical and mechanical aspects. Mechanical design of the chained form manipulator was established. On the contrary to the fact that the nonholonomic manipulator was designed focusing on the mechanical simplicity, the chained form manipulator was shown to satisfy control simplicity, which is especially useful for the high dimensional model. The presented simulation result showed that the chained form manipulator consists of large number of joint can be constructed.

In general, it is difficult to steer a chained form system in the region far from the origin. For a car with trailers, control is easier when they stand in a line. In the same sense, the chained form manipulator can be controlled easier when the joint angles are close to 0. Therefore, the chained form manipulator could find its practical application when it has a large number of joints. Since the control simplicity of the chained form manipulator remains even with the increase of number of joints, the chained form manipulator is useful for practical applications.

5

Prototyping and control of the chained form manipulator

5.1 Introduction

Since the chained form manipulator has a well-conditioned kinematic model, development of efficient control strategy for the chained form directly contributes to the improvement of actual performances in practice. In section 3.5, several control strategies were shown and verified with the prototype of the nonholonomic manipulator.

There are two major streams in control of nonholonomic systems. One is feedback control and the other is open loop motion planning. Feedback control is useful for stabilizing a system to the desired configuration and significant to deal with various uncertainties and disturbances. However, it has many disadvantages for the practical applications of nonholonomic systems. For example: 1. Overshoot cannot be reduced to a desired extent and transient behavior of the system cannot be controlled. 2. Although feedback control to the desired trajectory was achieved with a great success for a low dimensional system (*e.g.* single mobile robot), it is difficult to extend to high dimensional systems (*e.g.* a car with n trailers) 3. Stabilization to a nonzero configuration is difficult for the high dimensional system. For example, it is relatively easy to stabilize a car with trailers to a straight (aligned) shape. However, it is difficult to stabilize the system to arbitrary desired shapes. Because of these drawbacks, many researchers have focused on motion planning and open loop control.

A major advantage of open loop control is that it provides solutions for practical applications. For example, problems of avoiding obstacles or singularities can be solved. However, satisfactory performance is not guaranteed for open loop control if there exist disturbances or modelling uncertainties. Open and closed loop strategies have complementary advantages.

In this chapter, we discuss motion planning to obtain a motion that is insensitive to the initial-condition error when used for open loop control. First, a motion planning, which generates a path approximates given holonomic path, is presented by using sinusoidal inputs to the chained form. Using this scheme, any motion computed with conventional robot motion planner can be generated by the chained form manipulator The resultant motion is quite nice and easy to implement. Then, a new concept of motion planning is introduced through the analysis of the initial condition sensitivity of the planned motion. Initial condition sensitivity points out a new direction to improve actual performances. The result of the analysis provides a significant tool to plan the motion which is less sensitive to the error of the joint angles. Combining these two strategies, complete motion planning of the chained form manipulator is developed. The prototype of the chained form manipulator is introduced, then the proposed motion planning scheme is experimentally tested. The usefulness of the design of the chained form manipulator and the proposed control scheme is verified by presented experimental results.

5.2 Prototype of the chained form manipulator

For the experimental verification of the chained form manipulator, a prototype was fabricated as shown in fig. 5.1. It has a planar structure to avoid gravity. It can be designed with a spatial structure if joint actuating torques are sufficient enough to bear gravity. The prototype consists of four joint, whose kinematic model corresponds to the six dimensional chained form system including two control parameters.

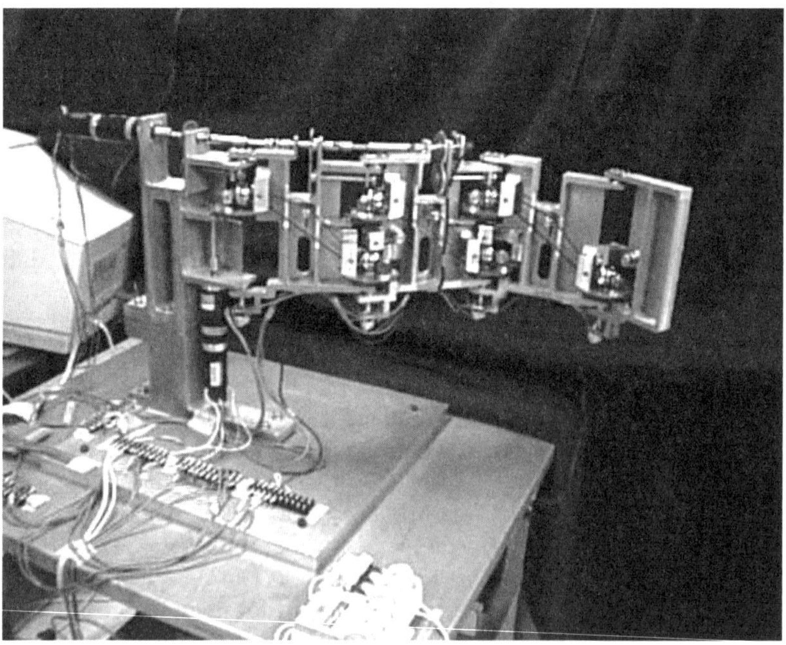

Fig. 5.1. Prototype of the chained form manipulator

Physical dimensions of the nonholonomic gears are the same as the case of the nonholonomic manipulator. Therefore, the mechanical analysis in section 3.3 can be applied in the same way.

Fig. 5.2 shows a joint structure, which is similar to fig. 4.10. In fig. 5.2, the upper and lower nonholonomic gears provide gear ratios of $sin\theta_i$ and $1/cos\theta_i$ respectively. The centers of the nonholonomic gears were precisely adjusted along the joint axis using a jig. Each joint is equipped with a potentiometer to measure the joint angle. The view from other side of the joint is shown in fig. 5.3.

The length of each link is $200mm$. The approximate weight of one link is $2.3kg$. Gear ratio k_g in eq. (4.8) is 1/10, which is created by a set of bevel gears (ratio of 1/2) and spur gears (ratio of 1/5) in sequence.

Fig. 5.2. Closeup view of joint 2 seen from one side.

Fig. 5.3. Joint 2 seen from the other side.

In fig. 5.4, the main power train illustrated in section 4.3 is presented. At the end of the power train, an encoder is placed to monitor the angular velocity of the shaft. By comparing the data from the encoder and the actual input u_1, it can be checked whether input u_1 is transmitted uniformly regardless of the joint motions. A problem of using universal joints is a backlash, which is inevitable to a certain extent. The encoder is used to see the effect of backlashes.

Fig. 5.5 shows the base of the prototype, where two actuators are placed to the base similarly to the nonholonomic manipulator.

The two actuators are brushless DC motors($100W$), which are equipped with planetary gear heads whose gear ratios are $1/216$ for the one that drives the first joint (u_2) and $1/72.4$ for the one that drives main power train (u_1) respectively. Velocity feedback control is adopted using their own controllers. The thickness of the base frame is $60mm$. Each link consists of two primary

Fig. 5.4. The main power train consisting of universal joints.

components. One is a frame ($t = 25mm$) and the other is a plate ($t = 4mm$). These two components are processed through the wire cutting. It contributes to reduce the fabrication cost of the link compared with the case of cutting down a material block by machining.

Fig. 5.5. The two actuators are placed to the manipulator base.

5.3 Motion planning

5.3.1 Related studies and unsolved problems

The nonholonomic motion planning problem has been one of the challenging topics and most of the strategies were developed for a driftless nonholonomic systems.

Lafferriere and Sussmann [26] proposed a motion planning scheme which gives exact solutions for nilpotent systems and approximate ones for non-nilpotent systems. The basic idea is to use an extended system by adding higher order Lie bracketing input vector fields. For the appropriate selection of the basis, P. Hall coordinates were used. Then, inputs for the extended system is obtained. Finally, inputs for the extended system are generated by the inputs of the original system. This procedure implies approximation of a given holonomic path by a feasible nonholonomic path. Although this approach is applicable quite generally, it is hard to implement because the concatenation of piecewise constant inputs was used to generate higher order inputs, which results in time-consuming motions.

Sussmann and Liu [81] used sinusoidal inputs of asymptotically high frequency and amplitude. As the frequency of the sinusoidal inputs goes to infinity, the generated path uniformly converges to the desired trajectory. Tilbury, Laumond, Murray, Sastry and Walsh [87] made a synthesis of [81] and [47] for an example of a mobile robot with two trailers. However, the resultant trajectory is highly oscillatory because of the use of high frequency inputs. Although this approach is applicable for a broad class of systems which are not necessarily nilpotent, it is still difficult to implement because the planned motion is highly oscillatory.

Many progresses were achieved for the mobile robots, by Laumond, Jacobs, Taïx and Murray [30], and Laumond, Sekhavat and M. Vaisset [31]. Since local planners are designed based on the geometric properties of mobile robots, these approaches cannot be directly applied to the motion planning of the chained form manipulator.

Tilbury, Murray and Sastry [89] proposed useful path planners for the chained form. Instead of time-consuming sinusoids in [47], sinusoidal inputs which can steer a system all at once were presented. Time polynomial inputs and piecewise constant inputs [44] can be alternative choices. Those strategies have a common feature that inputs can be computed by solving algebraic equations using the input parameterization.

Although there are many previous studies on the nonholonomic motion planning problems, it is still difficult to compute a path which satisfies practical needs, as we discuss further in this section. In section 3.5, it was shown that any existing controllers to the chained form can be applied to the control of the nonholonomic manipulator. The goal of this section is to develop an efficient motion planner which can be easily implemented for the chained form manipulator.

Two performance measures are proposed to achieve two major subgoals. The first subgoal is to compute the feasible nonholonomic path which approximates any given holonomic reference path. Then, it is shown that the computed path shows a nice performance in simulation. However, the actual performances are greatly affected by various errors, because it is an open loop control. In order to deal with such errors, the new performance measure is proposed. By taking the initial error of the configuration variables into account, less sensitive motion can be generated. By combining these two major approaches, the motion planning algorithm of the chained form manipulator is constructed.

5.3.2 Motion planning to approximate the given holonomic path

Although the controllable nonholonomic system can be exactly steered to the desired configurations, the resultant trajectory shows a detour to reach the goal. For example, a wheeled mobile robot cannot move sideways due to no-side-slip condition, as illustrated in fig. 5.6. In most cases, this fact results in an error between the holonomic, or the most preferred, path and a planned feasible path in the configuration space. The error can be reduced by using appropriate motion planning schemes, which implies approximating the holonomic path with the feasible nonholonomic path.

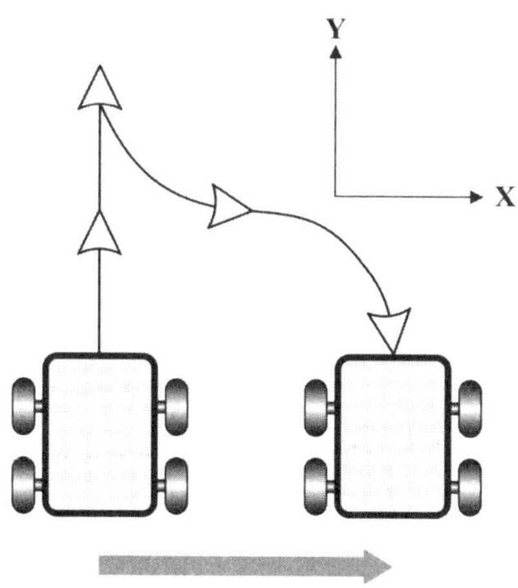

Fig. 5.6. A wheeled mobile robot making a detour to reach its desired configuration.

The motion planning scheme for the chained form manipulator is developed based on the sinusoidal inputs proposed in [89]. One of the requirements is that the motion planner should be easily extended to the high dimensional system with feasible computational cost. In addition, it is desirable for the resultant trajectory being smooth (not highly oscillatory) for practical implementations.

The three strategies proposed in [89] can be directly applied to steer the chained form manipulator. Major advantages are: 1. The resultant trajectories are not highly oscillatory. 2. Inputs are easily obtained by solving algebraic equations and this property is maintained for the high dimensional system. The authors in [89] suggested to apply the three schemes in appropriate ways, *e.g.* to avoid obstacles or singularities of transformation. Performances are dependent on the system and boundary conditions. In this section, sinusoidal inputs are selected.

Sinusoidal inputs are given as following equations:

$$v_1 = d_0 + d_1 sin(\omega t) \tag{5.1}$$

$$v_2 = e_0 + e_1 cos(\omega t) + \cdots + e_k cos(k\omega t) \tag{5.2}$$

$$k \in \{1, 2, \cdots, n-2\}$$

$d_1 \neq 0$ in eq. (5.1) is a parameter which can be chosen arbitrary and all the other coefficients are explicitly determined by solving algebraic equations under the given boundary conditions. For the case of the chained form manipulator, the holonomic reference path can be obtained by applying conventional robot motion planning techniques. Then, the path can be approximated by the feasible path generated by motion planner for the chained form, *i.e*, sinusoidal inputs in eqs. (5.1) and (5.2). Overparameterized d_1 is used to reduce the error of approximation.

The chained form manipulator has singularities at $\theta_i = \pm 90 deg$, where $i \in \{2, \ldots, n-2\}$, which is obvious from eq. (4.8). Trajectories in the z space should satisfy the condition $|z_i| < 1$, where $i \in \{2, \ldots, n-1\}$, which can be directly checked from eq. (4.11). Therefore, motion planning should be carefully carried out to avoid singularities. Since there are singularities also in the nonholonomic manipulator and the car with n trailer system, a motion planning scheme in this section can be applied to many systems which satisfies chained form convertibility. Furthermore, in a cluttered environment, a holonomic collision-free path can be approximated by the feasible path of the chained form manipulator.

The holonomic reference path is set to be a straight line connecting the initial and desired configuration in the configuration space. The index of the approximation error is defined by the maximum distance between the reference path and the points on the planned path. Therefore, the distance l is described as following equations.

$$l = \frac{(L_2^T L_2)(L_1^T L_1) - (L_1^T L_2)^2}{L_1^T L_1} \tag{5.3}$$

$$L_1 = P_2 - P_1 \quad L_2 = P_3 - P_1$$

where initial point is $P_1 \in \mathbb{R}^{n-2}$, desired point is $P_2 \in \mathbb{R}^{n-2}$ and points on the planned path is $P_3 \in \mathbb{R}^{n-2}$, where n is the dimension of the chained form

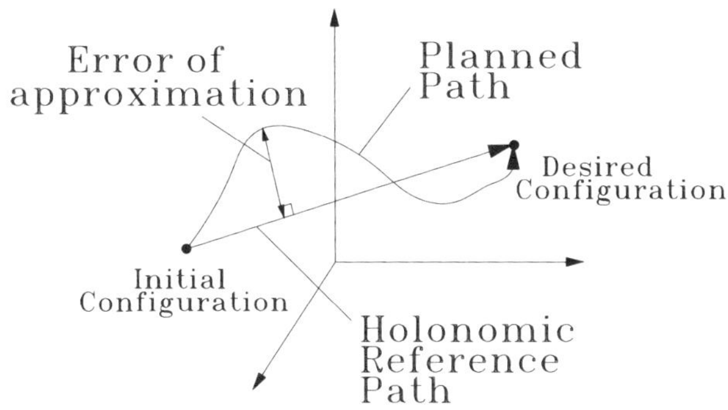

Fig. 5.7. Illustration of the index of the error of approximation.

manipulator and the number of joints is $n-2$. Conceptual illustration of the index in the configuration space is shown in fig. 5.7.

Since the trajectory is obtained explicitly, the error in eq. (5.3) is computed straightforward along the trajectory under a certain d_1. The parameter d_1 is used to minimize the error. Since this method implies one dimensional searching procedure in the parametric space, the computational cost is acceptably low. Although the resultant error depends on the boundary conditions, satisfactory motions were obtained in many cases.

To investigate the usefulness of the motion planning using sinusoidal inputs, a simulation is carried out for a four joint chained form manipulator, which is converted to the six dimensional chained form (simulation 5). The initial configuration $\theta(0)$ and the desired configuration θ^d are $\theta_i(0) = \varphi_2(0) = -30deg$, $\theta_i^d = \varphi_2^d = 30deg$ $(i \in \{1, \cdots, 4\})$. A parameter d_1, which was obtained by a parametric minimization, is -0.62 and the error of approximation is 0.025 (rad). Other parameters in eq. (5.2) are $[e_0, e_1, e_2, e_3, e_4] = [0.033, 0.044, 0.007, -0.015, -0.012]$. Total time to control is $T = 30(s)$. Boundary conditions of z_1 and z_n can be given arbitrary as explained in section 4.4. In a simulation, $z_1(0) = z_1(T) = 0$ were chosen to save a computational cost because it results in $d_0 = 0$.

Fig. 5.8 shows the computed trajectories of joint angles. It can be seen that all the joint angles started from the initial configurations are steered to their desired configurations smoothly. Input angular velocities of two actuators are plotted in fig. 5.9. If the required accelerations or velocities exceed the actuator limits, inputs can be modified by the time scaling technique explained in section 3.5. Therefore, planned motions can be always made physically feasible, which is common in the nonholonomic manipulator.

A manipulator motion in the plane is presented in fig. 5.10. The length of each manipulator link is 1. It is clear that the generated motion is quite smooth and similar to the motion of a conventional robot manipulator. The same motion is plotted versus time in fig. 5.11, where z axis represents time.

Fig. 5.12 and fig. 5.13 show the joint trajectories in the joint space. The trajectory is plotted in the three dimensional subspace. Three axes are θ_1, θ_2 and θ_3 in fig. 5.12 and θ_2, θ_3 and θ_4 in fig. 5.13. It can be seen that the computed trajectory approximates a holonomic reference path within a small error, *i.e.* the trajectory is almost straight.

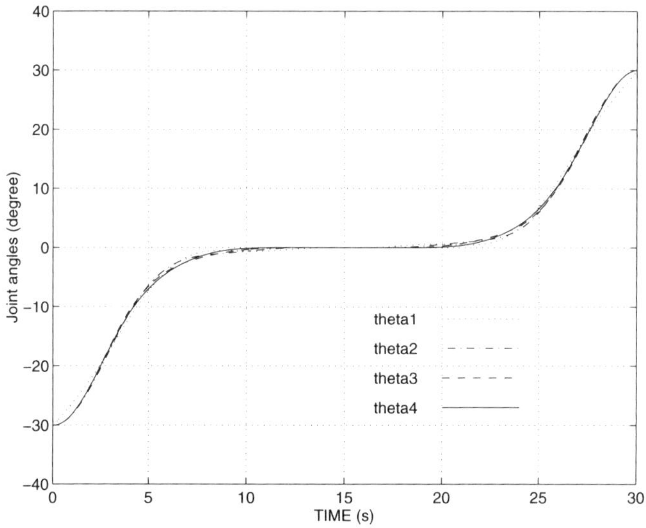

Fig. 5.8. Computed trajectories of joint angles using the sinusoidal inputs: simulation 5.

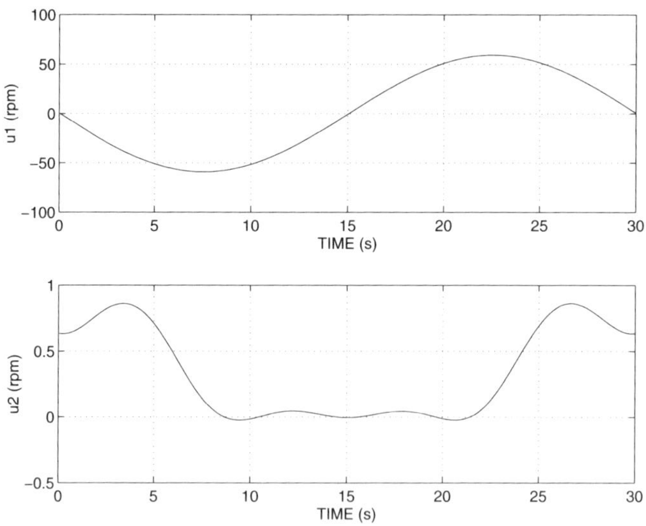

Fig. 5.9. Input angular velocities of actuators using the sinusoidal inputs: simulation 5.

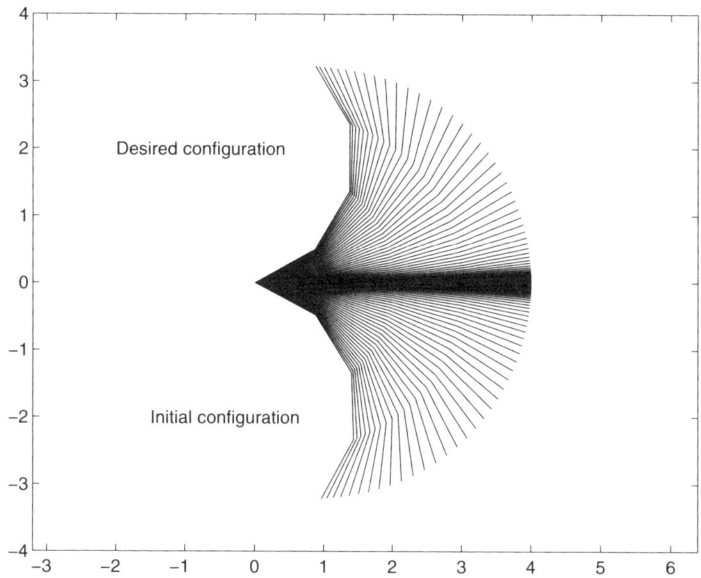

Fig. 5.10. Manipulator motion in $\mathbb{R}^2$ using sinusoidal inputs: simulation 5.

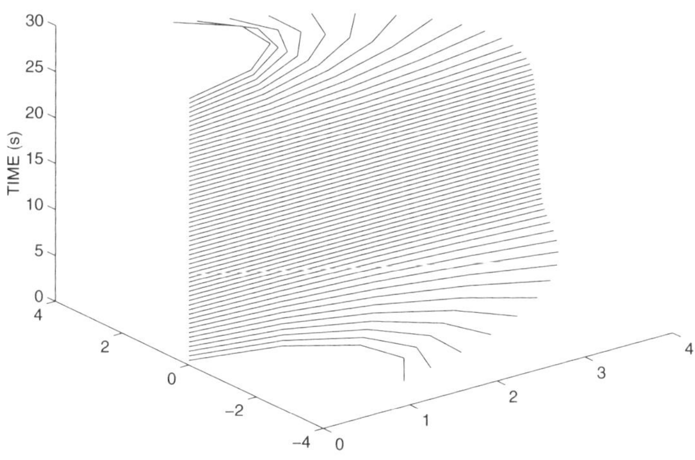

Fig. 5.11. Manipulator motion in $\mathbb{R}^2 \times t$ using sinusoidal inputs: simulation 5.

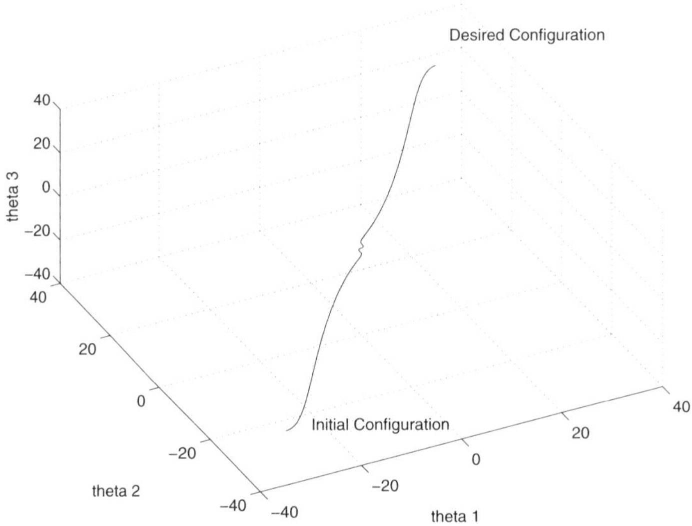

Fig. 5.12. A computed trajectory in the joint space where three axes are θ_1, θ_2 and θ_3: simulation 5.

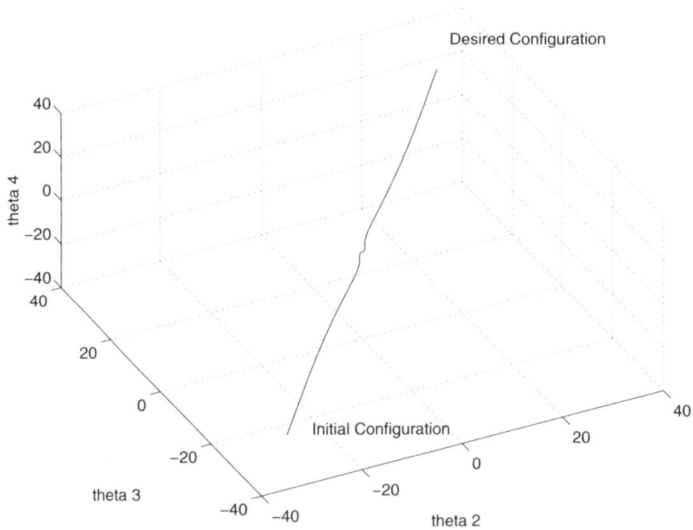

Fig. 5.13. A computed trajectory in the joint space where three axes are θ_2, θ_3 and θ_4: simulation 5.

In order to solve an obstacle avoidance problem, the resultant path should remain within the collision-free space. This fact implies the upper bound of the approximation error is given explicitly. The motion planning scheme presented in this section is an optimization and sometimes the optimized trajectory does not satisfy the condition. In such a case, the error reduced to the desired extent by setting appropriate intermediate points. Fig. 5.14 gives intuitive illustration of setting intermediate points.

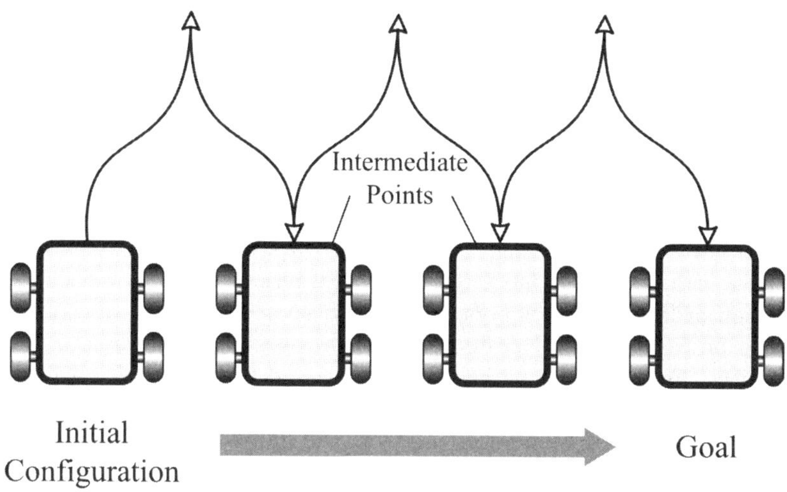

Fig. 5.14. The error of approximation can be reduced by setting appropriate intermediate points.

The motion planner should satisfy the topological property presented in [71]. The topological property implies that if the goal point comes close to the start point, the amount of detour (*i.e.* the error of approximation) is reduced. Since the local motion planner using the sinusoidal input satisfies the topological property, the upper bound of the error can be satisfied by setting appropriate intermediate points. It is one of the important reasons why sinusoidal inputs are selected in this section. However, the motion planning strategy proposed in [71] always requires a large number of intermediate point, which implies there are many switching stages of inputs. A strategy presented in this section generates a less switching trajectories in many cases, although it depends on the given boundary conditions.

5.4 Initial-condition sensitivity of planned motion

Although the motion planning scheme in the previous section shows satisfactory results in simulations, actual performances being used with open loop control are greatly affected by various errors. The velocity control performance of two actuators is a possible source of an error. Furthermore, there are various error sources in the mechanical components. Backlashes at gear trains and universal joints, adjustment of mechanical components and slip at the nonholonomic gear would be dominant error sources. In this section, a new performance measure is explored and defined to plan motions taking account of the initial-condition sensitivity.

In order to investigate the effect of the joint angle error, a simulation is carried out (simulation 6). Instead of the starting joint angle $\theta_i(0) = -30deg$, inputs are applied to the initial configuration of $[\theta_1(0), \theta_2(0), \theta_3(0), \theta_4(0)] = [-29, -31, -29, -31](deg)$, which implies each joint angle has an error of $1deg$. All the other parameters except for the initial joint angles are same to simulation 5. The resultant trajectory under the initial error of joint angles is shown in fig. 5.15. The effect of error is not negligible, which is obvious when fig. 5.15 is compared with fig. 5.8. The error of approximation in simulation 6 is 0.86 (rad), which is more than 30 times larger than that of simulation 5. Motion of fourth joint is poor because of the accumulation of errors. This fact can be easily checked in fig. 5.16, which represents the trajectory in the joint space. From the comparison of fig. 5.16 and fig. 5.13, it is obvious that the small error of joint angles greatly affects the actual performance.

In order to specify the effect of error, the motion of $z(t)$ is investigated under the existence of the initial error ϵ. When the sinusoidal inputs in eqs. (5.1) and (5.2) are applied to the chained form, $z(t)$ is obtained explicitly and it can be easily checked that $z = z(z(0), t, d_1, e_i, \omega)$, where $i \in \{1, \cdots, n-2\}$. For simplicity, free boundary conditions of z_1 are set as $z_1(0) = z_1(T) = 0$, where T represents a total time of control. Then, if $z(t)$ with the initial error is denoted by $z_{err}(t)$, the following relation is obtained by putting $z_{err}(0) = z(0) + \epsilon$.

$$z_{err} = z_{err}(z(0), \epsilon, t, d_1, e_i, \omega) \tag{5.4}$$

where $z_1(0) = z_1(T)$ is assumed. Then, the effect of the initial error on the resultant motion can be represented by the following sensitivity jacobian J_S.

$$J_S(\cdot) = \frac{\partial z_{err}}{\partial \epsilon} \tag{5.5}$$

The above equation is a new performance measure to estimate initial condition sensitivity of the planned motion. J_S is obtained explicitly by substituting inputs in eqs. (5.1) and (5.2) to the chained form and carrying out partial differentiations. Although the equations are quite complicated, the following relation is obtained finally by setting $\omega = \frac{2\pi}{T}$.

$$J_S = J_S(t, d_1) \tag{5.6}$$

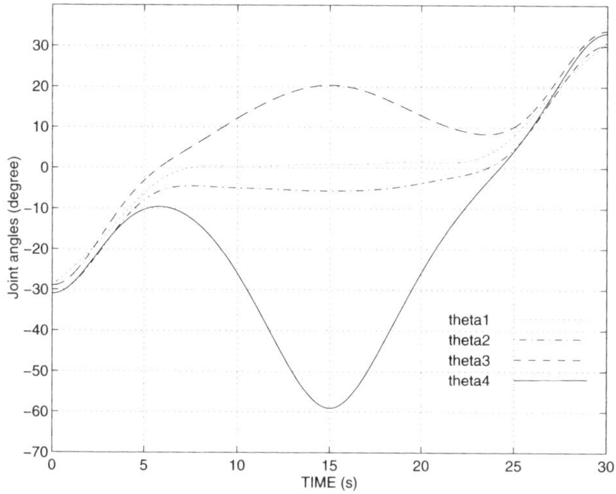

Fig. 5.15. Computed trajectories of joint angles under the initial error of joint angles: simulation 6.

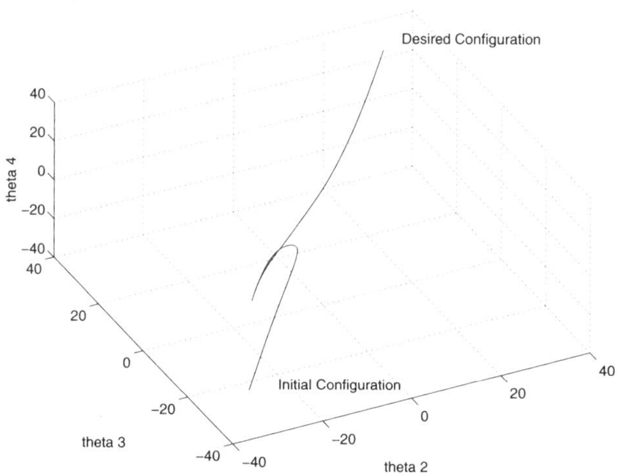

Fig. 5.16. A computed trajectory under the initial error where three axes are θ_2, θ_3 and θ_4: simulation 6.

where $J_S \in \mathbb{R}^{n-2}$ for the $n-2$ joint chained form manipulator. J_S is a lower triangular matrix and any of the diagonal elements is 1. Equation (5.6) is noteworthy and provides a significant tool of motion planning to deal with the error of the configuration variables. Since d_1 is a parameter which is independent on the given boundary conditions, it is clear that the sensitivity jacobian J_S is defined regardless of specific boundary conditions. The norm of J_S can be interpreted as the initial condition sensitivity of the planned motion.

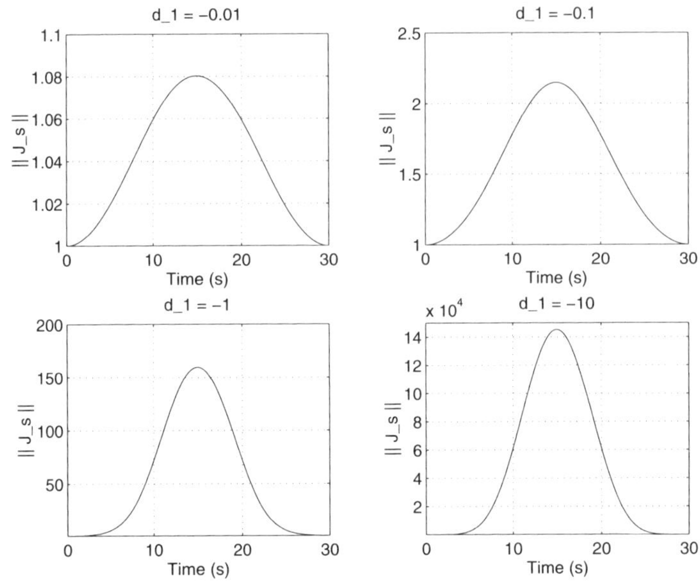

Fig. 5.17. $||J_S||$ shows maximum values always at $t = T/2$. Parameters are $d_1 = [-0.01, -0.1, -1, -10]$ respectively.

A computation of the norm of J_S is straightforward. $\frac{d||J_S||}{dt}$ is always zero at $t = T/2$ when $||J_S||$ is maximized. Since it is a little complicated and boring procedure to find out explicit solution of $\frac{d||J_S||}{dt} = 0$, numerical computation is carried out to show $||J_S||$ is maximum at $t = T/2$. Fig. 5.17 shows computed $||J_S||$ for various d_1 and it is clear that $||J_S||$ is maximum at $t = T/2$. Fig. 5.18 shows that $||J_S||$ is symmetric about the axis d_1 =0, which is a singularity of sinusoidal inputs.

Accordingly, $||J_S||$ is explicitly determined by taking the cross section of the surface presented in fig. 5.18 at $t = T/2$. Fig. 5.19 shows the computed result. It can be concluded that lower $|d_1|$ is preferable to minimize the effect of the initial error. In order to verify this conclusion, a simulation is carried out (simulation 7) for a different d_1 with the same boundary conditions as simulation 5. Fig. 5.20 shows a computed joint trajectories with $d_1 = -0.4$. The parameter d_1 which minimizes the error of approximation in simulation 5 was -0.62 and the error was 0.025 in the previous section. The error of approximation is 0.15 for $d_1 = -0.4$ in simulation 7.

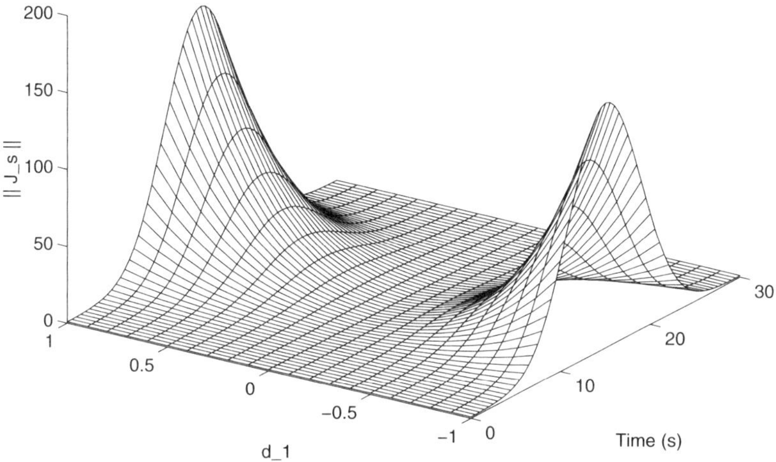

Fig. 5.18. $t \times d_1$ versus $||J_S||$ for $-1 \leq d_1 < 0$ and $0 < d_1 \leq 1$.

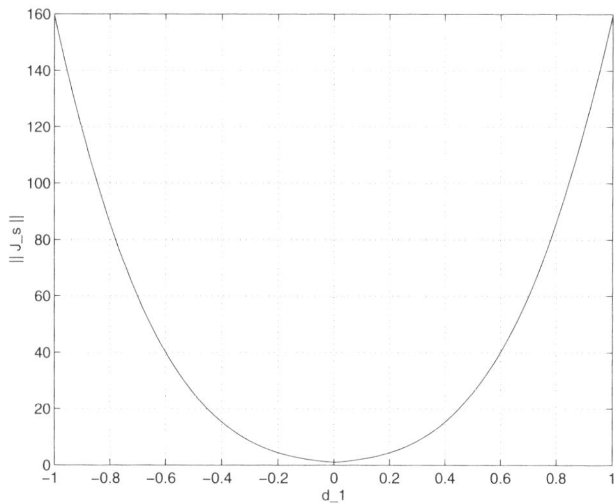

Fig. 5.19. d_1 versus $||J_S||$ for $-1 \leq d_1 < 0$ and $0 < d_1 \leq 1$ at $t = T/2$.

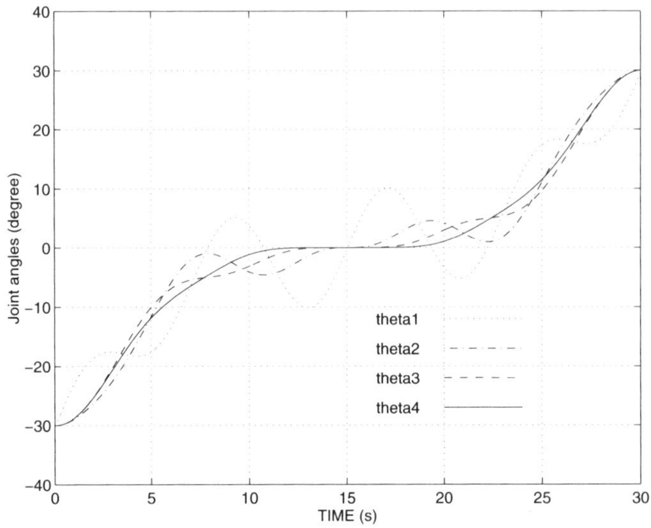

Fig. 5.20. Computed trajectories of joint angles ($d_1 = -0.4$): simulation 7.

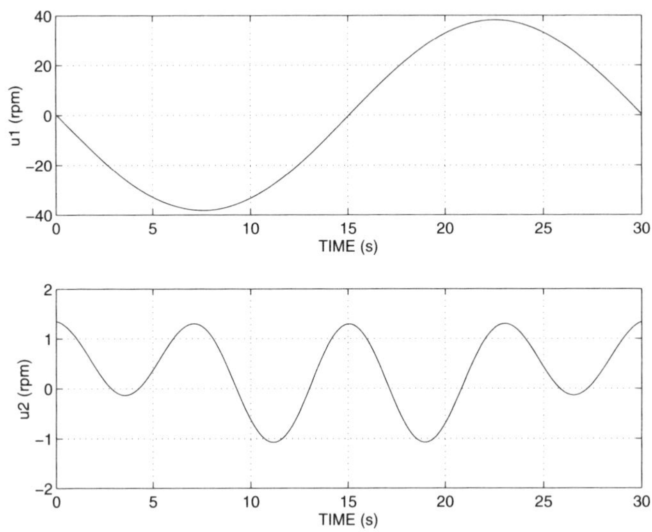

Fig. 5.21. Input angular velocities of actuators ($d_1 = -0.4$): simulation 7.

The aim of changing d_1 is to achieve the motion which is less sensitive to the error with the sacrifice of the error of approximation. From fig. 5.19, computed $||J_S||$'s are $||J_S|| = 43.7$ for $d_1 = -0.62$ and $||J_S|| = 15.3$ for $d_1 = -0.4$. Therefore, the effect of error is expected to be reduced for the case of $d_1 = -0.4$. Actuator inputs are plotted in fig. 5.21. u_1 is lower than simulation 5 and u_2 became a little oscillatory. Computed trajectories in the joint space are presented in fig. 5.22 and fig. 5.23. It can be seen that the error of approximation is a little higher than fig. 5.12 and fig. 5.13.

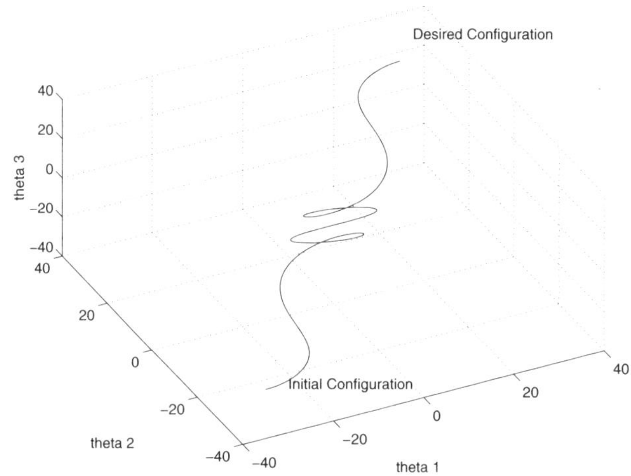

Fig. 5.22. A computed trajectory in the joint space where three axes are θ_1, θ_2 and θ_3 ($d_1 = -0.4$): simulation 7.

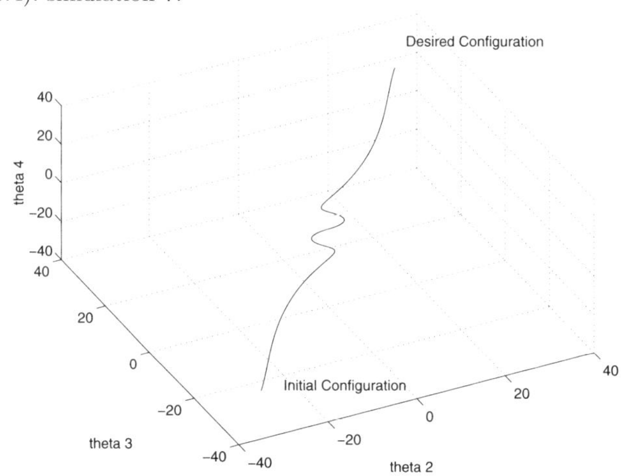

Fig. 5.23. A computed trajectory in the joint space where three axes are θ_2, θ_3 and θ_4 ($d_1 = -0.4$): simulation 7.

The simulation (simulation 8) shows an advantage of the less sensitive motion planning in simulation 7. Under the initial error of joint angles, joint angles are computed as shown in fig. 5.24. A starting point is $[-29, -31, -29, -31](deg)$, which implies each joint angle has an error of $1deg$. Although the joint trajectory is slightly different from fig. 5.20 because of the initial error, it is clear that the effect of error is acceptable compared with the unsatisfactory result of simulation 6. The error of approximation of simulation 8 is 0.38, which is increased by 2.5 times compared with simulation 6. Reminding the result of simulation 5, it is obvious that the motion ,which is less sensitive to the initial error of joint angles, is planned by setting $d_1 = -0.4$. Fig. 5.25 and fig. 5.26 show the trajectory in the joint space. It can be seen that the motion is similar to fig. 5.22 and fig. 5.23.

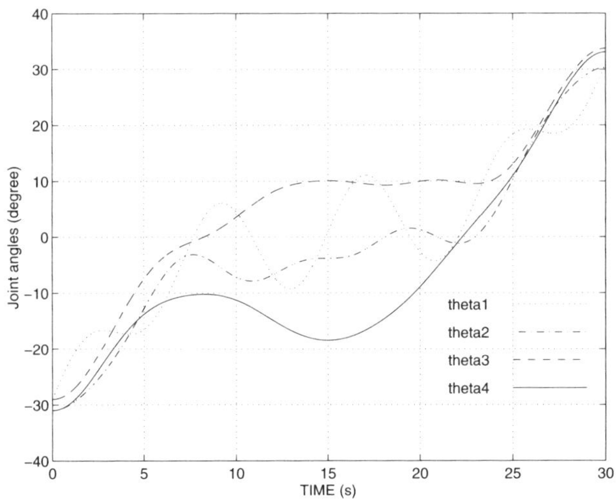

Fig. 5.24. Computed trajectories of joint angles with the initial error ($d_1 = -0.4$): simulation 8.

Since the initial condition sensitivity is dependent on the determination of d_1 only, it is not necessary to compute $||J_S||$ along the planned trajectory for every boundary conditions. Although joint errors cannot be specified, it can be assumed that a system may have a certain upper bound of the error. For the chained form manipulator, a dominant error source is backlashes at each joint and the error of joint angle is assumed to be bounded. Therefore, the sensitivity analysis in this section provides an estimation of the effect of error for a given system. Then, it implies that the upper bound of d_1 is estimated for the system. For the chained form manipulator, it can be concluded that it is safe to set the $|d_1| < 0.4$ from the experimental result. Of course, the specific upper bound of $|d_1|$ is dependent on the system ($e.g$ the accuracy of fabrication) and the environment ($e.g$ the source of disturbance). This concept provides a constraint in constructing the completed motion planning algorithm for the chained form manipulator. The algorithm can be interpreted as "*Minimizing the approximation error under the constraint of* $|d_1| < 0.4$."

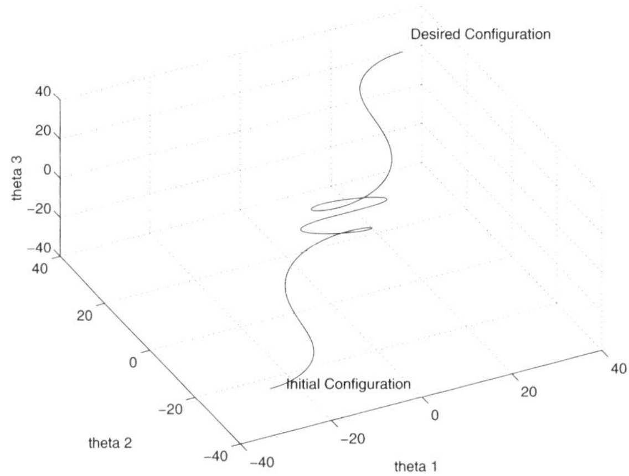

Fig. 5.25. A computed trajectory in the joint space with the initial error ($d_1 = -0.4$): simulation 8.

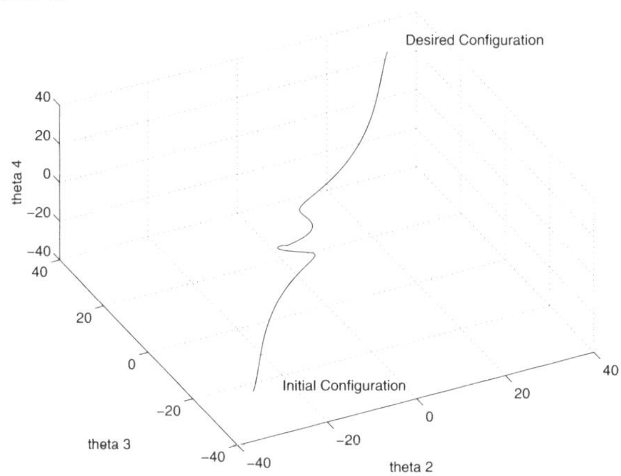

Fig. 5.26. A computed trajectory in the joint space with the initial error ($d_1 = -0.4$): simulation 8.

5.5 Experiments

5.5.1 Feedback control by pseudo-linearization

In this section, feedback control by pseudo-linearization in section 3.5.3 is experimentally tested with the prototype of thc chained form manipulator;

Table 5.1. Correspondence of simulations and experiments for the chained form manipulator.

—	Experiment D
Simulation 5 and 6	Experiment E
Simulation 7 and 8	Experiment F

experiment D. Feedback control of the chained form manipulator is adopted for precise positioning to the origin.

The initial configuration $\theta(0)$ is $\theta_i(0) = \varphi_2(0) = -40deg$, where $i \in \{1, \cdots, 4\}$. The goal configuration is the origin in the joint space. The parameter in eq. (3.15) is set as $a_0 = -0.3491$. We applied state feedback with feedback gains of $[g_1, g_2, g_3, g_4] = [-3.69, 4.63, -3.20, 1.00]$, which were computed by setting Q as an identity matrix and $R = 1$ in the quadratic form.

Fig. 5.27 shows an experimental result of joint trajectories. All the joint angles started from the initial configuration converged to the origin after 12.4(s). The condition for a termination of control was $E = 5deg$ in eq. (3.8) and resultant joint angles were $\theta^f = [1.3, 1.4, 1.0, 0.6](deg)$. Input trajectories of two actuators are plotted in fig. 5.28. Constant input u_1 was 33.3(r/min) and the velocity was controlled with high precision. A resultant motion of the prototype chained form manipulator is shown in fig. 5.29.

Although feedback control by pseudo-linearization is a powerful tool to stabilize the chained system, it is difficult to apply for the high dimensional system. A major drawback is that stabilizing a high dimensional chained system to the nonzero configuration is extremely difficult in practice. Although Sørdalen's coordinate transformation [79] which was used in section 3.5.3 can be applied, finding out appropriate control parameters, that results in reasonably small positioning error, is difficult. The author's experience is that feedback controllers which are designed based on the strict change of z_1 fails to reach their desired configuration in most cases. Examples of such controllers include Astolfi's exponentially convergent controller in [2, 3, 4] , Sampei's controller using time-state control form in [64, 67] and feedback control by pseudo-linearization presented in section 3.5.3. Sørdalen's controller in [73, 76] shows relatively good performance for stabilization to a nonzero configuration. However, it requires many switching stages and results in oscillatory motions in many cases.

Another drawback of feedback control is that the overshoot of the motion cannot be specified and maintained within the desired extent. Therefore, obstacle or singularity avoidance problem cannot be solved in the form of feedback. Because of these drawbacks, efficient motion planning is employed for the practical applications.

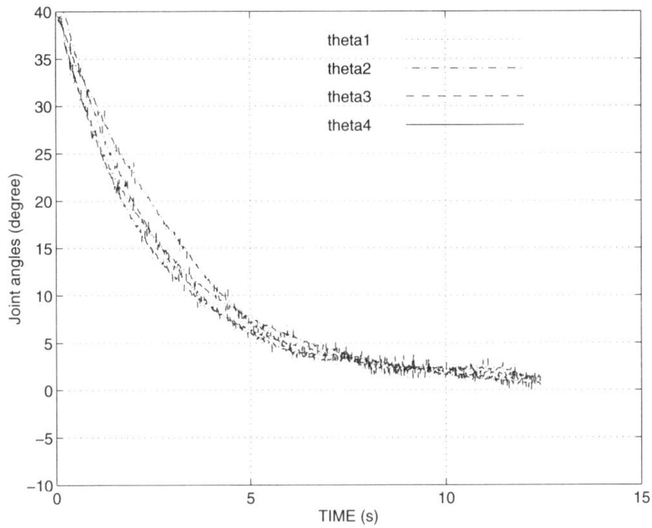

Fig. 5.27. Experimented trajectories of joint angles: experiment D.

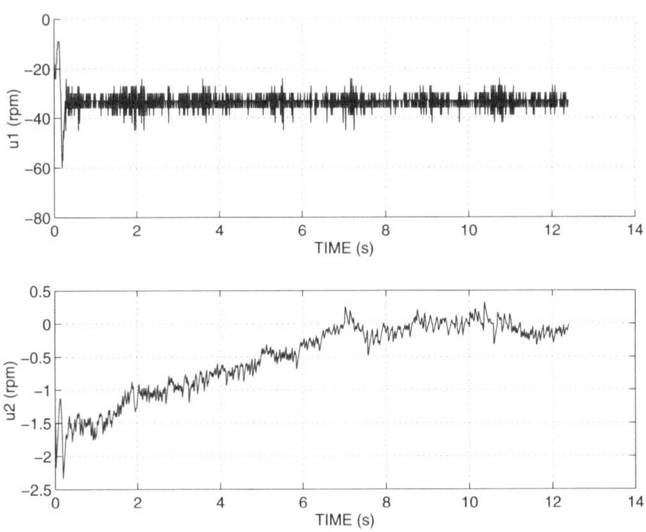

Fig. 5.28. Input angular velocities of actuators: experiment D.

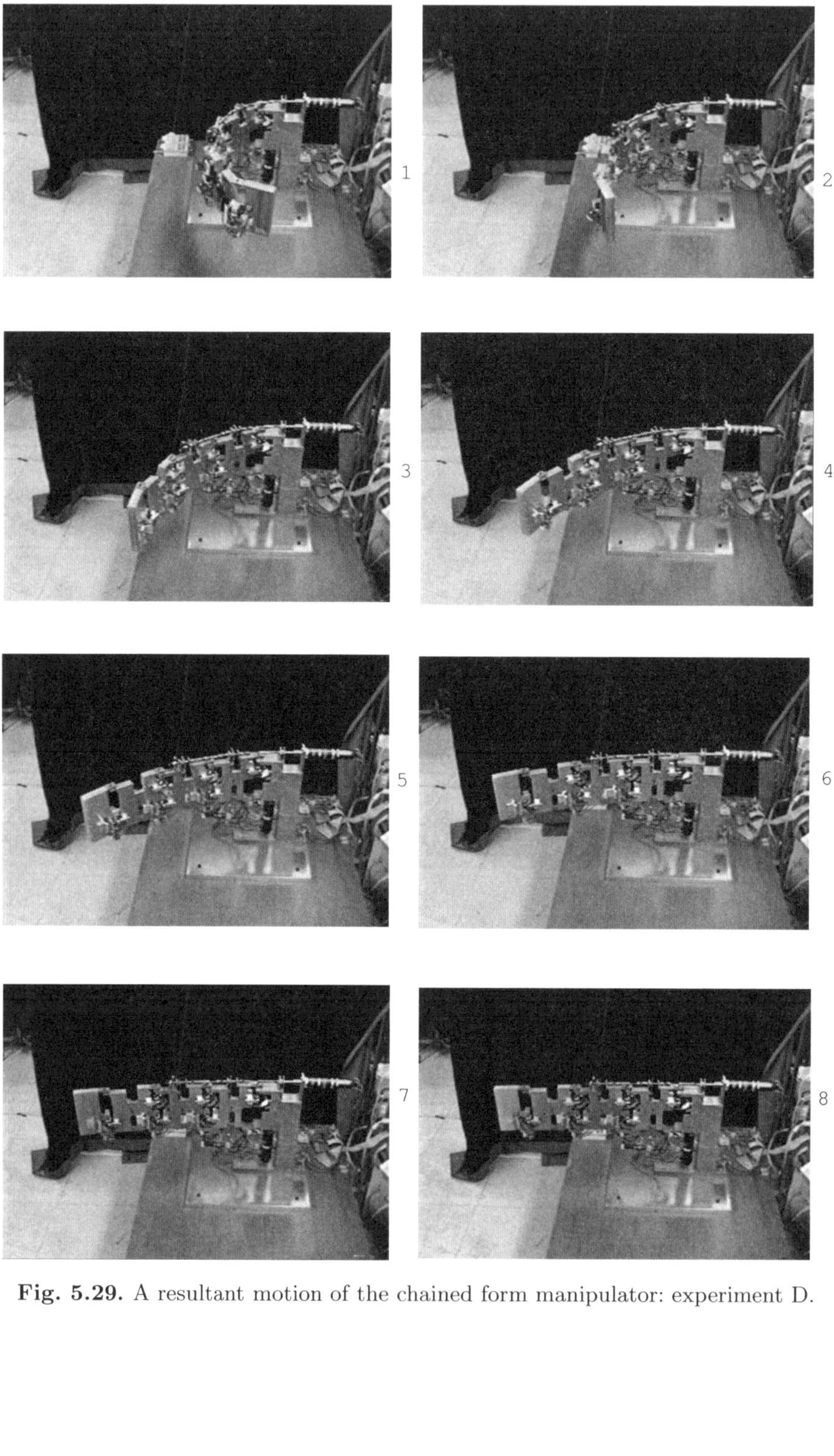

Fig. 5.29. A resultant motion of the chained form manipulator: experiment D.

5.5.2 Efficient motion planning using sinusoids

In this section, the usefulness of the motion planning scheme in section 5.3 is experimentally verified.

First experiment, experiment E is carried out under the same conditions with simulation 5. Fig. 5.30 shows experimental trajectory of the joint angles. The trajectory shows quite a large difference with fig. 5.8. Backlashes at each joint of the prototype are approximately 1 degree each. The unsatisfactory result in fig. 5.30 possibly resulted from the backlash. This fact can be predicted from the result in fig. 5.15. In addition, there are the other error sources in practice, for example, velocity control performance or slippage at the nonholonomic gear. Such small errors may cause seriously poor performances as a result. In fig. 5.30, θ_4 reached a singularity of $-90deg$ and the prototype failed to reach its desired configuration. Fig. 5.31 shows measured velocities of two actuators. A resultant motion of the prototype chained form manipulator is shown in fig. 5.32. Encoders are employed to read inputs. Since the velocity profiles are a little noisy. we use the rotation angles of the shafts $(s_i = \int_0^t u_i dt)$ to evaluate velocity control performance. Fig. 5.33 and Fig. 5.34 show measured angles and integrated values of inputs s_i in simulation 5 respectively. It can be seen that the input velocities are controlled precisely and the error of velocity control is less than 1.5%. Therefore, we can conclude that dominant error source may be backlashes at each manipulator joints.

From the result of the experiment E, it is clear that the initial condition sensitivity is extremely significant to achieve desirable performances. Experiment F is carried out under the same condition as in simulation 7. The motion now is less sensitive to the error of joint angles which was achieved by the change of d_1. Fig. 5.35 shows experimental joint trajectories. If there is no error, fig. 5.35 should be the same as fig. 5.20. However, fig. 5.35 is similar to fig. 5.24 because of backlashes. The performance is good because the planned motion is highly "robust" in some sense. Measured joint angles after completing control are $[\theta_1(30), \theta_2(30), \theta_3(30), \theta_4(30)] = [30.1, 27.8, 28.4, 29.7](deg)$. A resultant motion of the prototype chained form manipulator is shown in fig. 5.37.

It is important to check velocity transmission through the main power train illustrated in section 4.3.1. When the actuator at the base drives the power train with velocity u_1, it should be uniformly transmitted toward the end-effector regardless of the joint angles. However, universal joints combined with sliding link mechanism have some backlashes. Another problem is the precision of mechanisms of power train. This performance is checked by comparing the signals from two encoders. One is directly connected to the actuator shaft and the other is placed at the end of the power train. Fig. 5.38 shows measured angles from two encoders. It is clear that the input u_1 is transmitted with high precision under the joint motions. The error was less than 1.5%, which is acceptably low.

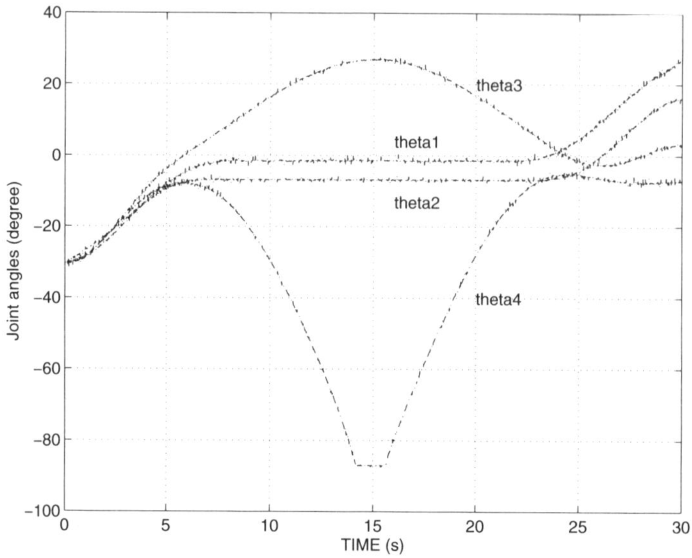

Fig. 5.30. Experimented trajectories of joint angles ($d_1 = -0.62$): experiment E.

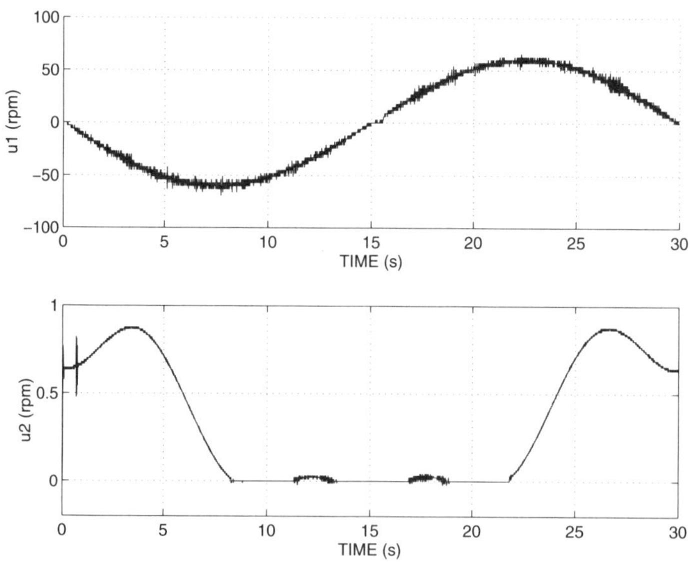

Fig. 5.31. Input angular velocities of actuators ($d_1 = -0.62$): experiment E.

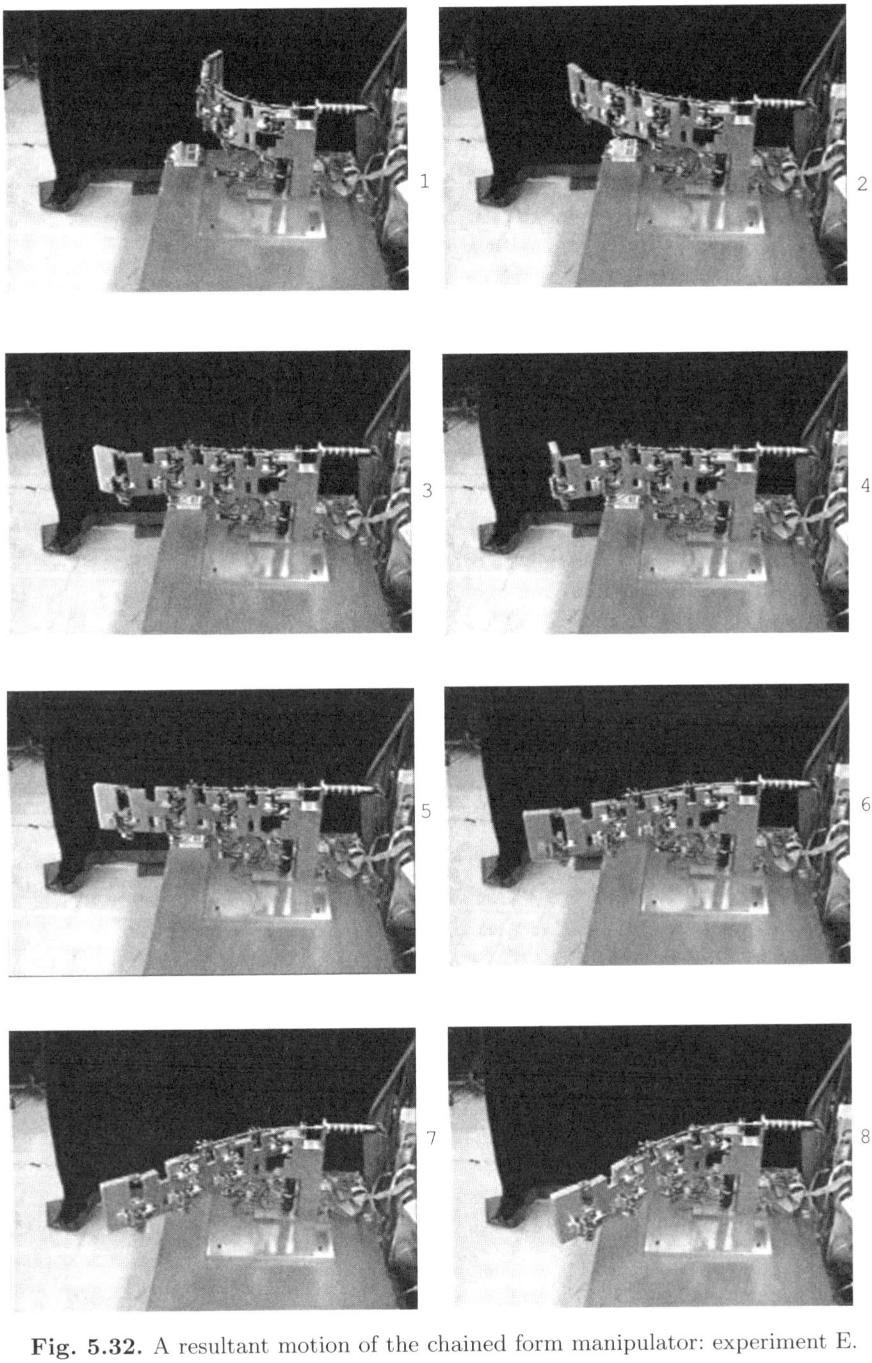

Fig. 5.32. A resultant motion of the chained form manipulator: experiment E.

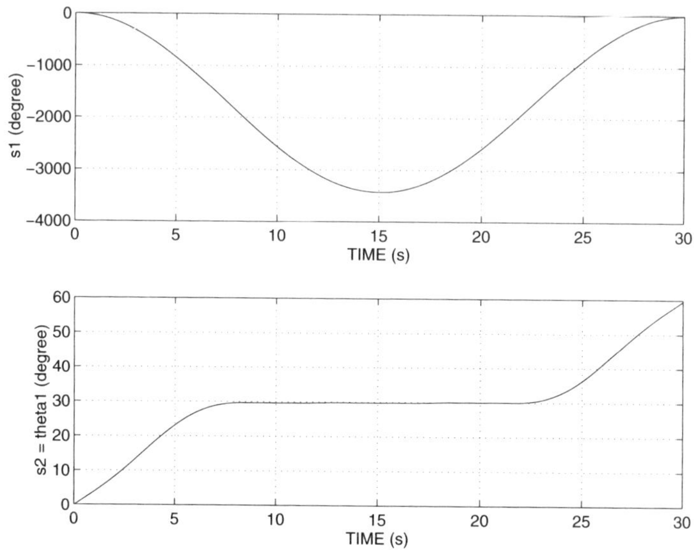

Fig. 5.33. Measured rotation angles of actuator shafts($d_1 = -0.62$): experiment E.

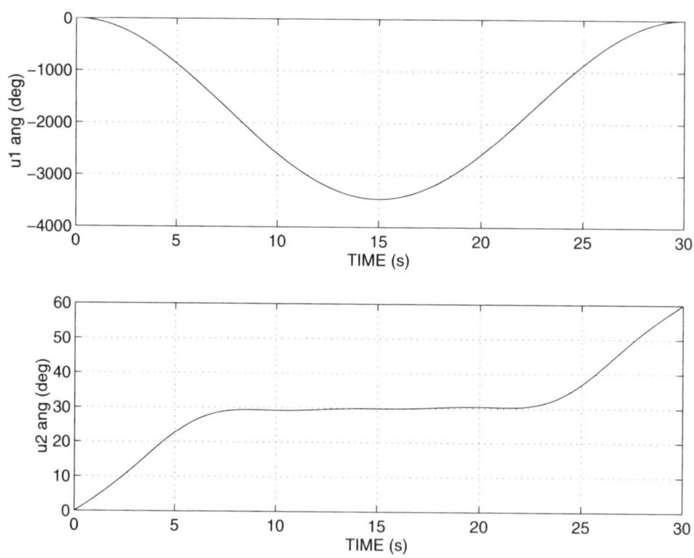

Fig. 5.34. Rotation angles computed by $s_i = \int_0^t u_i dt$ in simulation ($d_1 = -0.62$): simulation 5 which is compared with experiment E.

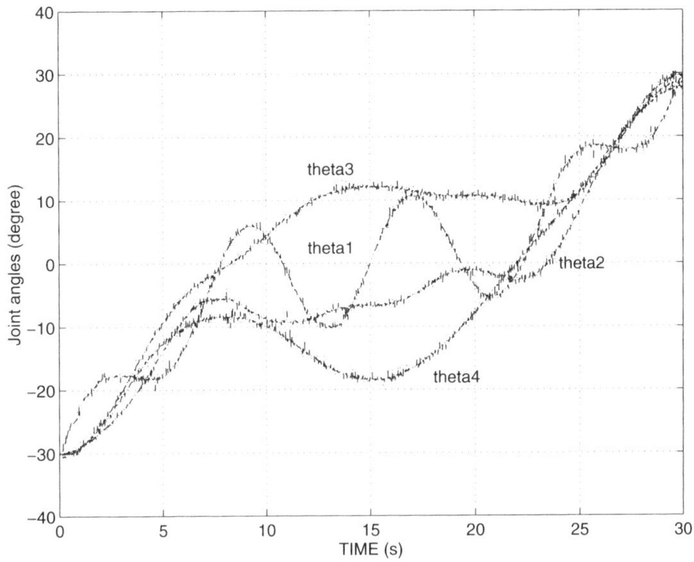

Fig. 5.35. Experimented trajectories of joint angles ($d_1 = -0.4$): experiment F.

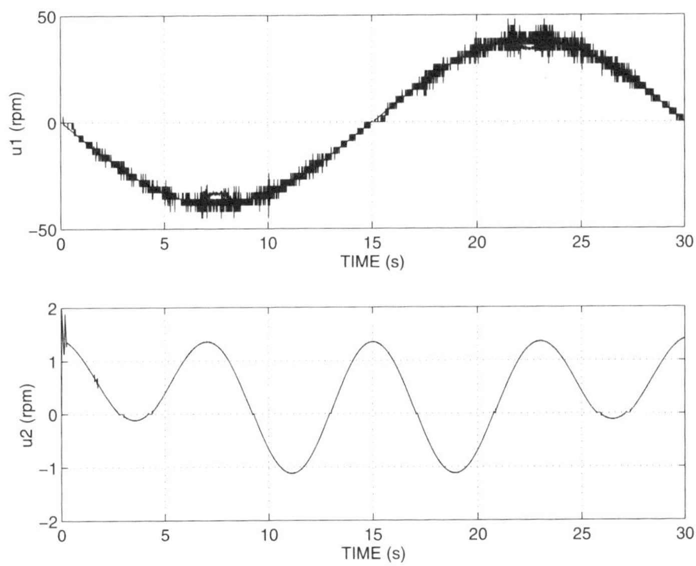

Fig. 5.36. Input angular velocities of actuators ($d_1 = -0.4$): experiment F.

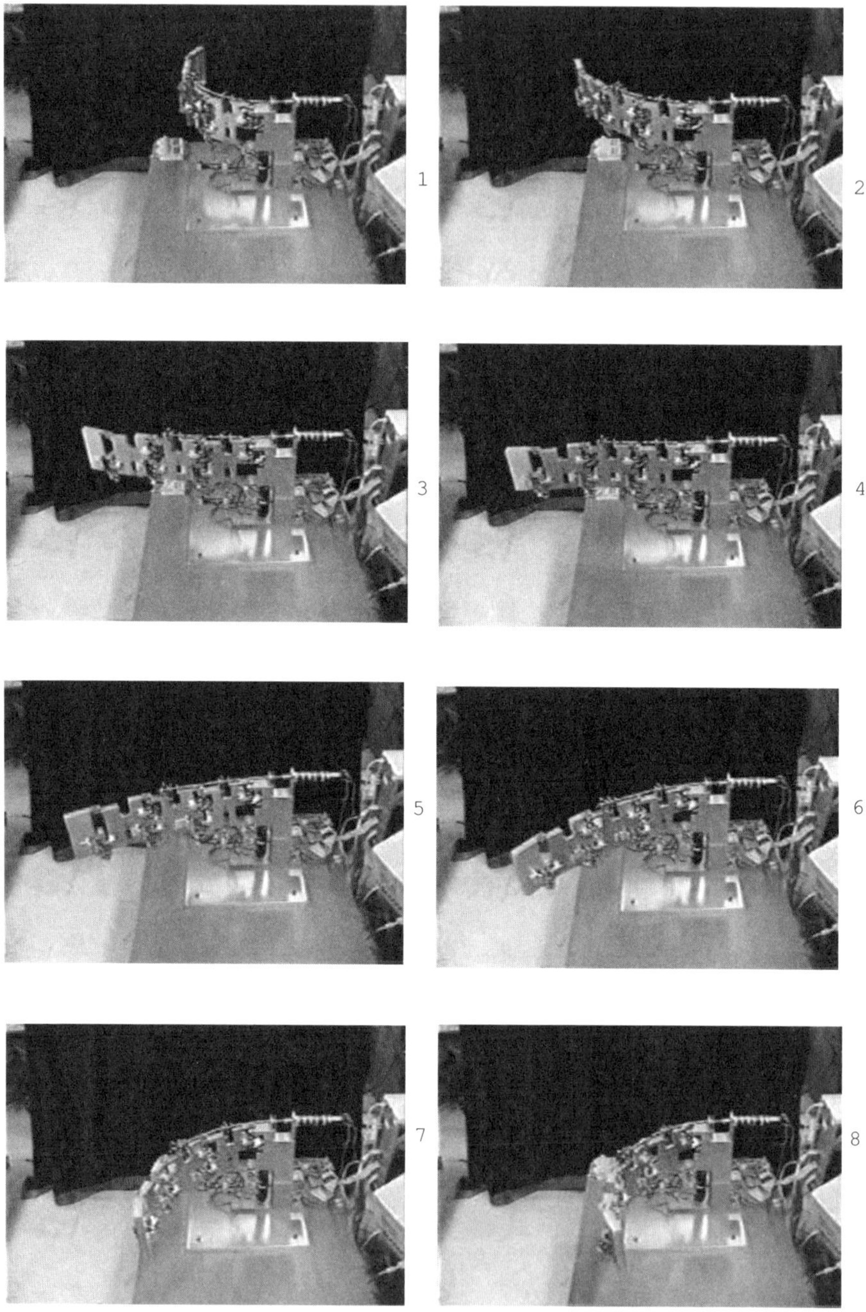

Fig. 5.37. A resultant motion of the chained form manipulator: experiment F.

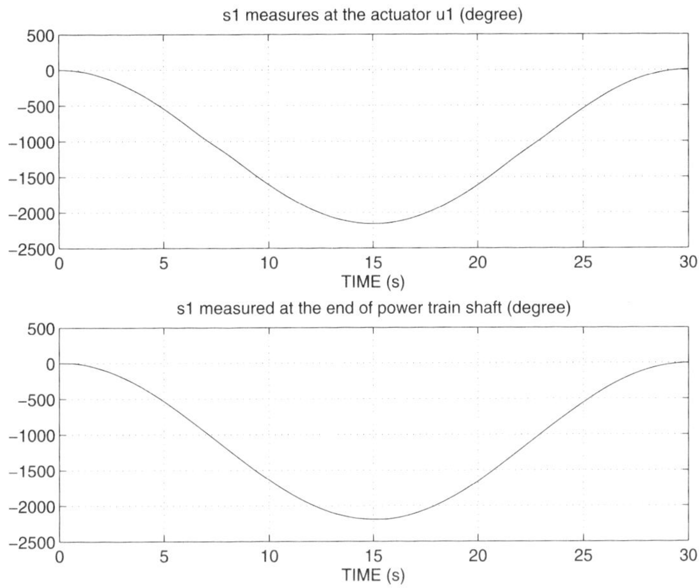

Fig. 5.38. s_1 measured from both sides of the main power train: experiment F.

5.6 Conclusion

A prototype of the chained form manipulator was fabricated for the experimental verification of the theoretical design and proposed control schemes for the chained form. An efficient motion planning strategy was developed and shown to generate trajectory which is easy to implement. The analysis on the initial-condition sensitivity was carried out under the use of sinusoidal inputs. It was verified that the less sensitive motion greatly contributes to reduce the effect of error when open loop control is applied.

Experiments were successfully carried out with the four joint model. Since the motion planning strategy can be easily extended to the higher dimension, there is no limitation of the number of joints from the control theoretic point of view. However, increasing the number of joint results in heavy weight and requires high actuating torques, which is a mechanical design problem.

We originally fabricated a six joint model, which was too heavy and required high torques. For experiments, we removed two joints to avoid the problem. Mechanical design to take account of the weight and torque would need to be carefully done for higher dimensional models.

6

Conclusion

Controllable underactuated manipulators were successfully developed based on the nonlinear control theoretic approach. The nonholonomic gear is designed to create nonlinear structure on the robot manipulator. The nonholonomic gear is a special type of velocity transmission consisting of a ball and wheels, and provides the kinematic nonholonomic constraint. The gear can be employed not only to build underactuated manipulators but also to be used in many other applications as a continuously variable transmission having a nonlinear gear ratio of trigonometric functions.

Theoretical design of the nonholonomic manipulator is established, then it was shown that the nonholonomic manipulator is controllable in higher dimension than the input space. The proof of the chained form convertibility guaranteed that various controllers to the chained form can be adopted to control the nonholonomic manipulator. Various issues of mechanical design were discussed and mechanical analysis was carried out for the fabrication of the prototype of the nonholonomic manipulator. In order to generate physically feasible motions of the nonholonomic manipulator to avoid actuator saturations and excessive joint accelerations, time scaling technique was proposed. Feedback control by pseudo-linearization was proposed for precise positioning. Presented control schemes were successfully experimented with the prototype. From the fact that the presented experimental results well matched with computed motions, the usefulness of theoretical design of the nonholonomic manipulator and applied control schemes was verified.

On the contrary to the fact that the nonholonomic manipulator was designed focusing on the simplicity of mechanical structure, the chained form manipulator was successfully designed to satisfy well-conditioned mathematical properties for the high dimensional model. Problems of the nonholonomic manipulator were clarified from mechanical and mathematical point of view. The chained form manipulator achieved an extremely simple kinematics for the high dimensional models with a cost of mechanical simplicity. For the experimental verification, we fabricated the prototype of the chained form manipulator. The efficient motion planning scheme is proposed. By applying the sinusoidal inputs to the chained form, a feasible nonholonomic motion was successfully planned with an acceptable computational cost. A new performance measure was introduced to reduce the effect of the initial error of joint angles. The initial condition sensitivity of planned motion under the various uncertainties is significant for the practical application. It is most important when the planned motion is used for open loop control. We proposed to plan motions taking account of the initial condition sensitivity. The result of analysis clearly pointed out a new direction of motion planning. The motion planning algorithm based on two performance measures was established. The usefulness of proposed motion planning strategy was verified by the experimental results.

References

1. P. K. Agarwal, J. C. Latombe, R. Motwani, and P. Raghavan. Nonholonomic path planning for pushing a disk among obstacles. In *Proc. 1997 IEEE International Conference on Robotics and Automation*, pages 3124–3129, Albuquerque, New Mexico, April 1997.
2. A. Astolfi. On the stabilization of non-holonomic systems. In *Proc. 33rd Conference on Decision and Control*, Lake Buena Vista, FL, December 1994. Invited Session: Discontinuous Stabilizing Control Design.
3. A. Astolfi. Exponential stabilization of a car-like vehicle. In *Proc. 1995 IEEE International Conference on Robotics and Automation*, pages 1391–1396, Nagoya, Japan, May 1995.
4. A. Astolfi. Exponential stabilization of nonholonomic systems via discontinuous control. In *Proceedings of IFAC Symp. Nonlinear Contr. Syst. Design (NOLCOS)*, pages 741–746, 1995.
5. A. Bicchi and R. Sorrentino. Dexterous manipulation through rolling. In *Proc. 1995 IEEE International Conference on Robotics and Automation*, pages 452–457, Nagoya, Japan, May 1995.
6. A. M. Bloch, M. Reyhanoglu, and N. H. McClamroch. Control and stabilization of nonholonomic dynamic systems. *IEEE Transactions on Robotics and Automation*, 37(11):1746–1757, 1992.
7. R. W. Brockett. Asymptotic stability and feedback stabilization. In R. W. Brockett, R. S. Millman, and H. J. Sussman, editors, *Differential Geometric Control Theory*, pages 181–208. Birkhauser, 1983.
8. L. Bushnell, D. Tilbury, and S. Sastry. Extended goursat normal forms with applications to nonholonomic motion planning. In *Proc. 32th Conference on Decision and Control*, pages 3447–3452, San Antonio, December 1993.
9. C. Canudas de Wit and O. J. Sørdalen. Exponential stabilization of mobile robots with nonholonomic constraints. *IEEE Transactions on Automatic Control*, 37(11):1791–1797, 1992.
10. V. H. L. Cheng. A direct way to stabilize continuous-time and discrete-time linear time-varying systems. *IEEE Transactions on Automatic Control*, 24(4):641–643, 1979.
11. J. E. Colgate, M. Peshkin, and W. Wannasuphoprasit. Nonholonomic haptic display. In *Proc. 1996 IEEE International Conference on Robotics and Automation*, pages 539–544, Minneapolis, Minnesota, April 1996.

12. J.-M. Coron. Global asymptotic stabilization for controllable systems without drift. *Mathematics of Control, Signals, and Systems*, 5:295–315, 1991.
13. S. H. Crandall, D. C. Karnopp, E. F. Kurtz Jr., and D. C. Pridmore-Brown. *Dynamics of mechanical and electomechanical systems.* Mcgraw-Hill Book co., – edition, 1968.
14. L. Dubins. On curves of minimal length with a constraint on average curvature and with prescribed initial and terminal positions and tangents. *American Journal of Mathematics*, 79:497–516, 1957.
15. Y. Kawada et al. *Handbook of the Strength of a Material.* Asakura Book co., 1st edition, 1966 (In Japanese).
16. P. Di Giamberardino, S. Monaco, and D. Normand-Cyrot. Digital control through finite feedback discretizability. In *Proc. 1996 IEEE International Conference on Robotics and Automation*, pages 3141–3146, Minneapolis, Minnesota, April 1996.
17. J. Godhavn and O. Egeland. A lyapunov approach to exponential stabilization of nonholonomic systems in power form. *IEEE Transactions on Automatic Control*, 42(7):1028–1032, 1997.
18. H. Goldstein. *Classical Mechanics.* Addison-Wesley, 2 edition, 1980.
19. Donald T. Greenwood. *Principles of dynamics.* Prentice hall, INC, 2nd edition, 1988.
20. L. Gurvits and Z. X. Li. Smooth time-periodic feedback solutions for nonholonomic motion planning. In Z. X. Li and J. Canny, editors, *Progress in Nonholonomic Motion Planning*, pages 53–108. Kluwer Academic Publisher, 1993.
21. J. Imura, K. Kobayashi, and T. Yoshikawa. Exponential stabilization problem of nonholonomic chained systems with specified transient response. In *Proc. 35th Conference on Decision and Control*, pages 4733–4738, Kobe, Japan, December 1996.
22. J. Imura, K. Kobayashi, and T. Yoshikawa. Exponential stabilization of nonholonomic chained systems with specified transient response. *Transactions of the Society of Instrument and Control Engineers*, 33(5):375–383, 1997.
23. S. Inada, M. Kubota, N. Hayashi, and K. Kitasato. *Kinematics.* Asakura Book co., 18th edition, 1977 (In Japanese).
24. K.Kato. Fundamentals and applications of friction drive. *Journal of the Japan Society for Precision Engineering*, 56(9):1602–1606, 1990 (In Japanese).
25. N. Kobayashi, K. Nonami, and I. Yasuzumi. Attitude control of free-flying space robot with discrete time sliding mode control algorithm. *Transactions of the Society of Instrument and Control Engineers*, 32(8):1206–1211, 1996.
26. G. Lafferriere and H. J. Sussmann. Motion planning for controllable systems without drift. In *Proc. 1991 IEEE International Conference on Robotics and Automation*, pages 1148–1153, Sacramento, California, April 1991.
27. J.-P. Laumond. Feasible trajectories for mobile robots with kinematic and environment constraints. In *Proc. Int. Conference on Intelligent Autonomous Systems*, pages 346–354, Amsterdam, The Netherlands, 1986.
28. J.-P. Laumond. Controllability of a multibody mobile robot. In *ICAR'91*, Pisa, Italy, June 1991.
29. J. P. Laumond. Controllability of a multibody mobile robot. *IEEE Transactions on Robotics and Automation*, 9(6):755–763, 1993.
30. J. P. Laumond, P. Jacobs, M. Taïx, and R. M. Murray. A motion planner for nonholonomic mobile robots. *IEEE Transactions on Robotics and Automation*, 10(5):577–593, 1994.

31. J. P. Laumond, S. Sekhavat, and M. Vaisset. Collision-free motion planning for a nonholonomic mobile robot with trailers. In *Preprints of the Fourth IFAC Symposium on Robot Control'94*, pages 171–177, Capri, Italy, September 1994.
32. J.-P. Laumond and T. Siméon. Motion planning for a two degrees of freedom mobile robot with towing. In *Proc. Int. Conference on Control and Applications*, Jerusalem, Israel, April 1989. IEEE.
33. J.-P. Laumond, M. Taïx, and P. Jacobs. A motion planner for car-like robots based on a mixed global/local approach. In *Proc. International Conference on Intelligent Robots and systems*, pages 765–773, Japan, 1990. IEEE.
34. Z. Li and J. Canny. Motion of two rigid bodies with rolling constraint. *IEEE Transactions on Robotics and Automation*, 6(1):62–72, 1990.
35. Z. Luo and M. Funaki. Design and control of a cartesian nonholonomic robot. *Journal of the Robotics Society of Japan*, 15(2):243–248, 1997(In Japanese).
36. K. M. Lynch. Locally controllable polygons by stable pushing. In *Proc. 1997 IEEE International Conference on Robotics and Automation*, pages 1442–1447, Albuquerque, New Mexico, April 1997.
37. K. M. Lynch and M. T. Mason. Controllability of pushing. In *Proc. 1995 IEEE International Conference on Robotics and Automation*, pages 112–119, Nagoya, Japan, May 1995.
38. H. Machida. Study of half toroidal continuously variable transmission. *Transactions of the Japan Society of Mechanical Engineers*, 57(533):271–276, 1991.
39. K. Machida, Z. Luo, and M. Funaki. Prototyping and control experiments of a cartesian nonholonomic robot. *Journal of the Robotics Society of Japan*, 13(7):1038–1043, 1995(In Japanese).
40. A. Marigo, Y. Chitour, and A. Bicchi. Manipulation of polyhedral parts by rolling. In *Proc. 1997 IEEE International Conference on Robotics and Automation*, pages 2992–2997, Albuquerque, New Mexico, April 1997.
41. R. T. M'Closkey and R. M. Murray. Convergence rates for nonholonomic systems in power form. In *American Control Conference*, pages 2967–2972, San Fransisco, California, June 1993.
42. R. T. M'Closkey and R. M. Murray. Exponential stabilization of driftless nonlinear control systems via time-varying, homogeneous feedback. In *Proc. 33rd Conference on Decision and Control*, pages 1317–1322, Lake Buena Vista, FL, December 1994.
43. R. T. M'Closkey and R. M. Murray. Exponential stabilization of driftless nonlinear control systems using homogeneous feedback. *IEEE Transactions on Automatic Control*, 42(5):614–628, 1997.
44. S. Monaco and D. Normand-Cyrot. An introduction to motion planning under multirate control. In *Proc. 31th Conference on Decision and Control*, pages 1780–1785, Tucson, Arizona, December 1992. IEEE.
45. R. M. Murray. Applications and extensions of goursat normal form to control of nonlinear systems. In *Proc. 32th Conference on Decision and Control*, pages 3425–3430, San Antonio, December 1993.
46. R. M. Murray and S. S. Sastry. Steering nonholonomic systems using sinusoids. In *Proc. 29th Conference on Decision and Control*, pages 2097–2101, Honolulu, Hawaii, December 1990.
47. R. M. Murray and S. S. Sastry. Nonholonomic motion planning: Steering using sinusoids. *IEEE Transactions on Automatic Control*, 38(5):700–716, 1993.

48. H. Nagata, A. Ishibashi, S. Hoyashita, and H. Takedomi. A variable-speed traction drive with internal spherical disks. *Transactions of the Japan Society of Mechanical Engineers*, 56(531):271–277, 1990.
49. Y. Nakamura and R. Mukherjee. Nonholonomic path planning of space robots via bi-directional approach. *IEEE Transactions on Robotics and Automation*, 7(4):500–514, 1991.
50. Y. Nakamura and R. Mukherjee. Exploiting nonholonomic redundancy of free-flying space robots. *IEEE Transactions on Robotics and Automation*, 9(4):499–506, 1993.
51. Y. Nakamura and S. Savant. Nonlinear tracking control of autonomous underwater vehicles. In *Proc. 1992 IEEE International Conference on Robotics and Automation*, pages A4–A9, Nice, France, May 1992. IEEE.
52. Y. Nakamura and T. Suzuki. "Planning Spiral Motions of Nonholonomic Free-Flying Space Robots". AIAA Journal of Spacecraft and Rockets, 34(1):137–143, 1997.
53. T. Narikiyo and S. Sugita. Exponential stabilization of nonholonomic systems described by chained form. *Transactions of the Society of Instrument and Control Engineers*, 32(8):1310–1312, 1996(In Japanese).
54. J. I. Neĭmark and N. A. Fufaev. *Dynamics of Nonholonomic Systems*, volume 33 of *Translations of Mathematical Monographs*. American Mathematical Society, 1972.
55. M. V. Nieustadt, M. Rathinam, and R. M. Murray. Differential flatness and absolute equivalence. In *Proc. 33rd Conference on Decision and Control*, pages 326–332, Lake Buena Vista, FL, December 1994.
56. H. Nijmeijer and A. J. van der Schaft. *Nonlinear Dynamical Control Systems*. Springer-Verlag, 1990.
57. N.Moronuki and Y.Furukawa. On the design of precise feed mechanism by friction drive. *Journal of the Japan Society for Precision Engineering*, 54(11):2113–2118, 1988 (In Japanese).
58. Takashige Noguchi. A new type of the harmonic analyzer. *Journal of the Japan Society of Mechanical Engineers*, 27(91):975–984, 1923 (In Japanese).
59. M. Peshkin, J. E. Colgate, and C. Moore. Passive robots and haptic displays based on nonholonomic elements. In *Proc. 1996 IEEE International Conference on Robotics and Automation*, pages 551–556, Minneapolis, Minnesota, April 1996.
60. J.-B. Pomet, B Thuilot, G. Bastin, and G. Campion. A hybrid strategy for the feedback stabilization of nonholonomic mobile robots. In *Proc. 1992 IEEE International Conference on Robotics and Automation*, pages 129–134, Nice, France, May 1992. IEEE.
61. J.-P. Pomet. Explicit design of time-varying stabilizing control laws for a class of controllable systems without drift. *System and Control Letters*, 18(2):147–158, 1992.
62. J. A. Reeds and L. A. Shepp. Optimal paths for a car that goes both forwards and backwards. *Pacific Journal of Mathematics*, 145(2):367–393, 1990.
63. P. Rouchon, M. Fliess, J. Lévine, and P. Martin. Flatness and motion planning: the car with n trailers. In *Proc. Second European Control Conference*, Groningen, The Netherlands, June 1993.
64. M. Sampei. A control strategy for a class of nonholonomic systems – time-state control form and its application –. In *Proc. 33rd Conference on Decision and Control*, pages 1120–1121, Lake Buena Vista, FL, December 1994.

65. M. Sampei. Feedback control of nonholonomic systems. *Journal of the Society of Instrument and Control Engineers*, 36(6):396–403, 1997 (In Japanese).
66. M. Sampei, M. Ishikawa, H. Kiyota, and M. Koga. Time-state control form and its application to non-holonomic systems. In *Proc. of the 12th Annual Conference of the Robotics Society of Japan*, pages 437–438, Japan, 1992 (In Japanese).
67. M. Sampei, H. Kiyota, M. Koga, and M. Suzuki. Necessary and sufficient conditions for transformation of nonholonomic system into time-state control form. In *Proc. 35th Conference on Decision and Control*, pages 4745–4746, Kobe, Japan, December 1996.
68. M. Sampei, S. Mizuno, M. Ishikawa, and M. Koga. Position control of a rigid ball between two parallel plates. *Journal of the Robotics Society of Japan*, 14(8):1237–1242, 1996 (In Japanese).
69. M. Sampei, T. Tamura, and T. Kobayashi. A control strategy for a class of nonlinear systems which cannot be stabilized with continuous state feedback. In *Preprints of the 36th Japan Joint Automatic Control Conference*, pages SS29–30, Japan (In Japanese), 1993.
70. C. Samson. Control of chained systems application to path following and time-varying point-stabilization of mobile robots. *IEEE Transactions on Automatic Control*, 40:64–77, 1995.
71. S. Sekhavat and J. P. Laumond. Topological property of trajectories computed from sinusoidal inputs for nonholonomic chained form systems. In *Proc. 1996 IEEE International Conference on Robotics and Automation*, pages 3383–3388, Minneapolis, Minnesota, April 1996.
72. O. J. Sørdalen. Conversion of the kinematics of a car with n trailers into a chained form. In *Proc. 1993 IEEE International Conference on Robotics and Automation*, pages 382–387, Atlanta, Georgia, May 1993.
73. O. J. Sørdalen. *Feedback Control of Nonholonomic Mobile Robots*. PhD thesis, The Norwegian Institute of Technology, February 1993. ITK-rapport 1993:5-W.
74. O. J. Sørdalen. On the global degree of nonholonomy of a car with n trailers. Technical Report UCB/ERL M93/39, University of California, Berkeley, May 1993.
75. O. J. Sørdalen and C. Canudas de Wit. Exponential control law for a mobil robot: Extension to path following. In *Proc. 1992 IEEE International Conference on Robotics and Automation*, pages 2158–2163, Nice, France, May 1992. IEEE.
76. O. J. Sørdalen and O. Egeland. Exponential stabilization of chained nonholonomic systems. In *Proc. Second European Control Conference*, Groningen, The Netherlands, June 1993.
77. O. J. Sørdalen and O. Egeland. Exponential stabilization of chained nonholonomic systems. *IEEE Transactions on Automatic Control*, 40(1):35–49, 1995.
78. O. J. Sørdalen, Y. Nakamura, and W. J. Chung. Control of a nonholonomic manipulator. In *Preprints of the Fourth IFAC Symposium on Robot Control'94*, pages 323–328, Capri, Italy, September 1994.
79. O. J. Sørdalen and K. Y. Wichlund. Exponential stabilization of a car with n trailers. In *Proc. 32th Conference on Decision and Control*, pages 978–983, San Antonio, December 1993.
80. S.Timoshenko and J.N.Goodier. *Theory of elasticity*. Macgraw-Hill Book co., 2nd edition, 1951.

81. H. J. Sussmann and W. Liu. Limits of highly oscillatory controls and the approximation of general paths by admissible trajectories. In *Proc. 30th Conference on Decision and Control*, pages 437–442, Brighton, England, December 1991. IEEE.
82. T. Suzuki and Y. Nakamura. "Planning Spiral Motion of Nonholonomic Space Robots". In 1996 IEEE Int. Conf. on Robotics and Automation, pages 718–725, Minneapolis, MN, April 1996.
83. M. Takahashi and J. Otsuka. Precise positioning using friction / traction drive. *Journal of the Japan Society for Precision Engineering*, 56(9):1611–1614, 1990 (In Japanese).
84. M. Takahashi and J. Otsuka. Study on precision positioning by friction drive (6th report) – ultra-precise and fast positioning. *Journal of the Japan Society for Precision Engineering*, 59(7):1103–1108, 1993 (In Japanese).
85. H. Tanaka and M. Eguchi. Speed ratio control of a half-toroidal traction drive cvt. *Transactions of the Japan Society of Mechanical Engineers*, 57(543):276–279, 1991.
86. A. R. Teel, R. M. Murray, and G. Walsh. Nonholonomic control systems: From steering to stabilization with sinusoids. In *Proc. 31th Conference on Decision and Control*, pages 1603–1609, Tucson, Arizona, December 1992.
87. D. Tilbury, J. P. Laumond, R. Murray, S. Sastry, and G. Walsh. Steering car-like systems with trailers using sinusoids. In *Proc. 1992 IEEE International Conference on Robotics and Automation*, pages 1993–1998, Nice, France, May 1992. IEEE.
88. D. Tilbury, R. Murray, and S. Sastry. Trajectory generation for the n-trailer problem using goursat normal form. Technical Report UCB/ERL M93/12, University of California at Berkeley, February 1993.
89. D. Tilbury, R. Murray, and S. Sastry. Trajectory generation for the n-trailer problem using goursat normal form. *IEEE Transactions on Automatic Control*, 40(5):802–819, 1995.
90. D. Tilbury, O. J. Sørdalen, L. Bushnell, and S. S. Sastry. A multisteering trailer system: Conversion into chained form using dynamic feedback. Technical Report UCB/ERL M93/55, University of California at Berkeley, July 1993.
91. D. Tilbury, O. J. Sørdalen, L. Bushnell, and S. S. Sastry. A multisteering trailer system: Conversion into chained form using dynamic feedback. *IEEE Transactions on Automatic Control*, 11(6):807–818, 1995.
92. W. Flűgge. *Handbook of Engineering mechanics.* Mcgraw-Hill Book co., 1st edition, 1962.
93. G. Walsh, D. Tilbury, S. Sastry, R. Murray, and J. P. Laumond. Stabilization of trajectories for systems with nonholonomic constraints. In *Proc. 1992 IEEE International Conference on Robotics and Automation*, pages 1999–2004, Nice, France, May 1992. IEEE.
94. G. Walsh, D. Tilbury, S. Sastry, R. Murray, and J. P. Laumond. Stabilization of trajectories for systems with nonholonomic constraints. *IEEE Transactions on Automatic Control*, 39(1):216–222, 1994.
95. W. Wannasuphoprasit, R. B. Gillespie, and J. E. Colgate M. A. Peshkin. Cobot control. In *Proc. 1997 IEEE International Conference on Robotics and Automation*, pages 3571–3576, Albuquerque, New Mexico, April 1997.

Springer Tracts in Advanced Robotics

Edited by B. Siciliano, O. Khatib, and F. Groen
Published Titles:

Vol. 1: Caccavale, F.; Villani, L. (Eds.)
Fault Diagnosis and Fault Tolerance for Mechatronic Systems: Recent Advances
191 p. 2002 [3-540-44159-X]

Vol. 2: Antonelli, G.
Underwater Robots: Motion and Force Control of Vehicle-Manipulator Systems
209 p. 2003 [3-540-00054-2]

Vol. 3: Natale, C.
Interaction Control of Robot Manipulators: Six-degrees-of-freedom Tasks
120 p. 2003 [3-540-00159-X]

Vol. 4: Bicchi, A.; Christensen, H.I.; Prattichizzo, D. (Eds.)
Control Problems in Robotics
296 p. 2003 [3-540-00251-0]

Vol. 5: Siciliano, B.; Dario, P. (Eds.)
Experimental Robotics VIII
685 p. 2003 [3-540-00305-3]

Vol. 6: Jarvis, R.A.; Zelinsky, A. (Eds.)
Robotics Research – The Tenth International Symposium
580 p. 2003 [3-540-00550-1]

Vol. 7: Boissonnat, J.-D.; Burdick, J.; Goldberg, K.; Hutchinson, S. (Eds.)
Algorithmic Foundations of Robotics V
577 p. 2004 [3-540-40476-7]

Vol. 8: Baeten, J.; De Schutter, J.
Integrated Visual Servoing and Force Control
198 p. 2004 [3-540-40475-9]

Vol. 9: Yamane, K.
Simulating and Generating Motions of Human Figures
176 p. 2004 [3-540-20317-6]

Vol. 10: Siciliano, B.; De Luca, A.; Melchiorri, C.; Casalino, G. (Eds.)
Advances in Control of Articulated and Mobile Robots
259 p. 2004 [3-540-20783-X]

Vol. 11: Kim, J.-H.; Kim, D.-H.; Kim, Y.-J.; Seow, K.-T.
Soccer Robotics
353 p. 2004 [3-540-21859-9]

Vol. 12: Iagnemma K.; Dubowsky, S.
Mobile Robots in Rough Terrain: Estimation, Motion Planning, and Control with Application to Planetary Rovers
123 p. 2004 [3-540-21968-4]

Printing: Mercedes-Druck, Berlin
Binding: Stein + Lehmann, Berlin